Lecture Notes in Computer Science 11425

Commenced Publication in 1973
Founding and Former Series Editors:
Gerhard Goos, Juris Hartmanis, and Jan van Leeuwen

Advanced Research in Computing and Software Science

Subline of Lecture Notes in Computer Science

More information about this series at http://www.springer.com/series/7407

Mikołaj Bojańczyk · Alex Simpson (Eds.)

Foundations of Software Science and Computation Structures

22nd International Conference, FOSSACS 2019
Held as Part of the European Joint Conferences on Theory and Practice of Software, ETAPS 2019
Prague, Czech Republic, April 6–11, 2019
Proceedings

Springer Open

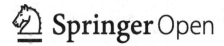

Editors
Mikołaj Bojańczyk
University of Warsaw
Warsaw, Poland

Alex Simpson
University of Ljubljana
Ljubljana, Slovenia

ISSN 0302-9743 ISSN 1611-3349 (electronic)
Lecture Notes in Computer Science
ISBN 978-3-030-17126-1 ISBN 978-3-030-17127-8 (eBook)
https://doi.org/10.1007/978-3-030-17127-8

Library of Congress Control Number: 2019936298

LNCS Sublibrary: SL1 – Theoretical Computer Science and General Issues

This Springer imprint is published by the registered company Springer Nature Switzerland AG
The registered company address is: Gewerbestrasse 11, 6330 Cham, Switzerland

ETAPS Foreword

Welcome to the 22nd ETAPS! This is the first time that ETAPS took place in the Czech Republic in its beautiful capital Prague.

ETAPS 2019 was the 22nd instance of the European Joint Conferences on Theory and Practice of Software. ETAPS is an annual federated conference established in 1998, and consists of five conferences: ESOP, FASE, FoSSaCS, TACAS, and POST. Each conference has its own Program Committee (PC) and its own Steering Committee (SC). The conferences cover various aspects of software systems, ranging from theoretical computer science to foundations to programming language developments, analysis tools, formal approaches to software engineering, and security.

Organizing these conferences in a coherent, highly synchronized conference program enables participation in an exciting event, offering the possibility to meet many researchers working in different directions in the field and to easily attend talks of different conferences. ETAPS 2019 featured a new program item: the Mentoring Workshop. This workshop is intended to help students early in the program with advice on research, career, and life in the fields of computing that are covered by the ETAPS conference. On the weekend before the main conference, numerous satellite workshops took place and attracted many researchers from all over the globe.

ETAPS 2019 received 436 submissions in total, 137 of which were accepted, yielding an overall acceptance rate of 31.4%. I thank all the authors for their interest in ETAPS, all the reviewers for their reviewing efforts, the PC members for their contributions, and in particular the PC (co-)chairs for their hard work in running this entire intensive process. Last but not least, my congratulations to all authors of the accepted papers!

ETAPS 2019 featured the unifying invited speakers Marsha Chechik (University of Toronto) and Kathleen Fisher (Tufts University) and the conference-specific invited speakers (FoSSaCS) Thomas Colcombet (IRIF, France) and (TACAS) Cormac Flanagan (University of California at Santa Cruz). Invited tutorials were provided by Dirk Beyer (Ludwig Maximilian University) on software verification and Cesare Tinelli (University of Iowa) on SMT and its applications. On behalf of the ETAPS 2019 attendants, I thank all the speakers for their inspiring and interesting talks!

ETAPS 2019 took place in Prague, Czech Republic, and was organized by Charles University. Charles University was founded in 1348 and was the first university in Central Europe. It currently hosts more than 50,000 students. ETAPS 2019 was further supported by the following associations and societies: ETAPS e.V., EATCS (European Association for Theoretical Computer Science), EAPLS (European Association for Programming Languages and Systems), and EASST (European Association of Software Science and Technology). The local organization team consisted of Jan Vitek and Jan Kofron (general chairs), Barbora Buhnova, Milan Ceska, Ryan Culpepper, Vojtech Horky, Paley Li, Petr Maj, Artem Pelenitsyn, and David Safranek.

The ETAPS SC consists of an Executive Board, and representatives of the individual ETAPS conferences, as well as representatives of EATCS, EAPLS, and EASST. The Executive Board consists of Gilles Barthe (Madrid), Holger Hermanns (Saarbrücken), Joost-Pieter Katoen (chair, Aachen and Twente), Gerald Lüttgen (Bamberg), Vladimiro Sassone (Southampton), Tarmo Uustalu (Reykjavik and Tallinn), and Lenore Zuck (Chicago). Other members of the SC are: Wil van der Aalst (Aachen), Dirk Beyer (Munich), Mikolaj Bojanczyk (Warsaw), Armin Biere (Linz), Luis Caires (Lisbon), Jordi Cabot (Barcelona), Jean Goubault-Larrecq (Cachan), Jurriaan Hage (Utrecht), Rainer Hähnle (Darmstadt), Reiko Heckel (Leicester), Panagiotis Katsaros (Thessaloniki), Barbara König (Duisburg), Kim G. Larsen (Aalborg), Matteo Maffei (Vienna), Tiziana Margaria (Limerick), Peter Müller (Zurich), Flemming Nielson (Copenhagen), Catuscia Palamidessi (Palaiseau), Dave Parker (Birmingham), Andrew M. Pitts (Cambridge), Dave Sands (Gothenburg), Don Sannella (Edinburgh), Alex Simpson (Ljubljana), Gabriele Taentzer (Marburg), Peter Thiemann (Freiburg), Jan Vitek (Prague), Tomas Vojnar (Brno), Heike Wehrheim (Paderborn), Anton Wijs (Eindhoven), and Lijun Zhang (Beijing).

I would like to take this opportunity to thank all speakers, attendants, organizers of the satellite workshops, and Springer for their support. I hope you all enjoy the proceedings of ETAPS 2019. Finally, a big thanks to Jan and Jan and their local organization team for all their enormous efforts enabling a fantastic ETAPS in Prague!

February 2019

Joost-Pieter Katoen
ETAPS SC Chair
ETAPS e.V. President

Preface

This volume contains the papers presented at the 22nd International Conference on Foundations of Software Science and Computation Structures (FoSSaCS), which took place in Prague during April 8–11, 2019. The conference is dedicated to foundational research with a clear significance for software science. It brings together research on theories and methods to support the analysis, integration, synthesis, transformation, and verification of programs and software systems.

The volume contains 29 contributed papers selected from 85 full paper submissions, and also a paper accompanying an invited talk by Thomas Colcombet (IRIF, France). Each submission was reviewed by at least three Program Committee members, with the help of external reviewers, and the final decisions took into account the feedback from a rebuttal phase. The conference submissions were managed using the EasyChair system, which was also used to assist with the compilation of the proceedings.

We wish to thank all the authors who submitted to FoSSaCS 2019, the Program Committee members, and the external reviewers. In addition, we would like to thank the ETAPS organization for providing an excellent environment for FoSSaCS alongside the other ETAPS conferences and workshops.

February 2019

Mikołaj Bojańczyk
Alex Simpson

Organization

Program Committee

Luca Aceto	Reykjavik University, Iceland
Achim Blumensath	Masaryk University, Brno, Czech Republic
Mikołaj Bojańczyk	University of Warsaw
Agata Ciabattoni	Vienna University of Technology, Austria
Flavio Corradini	University of Camerino, Italy
Nathanaël Fijalkow	CNRS, LaBRI, University of Bordeaux, France
Sergey Goncharov	FAU Erlangen-Nürnberg, Germany
Matthew Hague	Royal Holloway University of London, UK
Chris Heunen	The University of Edinburgh, UK
Patricia Johann	Appalachian State University, USA
Bartek Klin	University of Warsaw, Poland
Naoki Kobayashi	The University of Tokyo, Japan
Dexter Kozen	Cornell University, USA
Orna Kupferman	Hebrew University, Israel
Paul Blain Levy	University of Birmingham, UK
Peter Lefanu Lumsdaine	Stockholm University, Sweden
Radu Mardare	Aalborg University, Denmark
Angelo Montanari	University of Udine, Italy
Anca Muscholl	LaBRI, University of Bordeaux, France
Rasmus Ejlers Møgelberg	IT University of Copenhagen, Denmark
K. Narayan Kumar	Chennai Mathematical Institute, India
Dirk Pattinson	The Australian National University, Australia
Daniela Petrisan	Université Paris Diderot - Paris 7, France
Davide Sangiorgi	University of Bologna, Italy
Alex Simpson	University of Ljubljana, Slovenia
Ana Sokolova	University of Salzburg, Austria
James Worrell	University of Oxford, UK

Additional Reviewers

Achilleos, Antonis
Ahn, Ki Yung
Ahrens, Benedikt
Andres Martinez, Pablo
Atig, Mohamed Faouzi
Atkey, Robert
Bacci, Giorgio
Bacci, Giovanni
Bahr, Patrick
Bartocci, Ezio
Basold, Henning
Becker, Ruben
Benerecetti, Massimo
Bernardi, Giovanni
Blahoudek, František
Blondin, Michael

Bonchi, Filippo
Bresolin, Davide
Bruyère, Véronique
Cacciagrano, Diletta Romana
Cassar, Ian
Cerna, David
Chakraborty, Soham
Chen, Xiaohong
Clouston, Ranald
Dal Lago, Ugo
de Frutos Escrig, David
de Paiva, Valeria
Degorre, Aldric
Della Monica, Dario
Din, Crystal Chang
Dougherty, Daniel
Doumane, Amina
Dubut, Jérémy
Emmi, Michael
Enrique Moliner, Pau
Escardo, Martin
Faella, Marco
Ferreira, Carla
Furber, Robert
Fábregas, Ignacio
Gadducci, Fabio
Galesi, Nicola
García-Pérez, Álvaro
Gastin, Paul
Gavazzo, Francesco
Gorogiannis, Nikos
Goubault-Larrecq, Jean
Grädel, Erich
Haar, Stefan
Hamana, Makoto
Haselwarter, Philipp
Hasuo, Ichiro
Hausmann, Daniel
Heindel, Tobias
Herbreteau, Frédéric
Hoshino, Naohiko
Hosseini, Mehran
Hunt, Seb
Hyvernat, Pierre
Jaber, Guilhem
Jacq, Clément
Johnsen, Einar Broch
Kaarsgaard, Robin
Kaminski, Benjamin Lucien
Kammar, Ohad
Karvonen, Martti
Katsumata, Shin-Ya
Kerjean, Marie
Kop, Cynthia
Kurz, Alexander
Kuznets, Roman
Kučera, Antonín
Laird, James
Lefaucheux, Engel
Leitsch, Alexander
Leroux, Jérôme
Lhote, Nathan
Lindley, Sam
Loreti, Michele
Mamouras, Konstantinos
Marsden, Dan
Masini, Andrea
Mazowiecki, Filip
Mazza, Damiano
Mellies, Paul-Andre
Melliès, Paul-André
Merelli, Emanuela
Mostarda, Leonardo
Mukund, Madhavan
Neves, Renato
Norman, Gethin
North, Paige
Ohlmann, Pierre
Olarte, Carlos
Oortwijn, Wytse
Otop, Jan
Paquet, Hugo
Pedersen, Mathias Ruggaard
Perez, Guillermo
Peron, Adriano
Petrov, Tatjana
Pédrot, Pierre-Marie
Pérez, Jorge A.
Quaas, Karin
Ramanujam, R.
Rampersad, Narad
Rauch, Christoph

Re, Barbara
Rehak, Vojtech
Sala, Pietro
Schoepp, Ulrich
Schrijvers, Tom
Schröder, Lutz
Schwoon, Stefan
Sin'Ya, Ryoma
Sobocinski, Pawel
Sojakova, Kristina
Staton, Sam
Sumii, Eijiro
Sutre, Grégoire
Tang, Qiyi
Tesei, Luca
Thinnayam, Ramanathan
Tiezzi, Francesco
Tschaikowski, Max
Tsukada, Takeshi
Turrini, Andrea
Unno, Hiroshi
Uustalu, Tarmo
van Dijk, Tom
van Heerdt, Gerco
Vicary, Jamie
Vidal, German
Vignudelli, Valeria
Voigtländer, Janis
Wallbridge, James
Weil, Pascal
Winskel, Glynn
Wojtczak, Dominik
Wolter, Uwe
Ziemiański, Krzysztof

Contents

Universal Graphs and Good for Games Automata: New Tools for Infinite Duration Games

Thomas Colcombet[1(✉)] and Nathanaël Fijalkow[2]

[1] CNRS, IRIF, Université Paris-Diderot, Paris, France
thomas.colcombet@irif.fr
[2] CNRS, LaBRI, Université de Bordeaux, Bordeaux, France

Abstract. In this paper, we give a self contained presentation of a recent breakthrough in the theory of infinite duration games: the existence of a quasipolynomial time algorithm for solving parity games. We introduce for this purpose two new notions: good for small games automata and universal graphs.

The first object, good for small games automata, induces a generic algorithm for solving games by reduction to safety games. We show that it is in a strong sense equivalent to the second object, universal graphs, which is a combinatorial notion easier to reason with. Our equivalence result is very generic in that it holds for all existential memoryless winning conditions, not only for parity conditions.

1 Introduction

In this abstract, we are interested in the complexity of deciding the winner of finite turn-based perfect-information antagonistic two-player games. So typically, we are interested in parity games, or mean-payoff games, or Rabin games, etc...

In particular we revisit the recent advances showing that deciding the winner of parity games can be done in quasipolynomial time. Whether parity games can be solved in polynomial time is the main open question in this research area, and an efficient algorithm would have far-reaching consequences in verification, synthesis, logic, and optimisation. From a complexity-theoretic point of view, this is an intriguing puzzle: the decision problem is in **NP** and in **coNP**, implying that it is very unlikely to be **NP**-complete (otherwise **NP** = **coNP**). Yet no polynomial time algorithm has yet been constructed. For decades the best algorithms were exponential or mildly subexponential, most of them of the form $n^{O(d)}$, where n is the number of vertices and d the number of priorities (we refer to Section 2 for the role of these parameters).

Recently, Calude, Jain, Khoussainov, Li, and Stephan [CJK+17] constructed a quasipolynomial time algorithm for solving parity games, of complexity

This work was supported by the European Research Council (ERC) under the European Union's Horizon 2020 research and innovation programme (grant agreement No. 670624), and by the DeLTA ANR project (ANR-16-CE40-0007).

M. Bojańczyk and A. Simpson (Eds.): FOSSACS 2019, LNCS 11425, pp. 1–26, 2019.
https://doi.org/10.1007/978-3-030-17127-8_1

$n^{O(\log d)}$. Two subsequent algorithms with similar complexity were constructed by Jurdziński and Lazić [JL17], and by Lehtinen [Leh18].

Our aim in this paper is to understand these results through the prism of good for small games automata, which are used to construct generic reductions to solving safety games. A good for small games automaton can be understood as an approximation of the original winning condition which is correct for small games. The size of good for small games automata being critical in the complexity of these algorithms, we aim at understanding this parameter better.

A concrete instanciation of good for small games automata is the notion of separating automata, which was introduced by Bojańczyk and Czerwiński [BC18] to reformulate the first quasipolynomial time algorithm of [CJK+17]. Later Czerwiński, Daviaud, Fijalkow, Jurdziński, Lazić, and Parys [CDF+19] showed that the other two quasipolynomial time algorithms also can be understood as the construction of separating automata, and proved a quasipolynomial lower bound on the size of separating automata.

In this paper, we establish in particular Theorem 9 which states an equivalence between the size of good for small games automata, non-deterministic of separating automata, of deterministic separating automata and of universal graphs. This statement is generic in the sense that it holds for any winning condition which is memoryless for the existential player, hence in particular for parity conditions. At a technical level, the key notion that we introduce to show this equivalence is the combinatorial concept of universal graphs.

Our second contribution, Theorem 10, holds for the parity condition only, and is a new equivalence between universal trees and universal graphs. In particular we use a technique of saturation of graphs which simplifies greatly the arguments. The two theorems together give an alternative simpler proof of the result in [CDF+19].

Let us mention that the equivalence results have been very recently used to construct algorithms for mean-payoff games, leading to improvements over the best known algorithm [FGO18].

Structure of the paper In Section 2 we introduce the classical notions of games, automata, and good for games automata. In Section 3, we introduce the notion of good for small games automata, and show that in the context of memoryless for the existential player winning conditions these automata can be characterised in different ways, using in particular universal graphs (Theorem 9). In Section 4, we study more precisely the case of parity conditions.

2 Games and automata

We describe in this subsection classical material: arenas, games, strategies, automata and good for games automata. Section 2.1 introduces games, Section 2.2 the concept of memoryless strategy, and Section 2.3 the class of automata we use. Finally, Section 2.4 explains how automata can be used for solving games, and in particular defines the notion of automata that are good for games.

2.1 Games

We will consider several forms of graphs, which are all directed labelled graph with a root vertex. Let us fix the terminology now. Given a set X, an X-graph $H = (V, E, \mathrm{root}_H)$ has a set of vertices V, a set of X-labelled edges $E \subseteq V \times X \times V$, and a root vertex root_H. We write $x \xrightarrow{u}_H y$ if there exists a path from vertex x to vertex y labelled by the word $u \in X^*$. We write $x \xrightarrow{u}_H \infty$ if there exists an infinite path starting in vertex x labelled by the word $u \in X^\omega$. The graph is trimmed if all vertices are reachable from the root and have out-degree at least one. Note that as soon as a graph contains some infinite path starting from the root, it can be made trimmed by removing the bad vertices. A morphism of X-graphs from G to H is a map α from vertices of G to vertices of H, that sends the root of G to the root of H, and sends each edge of G to an edge of H, i.e., for all $a \in X$, $p \xrightarrow{a}_G q$ implies $\alpha(p) \xrightarrow{a}_H \alpha(q)$. A weak morphism of X-graphs is like a morphism but we lift the property that the root of G is sent to the root of H and instead require that if $\mathrm{root} \xrightarrow{a}_G x$ then $\mathrm{root} \xrightarrow{a}_H \alpha(x)$.

Definition 1. *Let C be a set (of colors). A C-arena A is a C-graph in which vertices are split into $V = V_\mathrm{E} \uplus V_\mathrm{A}$. The vertices are called positions. The positions in V_E are the positions owned by the existential player, and the ones in V_A are owned by the universal player. The root is the initial position. The edges are called moves. Infinite paths starting in the initial position are called plays. Finite paths starting in the initial position are called partial plays. The dual of an arena is obtained by swapping V_A and V_E, i.e., exchanging the ownership of the positions.*

A $\mathbb{W}$-game $\mathcal{G} = (A, \mathbb{W})$ consists of a C-arena A together with a set $\mathbb{W} \subseteq C^\omega$ called the winning condition.

For simplicity, we assume in this paper the following epsilon property[1]: there is a special color $\varepsilon \in C$ such that for all words $u, v \in C^\omega$, if u and v are equal after removing all the ε-letters, then $u \in \mathbb{W}$ if and only if $v \in \mathbb{W}$.

The dual of a game is obtained by dualising the arena, and complementing the winning condition.

If one compares with usual games – for instance checkers – then the arena represents the set of possible board configurations of the game (typically, the configuration of the board plus a bit telling whose turn to play it is). The configuration is an existential position if it is the first player's turn to play, otherwise it is a universal position. There is an edge from u to v if it is a valid move for the player to go from configuration u to configuration v. The interest of having

[1] This assumption is satisfied in an obvious way for all winning conditions seen in this paper. It could be avoided, but at the technical price of considering slightly different forms of games: games in which the moves are positive boolean combinations of pairs of colors and positions. Such 'move relations' form a joint generalisation of existential positions (which can be understood as logical disjunction) and universal position (which can be understood as logical conjunction).

colors and winning conditions may not appear clearly in this context, but the intent would be, for example, to tell who is the winner if the play is infinite.

Informally, the game is played as follows by two players: the existential player and the universal player[2]. At the beginning, a token is placed at the initial position of the game. Then the game proceeds in rounds. At each round, if the token is on an existential position then it is the existential player's turn to play, otherwise it is the universal player's turn. This player chooses an outgoing move from the position, and the token is pushed along this move. This interaction continues forever, inducing a play (defined as an infinite path in the arena) labelled by an infinite sequence of colors. If this infinite sequence belongs to the winning condition $\mathbb{W}$, then the existential player wins the play, otherwise, the universal player wins the play. It may happen that a player has to play but there is no move available from the current position: in this case the player immediately loses.

Classical winning conditions Before describing more precisely the semantics of games, let us recall what are the classical winning conditions considered in this context.

Definition 2. *We define the following classical winning conditions:*

safety condition *The safety condition is* $\texttt{Safety} = \{0\}^\omega$ *over the unique color* 0. *Expressed differently, all plays are winning. Note that the color* 0 *fulfills the requirement of the epsilon property.*

Muller condition *Given a finite set of colors C, a Muller condition is a Boolean combination of winning conditions of the form "the color c appears infinitely often". In general, no color fulfills the requirement of the epsilon property, but it is always possible to add an extra fresh color ε. The resulting condition satisfies the epsilon property.*

Rabin condition *Given a number p, we define the Rabin condition* $\texttt{Rabin}_p \subseteq \{\{1,2,3\}^p\}^\omega$ *by* $u \in \texttt{Rabin}_p$ *if there exists some $i \in 1,\ldots,p$ such that when projected on this component, 2 appears infinitely often in u, and 3 finitely often. Note that the constant vector* $\mathbf{1}$ *fulfills the epsilon property. The Rabin condition is a special case of Muller conditions.*

parity condition *Given a interval of integers $C = [i,j]$ (called priorities), a word $u = c_1c_2c_3\cdots \in C^\omega$ belongs to* $\texttt{Parity}_C$ *if the largest color appearing infinitely often in u is even.*

Büchi condition *The Büchi condition* $\texttt{Buchi}$ *is a parity condition over the restricted interval* $[1,2]$ *of priorities. A word belongs to* $\texttt{Buchi}$ *if it contains infinitely many occurrences of* 2.

coBüchi condition *The coBüchi condition* $\texttt{coBuchi}$ *is a parity condition over the restricted interval* $[0,1]$ *of priorities. A word belongs to* $\texttt{coBuchi}$ *if it it has only finitely many occurrences of* 1*'s.*

---

[2] In the literature, the players have many other names: 'Eve' and 'Adam', 'Eloise' and 'Abelard', 'Exist' and 'Forall', '0' and '1', or in specific contexts: 'Even' and 'Odd', 'Automaton' and 'Pathfinder', 'Duplicator' and 'Spoiler', ...

mean-payoff condition *Given a finite set $C \subseteq \mathbb{R}$, a word $u = c_1 c_2 c_3 \cdots \in C^\omega$ belongs to* $\mathtt{meanpayoff}_C$ *if*

$$\liminf_{n\to\infty} \frac{c_1 + c_2 + \cdots + c_n}{n} \geqslant 0 \ .$$

There are many variants of this definition (such as replacing lim inf *with* lim sup*), that all turn out to be equivalent on finite arenas.*

Strategies We describe now formally what it means to win a game. Let us take the point of view of the existential player. A strategy for the existential player is an object that describes how to play in every situation of the game that could be reached. It is a winning strategy if whenever these choices are respected during a play, the existential player wins this play. There are several ways one can define the notion of a strategy. Here we choose to describe a strategy as the set of partial plays that may be produced when it is used.

Definition 3. *A strategy s for the existential player s_E is a set of partial plays of the game that has the following properties:*

- *s_E is prefix-closed and non-empty,*
- *for all partial plays $\pi \in s_E$ ending in some $v \in V_E$, there exists exactly one partial play of length $|\pi| + 1$ in s_E that prolongs π,*
- *for all partial plays $\pi \in s_E$ ending in some $v \in V_A$, then all partial plays that prolong π of length $|\pi| + 1$ belong to s_E.*

A play is compatible with the strategy s_E if all its finite prefixes belong to s. A play is winning if it belongs to the winning condition $\mathbb{W}$. A game is won by the existential player if there exists a strategy for the existential player such that all plays compatible with it are won by the existential player. Such a strategy is called a winning strategy.

Symmetrically, a (winning) strategy for the universal player is a (winning) strategy for the existential player in the dual game. A game is won by the universal player if there exists a strategy for the universal player such that all infinite plays compatible with it are won by the universal player.

The idea behind this definition is that at any moment in the game, when following a strategy, a sequence of moves has already been played, yielding a partial play in the arena. The above definition guarantees that: 1. if a partial play belongs to the strategy, it is indeed reachable by a succession of moves that stay in the strategy, 2. if, while following the strategy, a partial play ends in a vertex owned by the existential player, there exists exactly one move that can be followed by the strategy at that moment, and 3. if, while following the strategy, a partial play ends in a vertex owned by the universal player, the strategy is able to face all possible choices of the opponent.

Remark 1. It is not possible that in a strategy defined in this way one reaches an existential position that would have no successor: indeed, 2. would not hold.

Remark 2. There are different ways to define a strategy in the literature. One is as a strategy tree: indeed one can see s_{E} as a set of nodes equipped with prefix ordering as the ancestor relation. Another way is to define a strategy as a partial map from paths to moves. All these definitions are equivalent. The literature also considers randomized strategies (in which the next move is chosen following a probability distribution): this is essential when the games are *concurrent* or *with partial information*, but not in the situation we consider in this paper.

Lemma 1 (at most one player wins). *It is not possible that both the existential player and the universal player win the same game.*

Of course, keeping the intuition of games in mind, one would expect also that one of the player wins. However, this is not necessarily the case. A game is called determined if either the existential or the universal player wins the game. The fact that a game is determined is referred to as its determinacy. A winning condition $\mathbb{W}$ is determined if all $\mathbb{W}$-games are determined. It happens that not all games are determined.

Theorem 1. *There exist winning conditions that are not determined (and it requires the axiom of choice to prove it).*

However, there are some situations in which games are determined. This is the case of finite duration games, of safety games, and more generally:

Theorem 2 (Martin's theorem of Borel determinacy [Mar75]). *Games with Borel winning conditions are determined.*

Defining the notion of Borel sets is beyond the scope of this paper. It suffices to know that this notion is sufficiently powerful for capturing a lot of natural winning conditions, and in particular all winning conditions in this paper are Borel; and thus determined.

2.2 Memory of strategies

A key insight in understanding a winning condition is to study the amount of memory required by winning strategies. To define the notion of memoryless strategies, we use an equivalent point of view on strategies, using strategy graphs.

Definition 4. *Given a C-arena A, an existential player strategy graph S_{E}, γ in A is a trimmed C-graph S_{E} together with a graph morphism γ from S_{E} to A such that for all vertices x in S_{E},*

- *if $\gamma(x)$ is an existential position, then there exists exactly one edge of the form (x, c, y) in S_{E},*
- *if $\gamma(x)$ is a universal position, then β induces a surjection between the edges originating from x in S_{E} and the moves originating from $\beta(x)$, i.e., for all moves of the form $(\beta(x), c, v)$, there exists an edge of the form (x, c, y) in S_{E} such that $\beta(y) = v$.*

The existential player strategy graph S_{E}, γ is memoryless if γ is injective. In general the memory of the strategy is the maximal cardinality of $\gamma^{-1}(v)$ for v ranging over all positions in the arena. For $\mathcal{G}$ a $\mathbb{W}$-game with $\mathbb{W} \subseteq C^\omega$, an existential player strategy graph S_{E} is winning if the labels of all its paths issued from the root belong to $\mathbb{W}$.

The (winning) universal player strategy graphs are defined as the (winning) existential player strategy graphs in the dual game.

The winning condition $\mathbb{W}$ is memoryless for the existential player if, whenever the existential player wins in a $\mathbb{W}$-game, there is a memoryless winning existential player strategy graph. It is memoryless for the existential player over finite arenas if this holds for finite $\mathbb{W}$-games only. The dual notion is the one of memoryless for the universal player winning condition.

Of course, as far as existence is concerned the two notions of strategy coincide:

Lemma 2. *There exists a winning existential player strategy graph if and only if there exists a winning strategy for the existential player.*

Proof. A strategy for the existential player s_{E} can be seen as a C-graph (in fact a tree) S_{E} of vertices s_{E}, of root ε, and with edges of the form $(\pi, a, \pi a)$ for all $\pi a \in s_{\mathrm{E}}$. If the strategy s_{E} is winning, then the strategy graph S_{E} is also winning. Conversely, given an existential player strategy graph S_{E}, the set s_{E} of its paths starting from the root is itself a strategy for the existential player. Again, the winning property is preserved. □

We list a number of important results stating that some winning conditions do not require memory.

Theorem 3 ([EJ91]). *The parity condition is memoryless for the existential player and for the universal player.*

Theorem 4 ([EM79, GKK88]). *The mean-payoff condition is memoryless for the existential player over finite arenas as well as for the universal player.*

Theorem 5 ([GH82]). *The Rabin condition is memoryless for the existential player, but not in general for the universal player.*

Theorem 6 ([McN93]). *Muller conditions are finite-memory for both players.*

Theorem 7 ([CFH14]). *Topologically closed conditions for which the residuals are totally ordered by inclusion are memoryless for the existential player.*

2.3 Automata

Definition 5 (automata over infinite words). *Let $\mathbb{W} \subseteq C^\omega$. A (non-deterministic) $\mathbb{W}$-automaton $\mathcal{A}$ over the alphabet A is a $(C \times A)$-graph. The convention is to call states its vertices, and transitions its edges. The root vertex is called the initial state. The set $\mathbb{W}$ is called the accepting condition (whereas it* *is the winning condition for games). The automaton $\mathcal{A}_p$ is obtained from $\mathcal{A}$ by setting the state p to be initial.*

A run of the automaton $\mathcal{A}$ over $u \in A^\omega$ is an infinite path in $\mathcal{A}$ that starts in the initial state and projects on its A-component to u. A run is accepting if it projects on its C-component to a word $v \in \mathbb{W}$. The language accepted by $\mathcal{A}$ is the set $\mathcal{L}(\mathcal{A})$ of infinite words $u \in A^\omega$ such that there exists an accepting run of $\mathcal{A}$ on u.

An automaton is deterministic (resp. complete) if for all states p and all letters $a \in A$, there exists at most one (resp. at least one) transition of the form $(p, (a, c), q)$. If the winning condition is parity, this is a parity automaton. If the winning condition is safety, this is a safety automaton, and we do not mention the C-component since there is only one color. I.e., the transitions form a subset of $Q \times A \times Q$, and the notion coincides with the one of a A-graph. For this reason, we may refer to the language $\mathcal{L}(H)$ accepted by an A-graph H: this is the set of labelling words of infinite paths starting in the root vertex of H.

Note that here we use non-deterministic automata for simplicity. However, the notions developed in this paper can be adapted to alternating automata.

The notion of ω-regularity. It is not the purpose of this paper to describe the rich theory of automata over infinite words. It suffices to say that a robust concept of ω-regular language emerges. These are the languages that are equivalently defined by means of Büchi automata, parity automata, Rabin automata, Muller automata, deterministic parity automata, deterministic Rabin automata, deterministic Muller automata, as well as many other formalisms (regular expressions, monadic second-order logic, ω-semigroup, alternating automata, ...). However, safety automata and deterministic Büchi automata define a subclass of ω-regular languages.

Note that the mean-payoff condition does not fall in this category, and automata defined with this condition do not recognize ω-regular languages in general.

2.4 Automata for solving games

There is a long tradition of using automata for solving games. The general principle is to use automata as reductions, i.e. starting from a $\mathbb{V}$-game $\mathcal{G}$ and a $\mathbb{W}$-automaton $\mathcal{A}$ that accepts the language $\mathbb{V}$, we construct a $\mathbb{W}$-game $\mathcal{G} \times \mathcal{A}$ called the product game that combines the two, and which is expected to have the same winner: this means that to solve the $\mathbb{V}$-game $\mathcal{G}$, it is enough to solve the $\mathbb{W}$-game $\mathcal{G} \times \mathcal{A}$. We shall see below that, unfortunately, this expected property does not always hold (Remark 4). The automata that guarantee the correction of the construction are called good for games, originally introduced by Henzinger and Piterman [HP06].

We begin our description by making precise the notion of product game. Informally, the new game requires the players to play like in the original game, and after each step, the existential player is required to provide a transition in the automaton that carries the same label.

Definition 6. *Let $\mathcal{D}$ be an arena over colors C, with positions P and moves M. Let also $\mathcal{A}$ be a $\mathbb{W}$-automaton over the alphabet C with states Q and transitions Δ. We construct the product arena $\mathcal{D} \times \mathcal{A}$ as follows:*

- *The set of positions in the product game is $(P \uplus M) \times Q$.*
- *The initial position is $(\mathrm{init}_{\mathcal{D}}, \mathrm{init}_{\mathcal{A}})$, in which $\mathrm{init}_{\mathcal{D}}$ is the initial position of $\mathcal{G}$, and $\mathrm{init}_{\mathcal{A}}$ is the initial state of $\mathcal{A}$.*
- *The positions of the form $(x,p) \in P \times Q$ are called game positions and are owned by the owner of x in $\mathcal{G}$. There is a move, called a game move, of the form $((x,p), \varepsilon, ((x,c,y),p))$ for all moves $(x,c,y) \in M$.*
- *The positions of the form $((x,c,y),p) \in M \times Q$ are called automaton positions and are owned by the existential player. There is a move, called an automaton move, of the form $(((x,c,y),p), d, (y,q))$ for all transitions of the form $(p,(c,d),q)$ in $\mathcal{A}$.*

Note that every game move $((x,p), \varepsilon, ((x,c,y),p))$ of $\mathcal{G} \times \mathcal{A}$ can be transformed into a move (x,c,y) of $\mathcal{G}$, called its game projection. Similarly every automaton move $(((x,c,y),p), d, (y,q))$ can be turned into a transition $(p,(c,d),q)$ of the automaton $\mathcal{A}$ called its automaton projection. Hence, every play π of the product game can be projected into the pair of a play π' in $\mathcal{G}$ of label u (called the game projection), and an infinite run ρ of the automaton over u (called the automaton projection). The product game is the game over the product arena, using the winning condition of the automaton.

Lemma 3 (folklore[3]). *Let $\mathcal{G}$ be a $\mathbb{V}$-game, and $\mathcal{A}$ be a $\mathbb{W}$-automaton that accepts a language $L \subseteq \mathbb{V}$, then if the existential player wins $\mathcal{G} \times Q_{\mathcal{A}}$, she wins $\mathcal{G}$.*

Proof. Assume that the existential player wins the game $\mathcal{G} \times \mathcal{A}$ using a strategy s_{E}. This strategy can be turned into a strategy for the existential player s'_{E} in $\mathcal{G}$ by performing a game projection. It is routine to check that this is a valid strategy.

Let us show that this strategy s'_{E} is $\mathbb{V}$-winning, and hence conclude that the existential player wins the game $\mathcal{G}$. Indeed, let π' be a play compatible with s'_{E}, say labelled by u. This play π' has been obtained by game projection of a play π compatible with s_{E} in $\mathcal{G} \times \mathcal{A}$. The automaton projection ρ of π is a run of $\mathcal{A}$ over u, and is accepting since s_{E} is a winning strategy. Hence, u is accepted by $\mathcal{A}$ and as a consequence belongs to $\mathbb{V}$. We have proved that s_{E} is winning. □

Corollary 1. *Let $\mathcal{G}$ be a $\mathbb{V}$-game, and $\mathcal{A}$ be a deterministic $\mathbb{W}$-automaton that accepts the language $\mathbb{V}$, then the games $\mathcal{G}$ and $\mathcal{G} \times \mathcal{A}$ have the same winner.*

Proof. We assume without loss of generality that $\mathcal{A}$ is deterministic and complete (note that this may require to slightly change the accepting condition, for instance in the case of safety). The results then follows from the application of Lemma 3 to the game $\mathcal{G}$ and its dual. □

[3] This technique of reduction is in fact more general, since the automaton may not be a safety automaton. Its use can be traced back, for instance, to the work of Büchi and Landweber [BL69].

The consequence of the above lemma is that when we know how to solve $\mathbb{W}$-games, and we have a deterministic $\mathbb{W}$-automaton $\mathcal{A}$ for a language $\mathbb{V}$, then we can decide the winner of $\mathbb{V}$-games by performing the product of the game with the automaton, and deciding the winner of the resulting game. Good for games automata are automata that need not be deterministic, but for which this kind of arguments still works.

Definition 7 (good for games automata [HP06]). *Let $\mathbb{V}$ be a language, and $\mathcal{A}$ be a $\mathbb{W}$-automaton. Then $\mathcal{A}$ is good for $\mathbb{V}$-games if for all $\mathbb{V}$-games $\mathcal{G}$, $\mathcal{G}$ and $\mathcal{G} \times \mathcal{A}$ have the same winner.*

Note that Lemma 1 says that deterministic automata are good for games automata.

Remark 3. It may seem strange, a priori, not to require in the definition that $\mathcal{L}(\mathcal{A}) = \mathbb{V}$. In fact, it holds anyway: if an automaton is good for $\mathbb{V}$-games, then it accepts the language $\mathbb{V}$. Indeed, let us assume that there exists a word $u \in \mathcal{L}(\mathcal{A}) \setminus \mathbb{V}$, then one can construct a game that has exactly one play, labelled u. This game is won by the universal player since $u \notin \mathbb{V}$, but the existential player wins $\mathcal{G} \times \mathcal{A}$. A contradiction. The same argument works if there is a word in $\mathbb{V} \setminus \mathcal{L}(\mathcal{A})$.

Examples of good for games automata can be found in [BKS17], together with a structural analysis of the extent to which they are non-deterministic.

Remark 4. We construct an automaton which is not good for games. The alphabet is $\{a, b\}$. The automaton $\mathcal{A}$ is a Büchi automaton: it has an initial state from which goes two ϵ-transitions: the first transition guesses that the word contains infinitely many a's, and the second transition guesses that the word contains infinitely many b's. Note that any infinite word contains either infinitely many a's or infinitely many b's, so the language $\mathbb{V}$ recognised by this automaton is the set of all words. However, this automaton requires a choice to be made at the very first step about which of the two alternatives hold. This makes it not good for games: indeed, consider a game $\mathcal{G}$ where the universal player picks any infinite word, letter by letter, and the winning condition is $\mathbb{V}$. It has only one position owned by the universal player. The existential player wins $\mathcal{G}$ because all plays are winning. However, the existential player loses $\mathcal{G} \times \mathcal{A}$, because in this game she has to declare at the first step whether there will be infinitely many a's or infinitely many b's, which the universal player can later contradict.

Let us conclude this part with Lemma 4, stating the possibility to compose good for games automata. We need before hand to defined the composition of automata.

Given $A \times B$-graph $\mathcal{A}$, and $B \times C$-graph $\mathcal{B}$, the composed graph $\mathcal{B} \circ \mathcal{A}$ has as states the product of the sets of states, as initial state the ordered pair of the initial states, and there is a transition $((p, q), (a, c), (p', q'))$ if there is a transition $(p, (a, b), p')$ in $\mathcal{A}$ and a transition $(q, (b, c), q')$. If $\mathcal{A}$ is in fact an automaton that uses the accepting condition $\mathbb{V}$, and $\mathcal{B}$ an automaton that uses the accepting condition $\mathbb{W}$, then the composed automaton $\mathcal{B} \circ \mathcal{A}$ uses has as underlying graph the composed graphs, and as accepting condition $\mathbb{W}$.

Lemma 4 (composition of good for games automata). *Let $\mathcal{A}$ be a good for games $\mathbb{W}$-automaton for the language $\mathbb{V}$, and $\mathcal{B}$ be good for games $\mathbb{V}$-automaton for the language L, then the composed automaton $\mathcal{A} \circ \mathcal{B}$ is a good for games $\mathbb{W}$-automaton for the language L.*

3 Efficiently solving games

From now on, graphs, games and automata are assumed to be finite.

We now present more recent material. We put forward the notion of good for n-games automata (good for small games) as a common explanation for the several recent algorithms for solving parity games 'efficiently'. After describing this notion in Section 3.1, we shall give more insight about it in the context of winning conditions that are memoryless for the existential player in Section 3.2

Much more can be said for parity games and good for small games safety automata: this will be the subject of Section 4.

3.1 Good for small games automata

We introduce the concept of (strongly) good for n-games automata (good for small games). The use of these automata is the same as for good for games automata, except that they are cannot be composed with any game, but only with small ones. In other words, a good for $(\mathbb{W}, n)$-game automaton yields a reduction for solving $\mathbb{W}$-games with at most n positions (Lemma 6). We shall see in Section 3.2 that as soon as the underlying winning condition is memoryless for the existential player, there are several characterisations for the smallest strongly good for n-games automata. It is good to keep in mind the definition of good for games automata (Definition 7) when reading the following one.

Definition 8. *Let $\mathbb{V}$ be a language, and $\mathcal{A}$ be a $\mathbb{W}$-automaton. Then $\mathcal{A}$ is good for $(\mathbb{V}, n)$-games if for all $\mathbb{V}$-games $\mathcal{G}$ with at most n positions, $\mathcal{G}$ and $\mathcal{G} \times \mathcal{A}$ have the same winner (we also write good for small games when there is no need for $\mathbb{V}$ and n to be explicit).*

It is strongly good for $(\mathbb{V}, n)$-games if it is good for $(\mathbb{V}, n)$-games and the language accepted by $\mathcal{A}$ is contained in $\mathbb{V}$.

Example 1 (automata that are good for small games). We have naturally the following chain of implications:

$$\text{good for games} \Longrightarrow \text{strongly good for } n\text{-games} \Longrightarrow \text{good for } n\text{-games}$$

The first implication is from Remark 3, and the second is by definition. Thus the first examples of automata that are strongly good for small games are the automata that are good for games.

Example 2. We consider the case of the coBüchi condition: recall that the set of colors is $\{0,1\}$ and the winning plays are the ones such that there ultimately contain only 0's. It can be shown that if the existential player wins in a coBüchi game with has at most n positions, then she also wins for the winning condition $L = (0^*(\varepsilon + 1))^n 0^\omega$, i.e., the words in which there is at most n occurrences of 1 (indeed, a winning memoryless strategy for the condition $\texttt{coBuchi}$ cannot contain a 1 in a cycle, and hence cannot contain more than n occurrences of 1 in the same play; thus the same strategy is also winning in the same game with the new winning condition L). As a consequence, a deterministic safety automaton that accepts the language $L \subseteq \texttt{coBuchi}$ (the minimal one has $n+1$ states) is good for $(\texttt{coBuchi}, n)$-games.

Mimicking Lemma 4 which states the closure under composition of good for games automata, we obtain the following variant for good for small games automata:

Lemma 5 (composition of good for small games automata). *Let $\mathcal{B}$ be a good for n-games $\mathbb{V}$-automaton for the language L with k states, and $\mathcal{A}$ be a good for kn-games $\mathbb{W}$-automaton for the language $\mathbb{V}$, then the composed automaton $\mathcal{A} \circ \mathcal{B}$ is a good for n-games $\mathbb{W}$-automaton for the language L.*

We also directly get an algorithm from such reductions.

Lemma 6. *Assume that there exists an algorithm for solving $\mathbb{W}$-games of size m in time $f(m)$. Let $\mathcal{G}$ be a $\mathbb{V}$-game with at most n positions and $\mathcal{A}$ be a good for $(\mathbb{V}, n)$-games $\mathbb{W}$-automaton of size k, there exists an algorithm for solving $\mathcal{G}$ of complexity $f(kn)$.*

Proof. Construct the game $\mathcal{G} \times \mathcal{A}$, and solve it. □

The third quasipolynomial time algorithm for solving parity games due to Lehtinen [Leh18] can be phrased using good for small games automata (note that it is not originally described in this form).

Theorem 8 ([Leh18, BL19]). *Given positive integers n, d, there exists a parity automaton with $n^{(\log d + O(1))}$ states and $1 + \lfloor \log n \rfloor$ priorities which is strongly good for n-games.*

Theorem 8 combined with Lemma 6 yields a quasipolynomial time algorithm for solving parity games. Indeed, consider a parity game $\mathcal{G}$ with n positions and d priorities. Let $\mathcal{A}$ be the good for n-games automaton constructed by Theorem 8. The game $\mathcal{G} \times \mathcal{A}$ is a parity game equivalent to $\mathcal{G}$, which has $m = n^{(\log d + O(1))}$ states and $d' = 1 + \lfloor \log n \rfloor$ priorities. Solving this parity game with a simple algorithm (of complexity $O(m^{d'})$) yields an algorithm of quasipolynomial complexity:

$$O(m^{d'}) = O(n^{(\log d + O(1)) d'}) = n^{O(\log(d) \log(n))}.$$

3.2 The case of memoryless winning conditions

In this section we fix a winning condition $\mathbb{W}$ which is memoryless for the existential player, and we establish several results characterising the smallest strongly good for small games automata in this case.

Our prime application is the case of parity conditions, that will be studied specifically in Section 4, but this part also applies to conditions such as mean-payoff or Rabin.

The goal is to establish the following theorem (the necessary definitions are introduced during the proof).

Theorem 9. *Let $\mathbb{W}$ be a winning condition which is memoryless for the existential player, then the following quantities coincide for all positive integers n:*

1. *the least number of states of a strongly $(\mathbb{W}, n)$-separating deterministic safety automaton,*
2. *the least number of states of a strongly good for $(\mathbb{W}, n)$-games safety automaton,*
3. *the least number of states of a strongly $(\mathbb{W}, n)$-separating safety automaton,*
4. *the least number of vertices of a $(\mathbb{W}, n)$-universal graph.*

The idea of separating automata[4] was introduced by Bojańczyk and Czerwiński [BC18] to reformulate the first quasipolynomial time algorithm [CJK+17]. Czerwiński, Daviaud, Fijalkow, Jurdziński, Lazić, and Parys [CDF+19] showed that the other two quasipolynomial time algorithms [JL17,Leh18] also can be understood as the construction of separating automata.

The proof of Theorem 9 spans over Sections 3.2 and 3.3. It it a consequence of Lemmas 7, 8, 11, and 12. We begin our proof of Theorem 9 by describing the notion of strongly separating automata.

Definition 9. *An automaton $\mathcal{A}$ is strongly $(\mathbb{W}, n)$-separating if*

$$\mathbb{W}|_n \subseteq \mathcal{L}(\mathcal{A}) \subseteq \mathbb{W} \ ,$$

in which $\mathbb{W}|_n$ is the union of all the languages accepted by safety automata with n states that accept sublanguages of $\mathbb{W}$.[5]

Lemma 7. *In the statement of Theorem 9, (1) $\Longrightarrow$ (2) $\Longrightarrow$ (3).*

Proof. Assume (1), i.e., there exists a strongly $(\mathbb{W}, n)$-separating deterministic safety automaton $\mathcal{A}$, then $\mathcal{L}(\mathcal{A}) \subseteq \mathbb{W}$. Let $\mathcal{G}$ be a $\mathbb{W}$-game with at most n positions. By Lemma 3, if the existential player wins $\mathcal{G} \times \mathcal{A}$, she wins the

[4] The definition used in [BC18] is not strictly equivalent to the one we use here: a separating automaton in [BC18] is a strongly separating automaton in our sense, but not conversely.

[5] Note that there is a natural, more symetric, notion of $(\mathbb{W}, n)$-separating automata in which the requested inclusions are $\mathbb{W}|_n \subseteq \mathcal{L}(\mathcal{A}) \subseteq \left(\mathbb{W}^{\mathtt{C}}\Big|_n\right)^{\mathtt{C}}$. However, nothing is known about this notion.

game $\mathcal{G}$. Conversely, assume that the existential player wins $\mathcal{G}$, then, by assumption she has a winning memoryless strategy graph $S_{\mathrm{E}}, \gamma\colon S_{\mathrm{E}} \to \mathcal{G}$, i.e., $\mathcal{L}(S_{\mathrm{E}}) \subseteq \mathbb{W}$ and γ is injective. By injectivity of γ, S_{E} has at most n vertices and hence $\mathcal{L}(S_{\mathrm{E}}) \subseteq \mathbb{W}|_n \subseteq \mathcal{L}(\mathcal{A})$. As a consequence, for every (partial) play π compatible with S_{E}, there exists a (partial) run of $\mathcal{A}$ over the labels of π (call this property $\star$). We construct a new strategy for the existential player in $\mathcal{G} \times \mathcal{A}$ as follows: When the token is in a game position, the existential player plays as in S_{E}; When the token is in an automaton position, the existential player plays the only available move (indeed, the move exists by property $\star$, and is unique by the determinism assumption). Since this is a safety game, the new strategy is winning. Hence the existential player wins $\mathcal{G} \times \mathcal{A}$, proving that $\mathcal{A}$ is good for $(\mathbb{W}, n)$-games. Item 2 is established.

Assume now (2), i.e., that $\mathcal{A}$ is some strongly good for $(\mathbb{W}, n)$-games automaton. Then by definition $\mathcal{L}(\mathcal{A}) \subseteq \mathbb{W}$. Now consider some word u in $\mathbb{W}|_n$. By definition, there exists some safety automaton $\mathcal{B}$ with at most n states such that $u \in \mathcal{L}(\mathcal{B}) \subseteq \mathbb{W}$. This automaton can be seen as a $\mathbb{W}$-game $\mathcal{G}$ in which all positions are owned by the universal player. Since $\mathcal{L}(\mathcal{B}) \subseteq \mathbb{W}$, the existential player wins the game $\mathcal{G}$. Since furthermore $\mathcal{A}$ is good for $(\mathbb{W}, n)$-games, the existential player has a winning strategy S_{E} in $\mathcal{G} \times \mathcal{A}$. Assume now that the universal player is playing the letters of u in the game $\mathcal{G} \times \mathcal{A}$, then the winning strategy S_{E} constructs an accepting run of $\mathcal{A}$ on u. Thus $u \in \mathcal{L}(\mathcal{A})$, and Item 3 is established. □

We continue our proof of Theorem 9 by introducing the notion of $(\mathbb{W}, n)$-universal graph.

Definition 10. *Given a winning condition $\mathbb{W} \subseteq C^\omega$ and a positive integer n, a C-graph U is $(\mathbb{W}, n)$-universal*[6] *if*

- *$\mathcal{L}(U) \subseteq \mathbb{W}$, and*
- *for all C-graphs H such that $\mathcal{L}(U) \subseteq \mathbb{W}$ and with at most n vertices, there is a weak graph morphism from H to U.*

We are now ready to prove one more implication of Theorem 9.

Lemma 8. *In the statement of Theorem 9, (4) $\Longrightarrow$ (1)*

Proof. Assume that there is a $(\mathbb{W}, n)$-universal graph U. We show that U seen as an safety automaton is strongly good for $(\mathbb{W}, n)$-games. One part is straightforward: $\mathcal{L}(U) \subseteq \mathbb{W}$ is by assumption. For the other part, consider a $\mathbb{W}$-game $\mathcal{G}$ with at most n positions. Assume that the existential player wins $\mathcal{G}$, this means that there exists a winning memoryless strategy for the existential player $S_{\mathrm{E}}, \gamma\colon S_{\mathrm{E}} \to \mathcal{G}$ in $\mathcal{G}$. We then construct a strategy for the existential player S'_{E} that maintains the property that the only game positions in $\mathcal{G} \times U$ that are met in S'_{E} are of the form $(x, \gamma(x))$. This is done as follows: when a game position is encountered, the existential player plays like the strategy S_{E}, and when an automaton position is encountered, the existential player plays in order to follow γ. This is possible since γ is a weak graph morphism. □

[6] Note that this is not the notion of (even weak) universality in categorical terms since U is not in general itself of size n.

3.3 Maximal graphs

In order to continue our proof of Theorem 9, more insight is needed: we have to understand what are the $\mathbb{W}$-maximal graphs. This is what we do now.

Definition 11. *A C-graph H is $\mathbb{W}$-maximal if $\mathcal{L}(H) \subseteq \mathbb{W}$ and if it is not possible to add a single edge to it without breaking this property, i.e., without producing an infinite path from the root vertex that does not belong to $\mathbb{W}$.*

Lemma 9. *For a winning condition $\mathbb{W} \subseteq C$ which is memoryless for the existential player, and H a $\mathbb{W}$-maximal graph, then the ε-edges in H form a transitive and total relation.*

Proof. Transitivity arises from the epsilon property of winning conditions (Definition 1): Consider three vertices x, y and z such that $\alpha = (x,\varepsilon,y)$ and $\beta = (y,\varepsilon,z)$ are edges of H. Let us add a new edge $\delta = (x,\varepsilon,y)$ yielding a new graph H'. Let us consider now any infinite path π in H' starting in the root (this path may contain finitely of infinitely many occurrences of δ, but not almost only δ's since $x \neq y$). Let π' be obtained from π by replacing each occurrence of δ by the sequence $\alpha\beta$. The resulting path π' belongs H, and thus its labelling belongs to $\mathbb{W}$. But since the labelings of π and π' agree after removing all the occurrences of ε, the epsilon property guarantees that the labelling of π belongs to $\mathbb{W}$. Since this holds for all choices of π, we obtain $\mathcal{L}(H') \subseteq \mathbb{W}$. Hence, by maximality, $\delta \in H$, which means that the ε-edges form a transitive relation.

Let us prove the totality. Let x and y be distinct vertices of H. We have to show that either $x \xrightarrow{\varepsilon} y$ or $y \xrightarrow{\varepsilon} x$. We can turn H into a game $\mathcal{G}$ as follows:

- all the vertices of H become positions that are owned by the universal player and we add a new position z owned by the existential player;
- all the edges of H that end in x or y become moves of $\mathcal{G}$ that now end in z,
- all the other edges of H become moves of $\mathcal{G}$ without change,
- and there are two new moves in $\mathcal{G}$, (z,ε,x) and (z,ε,y).

We claim first that the game $\mathcal{G}$ is won by the existential player. Let us construct a strategy s_{E} in $\mathcal{G}$ as follows. The only moment the existential player has a choice to make is when the play reaches the position z. This has to happen after a move of the form (t,a,z). This move originates either from an edge of the form (t,a,x), or from an edge of the form (t,a,y). In the first case the strategy s_{E} chooses the move (z,ε,x), and in the second case the move (z,ε,y). Let us consider a play π compatible with s_{E}, and let π' be obtained from π by replacing each occurrence of $(t,a,z)(z,\varepsilon,x)$ with (t,a,x) and each occurrence of $(t,a,z)(z,\varepsilon,y)$ with (t,a,y). The resulting π' is a path in H and hence its labeling belongs to $\mathbb{W}$. Since the labelings of π and π' are equivalent up to ε-letters, by the epsilon property, the labeling of π also belongs to $\mathbb{W}$. Hence the strategy s_{E} witnesses the victory of the existential player in $\mathcal{G}$. The claim is proved.

By assumption on $\mathbb{W}$, this means that there exists a winning memoryless strategy for the existential player $S_{\rm E}$ in $\mathcal{G}$. In this strategy, either the existential player always chooses (z, ε, x), or she always chooses (z, ε, y). Up to symmetry, we can assume the first case. Let now H' be the graph H to which a new edge $\delta = (y, \varepsilon, x)$ has been added. We aim that $\mathcal{L}(H') \subseteq \mathbb{W}$. Let π be an infinite path in H' starting from the root vertex. In this path, each occurrences of δ are preceded by an edge of the form (t, a, y). Thus, let π' be obtained from π by replacing each occurrence of a sequence of the form $(t, a, y)\delta$ by (t, a, y). The resulting path is a play compatible with $S_{\rm E}$. Hence the labeling of π' belongs to $\mathbb{W}$, and as a consequence, by the epsilon property, this is also the case for π. Since this holds for all choices of π, we obtain that $\mathcal{L}(H') \subseteq \mathbb{W}$. Hence, by $\mathbb{W}$-maximality assumption, (y, ε, x) is an edge of H.

Overall, the ε-edges form a total transitive relation. □

Let $\leqslant_\varepsilon$ be the least relation closed under reflexivity and that extends the ε-edge relation.

Lemma 10. *For a winning condition $\mathbb{W}$ which is memoryless for the existential player, and H a $\mathbb{W}$-maximal graph, then the following properties hold:*

- *The relation $\leqslant_\varepsilon$ is a total preorder.*
- *$x' \leqslant_\varepsilon x \xrightarrow{a}_H y \leqslant_\varepsilon y'$ implies $x' \xrightarrow{a}_H y'$, for all vertices x', x, y, y' and colors a.*
- *For all vertices p, q, $\mathcal{L}(H_p) \subseteq \mathcal{L}(H_q)$ if and only $q \leqslant_\varepsilon p$.*
- *for all vertices p, q and colors a, $a\mathcal{L}(H_q) \subseteq \mathcal{L}(H_p)$ if and only if $p \xrightarrow{a}_H q$.*

Proof. The first part is obvious from Lemma 9. For the second part, it is sufficient to prove that $x \xrightarrow{a} y \xrightarrow{\varepsilon} z$ implies $x \xrightarrow{a} y$ and that $x \xrightarrow{\varepsilon} y \xrightarrow{a} z$ implies $x \xrightarrow{a} y$. Both cases are are similar to the proof of transitivity in Lemma 9[7].

The two next items are almost the same. The difficult direction is to assume the language inclusion, and deduce the existence of an edge (left to right). Let us assume for an instant that H would be a finite word automaton, with all its states accepting. Then it is an obvious induction to show that if $a\mathcal{L}(H_q) \subseteq \mathcal{L}(H_p)$ (as languages of finite words), it is safe to add an ε-transitions from q to p without changing the language. The two above items are then obtained by limit passing (this is possible because the safety condition is topologically closed). □

We are now ready to provide the missing proofs for Theorem 9: from (3) to (4), and from (3) to (1). Both implications arise from Lemma 9.

Lemma 11. *In the statement of Theorem 9, (3) $\Longrightarrow$ (4).*

[7] This arises in fact from a more general simple phenomenon: if the sequence ab is 'indistinguishable in any context' from c (meaning that if one substitutes simultaneously infinitely many occurrences of ab with occurrences of c one does not change the membership to $\mathbb{W}$), then $x \xrightarrow{a} y \xrightarrow{b} z$ implies $x \xrightarrow{c} z$.

Proof. Let us start from a strongly $(\mathbb{W}, n)$-separating safety automaton $\mathcal{A}$. Without loss of generality, we can assume it is $\mathbb{W}$-maximal. We claim that it is $(\mathbb{W}, n)$-universal.

Let us define first for all languages $K \subseteq C^\omega$, its closure

$$\overline{K} = \bigcap_{\mathcal{L}(\mathcal{A}_s) \supseteq K} \mathcal{L}(\mathcal{A}_s)$$

(in case of an empty intersection, we assume C^ω). This is a closure operator: $K \subseteq K'$ implies $\overline{K} \subseteq \overline{K'}$, $K \subseteq \overline{K}$, and $\overline{\overline{K}} = \overline{K}$. Futhermore, $a\overline{K} \subseteq \overline{aK}$, for all letters $a \in C$. Let now H be a trimmed graph with at most n vertices such that $\mathcal{L}(H) \subseteq \mathbb{W}$. We have $\mathcal{L}(H) \subseteq \mathbb{W}|_n$ by definition of $\mathbb{W}|_n$.

We claim that for each vertex x of H, there is a state $\alpha(x)$ of $\mathcal{A}$ such that

$$\mathcal{L}(\mathcal{A}_{\alpha(x)}) = \overline{\mathcal{L}(H_x)}\ .$$

Indeed, note first that, since H is trimmed, there exists some word u such that $\text{root}_H \xrightarrow{u} x$. Hence, using the fact that $\mathcal{A}$ is strongly $(\mathbb{W}, n)$-separating, we get that for all $v \in \mathcal{L}(H_x)$, $uv \in \mathcal{L}(H) \subseteq \mathbb{W}|_n \subseteq \mathcal{L}(\mathcal{A})$. Let $\beta(v)$ be the state assumed after reading u by a run of $\mathcal{A}$ accepting uv. It is such that $v \in \mathcal{L}(\mathcal{A}_{\beta(v)})$. Since $\mathcal{A}$ is finite and its states are totally ordered under inclusion of residuals (Lemma 10), this means that there exists a state $\alpha(x)$ (namely the maximum over all the $\beta(w)$ for $w \in \mathcal{L}(H_x)$) such that $\mathcal{L}(\mathcal{A}_{\alpha(x)}) = \overline{\mathcal{L}(H_x)}$.

Let us show that α is a weak graph morphism[8] from H to $\mathcal{A}$. Consider some edge (x, a, y) of H. We have $a\mathcal{L}(H_y) \subseteq \mathcal{L}(H_x)$. Hence

$$a\mathcal{L}(\mathcal{A}_{\alpha(y)}) = a\overline{\mathcal{L}(H_y)} \subseteq \overline{a\mathcal{L}(H_y)} \subseteq \overline{\mathcal{L}(H_x)} = \mathcal{L}(\mathcal{A}_{\alpha(x)})\ ,$$

which implies by Lemma 10 that $\alpha(x) \xrightarrow{a}_{\mathcal{A}} \alpha(y)$. Let now $\text{root}_H \xrightarrow{a}_H x$ be some edge. By hypothesis, we have

$$a\mathcal{L}(H_x) \subseteq \mathcal{L}(H) \subseteq \mathbb{W}|_n \subseteq \mathcal{L}(\mathcal{A})\ .$$

Thus $\mathcal{L}(\mathcal{A}_{\alpha(x)}) = \overline{a\mathcal{L}(H_x)} \subseteq \overline{\mathcal{L}(\mathcal{A})} = \mathcal{L}(\mathcal{A}_{\text{root}_\mathcal{A}})$. We obtain $\text{root}_\mathcal{A} \xrightarrow{a}_\mathcal{A} \alpha(x)$ by Lemma 10. Hence, α is a weak graph morphism.

Since this holds for all choices of H, we have proved that $\mathcal{A}$ is a $(\mathbb{W}, n)$-universal graph. ☐

Lemma 12. *In the statement of Theorem 9, (3) $\Longrightarrow$ (1).*

Proof. Let us start from a strongly $(\mathbb{W}, n)$-separating safety automaton $\mathcal{A}$. Without loss of generality, we can assume it is maximal. Thus Lemma 10 holds.

[8] Note that in general that α is not a (non-weak) graph morphism, even for conditions like parity. Even more, such a graph morphism does not exist in general.

We now construct a deterministic safety automaton $\mathcal{D}$.

- the states of $\mathcal{D}$ are the same as the states of $\mathcal{A}$,
- the initial state of $\mathcal{D}$ is the initial state of $\mathcal{A}$,
- given a state $p \in \Delta_{\mathcal{D}}$ and a letter a, a transition of the form (p, a, q) exists if and only if there is some transition of the form (p, a, r) in $\mathcal{A}$, and q is chosen to be the least state r with this property.

We have to show that this deterministic safety automaton is strongly $(\mathbb{W}, n)$-separating. Note first that by definition $\mathcal{D}$ is obtained from $\mathcal{A}$ by removing transitions. Hence $\mathcal{L}(\mathcal{D}) \subseteq \mathcal{L}(\mathcal{A}) \subseteq \mathbb{W}$. Consider now some $u \in \mathbb{W}|_n$. By assumption, $u \in \mathcal{L}(\mathcal{A})$. Let $\rho = (p_0, u_1, p_1)(p_1, u_2, p_2)\cdots$ be the corresponding accepting run of $\mathcal{A}$. We construct by induction a (the) run of $\mathcal{D}$ $(q_0, u_1, q_1)(q_1, u_2, q_2)\cdots$ in such a way that $q_i \leqslant_\varepsilon p_i$. For the initial state, $p_0 = q_0$. Assume the run up to $q_i \leqslant_\varepsilon p_i$ has been constructed. By Lemma 10, (q_i, u_{i+1}, p_{i+1}) is a transition of $\mathcal{A}$. Hence the least r such that (q_i, u_{i+1}, r) is a transition of $\mathcal{A}$ does exist, and is $\leqslant_\varepsilon p_{i+1}$. Let us call it q_{i+1}; we indeed have that (q_i, u_{i+1}, q_{i+1}) is a transition of $\mathcal{D}$. Hence, u is accepted by $\mathcal{D}$. Thus $\mathbb{W}|_n \subseteq \mathcal{L}(\mathcal{D})$.

Overall $\mathcal{D}$ is a strongly $(\mathbb{W}, n)$-separating deterministic safety automaton that has at most as many states as $\mathcal{A}$. □

4 The case of parity conditions

We have seen above some general results on the notion of universal graphs, separating automata, and automata that are good for small games. In particular, we have seen Theorem 9 showing the equivalence of these objects for memoryless for the existential player winning conditions.

We are paying now a closer attention to the particular case of the parity condition. The technical developments that follow give an alternative proof of the equivalence results proved in [CDF+19] between strongly separating automata and universal trees.

4.1 Parity and cycles

We begin with a first classical lemma, which reduces the questions of satisfying a parity condition to checking the parity of cycles.

In a directed graph labelled by priorities, an even cycle is a cycle (all cycles are directed) such that the maximal priority occurring in it is even. Otherwise, it is an odd cycle. As usual, an elementary cycle is a cycle that does not meet twice the same vertex.

Lemma 13. *For a $[i, j]$-graph H that has all its vertices reachable from the root, the following properties are equivalent:*

- $\mathcal{L}(H) \subseteq \mathtt{Parity}_{[i,j]}$,
- *having all its cycles even,*
- *having all its elementary cycles even.*

Proof. Clearly, since all vertices are reachable, $\mathcal{L}(H) \subseteq \mathbb{W}$ implies that all the cycles are even. Also, if all cycles are even, then all elementary cycles also are. Finally assume that all the elementary cycles are even. Then we can consider H as a game, in which every positions is owned by the universal player. Assume that some infinite path from the root would not satisfy $\mathtt{Parity}_{[i,j]}$, then this path would be a winning strategy for the universal player in this game. Since $\mathtt{Parity}_{[i,j]}$ is a winning condition memoryless for the universal player, this means that the universal player has a winning memoryless strategy. But this winning memoryless strategy is nothing but a lasso, and thus contains an elementary cycle of maximal odd priority. □

4.2 The shape and size of universal graphs for parity games

We continue with a fixed d, and we consider parity conditions using priorities in $[0, 2d]$. More precisely, we relate the size of universal graphs for the parity condition with priorities $[0, 2d]$ to universal d-trees as defined now:

Definition 12. *A d-tree t is a balanced, unranked, ordered tree of height d (the root does not count: all branches contain exactly $d+1$ nodes). The order between nodes of same level is denoted $\leqslant_t$. Given a leaf x, and $i = 0 \dots i$, we denote $\mathtt{anc}^t_i(t)$ the ancestor at depth i of x (0 is the root, d is x).*

The d-tree t is n-universal if for all d-trees s with at most n nodes, there is a d-tree embedding of s into t, in which a d-tree embedding is an injective mapping from nodes of s to nodes of t that preserves the height of nodes, the ancestor relation, and the order of nodes. Said differently, s is obtained from t by pruning some subtrees (while keeping the structure of a d-tree).

Definition 13. *Given a d-tree t, $\mathtt{Graph}(t)$ is a $[0, 2d]$-graph with the following characteristics:*

- *the vertices are the leaves of t,*
- *for $0 \leqslant i \leqslant d$, $x \overset{2(d-i)}{\longrightarrow}_{\mathtt{Graph}(t)} y$ if $\mathtt{anc}^t_i(x) \leqslant_t \mathtt{anc}^t_i(y)$,*
- *for $0 < i \leqslant d$, $x \overset{2(d-i)+1}{\longrightarrow}_{\mathtt{Graph}(t)} y$ if $\mathtt{anc}^t_i(x) < \mathtt{anc}^t_i(y)$.*

Lemma 14. *For all d-trees t, $\mathcal{L}(\mathtt{Graph}(t)) \subseteq \mathtt{Parity}_{[0,2d]}$.*

Proof. Using Lemma 13, it is sufficient to prove that all cycle in $\mathtt{Graph}(t)$ are even. Thus, let us consider a cycle ρ. Assume that the highest priority occurring in α is $2(d-i)+1$. Note then that for all edges $\alpha = (x, k, y)$ occurring in ρ:

- $\mathtt{anc}^t_i(x) \leqslant_t \mathtt{anc}^t_i(y)$ since $k \leqslant i+1$,
- if $k = 2(d-i)+1$, $\mathtt{anc}^t_i(x) < \mathtt{anc}^t_i(y)$.

As a consequence, the first and last vertex of α cannot have the same ancestor at level i, and thus are different. □

Below, we develop sufficient results for establishing:

Theorem 10 ([CF18]). *For all positive integers d, n, the two following quantities are equal:*

- *the smallest number of leaves of an n-universal d-tree, and*
- *the smallest number of vertices of a $(\texttt{Parity}_{[0,2d]}, n)$-universal graph.*

Proof. We shall see below (Definition 14) a construction $\texttt{Tree}$ that maps all $\texttt{Parity}_{[0,2d]}$-maximal graphs G to a d-tree $\texttt{Tree}(G)$ of smaller or same size. Corollary 4 establishes that this construction is in some sense the converse of $\texttt{Tree}$ (in fact they form an adjunction). and that this correspondence preserves the notions of universality. This proves the above result: Given a n-universal d-tree t, then, by Corollary 4, $\texttt{Graph}(t)$ is a $(\texttt{Parity}_{[0,2d]}, n)$-universal graph that has as many vertices as leaves of graphs. Conversely, consider a $(\texttt{Parity}_{[0,2d]}, n)$-universal graph G. One can add to it edges until it becomes a $\texttt{Parity}_{[0,2d]}$-maximal graph G' with as many vertices. Then, by Corollary 4, $\texttt{Tree}(G')$ is an n-universal d-tree that has as much or less leaves than vertices of G'. □

Example 3. The complete d-tree t of degree n (that has n^d leaves) is n-universal. The $[0, 2d]$-graph $\texttt{Graph}(t)$ obtained in this way is used in the small progress measure algorithm [Jur00].

However, there exists n-universal d-trees that are much smaller than in the above example. The next theorem provides an upper and a lower bound.

Theorem 11 ([Fij18, CDF+19]). *Given positive integers n, d,*

- *there exists an n-universal d-tree with*

$$n \cdot \binom{\lceil \log(n) \rceil + d - 1}{\lceil \log(n) \rceil}$$

  leaves.
- *all n-universal d-trees have at least*

$$\binom{\lfloor \log(n) \rfloor + d - 1}{\lfloor \log(n) \rfloor}$$

  leaves.

Corollary 2. *The complexity of solving $\texttt{Parity}_{[0,d]}$-games with at most n-vertices is*

$$O\left(mn \log(n) \log(d) \cdot \binom{\lceil \log(n) \rceil + d/2 - 1}{\lceil \log(n) \rceil}\right).$$

and no algorithm based on good for small safety games can be faster than quasipolynomial time.

Maximal universal graphs for the parity condition We shall now analyse in detail the shape of $\texttt{Parity}_{[0,2d]}$-maximal graphs. This analysis culminates with the precise description of such graphs in Lemma 19, that essentially establishes a bijection with graphs of the form $\texttt{Graph}(t)$ (Corollary 4).

Let us note that, since the parity condition is memoryless for the existential player, using Lemma 10, and the fact that the parity condition is unchanged by modifying finite prefixes, we can always assume that the root vertex is the minimal one for the $\leqslant_\varepsilon$ ordering. Thus, from now, we do not have to pay attention to the root, in particular in weak graph morphisms. Thus, from now, we just mention the term morphism for weak graph morphisms.

Let us recall preference ordering $\sqsubseteq$ between the non-negative integers is defined as follows:

$$\cdots \sqsubset 2d+1 \sqsubset 2d-1 \sqsubset \cdots \sqsubset 3 \sqsubset 1 \sqsubset 0 \sqsubset 2 \sqsubset \cdots \sqsubset 2d-2 \sqsubset 2d \sqsubset \cdots$$

Fact 1. *Let $k \sqsubseteq \ell$ and u, v sequences of priorities. If the maximal priority occurring in ukv is even, then the maximal priority occurring in $u\ell v$ is also even.*

Lemma 15. *Let G be a $\texttt{Parity}_{[0,2d]}$-maximal graph and $k \sqsubseteq \ell$ be priorities in $[0, 2d]$. For all vertices x, y of G, $x \xrightarrow{k}_G y$ implies $x \xrightarrow{\ell}_G y$.*

Proof. Let us add (x, ℓ, y) to G. Let $u(x, \ell, y)v$ be some elementary cycle of the new graph involving the new edge (x, ℓ, y). By Lemma 13, $u(x, k, y)v$ is an even cycle in the original graph. Hence, by Fact 1, $u(x, \ell, y)v$ is also an even cycle. Thus, by Lemma 13, G with the newly added edge also satisfies $\mathcal{L}(G) \subseteq \texttt{Parity}_{[0,2d]}$. Using the maximality assumption for G, we obtain that (x, ℓ, y) was already present in G. □

Lemma 16. *Let G be a $\texttt{Parity}_{[0,2d]}$-maximal graph. For all vertices x, y, z of G, if $x \xrightarrow{k}_G y$ and $y \xrightarrow{\ell}_G z$, then $y \xrightarrow{\max(k,\ell)}_G z$.*

Proof. Let us add $(x, \max(k, \ell), z)$ to G. Let $u(x, \max(k, \ell), z)v$ be an elementary cycle in the new graph. By Lemma 13, $u(x, k, y)(y, \ell, z)v$, being a cycle of G, has to be even. Since, furthermore, the maximal priority that occurs in $u(x, k, y)(y, \ell, z)v$ is the same as the maximal one in $u(x, \max(k, \ell), z)v$, the cycle $u(x, \max(k, \ell), z)v$ is also even. Using the maximality assumption of G, we obtain that $(x, \max(k, \ell), z)$ was already present in G. □

Lemma 17. *Let G be a $\texttt{Parity}_{[0,2d]}$-maximal graph, and x, y be vertices, then $x \xrightarrow{0}_G x$, and $x \xrightarrow{2d}_G y$.*

Proof. For $x \xrightarrow{0}_G x$, it is sufficient to notice that adding the edge $(x, 0, x)$, if it was not present, simply creates one new elementary cycle to G, namely $(x, 0, x)$. Since it is an even cycle, by Lemma 13, the new graph also satisfies $\mathcal{L}(G) \subseteq \texttt{Parity}_{[0,2d]}$. Hence, by maximality assumption, the edge was already present in G before.

Consider the graph G with an extra edge $(x, 2d, y)$ added. Consider now an elementary cycle that contains $(x, 2d, y)$, i.e., of the form $u(x, 2d, y)v$. Its maximal priority is $2d$, and thus even. Hence by Lemma 13 and maximality assumption, the edge was already present in G. □

Lemma 18. *Let G be a $\mathtt{Parity}_{[0,2d]}$-maximal graph and $k = 0, 2, \ldots, 2d-2$. For all vertices x, y, $x \xrightarrow{k+1}_G y$ holds if and only if $y \xrightarrow{k}_G x$ does not hold.*

Proof. Assume first that $y \xrightarrow{k+1}_G x$ and $x \xrightarrow{k}_G y$ both holds. Then $y \xrightarrow{k+1}_G x \xrightarrow{k}_G y$ is an odd cycle contradicting Lemma 13.

Conversely, assume that adding the edge $x \xrightarrow{k+1}_G y$ would break the property $\mathcal{L}(G) \subseteq \mathtt{Parity}_{[0,2d]}$. This means that there is an elementary cycle of the form $u(x, k+1, y)v$ which is odd. Let ℓ be the maximal priority in vu. If $\ell \geqslant k+1$, then ℓ is odd, and thus $\ell \sqsubseteq k$, and we obtain $y \xrightarrow{k}_G x$ by Lemma 15. Otherwise, $\ell \leqslant k$, and again $\ell \sqsubseteq k$. Once more $y \xrightarrow{k}_G x$ holds by Lemma 15. □

Lemma 19. *A $[0, 2d]$-graph G is a $\mathtt{Parity}_{[0,2d]}$-maximal graph if and only if all the following properties hold:*

1. *$\xrightarrow{k}_G$ is a total preorder for all $k = 0, 2, \ldots, 2d$,*
2. *$\xrightarrow{k}_G \subseteq \xrightarrow{k+2}_G$ for all $k = 0, 2, \ldots, 2d-2$,*
3. *$\xrightarrow{2d}_G$ is the total equivalence relation,*
4. *$\xrightarrow{k+1}_G = (\xleftarrow{k}_G)^{\complement}$ for all $k = 0, 2, \ldots, 2d-2$.*[9]

Proof. First direction. Assume first that G is a $\mathtt{Parity}_{[0,2d]}$-maximal graph.

(1) Let $k = 0, 2, \ldots, 2d$; $\xrightarrow{k}_G$ is transitive by Lemma 16. Furthermore, by Lemma 17, $x \xrightarrow{0}_G x$ for all vertices x, and thus by Lemma 15, since $0 \sqsubseteq k$, $x \xrightarrow{k}_G x$. Hence $\xrightarrow{k}_G$ is also reflexive and hence a preorder. Consider now another vertex y. By Lemma 18, either $x \xrightarrow{k}_G y$ or $y \xrightarrow{k+1}_G x$. But by Lemma 15, $y \xrightarrow{k+1}_G x$ implies $y \xrightarrow{k}_G x$. Hence either $x \xrightarrow{k}_G y$ or $y \xrightarrow{k}_G k$. Thus $\xrightarrow{k}_G$ is a total preorder.

(2) For $k = 0, 2, \ldots, 2d-2$, since $k \sqsubseteq k+2$, by Lemma 15, $\xrightarrow{k}_G \subseteq \xrightarrow{k+2}_G$.

(3) $\xrightarrow{2d}_G$ is the maximal relation by Lemma 15.

(4) For $k = 0, 2, \ldots, 2d-2$ and x, y, we know that $y \xrightarrow{k}_G x$ holds if and only if $x \xrightarrow{k+1}_G y$ does not. This shows $\xrightarrow{k+1}_G = (\xleftarrow{k}_G)^{\complement}$.

Second direction. Assume now that G satisfies the conditions (1)-(4). Let us first show that $\mathcal{L}(G) \subseteq \mathtt{Parity}_{[0,2d]}$. For the sake of contradiction, consider an elementary cycle that would be odd. It can be written as $u(x, k, y)v$ with a

[9] Note that this also means, since $\xrightarrow{k}_G$ is a total preorder, that $\xrightarrow{k+1}_G = \xrightarrow{k}_G \setminus \xleftarrow{k}_G$.

maximal odd priority k. Note first that $\xrightarrow{\ell}\subseteq\xrightarrow{k-1}$ for all $\ell \leqslant k$: indeed, by (2), this is true if ℓ is even, and by (1) and (4), $\xrightarrow{j}\subseteq\xrightarrow{j-1}$ for all j odd. Also $\xrightarrow{k}_G$ is the strict version of the preorder $\xrightarrow{k-1}_G$. Hence, the path $u(x,k,y)v$ has to strictly advance with respect to the preorder $\xrightarrow{k-1}_G$: it cannot be a cycle.

Assume now that an edge (x,k,y) is not present in G. If k is even, since (x,k,y) is not present, by (4) this means that $(y,k+1,x)$ is present. Hence, adding the edge (x,k,y) would create the odd cycle $(x,k,y)(y,k+1,x)$. If k is odd, since (x,k,y) is not present, by (4) this means that $(y,k-1,x)$ is present. Hence, adding the edge (x,k,y) would create the odd cycle $(x,k,y)(y,k-1,x)$. Hence G is $\texttt{Parity}_{[0,2d]}$-maximal. □

Corollary 3. *Given a morphism α from a $\texttt{Parity}_{[0,2d]}$-maximal graph H to a $\texttt{Parity}_{[0,2d]}$-maximal graph G, then $x \xrightarrow{k}_H y$ if and only if $\alpha(x) \xrightarrow{k}_G \alpha(y)$, for all vertices x, y of H and integers k in $[0, 2d]$. Furthermore, if α is surjective, then every map β from G to H, such that $\alpha \circ \beta$ is the identity on G is an injective morphism.*

Proof. First part. From left to right, this is the definition of a morphism. The other direction is by (4) of Lemma 19: if $\alpha(x) \xrightarrow{k}_G \alpha(y)$ and k is odd, then $\alpha(x) \xrightarrow{k-1}_G \alpha(y)$ does not hold by (4), thus $x \xrightarrow{k-1}_H y$ does not hold by morphism, thus $x \xrightarrow{k}_H y$ holds by (4) again. The case of k even is similar (using $k+1$ this time).
For the second part, since $\alpha \circ \beta$ is the identity, β has to be injective. It is a morphism by the first part. □

The next definition, allowing to go from graphs to trees is shown meaningful by Lemma 19:

Definition 14. *Let G be a $\texttt{Parity}_{[0,2d]}$-maximal graph. The d-tree $\texttt{Tree}(G)$ is constructed as follows:*

- *the nodes of level $i = 0, \ldots, d$ are the pairs (i, C) for C ranging over the equivalence classes of $\xrightarrow{2(d-i)}_G \cap \xleftarrow{2(d-i)}_G$,*
- *a node (i, C) is an ancestor of (j, D) if $i \leqslant j$ and $D \subseteq C$,*
- *$(i, C) \leqslant_{\texttt{Tree}(G)} (i, D)$ if $x \xrightarrow{2(d-i)}_G x'$ for all $x \in C$ and $x' \in C'$.*

We shall see that $\texttt{Graph}$ and $\texttt{Tree}$ are almost the inverse one of the other. This is already transparent in the following lemma, which is just a reformulation of the definitions.

Lemma 20. *Let q be the quotient map from vertices of G to leaves of $\texttt{Tree}(G)$ that maps each vertex to its $(\xrightarrow{0}_G \cap \xleftarrow{0}_G)$-equivalence class. It has the following property for all vertices x, y of G:*

$$x \xrightarrow{2(d-i)}_G y \quad \textit{if and only if} \quad \mathtt{anc}_i^{\mathtt{Tree}(G)}(q(x)) \leqslant_{\mathtt{Tree}(G)} \mathtt{anc}_i^{\mathtt{Tree}(G)}(q(y)) \ ,$$

$$\textit{and} \quad x \xrightarrow{2(d-i)+1}_G y \quad \textit{if and only if} \quad \mathtt{anc}_i^{\mathtt{Tree}(G)}(q(x)) < [\mathtt{Tree}(G)]\mathtt{anc}_i^{\mathtt{Tree}(G)}(q(y)) \ .$$

The identity maps the vertices of $\mathtt{Graph}(t)$ *to the leaves of* t*, and has the property that for all vertices* x, y*:*

$$x \xrightarrow{2(d-i)}_{\mathtt{Graph}(t)} y \quad \textit{if and only if} \quad \mathtt{anc}_i^t(x) \leqslant_t \mathtt{anc}_i^t(y) \ ,$$

$$\textit{and} \quad x \xrightarrow{2(d-i)+1}_{\mathtt{Graph}(t)} y \quad \textit{if and only if} \quad \mathtt{anc}_i^t(x) < \mathtt{anc}_i^t(y) \ .$$

Corollary 4. [10]*For all* $\mathtt{Parity}_{[0,2d]}$*-maximal graphs* G, H*, all* d*-trees* t*, and all positive integers* n*,*

- $\mathtt{Graph}(\mathtt{Tree}(G))$ *is a quotient and an induced subgraph of* G*,*
- $\mathtt{Tree}(\mathtt{Graph}(t))$ *is isomorphic to* t*,*
- *there is a morphism from* H *to* $\mathtt{Graph}(t)$ *if and only if there is a tree embedding from* $\mathtt{Tree}(H)$ *to* t*,*
- $\mathtt{Tree}(G)$ *is* n*-universal if and only if* G *is* $(\mathtt{Parity}_{[0,2d]}, n)$*-universal,*
- $\mathtt{Graph}(t)$ *is* $(\mathtt{Parity}_{[0,2d]}, n)$*-universal if and only if* t *is* n*-universal.*

Proof. Let q be the quotient from Lemma 20. It can be seen as a surjective map from vertices of $\mathtt{Graph}(\mathtt{Tree}(G))$ to G. By Lemma 20 it is a morphism. By Corollary 3, $\mathtt{Graph}(\mathtt{Tree}(G))$ is also an induced subgraph of G.

The leaves of $\mathtt{Tree}(\mathtt{Graph}(t))$ are the singletons consisting of leaves of t. Hence, there is a bijective map from leaves of $\mathtt{Tree}(\mathtt{Graph}(t))$ to leaves of t that sends$\{\ell\}$ to ℓ. By Lemma 20, this is a morphism, and by Corollary 3 an isomorphism.

For the third item, assume first that there is a morphism from H to $\mathtt{Graph}(t)$. By the first point, there is an injective morphism from $\mathtt{Graph}(\mathtt{Tree}(H))$ to H. By composition, we obtain a morphism from $\mathtt{Graph}(\mathtt{Tree}(H))$ to $\mathtt{Graph}(t)$. By Lemma 20, it is also a tree embedding from $\mathtt{Tree}(H)$ to t. Conversely, assume that there exists an embedding from $\mathtt{Tree}(H)$ to t. It can be raised by Lemma 20 to a morphism from $\mathtt{Graph}(\mathtt{Tree}(H))$ to $\mathtt{Graph}(t)$. By the first point, there is a morphism from H to $\mathtt{Graph}(\mathtt{Tree}(H))$. By composition, we get a morphism from H to $\mathtt{Graph}(t)$.

The two last items are obvious from the one just before. □

Acknowledgements. We thank Pierre Ohlmann for many interesting discussions, and Marcin Jurdziński for his comments on an earlier draft of this paper.

[10] The careful reader will recognize $\mathtt{Tree}$ and $\mathtt{Graph}$ as left and right adjoints.

References

[BC18] Bojańczyk, M., Czerwiński, W.: An automata toolbox, February 2018. https://www.mimuw.edu.pl/~bojan/papers/toolbox-reduced-feb6.pdf

[BKS17] Boker, U., Kupferman, O., Skrzypczak, M.: How deterministic are good-for-games automata? In: FSTTCS, pp. 18:1–18:14 (2017)

[BL69] Büchi, J.R., Landweber, L.H.: Definability in the monadic second-order theory of successor. J. Symbolic Logic **34**(2), 166–170 (1969)

[BL19] Boker, U., Lehtinen, K.: Register games. Logical Methods Comput. Sci. (Submitted, 2019)

[CDF+19] Czerwiński, W., Daviaud, L., Fijalkow, N., Jurdziński, M., Lazić, R., Parys, P.: Universal trees grow inside separating automata: quasi-polynomial lower bounds for parity games. In: SODA, pp. 2333–2349 (2019)

[CF18] Colcombet, T., Fijalkow, N.: Parity games and universal graphs. CoRR, abs/1810.05106 (2018)

[CFH14] Colcombet, T., Fijalkow, N., Horn, F.: Playing safe. In: FSTTCS, pp. 379–390 (2014)

[CJK+17] Calude, C.S., Jain, S., Khoussainov, B., Li, W., Stephan, F.: Deciding parity games in quasipolynomial time. In: STOC, pp. 252–263 (2017)

[EJ91] Emerson, E.A., Jutla, C.S.: Tree automata, mu-calculus and determinacy (extended abstract). In: FOCS, pp. 368–377 (1991)

[EM79] Ehrenfeucht, A., Mycielski, J.: Positional strategies for mean payoff games. Int. J. Game Theory **109**(8), 109–113 (1979)

[FGO18] Fijalkow, N., Gawrychowski, P., Ohlmann, P.: The complexity of mean payoff games using universal graphs. CoRR, abs/1812.07072 (2018)

[Fij18] Fijalkow, N.: An optimal value iteration algorithm for parity games. CoRR, abs/1801.09618 (2018)

[GH82] Gurevich, Y., Harrington, L.: Trees, automata, and games. In: STOC, pp. 60–65 (1982)

[GKK88] Gurvich, V.A., Karzanov, A.V., Khachiyan, L.G.: Cyclic games and an algorithm to find minimax cycle means in directed graphs. USSR Comput. Math. Math. Phys. **28**, 85–91 (1988)

[HP06] Henzinger, T.A., Piterman, N.: Solving games without determinization. In: Ésik, Z. (ed.) CSL 2006. LNCS, vol. 4207, pp. 395–410. Springer, Heidelberg (2006). https://doi.org/10.1007/11874683_26

[JL17] Jurdziński, M., Lazić, R.: Succinct progress measures for solving parity games. In: LICS, pp. 1–9 (2017)

[Jur00] Jurdziński, M.: Small progress measures for solving parity games. In: Reichel, H., Tison, S. (eds.) STACS 2000. LNCS, vol. 1770, pp. 290–301. Springer, Heidelberg (2000). https://doi.org/10.1007/3-540-46541-3_24

[Leh18] Lehtinen, K.: A modal-μ perspective on solving parity games in quasi-polynomial time. In: LICS, pp. 639–648 (2018)

[Mar75] Martin, D.A.: Borel determinacy. Ann. Math. **102**(2), 363–371 (1975)

[McN93] McNaughton, R.: Infinite games played on finite graphs. Ann. Pure Appl. Logic **65**(2), 149–184 (1993)

Resource-Tracking Concurrent Games

Aurore Alcolei(✉), Pierre Clairambault, and Olivier Laurent

Université de Lyon, ENS de Lyon, CNRS, UCB Lyon 1, LIP, Lyon, France
{Aurore.Alcolei,Pierre.Clairambault,Olivier.Laurent}@ens-lyon.fr

Abstract. We present a framework for game semantics based on concurrent games, that keeps track of *resources* as data modified throughout execution but not affecting its control flow. Our leading example is *time*, yet the construction is in fact parametrized by a *resource bimonoid* $\mathcal{R}$, an algebraic structure expressing resources and the effect of their consumption either sequentially or in parallel. Relying on our construction, we give a sound resource-sensitive denotation to $\mathcal{R}$-IPA, an affine higher-order concurrent programming language with shared state and a primitive for resource consumption in $\mathcal{R}$. Compared with general operational semantics parametrized by $\mathcal{R}$, our resource analysis turns out to be finer, leading to non-adequacy. Yet, our model is not degenerate as adequacy holds for an operational semantics specialized to time.

In regard to earlier semantic frameworks for tracking resources, the main novelty of our work is that it is based on a non-interleaving semantics, and as such accounts for *parallel* use of resources accurately.

1 Introduction

Since its inception, *denotational semantics* has grown into a very wide subject. Its developments now cover numerous programming languages or paradigms, using approaches that range from the extensionality of *domain semantics* [24] (recording the input-output behaviour) to the intensionality of *game semantics* [1,17] (recording execution traces, formalized as *plays* in a 2-players game between the program ("Player") and its execution environment ("Opponent")). Denotational semantics has had significant influence on the theory of programming languages, with contributions ranging from program logics or reasoning principles to new language constructs and verification algorithms.

Most denotational models are *qualitative* in nature, meaning that they ignore efficiency of programs in terms of time, or other resources such as power or bandwith. To our knowledge, the first denotational model to cover time was Ghica's *slot games* [13], an extension of Ghica and Murawski's fully abstract model for a higher-order language with concurrency and shared state [14]. Slot games exploit the intensionality of game semantics and represent time via special

Supported by project Elica (ANR-14-CE25-0005) and Labex MiLyon (ANR-10-LABX-0070) of Université de Lyon, within the program "Investissements d'Avenir" (ANR-11-IDEX-0007), operated by the French National Research Agency (ANR).

M. Bojańczyk and A. Simpson (Eds.): FOSSACS 2019, LNCS 11425, pp. 27–44, 2019.
https://doi.org/10.1007/978-3-030-17127-8_2

moves called *tokens* matching the *ticks* of a clock. They are fully abstract *w.r.t.* the notion of observation in Sands' operational theory of *improvement* [26].

More recently, there has been a growing interest in capturing quantitative aspects denotationally. Laird *et al.* constructed [18] an enrichment of the relational model of Linear Logic [11], using weights from a *resource semiring* given as parameter. This way, they capture in a single framework several notions of resources for extensions of PCF, ranging from time to probabilistic weights. Two type systems with similar parametrizations were introduced simultaneously by, on the one hand, Ghica and Smith [15] and, on the other hand, Brunel, Gaboardi *et al.* [4]; the latter with a quantitative realizability denotational model.

In this paper, we give a resource-sensitive denotational model for $\mathcal{R}$-*IPA*, an affine higher-order programming language with concurrency, shared state, and with a primitive for resource consumption. With respect to slot games our model differs in that our resource analysis accounts for the fact that resource consumption may combine differently in parallel and sequentially – simply put, we mean to express that $\mathbf{wait}(1) \parallel \mathbf{wait}(1)$ may terminate in 1 s, rather than 2. We also take inspiration from weighted relational models [18] in that our construction is parametrized by an algebraic structure representing resources and their usage. Our *resource bimonoids* $\langle \mathcal{R}, 0, ;, \parallel, \leq \rangle$ differ however significantly from their resource semiring $\langle \mathcal{R}, 0, 1, +, \cdot \rangle$: while ; matches $\cdot$, $\parallel$ is a new operation expressing the consumption of resources in parallel. We have no counterpart for the $+$, which agglomerates distinct non-deterministically co-existing executions leading to the same value: instead our model keeps them separate.

Capturing parallel resource usage is technically challenging, as it can only be attempted relying on a representation of execution where parallelism is explicit. Accordingly, our model belongs to the family of *concurrent* or *asynchronous* game semantics pioneered by Abramsky and Melliès [2], pushed by Melliès [20] and later with Mimram [22], and by Faggian and Piccolo [12]; actively developed in the past 10 years prompted by the introduction of a more general framework by Rideau and Winskel [7,25]. In particular, our model is a refinement of the (qualitative) truly concurrent interpretation of *affine IPA* described in [5]. Our methodology to record resource usage is inspired by game semantics for first-order logic [3,19] where moves carry first-order terms from a signature – instead here they carry explicit *functions*, *i.e.* terms up to a congruence (it is also reminiscent of Melliès' construction of the free dialogue category over a category [21]).

As in [5] we chose to interpret an affine language: this lets us focus on the key phenomena which are already at play, avoiding the technical hindrance caused by replication. As suggested by recent experience with concurrent games [6,10], we expect the developments presented here to extend transparently in the presence of *symmetry* [8,9]; this would allow us to move to the general (non-affine) setting.

Outline. We start Sect. 2 by introducing the language $\mathcal{R}$-IPA. We equip it first with an interleaving semantics and sketch its interpretation in slot games. We then present resource bimonoids, give a new parallel operational semantics, and hint at our truly concurrent games model. In Sect. 3, we construct this model and prove its soundness. Finally in Sect. 4, we show adequacy for an operational semantics specialized to time, noting first that the general parallel operational semantics is too coarse *w.r.t.* our model.

2 From $\mathcal{R}$-IPA to $\mathcal{R}$-Strategies

2.1 Affine IPA

Terms and Types. We start by introducing the basic language under study, *affine Idealized Parallel Algol* (IPA). It is an affine variant of the language studied in [14], a call-by-name concurrent higher-order language with shared state. Its **types** are given by the following grammar:

$$A, B ::= \mathbf{com} \mid \mathbf{bool} \mid \mathbf{mem}_W \mid \mathbf{mem}_R \mid A \multimap B$$

Here, $\mathbf{mem}_W$ is the type of *writeable* references and $\mathbf{mem}_R$ is the type of *readable* references; the distinction is necessary in this affine setting as it allows to share accesses to a given state over subprocesses; this should make more sense in the next paragraph with the typing rules. In the sequel, non-functional types are called **ground types** (for which we use notation $\mathbb{X}$). We define terms directly along with their typing rules in Fig. 1. **Contexts** are simply lists $x_1 : A_1, \ldots, x_n : A_n$ of variable declarations (in which each variable occurs at most once), and the exchange rule is kept implicit. Weakening is not a rule but is admissible. We comment on a few aspects of these rules.

$$\frac{}{\Gamma \vdash \mathbf{skip} : \mathbf{com}} \qquad \frac{}{\Gamma \vdash \mathbf{tt} : \mathbf{bool}} \qquad \frac{}{\Gamma \vdash \mathbf{ff} : \mathbf{bool}} \qquad \frac{}{\Gamma \vdash \bot : \mathbb{X}} \qquad \frac{(x : A) \in \Gamma}{\Gamma \vdash x : A}$$

$$\frac{\Gamma, x : A \vdash M : B}{\Gamma \vdash \lambda x.\, M : A \multimap B} \qquad \frac{\Gamma \vdash M : A \multimap B \quad \Delta \vdash N : A}{\Gamma, \Delta \vdash M\, N : B} \qquad \frac{\Gamma \vdash M : \mathbf{mem}_R}{\Gamma \vdash !M : \mathbf{bool}}$$

$$\frac{\Gamma \vdash M : \mathbf{com} \quad \Delta \vdash N : \mathbb{X}}{\Gamma, \Delta \vdash M; N : \mathbb{X}} \qquad \frac{\Gamma \vdash M : \mathbf{com} \quad \Delta \vdash N : \mathbb{X}}{\Gamma, \Delta \vdash M \parallel N : \mathbb{X}} \qquad \frac{\Gamma \vdash M : \mathbf{mem}_W}{\Gamma \vdash M := \mathbf{tt} : \mathbf{com}}$$

$$\frac{\Gamma \vdash M : \mathbf{bool} \quad \Delta \vdash N_1 : \mathbb{X} \quad \Delta \vdash N_2 : \mathbb{X}}{\Gamma, \Delta \vdash \mathbf{if}\, M\, N_1\, N_2 : \mathbb{X}} \qquad \frac{\Gamma, x : \mathbf{mem}_W, y : \mathbf{mem}_R \vdash M : \mathbb{X}}{\Gamma \vdash \mathbf{new}\, x, y\, \mathbf{in}\, M : \mathbb{X}}$$

Fig. 1. Typing rules for affine IPA

Firstly, observe that the reference constructor $\mathbf{new}\, x, y\, \mathbf{in}\, M$ binds two variables x and y, one with a write permission and the other with a read permission. In this way, the permissions of a shared state can be distributed in different components of *e.g.* an application or a parallel composition, causing interferences despite the affine aspect of the language. Secondly, the assignment command, $M := \mathbf{tt}$, seems quite restrictive. Yet, the language is affine, so a variable can only be written to once, and, as we choose to initialize it to **ff**, the only useful thing to write is **tt**. Finally, many rules seem restrictive in that they apply only at ground type $\mathbb{X}$. More general rules can be defined as syntactic sugar; for instance we give (all other constructs extend similarly): $M;_{A \multimap B} N = \lambda x^A.\,(M;_B (N\,x))$.

Operational Semantics. We fix a countable set L of **memory locations**. Each location ℓ comes with two associated variable names ℓ_W and ℓ_R distinct from other variable names. Usually, stores are partial maps from L to $\{\mathbf{tt}, \mathbf{ff}\}$. Instead, we find it more convenient to introduce the notion of **state** of a memory location. A state corresponds to a history of memory actions (reads or writes) and follows the *state diagram* of Fig. 2 (ignoring for now the annotations with α, β). We write $(\mathsf{M}, \leq_{\mathsf{M}})$ for the induced set of states and accessibility relation on it. For each $m \in \mathsf{M}$, its set of **available actions** is $\mathrm{act}(m) = \{W, R\} \setminus m$ (the letters not occurring in m, annotations being ignored); and its **value** (in $\{\mathbf{tt}, \mathbf{ff}\}$) is $\mathrm{val}(m) = \mathbf{tt}$ iff W occurs in m.

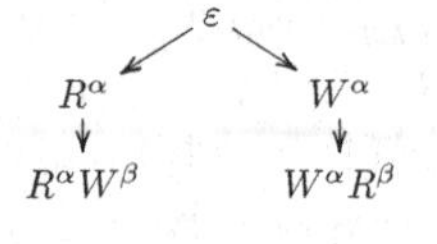

Fig. 2. State diagram

Finally, a **store** is a partial map $s : \mathsf{L} \to \mathsf{M}$ with finite domain, mapping each memory location to its current state. To each store corresponds a *typing context*

$$\Omega(s) = \{\ell_X : \mathbf{mem}_X \mid \ell \in \mathrm{dom}(s) \ \&\ X \in \mathrm{act}(s(\ell))\}.$$

The operational semantics operates on **configurations** defined as pairs $\langle M, s\rangle$ with s a store and $\Gamma \vdash M : A$ a term whose free variables are all memory locations with $\Gamma \subseteq \Omega(s)$. This property will be preserved by our rather standard small-step, call-by-name operational semantics. We refrain for now from giving the details, they will appear in Sect. 2.2 in the presence of resources.

2.2 Interleaving Cost Semantics, and $\mathcal{R}$-IPA

Ghica and Murawski [14] have constructed a *fully abstract*(for may-equivalence) model for (non-affine) IPA, relying on an extension of Hyland-Ong games [17].

Their model takes an *interleaving* view of the execution of concurrent programs: a program is represented by the set of all its possible executions, as decided non-deterministically by the scheduler. In game semantics, this is captured by lifting the standard requirement that the two players alternate. For instance, Fig. 3 shows a *play* in the interpretation of the program $x : \mathbf{com}, y : \mathbf{bool} \vdash x \parallel y : \mathbf{bool}$. The diagram is read from top to bottom, chronologically. Each line comprises one computational event ("move"), annotated with "−" if due to the execution environment ("Opponent") and with "+" if due to the program ("Player"); each move corresponds to a certain type component, under which it is placed. With the first move $\mathbf{q}^-$, the environment initiates the computation.

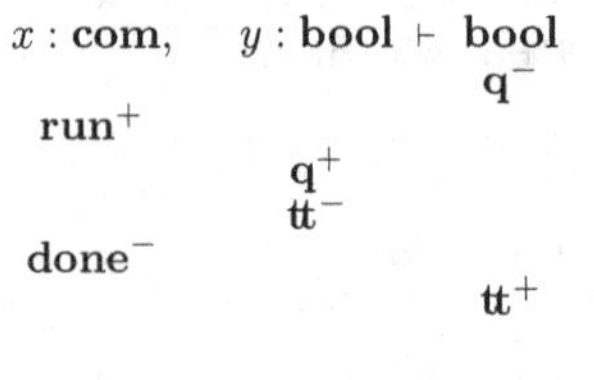

Fig. 3. A non-alternating play

Player then plays $\mathbf{run}^+$, triggering the evaluation of x. In standard game semantics, the control would then go back to the execution environment – Player would be stuck until Opponent plays. Here instead, due to parallelism Player can play a second move $\mathbf{q}^+$ immediately. At this point of execution, x and y are both running in parallel. Only when they have both returned (moves $\mathbf{done}^-$ and $\mathbf{tt}^-$) is Player able to respond $\mathbf{tt}^+$, terminating the computation. The full interpretation of $x : \mathbf{com}, y : \mathbf{bool} \vdash x \parallel y : \mathbf{bool}$, its *strategy*, comprises numerous plays like that, one for each interleaving.

As often in denotational semantics, Ghica and Murawski's model is invariant under reduction: if $\langle M, s\rangle \to \langle M', s'\rangle$, both have the same denotation. The model adequately describes the result of computation, but not its *cost* in terms, for instance, of time. Of course this cost is not yet specified: one must, for instance, define a *cost model* assigning a cost to all basic operations (*e.g.* memory operations, function calls, *etc*). In this paper we instead enrich the language with a primitive for *resource consumption* – cost models can then be captured by inserting this primitive concomitantly with the costly operations (see for example [18]).

$\mathcal{R}$-IPA. Consider a set $\mathcal{R}$ of **resources**. The language $\mathcal{R}$-IPA is obtained by adding to affine IPA a new construction, $\mathbf{consume}(\alpha)$, typed as in Fig. 4. When evaluated, $\mathbf{consume}(\alpha)$ triggers the consumption of resource $\mathcal{R}$. Time consumption will be a running example throughout the paper. In that case, we will consider the non-negative reals $\mathbb{R}_+$ as set $\mathcal{R}$, and for $t \in \mathbb{R}_+$ we will use $\mathbf{wait}(t)$ as a synonym for $\mathbf{consume}(t)$.

$$\frac{(\alpha \in \mathcal{R})}{\Gamma \vdash \mathbf{consume}(\alpha) : \mathbf{com}}$$

Fig. 4. Typing **consume**

$$\begin{array}{rlrl}
\langle \mathbf{skip};\, M, s, \alpha\rangle & \to \langle M, s, \alpha\rangle & \langle (\lambda x.\, M)\, N, s, \alpha\rangle & \to \langle M[N/x], s, \alpha\rangle \\
\langle \mathbf{skip} \parallel M, s, \alpha\rangle & \to \langle M, s, \alpha\rangle & \langle !\ell_R, s, \alpha\rangle & \to \langle \mathrm{val}(s(\ell)), s[\ell \mapsto s(\ell).R^\alpha], \alpha\rangle \\
\langle M \parallel \mathbf{skip}, s, \alpha\rangle & \to \langle M, s, \alpha\rangle & \langle \ell_W := \mathbf{tt}, s, \alpha\rangle & \to \langle \mathbf{skip}, s[\ell \mapsto s(\ell).W^\alpha], \alpha\rangle \\
\langle \mathbf{if}\, \mathbf{tt}\, N_1\, N_2, s, \alpha\rangle & \to \langle N_1, s, \alpha\rangle & \langle \mathbf{new}\, x, y\, \mathbf{in}\, M, s, \alpha\rangle & \to \langle M[\ell_W/x, \ell_R/y], s \uplus \{\ell \mapsto \varepsilon\}, \alpha\rangle \\
\langle \mathbf{if}\, \mathbf{ff}\, N_1\, N_2, s, \alpha\rangle & \to \langle N_2, s, \alpha\rangle & \langle \mathbf{consume}(\beta), s, \alpha\rangle & \to \langle \mathbf{skip}, s, \alpha; \beta\rangle
\end{array}$$

Fig. 5. Operational semantics: basic rules

To equip $\mathcal{R}$-IPA with an operational semantics we need operations on $\mathcal{R}$, they are introduced throughout this section. First we have $0 \in \mathcal{R}$, the null resource; if $\alpha, \beta \in \mathcal{R}$, we have some $\alpha; \beta \in \mathcal{R}$, the resource taken by consuming α, then β – for $\mathcal{R} = \mathbb{R}_+$, this is simply addition. To evaluate $\mathcal{R}$-IPA, the **configurations** are now triples $\langle M, s, \alpha\rangle$ with $\alpha \in \mathcal{R}$ tracking resources already spent. With that, we give in Fig. 5 the basic operational rules. The only rule affecting current resources is that for $\mathbf{consume}(\beta)$, the others leave it unchanged. However note that we store the current state of resources when performing memory operations, explaining the annotations in Fig. 2. These annotations do not impact the operational behaviour, but will be helpful in relating with the game semantics in Sect. 3. As usual, these rules apply within call-by-name evaluation contexts – we omit the details here but they will appear for our final operational semantics.

Slot Games. In [13], Ghica extends Ghica and Murawski's model to *slot games* in order to capture resource consumption. Slot games introduce a new action called a *token*, representing an atomic resource consumption, and written $\textcircled{\$}$ – writing $\textcircled{n}$ for n successive occurrences of $\textcircled{\$}$. A model of $\mathbb{N}_+$-IPA using slot games would have for instance the play in Fig. 6 in the interpretation of

$$H = (\mathbf{wait}(1);\, x;\, \mathbf{wait}(2)) \parallel (\mathbf{wait}(2);\, y;\, \mathbf{wait}(1))$$

in context $x : \mathbf{com}, y : \mathbf{bool}$, among with many others. Note, in examples, we use a more liberal typing rule for ';' allowing $y^{\mathbf{bool}};\, z^{\mathbf{com}} : \mathbf{bool}$ to avoid clutter: it can be encoded as $\mathbf{if}\, y\, (z;\, \mathtt{tt})\, (z;\, \mathtt{ff})$. Following the methodology of game semantics, the interpretation of $(\lambda xy.\, H)\, \mathbf{skip}\, \mathtt{tt}$ would yield, by composition, the strategy with only maximal play $\mathbf{q}^- \textcircled{6} \mathtt{tt}^+$, where $\textcircled{6}$ reflects the overall 6 time units (say "seconds") that have to pass in total before we see the result (3 in each thread). This seems wasteful, but it is indeed an adequate computational analysis, because both slot games and the operational semantics given so far implicitly assume a sequential operational model, *i.e.* that both threads compete to be scheduled on a *single* processor. Let us now question that assumption.

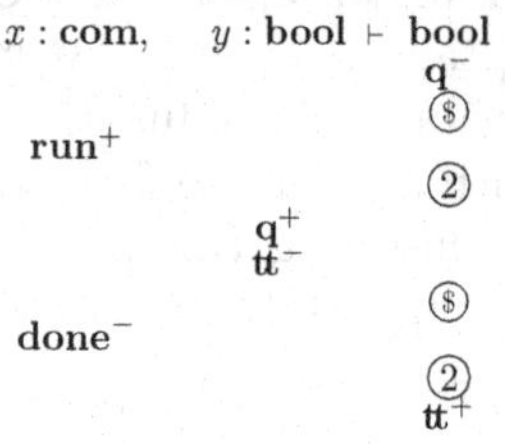

Fig. 6. A play with tokens

Parallel Resource Consumption. With a truly concurrent evaluation in mind, we should be able to prove that the program above may terminate in 3 s, rather than 6; as nothing prevents the threads from evaluating in parallel. Before we update the operational semantics to express that, we enrich our resource structure to allow it to express the effect of consuming resources in parallel.

We now introduce the full algebraic structure we require for resources.

Definition 1. *A* ***resource bimonoid*** *is* $\langle \mathcal{R}, 0, ;, \parallel, \leq \rangle$ *where* $\langle \mathcal{R}, 0, ;, \leq \rangle$ *is an ordered monoid,* $\langle \mathcal{R}, 0, \parallel, \leq \rangle$ *is an ordered commutative monoid, 0 is bottom for* $\leq$*, and* $\parallel$ *is* ***idempotent****,* i.e. *it satisfies* $\alpha \parallel \alpha = \alpha$.

A resource bimonoid is in particular a *concurrent monoid* in the sense of *e.g.* [16] (though we take $\leq$ in the opposite direction: we read $\alpha \leq_{\mathcal{R}} \alpha'$ as "α is *better/more efficient* than α'"). Our *Idempotence* assumption is rather strong as it entails that $\alpha \parallel \beta$ is the supremum of $\alpha, \beta \in \mathcal{R}$. This allows to recover a number of simple laws, *e.g.* $\alpha \parallel \beta \leq \alpha; \beta$, or the exchange rule $(\alpha; \beta) \parallel (\alpha'; \beta') \leq (\alpha \parallel \alpha'); (\beta \parallel \beta')$. Idempotence, which would not be needed for a purely functional language, is used crucially in our interpretation of state.

Our leading examples are $\langle \mathbb{N}_+, 0, +, \max, \leq \rangle$ and $\langle \mathbb{R}_+, 0, +, \max, \leq \rangle$ – we call the latter the *time bimonoid.* Others are the *permission bimonoid* $\langle \mathcal{P}(P), \emptyset, \cup, \cup, \subseteq \rangle$ for some set P of *permissions*: if reaching a state requires certain permissions, it does not matter whether these have been requested sequentially or in parallel; the bimonoid of *parametrized time* $\langle \mathcal{M}, 0, ;, \parallel, \leq \rangle$ with $\mathcal{M}$ the monotone functions from positive reals to positive reals, 0 the constant function, $\parallel$ the pointwise maximum, and $(f; g)(x) = f(x) + g(x + f(x))$: it tracks time consumption in a context where the time taken by **consume**(α) might grow over time.

Besides time-based bimonoids, it would be appealing to cover resources such as *power*, *bandwith* or *heapspace*. Those, however, clearly fail idempotence of $\parallel$, and are therefore not covered. It is not clear how to extend our model to those.

$$\frac{}{\langle M, s, \alpha\rangle \rightrightarrows \langle M, s, \alpha\rangle} \qquad \frac{\langle M, s, \alpha\rangle \to \langle M', s', \alpha'\rangle}{\langle M, s, \alpha\rangle \rightrightarrows \langle M', s', \alpha'\rangle} \qquad \frac{\langle M, s, \alpha\rangle \rightrightarrows \langle M', s', \alpha'\rangle}{\langle C[M], s, \alpha\rangle \rightrightarrows \langle C[M'], s', \alpha'\rangle}$$

$$\frac{\langle M, s, \alpha\rangle \rightrightarrows \langle M', s', \alpha'\rangle \qquad \langle M', s', \alpha'\rangle \rightrightarrows \langle M'', s'', \alpha''\rangle}{\langle M, s, \alpha\rangle \rightrightarrows \langle M'', s'', \alpha''\rangle} \qquad \frac{\langle M, s, \alpha\rangle \rightrightarrows \langle M', s', \alpha'\rangle \qquad \langle N, s, \alpha\rangle \rightrightarrows \langle N', s'', \alpha''\rangle}{\langle M \parallel N, s, \alpha\rangle \rightrightarrows \langle M' \parallel N', s' \uparrow s'', \alpha' \parallel \alpha''\rangle}$$

Fig. 7. Rules for parallel reduction

Parallel Operational Semantics. Let us fix a resource bimonoid $\mathcal{R}$. To express parallel resource consumption, we use the many-step *parallel reductions* defined in Fig. 7, with **call-by-name evaluation contexts** given by

$$C[] ::= [] \mid []\, N \mid [];\, N \mid \mathbf{if}\, []\, N_1\, N_2 \mid [] := \mathtt{tt} \mid ![] \mid ([] \parallel N) \mid (M \parallel [])$$

The rule for parallel composition carries some restrictions regarding memory: M and N can only reduce concurrently if they do not access the same memory cells. This is achieved by requiring that the *partial* operation $s \uparrow s'$ – that intuitively corresponds to "merging" two memory stores s and s' whenever there are no conflicts – is defined. More formally, the partial order $\leq_{\mathsf{M}}$ on memory states induces a partial order (also written $\leq_{\mathsf{M}}$) on stores, defined by $s \leq_{\mathsf{M}} s'$ iff $\mathrm{dom}(s) \subseteq \mathrm{dom}(s')$ and for all $\ell \in \mathrm{dom}(s)$ we have $s(\ell) \leq_{\mathsf{M}} s'(\ell)$. This order is a cpo in which s' and s'' are *compatible* (*i.e.* have an upper bound) iff for all $\ell \in \mathrm{dom}(s') \cap \mathrm{dom}(s'')$, $s'(\ell) \leq_{\mathsf{M}} s''(\ell)$ or $s''(\ell) \leq_{\mathsf{M}} s'(\ell)$ – so there has been no interference going to s' and s'' from their last common ancestor. When compatible, $s' \uparrow s''$ maps s' and s'' to their lub, and is undefined otherwise.

For $\vdash M : \mathbf{com}$, we set $M \Downarrow_\alpha$ if $\langle M, \emptyset, 0\rangle \rightrightarrows \langle \mathbf{skip}, s, \alpha\rangle$. For instance, instantiating the rules with the time bimonoid, we have

$$(\mathbf{wait}(1);\, \mathbf{wait}(2)) \parallel (\mathbf{wait}(2);\, \mathbf{wait}(1)) \Downarrow_3$$

2.3 Non-interleaving Semantics

To capture this parallel resource usage semantically, we build on the games model for affine IPA presented in [5]. Rather than presenting programs as collections of *sequences* of moves expressing all observable sequences of computational actions, this model adopts a *truly concurrent* view using collections of *partially ordered* plays. For each Player move, the order specifies its *causal dependencies*, *i.e.* the Opponent moves that need to have happened before. For instance, ignoring the subscripts, Fig. 8 displays a typical partially ordered play in the strategy for the term H of Sect. 2.2. One partially ordered play does not fully specify a sequential execution: that in Fig. 8 stands for *many* sequential executions, one of which is in Fig. 3. Behaviours expressed by partially ordered plays are deterministic *up to* choices of the scheduler irrelevant for the eventual result. Because $\mathcal{R}$-IPA is non-deterministic (via concurrency and shared state), our strategies will be *sets* of such partial orders.

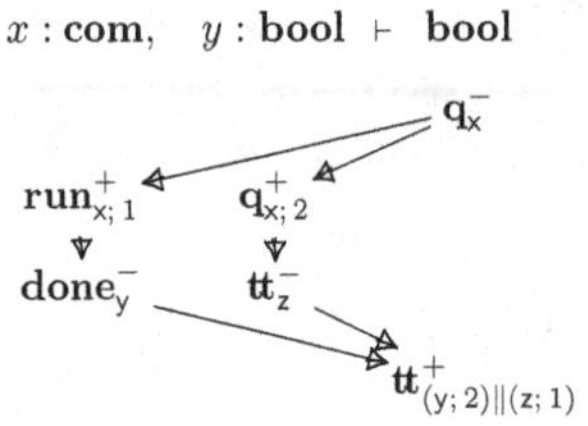

Fig. 8. A parallel $\mathcal{R}$-play

To express resources, we leverage the causal information and indicate, in each partially ordered play and for each positive move, an $\mathcal{R}$-expression representing its *additional cost* in function of the cost of its negative dependencies. Figure 8 displays such a $\mathcal{R}$-*play*: each Opponent move introduces a fresh variable, which can be used in annotations for Player moves. As we will see further on, once applied to strategies for values **skip** and **tt** (with no additional cost), this $\mathcal{R}$-play will answer to the initial Opponent move $\mathbf{q}_{\mathsf{x}}^-$ with $\mathbf{tt}^+_{\mathsf{x};\,\alpha}$ where $\alpha = (1;\,2) \parallel (2;\,1) =_{\mathbb{R}_+} 3$, as prescribed by the more efficient parallel operational semantics.

We now go on to define formally our semantics.

3 Concurrent Game Semantics of IPA

3.1 Arenas and $\mathcal{R}$-Strategies

Arenas. We first introduce *arenas*, the semantic representation of types in our model. As in [5], an arena will be a certain kind of *event structure* [27].

Definition 2. *An* ***event structure*** *comprises* $(E, \leq_E, \#_E)$ *where* E *is a set of* events, $\leq_E$ *is a partial order called* causal dependency, *and* $\#_E$ *is an irreflexive symmetric binary relation called* conflict, *subject to the two axioms:*

$$\forall e \in E, [e]_E = \{e' \in E \mid e' \leq_E e\} \textit{ is finite}$$
$$\forall e_1 \;\#_E\; e_2, \forall e_1 \leq_E e_1', e_1' \;\#_E\; e_2$$

We will use some vocabulary and notations from event structures. A **configuration** $x \subseteq E$ is a down-closed, consistent (*i.e.* for all $e, e' \in x$, $\neg(e \;\#_E\; e')$) finite set of events. We write $\mathscr{C}(E)$ for the set of configurations of E. We write $\rightarrowtriangle_E$ for **immediate causality**, *i.e.* $e \rightarrowtriangle_E e'$ iff $e <_E e'$ with nothing in between – this is the relation represented in diagrams such as Fig. 8. A conflict $e_1 \#_E e_2$ is **minimal** if for all $e_1' <_E e_1$, $\neg(e_1' \#_E e_2)$ and symmetrically. We write $e_1 \sim_E e_2$ to indicate that e_1 and e_2 are in minimal conflict.

With this, we now define arenas.

Definition 3. *An* ***arena*** *is* $(A, \leq_A, \#_A, \mathrm{pol}_A)$, *an event structure along with a* ***polarity function*** $\mathrm{pol}_A : A \longrightarrow \{-,+\}$ *subject to: (1)* $\leq_A$ *is forest-shaped, (2)* $\rightarrowtriangle_A$ *is alternating: if* $a_1 \rightarrowtriangle_A a_2$, *then* $\mathrm{pol}_A(a_1) \neq \mathrm{pol}_A(a_2)$, *and (3) it is race-free,* i.e. *if* $a_1 \sim_A a_2$, *then* $\mathrm{pol}_A(a_1) = \mathrm{pol}_A(a_2)$.

Arenas present the computational actions available on a type, following a call-by-name evaluation strategy. For instance, the observable actions of a closed term on **com** are that it can be ran, and it may terminate, leading to the arena **com** = $\mathbf{run}^- \rightarrowtriangle \mathbf{done}^+$. Likewise, a boolean can be evaluated, and can terminate on **tt** or **ff**, yielding the arena on the right of Fig. 9 (when drawing arenas, immediate causality is written with a dotted line, from top to bottom). We present some simple arena constructions. The **empty arena**, written 1, has no events. If A is an arena, then its **dual** $A^\perp$ has the same components, but polarity reversed. The **parallel composition** of A and B, written $A \parallel B$, has as events the tagged disjoint union $\{1\} \times A \cup \{2\} \times B$, and all other components inherited. For $x_A \in \mathscr{C}(A)$ and $x_B \in \mathscr{C}(B)$, we also write $x_A \parallel x_B \in \mathscr{C}(A \parallel B)$. Figure 9 displays the arena $\mathbf{com}^\perp \parallel \mathbf{bool}^\perp \parallel \mathbf{bool}$.

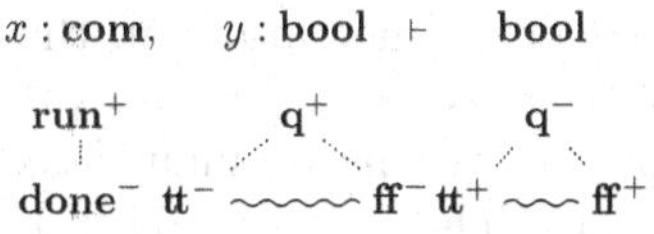

Fig. 9. An arena for a sequent

$\mathcal{R}$-Augmentations. As hinted before, $\mathcal{R}$-strategies will be collections of partially ordered plays with resource annotations in $\mathcal{R}$, called *$\mathcal{R}$-augmentations.*

Definition 4. *An* ***augmentation*** *[5] on arena A is a finite partial order* $\mathbb{q} = (|\mathbb{q}|, \leq_{\mathbb{q}})$ *such that* $\mathscr{C}(\mathbb{q}) \subseteq \mathscr{C}(A)$ *(concerning configurations, augmentations are considered as event structures with empty conflict), which is* ***courteous****, in the sense that for all* $a_1 \rightarrowtriangle_{\mathbb{q}} a_2$, *if* $\mathrm{pol}_A(a_1) = +$ *or* $\mathrm{pol}_A(a_2) = -$, *then* $a_1 \rightarrowtriangle_A a_2$.

A ***$\mathcal{R}$-augmentation*** *also has (with* $[a]_{\mathbb{q}}^- = \{a' \leq_{\mathbb{q}} a \mid \mathrm{pol}_A(a') = -\}$*)*

$$\lambda_{\mathbb{q}} : (a \in |\mathbb{q}|) \quad \longrightarrow \quad \left(\mathcal{R}^{[a]_{\mathbb{q}}^-} \to \mathcal{R}\right)$$

such that if $\mathrm{pol}_A(a) = -$, *then* $\lambda_{\mathbb{q}}(a)(\rho) = \rho_a$, *the projection on a of* $\rho \in \mathcal{R}^{[a]_{\mathbb{q}}^-}$, *and for all* $a \in |\mathbb{q}|$, $\lambda_{\mathbb{q}}(a)$ *is monotone* w.r.t. *all of its variables.*

We write $\mathcal{R}$-Aug(A) *for the set of* $\mathcal{R}$*-augmentations on A.*

If $\mathbb{q}, \mathbb{q}' \in \mathcal{R}$-Aug$(A)$, $\mathbb{q}$ is **rigidly embedded** in $\mathbb{q}'$, or a **prefix** of $\mathbb{q}'$, written $\mathbb{q} \hookrightarrow \mathbb{q}'$, if $|\mathbb{q}| \in \mathscr{C}(\mathbb{q}')$, for all $a, a' \in |\mathbb{q}|$, $a \leq_{\mathbb{q}} a'$ iff $a \leq_{\mathbb{q}'} a'$, and for all $a \in |\mathbb{q}|$, $\lambda_{\mathbb{q}}(a) = \lambda_{\mathbb{q}'}(a)$. The *$\mathcal{R}$-plays* of Sect. 2.3 are formalized as $\mathcal{R}$-augmentations: Fig. 8 presents an $\mathcal{R}$-augmentation on the arena of Fig. 9. The functional dependency in the annotation of positive events is represented by using the free variables introduced alongside negative events, however this is only a symbolic representation: the formal annotation is a function for each positive event. In the model of $\mathcal{R}$-IPA, we will only use the particular case where the annotations of positive events only depend on the annotations of their immediate predecessors.

$\mathcal{R}$-Strategies. We start by defining $\mathcal{R}$-strategies on arenas.

Definition 5. *A* ***$\mathcal{R}$-strategy*** *on A is a non-empty prefix-closed set of $\mathcal{R}$-augmentations $\sigma \subseteq \mathcal{R}\text{-Aug}(A)$ which is* ***receptive*** *[5]: for $\mathfrak{q} \in \sigma$ such that $|\mathfrak{q}|$ extends with $a^- \in A$ (i.e. $\mathrm{pol}(a) = -$, $a \notin |\mathfrak{q}|$, and $|\mathfrak{q}| \cup \{a\} \in \mathscr{C}(A)$), there is $\mathfrak{q} \hookrightarrow \mathfrak{q}' \in \sigma$ such that $|\mathfrak{q}'| = |\mathfrak{q}| \cup \{a\}$.*

If σ is a $\mathcal{R}$-strategy on arena A, we write $\sigma : A$.

Observe that $\mathcal{R}$-strategies are fully described by their *maximal* augmentations, *i.e.* augmentations that are the prefix of no other augmentations in the strategy. Our interpretation of new will use the $\mathcal{R}$-strategy cell : $[\![\mathbf{mem}_W]\!] \parallel [\![\mathbf{mem}_R]\!]$ (with arenas presented in Fig. 10), comprising all the $\mathcal{R}$-augmentations rigidly included in either of the two from Fig. 11. These two match the race when reading and writing simultaneously: if both $\mathbf{wtt}^-$ and $\mathbf{r}^-$ are played the read may return $\mathbf{tt}^+$ or $\mathbf{ff}^+$, but it can only return $\mathbf{tt}^+$ in the presence of $\mathbf{wtt}^-$.

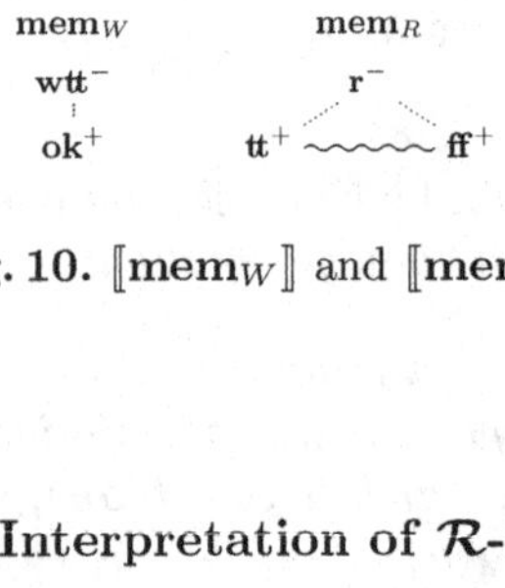

Fig. 10. $[\![\mathbf{mem}_W]\!]$ and $[\![\mathbf{mem}_R]\!]$

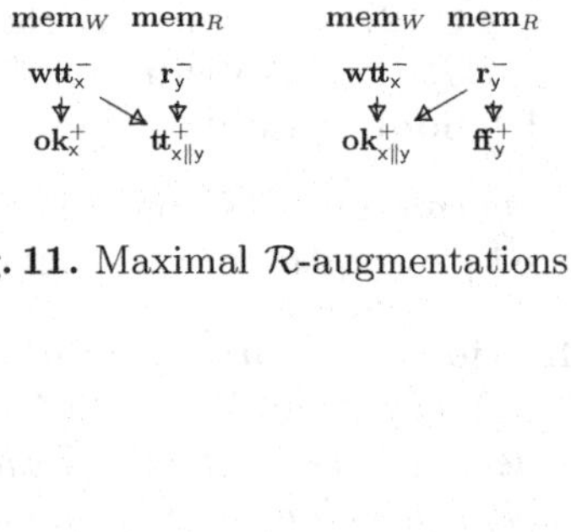

Fig. 11. Maximal $\mathcal{R}$-augmentations of cell

3.2 Interpretation of $\mathcal{R}$-IPA

Categorical Structure. In order to define the interpretation of terms of $\mathcal{R}$-IPA as $\mathcal{R}$-strategies, a key step is to show how to form a *category* of $\mathcal{R}$-strategies. To do that we follow the standard idea of considering **$\mathcal{R}$-strategies from A to B** to be simply $\mathcal{R}$-strategies on the compound arena $A^\perp \parallel B$. As usual, our first example of a $\mathcal{R}$-strategy between arenas is the *copycat $\mathcal{R}$-strategy*.

Definition 6. *Let A be an arena. We define a partial order $\leq_{\mathbb{C}_A}$ on $A^\perp \parallel A$:*

$$\begin{aligned}\leq_{\mathbb{C}_A} = (&\{((1,a),(1,a')) \mid a \leq_A a'\} \cup \{((2,a),(2,a')) \mid a \leq_A a'\} \cup \\ &\{((1,a),(2,a)) \mid \mathrm{pol}_A(a) = +\} \cup \{((2,a),(1,a)) \mid \mathrm{pol}_A(a) = -\})^+\end{aligned}$$

where $(-)^+$ denotes the transitive closure of a relation. Note that if $a \in A^\perp \parallel A$ is positive, it has a unique immediate predecessor $\mathrm{pred}(a) \in A^\perp \parallel A$ *for $\leq_{\mathbb{C}_A}$.*

If $x \parallel y \in \mathscr{C}(A^\perp \parallel A)$ is down-closed for $\leq_{\text{Ⅽ}_A}$ (write $\leq_{x,y}$ for the restriction of $\leq_{\text{Ⅽ}_A}$ to $x \parallel y$), we define an $\mathcal{R}$-augmentation $\mathbb{q}_{x,y} = (x \parallel y, \leq_{x,y}, \lambda_{x,y})$ where

$$\lambda_{x,y} : (a \in x \parallel y) \quad \longrightarrow \quad \left(\mathcal{R}^{[a]^-_{x \parallel y}} \to \mathcal{R}\right)$$

with $\lambda_{x,y}(a^-)(\rho) = \rho_a$, and $\lambda_{x,y}(a^+)(\rho) = \rho_{\text{pred}(a)}$. Then, Ⅽ_A is the $\mathcal{R}$-strategy comprising all $\mathbb{q}_{x,y}$ for $x \parallel y \in \mathscr{C}(A^\perp \parallel A)$ down-closed in A.

We first define *interactions* of $\mathcal{R}$-augmentations, extending [5].

Definition 7. *We say that $\mathbb{q} \in \mathcal{R}\text{-Aug}(A^\perp \parallel B)$, and $\mathbb{p} \in \mathcal{R}\text{-Aug}(B^\perp \parallel C)$ are* ***causally compatible*** *if $|\mathbb{q}| = x_A \parallel x_B$, $|\mathbb{p}| = x_B \parallel x_C$, and the preorder $\leq_{\mathbb{p} \circledast \mathbb{q}}$ on $x_A \parallel x_B \parallel x_C$ defined as $(\leq_\mathbb{q} \cup \leq_\mathbb{p})^+$ is a partial order.*

Say $e \in x_A \parallel x_B \parallel x_C$ is negative *if it is negative in $A^\perp \parallel C$. We define*

$$\lambda_{\mathbb{p} \circledast \mathbb{q}} : (e \in x_A \parallel x_B \parallel x_C) \quad \longrightarrow \quad \left(\mathcal{R}^{[e]^-_{\mathbb{p} \circledast \mathbb{q}}} \to \mathcal{R}\right)$$

as follows, by well-founded induction on $<_{\mathbb{p} \circledast \mathbb{q}}$, for $\rho \in \mathcal{R}^{[e]^-_{\mathbb{p} \circledast \mathbb{q}}}$:

$$\lambda_{\mathbb{p} \circledast \mathbb{q}}(e)(\rho) = \begin{cases} \lambda_\mathbb{p}(e)\left(\langle \lambda_{\mathbb{p} \circledast \mathbb{q}}(e')(\rho) \mid e' \in [e]^-_\mathbb{p} \rangle\right) & \text{if } \mathrm{pol}_{B^\perp \parallel C}(e) = +, \\ \lambda_\mathbb{q}(e)\left(\langle \lambda_{\mathbb{p} \circledast \mathbb{q}}(e')(\rho) \mid e' \in [e]^-_\mathbb{q} \rangle\right) & \text{if } \mathrm{pol}_{A^\perp \parallel B}(e) = +, \\ \rho_e & \text{otherwise, i.e. } e \text{ negative} \end{cases}$$

The ***interaction*** *$\mathbb{p} \circledast \mathbb{q}$ of compatible $\mathbb{q}, \mathbb{p}$ is $(x_A \parallel x_B \parallel x_C, \leq_{\mathbb{p} \circledast \mathbb{q}}, \lambda_{\mathbb{p} \circledast \mathbb{q}})$.*

If $\sigma : A^\perp \parallel B$ and $\tau : B^\perp \parallel C$, we write $\tau \circledast \sigma$ for the set comprising all $\mathbb{p} \circledast \mathbb{q}$ such that $\mathbb{p} \in \tau$ and $\mathbb{q} \in \sigma$ are causally compatible. For $\mathbb{q} \in \sigma$ and $\mathbb{p} \in \tau$ causally compatible with $|\mathbb{p} \circledast \mathbb{q}| = x_A \parallel x_B \parallel x_C$, their **composition** is $\mathbb{p} \odot \mathbb{q} = (x_A \parallel x_C, \leq_{\mathbb{p} \odot \mathbb{q}}, \lambda_{\mathbb{p} \odot \mathbb{q}})$ where $\leq_{\mathbb{p} \odot \mathbb{q}}$ and $\lambda_{\mathbb{p} \odot \mathbb{q}}$ are the restrictions of $\leq_{\mathbb{p} \circledast \mathbb{q}}$ and $\lambda_{\mathbb{p} \circledast \mathbb{q}}$. Finally, the **composition** of $\sigma : A^\perp \parallel B$ and $\tau : B^\perp \parallel C$ is the set comprising all $\mathbb{p} \odot \mathbb{q}$ for $\mathbb{q} \in \sigma$ and $\mathbb{p} \in \tau$ causally compatible.

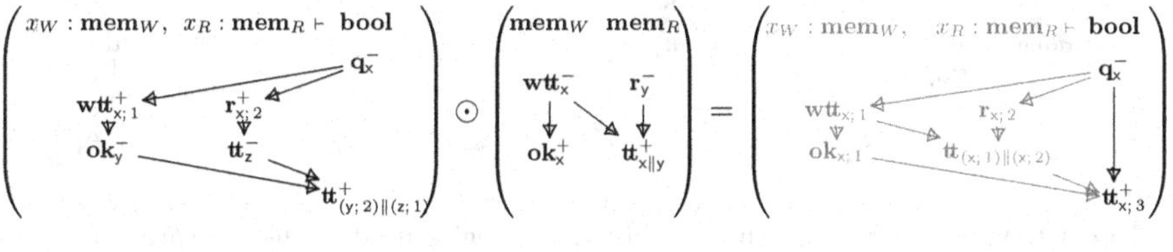

Fig. 12. Example of interaction and composition between $\mathbb{R}_+$-augmentations

In Fig. 12, we display an example composition between $\mathbb{R}_+$-augmentations – with also in gray the underlying interaction. The reader may check that the variant of the left $\mathbb{R}_+$-augmentation with $\mathbf{tt}$ replaced with $\mathbf{ff}$ is causally compatible with the other augmentation in Fig. 11, with composition $\mathbf{q}^-_\mathsf{x} \rightarrow \mathbf{ff}^+_{\mathsf{x};\,4}$.

We also have a tensor operation: on arenas, $A \otimes B$ is simply a synonym for $A \parallel B$. If $\mathfrak{q}_1 \in \mathcal{R}\text{-Aug}(A_1^\perp \parallel B_1)$ and $\mathfrak{q}_2 \in \mathcal{R}\text{-Aug}(A_2^\perp \parallel B_2)$, their **tensor product** $\mathfrak{q}_1 \otimes \mathfrak{q}_2 \in \mathcal{R}\text{-Aug}((A_1 \otimes A_2)^\perp \parallel (B_1 \otimes B_2))$ is defined in the obvious way. This is lifted to $\mathcal{R}$-strategies element-wise. As is common when constructing basic categories of games and strategies, we have:

Proposition 1. *There is a compact closed category* $\mathcal{R}$-Strat *having arenas as objects, and as morphisms,* $\mathcal{R}$*-strategies between them.*

Negative Arenas and $\mathcal{R}$*-Strategies.* As a compact closed category, $\mathcal{R}$-Strat is a model of the linear λ-calculus. However, we will (as usual for call-by-name) instead interpret $\mathcal{R}$-IPA in a sub-category of *negative* arenas and strategies, in which the empty arena 1 is terminal, providing the interpretation of weakening. We will stay very brief here, as this proceeds exactly as in [5].

A partial order with polarities is **negative** if all its minimal events are. This applies in particular to arenas, and $\mathcal{R}$-augmentations. A $\mathcal{R}$-strategy is **negative** if all its $\mathcal{R}$-augmentations are. A negative $\mathcal{R}$-augmentation $\mathfrak{q} \in \mathcal{R}\text{-Aug}(A)$ is **well-threaded** if for all $a \in |\mathfrak{q}|$, $[a]_\mathfrak{q}$ has exactly one minimal event; a $\mathcal{R}$-strategy is **well-threaded** iff all its $\mathcal{R}$-augmentations are. We have:

Proposition 2. Negative arenas *and* negative well-threaded $\mathcal{R}$*-strategies form a cartesian symmetric monoidal closed category* $\mathcal{R}$-Strat$_-$*, with* 1 *terminal.*

We also write $\sigma : A \stackrel{+}{\twoheadrightarrow} B$ *for morphisms in* $\mathcal{R}$-Strat$_-$*.*

The closure of $\mathcal{R}$-Strat does not transport to $\mathcal{R}$-Strat$_-$ as $A^\perp \parallel B$ is never negative if A is non-empty, thus we replace it with a negative version. Here we describe only a restricted case of the general construction in [5], which is however sufficient for the types of $\mathcal{R}$-IPA. If A, B are negative arenas and B is **well-opened**, *i.e.* it has exactly one minimal event b, we form $A \multimap B$ as having all components as in $A^\perp \parallel B$, with additional dependencies $\{((2,b),(1,a)) \mid a \in A\}$.

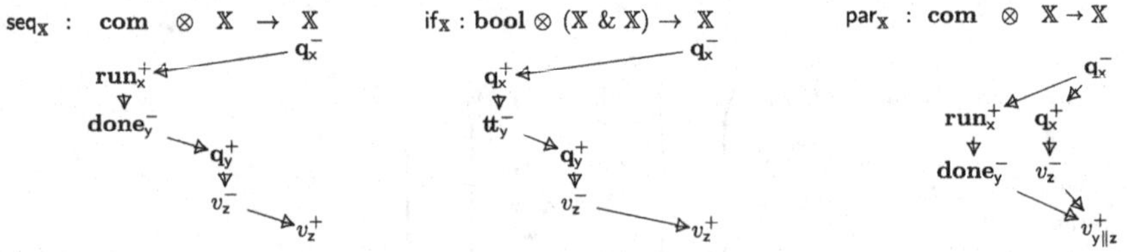

Fig. 13. Maximal $\mathcal{R}$-augmentations of $\mathcal{R}$-strategies used in the interpretation

Using the compact closed structure of $\mathcal{R}$-Strat it is easy to build a copycat $\mathcal{R}$-strategy $\mathsf{ev}_{A,B} : (A \multimap B) \otimes A \stackrel{+}{\twoheadrightarrow} B$, and to associate to any $\sigma : C \otimes A \stackrel{+}{\twoheadrightarrow} B$ some $\Lambda(\sigma) : C \stackrel{+}{\twoheadrightarrow} A \multimap B$ providing the monoidal closure. The cartesian product of A and B is $A \,\&\, B$ with components the same as $A \parallel B$, except for $(1,a) \# (2,b)$ for all $a \in A, b \in B$. We write $\pi_i : A_1 \,\&\, A_2 \stackrel{+}{\twoheadrightarrow} A_i$ for the projections, and $\langle \sigma, \tau \rangle : A \stackrel{+}{\twoheadrightarrow} B \,\&\, C$ for the pairing of $\sigma : A \stackrel{+}{\twoheadrightarrow} B$, and $\tau : A \stackrel{+}{\twoheadrightarrow} C$.

Interpretation of $\mathcal{R}$-IPA. We set $[\![\mathbf{com}]\!] = \mathbf{run}^- \rightarrow \mathbf{done}^+$, $[\![\mathbf{bool}]\!]$ as in the right-hand side of Fig. 9, $[\![\mathbf{mem}_W]\!]$ and $[\![\mathbf{mem}_R]\!]$ as in Fig. 10, and $[\![A \multimap B]\!] = [\![A]\!] \multimap [\![B]\!]$ as expected. Contexts $\Gamma = x_1 : A_1, \ldots, x_n : A_n$ are interpreted as $[\![\Gamma]\!] = \otimes_{1 \leq i \leq n}[\![A_i]\!]$. Terms $\Gamma \vdash M : A$ are interpreted as $[\![t]\!] : [\![\Gamma]\!] \twoheadrightarrow [\![A]\!]$ as follows: $[\![\bot]\!]$ is the diverging $\mathcal{R}$-strategy (no player move), $[\![\mathbf{consume}(\alpha)]\!]$ has only maximal $\mathcal{R}$-augmentation $\mathbf{run}_{\mathsf{x}}^- \rightarrow \mathbf{done}_{\mathsf{x};\alpha}^+$, $[\![\mathbf{skip}]\!]$ is $[\![\mathbf{consume}(0)]\!]$, and $\mathbf{tt}$ and $\mathbf{ff}$ are interpreted similarly with the adequate constant $\mathcal{R}$-strategies. The rest of the interpretation is given on the left, using the two obvious isos $\mathsf{deref} : [\![\mathbf{mem}_R]\!] \twoheadrightarrow [\![\mathbf{bool}]\!]$ and $\mathsf{assign} : [\![\mathbf{mem}_W]\!] \twoheadrightarrow [\![\mathbf{com}]\!]$; the $\mathcal{R}$-strategy cell introduced in Fig. 11; and additional $\mathcal{R}$-strategies with typical $\mathcal{R}$-augmentations in Fig. 13. We omit the (standard) clauses for the λ-calculus.

$$[\![M; N : \mathbb{X}]\!] = \mathsf{seq}_{\mathbb{X}} \odot ([\![M]\!] \otimes [\![N]\!])$$

$$[\![M \parallel N : \mathbb{X}]\!] = \mathsf{par}_{\mathbb{X}} \odot ([\![M]\!] \otimes [\![N]\!])$$

$$[\![\mathbf{if}\, M\, N_1\, N_2 : \mathbb{X}]\!] = \mathsf{if}_{\mathbb{X}} \odot ([\![M]\!] \otimes \langle [\![N_1]\!], [\![N_2]\!] \rangle)$$

$$[\![!M : \mathbf{bool}]\!] = \mathsf{deref} \odot [\![M]\!]$$

$$[\![M := \mathbf{tt} : \mathbf{com}]\!] = \mathsf{assign} \odot [\![M]\!]$$

$$[\![\mathbf{new}\, x, y\, \mathbf{in}\, M : \mathbb{X}]\!] = [\![M]\!] \odot ([\![\Gamma]\!] \otimes \mathsf{cell})$$

3.3 Soundness

Now that we have defined the game semantics of $\mathcal{R}$-IPA, we set to prove that it is sound with respect to the operational semantics given in Sect. 2.2.

We first introduce a useful notation. For any type A, $[\![A]\!]$ has a unique minimal event; write $(\!|A|\!)$ for the arena without this minimal event. Likewise, if $\Gamma \vdash M : A$, then by construction, $[\![M]\!] : [\![\Gamma]\!]^\perp \parallel [\![A]\!]$ is a negative $\mathcal{R}$-strategy whose augmentations all share the same minimal event $\mathbf{q}_{\mathsf{x}}^-$ where $\mathbf{q}^-$ is minimal in A. For $\alpha \in \mathcal{R}$, write $(\!|M|\!)_\alpha$ for $[\![M]\!]$ without $\mathbf{q}_{\mathsf{x}}^-$, with x replaced by α. Then we have $(\!|M|\!)_\alpha : [\![\Gamma]\!]^\perp \parallel (\!|A|\!)$ – one may think of $(\!|M|\!)_\alpha$ as "M started with consumed resource α".

Naively, one may expect soundness to state that for all $\vdash M : \mathbf{com}$, if $M \Downarrow_\alpha$, then $(\!|M|\!)_0 = \mathbf{done}_\alpha^+$. However, whereas the resource annotations in the semantics are always as good as permitted by the causal constraints, derivations in the operational semantics may be sub-optimal. For instance, we may derive $M \Downarrow_\alpha$ not using the parallel rule at all. So our statement is:

Theorem 1. *If* $\vdash M : \mathbf{com}$ *with* $M \Downarrow_\alpha$, *there is* $\beta \leq_{\mathcal{R}} \alpha$ s.t. $(\!|M|\!)_0 = \mathbf{done}_\beta^+$.

Our proof methodology is standard: we replay operational derivations as augmentations in the denotational semantics. Stating the invariant successfully proved by induction on operational derivations requires some technology.

If s is a store, then write $\mathsf{cell}_s : [\![\Omega(s)]\!]$ for the memory strategy for store s. It is defined as $\otimes_{\ell \in \mathrm{dom}(s)} \mathsf{cell}_{s(\ell)}$ where $\mathsf{cell}_\varepsilon = \mathsf{cell}$, cell_{R^α} is the $\mathcal{R}$-strategy with only maximal $\mathcal{R}$-augmentation $\mathbf{wtt}_{\mathsf{x}}^- \rightarrow \mathbf{ok}_{\mathsf{x} \parallel \alpha}^+$, cell_{W^α} has maximal $\mathcal{R}$-augmentation $\mathbf{r}_{\mathsf{y}}^- \rightarrow \mathbf{tt}_{\alpha \parallel \mathsf{y}}^+$, and the empty $\mathcal{R}$-strategy for the other cases. If $s \leq_{\mathsf{M}} s'$, then

s' can be obtained from s using memory operations and there is a matching $\mathcal{R}$-augmentation $\mathbb{q}_{s \triangleright s'} \in \mathsf{cell}_s$ defined location-wise in the obvious way.

Now, if $\sigma : [\![\Omega(s)]\!]^\perp \parallel (\!|A|\!)$ is a $\mathcal{R}$-strategy and $\mathbb{q} \in \sigma$ with moves only in $[\![\Omega(s)]\!]^\perp$ is causally compatible with $\mathbb{q}_{s \triangleright s'}$, we define the **residual** of σ after $\mathbb{q}$:

$$\sigma/(\mathbb{q} \circledast \mathbb{q}_{s \triangleright s'}) : [\![\Omega(s')]\!]^\perp \parallel (\!|A|\!)$$

If $\mathbb{p} \in \sigma$ with $\mathbb{q} \hookrightarrow \mathbb{p}$, we write first $\mathbb{p}' = \mathbb{p}/(\mathbb{q} \circledast \mathbb{q}_{s \triangleright s'})$ the $\mathcal{R}$-augmentation with $|\mathbb{p}'| = |\mathbb{p}| \setminus |\mathbb{q}|$, and with causal order the restriction of that of $\mathbb{p}$. For $e \in |\mathbb{p}'|$, we set $\lambda_{\mathbb{p}'}(e)$ to be $\lambda_{\mathbb{p}}(e)$ whose arguments corresponding to negative events e' in $\mathbb{q}$ are instantiated with $\lambda_{\mathbb{q} \circledast \mathbb{q}_{s \triangleright s'}}(e') \in \mathcal{R}$. With that, we set $\sigma/(\mathbb{q} \circledast \mathbb{q}_{s \triangleright s'})$ as comprising all $\mathbb{p}/(\mathbb{q} \circledast \mathbb{q}_{s \triangleright s'})$ for $\mathbb{p} \in \sigma$ with $\mathbb{q} \hookrightarrow \mathbb{p}$.

Informally, this means that, considering some $\mathbb{q}$ which represents a scheduling of the memory operations turning s into s', we extract from σ its behavior after the execution of these memory operations. Finally, we generalize $\leq_{\mathcal{R}}$ to $\mathcal{R}$-augmentations by setting $\mathbb{q} \leq_{\mathcal{R}} \mathbb{q}'$ iff they have the same underlying partial order and for all $e \in |\mathbb{q}|$, $\lambda_{\mathbb{q}}(e) \leq_{\mathcal{R}} \lambda_{\mathbb{q}'}(e)$. With that, we can finally state:

Lemma 1. *Let* $\Omega(s) \vdash M : A$, $\langle M, s_1, \alpha \rangle \rightrightarrows \langle M', s_1' \uplus s_2', \alpha' \rangle$ *with* $\mathrm{dom}(s_1) = \mathrm{dom}(s_1')$, *and all resource annotations in* s_1 *lower than* α. *Then, there is* $\mathbb{q} \in (\!|M|\!)_\alpha$ *with events in* $[\![\Omega(s)]\!]$, *causally compatible with* $\mathbb{q}_{s_1 \triangleright s_1'}$, *and a function*

$$\varphi : (\!|M'|\!)_{\alpha'} \circledast \mathsf{cell}_{s_2'} \quad \longrightarrow \quad (\!|M|\!)_\alpha/(\mathbb{q} \circledast \mathbb{q}_{s_1 \triangleright s_1'})$$

preserving $\hookrightarrow$ *and* s.t. *for all* $\mathbb{p} \circledast \mathbb{q}_{s_2'} \in (\!|M'|\!)_{\alpha'} \circledast \mathsf{cell}_{s_2'}$, $\varphi(\mathbb{p} \circledast \mathbb{q}_{s_2'}) \leq_{\mathcal{R}} \mathbb{p} \odot \mathbb{q}_{s_2'}$.

This is proved by induction on the operational semantics – the critical cases are: assignment and dereferenciation exploiting that if $\alpha \leq_{\mathcal{R}} \beta$, then $\alpha \parallel \beta = \beta$ (which boils down to idempotence); and parallel composition where compatibility of s' and s'' entails that the corresponding augmentations of cell_s are compatible.

Lemma 1, instantiated with $\langle M, \emptyset, 0 \rangle \rightrightarrows \langle \mathbf{skip}, s, \alpha \rangle$, yields soundness.

Non-adequacy. Our model is not adequate. To see why, consider:

$$\vdash \mathbf{new}\, x_W, x_R\, \mathbf{in} \left(\begin{array}{r} \mathbf{wait}(1); \\ x_W := \mathtt{tt}; \\ \mathbf{wait}(2) \end{array} \middle\| \begin{array}{l} \mathbf{wait}(2); \\ !x_R; \\ \mathbf{wait}(1) \end{array} \right) : \mathbf{bool}$$

Our model predicts that this may evaluate to $\mathtt{tt}$ in 3 s (see Fig. 12) and to $\mathtt{ff}$ in 4 s. However, the operational semantics can only evaluate it (both to $\mathtt{tt}$ and $\mathtt{ff}$) in 4 s. Intuitively, the reason is that the causal shapes implicit in the reduction $\rightrightarrows$ are all series-parallel (generated with sequential and parallel composition), whereas the interaction in Fig. 12 is not.

Our causal semantic approach yields a finer resource analysis than achieved by the parallel operational semantics. The operational semantics, rather than our model, is to blame for non-adequacy: indeed, we now show that for $\mathcal{R} = \mathbb{R}_+$ our model is adequate *w.r.t.* an operational semantics specialized for time.

4 Adequacy for Time

For time, we may refine the operational semantics by adding the following rule

$$\langle \mathbf{wait}(t_1 + t_2), s, t_0 \rangle \to \langle \mathbf{wait}(t_2), s, t_0 + t_1 \rangle$$

using which the program above evaluates to $\mathbf{tt}$ in 3 s. It is clear that the soundness theorem of the previous section is retained.

We first focus on adequacy for first-order programs without abstraction or application, written $\Omega(s) \vdash_1 M : \mathbf{com}$. For any $t_0 \in \mathbb{R}_+$ there is $\langle M, s, t_0 \rangle \rightrightarrows \langle M', s \uplus s', t_0 \rangle$ where $(\!|M|\!)_{t_0} = (\!|M'|\!)_{t_0} \odot \mathsf{cell}_{s'}$ and M' is in **canonical form**: it cannot be decomposed as $C[\mathbf{skip}; N]$, $C[\mathbf{skip} \parallel N]$, $C[N \parallel \mathbf{skip}]$, $C[\mathbf{if}\,\mathbf{tt}\,N_1\,N_2]$, $C[\mathbf{if}\,\mathbf{ff}\,N_1\,N_2]$, $C[\mathbf{wait}(0)]$ and $C[\mathbf{new}\,x, y\,\mathbf{in}\,N]$ for $C[]$ an evaluation context.

Consider $\Omega(s) \vdash_1 M : \mathbf{com}$, and $\mathfrak{q} \in (\!|M|\!)_{t_0} \circledast \mathsf{cell}_s$ with a top element $\mathbf{done}^+_{t_f}$ in $(\!|\mathbf{com}|\!)$, the **result** – *i.e.* $\mathfrak{q}$ describes an interaction between $(\!|M|\!)_{t_0}$ and the memory leading to a successful evaluation to **done** at time t_f. To prove adequacy, we must extract from it a derivation from $\langle M, s, t_0 \rangle$, at time t_f.

Apart from the top $\mathbf{done}^+_{t_f}$, $\mathfrak{q}$ only records memory operations, which we must replicate operationally in the adequate order. A **minimal operation with timing** t is either the top $\mathbf{done}^+_t$ if it is the only event in $\mathfrak{q}$, or a prefix $(m_t \rightarrowtriangle n_t) \hookrightarrow \mathfrak{q}$ corresponding to a memory operation (for instance, in augmentations of Fig. 14, the only minimal operation has timing 2). If $t = t_0$, this operation should be performed immediately. If $t > t_0$ we need to spend time to trigger it – it is then critical to spend time on *all available* **wait***s in parallel*:

Lemma 2. *For* $\Omega(s) \vdash_1 M : \mathbf{com}$ *in canonical form,* $t_0 \in \mathbb{R}_+$, $\mathfrak{q} \in (\!|M|\!)_{t_0} \circledast \mathsf{cell}_s$ *with result* $\mathbf{done}^+_{t_f}$*, if all minimal operations have timing strictly greater than* t_0*,*

$$\langle M, s, t_0 \rangle \rightrightarrows \langle M', s, t_0 + t \rangle$$

for some $t > 0$ *and* M' *only differing from* M *by having smaller annotations in* **wait** *commands and at least one* **wait** *changed to* **skip***.*

Furthermore, there is $\mathfrak{q} \leq_{\mathcal{R}} \mathfrak{q}'$ *with* $\mathfrak{q}' \in (\!|M'|\!)_{t_0+t} \circledast \mathsf{cell}_s$ *with result* $\mathbf{done}^+_{t_f}$.

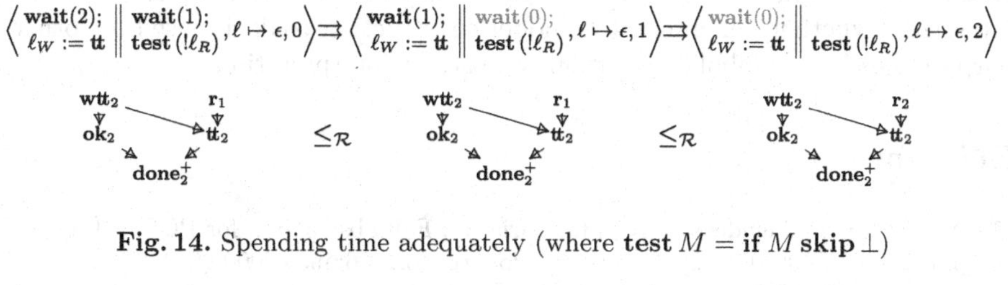

Fig. 14. Spending time adequately (where $\mathbf{test}\,M = \mathbf{if}\,M\,\mathbf{skip}\,\bot$)

Proof. As M is in canonical form, all delays in minimal operations are impacted by **wait**(t) commands in head position (*i.e.* such that $M = C[\mathbf{wait}(t)]$). Let $t_{\min}$ be the minimal time appearing in those **wait**$(-)$ commands in head position. Using our new rule and parallel composition, we remove $t_{\min}$ to all such instances of **wait**$(-)$; then transform the resulting occurrences of **wait**(0) to **skip**.

A representative example is displayed in Fig. 14. In the second step, though $!\ell_R$ is available immediately, we must wait to get the right result.

With that we can prove the key lemma towards adequacy.

Lemma 3. *Let* $\Omega(s) \vdash_1 M : \mathbf{com}$*,* $t_0 \in \mathbb{R}_+$*, and* $\mathbb{q} \in (\!|M|\!)_{t_0} \circledast \mathsf{cell}_s$ *with result* $\mathbf{done}^+_{t_f}$ *in* $(\!|\mathbf{com}|\!)$*. Then, there is* $\langle M, s, t_0\rangle \rightrightarrows \langle \mathbf{skip}, -, t_f\rangle$*.*

Proof. By induction on the size of M. First, we convert M to canonical form. If all minimal operations in $\mathbb{q} \in (\!|M|\!)_{t_0}$ have timing strictly greater than t_0, we apply Lemma 2 and conclude by induction hypothesis.

Otherwise, at least one minimal operation has timing t_0. If it is the result $\mathbf{done}^+_{t_0}$ in $(\!|\mathbb{X}|\!)$, then M is the constant **skip**. Otherwise, it is a memory operation, say $\mathbb{p} \hookrightarrow \mathbb{q}$ with $\mathbb{p} = (\mathbf{r}_{t_0} \twoheadrightarrow b_{t_0})$ and write also $s' = s[\ell \mapsto s(\ell).R^{t_0}]$. It follows then by an induction on M that $M = C[!\ell_R]$ for some $C[]$, with

$$\mathbb{q}/(\mathbb{p} \circledast \mathbb{q}_{s \rhd s'}) \in (\!|C[b]|\!)_{t_0} \circledast \mathsf{cell}_s$$

so $\langle M, s, t_0\rangle \rightrightarrows \langle C[b], s', t_0\rangle \rightrightarrows \langle \mathbf{skip}, -, t_f\rangle$ by induction hypothesis.

Adequacy follows for higher-order programs: in general, any $\vdash M : \mathbf{com}$ can be β-reduced to first-order M', leaving the interpretation unchanged. By Church-Rosser, M' behaves like M operationally, up to weak bisimulation. Hence:

Theorem 2. *Let* $\vdash M : \mathbf{com}$*. For any* $t \in \mathbb{R}_+$*, if* $\mathbf{done}^+_t \in (\!|M|\!)_0$ *then* $M \Downarrow_t$*.*

5 Conclusion

It would be interesting to compare our model with structures used in timing analysis, for instance [23] relies on a concurrent generalization of control flow graphs that is reminiscent of event structures. In future work we also plan to investigate whether our annotated model construction could be used for other purposes, such as symbolic execution or abstract interpretation.

References

1. Abramsky, S., Jagadeesan, R., Malacaria, P.: Full abstraction for PCF. Inf. Comput. **163**(2), 409–470 (2000). https://doi.org/10.1006/inco.2000.2930
2. Abramsky, S., Melliès, P.: Concurrent games and full completeness. In: 14th Annual IEEE Symposium on Logic in Computer Science, Trento, Italy, 2–5 July 1999, pp. 431–442 (1999). https://doi.org/10.1109/LICS.1999.782638

3. Alcolei, A., Clairambault, P., Hyland, M., Winskel, G.: The true concurrency of Herbrand's theorem. In: 27th EACSL Annual Conference on Computer Science Logic, CSL 2018, Birmingham, UK, 4–7 September 2018, pp. 5:1–5:22 (2018). https://doi.org/10.4230/LIPIcs.CSL.2018.5
4. Brunel, A., Gaboardi, M., Mazza, D., Zdancewic, S.: A core quantitative coeffect calculus. In: Shao, Z. (ed.) ESOP 2014. LNCS, vol. 8410, pp. 351–370. Springer, Heidelberg (2014). https://doi.org/10.1007/978-3-642-54833-8_19
5. Castellan, S., Clairambault, P.: Causality vs. interleavings in concurrent game semantics. In: Desharnais, J., Jagadeesan, R. (eds.) 27th International Conference on Concurrency Theory, CONCUR 2016, Québec City, Canada, 23–26 August 2016. LIPIcs, vol. 59, pp. 32:1–32:14. Schloss Dagstuhl - Leibniz-Zentrum fuer Informatik (2016). https://doi.org/10.4230/LIPIcs.CONCUR.2016.32
6. Castellan, S., Clairambault, P., Paquet, H., Winskel, G.: The concurrent game semantics of probabilistic PCF. In: Proceedings of the 33rd Annual ACM/IEEE Symposium on Logic in Computer Science, LICS 2018, Oxford, UK, 09–12 July 2018, pp. 215–224 (2018). https://doi.org/10.1145/3209108.3209187
7. Castellan, S., Clairambault, P., Rideau, S., Winskel, G.: Games and strategies as event structures. Logical Methods Comput. Sci. **13**(3) (2017). https://doi.org/10.23638/LMCS-13(3:35)2017
8. Castellan, S., Clairambault, P., Winskel, G.: The parallel intensionally fully abstract games model of PCF. In: 30th Annual ACM/IEEE Symposium on Logic in Computer Science, LICS 2015, Kyoto, Japan, 6–10 July 2015, pp. 232–243 (2015). https://doi.org/10.1109/LICS.2015.31
9. Castellan, S., Clairambault, P., Winskel, G.: Thin games with symmetry and concurrent hyland-ong games. Logical Methods Comput. Sci. (to appear, 2019)
10. Clairambault, P., de Visme, M., Winskel, G.: Game semantics for quantum programming. PACMPL **3**(POPL), 32:1–32:29 (2019). https://doi.org/10.1145/3290345
11. Ehrhard, T.: The Scott model of linear logic is the extensional collapse of its relational model. Theor. Comput. Sci. **424**, 20–45 (2012). https://doi.org/10.1016/j.tcs.2011.11.027
12. Faggian, C., Piccolo, M.: Partial orders, event structures and linear strategies. In: Curien, P.-L. (ed.) TLCA 2009. LNCS, vol. 5608, pp. 95–111. Springer, Heidelberg (2009). https://doi.org/10.1007/978-3-642-02273-9_9
13. Ghica, D.R.: Slot games: a quantitative model of computation. In: Proceedings of the 32nd ACM SIGPLAN-SIGACT Symposium on Principles of Programming Languages, POPL 2005, Long Beach, California, USA, 12–14 January 2005, pp. 85–97 (2005). https://doi.org/10.1145/1040305.1040313
14. Ghica, D.R., Murawski, A.S.: Angelic semantics of fine-grained concurrency. Ann. Pure Appl. Logic **151**(2–3), 89–114 (2008). https://doi.org/10.1016/j.apal.2007.10.005
15. Ghica, D.R., Smith, A.I.: Bounded linear types in a resource semiring. In: Shao, Z. (ed.) ESOP 2014. LNCS, vol. 8410, pp. 331–350. Springer, Heidelberg (2014). https://doi.org/10.1007/978-3-642-54833-8_18
16. Hoare, T., Möller, B., Struth, G., Wehrman, I.: Concurrent Kleene algebra and its foundations. J. Log. Algebr. Program. **80**(6), 266–296 (2011). https://doi.org/10.1016/j.jlap.2011.04.005
17. Hyland, J.M.E., Ong, C.L.: On full abstraction for PCF: I, II, and III. Inf. Comput. **163**(2), 285–408 (2000). https://doi.org/10.1006/inco.2000.2917

18. Laird, J., Manzonetto, G., McCusker, G., Pagani, M.: Weighted relational models of typed lambda-calculi. In: 28th Annual ACM/IEEE Symposium on Logic in Computer Science (LICS), New Orleans, USA, Proceedings, pp. 301–310 (2013)
19. Laurent, O.: Game semantics for first-order logic. Logical Methods Comput. Sci. **6**(4) (2010). https://doi.org/10.2168/LMCS-6(4:3)2010
20. Melliès, P.: Asynchronous games 4: a fully complete model of propositional linear logic. In: 20th IEEE Symposium on Logic in Computer Science (LICS 2005), Chicago, IL, USA, 26–29 June 2005, Proceedings, pp. 386–395 (2005). https://doi.org/10.1109/LICS.2005.6
21. Melliès, P.: Game semantics in string diagrams. In: Proceedings of the 27th Annual IEEE Symposium on Logic in Computer Science, LICS 2012, Dubrovnik, Croatia, 25–28 June 2012, pp. 481–490 (2012). https://doi.org/10.1109/LICS.2012.58
22. Melliès, P.-A., Mimram, S.: Asynchronous games: innocence without alternation. In: Caires, L., Vasconcelos, V.T. (eds.) CONCUR 2007. LNCS, vol. 4703, pp. 395–411. Springer, Heidelberg (2007). https://doi.org/10.1007/978-3-540-74407-8_27
23. Mittermayr, R., Blieberger, J.: Timing analysis of concurrent programs. In: Vardanega, T. (ed.) 12th International Workshop on Worst-Case Execution Time Analysis, WCET 2012, Pisa, Italy, 10 July 2012. OASICS, vol. 23, pp. 59–68. Schloss Dagstuhl - Leibniz-Zentrum fuer Informatik (2012). https://doi.org/10.4230/OASIcs.WCET.2012.59
24. Plotkin, G.D.: Post-graduate lecture notes in advanced domain theory (incorporating the "Pisa notes"). Department of Computer Science, University of Edinburgh (1981)
25. Rideau, S., Winskel, G.: Concurrent strategies. In: Proceedings of the 26th Annual IEEE Symposium on Logic in Computer Science, LICS 2011, Toronto, Ontario, Canada, 21–24 June 2011, pp. 409–418 (2011). https://doi.org/10.1109/LICS.2011.13
26. Sands, D.: Operational theories of improvement in functional languages (extended abstract). In: Heldal, R., Holst, C.K., Wadler, P. (eds.) Functional Programming, Glasgow 1991, pp. 298–311. Springer, London (1991). https://doi.org/10.1007/978-1-4471-3196-0_24
27. Winskel, G.: Event structures. In: Brauer, W., Reisig, W., Rozenberg, G. (eds.) ACPN 1986. LNCS, vol. 255, pp. 325–392. Springer, Heidelberg (1987). https://doi.org/10.1007/3-540-17906-2_31

Change Actions: Models of Generalised Differentiation

Mario Alvarez-Picallo(✉) and C.-H. Luke Ong(✉)

University of Oxford, Oxford, UK
{mario.alvarez-picallo,luke.ong}@cs.ox.ac.uk

Abstract. Change structures, introduced by Cai et al., have recently been proposed as a semantic framework for incremental computation. We generalise change actions, an alternative to change structures, to arbitrary cartesian categories and propose the notion of *change action model* as a categorical model for (higher-order) generalised differentiation. Change action models naturally arise from many geometric and computational settings, such as (generalised) cartesian differential categories, group models of discrete calculus, and Kleene algebra of regular expressions. We show how to build canonical change action models on arbitrary cartesian categories, reminiscent of the Fàa di Bruno construction.

1 Introduction

Incremental computation is the process of incrementally updating the output of some given function as the input is gradually changed, without recomputing the entire function from scratch. Recently, Cai et al. [6] introduced the notion of change structure to give a semantic account of incremental computation. Change structures have subsequently been generalised to *change actions* [2], and proposed as a model for automatic differentiation [16]. These developments raise a number of questions about the structure of change actions themselves and how they relate to more traditional notions of differentiation.

A *change action* $A = (|A|, \Delta A, \oplus_A, +_A, 0)$ is a set $|A|$ equipped with a monoid $(\Delta A, +_A, 0_A)$ acting on it, via action $\oplus_A : |A| \times \Delta A \to |A|$. For example, every monoid $(S, +, 0)$ gives rise to a (so-called *monoidal*) change action $(S, S, +, +, 0)$. Given change actions A and B, consider functions $f : |A| \to |B|$. A *derivative* of f is a function $\partial f : |A| \times \Delta A \to \Delta B$ such that for all $a \in |A|, \delta a \in \Delta A$, $f(a \oplus_A \delta a) = f(a) \oplus_B \partial f(a, \delta a)$. Change actions and differentiable functions (i.e. functions that have a regular derivative) organise themselves into categories (and indeed 2-categories) with finite (co)products, whereby morphisms are composed via the chain rule.

The definition of change actions (and derivatives of functions) makes no use of properties of **Set** beyond the existence of products. We develop the theory of change actions on arbitrary cartesian categories and study their properties.

M. Bojańczyk and A. Simpson (Eds.): FOSSACS 2019, LNCS 11425, pp. 45–61, 2019.
https://doi.org/10.1007/978-3-030-17127-8_3

A first contribution is the notion of a *change action model*, which is defined to be a coalgebra for a certain (copointed) endofunctor CAct on the category $\mathbf{Cat}_\times$ of (small) cartesian categories. The functor CAct sends a category $\mathbf{C}$ to the category $\mathrm{CAct}(\mathbf{C})$ of (internal) change actions and differential maps on $\mathbf{C}$.

There is a natural, extrinsic, notion of higher-order derivative in change action models. In such a model $\alpha : \mathbf{C} \to \mathrm{CAct}(\mathbf{C})$, a $\mathbf{C}$-object A is associated (via α) with a change action, the carrier object of whose monoid is in turn associated with a change action, and so on *ad infinitum*. We construct a "canonical" change action model, $\mathrm{CAct}_\omega(\mathbf{C})$, that internalises such ω-sequences that exhibit higher-order differentiation. Objects of $\mathrm{CAct}_\omega(\mathbf{C})$ are ω-sequences of "contiguously compatible" change actions; and morphisms are corresponding ω-sequences of differential maps, each map being the canonical (via α) derivative of the preceding in the ω-sequence. We show that $\mathrm{CAct}_\omega(\mathbf{C})$ is the final CAct-coalgebra (relativised to change action models on $\mathbf{C}$). The category $\mathrm{CAct}_\omega(\mathbf{C})$ may be viewed as a kind of Faà di Bruno construction [8,10] in the more general setting of change action models.

Change action models capture many versions of differentiation that arise in mathematics and computer science. We illustrate their generality via three examples. The first, *(generalised) cartesian differential categories* (GCDC) [4, 10], are themselves an axiomatisation of the essential properties of the derivative. We show that a GCDC $\mathbf{C}$—which by definition associates every object A with a monoid $L(A) = (L_0(A), +_A, 0_A)$—gives rise to change action models in various non-trivial ways.

Secondly we show how discrete differentiation in both the *calculus of finite differences* [15] and *Boolean differential calculus* [22,23] can be modelled using the full subcategory $\mathbf{Grp}_{\mathbf{Set}}$ of $\mathbf{Set}$ whose objects are groups. Our unifying formulation generalises these discrete calculi to arbitrary groups, and gives an account of the chain rule in these settings.

Our third example is differentiation of regular expressions. Recall that Kleene algebra $\mathbb{K}$ is the algebra of regular expressions. We show that the algebra of polynomials over a commutative Kleene algebra is a change action model.

Outline. In Sect. 2 we present the basic definitions of change actions and differential maps, and show how they can be organised into categories. The theory of change action is extended to arbitrary cartesian categories $\mathbf{C}$ in Sect. 3: we introduce the category $\mathrm{CAct}(\mathbf{C})$ of internal change actions on $\mathbf{C}$. In Sect. 4 we present change action models, and properties of the tangent bundle functors. In Sect. 5 we illustrate the unifying power of change action models via three examples. In Sect. 6, we study the category $\mathrm{CAct}_\omega(\mathbf{C})$ of ω-change actions and ω-differential maps. Missing proofs are provided in an extended version of the present paper [1].

2 Change Actions

A *change action* is a tuple $A = (|A|, \Delta A, \oplus_A, +_A, 0_A)$ where $|A|$ and ΔA are sets, $(\Delta A, +_A, 0_A)$ is a monoid, and $\oplus_A : |A| \times \Delta A \to |A|$ is an action of the monoid on $|A|$.[1] We omit the subscript from $\oplus_A, +_A$ and 0_A whenever we can.

Definition 1 (Derivative condition). Let A and B be change actions. A function $f : |A| \to |B|$ is *differentiable* if there is a function $\partial f : |A| \times \Delta A \to \Delta B$ satisfying $f(a \oplus_A \delta a) = f(a) \oplus_B \partial f(a, \delta a)$, for all $a \in |A|, \delta a \in \Delta A$. We call ∂f a *derivative* for f, and write $f : A \to B$ whenever f is differentiable.

Lemma 1 (Chain rule). *Given $f : A \to B$ and $g : B \to C$ with derivatives ∂f and ∂g respectively, the function $\partial(g \circ f) : |A| \times \Delta A \to \Delta C$ defined by $\partial(g \circ f)(a, \delta a) := \partial g(f(a), \partial f(a, \delta a))$ is a derivative for $g \circ f : |A| \to |C|$.*

Proof. Unpacking the definition, we have $(g \circ f)(a) \oplus_C \partial(g \circ f)(a, \delta a) = g(f(a)) \oplus_C \partial g(f(a), \partial f(a, \delta a)) = g(f(a) \oplus_B \partial f(a, \delta a)) = g(f(a \oplus_A \delta a))$, as desired. □

Example 1 (Some useful change actions).

1. If $(A, +, 0)$ is a monoid, $(A, A, +, +, 0)$ is a change action (called *monoidal*).
2. For any set A, $A_\star := (A, \{\star\}, \pi_1, \pi_1, \star)$ is a (trivial) change action.
3. Let $A \Rightarrow B$ be the set of functions from A from B, and $\mathrm{ev}_{A,B} : A \times (A \Rightarrow B) \to B$ be the usual evaluation map. Then $(A, A \Rightarrow A, \mathrm{ev}_{A,A}, \circ, \mathrm{Id}_A)$ is a change action. If $U \subseteq (A \Rightarrow A)$ contains the identity map and is closed under composition, $(A, U, \mathrm{ev}_{A,A} \restriction_{A \times U}, \circ \restriction_{U \times U}, \mathrm{Id}_U)$ is a change action.

Regular Derivatives. The preceding definitions neither assume nor guarantee a derivative to be additive (i.e. they may not satisfy $\partial f(x, \Delta a + \Delta b) = \partial f(x, \Delta a) + \partial f(x, \Delta b)$), as they are in standard differential calculus. A strictly weaker condition that we will now require is *regularity*: if a derivative is additive in its second argument then it is regular, but not vice versa. Under some conditions, the converse is also true.

Definition 2. Given a differentiable map $f : A \to B$, a derivative ∂f for f is *regular* if, for all $a \in |A|$ and $\delta a, \delta b \in \Delta A$, we have $f(a, 0_A) = 0_B$ and $\partial f(a, \delta a +_A \delta b) = \partial f(a, \delta a) +_B \partial f(a \oplus_A \delta a, \delta b)$.

Proposition 1. *Whenever $f : A \to B$ is differentiable and has a unique derivative ∂f, this derivative is regular.*

Proposition 2. *Given $f : A \to B$ and $g : B \to C$ with regular derivatives ∂f and ∂g respectively, the derivative $\partial(g \circ f) = \partial g \circ \langle f \circ \pi_1, \partial f \rangle$ is regular.*

---

[1] Change actions are closely related to the notion of *change structures* introduced in [6] but differ from the latter in not being dependently typed or assuming the existence of an $\ominus$ operator, and requiring ΔA to have a monoid structure compatible with the map $\oplus$.

Two Categories of Change Actions. The study of change actions can be undertaken in two ways: one can consider functions that are differentiable (without choosing a derivative); alternatively, the derivative itself can be considered part of the morphism. The former leads to the category $\mathbf{CAct}^-$, whose objects are change actions and morphisms are the differentiable maps.

The category $\mathbf{CAct}^-$ was the category we originally proposed [2]. It is well-behaved, possessing limits, colimits, and exponentials, which is a trivial corollary of the following result:

Theorem 1. *The category* $\mathbf{CAct}^-$ *of change actions and differentiable morphisms is equivalent to* **PreOrd**, *the category of preorders and monotone maps.*

The actual structure of the limits and colimits in $\mathbf{CAct}^-$ is, however, not so satisfactory. One can, for example, obtain the product of two change actions A and B by taking their product in **PreOrd** and turning it into a change action, but the corresponding monoid action map $\oplus$ is not, in general, easily expressible, even if those for A and B are. Derivatives of morphisms in $\mathbf{CAct}^-$ can also be hard to obtain, as exhibiting f as a morphism in $\mathbf{CAct}^-$ merely proves it is differentiable but gives no clue as to how a derivative might be constructed.

A more constructive approach is to consider morphism as a function together with a choice of a derivative for it.

Definition 3. Given change actions A and B, a *differential map* $f : A \to B$ is a pair $(|f|, \partial f)$ where $|f| : |A| \to |B|$ is a function, and $\partial f : |A| \times \Delta A \to \Delta B$ is a regular derivative for $|f|$.

The category **CAct** has change actions as objects and differential maps as morphisms. The identity morphisms are (Id_A, π_1); given morphisms $f : A \to B$ and $g : B \to C$, define the composite $g \circ f := (|g| \circ |f|, \partial g \circ \langle |f| \circ \pi_1, \partial f\rangle) : A \to C$.

Finite products and coproducts exist in **CAct** (see Theorems 2 and 4 for a more general statement). Whether limits and colimits exist in **CAct** beyond products and coproducts is open.

Remark 1. If one thinks of changes (i.e. elements of ΔA) as morphisms between elements of $|A|$, then regularity resembles functoriality. This intuition is explored in [1, Appendix F], where we show that categories of change actions organise themselves into 2-categories.

3 Change Actions on Arbitrary Categories

The definition of change actions makes no use of any properties of **Set** beyond the existence of products. Indeed, change actions can be characterised as just a kind of multi-sorted algebra, which is definable in any category with products.

The Category CAct(C). Consider the category $\mathbf{Cat}_\times$ of (small) cartesian categories (i.e. categories with chosen finite products) and product-preserving functors. We can define an endofunctor $\mathrm{CAct} : \mathbf{Cat}_\times \to \mathbf{Cat}_\times$ sending a category $\mathbf{C}$ to the category of (internal) change actions on $\mathbf{C}$.

The objects of $\mathrm{CAct}(\mathbf{C})$ are tuples $A = (|A|, \Delta A, \oplus_A, +_A, 0_A)$ where $|A|$ and ΔA are (arbitrary) objects in $\mathbf{C}$, $(\Delta A, +_A, 0_A)$ is a monoid object in $\mathbf{C}$, and $\oplus_A : |A| \times \Delta A \to |A|$ is a monoid action in $\mathbf{C}$, i.e. a $\mathbf{C}$-morphism satisfying, for all $a : C \to |A|, \delta_1 a, \delta_2 a : C \to \Delta A$:

$$\oplus_A \circ \langle a, 0_A \circ ! \rangle = a$$
$$\oplus_A \circ \langle a, +_A \circ \langle \delta_1 a, \delta_2 a \rangle \rangle = \oplus_A \circ \langle \oplus_A \circ \langle a, \delta_1 a \rangle, \delta_2 a \rangle$$

Given objects A, B in $\mathrm{CAct}(\mathbf{C})$, the morphisms of $\mathrm{CAct}(A, B)$ are pairs $f = (|f|, \partial f)$ where $|f| : |A| \to |B|$ and $\partial f : |A| \times \Delta A \to \Delta B$ are morphisms in $\mathbf{C}$, satisfying a diagrammatic version of the derivative condition:

$$
\begin{array}{ccc}
|A| \times \Delta A & \xrightarrow{\langle |f| \circ \pi_1, \partial f \rangle} & |B| \times \Delta B \\
\oplus_A \downarrow & & \downarrow \oplus_B \\
|A| & \xrightarrow{\quad |f| \quad} & |B|
\end{array}
$$

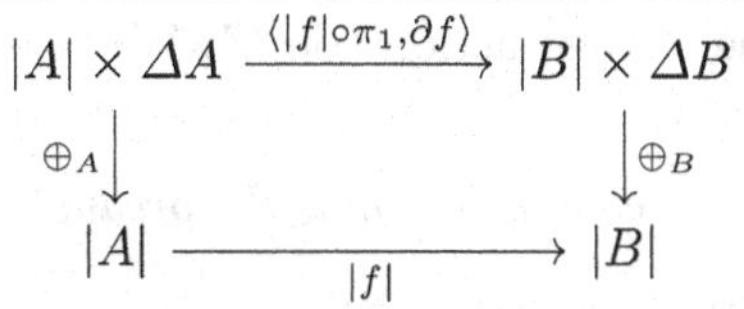

Additionally, we require our derivatives to be regular, as in Definition 2, i.e. for all morphisms $a : C \to |A|, \delta_1 a, \delta_2 a : C \to \Delta A$, the following equations hold:

$$\partial f \circ \langle a, 0_A \circ ! \rangle = 0_B$$
$$\partial f \circ \langle a, +_A \circ \langle \delta_1 a, \delta_2 a \rangle \rangle = +_A \circ \langle \partial f \circ \langle a, \delta_1 a \rangle, \partial f \circ \langle +_A \circ \langle a, \delta_1 a \rangle, \delta_2 a \rangle \rangle$$

The chain rule can then be expressed naturally by pasting two instances of the previous diagram together:

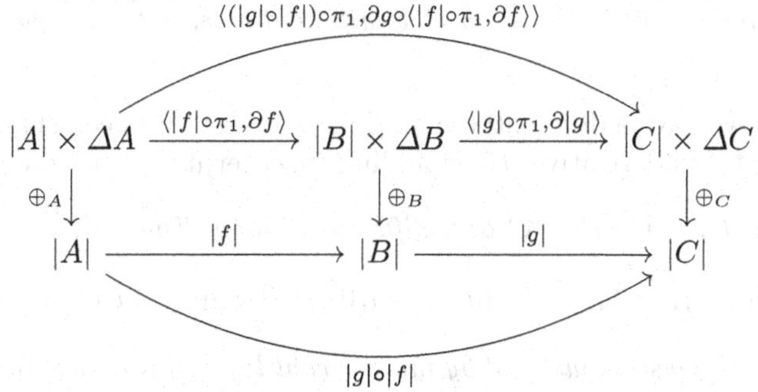

Hence $f \circ g = \langle (|g| \circ |f|) \circ \pi_1, \partial g \circ \langle |f| \circ \pi_1, \partial f \rangle \rangle$.

Now, given a product-preserving functor $\mathrm{F} : \mathbf{C} \to \mathbf{D}$, there is a corresponding functor $\mathrm{CAct}(\mathrm{F}) : \mathrm{CAct}(\mathbf{C}) \to \mathrm{CAct}(\mathbf{D})$ given by:

$$\mathrm{CAct}(\mathrm{F})(|A|, \Delta A, \oplus_A, +_A, 0_A) := (\mathrm{F}(|A|), \mathrm{F}(\Delta A), \mathrm{F}(\oplus_A), \mathrm{F}(+_A), \mathrm{F}(0_A))$$
$$\mathrm{CAct}(\mathrm{F})(|f|, \partial f) := (\mathrm{F}(|f|), \mathrm{F}(\partial f))$$

We can embed $\mathbf{C}$ fully and faithfully into $\mathrm{CAct}(\mathbf{C})$ via the functor $\eta_{\mathbf{C}}$ which sends an object A of $\mathbf{C}$ to the "trivial" change action $A_\star = (A, \top, \pi_1, !, !)$ and every morphism $f : A \to B$ of $\mathbf{C}$ to the morphism $(f, !)$. As before, this functor extends to a natural transformation from the identity functor to CAct.

Additionally, there is an obvious forgetful functor $\varepsilon_{\mathbf{C}} : \mathrm{CAct}(\mathbf{C}) \to \mathbf{C}$, which defines the components of a natural transformation ε from the functor CAct to the identity endofunctor Id.

Given $\mathbf{C}$, we write $\xi_{\mathbf{C}}$ for the functor $\mathrm{CAct}(\varepsilon_{\mathbf{C}}) : \mathrm{CAct}(\mathrm{CAct}(\mathbf{C})) \to \mathrm{CAct}(\mathbf{C})$.[2] Explicitly, this functor maps an object $(A, B, \oplus, +, 0)$ in $\mathrm{CAct}(\mathrm{CAct}(\mathbf{C}))$ to the object $(|A|, |B|, |\oplus|, |+|, |0|)$. Intuitively, $\varepsilon_{\mathrm{CAct}(\mathbf{C})}$ prefers the "original" structure on objects, whereas $\xi_{\mathbf{C}}$ prefers the "higher" structure. The equaliser of these two functors is precisely the category of change actions whose higher structure is the original structure.

Products and Coproducts in CAct(C). We have defined CAct as an endofunctor on cartesian categories. This is well-defined: if $\mathbf{C}$ has all finite (co)products, so does $\mathrm{CAct}(\mathbf{C})$. Let $A = (|A|, \Delta A, \oplus_A, +_A, 0_A)$ and $B = (|B|, \Delta B, \oplus_B, +_B, 0_B)$ be change actions on $\mathbf{C}$. We present their product and coproducts as follows.

Theorem 2. *The following change action is the product of A and B in* $\mathrm{CAct}(\mathbf{C})$

$$A \times B := (|A| \times |B|, \Delta A \times \Delta B, \oplus_{A\times B}, +_{A\times B}, \langle 0_A, 0_B \rangle)$$

where $\oplus_{A\times B} := \langle \oplus_A \circ (\pi_1 \times \pi_1), \oplus_B \circ (\pi_2 \times \pi_2) \rangle$ *and* $+_{A\times B} := \langle +_A \circ (\pi_1 \times \pi_1), +_B \circ (\pi_2 \times \pi_2) \rangle$. *The projections are* $\overline{\pi_1} = (\pi_1, \pi_1 \circ \pi_2)$ *and* $\overline{\pi_2} = (\pi_2, \pi_2 \circ \pi_2)$, *writing $\overline{f}$ for maps f in* CAct *to distinguish them from* $\mathbf{C}$*-maps.*

Theorem 3. *The change action $\overline{\top} = (\top, \top, \pi_1, \pi_1, \mathrm{Id}_\top)$ is the terminal object in* $\mathrm{CAct}(\mathbf{C})$*, where $\top$ is the terminal object of* $\mathbf{C}$*. Furthermore, if A is a change action every point $|f| : \top \to |A|$ in* $\mathbf{C}$ *is differentiable, with (unique) derivative* 0_A.

Whenever we have a differential map $f : A \times B \to C$ between change actions, we can compute its derivative ∂f by adding together its "partial" derivatives:[3].

Lemma 2. *Let $f : A \times B \to C$ be a differential map. Then*

$$\partial f((a,b),(\delta a, \delta b)) = +_C \circ \langle \partial f((a,b),(\delta a, 0_B)), \partial f((\oplus_A \circ \langle a, \delta a \rangle, b), (0_A, \delta b)) \rangle$$

(The notational abuse is justified by the internal logic of a cartesian category.)

Theorem 4. *If* $\mathbf{C}$ *is distributive, with law $\delta_{A,B,C} : (A \sqcup B) \times C \to (A \times C) \sqcup (B \times C)$, the following change action is the coproduct of A and B in* $\mathrm{CAct}(\mathbf{C})$

$$A \sqcup B := (|A| \sqcup |B|, \Delta A \times \Delta B, \oplus_{A\sqcup B}, +_{A\sqcup B}, \langle 0_A, 0_B \rangle)$$

where $\oplus_{A\sqcup B} := [\oplus_A \circ (\mathrm{Id}_A \times \pi_1), \oplus_B \circ (\mathrm{Id}_B \times \pi_2)] \circ \delta_{A,B,C}$*, and* $+_{A\sqcup B} := \langle +_A \circ (\pi_1 \times \pi_1), +_B \circ (\pi_2 \times \pi_2) \rangle$. *The injections are* $\overline{\iota_1} = (\iota_1, \langle \pi_2, 0_B \rangle)$ *and* $\overline{\iota_2} = (\iota_2, \langle 0_A, \pi_2 \rangle)$.

[2] One might expect CAct to be a comonad with ε as a counit. But if this were the case, we would have $\xi_{\mathbf{C}} = \varepsilon_{\mathrm{CAct}(\mathbf{C})}$, which is, in general, not true.

[3] Alternatively, one can define the (first) partial derivative of a map $f(x, y)$ as a map $\delta_1 f$ such that $f(x \oplus \delta x, y) = f(x, y) \oplus \delta_1(x, y, \delta x)$. It can be shown that a map is differentiable iff its first and second derivatives exist.

Stable Derivatives and Additivity. We do not require derivatives to be additive in their second argument; indeed in many cases they are not. Under some simple conditions, however, (regular) derivatives can be shown to be additive.

Definition 4. Given a (internal) change action A and objects $|B|, |C|$ in a cartesian category $\mathbf{C}$, a morphism $u : |A| \times |B| \to |C|$ is *stable* whenever the diagram commutes:

$$\begin{array}{ccc} (|A| \times \Delta A) \times |B| & \xrightarrow{\oplus_A \times \mathrm{Id}} & |A| \times |B| \\ {\scriptstyle \pi_1 \times \mathrm{Id}} \downarrow & & \downarrow {\scriptstyle u} \\ |A| \times |B| & \xrightarrow{\quad u \quad} & |C| \end{array}$$

If one thinks of ΔA as the object of "infinitesimal" transformations on $|A|$, then the preceding definition says that a morphism $u : |A| \times |B| \to |C|$ is stable whenever infinitesimal changes on the input A do not affect its output.

Lemma 3. *Let $f = (|f|, \partial f)$ be a differential map in* $\mathrm{CAct}(\mathbf{C})$. *If ∂f is stable, then it is additive in its second argument*[4], *i.e. for all $x, \delta_1 x, \delta_2 x$ we have:*

$$\partial f \circ \langle x, +_A \circ \langle \delta_1 x, \delta_2 x \rangle \rangle = + \circ \langle \partial f \circ \langle x, \delta_1 x \rangle, \partial f \circ \langle x, \delta_2 x \rangle \rangle$$

Lemma 4. *Let $f = (|f|, \partial f)$ and $g = (|g|, \partial g)$ be differential maps, with ∂g stable. Then $\partial(g \circ f)$ is stable.*

It is straightforward to see that the category $\mathrm{Stab}(\mathbf{C})$ of change actions and differential maps with stable derivatives is a subcategory of $\mathrm{CAct}(\mathbf{C})$.

4 Higher-Order Derivatives: The Extrinsic View

In this section we study categories in which every object is equipped with a change action, and every morphism specifies a corresponding differential map. This provides a simple way of characterising categories which are models of higher-order differentiation purely in terms of change actions.

Change Action Models. Recall that a *copointed endofunctor* is a pair (F, σ) where the endofunctor $\mathrm{F} : \mathbf{C} \to \mathbf{C}$ is equipped with a natural transformation $\sigma : \mathrm{F} \stackrel{\cdot}{\to} \mathrm{Id}$. A *coalgebra of a copointed endofunctor* (F, σ) is an object A of $\mathbf{C}$ together with a morphism $\alpha : A \to \mathrm{F}A$ such that $\sigma_A \circ \alpha = \mathrm{Id}_A$.

Definition 5. We call a coalgebra $\alpha : \mathbf{C} \to \mathrm{CAct}(\mathbf{C})$ of the copointed endofunctor $(\mathrm{CAct}, \varepsilon)$ a *change action model* (on $\mathbf{C}$).

Assumption. Throughout Sect. 4, we fix a change action model $\alpha : \mathbf{C} \to \mathrm{CAct}(\mathbf{C})$.

Given an object A of $\mathbf{C}$, the coalgebra α specifies a (internal) change action $\alpha(A) = (A, \Delta A, \oplus_A, +_A, 0_A)$ in $\mathrm{CAct}(\mathbf{C})$. (We abuse notation and write ΔA for the carrier object of the monoid specified in $\alpha(A)$; similarly for $+_A, \oplus_A$, and 0_A.) Given a morphism $f : A \to B$ in $\mathbf{C}$, there is an associated differential map

[4] Note that the converse is not the case, i.e. a derivative can be additive but not stable.

$\alpha(f) = (f, \partial f) : \alpha(A) \to \alpha(B)$. Since $\partial f : A \times \Delta A \to \Delta B$ is also a $\mathbf{C}$-morphism, there is a corresponding differential map $\alpha(\partial f) = (\partial f, \partial^2 f)$ in $\mathrm{CAct}(\mathbf{C})$, where $\partial^2 f : (A \times \Delta A) \times (\Delta A \times \Delta^2 A) \to \Delta^2 B$ is a second derivative for f. Iterating this process, we obtain an n-th derivative $\partial^n f$ for every $\mathbf{C}$-morphism f. Thus change action models offer a setting for reasoning about higher-order differentiation.

Tangent Bundles in Change Action Models. In differential geometry the tangent bundle functor, which maps every manifold to its tangent bundle, is an important construction. There is an endofunctor on change action models reminiscent of the tangent bundle functor, with analogous properties.

Definition 6. The *tangent bundle functor* $\mathrm{T} : \mathbf{C} \to \mathbf{C}$ is defined as $\mathrm{T}A := A \times \Delta A$ and $\mathrm{T}f := \langle f \circ \pi_1, \partial f \rangle$.

Notation. We use shorthand $\pi_{ij} := \pi_i \circ \pi_j$.

The tangent bundle functor T preserves products up to isomorphism, i.e. for all objects A, B of $\mathbf{C}$, we have $\mathrm{T}(A \times B) \cong \mathrm{T}A \times \mathrm{T}B$ and $\mathrm{T}1 \cong 1$. In particular, $\phi_{A,B} := \langle \langle \pi_{11}, \pi_{12} \rangle, \langle \pi_{21}, \pi_{22} \rangle \rangle : \mathrm{T}A \times \mathrm{T}B \to \mathrm{T}(A \times B)$ is an isomorphism. Consequently, given maps $f : A \to B$ and $g : A \to C$, then, up to the previous isomorphism, $\mathrm{T}\langle f, g \rangle = \langle \mathrm{T}f, \mathrm{T}g \rangle$.

A consequence of the structure of products in $\mathrm{CAct}(\mathbf{C})$ is that the map $\oplus_{A \times B}$ inherits the pointwise structure in the following sense:

Lemma 5. *Let* $\phi_{A,B} : \mathrm{T}A \times \mathrm{T}B \to \mathrm{T}(A \times B)$ *be the canonical isomorphism described above. Then* $\oplus_{A \times B} \circ \phi_{A,B} = \oplus_A \times \oplus_B$.

It will often be convenient to operate directly on the functor T, rather than on the underlying derivatives. For these, the following results are useful:

Lemma 6. *The following families of morphisms are natural transformations:* $\pi_1, \oplus_A : \mathrm{T}(A) \to A$, $\mathrm{z} := \langle \mathrm{Id}, 0 \rangle : A \to \mathrm{T}(A)$ $\mathrm{l} := \langle \langle \pi_1, 0 \rangle, \langle \pi_2, 0 \rangle \rangle : \mathrm{T}(A) \to \mathrm{T}^2(A)$. *Additionally, the triple* $(\mathrm{T}, \mathrm{z}, \mathrm{T}\oplus)$ *defines a monad on* $\mathbf{C}$.

A particularly interesting class of change action models are those that are also cartesian closed. Surprisingly, this has as an immediate consequence that differentiation is itself internal to the category.

Lemma 7 (Internalisation of derivatives). *Whenever* $\mathbf{C}$ *is cartesian closed, there is a morphism* $\mathrm{d}_{A,B} : (A \Rightarrow B) \to (A \times \Delta A) \Rightarrow \Delta B$ *such that, for any morphism* $f : 1 \times A \to B$, $\mathrm{d}_{A,B} \circ \Lambda f = \Lambda(\partial f \circ \langle \langle \pi_1, \pi_{12} \rangle, \langle \pi_1, \pi_{22} \rangle \rangle)$.

Under some conditions, we can classify the structure of the exponentials in $(\mathrm{CAct}, \varepsilon)$-coalgebras. This requires the existence of an infinitesimal object.[5]

[5] The concept of "infinitesimal object" is borrowed from synthetic differential geometry [18]. However, there is nothing intrinsically "infinitesimal" about such objects here.

Definition 7. If $\mathbf{C}$ is cartesian closed, an *infinitesimal object* D is an object of $\mathbf{C}$ such that the tangent bundle functor T is represented by the covariant Hom-functor $D \Rightarrow (\cdot)$, i.e. there is a natural isomorphism $\phi : (D \Rightarrow (\cdot)) \dot{\rightarrow} \mathrm{T}$.

Lemma 8. *Whenever there is an infinitesimal object in* $\mathbf{C}$*, the tangent bundle* $\mathrm{T}(A \Rightarrow B)$ *is naturally isomorphic to* $A \Rightarrow \mathrm{T}B$.

We would like the tangent bundle functor to preserve the exponential structure; in particular we would expect a result of the form $\frac{\partial(\lambda y.t)}{\partial x} = \lambda y.\frac{\partial t}{\partial x}$, which is true in differential λ-calculus [11]. Unfortunately it seems impossible to prove in general that this equation holds, although weaker results are available. If the tangent bundle functor is representable, however, additional structure is preserved.

Theorem 5. *The isomorphism between the functors* $\mathrm{T}(A \Rightarrow (\cdot))$ *and* $A \Rightarrow \mathrm{T}(\cdot)$ *respects the structure of* T*, in the sense that the diagram commutes.*

$$\begin{array}{ccc} \mathrm{T}(A \Rightarrow B) & \xrightarrow{\cong} & A \Rightarrow \mathrm{T}(B) \\ \downarrow{\scriptstyle \oplus_{A \Rightarrow B}} & \swarrow{\scriptstyle \mathrm{Id}_A \Rightarrow \oplus_B} & \\ A \Rightarrow B & & \end{array}$$

5 Examples of Change Action Models

Generalised Cartesian Differential Categories. *Generalised cartesian differential categories* (GCDC) [10]—a recent generalisation of cartesian differential categories [4]—are models of differential calculi. We show that change action models generalise GCDC in that GCDCs give rise to change action models in three[6] different (non-trivial) ways. In this subsection let $\mathbf{C}$ be a GCDC (we assume familiarity with the definitions and notations in [10]).

1. The Flat Model. Define the functor $\alpha : \mathbf{C} \rightarrow \mathrm{CAct}(\mathbf{C})$ as follows. Let $f : A \rightarrow B$ be a $\mathbf{C}$-morphism. Then $\alpha(A) := (A, L_0(A), \pi_1, +_A, 0_A)$ and $\alpha(f) := (f, \mathrm{D}[f])$.

Theorem 6. *The functor* α *is a change action model.*

2. The Kleisli Model. GCDCs admit a tangent bundle functor, defined analogously to the standard notion in differential geometry. Let $f : A \rightarrow B$ be a $\mathbf{C}$-morphism. Define the *tangent bundle functor* $\mathrm{T} : \mathbf{C} \rightarrow \mathbf{C}$ as: $\mathrm{T}A := A \times L_0(A)$, and $\mathrm{T}f := \langle f \circ \pi_1, \mathrm{D}[f] \rangle$. The functor T is in fact a monad, with unit $\eta = \langle \mathrm{Id}, 0_A \rangle : A \rightarrow A \times L_0(A)$ and multiplication $\mu : (A \times L_0(A)) \times L_0(A)^2 \rightarrow A \times L_0(A)$ defined by the composite:

$$(A \times L_0(A)) \times L_0(A)^2 \xrightarrow{\langle \pi_1 \circ \pi_1, \langle \pi_2 \circ \pi_1, \pi_1 \circ \pi_2 \rangle \rangle} A \times L_0(A)^2 \xrightarrow{\mathrm{Id} \times +_A} A \times L_0(A)$$

Thus we can define the Kleisli category of this functor by $\mathbf{C}_{\mathrm{T}}$ which has geometric significance as a category of generalised vector fields.

[6] The third, the Eilenberg-Moore model, is presented in [1, Appendix D].

We define the functor $\alpha_{\mathrm{T}} : \mathbf{C}_{\mathrm{T}} \to \mathrm{CAct}(\mathbf{C}_{\mathrm{T}})$: given a $\mathbf{C}_{\mathrm{T}}$-morphism $f : A \to B$, set $\alpha_{\mathrm{T}}(A) := (A, L_0(A), \mathrm{Id}_A \times \mathrm{Id}_{L_0(A)}, \eta \circ +_A, \eta \circ 0_A)$ and $\alpha_{\mathrm{T}}(f) := (f, \mathrm{D}[f])$.

Lemma 9. *α_{T} is a change action model.*

Remark 2. The converse is not true: in general the existence of a change action model on $\mathbf{C}$ does not imply that $\mathbf{C}$ satisfies the GCDC axioms. However, if one requires, additionally, $(\Delta A, +_A, 0_A)$ to be commutative, with $\Delta(\Delta A) = \Delta A$ and $\oplus_{\Delta A} = +_A$ for all objects A, and some technical conditions (stability and uniqueness of derivatives), then it can be shown that $\mathbf{C}$ is indeed a GCDC.

Difference Calculus and Boolean Differential Calculus. Consider the full subcategory $\mathbf{Grp}_{\mathbf{Set}}$ of $\mathbf{Set}$ whose objects are all the groups[7]. This is a cartesian closed category which can be endowed with the structure of a $(\mathrm{CAct}, \varepsilon)$-coalgebra α in a straightforward way.

Given a group $A = (A, +, 0, -)$, define change action $\alpha(A) := (A, A, +, +, 0)$ Given a function $f : A \to B$, define differential map $\alpha(f) := (f, \partial f)$ where $\partial f(x, \delta x) := -f(x) + f(x \oplus \delta x)$. Notice $f(x) \oplus \partial f(x, \delta x) = f(x) + (-f(x) + f(x + \delta x)) = f(x + \delta x) = f(x \oplus \delta x)$; hence ∂f is a derivative for f which is regular (but not necessarily additive), and $\alpha(f)$ a map in $\mathrm{CAct}(\mathbf{Grp}_{\mathbf{Set}})$. The following result is then immediate.

Lemma 10. *$\alpha : \mathbf{Grp}_{\mathbf{Set}} \to \mathrm{CAct}(\mathbf{Grp}_{\mathbf{Set}})$ defines a change action model.*

This result is significant: in the calculus of finite differences [15], the *discrete derivative* (or *discrete difference operator*) of a function $f : \mathbb{Z} \to \mathbb{Z}$ is defined as $\delta f(x) := f(x+1) - f(x)$. In fact the discrete derivative δf is (an instance of) the derivative of f *qua* morphism in $\mathbf{Grp}_{\mathbf{Set}}$, i.e. $\delta f(x) = \partial f(x, 1)$.

Finite difference calculus [13,15] has found applications in combinatorics and numerical computation. Our formulation via change action model over $\mathbf{Grp}_{\mathbf{Set}}$ has several advantages. First it justifies the chain rule, which seems new. Secondly, it generalises the calculus to arbitrary groups. To illustrate this, consider *Boolean differential calculus* [22,23], a technique that applies methods from calculus to the space $\mathbb{B}^n$ of vectors of elements of some Boolean algebra $\mathbb{B}$.

Definition 8. Given a Boolean algebra $\mathbb{B}$ and function $f : \mathbb{B}^n \to \mathbb{B}^m$, the *i-th Boolean derivative of f at $(u_1, \ldots, u_n) \in \mathbb{B}^n$* is the value $\frac{\partial f}{\partial x_i}(u_1, \ldots, u_n) := f(u_1, \ldots, u_n) \not\leftrightarrow f(u_1, \ldots, \neg u_i, \ldots, u_n)$ writing $u \not\leftrightarrow v := (u \wedge \neg v) \vee (\neg u \wedge v)$ for exclusive-or.

Now $\mathbb{B}^n$ is a $\mathbf{Grp}_{\mathbf{Set}}$-object. Set $\top_i := (\bot, \overset{i-1}{\ldots}, \bot, \top, \bot, \overset{n-i}{\ldots}, \bot) \in \mathbb{B}^n$.

Lemma 11. *The Boolean derivative of $f : \mathbb{B}^n \to \mathbb{B}^m$ coincides with its derivative qua morphism in $\mathbf{Grp}_{\mathbf{Set}}$: $\frac{\partial f}{\partial x_i}(u_1, \ldots, u_n) = \partial f((u_1, \ldots, u_n), \top_i)$.*

[7] We consider arbitrary functions, rather than group homomorphisms, since, according to this change action structure, every function between groups is differentiable.

Polynomials over Commutative Kleene Algebras. The algebra of polynomials over a commutative Kleene algebra [14,17] (see [12,21] for work of a similar vein) is a change action model. Recall that Kleene algebra is the algebra of regular expressions [5,9]. Formally a *Kleene algebra* $\mathbb{K}$ is a tuple $(K, +, \cdot, ^\star, 0, 1)$ such that $(K, +, \cdot, 0, 1)$ is an idempotent semiring under $+$ satisfying, for all $a, b, c \in K$:

$$1 + a\,a^\star = a^\star \quad 1 + a^\star a = a^\star \quad b + a\,c \leq c \rightarrow a^\star\,b \leq c \quad b + c\,a \leq c \rightarrow b\,a^\star \leq c$$

where $a \leq b := a + b = b$. A Kleene algebra is *commutative* whenever $\cdot$ is.

Henceforth fix a commutative Kleene algebra $\mathbb{K}$. Define the *algebra of polynomials* $\mathbb{K}[\overline{x}]$ as the free extension of the algebra $\mathbb{K}$ with elements $\overline{x} = x_1, \ldots, x_n$. We write $p(\overline{a})$ for the value of $p(\overline{x})$ evaluated at $\overline{x} \mapsto \overline{a}$. Polynomials, viewed as functions, are closed under composition: when $p \in \mathbb{K}[\overline{x}], q_1, \ldots, q_n \in \mathbb{K}[\overline{y}]$ are polynomials, so is the composite $p(q_1(\overline{y}), \ldots, q_n(\overline{y}))$.

Given a polynomial $p = p(\overline{x})$, we define its *i-th derivative* $\frac{\partial p}{\partial x_i}(\overline{x}) \in \mathbb{K}[\overline{x}]$:

$$\frac{\partial\, a}{\partial x_i}(\overline{x}) = 0 \qquad \frac{\partial\, p^\star}{\partial x_i}(\overline{x}) = p^\star(\overline{x})\,\frac{\partial\, p}{\partial x_i}(\overline{x}) \qquad \frac{\partial\, x_j}{\partial x_i}(\overline{x}) = \begin{cases} 1 \text{ if } i = j \\ 0 \text{ otherwise} \end{cases}$$

$$\frac{\partial\,(p+q)}{\partial x_i}(\overline{x}) = \frac{\partial\, p}{\partial x_i}(\overline{x}) + \frac{\partial\, q}{\partial x_i}(\overline{x}) \qquad \frac{\partial\,(p\,q)}{\partial x_i}(\overline{x}) = p(\overline{x})\frac{\partial\, q}{\partial x_i}(\overline{x}) + q(\overline{x})\frac{\partial\, p}{\partial x_i}(\overline{x})$$

Write $\frac{\partial p}{\partial x_i}(\overline{e})$ to mean the result of evaluating the polynomial $\frac{\partial p}{\partial x_i}(\overline{x})$ at $\overline{x} \mapsto \overline{e}$.

Theorem 7 (Taylor's formula [14]). *Let $p(x) \in \mathbb{K}[x]$. For all $a, b \in \mathbb{K}[x]$, we have $p(a + b) = p(a) + b \cdot \frac{\partial p}{\partial x}(a + b)$.*

The category of finite powers of $\mathbb{K}$, $\mathbb{K}_\times$, has all natural numbers n as objects. The morphisms $\mathbb{K}_\times[m, n]$ are n-tuples of polynomials $(p_1, \ldots, p_n)$ where $p_1, \ldots, p_n \in \mathbb{K}[x_1, \ldots, x_m]$. Composition of morphisms is the usual composition of polynomials.

Lemma 12. *The category $\mathbb{K}_\times$ is a cartesian category, endowed with a change action model $\alpha : \mathbb{K}_\times \rightarrow \mathrm{CAct}(\mathbb{K}_\times)$ whereby $\alpha(\mathbb{K}) := (\mathbb{K}, \mathbb{K}, +, +, 0)$, $\alpha(\mathbb{K}^i) := \alpha(\mathbb{K})^i$; for $\overline{p} = (p_1(\overline{x}), \ldots, p_n(\overline{x})) : \mathbb{K}^m \rightarrow \mathbb{K}^n$, $\alpha(\overline{p}) := (\overline{p}, (p'_1, \ldots, p'_n))$, where $(p'_i = p'_i(x_1, \ldots, x_m, y_1, \ldots, y_m) := \sum_{j=1}^{n} y_j \cdot \frac{\partial p_i}{\partial x_j}(x_1 + y_1, \ldots, x_m + y_m)$.*

Remark 3. Interestingly derivatives are not additive in the second argument. Take $p(x) = x^2$. Then $\partial p(a, b + c) > \partial p(a, b) + \partial p(a, c)$. It follows that $\mathbb{K}[\overline{x}]$ cannot be modelled by GCDC (because of axiom [CD.2]).

6 ω-Change Actions and ω-Differential Maps

A change action model $\alpha : \mathbf{C} \rightarrow \mathrm{CAct}(\mathbf{C})$ is a category that supports higher-order differentials: each $\mathbf{C}$-object A is associated with an ω-sequence of change actions—$\alpha(A), \alpha(\Delta A), \alpha(\Delta^2 A), \ldots$—in which every change action is compatible with the neighbouring change actions. We introduce *ω-change actions* as a means of constructing change action models "freely": given a cartesian category $\mathbf{C}$, the objects of the category $\mathrm{CAct}_\omega(\mathbf{C})$ are all ω-sequences of "contiguously compatible" change actions.

We work with ω-sequences $[A_i]_{i\in\omega}$ and $[f_i]_{i\in\omega}$ of objects and morphisms in $\mathbf{C}$. We write $\mathsf{p}_k([A_i]_{i\in\omega}) := A_k$ for the k-th element of the ω-sequence (similarly for $\mathsf{p}_k([f_i]_{i\in\omega})$), and omit the subscript '$i \in \omega$' from $[A_i]_{i\in\omega}$ to reduce clutter. Given ω-sequences $[A_i]$ and $[B_i]$ of objects of a cartesian category $\mathbf{C}$, define ω-sequences, *product* $[A_i] \times [B_i]$, *left shift* $\Pi[A_i]$ and *derivative space* $\mathbf{D}[A_i]$, by:

$$\mathsf{p}_j([A_i] \times [B_i]) := A_j \times B_j \qquad \mathsf{p}_j(\Pi[A_i]) := A_{j+1}$$
$$\mathsf{p}_0(\mathbf{D}[A_i]) := A_0 \qquad \mathsf{p}_{j+1}\mathbf{D}[A_i] := \mathsf{p}_j\mathbf{D}[A_i] \times \mathsf{p}_j\mathbf{D}(\Pi[A_i])$$

Example 2. Given an ω-sequence $[A_i]$, the first few terms of $\mathbf{D}[A_i]$ are:

$$\mathsf{p}_0\mathbf{D}[A_i] = A_0 \quad \mathsf{p}_1\mathbf{D}[A_i] = A_0 \times A_1 \quad \mathsf{p}_2\mathbf{D}[A_i] = (A_0 \times A_1) \times (A_1 \times A_2)$$
$$\mathsf{p}_3\mathbf{D}[A_i] = ((A_0 \times A_1) \times (A_1 \times A_2)) \times ((A_1 \times A_2) \times (A_2 \times A_3))$$

Definition 9. Given ω-sequences $[A_i]$ and $[B_i]$, a *pre-ω-differential map* between them, written $[f_i] : [A_i] \to [B_i]$, is an ω-sequence $[f_i]$ such that for each j, $f_j : \mathsf{p}_j\mathbf{D}[A_i] \to B_j$ is a $\mathbf{C}$-morphism.

We explain the intuition behind the derivative space $\mathbf{D}[A_i]$. Take a morphism $f : A \to B$, and set $A_i = \Delta^i A$ (where $\Delta^0 := A$ and $\Delta^{n+1}A := \Delta(\Delta^n A)$). Since Δ distributes over product, the domain of the n-th derivative of f is $\mathsf{p}_n\mathbf{D}[A_i]$.

Notation. Define $\pi_1^{\langle 0\rangle} := \pi_1$ and $\pi_1^{\langle j+1\rangle} := \pi_1^{\langle j\rangle} \times \pi_1^{\langle j\rangle}$; and define $\pi_2^{(0)} := \mathrm{Id}$ and $\pi_2^{(j+1)} := \pi_2 \circ \pi_2^{(j)}$.

Definition 10. Let $[f_i] : [A_i] \to [B_i]$ and $[g_i] : [B_i] \to [C_i]$ be pre-ω-differential maps. The *derivative sequence* $\mathbf{D}[f_i]$ is the ω-sequence defined by:

$$\mathsf{p}_j\mathbf{D}[f_i] := \langle f_j \circ \pi_1^{\langle j\rangle}, f_{j+1}\rangle : \mathsf{p}_{j+1}\mathbf{D}[A_i] \to B_j \times B_{j+1}$$

Using the shorthand $\mathbf{D}^n[f_i] := \underbrace{\mathbf{D}(\ldots(\mathbf{D}}_{n \text{ times}}[f_i]))$, the *composite* $[g_i] \circ [f_i] : [A_i] \to [C_i]$ is the pre-ω-differential map given by $\mathsf{p}_j([g_i] \circ [f_i]) = g_j \circ \mathsf{p}_0(\mathbf{D}^j[f_i])$. The *identity* pre-$\omega$-differential map $\mathrm{Id} : [A_i] \to [A_i]$ is defined as: $\mathsf{p}_j\mathrm{Id} := \pi_2^{(j)} : \mathsf{p}_j\mathbf{D}[A_i] \to A_j$.

Example 3. Consider ω-sequences $[f_i]$ and $[g_i]$ as above. Then:

$$\mathsf{p}_0\mathbf{D}[f_i] = \langle f_0 \circ \pi_1^{\langle 0\rangle}, f_1\rangle \qquad \mathsf{p}_1\mathbf{D}[f_i] = \langle f_1 \circ \pi_1^{\langle 1\rangle}, f_2\rangle$$
$$\mathsf{p}_0\mathbf{D}^2[f_i] = \langle\langle f_0 \circ \pi_1^{\langle 0\rangle}, f_1\rangle \circ \pi_1, \langle f_1 \circ \pi_1^{\langle 1\rangle}, f_2\rangle\rangle$$
$$\mathsf{p}_1\mathbf{D}^2[f_i] = \langle\langle f_1 \circ \pi_1^{\langle 1\rangle}, f_2\rangle \circ \pi_1^{\langle 1\rangle}, \langle f_2 \circ \pi_1^{\langle 2\rangle}, f_3\rangle\rangle$$
$$\mathsf{p}_0\mathbf{D}^3[f_i] = \langle \mathsf{p}_0\mathbf{D}^2[f_i] \circ \pi_1^{\langle 0\rangle}, \langle\langle f_1 \circ \pi_1^{\langle 1\rangle}, f_2\rangle \circ \pi_1^{\langle 1\rangle}, \langle f_2 \circ \pi_1^{\langle 2\rangle}, f_3\rangle\rangle\rangle$$

It follows that the first few terms of the composite $[g_i] \circ [f_i]$ are:

$$\mathsf{p}_0([g_i] \circ [f_i]) = g_0 \circ f_0 \qquad \mathsf{p}_1([g_i] \circ [f_i]) = g_1 \circ \langle f_0 \circ \pi_1^{\langle 0 \rangle}, f_1 \rangle$$
$$\mathsf{p}_2([g_i] \circ [f_i]) = g_2 \circ \langle \langle f_0 \circ \pi_1, f_1 \rangle \circ \pi_1^{\langle 0 \rangle}, \langle f_1 \circ \pi_1^{\langle 1 \rangle}, f_2 \rangle \rangle$$

Notice that these correspond to iterations of the chain rule, assuming $f_{i+1} = \partial f_i$ and $g_{i+1} = \partial g_i$.

Proposition 3. *For any pre-ω-differential map $[f_i]$, $\mathrm{Id} \circ [f_i] = [f_i] \circ \mathrm{Id} = [f_i]$.*

Proposition 4. *Composition of pre-ω-differential maps is associative: given pre-ω-differential maps $[f_i] : [A_i] \to [B_i]$, $[g_i] : [B_i] \to [C_i]$ and $[h_i] : [C_i] \to [D_i]$, then for all $n \geq 0$, $h_n \circ \mathsf{p}_0\mathbf{D}^n([g_i] \circ [f_i]) = (h_n \circ \mathsf{p}_0\mathbf{D}^n[g_i]) \circ \mathsf{p}_0\mathbf{D}^n[f_i]$.*

Definition 11. Given pre-ω-differential maps $[f_i] : [A_i] \to [B_i], [g_i] : [A_i] \to [C_i]$, the *pairing* $\langle [f_i], [g_i] \rangle : [A_i] \to [B_i] \times [C_i]$ is the pre-ω-differential map defined by: $\mathsf{p}_j \langle [f_i], [g_i] \rangle = \langle f_j, g_j \rangle$. Define pre-$\omega$-differential maps $\pi_\mathbf{1} := [\pi_{\mathbf{1}i}] : [A_i] \times [B_i] \to [A_i]$ by $\mathsf{p}_j[\pi_{\mathbf{1}i}] := \pi_1 \circ \pi_2^{(j)}$, and $\pi_\mathbf{2} := [\pi_{\mathbf{2}i}] : [A_i] \times [B_i] \to [B_i]$ by $\mathsf{p}_j[\pi_{\mathbf{2}i}] := \pi_2 \circ \pi_2^{(j)}$.

Definition 12. A *pre-ω-change action* on a cartesian category $\mathbf{C}$ is a quadruple $\widehat{A} = ([A_i], [\widehat{\oplus^A}_i], [\widehat{+^A}_i], [0_i^A])$ where $[A_i]$ is an ω-sequence of $\mathbf{C}$-objects, and for each $j \geq 0$, $\widehat{\oplus^A}_j$ and $\widehat{+^A}_j$ are ω-sequences, satisfying

1. $\widehat{\oplus^A}_j : \Pi^j[A_i] \times \Pi^{j+1}[A_i] \to \Pi^j[A_i]$ is a pre-ω-differential map.
2. $\widehat{+^A}_j : \Pi^{j+1}[A_i] \times \Pi^{j+1}[A_i] \to \Pi^{j+1}[A_i]$ is a pre-ω-differential map.
3. $0_j^A : \top \to A_{j+1}$ is a $\mathbf{C}$-morphism.
4. $\Delta(\widehat{A}, j) := (A_j, A_{j+1}, \mathsf{p}_0\widehat{\oplus^A}_j, \mathsf{p}_0\widehat{+^A}_j, 0_j^A)$ is a change action in $\mathbf{C}$.

We extend the left-shift operation to pre-ω-change actions by defining $\Pi\widehat{A} := (\Pi[A_i], \Pi[\widehat{\oplus^A}_i], \Pi[\widehat{+^A}_i], [0_i^A])$. Then we define the change actions $\mathbf{D}(\widehat{A}, j)$ inductively by: $\mathbf{D}(\widehat{A}, 0) := \Delta(\widehat{A}, 0)$ and $\mathbf{D}(\widehat{A}, j+1) := \Delta(\widehat{A}, j) \times \Delta(\Pi\widehat{A}, j)$. Notice that the carrier object of $\mathbf{D}(\widehat{A}, j)$ is the j-th element of the ω-sequence $\mathbf{D}[A_i]$.

Definition 13. Given pre-ω-change actions $\widehat{A}$ and $\widehat{B}$ (using the preceding notation), a pre-ω-differential map $[f_i] : [A_i] \to [B_i]$ is *ω-differential* if, for each $j \geq 0$, (f_j, f_{j+1}) is a differential map from the change action $\mathbf{D}(\widehat{A}, j)$ to $\Delta(\widehat{B}, j)$. Whenever $[f_i]$ is an ω-differential map, we write $\widehat{f} : \widehat{A} \to \widehat{B}$.

We say that a pre-ω-change action $\widehat{A}$ is an *ω-change action* if, for each $i \geq 0$, $\widehat{\oplus^A}_i$ and $\widehat{+^A}_i$ are ω-differential maps.[8]

[8] It is important to sequence the definitions appropriately. Notice that we only define ω-differential maps once there is a notion of pre-ω-change action, but pre-ω-change actions need pre-ω-differential maps to make sense of the monoidal sum $\widehat{+}_j$ and action $\widehat{\oplus}_j$.

Remark 4. The reason for requiring each $\widehat{\oplus^A}_i$ and $\widehat{+^A}_i$ in an ω-change object $\widehat{A}$ to be ω-differential is so that $\widehat{A}$ is *internally* a change action in $\mathrm{CAct}_\omega(\mathbf{C})$ (see Definition 15).

Lemma 13. *Let $\widehat{f} : \widehat{A} \to \widehat{B}$ and $\widehat{g} : \widehat{B} \to \widehat{C}$ be ω-differential maps. Qua pre-ω-differential maps, their composite $[g_i] \circ [f_i]$ is ω-differential. Setting $\widehat{g} \circ \widehat{f} := [g_i] \circ [f_i] : \widehat{A} \to \widehat{C}$, it follows that composition of ω-differential maps is associative.*

Lemma 14. *For any ω-change action $\widehat{A}$, the pre-ω-differential map $\mathrm{Id} : [A_i] \to [A_i]$ is ω-differential. Hence $\widehat{\mathrm{Id}} := \mathrm{Id} : \widehat{A} \to \widehat{A}$ satisfies the identity laws.*

Definition 14. Given ω-change actions $\widehat{A}$ and $\widehat{B}$, we define the *product ω-change action* by: $(\widehat{A} \times \widehat{B} := ([A_i \times B_i], [\widehat{\oplus'}_i], [\widehat{+'}_i], [0'_i])$ where

1. $\widehat{\oplus'}_j := \langle \widehat{\oplus^A}_j, \widehat{\oplus^B}_j \rangle \circ \langle \langle \widehat{\pi_{11}}, \widehat{\pi_{12}} \rangle, \langle \widehat{\pi_{21}}, \widehat{\pi_{22}} \rangle \rangle$
2. $\widehat{+'}_j := \langle \widehat{+^A}_j, \widehat{+^B}_j \rangle \circ \langle \langle \widehat{\pi_{11}}, \widehat{\pi_{12}} \rangle, \langle \widehat{\pi_{21}}, \widehat{\pi_{22}} \rangle \rangle$
3. $0'_j := \langle 0^A_j, 0^B_j \rangle$

Notice that $\Delta(\widehat{A} \times \widehat{B}, j) := (A_j \times B_j, A_{j+1} \times B_{j+1}, \mathsf{p}_0 \widehat{\oplus'}_j, \mathsf{p}_0 \widehat{+'}_j, 0'_j)$ is a change action in $\mathbf{C}$ by construction.

Lemma 15. *The pre-ω-differential maps π_1, π_2 are ω-differential. Moreover, for any ω-differential maps $\widehat{f} : \widehat{A} \to \widehat{B}$ and $\widehat{g} : \widehat{A} \to \widehat{C}$, the map $\langle \widehat{f}, \widehat{g} \rangle := \langle [f_i], [g_i] \rangle$ is ω-differential, satisfying $\widehat{\pi_1} \circ \langle \widehat{f}, \widehat{g} \rangle = \widehat{f}$ and $\widehat{\pi_2} \circ \langle \widehat{f}, \widehat{g} \rangle = \widehat{g}$.*

Definition 15. Define the functor $\mathrm{CAct}_\omega : \mathbf{Cat}_\times \to \mathbf{Cat}_\times$ as follows.

- $\mathrm{CAct}_\omega(\mathbf{C})$ is the category whose objects are the ω-change actions over $\mathbf{C}$ and whose morphisms are the ω-differential maps.
- If $\mathrm{F} : \mathbf{C} \to \mathbf{D}$ is a (product-preserving) functor, then $\mathrm{CAct}_\omega(\mathrm{F}) : \mathrm{CAct}_\omega(\mathbf{C}) \to \mathrm{CAct}_\omega(\mathbf{C})$ is the functor mapping the ω-change action $([A_i], [[\oplus_i]_j], [[+_i]_j], [0_j])$ to $([\mathrm{F}A_i], [[\mathrm{F}\oplus_i]_j], [[\mathrm{F}+_i]_j], [\mathrm{F}0_j])$; and the ω-differential map $[f_i]$ to $[\mathrm{F}f_i]$.

Theorem 8. *The category $\mathrm{CAct}_\omega(\mathbf{C})$ is cartesian, with product given in Definition 14. Moreover if $\mathbf{C}$ is closed and has countable limits, $\mathrm{CAct}_\omega(\mathbf{C})$ is cartesian closed.*

Theorem 9. *The category $\mathrm{CAct}_\omega(\mathbf{C})$ is equipped with a canonical change action model: $\gamma : \mathrm{CAct}_\omega(\mathbf{C}) \to \mathrm{CAct}(\mathrm{CAct}_\omega(\mathbf{C}))$.*

Theorem 10 (Relativised final coalgebra). *Let $\mathbf{C}$ be a change action model. The canonical change action model $\gamma : \mathrm{CAct}_\omega(\mathbf{C}) \to \mathrm{CAct}(\mathrm{CAct}_\omega(\mathbf{C}))$ is a relativised*[9] *final coalgebra of $(\mathrm{CAct}, \varepsilon)$.*
i.e. for all change action models on $\mathbf{C}$, $\alpha : \mathbf{C} \to \mathrm{CAct}(\mathbf{C})$, there is a unique coalgebra homomorphism $\alpha_\omega : \mathbf{C} \to \mathrm{CAct}_\omega(\mathbf{C})$, as witnessed by the commuting diagram:

$$\begin{array}{ccc} \mathbf{C} & \xrightarrow{\alpha} & \mathrm{CAct}(\mathbf{C}) \\ {\scriptstyle \exists!\,\alpha_\omega}\downarrow & & \downarrow{\scriptstyle \mathrm{CAct}(\alpha_\omega)} \\ \mathrm{CAct}_\omega(\mathbf{C}) & \xrightarrow{\gamma} & \mathrm{CAct}(\mathrm{CAct}_\omega(\mathbf{C})) \end{array}$$

[9] Here CAct is restricted to the full subcategory of $\mathbf{Cat}_\times$ with $\mathbf{C}$ as the only object.

Proof. We first exhibit the functor $\alpha_\omega : \mathbf{C} \to \mathrm{CAct}_\omega(\mathbf{C})$.

Take a $\mathbf{C}$-morphism $f : A \to B$. We define the ω-differential map $\alpha_\omega(f) := \widehat{f} : \widehat{A} \to \widehat{B}$, where $\widehat{A} := \big([A_i], [\widehat{\oplus}_i], [\widehat{+}_i], [0_i]\big)$ is the ω-change action determined by A under *iterative actions of* α. I.e. for each $i \geq 0$: $A_i := \Delta^i A$ (by abuse of notation, we write $\Delta A'$ to mean the carrier object of the monoid of the internal change action $\alpha(A')$, for any $\mathbf{C}$-object A'); $\widehat{\oplus}_j : \Pi^j[A_i] \times \Pi^{j+1}[A_i] \to \Pi^j[A_i]$ is specified by: $\mathsf{p}_k \widehat{\oplus}_j$ is the monoid action morphism of $\alpha(A_{j+k})$; $\widehat{+}_j : \Pi^{j+1}[A_i] \times \Pi^{j+1}[A_i] \to \Pi^{j+1}[A_i]$ is specified by: $\mathsf{p}_k \widehat{\oplus}_j$ is the monoid sum morphism of $\alpha(A_{j+k})$; 0_i is the zero object of $\alpha(A_i)$.

The ω-sequence $\widehat{f} := [f_i]$ is defined by induction: $f_0 := f$; assume $f_n : (\mathbf{D}\widehat{A})_n \to B_n$ is defined and suppose $\alpha(f_n) = (f_n, \partial f_n)$ then define $f_{n+1} := \partial f_n$.

To see that the diagram commutes, notice that $\gamma(\widehat{f}) = (\widehat{f}, \Pi \widehat{f})$ and $\mathrm{CAct}(\alpha_\omega)$ maps $\alpha(f) = (f, \partial f)$ to $(\widehat{f}, \widehat{\partial f})$; then observe that $\Pi \widehat{f} = \widehat{\partial f}$ follows from the construction of $\widehat{f}$.

Finally to see that the functor α_ω is unique, consider the $\mathbf{C}$-morphisms $\partial^n f$ ($n = 0, 1, 2, \cdots$) where $\alpha(\partial^n f) = (\partial^n f, \partial^{n+1} f)$. Suppose $\beta : \mathbf{C} \to \mathrm{CAct}_\omega(\mathbf{C})$ is another homomorphism. Thanks to the commuting diagram, we must have $\Pi^n \beta(f) = \beta(\partial^n f)$, and so, in particular $(\beta(f))_n = (\Pi^n \beta(f))_0 = (\beta(\partial^n f))_0 = \partial^n f$, for each $n \geq 0$. Thus $\widehat{f} = \beta(f)$ as desired. □

Intuitively any change action model on $\mathbf{C}$ is always a "subset" of the change action model on $\mathrm{CAct}_\omega(\mathbf{C})$.

Theorem 11. *The category* $\mathrm{CAct}_\omega(\mathbf{C})$ *is the limit in* $\mathbf{Cat}_\times$ *of the diagram.*

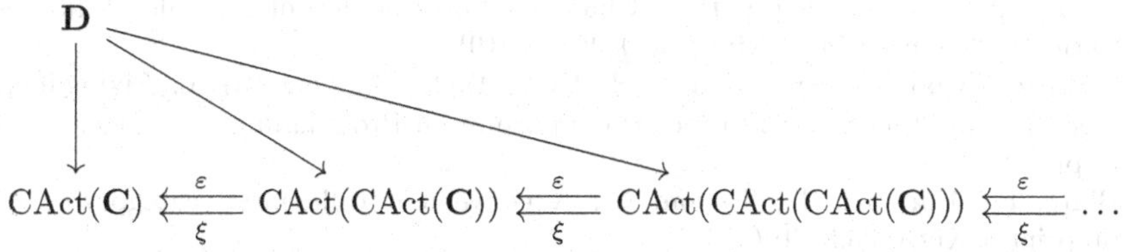

7 Related Work, Future Directions and Conclusions

The present work directly expands upon work by the authors and others in [2], where the notion of change action was developed in the context of the incremental evaluation of Datalog programs. This work generalizes some results in [2] and addresses two significant questions that had been left open, namely: how to construct cartesian closed categories of change actions and how to formalize higher-order derivatives.

Our work is also closely related to Cockett, Seely and Cruttwell's work on cartesian differential categories [3,4,7] and Cruttwell's more recent work on generalised cartesian differential categories [10]. Both cartesian differential categories and change action models aim to provide a setting for differentiation, and the construction of ω-change actions resembles the Faà di Bruno construction

[8,10] (especially its recent reformulation by Lemay [20]) which, given an arbitrary category **C**, builds a cofree cartesian differential category for it. The main difference between these two settings lies in the specific axioms required (change action models are significantly weaker: see Remark 2).

In this sense, the derivative condition is close to the Kock-Lawvere axiom from synthetic differential geometry [18,19], which has provided much of the driving intuition behind this work, and making this connection precise is the subject of ongoing research.

In a different direction, the simplicity of products and exponentials in closed change action models (see Theorem 5) suggests that there should be a reasonable calculus for change action models. Exploring such a calculus and its connections to the differential λ-calculus [11] could lead to practical applications to languages for incremental computation or higher-order automatic differentiation [16].

In conclusion, change actions and change action models constitute a new setting for reasoning about differentiation that is able to unify "discrete" and "continuous" models, as well as higher-order functions. Change actions are remarkably well-behaved and show tantalising connections with geometry and 2-categories. We believe that most ad hoc notions of derivatives found in disparate subjects can be elegantly integrated into the framework of change action models. We therefore expect any further work in this area to have the potential of benefiting these notions of derivatives.

References

1. Alvarez-Picallo, M., Ong, C.H.L.: Change actions: models of generalised differentiation. arXiv preprint arXiv:1902.05465 (2019)
2. Alvarez-Picallo, M., Peyton-Jones, M., Eyers-Taylor, A., Ong, C.H.L.: Fixing incremental computation. In: European Symposium on Programming. Springer (2019, in press)
3. Blute, R., Ehrhard, T., Tasson, C.: A convenient differential category. arXiv preprint arXiv:1006.3140 (2010)
4. Blute, R.F., Cockett, J.R.B., Seely, R.A.: Cartesian differential categories. Theory Appl. Categories **22**(23), 622–672 (2009)
5. Brzozowski, J.A.: Derivatives of regular expressions. J. ACM **11**(4), 481–494 (1964). https://doi.org/10.1145/321239.321249
6. Cai, Y., Giarrusso, P.G., Rendel, T., Ostermann, K.: A theory of changes for higher-order languages: incrementalizing λ-calculi by static differentiation. ACM SIGPLAN Not. **49**, 145–155 (2014)
7. Cockett, J.R.B., Cruttwell, G.S.H.: Differential structure, tangent structure, and SDG. Appl. Categorical Struct. **22**(2), 331–417 (2014)
8. Cockett, J.R.B., Seely, R.A.G.: The Faà di Bruno construction. Theory Appl. Categories **25**, 393–425 (2011)
9. Conway, J.H.: Regular Algebra and Finite Machines. Chapman and Hall, London (1971)
10. Cruttwell, G.S.: Cartesian differential categories revisited. Math. Struct. Comput. Sci. **27**(1), 70–91 (2017)
11. Ehrhard, T., Regnier, L.: The differential lambda-calculus. Theor. Comput. Sci. **309**(1–3), 1–41 (2003). https://doi.org/10.1016/S0304-3975(03)00392-X

12. Esparza, J., Kiefer, S., Luttenberger, M.: Newtonian program analysis. J. ACM **57**(6), 33:1–33:47 (2010). https://doi.org/10.1145/1857914.1857917
13. Gleich, D.: Finite calculus: a tutorial for solving nasty sums. Stanford University (2005)
14. Hopkins, M.W., Kozen, D.: Parikh's theorem in commutative Kleene algebra. In: 14th Annual IEEE Symposium on Logic in Computer Science, Trento, Italy, 2–5 July 1999, pp. 394–401 (1999). https://doi.org/10.1109/LICS.1999.782634
15. Jordan, C.: Calculus of Finite Differences, vol. 33. American Mathematical Society, New York (1965)
16. Kelly, R., Pearlmutter, B.A., Siskind, J.M.: Evolving the incremental lambda-calculus into a model of forward automatic differentiation (ad). arXiv preprint arXiv:1611.03429 (2016)
17. Kleene, S.C.: Representation of events in nerve nets and finite automata. In: Shannon, C.E., McCarthy, J. (eds.) Automata Studies, pp. 3–41. Princeton University Press, Princeton (1956)
18. Kock, A.: Synthetic Differential Geometry, 2nd edn. Cambridge University Press, Cambridge (2006)
19. Lavendhomme, R.: Basic Concepts of Synthetic Differential Geometry, vol. 13. Springer, Boston (2018). https://doi.org/10.1007/978-1-4757-4588-7
20. Lemay, J.S.: A tangent category alternative to the Faà di Bruno construction. arXiv preprint arXiv:1805.01774v1 (2018)
21. Lombardy, S., Sakarovitch, J.: How expressions can code for automata. In: 6th Latin American Symposium on Theoretical Informatics, LATIN 2004, Buenos Aires, Argentina, 5–8 April 2004, Proceedings, pp. 242–251 (2004). https://doi.org/10.1007/978-3-540-24698-5_28
22. Steinbach, B., Posthoff, C.: Boolean differential calculus. Synth. Lect. Digit. Circ. Syst. **12**(1), 1–215 (2017)
23. Thayse, A. (ed.): Boolean Calculus of Differences. LNCS, vol. 101. Springer, Heidelberg (1981). https://doi.org/10.1007/3-540-10286-8

Coalgebra Learning via Duality

Simone Barlocco[1], Clemens Kupke[1(✉)], and Jurriaan Rot[2]

[1] University of Strathclyde, Glasgow, UK
{simone.barlocco,clemens.kupke}@strath.ac.uk
[2] Radboud University, Nijmegen, Netherlands
j.rot@cs.ru.nl

Abstract. Automata learning is a popular technique for inferring minimal automata through membership and equivalence queries. In this paper, we generalise learning to the theory of coalgebras. The approach relies on the use of logical formulas as tests, based on a dual adjunction between states and logical theories. This allows us to learn, e.g., labelled transition systems, using Hennessy-Milner logic. Our main contribution is an abstract learning algorithm, together with a proof of correctness and termination.

1 Introduction

In recent years, automata learning is applied with considerable success to infer models of systems and in order to analyse and verify them. Most current approaches to active automata learning are ultimately based on the original algorithm due to Angluin [4], although numerous improvements have been made, in practical performance and in extending the techniques to different models [30].

Our aim is to move from automata to *coalgebras* [14,26], providing a generalisation of learning to a wide range of state-based systems. The key insight underlying our work is that dual adjunctions connecting coalgebras and tailor-made logical languages [12,19,21,22,26] allow us to devise a generic learning algorithm for coalgebras that is parametric in the type of system under consideration. Our approach gives rise to a fundamental distinction between *states* of the learned system and *tests*, modelled as logical formulas. This distinction is blurred in the classical DFA algorithm, where tests are also used to specify the (reachable) states. It is precisely the distinction between tests and states which allows us to move beyond classical automata, and use, for instance, Hennessy-Milner logic to learn bisimilarity quotients of labelled transition systems.

To present learning via duality we need to introduce new notions and refine existing ones. First, in the setting of coalgebraic modal logic, we introduce the new notion of *sub-formula closed* collections of formulas, generalising suffix-closed sets of words in Angluin's algorithm (Sect. 4). Second, we import the abstract notion of *base* of a functor from [8], which allows us to speak about

C. Kupke—Partially supported by EPSRC grant EP/N015843/1.

M. Bojańczyk and A. Simpson (Eds.): FOSSACS 2019, LNCS 11425, pp. 62–79, 2019.
https://doi.org/10.1007/978-3-030-17127-8_4

'successor states' (Sect. 5). In particular, the base allows us to characterise *reachability* of coalgebras in a clear and concise way. This yields a canonical procedure for computing the reachable part from a given initial state in a coalgebra, thus generalising the notion of a generated subframe from modal logic.

We then rephrase *coalgebra learning* as the problem of inferring a coalgebra which is reachable, minimal and which cannot be distinguished from the original coalgebra held by the teacher using tests. This requires suitably adapting the computation of the reachable part to incorporate tests, and only learn 'up to logical equivalence'. We formulate the notion of *closed table*, and an associated procedure to close tables. With all these notions in place, we can finally define our abstract algorithm for coalgebra learning, together with a proof of correctness and termination (Sect. 6). Overall, we consider this correctness and termination proof as the main contribution of the paper; other contributions are the computation of reachability via the base and the notion of sub-formula closedness. At a more conceptual level, our paper shows how states and tests interact in automata learning, by rephrasing it in the context of a dual adjunction connecting coalgebra (systems) and algebra (logical theories). As such, we provide a new foundation of learning state-based systems.

Related Work. The idea that tests in the learning algorithm should be formulas of a distinct logical language was proposed first in [6]. However, the work in *loc. cit.* is quite ad-hoc, confined to Boolean-valued modal logics, and did not explicitly use duality. This paper is a significant improvement: the dual adjunction framework and the definition of the base [8] enables us to present a description of Angluin's algorithm in purely categorical terms, including a proof of correctness and, crucially, termination. Our abstract notion of logic also enables us to recover *exactly* the standard DFA algorithm (where tests are words) and the algorithm for learning Mealy machines (where test are many-valued), something that is not possible in [6] where tests are modal formulas. Closely related to our work is also the line of research initiated by [15] and followed up within the CALF project [11–13] which applies ideas from category theory to automata learning. Our approach is orthogonal to CALF: the latter focuses on learning a general version of *automata*, whereas our work is geared towards learning bisimilarity quotients of state-based transition systems. While CALF lends itself to studying automata in a large variety of base categories, our work thus far is concerned with varying the type of transition structures.

2 Learning by Example

The aim of this section is twofold: (i) to remind the reader of the key elements of Angluin's L* algorithm [4] and (ii) to motivate and outline our generalisation.

In the classical L* algorithm, the learner tries to learn a regular language $\mathcal{L}$ over some alphabet A or, equivalently, a DFA $\mathcal{A}$ accepting that language. Learning proceeds by asking queries to a teacher who has access to this automaton. *Membership queries* allow the learner to test whether a given word is in the language, and *equivalence queries* to test whether the correct DFA has been learned already. The algorithm constructs so-called tables (S, E) where $S, E \subseteq A^*$ are the rows and columns of the table, respectively. The value at position (s, e) of the table is the answer to the membership query "$se \in \mathcal{L}$?".

Words play a double role: On the one hand, a word $w \in S$ represents the state which is reached when reading w at the initial state. On the other hand, the set E represents the set of membership queries that the learner is asking about the states in S. A table is *closed* if for all $w \in S$ and all $a \in A$ either $wa \in S$ or there is a state $v \in S$ such that wa is equivalent to v w.r.t. membership queries of words in E. If a table is not closed we extend S by adding words of the form wa for $w \in S$ and $a \in A$. Once it is closed, one can define a *conjecture*,[1] i.e., a DFA with states in S. The learner now asks the teacher whether the conjecture is correct. If it is, the algorithm terminates. Otherwise the teacher provides a *counterexample*: a word on which the conjecture is incorrect. The table is now extended using the counterexample. As a result, the table is not closed anymore and the algorithm continues again by closing the table.

Our version of L* introduces some key conceptual differences: tables are pairs (S, Ψ) such that S (set of rows) is a selection of states of $\mathcal{A}$ and Ψ (set of columns) is a collection of tests/formulas. Membership queries become checks of tests in Ψ at states in S and equivalence queries verify whether or not the learned structure is logically equivalent to the original one. A table (S, Ψ) is closed if for all successors x' of elements of S there exists an $x \in S$ such that x and x' are equivalent w.r.t. formulas in Ψ. The clear distinction between states and tests in our algorithm means that counterexamples are formulas that have to be added to Ψ. Crucially, the move from words to formulas allows us to use the rich theory of coalgebra and coalgebraic logic to devise a generic algorithm.

We consider two examples within our generic framework: classical DFAs, yielding essentially the L* algorithm, and labelled transition systems, which is to the best of our knowledge not covered by standard automata learning algorithms.

For the DFA case, let $L = \{u \in \{a, b\}^* \mid \text{number of } a\text{'s mod } 3 = 0\}$ and assume that the teacher uses the following (infinite) automaton describing L:

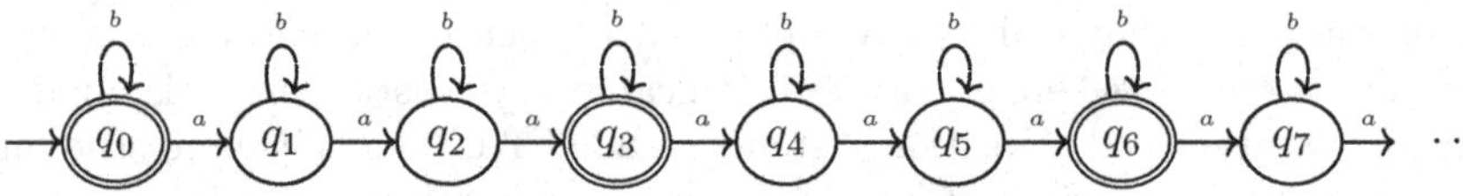

As outlined above, the learner starts to construct tables (S, Ψ) where S is a selection of states of the automaton and Ψ are formulas. For DFAs we will see (Example 1) that our formulas are just words in $\{a, b\}^*$. Our starting table is $(\{q_0\}, \emptyset)$, i.e., we select the initial state and do not check any logical properties. This table is trivially closed, as all states are equivalent w.r.t. $\emptyset$. The first conjecture is the automaton consisting of one accepting state q_0 with a- and b-loops, whose language is $\{a, b\}^*$. This is incorrect and the teacher provides, e.g., aa as counterexample. The resulting table is $(\{q_0\}, \{\varepsilon, a, aa\})$ where the

[1] The algorithm additionally requires *consistency*, but this is not needed if counterexamples are added to E. This idea goes back to [22].

second component was generated by closing $\{aa\}$ under suffixes. Suffix closedness features both in the original L* algorithm and in our framework (Sect. 4). The table $(\{q_0\}, \{\varepsilon, a, aa\})$ is not closed as q_1, the a-successor of q_0, does not accept ε whereas q_0 does. Therefore we extend the table to $(\{q_0, q_1\}, \{\varepsilon, a, aa\})$. Note that, unlike in the classical setting, exploring successors of already selected states cannot be achieved by appending letters to words, but we need to *locally* employ the transition structure on the automaton $\mathcal{A}$ instead. A similar argument shows that we need to extend the table further to $(\{q_0, q_1, q_2\}, \{\varepsilon, a, aa\})$ which is closed. This leads to the (correct) conjecture depicted on the right below. The acceptance condition and transition structure has been read off from the original automaton, where the transition from q_2 to q_0 is obtained by realising that q_2's successor q_3 is represented by the equivalent state $q_0 \in S$.

A key feature of our work is that the L* algorithm can be systematically generalised to new settings, in particular, to the learning of bisimulation quotients of transition systems. Consider the following labelled transition system (LTS). We would like to learn its minimal representation, i.e., its quotient modulo bisimulation.

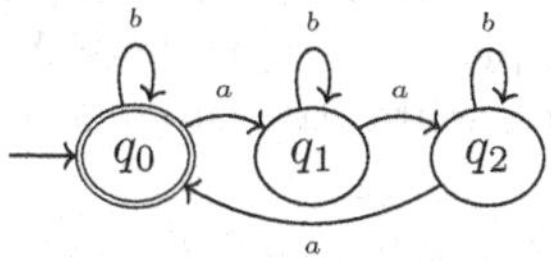

Our setting allows us to choose a suitable logical language. For LTSs, the language consists of the formulas of standard multi-modal logic (cf. Example 3). The semantics is as usual where $\langle a \rangle \phi$ holds at a state if it has an a-successor that makes ϕ true.

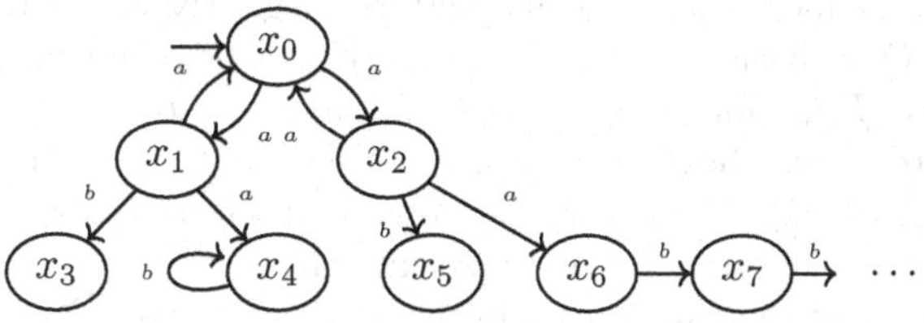

As above, the algorithm constructs tables, starting with $(S = \{x_0\}, \Psi = \emptyset)$. The table is closed, so the first conjecture is a single state with an a-loop with no proposition letter true (note that x_0 has no b or c successor and no proposition is true at x_0). It is, however, easy for the teacher to find a counterexample. For example, the formula $\langle a \rangle \langle b \rangle \top$ is true at the root of the original LTS but false in the conjecture. We add the counterexample and all its subformulas to Ψ and obtain a new table $(\{x_0\}, \Psi'\}$ with $\Psi' = \{\langle a \rangle \langle b \rangle \top, \langle b \rangle \top, \top\}$. Now, the table is not closed, as x_0 has successor x_1 that satisfies $\langle b \rangle \top$ whereas x_0 does not satisfy $\langle b \rangle \top$. Therefore we add x_1 to the table to obtain $(\{x_0, x_1\}, \Psi')$. Similar arguments will lead to the closed table $(\{x_0, x_1, x_3, x_4\}, \Psi')$ which also yields the correct conjecture. Note that the state x_2 does not get added to the table as it is equivalent to x_1 and thus already represented. This demonstrates a remarkable fact: we computed the bisimulation quotient of the LTS without inspecting the (infinite) right-hand side of the LTS.

Another important example that fits smoothly into our framework is the well-known variant of Angluin's algorithm to learn Mealy machines (Example 2). Thanks to our general notion of logic, our framework allows to use an intuitive language, where a formula is simply an input word w whose truth value at a state

x is the observed output after entering w at x. This is in contrast to [6] where formulas had to be Boolean valued. Multi-valued logics fit naturally in our setting; this is expected to be useful to deal with systems with quantitative information.

3 Preliminaries

The general learning algorithm in this paper is based on the theory of *coalgebras*, which provides an abstract framework for representing state-based transition systems. In what follows we assume that the reader is familiar with basic notions of category theory and coalgebras [14,26]. We briefly recall the notion of pointed coalgebra, modelling a coalgebra with an initial state. Let $\mathcal{C}$ be a category with a terminal object 1 and let $B\colon \mathcal{C} \to \mathcal{C}$ be a functor. A pointed B-coalgebra is a triple (X, γ, x_0) where $X \in \mathcal{C}$ and $\gamma\colon X \to BX$ and $x_0\colon 1 \to X$, specifying the coalgebra structure and the point ("initial state") of the coalgebra, respectively.

Coalgebraic Modal Logic. Modal logics are used to describe properties of state-based systems, modelled here as coalgebras. The close relationship between coalgebras and their logics is described elegantly via dual adjunctions [18,20,21,24].

Our basic setting consists of two categories $\mathcal{C}, \mathcal{D}$ connected by functors P, Q forming a dual adjunction $P \dashv Q\colon \mathcal{C} \leftrightarrows \mathcal{D}^{\mathsf{op}}$. In other words, we have a natural bijection $\mathcal{C}(X, Q\Delta) \cong \mathcal{D}(\Delta, PX)$ for $X \in \mathcal{C}, \Delta \in \mathcal{D}$. Moreover, we assume two functors, $B\colon \mathcal{C} \to \mathcal{C}, L\colon \mathcal{D} \to \mathcal{D}$, see (1). The functor L represents the syntax of the (modalities in the) logic: assuming that L has an initial algebra $\alpha\colon L\Phi \to \Phi$ we think of Φ as the collection of formulas, or tests. In this logical perspective, the functor P maps an object X of $\mathcal{C}$ to the collection of predicates and the functor Q maps an object Δ of $\mathcal{D}$ to the collection $Q\Delta$ of Δ-theories.

$$B \circlearrowright \mathcal{C} \underset{Q}{\overset{P}{\rightleftarrows}} \mathcal{D}^{\mathsf{op}} \circlearrowleft L \qquad (\text{with } P \dashv Q) \tag{1}$$

The connection between coalgebras and their logics is specified via a natural transformation $\delta\colon LP \Rightarrow PB$, sometimes referred to as the one-step semantics of the logic. The δ is used to define the semantics of the logic on a B-coalgebra (X, γ) by initiality, as in (2). Furthermore, using the bijective correspondence of the dual adjunction between P and Q, the map $[\![\_]\!]$ corresponds to a map $th^\gamma\colon X \to Q\Phi$ that we will refer to as the theory map of (X, γ).

$$\begin{array}{ccccc} L\Phi & \xrightarrow{L[\![\_]\!]} & LPX & \xrightarrow{\delta_X} & PBX \\ {\scriptstyle\alpha}\downarrow & & & & \downarrow{\scriptstyle P\gamma} \\ \Phi & & \xrightarrow{\exists! [\![\_]\!]} & & PX \end{array} \tag{2}$$

The theory map can be expressed directly via a universal property, by making use of the so-called *mate* $\delta^\flat\colon BQ \Rightarrow QL$ of the one-step semantics δ (cf. [18,24]). More precisely, we have $\delta^\flat = QL\varepsilon \circ Q\delta Q \circ \eta BQ$, where η, ε are the unit and counit of the adjunction. Then $th^\gamma\colon X \to Q\Phi$ is the unique morphism making (3) commute.

$$\begin{array}{ccccc} BX & \xrightarrow{Bth^\gamma} & BQ\Phi & \xrightarrow{\delta^\flat_\Phi} & QL\Phi \\ {\scriptstyle\gamma}\uparrow & & & & \uparrow{\scriptstyle Q\alpha} \\ X & & \xrightarrow{\exists! th^\gamma} & & Q\Phi \end{array} \tag{3}$$

Example 1. Let $\mathcal{C} = \mathcal{D} = \mathsf{Set}$, $P = Q = 2^-$ the contravariant power set functor, $B = 2 \times -^A$ and $L = 1 + A \times -$. In this case B-coalgebras can be thought of as deterministic automata with input alphabet A (e.g., [25]). It is well-known that the initial L-algebra is $\Phi = A^*$ with structure $\alpha = [\varepsilon, \text{cons}]: 1 + A \times A^* \to A^*$ where ε selects the empty word and cons maps a pair $(a, w) \in A \times A^*$ to the word $aw \in A^*$, i.e., in this example our tests are words with the intuitive meaning that a test succeeds if the word is accepted by the given automaton. For $X \in \mathcal{C}$, the X-component of the (one-step) semantics $\delta: LP \Rightarrow PB$ is defined as follows: $\delta_X(*) = \{(i, f) \in 2 \times X^A \mid i = 1\}$, and $\delta_X(a, U) = \{(i, f) \in 2 \times X^A \mid f(a) \in U\}$. It is matter of routine checking that the semantics of tests in Φ on a B-coalgebra (X, γ) is as follows: we have $[\![\varepsilon]\!] = \{x \in X \mid \pi_1(\gamma(x)) = 1\}$ and $[\![aw]\!] = \{x \in X \mid \pi_2(\gamma(x))(a) \in [\![w]\!]\}$, where π_1 and π_2 are the projection maps. The theory map th^γ sends a state to the language accepted by that state in the usual way.

Example 2. Again let $\mathcal{C} = \mathcal{D} = \mathsf{Set}$ and consider the functors $P = Q = O^-$, $B = (O \times -)^A$ and $L = A \times (1 + -)$, where A and O are fixed sets, thought of as input and output alphabet, respectively. Then B-coalgebras are Mealy machines and the initial L-algebra is given by the set A^+ of finite non-empty words over A. For $X \in \mathcal{C}$, the one-step semantics $\delta_X: A \times (1 + O^X) \to O^{BX}$ is defined by $\delta_X(a, \text{inl}(*)) = \lambda f.\pi_1(f(a))$ and $\delta_X(a, \text{inr}(g)) = \lambda f.g(\pi_2(f(a)))$. Concretely, formulas are words in A^+; the (O-valued) semantics of $w \in A^+$ at state x is the output $o \in O$ that is produced after processing the input w from state x.

Example 3. Let $\mathcal{C} = \mathsf{Set}$ and $\mathcal{D} = \mathsf{BA}$, where the latter denotes the category of Boolean algebras. Again $P = 2^-$, but this time 2^X is interpreted as a Boolean algebra. The functor Q maps a Boolean algebra to the collection of ultrafilters over it [7]. Furthermore $B = (\mathcal{P}-)^A$ where $\mathcal{P}$ denotes covariant power set and A a set of actions. Coalgebras for this functor correspond to labelled transition systems, where a state has a set of successors that depends on the action/input from A. The dual functor $L: \mathsf{BA} \to \mathsf{BA}$ is defined as $LY := F_{\mathsf{BA}}(\{\langle a \rangle y \mid a \in A, y \in Y\})/\equiv$ where $F_{\mathsf{BA}}: \mathsf{Set} \to \mathsf{BA}$ denotes the free Boolean algebra functor and where, roughly speaking, $\equiv$ is the congruence generated from the axioms $\langle a \rangle \bot \equiv \bot$ and $\langle a \rangle (y_1 \vee y_2) \equiv \langle a \rangle (y_1) \vee \langle a \rangle (y_2)$ for each $a \in A$. This is explained in more detail in [21]. The initial algebra for this functor is the so-called Lindenbaum-Tarski algebra [7] of modal formulas $(\phi ::= \bot \mid \phi \vee \phi \mid \neg\phi \mid \langle a \rangle \phi)$ quotiented by logical equivalence. The definition of an appropriate δ can be found in, e.g., [21]—the semantics $[\![\_]\!]$ of a formula then amounts to the standard one [7].

Different types of probabilistic transition systems also fit into the dual adjunction framework, see, e.g, [17].

Subobjects and Intersection-Preserving Functors. We denote by $\mathsf{Sub}(X)$ the collection of subobjects of an object $X \in \mathcal{C}$. Let $\leq$ be the order on subobjects $s: S \rightarrowtail X$, $s': S' \rightarrowtail X$ given by $s \leq s'$ iff there is $m: S \to S'$ s.t. $s = s' \circ m$. The *intersection* $\bigwedge J \rightarrowtail X$ of a family $J = \{s_i: S_i \to X\}_{i \in I}$ is defined as the greatest lower bound w.r.t. the order $\leq$. In a complete category, it can be computed by (wide) pullback. We denote the maps in the limiting cone by $x_i \colon \bigwedge J \rightarrowtail S_i$.

For a functor $B \colon \mathcal{C} \to \mathcal{D}$, we say B *preserves (wide) intersections* if it preserves these wide pullbacks, i.e., if $(B(\bigwedge J), \{Bx_i\}_{i \in I})$ is the pullback of $\{Bs_i \colon BS_i \to BX\}_{i \in I}$. By [2, Lemma 3.53] (building on [29]), *finitary* functors on Set 'almost' preserve wide intersections: for every such functor B there is a functor B' which preserves wide intersections and agrees with B on all non-empty sets. Finally, if B preserves intersections, then it preserves monos.

Minimality Notions. The algorithm that we will describe in this paper learns a minimal and reachable representation of an object. The intuitive notions of minimality and reachability are formalised as follows.

Definition 4. *We call a B-coalgebra (X, γ)* minimal w.r.t. logical equivalence *if the theory map $th^{\gamma} \colon X \to Q\Phi$ is a monomorphism.*

Definition 5. *We call a pointed B-coalgebra (X, γ, x_0)* reachable *if for any subobject $s \colon S \to X$ and $s_0 \colon 1 \to S$ with $x_0 = s \circ s_0$: if S is a subcoalgebra of (X, γ) then s is an isomorphism.*

For expressive logics [27], behavioural equivalence coincides with logical equivalence. Hence, in that case, our algorithm learns a "well-pointed coalgebra" in the terminology of [2], i.e., a pointed coalgebra that is reachable and minimal w.r.t. behavioural equivalence. All logics appearing in this paper are expressive.

Assumption on $\mathcal{C}$ and Factorisation System. Throughout the paper we will assume that $\mathcal{C}$ is a complete and well-powered category. Well-powered means that for each $X \in \mathcal{C}$ the collection $\mathsf{Sub}(X)$ of subobjects of a given object forms a set. Our assumptions imply [10, Proposition 4.4.3] that every morphism f in $\mathcal{C}$ factors uniquely (up to isomorphism) as $f = m \circ e$ with m a mono and e a strong epi. Recall that an epimorphism $e \colon X \to Y$ is strong if for every commutative square in (4) where the bottom arrow is a monomorphism, there exists a unique diagonal morphism d such that the entire diagram commutes.

$$\begin{array}{ccc} X & \stackrel{e}{\twoheadrightarrow} & Y \\ {\scriptstyle h}\downarrow & \swarrow{\scriptstyle d} & \downarrow{\scriptstyle g} \\ U & \underset{m}{\rightarrowtail} & Z \end{array} \tag{4}$$

4 Subformula Closed Collections of Formulas

Our learning algorithm will construct conjectures that are "partially" correct, i.e., correct with respect to a subobject of the collection of all formulas/tests. Recall this collection of all tests are formalised in our setting as the initial L-algebra $(\Phi, \alpha \colon L\Phi \to \Phi)$. To define a notion of partial correctness we need to consider subobjects of Φ to which we can restrict the theory map. This is formalised via the notion of "subformula closed" subobject of Φ.

The definition of such subobjects is based on the notion of *recursive coalgebra*. For $L\colon \mathcal{D} \to \mathcal{D}$ an endofunctor, a coalgebra $f\colon X \to LX$ is called *recursive* if for every L-algebra $g\colon LY \to Y$ there is a unique 'coalgebra-to-algebra' map $g^\dagger$ making (5) commute.

$$\begin{array}{ccc} LX & \xrightarrow{Lg^\dagger} & LY \\ {\scriptstyle f}\uparrow & & \downarrow{\scriptstyle g} \\ X & \xrightarrow{g^\dagger} & Y \end{array} \qquad (5)$$

Definition 6. *A subobject $j\colon \Psi \to \Phi$ is called a* subformula closed collection *(of formulas) if there is a unique L-coalgebra structure $\sigma\colon \Psi \to L\Psi$ such that (Ψ, σ) is a recursive L-coalgebra and j is the (necessarily unique) coalgebra-to-algebra map from (Ψ, σ) to the initial algebra (Φ, α).*

Remark 7. The uniqueness of σ in Definition 6 is implied if L preserves monomorphisms. This is the case in our examples. The notion of recursive coalgebra goes back to [23,28]. The paper [1] contains a claim that the first item of our definition of subformula closed collection is implied by the second one if L preserves preimages. In our examples both properties of (Ψ, σ) are verified directly, rather than by relying on general categorical results.

Example 8. In the setting of Example 1, where the initial L-algebra is based on the set A^* of words over the set (of inputs) A, a subset $\Psi \subseteq A^*$ is subformula-closed if it is suffix-closed, i.e., if for all $aw \in \Psi$ we have $w \in \Psi$ as well.

Example 9. In the setting that $B = (\mathcal{P}-)^A$ for some set of actions A, $\mathcal{C} = \mathsf{Set}$ and $\mathcal{D} = \mathsf{BA}$, the logic is given as a functor L on Boolean algebras as discussed in Example 3. As a subformula closed collection is an object in Ψ, we are not simply dealing with a set of formulas, but with a Boolean algebra. The connection to the standard notion of being closed under taking subformulas in modal logic [7] can be sketched as follows: given a set Δ of modal formulas that is closed under taking subformulas, we define a Boolean algebra $\Psi_\Delta \subseteq \Phi$ as the smallest Boolean subalgebra of Φ that is generated by the set $\hat{\Delta} = \{[\phi]_\Phi \mid \phi \in \Delta\}$ where for a formula ϕ we let $[\phi]_\Phi \in \Phi$ denote its equivalence class in Φ.

It is then not difficult to define a suitable $\sigma\colon \Psi_\Delta \to L\Psi_\Delta$. As Ψ_Δ is generated by closing $\hat{\Delta}$ under Boolean operations, any two states x_1, x_2 in a given coalgebra (X, γ) satisfy $(\forall b \in \Psi_\Delta . x_1 \in [\![b]\!] \Leftrightarrow x_2 \in [\![b]\!])$ iff $\left(\forall b \in \hat{\Delta} . x_1 \in [\![b]\!] \Leftrightarrow x_2 \in [\![b]\!]\right)$. In other words, equivalence w.r.t. Ψ_Δ coincides with equivalence w.r.t. the *set* of formulas Δ. This explains why in the concrete algorithm, we do not deal with Boolean algebras explicitly, but with subformula closed sets of formulas instead.

The key property of subformula closed collections Ψ is that we can restrict our attention to the so-called Ψ-theory map. Intuitively, subformula closedness is what allows us to define this theory map inductively.

$$\begin{array}{ccccc} X & & \xrightarrow{\quad th^\gamma_\Psi \quad} & & Q\Psi \\ {\scriptstyle \gamma}\downarrow & & & & \uparrow{\scriptstyle Q\sigma} \\ BX & \xrightarrow{Bth^\gamma_\Psi} & BQ\Psi & \xrightarrow{\delta^\flat_\Psi} & QL\Psi \end{array} \qquad (6)$$

Lemma 10. *Let $\Psi \overset{j}{\rightarrowtail} \Phi$ be a sub-formula closed collection, with coalgebra structure $\sigma \colon \Psi \to L\Psi$. Then $th^{\gamma}_{\Psi} = Qj \circ th^{\gamma}_{\Phi}$ is the unique map making* (6) *commute. We call th^{γ}_{Ψ} the Ψ-theory map, and omit the Ψ if it is clear from the context.*

5 Reachability and the Base

In this section, we define the notion of *base* of an endofunctor, taken from [8]. This allows us to speak about the (direct) successors of states in a coalgebra, and about reachability, which are essential ingredients of the learning algorithm.

Definition 11. *Let $B \colon \mathcal{C} \to \mathcal{C}$ be an endofunctor. We say B* has a base *if for every arrow $f \colon X \to BY$ there exist $g \colon X \to BZ$ and $m \colon Z \rightarrowtail Y$ with m a monomorphism such that $f = Bm \circ g$, and for any pair $g' \colon X \to BZ', m' \colon Z' \rightarrowtail Y$ with $Bm' \circ g' = f$ and m' a monomorphism there is a unique arrow $h \colon Z \to Z'$ such that $Bh \circ g = g'$ and $m' \circ h = m$, see Diagram* (7)*. We call (Z, g, m) the (B)-*base *of the morphism f.*

We sometimes refer to $m \colon Z \rightarrowtail Y$ as the base of f, omitting the g when it is irrelevant, or clear from the context. Note that the terminology 'the' base is justified, as it is easily seen to be unique up to isomorphism.

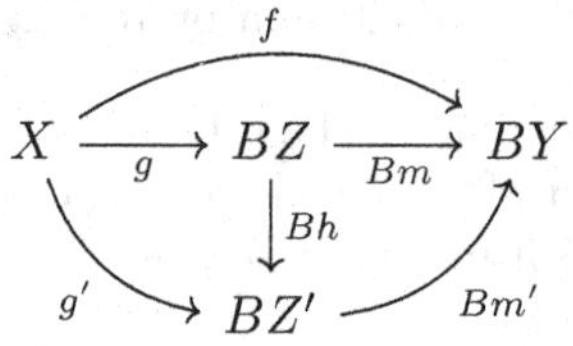

(7)

For example, let $B \colon \mathsf{Set} \to \mathsf{Set}$, $BX = 2 \times X^A$. The base of a map $f \colon X \to BY$ is given by $m \colon Z \rightarrowtail Y$, where $Z = \{(\pi_2 \circ f)(x)(a) \mid x \in X, a \in A\}$, and m is the inclusion. The associated $g \colon X \to BZ$ is the corestriction of f to BZ.

For $B = (\mathcal{P}-)^A \colon \mathsf{Set} \to \mathsf{Set}$, the B-*base* of $f \colon X \to Y$ is given by the inclusion $m \colon Z \rightarrowtail Y$, where $Z = \{y \in Y \mid \exists x \in X, \exists a \in A \text{ s.t. } y \in f(x)(a)\}$.

Proposition 12. *Suppose $\mathcal{C}$ is complete and well-powered, and $B \colon \mathcal{C} \to \mathcal{C}$ preserves (wide) intersections. Then B has a base.*

If $\mathcal{C}$ is a locally presentable category, then it is complete and well-powered [3, Remark 1.56]. Hence, in that case, any functor $B \colon \mathcal{C} \to \mathcal{C}$ which preserves intersections has a base. The following lemma will be useful in proofs.

Lemma 13. *Let $B \colon \mathcal{C} \to \mathcal{C}$ be a functor that has a base and that preserves preimages. Let $f \colon S \to BX$ and $h \colon X \to Y$ be morphisms, let (Z, g, m) be the base of f and let $e \colon Z \to W, m' \colon W \to Y$ be the (strong epi, mono)-factorisation of $h \circ m$. Then $(W, Be \circ g, m')$ is the base of $Bh \circ f$.*

The B-base provides an elegant way to relate reachability within a coalgebra to a monotone operator on the (complete) lattice of subobjects of the carrier of the coalgebra. Moreover, we will see that the least subcoalgebra that contains a given subobject of the carrier can be obtained via a standard least fixpoint construction. Finally, we will introduce the notion of prefix closed subobject of a coalgebra, generalising the prefix closedness condition from Angluin's algorithm.

By our assumption on $\mathcal{C}$ at the end of Sect. 3, the collection of subobjects $(\mathsf{Sub}(X), \leq)$ ordered as usual (cf. Section 3) forms a complete lattice. Recall that the meet on $\mathsf{Sub}(X)$ (intersection) is defined via pullbacks. In categories with coproducts, the join $s_1 \vee s_2$ of subobjects $s_1, s_2 \in \mathsf{Sub}(X)$ is defined as the mono part of the factorisation of the map $[s_1, s_2]\colon S_1 + S_2 \to X$, i.e., $[s_1, s_2] = (s_1 \vee s_2) \circ e$ for a strong epi e. In Set, this amounts to taking the union of subsets.

For a binary join $s_1 \vee s_2$ we denote by $\mathit{inl}_\vee\colon S_1 \to (S_1 \vee S_2)$ and $\mathit{inr}_\vee\colon S_2 \to (S_1 \vee S_2)$ the embeddings that exist by $s_i \leq s_1 \vee s_2$ for $i = \{1, 2\}$. Let us now define the key operator of this section.

$$\begin{array}{ccc} S & \xrightarrow{\;s\;} & X \\ {\scriptstyle g}\downarrow & & \downarrow{\scriptstyle \gamma} \\ B\Gamma(S) & \xrightarrow{B\Gamma^B_\gamma(s)} & BX \end{array} \qquad (8)$$

Definition 14. *Let B be a functor that has a base, $s\colon S \rightarrowtail X$ a subobject of some $X \in \mathcal{C}$ and let (X, γ) be a B-coalgebra. Let $(\Gamma(S), g, \Gamma^B_\gamma(s))$ be the B-base of $\gamma \circ s$, see Diagram (8). Whenever B and γ are clear from the context, we write $\Gamma(s)$ instead of $\Gamma^B_\gamma(s)$.*

Lemma 15. *Let $B\colon \mathcal{C} \to \mathcal{C}$ be a functor with a base and let (X, γ) be a B-coalgebra. The operator $\Gamma\colon \mathsf{Sub}(X) \to \mathsf{Sub}(X)$ defined by $s \mapsto \Gamma(s)$ is monotone.*

Intuitively, Γ computes for a given set of states S the set of "immediate successors", i.e., the set of states that can be reached by applying γ to an element of S. We will see that pre-fixpoints of Γ correspond to subcoalgebras. Furthermore, Γ is the key to formulate our notion of closed table in the learning algorithm.

Proposition 16. *Let $s\colon S \rightarrowtail X$ be a subobject and $(X, \gamma) \in \mathsf{Coalg}(B)$ for $X \in \mathcal{C}$ and $B\colon \mathcal{C} \to \mathcal{C}$ a functor that has a base. Then s is a subcoalgebra of (X, γ) if and only if $\Gamma(s) \leq s$. Consequently, the collection of subcoalgebras of a given B-coalgebra forms a complete lattice.*

Using this connection, reachability of a pointed coalgebra (Definition 5) can be expressed in terms of the least fixpoint lfp of an operator defined in terms of Γ.

Theorem 17. *Let $B\colon \mathcal{C} \to \mathcal{C}$ be a functor that has a base. A pointed B-coalgebra (X, γ, x_0) is reachable iff $X \cong \mathsf{lfp}(\Gamma \vee x_0)$ (isomorphic as subobjects of X, i.e., equal).*

This justifies defining the reachable part from an initial state $x_0\colon 1 \rightarrowtail X$ as the least fixpoint of the monotone operator $\Gamma \vee x_0$. Standard means of computing the least fixpoint by iterating this operator then give us a way to compute this subcoalgebra. Further, Γ provides a way to generalise the notion of "prefixed closedness" from Angluin's L* algorithm to our categorical setting.

Definition 18. *Let $s_0, s \in \mathsf{Sub}(X)$ for some $X \in \mathcal{C}$ and let (X, γ) be a B-coalgebra. We call s s_0-*prefix closed w.r.t.* γ if $s = \bigvee_{i=0}^{n} s_i$ for some $n \geq 0$ and a collection $\{s_i \mid i = 1, \ldots, n\}$ with $s_{j+1} \leq \Gamma(\bigvee_{i=0}^{j} s_i)$ for all j with $0 \leq j < n$.*

6 Learning Algorithm

We define a general learning algorithm for B-coalgebras. First, we describe the setting, in general and slightly informal terms. The teacher has a pointed B-coalgebra (X, γ, s_0). Our task is to 'learn' a pointed B-coalgebra $(S, \hat{\gamma}, \hat{s}_0)$ s.t.:

- $(S, \hat{\gamma}, \hat{s}_0)$ is *correct* w.r.t. the collection Φ of all tests, i.e., the theory of (X, γ) and $(S, \hat{\gamma})$ coincide on the initial states s_0 and $\hat{s}_0$, (Definition 25);
- $(S, \hat{\gamma}, \hat{s}_0)$ is minimal w.r.t. logical equivalence;
- $(S, \hat{\gamma}, \hat{s}_0)$ is reachable.

The first point means that the learned coalgebra is 'correct', that is, it agrees with the coalgebra of the teacher on all possible tests from the initial state. For instance, in case of deterministic automata and their logic in Example 1, this just means that the language of the learned automaton is the correct one.

In the learning game, we are only provided limited access to the coalgebra $\gamma \colon X \to BX$. Concretely, the teacher gives us:

- for any subobject $S \rightarrowtail X$ and sub-formula closed subobject Ψ of Φ, the composite theory map $S \rightarrowtail X \xrightarrow{th^{\gamma}_{\Psi}} Q\Psi$;
- for $(S, \hat{\gamma}, \hat{s}_0)$ a pointed coalgebra, whether or not it is correct w.r.t. the collection Φ of all tests;
- in case of a negative answer to the previous question, a *counterexample*, which essentially is a subobject Ψ' of Φ representing some tests on which the learned coalgebra is wrong (defined more precisely below);
- for a given subobject S of X, the 'next states'; formally, the computation of the B-base of the composite arrow $S \rightarrowtail X \xrightarrow{\gamma} BX$.

The first three points correspond respectively to the standard notions of membership query ('filling in' the table with rows S and columns Ψ), equivalence query and counterexample generation. The last point, about the base, is more unusual: it does not occur in the standard algorithm, since there a canonical choice of (X, γ) is used, which allows to represent next states in a fixed manner. It is required in our setting of an arbitrary coalgebra (X, γ).

In the remainder of this section, we describe the abstract learning algorithm and its correctness. First, we describe the basic ingredients needed for the algorithm: tables, closedness, counterexamples and a procedure to close a given table (Sect. 6.1). Based on these notions, the actual algorithm is presented (Sect. 6.2), followed by proofs of correctness and termination (Sect. 6.3).

Assumption 19. *Throughout this section, we assume*

- *that we deal with coalgebras over the base category* $\mathcal{C} = \mathsf{Set}$;
- *a functor* $B \colon \mathcal{C} \to \mathcal{C}$ *that preserves pre-images and wide intersections;*
- *a category* $\mathcal{D}$ *with an initial object* 0 *s.t. arrows with domain* 0 *are monic;*
- *a functor* $L \colon \mathcal{D} \to \mathcal{D}$ *with an initial algebra* $L\Phi \xrightarrow{\cong} \Phi$;
- *an adjunction* $P \dashv Q \colon \mathcal{C} \leftrightarrows \mathcal{D}^{\mathsf{op}}$, *and a logic* $\delta \colon LP \Rightarrow PB$.

Moreover, we assume a pointed B-coalgebra (X, γ, s_0).

Remark 20. We restrict to $\mathcal{C} = \mathsf{Set}$, but see it as a key contribution to state the algorithm in categorical terms: the assumptions cover a wide class of functors on Set, which is the main direction of generalisation. Further, the categorical approach will enable future generalisations. The assumptions on the category $\mathcal{C}$ are: it is complete, well-powered and satisfies that for all (strong) epis $q\colon S \to \overline{S} \in \mathcal{C}$ and all monos $i\colon S' \to S$ such that $q \circ i$ is mono there is a morphism $q^{-1}\colon \overline{S} \to S$ such that (i) $q \circ q^{-1} = \mathsf{id}$ and $q^{-1} \circ q \circ i = i$.

6.1 Tables and Counterexamples

Definition 21. *A* table *is a pair* $(S \overset{s}{\rightarrowtail} X, \Psi \overset{i}{\rightarrowtail} \Phi)$ *consisting of a subobject s of X and a subformula-closed subobject i of Φ.*

To make the notation a bit lighter, we sometimes refer to a table by (S, Ψ), using s and i respectively to refer to the actual subobjects. The pair (S, Ψ) represents 'rows' and 'columns' respectively, in the table; the 'elements' of the table are given abstractly by the map $th^{\gamma}_{\Psi} \circ s$. In particular, if $\mathcal{C} = \mathcal{D} = \mathsf{Set}$ and $Q = 2^{-}$, then this is a map $S \to 2^{\Psi}$, assigning a Boolean value to every pair of a row (state) and a column (formula).

For the definition of closedness, we use the operator $\Gamma(S)$ from Definition 14, which characterises the successors of a subobject $S \rightarrowtail X$.

$$\begin{array}{ccccc} S & \overset{s}{\rightarrowtail} & X & \xrightarrow{th^{\gamma}} & Q\Psi \\ {\scriptstyle k}\uparrow & & & \nearrow{\scriptstyle th^{\gamma}} & \\ \Gamma(S) & \xrightarrow[\Gamma(s)]{} & X & & \end{array} \tag{9}$$

Definition 22. *A table (S, Ψ) is* closed *if there exists a map $k\colon \Gamma(S) \to S$ such that Diagram (9) commutes. A table (S, Ψ) is* sharp *if the composite map $S \xrightarrow{s} X \xrightarrow{th^{\gamma}} Q\Psi$ is monic.*

Thus, a table (S, Ψ) is closed if all the successors of states (elements of $\Gamma(S)$) are already represented in S, up to equivalence w.r.t. the tests in Ψ. In other terms, the rows corresponding to successors of existing rows are already in the table. Sharpness amounts to minimality w.r.t. logical equivalence: every row has a unique value. The latter will be an invariant of the algorithm (Theorem 32).

A *conjecture* is a coalgebra on S, which is not quite a subcoalgebra of X: instead, it is a subcoalgebra 'up to equivalence w.r.t. Ψ', that is, the successors agree up to logical equivalence.

$$\begin{array}{ccccc} S & \overset{s}{\rightarrowtail} & X & \xrightarrow{\gamma} & BX \\ {\scriptstyle \hat{\gamma}}\downarrow & & & & \downarrow{\scriptstyle Bth^{\gamma}} \\ BS & \xrightarrow[Bs]{} & BX & \xrightarrow[Bth^{\gamma}]{} & BQ\Psi \end{array} \tag{10}$$

Definition 23. *Let (S, Ψ) be a table. A coalgebra structure $\hat{\gamma}\colon S \to BS$ is called a* conjecture *(for (S, Ψ)) if Diagram (10) commutes.*

It is essential to be able to construct a conjecture from a closed table. The following, stronger result is a variation of Proposition 16.

Theorem 24. *A sharp table is closed iff there exists a conjecture for it. Moreover, if the table is sharp and B preserves monos, then this conjecture is unique.*

Our goal is to learn a pointed coalgebra which is correct w.r.t. all formulas. To this aim we ensure correctness w.r.t. an increasing sequence of subformula closed collections Ψ.

$$\begin{array}{ccccc} & & X & & \\ & \overset{s_0}{\nearrow} & & \overset{th^{\gamma}}{\searrow} & \\ 1 & \underset{\hat{s}_0}{\rightarrowtail} & S & \xrightarrow[th^{\hat{\gamma}}]{} & Q\Psi \end{array} \qquad (11)$$

Definition 25. *Let (S, Ψ) be a table, and let $(S, \hat{\gamma}, \hat{s}_0)$ be a pointed B-coalgebra on S. We say $(S, \hat{\gamma}, \hat{s}_0)$ is* correct *w.r.t. Ψ if Diagram* (11) *commutes.*

All conjectures constructed during the learning algorithm will be correct w.r.t. the subformula closed collection Ψ of formulas under consideration.

Lemma 26. *Suppose (S, Ψ) is closed, and $\hat{\gamma}$ is a conjecture. Then $th^{\gamma}_{\Psi} \circ s = th^{\hat{\gamma}}_{\Psi} \colon S \to Q\Psi$. If $\hat{s}_0 \colon 1 \to S$ satisfies $s \circ \hat{s}_0 = s_0$ then $(S, \hat{\gamma}, \hat{s}_0)$ is correct w.r.t. Ψ.*

We next define the crucial notion of *counterexample* to a pointed coalgebra: a subobject Ψ' of Ψ on which it is 'incorrect'.

Definition 27. *Let (S, Ψ) be a table, and let $(S, \hat{\gamma}, \hat{s}_0)$ be a pointed B-coalgebra on S. Let Ψ' be a subformula closed subobject of Φ, such that Ψ is a subcoalgebra of Ψ'. We say Ψ' is a* counterexample *(for $(S, \hat{\gamma}, \hat{s}_0)$,* extending *$\Psi$) if $(S, \hat{\gamma}, \hat{s}_0)$ is* not *correct w.r.t. Ψ'.*

The following elementary lemma states that if there are no more counterexamples for a coalgebra, then it is correct w.r.t. the object Φ of all formulas.

Lemma 28. *Let (S, Ψ) be a table, and let $(S, \hat{\gamma}, \hat{s}_0)$ be a pointed B-coalgebra on S. Suppose that there are no counterexamples for $(S, \hat{\gamma}, \hat{s}_0)$ extending Ψ. Then $(S, \hat{\gamma}, \hat{s}_0)$ is correct w.r.t. Φ.*

The following describes, for a given table, how to extend it with the successors (in X) of all states in S. As we will see below, by repeatedly applying this construction, one eventually obtains a closed table.

Definition 29. *Let (S, Ψ) be a sharp table. Let $(\overline{S}, q, r)$ be the (strong epi, mono)-factorisation of the map $th^{\gamma} \circ (s \vee \Gamma(s))$, as in the diagram:*

$$\begin{array}{ccccc} S \vee \Gamma(S) & \xrightarrow{s \vee \Gamma(s)} & X & \xrightarrow{th^{\gamma}} & Q\Psi \\ & \underset{q}{\searrow\!\!\!\searrow} & & \underset{r}{\nearrow} & \\ & & \overline{S} & & \end{array}$$

We define $\mathsf{close}(S, \Psi) := \{\overline{s} \colon \overline{S} \rightarrowtail X \mid th^{\gamma} \circ \overline{s} = r, s \leq \overline{s} \leq s \vee \Gamma(s)\}$. *For each $\overline{s} \in \mathsf{close}(S, \Psi)$ we have $s \leq \overline{s}$ and thus $s = \overline{s} \circ \kappa$ for some $\kappa \colon S \to \overline{S}$.*

Lemma 30. *In Definition 29, for each $\overline{s} \in \mathsf{close}(S, \Psi)$, we have $\kappa = q \circ inl_{\vee}$.*

We will refer to $\kappa = q \circ inl_{\vee}$ as the connecting map from s to $\overline{s}$.

Lemma 31. *In Definition 29, if there exists $q^{-1} \colon \overline{S} \to S \vee \Gamma(S)$ such that $q \circ q^{-1} = \mathsf{id}$ and $q^{-1} \circ q \circ inl_{\vee} = inl_{\vee}$, then $\mathsf{close}(S, \Psi)$ is non-empty.*

By our assumptions, the hypothesis of Lemma 31 is satisfied (Remark 20), hence $\mathsf{close}(S, \Psi)$ is non-empty. It is precisely (and only) at this point that we need the strong condition about existence of right inverses to epimorphisms.

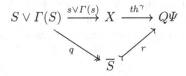

6.2 The Algorithm

Having defined closedness, counterexamples and a procedure for closing a table, we are ready to define the abstract algorithm. In the algorithm, the teacher has access to a function $\mathsf{counter}((S, \hat{\gamma}, \hat{s}_0), \Psi)$, which returns the set of all counterexamples (extending Ψ) for the conjecture $(S, \hat{\gamma}, \hat{s}_0)$. If this set is empty, the coalgebra $(S, \hat{\gamma}, \hat{s}_0)$ is correct (see Lemma 28), otherwise the teacher picks one of its elements Ψ'. We also make use of $\mathsf{close}(S, \Psi)$, as given in Definition 29.

Algorithm 1. Abstract learning algorithm

1: $(S \overset{s}{\rightarrowtail} X) \leftarrow (1 \overset{s_0}{\rightarrowtail} X)$
2: $\hat{s}_0 \leftarrow \mathsf{id}_1$
3: $\Psi \leftarrow 0$
4: **while true do**
5:     **while** $(S \overset{s}{\rightarrowtail} X, \Psi)$ is not closed **do**
6:         let $(\overline{S} \overset{\overline{s}}{\rightarrowtail} X) \in \mathsf{close}(S, \Psi)$, with connecting map $\kappa \colon S \rightarrowtail \overline{S}$
7:         $(S \overset{s}{\rightarrowtail} X) \leftarrow (\overline{S} \overset{\overline{s}}{\rightarrowtail} X)$
8:         $\hat{s}_0 \leftarrow \kappa \circ \hat{s}_0$
9:     **end while**
10:     let $(S, \hat{\gamma})$ be a conjecture for (S, Ψ)
11:     **if** $\mathsf{counter}((S, \hat{\gamma}, \hat{s}_0), \Psi) = \emptyset$ **then**
12:         **return** $(S, \hat{\gamma}, \hat{s}_0)$
13:     **else**
14:         $\Psi \leftarrow \Psi'$ for some $\Psi' \in \mathsf{counter}((S, \hat{\gamma}, \hat{s}_0), \Psi)$
15:     **end if**
16: **end while**

The algorithm takes as input the coalgebra (X, γ, s_0) (which we fixed throughout this section). In every iteration of the outside loop, the table is first closed by repeatedly applying the procedure in Definition 29. Then, if the conjecture corresponding to the closed table is correct, the algorithm returns it (Line 12). Otherwise, a counterexample is chosen (Line 14), and the algorithm continues.

6.3 Correctness and Termination

Correctness is stated in Theorem 33. It relies on establishing loop invariants:

Theorem 32. *The following is an invariant of both loops in Algorithm 1 in Sect. 6.2: 1. (S, Ψ) is sharp, 2. $s \circ \hat{s}_0 = s_0$, and 3. s is s_0-prefix closed w.r.t. γ.*

Theorem 33. *If Algorithm 1 in Sect. 6.2 terminates, then it returns a pointed coalgebra $(S, \hat{\gamma}, \hat{s}_0)$ which is minimal w.r.t. logical equivalence, reachable and correct w.r.t. Φ.*

In our termination arguments, we have to make an assumption about the coalgebra which is to be learned. It does not need to be finite itself, but it should be finite up to logical equivalence—in the case of deterministic automata, for instance, this means the teacher has a (possibly infinite) automaton representing a regular language. To speak about this precisely, let Ψ be a subobject of Φ. We take a (strong epi, mono)-factorisation of the theory map, i.e., $th^{\gamma}_{\Psi} = \left(X \xrightarrow{e_{\Psi}} |X|_{\Psi} \xrightarrow{m_{\Psi}} Q\Psi \right)$ for some strong epi e and mono m. We call the object $|X|_{\Psi}$ in the middle the *Ψ-logical quotient*. For the termination result (Theorem 37), $|X|_{\Phi}$ is assumed to have finitely many quotients and subobjects, which just amounts to finiteness, in Set.

We start with termination of the inner while loop (Corollary 36). This relies on two results: first, that once the connecting map κ is an iso, the table is closed, and second, that—under a suitable assumption on the coalgebra (X, γ)—during execution of the inner while loop, the map κ will eventually be an iso.

Theorem 34. *Let (S, Ψ) be a sharp table, let $\overline{S} \in \mathsf{close}(S, \Psi)$ and let $\kappa \colon S \to \overline{S}$ be the connecting map. If κ is an isomorphism, then (S, Ψ) is closed.*

Lemma 35. *Consider a sequence of sharp tables $(S_i \overset{s_i}{\rightarrowtail} X, \Psi)_{i \in \mathbb{N}}$ such that $s_{i+1} \in \mathsf{close}(S_i, \Psi)$ for all i. Moreover, let $(\kappa_i \colon S_i \to S_{i+1})_{i \in \mathbb{N}}$ be the connecting maps (Definition 29). If the logical quotient $|X|_{\Phi}$ of X has finitely many subobjects, then κ_i is an isomorphism for some $i \in \mathbb{N}$.*

Corollary 36. *If the Φ-logical quotient $|X|_{\Phi}$ has finitely many subobjects, then the inner while loop of Algorithm 1 terminates.*

For the outer loop, we assume that $|X|_{\Phi}$ has finitely many quotients, ensuring that every sequence of counterexamples proposed by the teacher is finite.

Theorem 37. *If the Φ-logical quotient $|X|_{\Phi}$ has finitely many quotients and finitely many subobjects, then Algorithm 1 terminates.*

7 Future Work

We showed how duality plays a natural role in automata learning, through the central connection between states and tests. Based on this foundation, we proved correctness and termination of an abstract algorithm for coalgebra learning. The generality is not so much in the base category (which, for the algorithm, we take to be Set) but rather in the functor used; we only require a few mild conditions on the functor, and make no assumptions about its shape. The approach is thus considered *coalgebra learning* rather than automata learning.

Returning to automata, an interesting direction is to extend the present work to cover learning of, e.g., non-deterministic or alternating automata [5,9] for a regular language. This would require explicitly handling branching in the type of coalgebra. One promising direction would be to incorporate the forgetful logics of [19], which are defined within the same framework of coalgebraic logic as the current work. It is not difficult to define in this setting what it means for a table to be closed 'up to the branching part', stating, e.g., that even though the table is not closed, all the successors of rows are present as combinations of other rows.

Another approach would be to integrate monads into our framework, which are also used to handle branching within the theory of coalgebras [16]. It is an intriguing question whether the current approach, which allows to move beyond automata-like examples, can be combined with the CALF framework [13], which is very far in handling branching occurring in various kinds of automata.

Acknowledgments. We are grateful to Joshua Moerman, Nick Bezhanishvili, Gerco van Heerdt, Aleks Kissinger and Stefan Milius for valuable discussions and suggestions.

References

1. Adámek, J., Lücke, D., Milius, S.: Recursive coalgebras of finitary functors. ITA **41**(4), 447–462 (2007)
2. Adámek, J., Milius, S., Moss, L.S., Sousa, L.: Well-pointed coalgebras. Logical Methods Comput. Sci. **9**(3) (2013)
3. Adámek, J., Rosický, J.: Locally Presentable and Accessible Categories. Cambridge Tracts in Mathematics. Cambridge University Press, Cambridge (1994)
4. Angluin, D.: Learning regular sets from queries and counterexamples. Inf. Comput. **75**(2), 87–106 (1987)
5. Angluin, D., Eisenstat, S., Fisman, D.: Learning regular languages via alternating automata. In: Yang, Q., Wooldridge, M. (eds.) IJCAI 2015, pp. 3308–3314. AAAI Press (2015)
6. Barlocco, S., Kupke, C.: Angluin learning via logic. In: Artemov, S., Nerode, A. (eds.) LFCS 2018. LNCS, vol. 10703, pp. 72–90. Springer, Cham (2018). https://doi.org/10.1007/978-3-319-72056-2_5
7. Blackburn, P., de Rijke, M., Venema, Y.: Modal Logic. Cambridge Tracts in Theoretical Computer Science, vol. 53. Cambridge University Press, Cambridge (2001)
8. Blok, A.: Interaction, observation and denotation. Master's thesis, ILLC Amsterdam (2012)
9. Bollig, B., Habermehl, P., Kern, C., Leucker, M.: Angluin-style learning of NFA. In: Boutilier, C. (ed.) Proceedings of the 21st International Joint Conference on Artificial Intelligence, IJCAI 2009, pp. 1004–1009 (2009)
10. Borceux, F.: Handbook of Categorical Algebra. Encyclopedia of Mathematics and its Applications, vol. 1. Cambridge University Press, Cambridge (1994)
11. van Heerdt, G.: An abstract automata learning framework. Master's thesis, Radboud Universiteit Nijmegen (2016)
12. van Heerdt, G., Sammartino, M., Silva, A.: CALF: categorical automata learning framework. In: Goranko, V., Dam, M. (eds.) 26th EACSL Annual Conference on Computer Science Logic, CSL 2017. LIPIcs, vol. 2, pp. 29:1–29:24. Schloss Dagstuhl - Leibniz-Zentrum fuer Informatik (2017)
13. van Heerdt, G., Sammartino, M., Silva, A.: Learning automata with side-effects. CoRR, abs/1704.08055 (2017)
14. Jacobs, B.: Introduction to Coalgebra: Towards Mathematics of States and Observation. Cambridge Tracts in Theoretical Computer Science, vol. 59. Cambridge University Press, Cambridge (2016)

15. Jacobs, B., Silva, A.: Automata learning: a categorical perspective. In: van Breugel, F., Kashefi, E., Palamidessi, C., Rutten, J. (eds.) Panangaden Festschrift. LNCS, vol. 8464, pp. 384–406. Springer, Cham (2014). https://doi.org/10.1007/978-3-319-06880-0_20
16. Jacobs, B., Silva, A., Sokolova, A.: Trace semantics via determinization. J. Comput. Syst. Sci. **81**(5), 859–879 (2015)
17. Jacobs, B., Sokolova, A.: Exemplaric expressivity of modal logics. J. Logic Comput. **20**(5), 1041–1068 (2009)
18. Klin, B.: Coalgebraic modal logic beyond sets. Electr. Notes Theor. Comput. Sci. **173**, 177–201 (2007)
19. Klin, B., Rot, J.: Coalgebraic trace semantics via forgetful logics. Logical Methods Comput. Sci. **12**(4) (2016)
20. Kupke, C., Kurz, A., Pattinson, D.: Algebraic semantics for coalgebraic logics. Electr. Notes Theor. Comput. Sci. **106**, 219–241 (2004)
21. Kupke, C., Pattinson, D.: Coalgebraic semantics of modal logics: an overview. Theor. Comput. Sci. **412**(38), 5070–5094 (2011)
22. Maler, O., Pnueli, A.: On the learnability of infinitary regular sets. Inf. Comput. **118**(2), 316–326 (1995)
23. Osius, G.: Categorical set theory: a characterization of the category of sets. J. Pure Appl. Algebra **4**(1), 79–119 (1974)
24. Pavlovic, D., Mislove, M., Worrell, J.B.: Testing semantics: connecting processes and process logics. In: Johnson, M., Vene, V. (eds.) AMAST 2006. LNCS, vol. 4019, pp. 308–322. Springer, Heidelberg (2006). https://doi.org/10.1007/11784180_24
25. Rutten, J.J.M.M.: Automata and coinduction (an exercise in coalgebra). In: Sangiorgi, D., de Simone, R. (eds.) CONCUR 1998. LNCS, vol. 1466, pp. 194–218. Springer, Heidelberg (1998). https://doi.org/10.1007/BFb0055624
26. Rutten, J.J.M.M.: Universal coalgebra: a theory of systems. Theor. Comput. Sci. **249**(1), 3–80 (2000)
27. Schröder, L.: Expressivity of coalgebraic modal logic: the limits and beyond. Theor. Comput. Sci. **390**(2–3), 230–247 (2008)
28. Taylor, P.: Practical Foundations of Mathematics. Cambridge University Press, Cambridge (1999)
29. Trnková, V.: On descriptive classification of set-functors. I. Comment. Math. Univ. Carolinae **12**(1), 143–174 (1971)
30. Vaandrager, F.W.: Model learning. Commun. ACM **60**(2), 86–95 (2017)

Tight Worst-Case Bounds for Polynomial Loop Programs

Amir M. Ben-Amram[1] and Geoff W. Hamilton[2(✉)]

[1] School of Computer Science, Tel-Aviv Academic College, Tel Aviv, Israel
amirben@mta.ac.il
[2] School of Computing, Dublin City University, Dublin 9, Ireland
hamilton@computing.dcu.ie

Abstract. In 2008, Ben-Amram, Jones and Kristiansen showed that for a simple programming language—representing non-deterministic imperative programs with bounded loops, and arithmetics limited to addition and multiplication—it is possible to decide precisely whether a program has certain growth-rate properties, in particular whether a computed value, or the program's running time, has a polynomial growth rate.

A natural and intriguing problem was to improve the precision of the information obtained. This paper shows how to obtain asymptotically-tight *multivariate* polynomial bounds for this class of programs. This is a complete solution: whenever a polynomial bound exists it will be found.

1 Introduction

One of the most important properties we would like to know about programs is their *resource usage*, i.e., the amount of resources (such as time, memory and energy) required for their execution. This information is useful during development, when performance bugs and security vulnerabilities exploiting performance issues can be avoided. It is also particularly relevant for mobile applications, where resources are limited, and for cloud services, where resource usage is a major cost factor.

In the literature, a lot of different "cost analysis" problems (also called "resource bound analysis," etc.) have been studied (e.g. [1,11,13,18,19,24,26,27]); several of them may be grouped under the following general definition. The *countable resource problem* asks about the maximum usage of a "resource" that accumulates during execution, and which one can explicitly count, by instrumenting the program with an accumulator variable and instructions to increment it where necessary. For example, we can estimate the *execution time* of a program by counting certain "basic steps". Another example is counting the number of visits to designated program locations. Realistic problems of this type include bounding the number of calls to specific functions, perhaps to system services; the number of I/O operations; number of accesses to memory, etc. The consumption of resources such as *energy* suits our problem formulation as long as such explicit bookkeeping is possible (we have to assume that the increments, if not constant, are given by a monotone polynomial expression).

M. Bojańczyk and A. Simpson (Eds.): FOSSACS 2019, LNCS 11425, pp. 80–97, 2019.
https://doi.org/10.1007/978-3-030-17127-8_5

In this paper we solve the *bound analysis problem* for a particular class of programs, defined in [7]. The bound analysis problem is to find symbolic bounds on the maximal possible value of an integer variable at the end of the program, in terms of some integer-valued variables that appear in the initial state of a computation. Thus, a solution to this problem might be used for any of the resource-bound analyses above. In this work we focus on values that grow polynomially (in the sense of being bounded by a polynomial), and our goal is to find polynomial bounds that are tight, in the sense of being precise up to a constant factor.

The programs we study are expressed by the so-called *core language*. It is imperative, including bounded loops, non-deterministic branches and restricted arithmetic expressions; the syntax is shown in Fig. 1. Semantics is explained and motivated below, but is largely intuitive; see also the illustrative example in Fig. 2. In 2008, it was proved [7] that for this language it is decidable whether a computed result is polynomially bounded or not. This makes the language an attractive target for work on the problem of computing tight bounds. However, for the past ten years there has been no improvement on [7]. We now present an algorithm to compute, for every program in the language, and every variable in the program which has a polynomial upper bound (in terms of input values), a tight polynomial bound on its largest attainable value (informally, "the worst-case value") as a function of the input values. The bound is guaranteed to be tight up to a multiplicative constant factor but constants are left implicit (for example a bound quadratic in n will always be represented as n^2). The algorithm could be extended to compute upper and lower bounds with explicit constant factors, but choosing to ignore coefficients simplifies the algorithm considerably. In fact, we have striven for a simple, comprehensible algorithm, and we believe that the algorithm we present is sufficiently simple that, beyond being comprehensible, offers insight into the structure of computations in this model.

1.1 The Core Language

Data. It is convenient to assume (without loss of generality) that the only type of data is non-negative integers. Note that a realistic (not "core") program may include many statements that manipulate non-integer data that are not relevant to loop control—so in a complexity analysis, we may be able to abstract these parts away and still analyze the variables of interest. In other cases, it is

$$
\begin{array}{rcl}
\texttt{X} \in \text{Variable} & ::= & \texttt{X}_1 \mid \texttt{X}_2 \mid \texttt{X}_3 \mid \ldots \mid \texttt{X}_n \\
\texttt{E} \in \text{Expression} & ::= & \texttt{X} \mid \texttt{E + E} \mid \texttt{E * E} \\
\texttt{C} \in \text{Command} & ::= & \texttt{skip} \mid \texttt{X:=E} \mid \texttt{C}_1\texttt{;C}_2 \mid \texttt{loop E \{C\}} \\
 & \mid & \texttt{choose C}_1 \texttt{ or C}_2
\end{array}
$$

Fig. 1. Syntax of the core language.

possible to preprocess a program to replace complex data values with their size (or "norm"), which is the quantity of importance for loop control. Methods for this process have been widely studied in conjunction with termination and cost analysis.

Command Semantics. The core language is inherently non-deterministic. The `choose` command represents a non-deterministic choice, and can be used to abstract any concrete conditional command by simply ignoring the condition; this is necessary to ensure that our analysis problem is decidable. Note that what we ignore is branches within a loop body and not branches that implement the loop control, which we represent by a dedicated loop command. The command `loop E {C}` repeats `C` a (non-deterministic) number of times bounded by the value of `E`, which is evaluated just before the loop is entered. Thus, as a conservative abstraction, it may be used to model different forms of loops (for-loops, while-loops) as long as a bound on the number of iterations, as a function of the program state on loop initiation, can be determined and expressed in the language. There is an ample body of research on analysing programs to find such bounds where they are not explicitly given by the programmer; in particular, bounds can be obtained from a *ranking function* for the loop [2,3,5,6,23]. Note that the arithmetic in our language is too restricted to allow for the maintenance of counters and the creation of *while* loops, as there is no subtraction, no explicit constants and no tests. Thus, for realistic "concrete" programs which use such devices, loop-bound analysis is supposed to be performed *on the concrete program* as part of the process of abstracting it to the core language. This process is illustrated in [9, Sect. 2].

From a computability viewpoint, the use of bounded loops restricts the programs that can be represented to such that compute primitive recursive functions; this is a rich enough class to cover a lot of useful algorithms and make the analysis problem challenging. In fact, our language resembles a weakened version of Meyer and Ritchie's LOOP language [20], which computes all the primitive recursive functions, and where behavioral questions like "is the result linearly bounded" are undecidable.

```
loop X1 {
   loop X2 + X3 { choose { X3:= X1; X2:= X4 } or { X3:= X4; X2:= X1 } };
   X4:= X2 + X3
};
loop X4 { choose  { X3:= X1 + X2 + X3 } or { X3:= X2;  X2:= X1 } }
```

Fig. 2. A core-language program. `loop` n `C` means "do `C` at most n times."

1.2 The Algorithm

Consider the program in Fig. 2. Suppose that it is started with the values of the variables $X_1, X_2, \ldots$ being $x_1, x_2, \ldots$. Our purpose is to bound the values of all variables at the conclusion of the program in terms of those initial values. Indeed, they are all polynomially bounded, and our algorithm provides tight bounds. For instance, it establishes that the final value of $\mathtt{X}_3$ is tightly bounded (up to a constant factor) by $\max(x_4(x_4+x_1^2), x_4(x_2+x_3+x_1^2))$.

In fact, it produces information in a more precise form, as *a disjunction of simultaneous bounds*. This means that it generates vectors, called *multi-polynomials*, that give simultaneous bounds on all variables; for example, with the program in Fig. 2, one such multi-polynomial is $\langle x_1, x_2, x_3, x_4\rangle$ (this is the result of all loops taking a very early exit). This form is important in the context of a compositional analysis. To see why, suppose that we provide, for a command with variables $\mathtt{X}, \mathtt{Y}$, the bounds $\langle x, y\rangle$ and $\langle y, x\rangle$. Then we know that the *sum* of their values is always bounded by $x + y$, a result that would have not been deduced had we given the bound $\max(x, y)$ on each of the variables. The difference may be critical for the success of analyzing an enclosing or subsequent command.

Multivariate bounds are often of interest, and perhaps require no justification, but let us point out that multivariate polynomials are necessary even if we're ultimately interested in a univariate bound, in terms of some single initial value, say n. This is, again, due to the analysis being compositional. When we analyze an internal command that uses variables $\mathtt{X}, \mathtt{Y}, \ldots$ we do not know in what possible contexts the command will be executed and how the values of these variables will be related to n.

Some highlights of our solution are as follows.

- We reduce the problem of analyzing any core-language program to the problem of analyzing a single loop, whose body is already processed, and therefore presented as a collection of multi-polynomials. This is typical of algorithms that analyze a structured imperative language and do so compositionally.
- Since we are computing bounds only up to a constant factor, we work with *abstract* polynomials, that have no numeric coefficients.
- We further introduce *τ-polynomials*, to describe the evolution of values in a loop. These have an additional parameter τ (for "time"; more precisely, number of iterations). Introducing τ-polynomials was a key step in the solution.
- The analysis of a loop is simply a closure computation under two operations: ordinary composition, and *generalization* which is the operation that predicts the evolution of values by judiciously adding τ's to *idempotent* abstract multi-polynomials.

The remainder of this paper is structured as follows. In Sect. 2 we give some definitions and state our main result. In Sects. 3, 4 and 5 we present our algorithm. In Sect. 6, we outline the correctness proofs. Section 7 considers related work, and Sect. 8 concludes and discusses ideas for further work.

2 Preliminaries

In this section, we give some basic definitions, complete the presentation of our programming language and precisely state the main result.

2.1 Some Notation and Terminology

The Language. We remark that in our language syntax there is no special form for a "program unit"; in the text we sometimes use "program" for the subject of our analysis, yet syntactically it's just a command.

Polynomials and Multi-polynomials. We work throughout this article with multivariate polynomials in $x_1, \dots, x_n$ that have non-negative integer coefficients and no variables other than $x_1, \dots, x_n$; when we speak of a polynomial we always mean one of this kind. Note that over the non-negative integers, such polynomials are monotonically (weakly) increasing in all variables.

The post-fix substitution operator $[a/b]$ may be applied to any sort of expression containing a variable b, to substitute a instead; e.g., $(x^2 + yx + y)[2z/y] = x^2 + 2zx + 2z$.

When discussing a command, state-transition, or program trace, with a variable $\mathtt{X}_i$, x_i will denote, as a rule, the initial value of this variable, and x'_i its final value. Thus we distinguish the syntactic entity by the typewriter font. We write the polynomials manipulated by our algorithms using the variable names x_i. We presume that an implementation of the algorithm represents polynomials concretely so that ordinary operations such as composition can be applied, but otherwise we do not concern ourselves much with representation.

The parameter n always refers to the number of variables in the subject program. The set $[n]$ is $\{1, \dots, n\}$. For a set S an n-tuple over S is a mapping from $[n]$ to S. The set of these tuples is denoted by S^n. Throughout the paper, various natural liftings of operators to collections of objects is tacitly assumed, e.g., if S is a set of integers then $S + 1$ is the set $\{s + 1 \mid s \in S\}$ and $S + S$ is $\{s + t \mid s, t \in S\}$. We use such lifting with sets as well as with tuples. If S is ordered, we extend the ordering to S^n by comparing tuples element-wise (this leads to a partial order, in general, e.g., with natural numbers, $\langle 1, 3\rangle$ and $\langle 2, 2\rangle$ are incomparable).

Definition 1. *A* polynomial transition (PT) *represents a mapping of an "input" state* $\mathbf{x} = \langle x_1, \dots, x_n\rangle$ *to a "result" state* $\mathbf{x}' = \langle x'_1, \dots, x'_n\rangle = \mathbf{p}(\boldsymbol{x})$ *where* $\mathbf{p} = \langle \mathbf{p}[1], \dots, \mathbf{p}[n]\rangle$ *is an n-tuple of polynomials. Such a* $\mathbf{p}$ *is called a* a multi-polynomial (MP)*; we denote by* `MPol` *the set of multi-polynomials, where the number of variables n is fixed by context.*

Multi-polynomials are used in this work to represent the effect of a command. Various operations will be applied to MPs, mostly obvious—in particular, composition (which corresponds to sequential application of the transitions). Note that composition of multi-polynomials, $\mathbf{q} \circ \mathbf{p}$, is naturally defined since $\mathbf{p}$ supplies n values for the n variables of $\mathbf{q}$ (in other words, they are composed as functions in $\mathbb{N}^n \to \mathbb{N}^n$). We define *Id* to be the identity transformation, $\mathbf{x}' = \mathbf{x}$ (in MP notation: $\mathbf{p}[i] = x_i$ for $i = 1, \dots, n$).

2.2 Formal Semantics of the Core Language

The semantics associates with every command C over variables $\texttt{X}_1, \dots, \texttt{X}_n$ a relation $[\![\texttt{C}]\!] \subseteq \mathbb{N}^n \times \mathbb{N}^n$. In the expression $\boldsymbol{x}[\![\texttt{C}]\!]\boldsymbol{y}$, vector $\boldsymbol{x}$ (respectively $\boldsymbol{y}$) is the store before (after) the execution of C.

The semantics of `skip` is the identity. The semantics of an assignment $\texttt{X}_i\texttt{:=E}$ associates to each store $\boldsymbol{x}$ a new store $\boldsymbol{y}$ obtained by replacing the component x_i by the value of the expression E when evaluated over store $\boldsymbol{x}$. This is defined in the natural way (details omitted), and is denoted by $[\![\texttt{E}]\!]\boldsymbol{x}$. Composite commands are described by the straight-forward equations:

$$
\begin{aligned}
[\![\texttt{C}_1;\texttt{C}_2]\!] &= [\![\texttt{C}_2]\!] \circ [\![\texttt{C}_1]\!] \\
[\![\texttt{choose C}_1 \texttt{ or C}_2]\!] &= [\![\texttt{C}_1]\!] \cup [\![\texttt{C}_2]\!] \\
[\![\texttt{loop E \{C\}}]\!] &= \{(\boldsymbol{x}, \boldsymbol{y}) \mid \exists i \le [\![\texttt{E}]\!]\boldsymbol{x} : \boldsymbol{x}[\![\texttt{C}]\!]^i \boldsymbol{y}\}
\end{aligned}
$$

where $[\![\texttt{C}]\!]^i$ represents $[\![\texttt{C}]\!] \circ \dots \circ [\![\texttt{C}]\!]$ (i occurrences of $[\![\texttt{C}]\!]$); and $[\![\texttt{C}]\!]^0 = Id$.

Remarks. The following two changes may enhance the applicability of the core language for simulating certain concrete programs; we include them as "options" because they do not affect the validity of our proofs.

1. The semantics of an assignment operation may be non-deterministic: `X:=E` assigns to X some non-negative value *bounded* by E. This is useful to abstract expressions which are not in the core language, and also to use the results of size analysis of subprograms. Such an analysis may determine invariants such as "the value of `f(X,Y)` is at most the sum of X and Y."
2. The domain of the integer variables may be extended to $\mathbb{Z}$. In this case the bounds that we seek are on the absolute value of the output in terms of absolute values of the inputs. This change does not affect our conclusions because of the facts $|xy| = |x| \cdot |y|$ and $|x + y| \le |x| + |y|$. The semantics of the loop command may be defined either as doing nothing if the loop bound is not positive, or using the absolute value as a bound.

2.3 Detailed Statement of the Main Result

The *polynomial-bound analysis problem* is to find, for any given command, which output variables are bounded by a polynomial in the input values (which are simply the values of all variables upon commencement of the program), and to bound these output values tightly (up to constant factors). The problem of *identifying* the polynomially-bounded variables is completely solved by [7]. We rely on that algorithm, which is polynomial-time, to do this for us (as further explained below).

Our main result is thus stated as follows.

Theorem 1. *There is an algorithm which, for a command* C, *over variables* $\texttt{X}_1$ *through* $\texttt{X}_n$, *outputs a set* $\mathcal{B}$ *of multi-polynomials, such that the following hold, where* PB *is the set of indices* i *of variables* $\texttt{X}_i$ *which are polynomially bounded under* $[\![\texttt{C}]\!]$.

1. *(Bounding) There is a constant* $c_{\mathbf{p}}$ *associated with each* $\mathbf{p} \in \mathcal{B}$*, such that*

$$\forall \boldsymbol{x}, \boldsymbol{y} \,.\, \boldsymbol{x}[\![\mathtt{C}]\!]\boldsymbol{y} \Longrightarrow \exists \mathbf{p} \in \mathcal{B} \,.\forall i \in PB \,.\, y_i \leq c_{\mathbf{p}}\mathbf{p}[i](\boldsymbol{x})$$

2. *(Tightness) For every* $\mathbf{p} \in \mathcal{B}$ *there are constants* $d_{\mathbf{p}} > 0$, $\boldsymbol{x}_0$ *such that for all* $\boldsymbol{x} \geq \boldsymbol{x}_0$ *there is a* $\boldsymbol{y}$ *such that*

$$\boldsymbol{x}[\![\mathtt{C}]\!]\boldsymbol{y} \text{ and } \forall i \in PB \,.\, y_i \geq d_{\mathbf{p}}\mathbf{p}[i](\boldsymbol{x}).$$

3 Analysis Algorithm: First Concepts

The following sections describe our analysis algorithm. Naturally, the most intricate part of the analysis concerns loops. In fact we break the description into stages: first we reduce the problem of analyzing any program to that of analyzing *simple disjunctive loops*, defined next. Then, we approach the analysis of such loops, which is the main effort in this work.

Definition 2. *A* simple disjunctive loop (SDL) *is a finite set of PTs.*

The loop is "disjunctive" because its meaning is that in every iteration, any of the given transitions may be applied. The semantics is formalized by *traces* (Definition 4). A SDL does not specify the number of iterations; our analysis generates polynomials which depend on the number of iterations as well as the initial state. For this purpose, we now introduce τ-polynomials where τ represents the number of iterations.

Definition 3. *τ-polynomials are polynomials in* $x_1, \ldots, x_n$ *and* τ.

τ has a special status and does not have a separate component in the polynomial giving its value. If p is a τ-polynomial, then $p(v_1, \ldots, v_n)$ is the result of substituting each v_i for the respective x_i; and we also write $p(v_1, \ldots, v_n, t)$ for the result of substituting t for τ as well. The set of τ-polynomials in n variables (n known from context) is denoted $\tau\mathtt{Pol}$.

Multi-polynomials and polynomial transitions are formed from τ-polynomials just as previously defined and are used to represent the effect of a variable number of iterations. For example, the τ-polynomial transition $\langle x_1', x_2' \rangle = \langle x_1,\ x_2 + \tau x_1 \rangle$ represents the effect of repeating (τ times) the assignment $\mathtt{X}_2 \mathtt{:=}\ \mathtt{X}_2 + \mathtt{X}_1$. The effect of iterating the composite command: $\mathtt{X}_2 \mathtt{:=}\ \mathtt{X}_2 + \mathtt{X}_1\mathtt{;}\ \mathtt{X}_3 \mathtt{:=}\ \mathtt{X}_3 + \mathtt{X}_2$ has an effect described by $\mathbf{x}' = \langle x_1,\ x_2 + \tau x_1,\ x_3 + \tau x_2 + \tau^2 x_1 \rangle$ (here we already have an upper bound which is not reached precisely, but is correct up to a constant factor). We denote the set of τ-polynomial transitions by τ**MPol**. We should note that composition $\mathbf{q} \circ \mathbf{p}$ over τ**MPol** is performed by substituting $\mathbf{p}[i]$ for each occurrence of x_i in $\mathbf{q}$. Occurrences of τ are unaffected (since τ is not part of the state). We make a couple of preliminary definitions before reaching our goal which is the definition of the *simple disjunctive loop problem* (Definition 6).

Definition 4. *Let* $\mathcal{S}$ *be a set of polynomial transitions. An* (abstract) trace *over* $\mathcal{S}$ *is a finite sequence* $\mathbf{p}_1; \ldots; \mathbf{p}_{|\sigma|}$ *of elements of* $\mathcal{S}$*. Thus* $|\sigma|$ *denotes the* length *of the trace. The set of all traces is denoted* $\mathcal{S}^*$*. We write* $[\![\sigma]\!]$ *for the composed relation* $\mathbf{p}_{|\sigma|} \circ \cdots \circ \mathbf{p}_1$ *(for the empty trace,* ε*, we have* $[\![\varepsilon]\!] = Id$*).*

Definition 5. *Let $p(\mathbf{x})$ be a (concrete or abstract) τ-polynomial. We write $\dot{p}$ for the sum of* linear monomials *of p, namely any one of the form ax_i with constant coefficient a. We write $\ddot{p}$ for the rest. Thus $p = \dot{p} + \ddot{p}$.*

Definition 6 (Simple disjunctive loop problem). *The* simple disjunctive loop problem *is: given the set $\mathcal{S}$, find (if possible) a finite set $\mathcal{B}$ of τ-polynomial transitions which* tightly bound *all traces over $\mathcal{S}$. More precisely, we require:*

1. *(Bounding) There is a constant $c_{\mathbf{p}} > 0$ associated with each $\mathbf{p} \in \mathcal{B}$, such that*

$$\forall \boldsymbol{x}, \boldsymbol{y}, \sigma \;.\; \boldsymbol{x}[\![\sigma]\!]\boldsymbol{y} \Longrightarrow \exists \mathbf{p} \in \mathcal{B} \,.\, \boldsymbol{y} \le c_{\mathbf{p}}\mathbf{p}(\boldsymbol{x}, |\sigma|)$$

2. *(Tightness) For every $\mathbf{p} \in \mathcal{B}$ there are constants $d_{\mathbf{p}} > 0$, $\boldsymbol{x}_0$ such that for all $\boldsymbol{x} \ge \boldsymbol{x}_0$ there are a trace σ and a state vector $\boldsymbol{y}$ such that*

$$\boldsymbol{x}[\![\sigma]\!]\boldsymbol{y} \wedge \boldsymbol{y} \ge \dot{\mathbf{p}}(\boldsymbol{x}, |\sigma|) + d_{\mathbf{p}}\ddot{\mathbf{p}}(\boldsymbol{x}, |\sigma|)\,.$$

Note that in the lower-bound clause (2), the linear monomials of p are not multiplied, in the left-hand side, by the coefficient $d_{\mathbf{p}}$; this sets, in a sense, a stricter requirement for them: if the trace maps x to x^2 then the bound $2x^2$ is acceptable, but if it maps x to x, the bound $2x$ is not accepted. The reader may understand this technicality by considering the effect of iteration: it is important to distinguish the transition $x_1' = x_1$, which can be iterated ad libitum, from the transition $x_1' = 2x_1$, which produces exponential growth on iteration. Distinguishing $x_1' = x_1^2$ from $x_1' = 2x_1^2$ is not as important. The result set $\mathcal{B}$ above is sometimes called a *loop summary*. We remark that Definition 6 implies that the **max** of all these polynomials provides a "big Theta" bound for the worst-case (namely biggest) results of the loop's computation. We prefer, however, to work with sets of polynomials. Another technical remark is that $c_{\mathbf{p}}, d_{\mathbf{p}}$ range over real numbers. However, our data and the coefficients of polynomials remain integers, it is only such comparisons that are performed with real numbers (specifically, to allow $c_{\mathbf{p}}$ to be smaller than one).

4 Reduction to Simple Disjunctive Loops

We show how to reduce the problem of analysing core-language programs to the analysis of polynomially-bounded simple disjunctive loops.

4.1 Symbolic Evaluation of Straight-Line Code

Straight-line code consists of atomic commands—namely assignments (or `skip`, equivalent to `X`$_1$ `:=` `X`$_1$), composed sequentially. It is obvious that symbolic evaluation of such code leads to polynomial transitions.

Example 1. `X`$_2$`:= X`$_1$`; X`$_4$`:= X`$_2$` + X`$_3$`; X`$_1$`:= X`$_2$` * X`$_3$ is precisely represented by the transition $\langle x_1, x_2, x_3\rangle' = \langle x_1x_3,\; x_1,\; x_3,\; x_1 + x_3\rangle$.

4.2 Evaluation of Non-deterministic Choice

Evaluation of the command `choose` C_1 `or` C_2 yields a set of possible outcomes. Hence, the result of analyzing a command will be a *set* of multi-polynomial transitions. We express this in the common notation of abstract semantics:

$$[\![\mathtt{C}]\!]^S \in \wp(\mathtt{MPol})\,.$$

For uniformity, we consider $[\![\mathtt{C}]\!]^S$ for an atomic command to be a singleton in $\wp(\mathtt{MPol})$ (this means that we represent a transition $\boldsymbol{x}' = \mathbf{p}(\boldsymbol{x})$ by $\{\mathbf{p}\}$). Composition is naturally extended to sets, and the semantics of a choice command is now simply set union, so we have:

$$[\![\mathtt{C}_1;\mathtt{C}_2]\!]^S = [\![\mathtt{C}_2]\!]^S \circ [\![\mathtt{C}_1]\!]^S$$
$$[\![\mathtt{choose}\ \mathtt{C}_1\ \mathtt{or}\ \mathtt{C}_2]\!]^S = [\![\mathtt{C}_1]\!]^S \cup [\![\mathtt{C}_2]\!]^S$$

Example 2. `X`$_2$`:= X`$_1$`; choose { X`$_4$`:= X`$_2$` + X`$_3$` } or { X`$_1$`:= X`$_2$` * X`$_3$` }` is represented by the set $\{\langle x_1, x_1, x_3, x_1 + x_3\rangle,\ \langle x_1 x_3, x_1, x_3, x_4\rangle\}$.

4.3 Handling Loops

The above shows that any loop-free command in our language can be precisely represented by a finite set of PTs. Consequently, the problem of analyzing *any* command is reduced to the analysis of simple disjunctive loops.

Suppose that we have an algorithm SOLVE that takes a simple disjunctive loop and computes tight bounds for it (see Definition 6). We use it to complete the analysis of any program by the following definition:

$$[\![\mathtt{loop}\ \mathtt{E}\ \{\mathtt{C}\}]\!]^S = (\textsc{Solve}([\![\mathtt{C}]\!]^S)[E/\tau]\,.$$

Thus, the whole solution is constructed as an ordinary abstract interpretation, following the semantics of the language, except for procedure SOLVE, described below.

Example 3. `X`$_4$`:= X`$_1$`; loop X`$_4$` { X`$_2$`:= X`$_1$` + X`$_2$`; X`$_3$`:= X`$_2$` }`.
The loop includes just one PT. Solving the loop yields a set $\mathcal{L} = \{\langle x_1, x_2, x_3, x_4\rangle, \langle x_1, x_2 + \tau x_1, x_2 + \tau x_1, x_4\rangle\}$ (the first MP accounts for zero iterations, the second covers any positive number of iterations). We can now compute the effect of the given command as

$$\begin{aligned}\mathcal{L}[x_4/\tau] \circ [\![\mathtt{X}_4\ \mathtt{:=}\ \mathtt{X}_1]\!]^S &= \mathcal{L}[x_4/\tau] \circ \{\langle x_1, x_2, x_3, x_1\rangle\}\\ &= \{\langle x_1, x_2, x_3, x_1\rangle, \langle x_1, x_2 + x_1^2, x_2 + x_1^2, x_1\rangle\}.\end{aligned}$$

The next section describes procedure Solve, and operates under the assumption that all variables are polynomially bounded in the loop. However, a loop can generate exponential growth. To cover this eventuality, we first apply the algorithm of [7] which identifies which variables are polynomially bounded. If some $\texttt{X}_i$ is *not* polynomially bounded we replace the ith component of all the loop transitions with x_n (where we assume x_n to be a dedicated, unmodified variable). Clearly, after this change, all variables are polynomially bounded; moreover, variables which are genuinely polynomial are unaffected, because they cannot depend on a super-exponential quantity (given the restricted arithmetics in our language). In reporting the results of the algorithm, we should display "`super-polynomial`" instead of all bounds that depend on x_n.

5 Simple Disjunctive Loop Analysis Algorithm

Intuitively, evaluating `loop E {C}` abstractly consists of simulating any finite number of iterations, i.e., computing

$$Q_i = \{Id\} \cup P \cup (P \circ P) \cup \cdots \cup P^{(i)} \tag{1}$$

where $P = [\![\texttt{C}]\!]^S \in \wp(\texttt{MPol})$. The question now is whether the sequence (1) reaches a fixed point. In fact, it often doesn't. However, it is quite easy to see that in the *multiplicative fragment* of the language, that is, where the addition operator is not used, such non-convergence is associated with exponential growth. Indeed, since there is no addition, all our polynomials are monomials with a leading coefficient of 1 (*monic monomials*)—this is easy to verify. It follows that if the sequence (1) does not converge, higher and higher exponents must appear, which indicates that some variable cannot be bounded polynomially. Taking the contrapositive, we conclude that if all variables are known to be polynomially bounded the sequence will converge. Thus we have the following easy (and not so satisfying) result:

Observation 2. *For a SDL that does not use addition, the sequence Q_i as in* (1) *reaches a fixed point, and the fixed point provides tight bounds for all the polynomially-bounded variables.*

When we have addition, we find that knowing that all variables are polynomially bounded does not imply convergence of the sequence (1). An example is: `loop` $\texttt{X}_3$ `{` $\texttt{X}_1$`:=` $\texttt{X}_1$ `+` $\texttt{X}_2$ `}` yielding the infinite sequence of MPs $\langle x_1, x_2, x_3\rangle$, $\langle x_1 + x_2, x_2, x_3\rangle$, $\langle x_1 + 2x_2, x_2, x_3\rangle$, ... Our solution employs two means. One is the introduction of τ-polynomials, already presented. The other is a kind of *abstraction*—intuitively, ignoring the concrete values of (non-zero) coefficients. Let us first define this abstraction:

Definition 7. `APol`*, the set of abstract polynomials, consists of formal sums of distinct monomials over $x_1, \ldots, x_n$, where the coefficient of every monomial included is* 1. *We extend the definition to an abstraction of τ-polynomials, denoted τ*`APol`.

The meaning of abstract polynomials is given by the following rules:

1. The abstraction of a polynomial p, $\alpha(p)$, is obtained by modifying all (non-zero) coefficients to 1.
2. Addition and multiplication in τAPol is defined in a natural way so that $\alpha(p)+\alpha(q) = \alpha(p+q)$ and $\alpha(p) \cdot \alpha(q) = \alpha(p \cdot q)$ (to carry these operations out, you just go through the motions of adding or multiplying ordinary polynomials, ignoring the coefficient values).
3. The *canonical concretization* of an abstract polynomial, $\gamma(\mathbf{p})$ is obtained by simply regarding it as an ordinary polynomial.
4. These definitions extend naturally to tuples of (abstract) polynomials.
5. The set of abstract multi-polynomials AMPol and their extension with τ (τAMPol) are defined as n-tuples over APol (respectively, τAPol). We use AMP as an abbreviation for abstract multi-polynomial.
6. Composition $\mathbf{p} \bullet \mathbf{q}$, for $\mathbf{p}, \mathbf{q} \in$ AMPol (or τAMPol) is defined as $\alpha(\gamma(\mathbf{p}) \circ \gamma(\mathbf{q}))$; it is easy to see that one can perform the calculation without the detour through polynomials with coefficients. The different operator symbol ("•" versus "∘") helps in disambiguating expressions.

Analysing a SDL. To analyse a SDL specified by a set of MPs $\mathcal{S}$, we start by computing $\alpha(\mathcal{S})$. The rest of the algorithm computes within τAMPol. We define two operations that are combined in the analysis of loops. The first, which we call *closure*, is simply the fixed point of accumulated iterations as in the multiplicative case. It is introduced by the following two definitions.

Definition 8 (iterated composition). *Let* $\mathbf{t}$ *be any abstract* τ*-MP. We define* $\mathbf{t}^{\bullet(n)}$*, for* $n \geq 0$*, by:*

$$\mathbf{t}^{\bullet(0)} = Id$$
$$\mathbf{t}^{\bullet(n+1)} = \mathbf{t} \bullet \mathbf{t}^{\bullet(n)}.$$

For a set $\mathcal{T}$ *of abstract* τ*-MPs, we define, for* $n \geq 0$*:*

$$\mathcal{T}^{\bullet(0)} = \{Id\}$$
$$\mathcal{T}^{\bullet(n+1)} = \mathcal{T}^{\bullet(n)} \cup \bigcup_{\mathbf{q} \in \mathcal{T},\ \mathbf{p} \in \mathcal{T}^{\bullet(n)}} \mathbf{q} \bullet \mathbf{p}.$$

Note that $\mathbf{t}^{\bullet(n)} = \alpha(\gamma(\mathbf{t})^{(n)})$, where $\mathbf{p}^{(n)}$ is defined using ordinary composition.

Definition 9 (abstract closure). *For finite* $P \subset \tau$AMPol*, we define:*

$$Cl(P) = \bigcup_{i=0}^{\infty} P^{\bullet(i)}.$$

In the correctness proof, we argue that when all variables are polynomially bounded in a loop $\mathcal{S}$, the closure of $\alpha(\mathcal{S})$ can be computed in finite time; equivalently, it equals $\bigcup_{i=0}^{k}(\alpha(\mathcal{S}))^{\bullet(i)}$ for some k. The argument is essentially the same as in the multiplicative case.

The second operation is called *generalization* and its role is to capture the behaviour of accumulator variables, meaning variables that grow by accumulating increments in the loop, and make explicit the dependence on the number of iterations. The identification of which additive terms in a MP should be considered as increments that accumulate is at the heart of our problem, and is greatly simplified by concentrating on idempotent AMPs.

Definition 10. $\mathbf{p} \in \tau\texttt{AMPol}$ *is called* idempotent *if* $\mathbf{p} \bullet \mathbf{p} = \mathbf{p}$.

Note that this is composition in the abstract domain. So, for instance, $\langle x_1, x_2 \rangle$ is idempotent, and so is $\langle x_1 + x_2, x_2 \rangle$, while $\langle x_1 x_2, x_2 \rangle$ and $\langle x_1 + x_2, x_1 \rangle$ are not.

Definition 11. *For* $\mathbf{p}$ *an (abstract) multi-polynomial, we say that* x_i *is* self-dependent *in* $\mathbf{p}$ *if* $\mathbf{p}[i]$ *depends on* x_i. *We call a monomial self-dependent if all the variables appearing in it are.*

Definition 12. *We define a notational convention for* τ*-MPs. Assuming that* $\mathbf{p}[i]$ *depends on* x_i, *we write*

$$\mathbf{p}[i] = x_i + \tau\mathbf{p}[i]' + \mathbf{p}[i]'' + \mathbf{p}[i]''' ,$$

where $\mathbf{p}[i]'''$ *includes all the non-self-dependent monomials of* $\mathbf{p}[i]$, *while the self-dependent monomials (other than* x_i*) are grouped into two sums:* $\tau\mathbf{p}[i]'$, *including all monomials with a positive degree of* τ, *and* $\mathbf{p}[i]''$ *which includes all the* τ*-free monomials.*

Example 4. Let $\mathbf{p} = \langle x_1 + \tau x_2 + \tau x_3 + x_3 x_4,\ x_3,\ x_3,\ x_4 \rangle$. The self-dependent variables are all but x_2. Since x_1 is self-dependent, we will apply the above definition to $\mathbf{p}[1]$, so that $\mathbf{p}[1]' = x_3$, $\mathbf{p}[1]'' = x_3 x_4$ and $\mathbf{p}[1]''' = \tau x_2$. Note that a factor of τ is stripped in $\mathbf{p}[1]'$. Had the monomial been $\tau^2 x_3$, we would have $\mathbf{p}[1]' = \tau x_3$.

Definition 13 (generalization). *Let* $\mathbf{p}$ *be idempotent in* $\tau\texttt{AMPol}$; *define* $\mathbf{p}^\tau$ *by*

$$\mathbf{p}^\tau[i] = \begin{cases} x_i + \tau\mathbf{p}[i]' + \tau\mathbf{p}[i]'' + \mathbf{p}[i]''' & \textit{if } \mathbf{p}[i] \textit{ depends on } x_i \\ \mathbf{p}[i] & \textit{otherwise.} \end{cases}$$

Note that the arithmetic here is abstract (see examples below). Note also that in the term $\tau\mathbf{p}[i]'$ the τ is already present in $\mathbf{p}$, while in $\tau\mathbf{p}[i]''$ it is added to existing monomials. In this definition, the monomials of $\mathbf{p}[i]'''$ are treated like those of $\tau\mathbf{p}[i]'$; however, in certain steps of the proofs we treat them differently, which is why the notation separates them.

Example 5. Let $\mathbf{p} = \langle x_1 + x_3,\ x_2 + x_3 + x_4,\ x_3,\ x_3 \rangle$.

Note that $\mathbf{p} \bullet \mathbf{p} = \mathbf{p}$. We have $\mathbf{p}^\tau = \langle x_1 + \tau x_3,\ x_2 + \tau x_3 + x_4,\ x_3,\ x_3 \rangle$.

Example 6. Let $\mathbf{p} = \langle x_1 + \tau x_2 + \tau x_3 + \tau x_3 x_4,\ x_3,\ x_3,\ x_4\rangle$.

Note that $\mathbf{p} \bullet \mathbf{p} = \mathbf{p}$. The self-dependent variables are all but x_2.

We have $\mathbf{p}^\tau = \langle x_1 + \tau x_2 + \tau x_3 + \tau x_3 x_4,\ x_3,\ x_3,\ x_4\rangle = \mathbf{p}$.

Finally we can present the analysis of the loop command.

Algorithm SOLVE($\mathcal{S}$)
Input: $\mathcal{S}$, a polynomially-bounded disjunctive simple loop
Output: a set of τ-MPs which tightly approximates the effect of all $\mathcal{S}$-traces.

1. Set $T = \alpha(\mathcal{S})$.
2. Repeat the following steps until T remains fixed:
   (a) Closure: Set T to $Cl(T)$.
   (b) Generalization: For all $\mathbf{p} \in T$ such that $\mathbf{p} \bullet \mathbf{p} = \mathbf{p}$, add $\mathbf{p}^\tau$ to T.

Example 7. `loop X3 { X1:= X1 + X2; X2:= X2 + X3; X4:= X3 }`
The body of the loop is evaluated symbolically and yields the multi-polynomial:

$$\mathbf{p} = \langle x_1 + x_2,\ x_2 + x_3,\ x_3,\ x_3\rangle$$

Now, computing within `AMPol`,

$$\begin{aligned}\alpha(\mathbf{p})^{\bullet(2)} &= \alpha(\mathbf{p}) \bullet \alpha(\mathbf{p}) = \langle x_1 + x_2 + x_3,\ x_2 + x_3,\ x_3,\ x_3\rangle;\\ \alpha(\mathbf{p})^{\bullet(3)} &= \alpha(\mathbf{p})^{\bullet(2)}.\end{aligned}$$

Here the closure computation stops. Since $\alpha(\mathbf{p}^{\bullet(2)})$ is idempotent, we compute

$$\mathbf{q} = (\alpha(\mathbf{p})^{\bullet(2)})^\tau = \langle x_1 + \tau x_2 + \tau x_3,\ x_2 + \tau x_3,\ x_3,\ x_3\rangle$$

and applying closure again, we obtain some additional results:

$$\begin{aligned}\mathbf{q} \bullet \alpha(\mathbf{p}) &= \langle x_1 + x_2 + x_3 + \tau x_2 + \tau x_3,\ x_2 + x_3 + \tau x_3,\ x_3,\ x_3\rangle\\ (\mathbf{q})^{\bullet(2)} &= \langle x_1 + \tau x_2 + \tau x_3 + \tau^2 x_3,\ x_2 + \tau x_3,\ x_3,\ x_3\rangle\\ (\mathbf{q})^{\bullet(2)} \bullet \alpha(\mathbf{p}) &= \langle x_1 + x_2 + x_3 + \tau x_2 + \tau x_3 + \tau^2 x_3,\ x_2 + x_3 + \tau x_3,\ x_3,\ x_3\rangle\end{aligned}$$

The last element is idempotent but applying generalization does not generate anything new. Thus the algorithm ends. The reader may reconsider the source code to verify that we have indeed obtained tight bounds for the loop.

6 Correctness

We claim that our algorithm obtains a description of the worst-case results of the program that is precise up to constant factors. That is, we claim that the set of MPs returned provides an upper bound (on all executions) which is also tight; tightness means that every MP returned is also a lower bound (up to a constant factor) on an infinite sequence of possible executions. Unfortunately, due to space constraints, we are not able to give full details of the proofs here; however, we give the main highlights. Intuitively, what we want to prove is that the multi-polynomials we compute cover all "behaviors" of the loop. More precisely, in the upper-bound part of the proof we want to cover all behaviors: upper-bounding is a universal statement. To prove that bounds are tight, we show that each such bound constitutes a *lower bound* on a certain "worst-case behavior": tightness is an existential statement. The main aspects of these proofs are as follows:

- A key notion in our proofs is that of *realizability*. Intuitively, when we come up with a bound, we want to show that there are traces that achieve (realize) this bound for arbitrarily large input values.
- In the lower-bound proof, we describe a "behavior" by a *pattern*. A pattern is constructed like a regular expression with concatenation and Kleene-star. However, they allow no nested iteration constructs, and the starred sub-expressions have to be repeated the same number of times; for example, the pattern $\mathbf{p}^*\mathbf{q}^*$ generates the traces $\{\mathbf{p}^t\mathbf{q}^t,\ t \geq 0\}$. The proof constructs a pattern for every multi-polynomial computed, showing it is realizable. It is interesting that such simple patterns suffice to establish tight lower bounds for all our programs.
- In the upper-bound proof, we describe all "behaviors" by a finite set of *well-typed regular expressions* [10]. This elegant tool channels the power of the Factorization Forest Theorem [25]; this brings out the role of idempotent elements, which is key in our algorithm.
- Interestingly, the lower-bound proof not only justifies the tightness of our upper bounds, it also justifies the termination of the algorithm and the application of the Factorization Forest Theorem in the upper-bound proof, because it shows that our abstract multi-polynomials generate a finite monoid.

7 Related Work

Bound analysis, in the sense of finding symbolic bounds for data values, iteration bounds and related quantities, is a classic field of program analysis [18,24,27]. It is also an area of active research, with tools being currently (or recently) developed including COSTA [1], APROVE [13], CIAOPP [19], C^4B [11], LOOPUS [26]—all for imperative programs. There is also work on functional and logic programs, term rewriting systems, recurrence relations, etc. which we cannot attempt to survey here. In the rest of this section we survey work which is more directly related to ours, and has even inspired it.

The LOOP language is due to Meyer and Ritchie [20], who note that it computes only primitive recursive functions, but complexity can rise very fast, even for programs with nesting-depth 2. Subsequent work [15–17,22] concerning similar languages attempted to analyze such programs more precisely; most of them proposed syntactic criteria, or analysis algorithms, that are sufficient for ensuring that the program lies in a desired class (often, polynomial-time programs), but are not both necessary and sufficient: thus, they do not prove decidability (the exception is [17] which has a decidability result for a weak "core" language). The core language we use in this paper is from Ben-Amram et al. [7], who observed that by introducing weak bounded loops instead of concrete loop commands and non-deterministic branching instead of "`if`", we have weakened the semantics just enough to obtain decidability of polynomial growth-rate. Justifying the necessity of these relaxations, [8] showed undecidability for a language that can only do addition and definite loops (that cannot exit early).

In the vast literature on bound analysis in various forms, there are a few other works that give a complete solution for a weak language. *Size-change programs* are considered by [12,28]. Size-change programs abstract away nearly everything in the program, leaving a control-flow graph annotated with assertions about variables which decrease (or do not increase) in a transition. Thus, it does not assume structured and explicit loops, and it cannot express information about values which increase. Both works yield tight bounds on the number of transitions until termination.

Dealing with a somewhat different problem, [14,21] both check, or find, *invariants* in the form of polynomial equations. We find it remarkable that they give complete solutions for weak languages, where the weakness lies in the non-deterministic control-flow, as in our language. If one could give a complete solution for polynomial *inequalities*, this would have implied a solution to our problem as well.

8 Conclusion and Further Work

We have solved an open problem in the area of analyzing programs in a simple language with bounded loops. For our language, it has been previously shown that it is possible to decide whether a variable's value, number of steps in the program, etc. are polynomially bounded or not. Now, we have an algorithm that computes tight polynomial bounds on the final values of variables in terms of initial values. The bounds are tight up to constant factors (suitable constants are also computable). This result improves our understanding of what is computable by, and about, programs of this form. An interesting corollary of our algorithm is that as long as variables are *polynomially bounded*, their worst-case bounds are described tightly by (multivariate) *polynomials*. This is, of course, not true for common Turing-complete languages. Another interesting corollary of the *proofs* is the definition of a simple class of patterns that suffice to realize the worst-case behaviors. This will appear in a planned extended version of this paper.

There are a number of possible directions for further work. We would like to look for decidability results for richer (yet, obviously, sub-recursive) languages. Some possible language extensions include deterministic loops, variable resets (cf. [4]), explicit constants, and procedures. The inclusion of explicit constants is a particularly challenging open problem.

Rather than extending the language, we could extend the range of bounds that we can compute. In light of the results in [17], it seems plausible that the approach can be extended to classify the Grzegorczyk-degree of the growth rate of variables when they are super-polynomial. There may also be room for progress regarding precise bounds of the form 2^{poly}.

In terms of time complexity, our algorithm is polynomial in the size of the program times n^{nd}, where d is the highest degree of any MP computed. Such exponential behavior is to be expected, since a program can be easily written to compute a multivariate polynomial that is exponentially long to write. But there is still room for finer investigation of this issue.

References

1. Albert, E., Arenas, P., Genaim, S., Puebla, G., Zanardini, D.: Cost analysis of object-oriented bytecode programs. Theor. Comput. Sci. **413**(1), 142–159 (2012). https://doi.org/10.1016/j.tcs.2011.07.009
2. Alias, C., Darte, A., Feautrier, P., Gonnord, L.: Multi-dimensional rankings, program termination, and complexity bounds of flowchart programs. In: Cousot, R., Martel, M. (eds.) SAS 2010. LNCS, vol. 6337, pp. 117–133. Springer, Heidelberg (2010). https://doi.org/10.1007/978-3-642-15769-1_8
3. Bagnara, R., Hill, P.M., Zaffanella, E.: The parma polyhedra library: toward a complete set of numerical abstractions for the analysis and verification of hardware and software systems. Sci. Comput. Program. **72**(1–2), 3–21 (2008)
4. Ben-Amram, A.M.: On decidable growth-rate properties of imperative programs. In: Baillot, P. (ed.) International Workshop on Developments in Implicit Computational complExity (DICE 2010). EPTCS, vol. 23, pp. 1–14 (2010). https://doi.org/10.4204/EPTCS.23.1
5. Ben-Amram, A.M., Genaim, S.: Ranking functions for linear-constraint loops. J. ACM **61**(4), 26:1–26:55 (2014). https://doi.org/10.1145/2629488
6. Ben-Amram, A.M., Genaim, S.: On multiphase-linear ranking functions. In: Majumdar, R., Kunčak, V. (eds.) CAV 2017. LNCS, vol. 10427, pp. 601–620. Springer, Cham (2017). https://doi.org/10.1007/978-3-319-63390-9_32
7. Ben-Amram, A.M., Jones, N.D., Kristiansen, L.: Linear, polynomial or exponential? Complexity inference in polynomial time. In: Beckmann, A., Dimitracopoulos, C., Löwe, B. (eds.) CiE 2008. LNCS, vol. 5028, pp. 67–76. Springer, Heidelberg (2008). https://doi.org/10.1007/978-3-540-69407-6_7
8. Ben-Amram, A.M., Kristiansen, L.: On the edge of decidability in complexity analysis of loop programs. Int. J. Found. Comput. Sci. **23**(7), 1451–1464 (2012). https://doi.org/10.1142/S0129054112400588
9. Ben-Amram, A.M., Pineles, A.: Flowchart programs, regular expressions, and decidability of polynomial growth-rate. In: Hamilton, G., Lisitsa, A., Nemytykh, A.P. (eds.) Proceedings of the Fourth International Workshop on Verification and Program Transformation (VPT). EPTCS, vol. 216, pp. 24–49 (2016). https://doi.org/10.4204/EPTCS.216.2
10. Bojańczyk, M.: Factorization forests. In: Diekert, V., Nowotka, D. (eds.) DLT 2009. LNCS, vol. 5583, pp. 1–17. Springer, Heidelberg (2009). https://doi.org/10.1007/978-3-642-02737-6_1
11. Carbonneaux, Q., Hoffmann, J., Shao, Z.: Compositional certified resource bounds. In: Proceedings of the ACM SIGPLAN 2015 Conference on Programming Language Design and Implementation (PLDI). ACM (2015)

12. Colcombet, T., Daviaud, L., Zuleger, F.: Size-change abstraction and max-plus automata. In: Csuhaj-Varjú, E., Dietzfelbinger, M., Ésik, Z. (eds.) MFCS 2014. LNCS, vol. 8634, pp. 208–219. Springer, Heidelberg (2014). https://doi.org/10.1007/978-3-662-44522-8_18
13. Giesl, J., et al.: Analyzing program termination and complexity automatically with AProVE. J. Autom. Reasoning **58**(1), 3–31 (2017). https://doi.org/10.1007/s10817-016-9388-y
14. Hrushovski, E., Ouaknine, J., Pouly, A., Worrell, J.: Polynomial invariants for affine programs. In: Proceedings of the 33rd Annual ACM/IEEE Symposium on Logic in Computer Science, LICS 2018, pp. 530–539. ACM, New York (2018). https://doi.org/10.1145/3209108.3209142
15. Jones, N.D., Kristiansen, L.: A flow calculus of mwp-bounds for complexity analysis. ACM Trans. Comput. Logic **10**(4), 1–41 (2009). https://doi.org/10.1145/1555746.1555752
16. Kasai, T., Adachi, A.: A characterization of time complexity by simple loop programs. J. Comput. Syst. Sci. **20**(1), 1–17 (1980). https://doi.org/10.1016/0022-0000(80)90001-X
17. Kristiansen, L., Niggl, K.H.: On the computational complexity of imperative programming languages. Theor. Comput. Sci. **318**(1–2), 139–161 (2004). https://doi.org/10.1016/j.tcs.2003.10.016
18. Le Métayer, D.: ACE: an automatic complexity evaluator. ACM Trans. Program. Lang. Syst. **10**(2), 248–266 (1988). https://doi.org/10.1145/42190.42347
19. López-García, P., Darmawan, L., Klemen, M., Liqat, U., Bueno, F., Hermengildo, M.V.: Interval-based resource usage verification by translation into Horn clauses and an application to energy consumption. Theory Pract. Logic Program. **18**(2), 167–223 (2018)
20. Meyer, A.R., Ritchie, D.M.: The complexity of loop programs. In: Proceedings of the 22nd ACM National Conference, Washington, DC, pp. 465–469 (1967)
21. Müller-Olm, M., Seidl, H.: Computing polynomial program invariants. Inf. Process. Lett. **91**(5), 233–244 (2004). https://doi.org/10.1016/j.ipl.2004.05.004
22. Niggl, K.H., Wunderlich, H.: Certifying polynomial time and linear/polynomial space for imperative programs. SIAM J. Comput. **35**(5), 1122–1147 (2006). https://doi.org/10.1137/S0097539704445597
23. Podelski, A., Rybalchenko, A.: A complete method for the synthesis of linear ranking functions. In: Steffen, B., Levi, G. (eds.) VMCAI 2004. LNCS, vol. 2937, pp. 239–251. Springer, Heidelberg (2004). https://doi.org/10.1007/978-3-540-24622-0_20
24. Rosendahl, M.: Automatic complexity analysis. In: Proceedings of the Conference on Functional Programming Languages and Computer Architecture, FPCA 1989, pp. 144–156. ACM (1989). https://doi.org/10.1145/99370.99381
25. Simon, I.: Factorization forests of finite height. Theor. Comput. Sci. **72**(1), 65–94 (1990). https://doi.org/10.1016/0304-3975(90)90047-L
26. Sinn, M., Zuleger, F., Veith, H.: Complexity and resource bound analysis of imperative programs using difference constraints. J. Autom. Reasoning **59**(1), 3–45 (2017). https://doi.org/10.1007/s10817-016-9402-4
27. Wegbreit, B.: Mechanical program analysis. Commun. ACM **18**(9), 528–539 (1975). https://doi.org/10.1145/361002.361016
28. Zuleger, F.: Asymptotically precise ranking functions for deterministic size-change systems. In: Beklemishev, L.D., Musatov, D.V. (eds.) CSR 2015. LNCS, vol. 9139, pp. 426–442. Springer, Cham (2015). https://doi.org/10.1007/978-3-319-20297-6_27

A Complete Normal-Form Bisimilarity for State

Dariusz Biernacki[1], Sergueï Lenglet[2(✉)], and Piotr Polesiuk[1]

[1] University of Wrocław, Wrocław, Poland
{dabi,ppolesiuk}@cs.uni.wroc.pl
[2] Université de Lorraine, Nancy, France
serguei.lenglet@univ-lorraine.fr

Abstract. We present a sound and complete bisimilarity for an untyped λ-calculus with higher-order local references. Our relation compares values by applying them to a fresh variable, like normal-form bisimilarity, and it uses environments to account for the evolving store. We achieve completeness by a careful treatment of evaluation contexts comprising open stuck terms. This work improves over Støvring and Lassen's incomplete environment-based normal-form bisimilarity for the $\lambda\rho$-calculus, and confirms, in relatively elementary terms, Jaber and Tabareau's result, that the state construct is discriminative enough to be characterized with a bisimilarity without any quantification over testing arguments.

1 Introduction

Two terms are contextually equivalent if replacing one by the other in a bigger program does not change the behavior of the program. The quantification over program contexts makes contextual equivalence hard to use in practice and it is therefore common to look for more effective characterizations of this relation. In a calculus with local state, such a characterization has been achieved either through *logical relations* [1,5,15], which rely on types, denotational models [6,10,13], or coinductively defined *bisimilarities* [9,12,17–19].

Koutavas et al. [8] argue that to be sound w.r.t. contextual equivalence, a bisimilarity for state should accumulate the tested terms in an environment to be able to try them again as the store evolves. Such *environmental bisimilarities* usually compare terms by applying them to arguments built from the environment [12,17,19], and therefore still rely on some universal quantification over testing arguments. An exception is Støvring and Lassen's bisimilarity [18], which compares terms by applying them to a fresh variable, like one would do with a *normal-form* (or *open*) bisimilarity [11,16]. Their bisimilarity characterizes contextual equivalence in a calculus with control and state, but is not *complete* in a calculus with state only: there exist equivalent terms that are not related by the bisimilarity. Jaber and Tabareau [6] go further and propose a sound and complete *Kripke Open Bisimilarity* for a calculus with local state, which also compares terms by applying them to a fresh variable, but uses notions from Kripke logical relations, namely transition systems of invariants, to reason about heaps.

M. Bojańczyk and A. Simpson (Eds.): FOSSACS 2019, LNCS 11425, pp. 98–114, 2019.
https://doi.org/10.1007/978-3-030-17127-8_6

In this paper, we propose a sound and complete normal-form bisimilarity for a call-by-value λ-calculus with local references which relies on environments to handle heaps. We therefore improve over Støvring and Lassen's work, since our relation is complete, by following a different, potentially simpler, path than Jaber and Tabareau, since we use environments to represent possible worlds and do not rely on any external structures such as transition systems of invariants. Moreover, we do not need types and define our relation in an untyped calculus.

We obtain completeness by treating carefully normal forms that are not values, i.e., open stuck terms of the form $E[x\,v]$. First, we distinguish in the environment the terms which should be tested multiple times from the ones that should be run only once, namely the evaluation contexts like E in the above term. The latter are kept in a separate environment that takes the form of a stack, according to the idea presented by Laird [10] and by Jagadeesan et al. [7]. Second, we relate the so-called *deferred diverging* terms [5,6], i.e., open stuck terms which hide a diverging behavior in the evaluation context E, with the regular diverging terms.

It may be worth stressing that our congruence proof is based on the machinery we have developed before [3] and is simpler than Støvring and Lassen's one, in particular in how it accounts for the extensionality of functions.

We believe that this work makes a contribution to the understanding of how one should adjust the normal-form bisimulation proof principle when the calculus under consideration becomes less discriminative, assuming that one wishes to preserve completeness of the theory. In particular, it is quite straightforward to define a complete normal-form bisimilarity for the λ-calculus with first-class continuations and global store, with no need to refer to other notions than the ones already present in the reduction semantics. Similarly, in the $\lambda\mu\rho$-calculus (continuations and local references), one only needs to introduce environments to ensure soundness of the theory, but essentially nothing more is required to obtain completeness [18]. In this article we show which new ingredients are needed when moving from these two highly expressive calculi to the corresponding, less discriminative ones—with global or local references only—that do not offer access to the current continuation.

The rest of this paper is as follows. In Sect. 2, we study a simple calculus with global store to see how to reach completeness in that case. In particular, we show in Sect. 2.2 how we deal with deferred diverging terms. We remind in Sect. 2.3 the notion of *diacritical progress* [3] and the framework our bisimilarity and its proof of soundness are based upon. We sketch the completeness proof in Sect. 2.4. Section 2 paves the way for the main result of the paper, described in Sect. 3, where we turn to the calculus with local store. We define the bisimilarity in Sect. 3.2, prove its soundness and completeness in Sect. 3.3, and use it in Sect. 3.4 on examples taken from the literature. We conclude in Sect. 4, where we discuss related work and in particular compare our work to Jaber and Tabareau's. A companion report expands on the proofs [4].

2 Global Store

We first consider a calculus where terms share a global store and present how we deal with deferred diverging terms to get a complete bisimilarity.

2.1 Syntax, Semantics, and Contextual Equivalence

We extend the call-by-value λ-calculus with the ability to read and write a global memory. We let x, y, ... range over term variables and l range over references. A *store*, denoted by h, g, is a finite map from references to values; we write $\mathsf{dom}(h)$ for the domain of h, i.e., the set of references on which h is defined. We write $\emptyset$ for the empty store, $h \uplus g$ for the union of two stores, assuming $\mathsf{dom}(h) \cap \mathsf{dom}(g) = \emptyset$. The syntax of terms and contexts is defined as follows.

$$\begin{array}{ll} \text{Terms:} & t, s ::= v \mid t\,t \mid l := t; t \mid !l \\ \text{Values:} & v, w ::= x \mid \lambda x.t \\ \text{Evaluation contexts:} & E, F ::= \square \mid E\,t \mid v\,E \mid l := E; t \end{array}$$

The term $l := t; s$ evaluates t (if possible) and stores the resulting value in l before continuing as s, while $!l$ reads the value kept in l. When writing examples and in the completeness proofs, we use natural numbers, booleans, the conditional if ... then ... else ..., local definitions $\mathsf{let} \ldots \mathsf{in} \ldots$, sequence ;, and unit () assuming the usual call-by-value encodings for these constructs.

A λ-abstraction $\lambda x.t$ binds x in t; we write $\mathsf{fv}(t)$ (respectively $\mathsf{fv}(E)$) for the set of free variables of t (respectively E). We identify terms up to α-conversion of their bound variables. A variable or reference is *fresh* if it does not occur in any other entities under consideration, and a store is fresh if it maps references to pairwise distinct fresh variables. A term or context is *closed* if it has no free variables. We write $\mathsf{fr}(t)$ for the set of references that occur in t.

The call-by-value semantics of the calculus is defined on *configurations* $\langle h \mid t \rangle$ such that $\mathsf{fr}(t) \subseteq \mathsf{dom}(h)$ and for all $l \in \mathsf{dom}(h)$, $\mathsf{fr}(h(l)) \subseteq \mathsf{dom}(h)$. We let c and d range over configurations. We write $t\{v/x\}$ for the usual capture-avoiding substitution of x by v in t, and we let $\int$ range over simultaneous substitutions $.\{v_1/x_1\} \ldots \{v_n/x_n\}$. We write $h[l := v]$ for the operation updating the value of l to v. The reduction semantics $\rightarrow$ is defined by the following rules.

$$\begin{array}{ll} \langle h \mid (\lambda x.t)\, v \rangle \rightarrow \langle h \mid t\{v/x\} \rangle & \langle h \mid !l \rangle \rightarrow \langle h \mid h(l) \rangle \\ \langle h \mid l := v; t \rangle \rightarrow \langle h[l := v] \mid t \rangle & \langle h \mid E[t] \rangle \rightarrow \langle g \mid E[s] \rangle \text{ if } \langle h \mid t \rangle \rightarrow \langle g \mid s \rangle \end{array}$$

The well-formedness condition on configurations ensures that a read operation $!l$ cannot fail. We write $\rightarrow^*$ for the reflexive and transitive closure of $\rightarrow$.

A term t of a configuration $\langle h \mid t \rangle$ which cannot reduce further is called a *normal form*. Normal forms are either values or *open-stuck terms* of the form $E[x\, v]$; closed normal forms can only be λ-abstractions. A configuration *terminates*, written $c \Downarrow$ if it reduces to a normal-form configuration; otherwise it *diverges*, written $c \Uparrow$, like configurations running $\Omega \stackrel{\mathsf{def}}{=} (\lambda x.x\, x)\, (\lambda x.x\, x)$.

Contextual equivalence equates terms behaving the same in all contexts. A substitution $\int$ closes a term t if $t\int$ is closed; it closes a configuration $\langle h \mid t \rangle$ if it closes t and the values in h.

Definition 1. *t and s are contextually equivalent, written $t \equiv s$, if for all contexts E, fresh stores h, and closing substitutions $\int$, $\langle h \mid E[t] \rangle \int \Downarrow$ iff $\langle h \mid E[s] \rangle \int \Downarrow$.*

Testing only evaluation contexts is not a restriction, as it implies the equivalence w.r.t. all contexts $\equiv_C$: one can show that $t \equiv_C s$ iff $\lambda x.t \equiv_C \lambda x.s$ iff $\lambda x.t \equiv \lambda x.s$.

2.2 Normal-Form Bisimulation

Informal Presentation. Two open terms are normal-form bisimilar if their normal forms can be decomposed into bisimilar subterms. For example in the plain λ-calculus, a stuck term $E[x\ v]$ is bisimilar to t if t reduces to a stuck term $F[xw]$ so that respectively E, F and v, w are bisimilar when they are respectively plugged with and applied to a fresh variable.

Such a requirement is too discriminating for many languages, as it distinguishes terms that should be equivalent. For instance in plain λ-calculus, given a closed value v, $t \stackrel{\mathsf{def}}{=} x\ v$ is not normal form bisimilar to $s \stackrel{\mathsf{def}}{=} (\lambda y.x\ v)\ (x\ v)$. Indeed, $\Box$ is not bisimilar to $(\lambda y.x\ v)\ \Box$ when plugged with a fresh z: the former produces a value z while the latter reduces to a stuck term $x\ v$. However, t and s are contextually equivalent, as for all closed value w, $t\{w/x\}$ and $s\{w/x\}$ behave like $w\ v$: if $w\ v$ diverges, then they both diverges, and if $w\ v$ evaluates to some value w', then they also evaluates to w'. Similarly, $x\ v\ \Omega$ and Ω are not normal-form bisimilar (one is a stuck term while the other is diverging), but they are contextually equivalent by the same reasoning.

The terms t and s are no longer contextually equivalent in a λ-calculus with store, since a function can count how many times it is applied and change its behavior accordingly. More precisely, t and s are distinguished by the context $l := 0; (\lambda x.\Box)\ \lambda z.l := !l + 1; \mathsf{if}\ !l = 1\ \mathsf{then}\ 0\ \mathsf{else}\ \Omega$. But this counting trick is not enough to discriminate $x\ v\ \Omega$ and Ω, as they are still equivalent in a λ-calculus with store. Although $x\ v\ \Omega$ is a normal form, it is in fact always diverging when we replace x by an arbitrary closed value w, either because $w\ v$ itself diverges, or it evaluates to some w' and then $w'\ \Omega$ diverges. A stuck term which hides a diverging behavior has been called *deferred diverging* in the literature [5,6].

It turns out that being able to relate a diverging term to a deferred diverging term is all we need to change from the plain λ-calculus normal-form bisimilarity to get a complete equivalence when we add global store. We do so by distinguishing two cases in the clause for open-stuck terms: a configuration $\langle h \mid E[x\ v] \rangle$ is related to c either if c can reduce to a stuck configuration with related subterms, or if E is a diverging context, and we do not require anything of c. The resulting simulation is not symmetric as it relates a deferred diverging configuration with any configuration c (even converging one), but the corresponding notion of bisimulation equates such configuration only to either a configuration of the same kind or a diverging configuration such as $\langle h \mid \Omega \rangle$.

Progress. We define simulation using the notion of *diacritical progress* we developed in a previous work [2,3], which distinguishes between *active* and *passive* clauses. Roughly, passive clauses are between simulation states which should be considered equal, while active clauses are between states where actual progress is taking place. This distinction does not change the notions of bisimulation or bisimilarity, but it simplifies the soundness proof of the bisimilarity. It also allows for the definition of powerful *up-to techniques*, relations that are easier to use than bisimulations but still imply bisimilarity. For normal-form bisimilarity, our framework enables up-to techniques which respects η-expansion [3].

Progress is defined between objects called *candidate relations*, denoted by $\mathcal{R}$, $\mathcal{S}$, $\mathcal{T}$. A candidate relation $\mathcal{R}$ contains pairs of configurations, and a set of configurations written $\mathcal{R}\!\uparrow$, which we expect to be composed of diverging or deferred diverging configurations (for such relations we take $\mathcal{R}^{-1}\!\uparrow$ to be $\mathcal{R}\!\uparrow$). We extend $\mathcal{R}$ to stores, terms, values, and contexts with the following definitions.

$$\frac{\mathsf{dom}(h) = \mathsf{dom}(g) \qquad \forall l, h(l)\ \mathcal{R}^{\mathsf{v}}\ g(l)}{h\ \mathcal{R}^{\mathsf{h}}\ g} \qquad \frac{\langle h \mid t\rangle\ \mathcal{R}\ \langle h \mid s\rangle \qquad h \text{ fresh}}{t\ \mathcal{R}^{\mathsf{t}}\ s}$$

$$\frac{v\,x\ \mathcal{R}^{\mathsf{t}}\ w\,x \qquad x \text{ fresh}}{v\ \mathcal{R}^{\mathsf{v}}\ w} \qquad \frac{E[x]\ \mathcal{R}^{\mathsf{t}}\ F[x] \qquad x \text{ fresh}}{E\ \mathcal{R}^{\mathsf{c}}\ F} \qquad \frac{\langle h \mid E[x]\rangle \in \mathcal{R}\!\uparrow \qquad x, h \text{ fresh}}{E \in \mathcal{R}\!\uparrow^{\mathsf{c}}}$$

We use these extensions to define progress as follows.

Definition 2. *A candidate relation $\mathcal{R}$ progresses to $\mathcal{S}$, $\mathcal{T}$ written $\mathcal{R} \rightarrowtail \mathcal{S}, \mathcal{T}$, if $\mathcal{R} \subseteq \mathcal{S}$, $\mathcal{S} \subseteq \mathcal{T}$, and*

1. *$c\ \mathcal{R}\ d$ implies*
   - *if $c \to c'$, then $d \to^* d'$ and $c'\ \mathcal{T}\ d'$;*
   - *if $c = \langle h \mid v\rangle$, then $d \to^* \langle g \mid w\rangle$, $h\ \mathcal{S}^{\mathsf{h}}\ g$, and $v\ \mathcal{S}^{\mathsf{v}}\ w$;*
   - *if $c = \langle h \mid E[x\,v]\rangle$, then either*
     - *$d \to^* \langle g \mid F[x\,w]\rangle$, $h\ \mathcal{T}^{\mathsf{h}}\ g$, $E\ \mathcal{T}^{\mathsf{c}}\ F$, and $v\ \mathcal{T}^{\mathsf{v}}\ w$, or*
     - *$E \in \mathcal{T}\!\uparrow^{\mathsf{c}}$.*
2. *$c \in \mathcal{R}\!\uparrow$ implies $c \neq \langle h \mid v\rangle$ for all h and v and*
   - *if $c \to c'$, then $c' \in \mathcal{T}\!\uparrow$;*
   - *if $c = \langle h \mid E[x\,v]\rangle$, then $E \in \mathcal{T}\!\uparrow^{\mathsf{c}}$.*

A normal-form simulation is a candidate relation $\mathcal{R}$ such that $\mathcal{R} \rightarrowtail \mathcal{R}, \mathcal{R}$, and a bisimulation is a candidate relation $\mathcal{R}$ such that $\mathcal{R}$ and $\mathcal{R}^{-1}$ are simulations. Normal-form bisimilarity $\approx$ is the union of all normal-form bisimulations.

We test values and contexts by applying or plugging them with a fresh variable x, and running them in a fresh store; with a global memory, the value represented by x may access any reference and assign it an arbitrary value, hence the need for a fresh store. The stores of two bisimilar value configurations must have the same domain, as it would be easy to distinguish them otherwise by testing the content of the references that would be in one store but not in the other.

The main novelty compared to usual definitions of normal-form bisimilarity [3,11] is the set of (deferred) diverging configurations used in the stuck terms clause. We detect that E in a configuration $\langle h \mid E[x\,v]\rangle$ is (deferred) diverging by running $\langle h' \mid E[y]\rangle$ where y and h' are fresh; this configuration may then diverge or evaluate to an other deferred diverging configuration $\langle h \mid E'[x\,v]\rangle$.

Like in the plain λ-calculus [3], $\mathcal{R}$ progresses towards $\mathcal{S}$ in the value clause and $\mathcal{T}$ in the others; the former is passive while the others are active. Our framework prevents some up-to techniques from being applied after a passive transition. In particular, we want to forbid the application of bisimulation up to context as it would be unsound: we could deduce that $v\,x$ and $w\,x$ are equivalent for all v and w just by building a candidate relation containing v and w.

Example 1. To prove that $\langle h \mid x\ v\ \Omega\rangle \approx \langle h \mid \Omega\rangle$ holds for all v and h, we prove that $\mathcal{R} \stackrel{\mathsf{def}}{=} \{(\langle h \mid x\ v\ \Omega\rangle, \langle h \mid \Omega\rangle), \{\langle g \mid y\ \Omega\rangle \mid y, g \text{ fresh}\}\}$ is a bisimulation. Indeed, $\langle h \mid x\ v\ \Omega\rangle$ is stuck with $\langle g \mid y\ \Omega\rangle \in \mathcal{R}{\uparrow}$ for fresh y and g, and we have $\langle g \mid y\ \Omega\rangle \rightarrow \langle g \mid y\ \Omega\rangle$. Conversely, the transition $\langle h \mid \Omega\rangle \rightarrow \langle h \mid \Omega\rangle$ is matched by $\langle h \mid x\ v\ \Omega\rangle \rightarrow^* \langle h \mid x\ v\ \Omega\rangle$ and the resulting terms are in $\mathcal{R}$.

2.3 Soundness

In this framework, proving that $\approx$ is sound is a consequence that a form of bisimulation up to context is valid, a result which itself may require to prove that other up-to techniques are valid. We distinguish the techniques which can be used in passive clauses (called *strong* up-to techniques), from the ones which cannot. An up-to technique (resp. strong up-to technique) is a function f such that $\mathcal{R} \rightarrowtail \mathcal{R}, f(\mathcal{R})$ (resp. $\mathcal{R} \rightarrowtail f(\mathcal{R}), f(\mathcal{R})$) implies $\mathcal{R} \subseteq \approx$. To show that a given f is an up-to technique, we rely on a notion of *respectfulness*, which is simpler to prove and gives sufficient conditions for f to be an up-to technique.

We briefly recall the notions we need from our previous work [2]. We extend $\subseteq$ and $\cup$ to functions argument-wise (e.g., $(f \cup g)(\mathcal{R}) = f(\mathcal{R}) \cup g(\mathcal{R})$), and given a set $\mathfrak{F}$ of functions, we also write $\mathfrak{F}$ for the function defined as $\bigcup_{f \in \mathfrak{F}} f$. We define f^ω as $\bigcup_{n \in \mathbb{N}} f^n$. We write id for the identity function on relations, and $\widehat{f}$ for $f \cup \mathsf{id}$. A function f is monotone if $\mathcal{R} \subseteq \mathcal{S}$ implies $f(\mathcal{R}) \subseteq f(\mathcal{S})$. We write $\mathcal{P}_{\mathit{fin}}(\mathcal{R})$ for the set of finite subsets of $\mathcal{R}$, and we say f is continuous if it can be defined by its image on these finite subsets, i.e., if $f(\mathcal{R}) \subseteq \bigcup_{\mathcal{S} \in \mathcal{P}_{\mathit{fin}}(\mathcal{R})} f(\mathcal{S})$. The up-to techniques we use are defined by inference rules with a finite number of premises, so they are trivially continuous.

Definition 3. *A function f evolves to g, h, written $f \rightsquigarrow g, h$, if for all $\mathcal{R}$ and $\mathcal{T}$, $\mathcal{R} \rightarrowtail \mathcal{R}, \mathcal{T}$ implies $f(\mathcal{R}) \rightarrowtail g(\mathcal{R}), h(\mathcal{T})$. A function f* strongly *evolves to g, h, written $f \rightsquigarrow_{\mathsf{s}} g, h$, if for all $\mathcal{R}$, $\mathcal{S}$, and $\mathcal{T}$, $\mathcal{R} \rightarrowtail \mathcal{S}, \mathcal{T}$ implies $f(\mathcal{R}) \rightarrowtail g(\mathcal{S}), h(\mathcal{T})$.*

Evolution can be seen as progress for functions on relations. Evolution is more restrictive than strong evolution, as it requires $\mathcal{R}$ such that $\mathcal{R} \rightarrowtail \mathcal{R}, \mathcal{T}$.

Definition 4. *A set $\mathfrak{F}$ of continuous functions is* respectful *if there exists $\mathfrak{S}$ such that $\mathfrak{S} \subseteq \mathfrak{F}$ and*

- *for all $f \in \mathfrak{S}$, we have $f \rightsquigarrow_{\mathsf{s}} \widehat{\mathfrak{S}}^\omega, \widehat{\mathfrak{F}}^\omega$;*
- *for all $f \in \mathfrak{F}$, we have $f \rightsquigarrow \widehat{\mathfrak{S}}^\omega \circ \widehat{\mathfrak{F}} \circ \widehat{\mathfrak{S}}^\omega, \widehat{\mathfrak{F}}^\omega$.*

$$\frac{c \mathrel{\mathcal{R}} d \qquad v \mathrel{\mathcal{R}^{\mathsf{v}}} w}{c\{v/x\} \mathrel{\mathsf{subst}(\mathcal{R})} d\{w/x\}} \qquad \frac{c \in \mathcal{R}\!\uparrow}{c\{v/x\} \in \mathsf{subst}(\mathcal{R})\!\uparrow} \qquad \frac{\langle h \mid t\rangle \mathrel{\mathcal{R}} \langle g \mid s\rangle \qquad E \mathrel{\mathcal{R}^{\mathsf{c}}} F}{\langle h \mid E[t]\rangle \mathrel{\mathsf{plug}_{\mathsf{c}}(\mathcal{R})} \langle g \mid F[s]\rangle}$$

$$\frac{\langle h \mid t\rangle \in \mathcal{R}\!\uparrow}{\langle h \mid E[t]\rangle \in \mathsf{plug}_{\uparrow}(\mathcal{R})\!\uparrow} \qquad \frac{c \to^* c' \qquad d \to^* d' \qquad c' \mathrel{\mathcal{R}} d'}{c \mathrel{\mathsf{red}(\mathcal{R})} d}$$

$$\frac{c \in \mathcal{R}\!\uparrow}{c \mathrel{\mathsf{div}(\mathcal{R})} d} \qquad \frac{E \in \mathcal{R}\!\uparrow^{\mathsf{c}}}{\langle h \mid E[t]\rangle \in \mathsf{plugdiv}(\mathcal{R})\!\uparrow}$$

Fig. 1. Up-to techniques for the calculus with global store

In words, a function is in a respectful set $\mathfrak{F}$ if it evolves towards a combination of functions in $\mathfrak{F}$ after active clauses, and in $\mathfrak{S}$ after passive ones. When checking that f is regular (second case), we can use a regular function at most once after a passive clause. The (possibly empty) subset $\mathfrak{S}$ intuitively represents the strong up-to techniques of $\mathfrak{F}$. If $\mathfrak{S}_1$ and $\mathfrak{S}_2$ are subsets of $\mathfrak{F}$ which verify the conditions of the definition, then $\mathfrak{S}_1 \cup \mathfrak{S}_2$ also does, so there exists the largest subset of $\mathfrak{F}$ which satisfies the conditions, written $\mathsf{strong}(\mathfrak{F})$.

Lemma 1. *Let $\mathfrak{F}$ be a respectful set.*

- *If $f \in \mathfrak{F}$, then f is an up-to technique. If $f \in \mathsf{strong}(\mathfrak{F})$, then f is a strong up-to technique.*
- *For all $f \in \mathfrak{F}$, we have $f(\approx) \subseteq \approx$.*

Showing that f is in a respectful set $\mathfrak{F}$ is easier than proving it is an up-to technique. Besides, proving that a bisimulation up to context is respectful implies that $\approx$ is preserved by contexts thanks to the last property of Lemma 1.

The up-to techniques for the calculus with global store are given in Fig. 1. The techniques subst and plug allow to prove that $\approx$ is preserved by substitution and by evaluation contexts. The remaining ones are auxiliary techniques which are used in the respectfulness proof: red relies on the fact that the calculus is deterministic to relate terms up to reduction steps. The technique div allows to relate a diverging configuration to any other configuration, while plugdiv states that if E is a diverging context, then $\langle h \mid E[t]\rangle$ is a diverging configuration for all h and t. We distinguish the technique $\mathsf{plug}_{\mathsf{c}}$ from $\mathsf{plug}_{\uparrow}$ to get a more fine-grained classification, as $\mathsf{plug}_{\mathsf{c}}$ is the only one which is not strong.

Lemma 2. *The set $\mathfrak{F} \stackrel{\mathsf{def}}{=} \{\mathsf{subst}, \mathsf{plug}_m, \mathsf{red}, \mathsf{div}, \mathsf{plugdiv} \mid m \in \{\mathsf{c}, \uparrow\}\}$ is respectful, with $\mathsf{strong}(\mathfrak{F}) = \mathfrak{F} \setminus \{\mathsf{plug}_{\mathsf{c}}\}$.*

We omit the proof, as it is similar but much simpler than for the calculus with local store of Sect. 3. We deduce that $\approx$ is sound using Lemma 1.

Theorem 1. *For all t, s, and fresh store h, if $\langle h \mid t\rangle \approx \langle h \mid s\rangle$, then $t \equiv s$.*

2.4 Completeness

We prove the reverse implication by building a bisimulation which contains $\equiv$.

Theorem 2. *For all t, s, if $t \equiv s$, then for all fresh stores h, $\langle h \mid t\rangle \approx \langle h \mid s\rangle$.*

Proof (Sketch). It suffices to show that the candidate $\mathcal{R}$ defined as

$$\{(\langle h \mid t\rangle, \langle g \mid s\rangle) \mid \forall E, h_E, \text{ closing } \int, \langle h \uplus h_E \mid E[t]\rangle \int \Downarrow \;\Rightarrow\; \langle g \uplus h_E \mid E[s]\rangle \int \Downarrow\}$$
$$\cup \{\langle h \mid t\rangle \mid \forall E, h_E, \text{ closing } \int, \langle h \uplus h_E \mid E[t]\rangle \int \Uparrow\}$$

is a simulation. We proceed by case analysis on the behavior of $\langle h \mid t\rangle$. The details are in the report [4]; we sketch the proof in the case when $\langle h \mid t\rangle \mathrel{\mathcal{R}} \langle g \mid s\rangle$, $t = E[x\, v]$, and E is not deferred diverging.

A first step is to show that $\langle g \mid s\rangle$ also evaluates to an open-stuck configuration with x in function position. To do so, we consider a fresh l and we define $\int$ such that $\int(y)$ sets l at 1 when it is first applied if $y = x$, and at 2 if $y \neq x$. Then $\langle h \uplus l := 0 \mid t\rangle \int$ sets l at 1, which should also be the case of $\langle g \uplus l := 0 \mid s\rangle \int$, and it is possible only if $\langle g \mid s\rangle \rightarrow^* \langle g' \mid F[x\, w]\rangle$ for some g', F, and w.

We then have to show that $E \mathrel{\mathcal{R}^{\mathsf{c}}} F$, $v \mathrel{\mathcal{R}^{\mathsf{v}}} w$, and $h \mathrel{\mathcal{R}^{\mathsf{h}}} g'$. We sketch the proof for the contexts, as the proofs for the values and the stores are similar. Given h_f a fresh store, y a fresh variable, E' a context, $h_{E'}$ a store, $\int$ a closing substitution, we want $\langle h_f \uplus h_{E'} \mid E'[E[y]]\rangle \int \Downarrow$ iff $\langle h_f \uplus h_{E'} \mid E'[F[y]]\rangle \int \Downarrow$.

Let l be a fresh reference. Assuming $\mathsf{dom}(h) = \{l_1 \dots l_n\}$, given a term t, we write $\bigcup_i l_i := h; t$ for $l_1 := h(l_1); \dots l_n := h(l_n); t$. We define

$$\int_x \stackrel{\mathsf{def}}{=} \begin{cases} x \mapsto \lambda a.\mathsf{if}\ !l = 0\ \mathsf{then}\ l := 1; \bigcup_i l_i := h_f \uplus h_{E'}; \int(y)\ \mathsf{else}\ \int(x)\, a \\ z \mapsto \int'(z) \quad \text{if } z \neq x \end{cases}$$

The substitution $\int_x$ behaves like $\int$ except that when $\int_x(x)$ is applied for the first time, it replaces its argument by $\int(y)$ and sets the store to $h_f \uplus h_{E'}$. Therefore $\langle h \uplus l := 0 \mid E'[t]\rangle \int_x \rightarrow^* \langle h_f \uplus h_{E'} \uplus l := 1 \mid E'[E[y]]\rangle \int_x$, but this configuration then behaves like $\langle h_f \uplus h_{E'} \mid E'[E[y]]\rangle \int$. Similarly, $\langle g \uplus l := 0 \mid E'[s]\rangle \int_x$ evaluates to a configuration equivalent to $\langle h_f \uplus h_{E'} \mid E'[F[y]]\rangle \int$, and since $\langle h \uplus l := 0 \mid E'[t]\rangle \int_x \Downarrow$ implies $\langle g \uplus l := 0 \mid E'[s]\rangle \int_x \Downarrow$, we can conclude from there.

3 Local Store

We adapt the ideas of the previous section to a calculus where terms create their own local store. To be able to deal with local resources, the relation we define mixes principles from normal-form and environmental bisimilarities.

3.1 Syntax, Semantics, and Contextual Equivalence

In this section, the terms no longer share a global store, but instead must create local references before storing values. We extend the syntax of Sect. 2 with a construct to create a new reference.

Terms: $t, s ::= \ldots \mid \mathsf{new}\ l := v\ \mathsf{in}\ t$

Reference creation $\mathsf{new}\ l := v\ \mathsf{in}\ t$ binds l in t; we identify terms up to α-conversion of their references. We write $\mathsf{fr}(t)$ and $\mathsf{fr}(E)$ for the set of free references of t or E, and a term or context is *reference-closed* if its set of free references is empty. Following [18] and in contrast with [5,6], references are not values, but we can still give access to a reference l by passing $\lambda x.!l$ and $\lambda x.l := x; \lambda y.y$.

As before, the semantics is defined on configurations $\langle h \mid t \rangle$ verifying $\mathsf{fr}(t) \subseteq \mathsf{dom}(h)$ and for all $l \in \mathsf{dom}(h)$, $\mathsf{fr}(h(l)) \subseteq \mathsf{dom}(h)$. We add to the rules of Sect. 2 the following one for reference creation.

$$\langle h \mid \mathsf{new}\ l := v\ \mathsf{in}\ t \rangle \to \langle h \uplus l := v \mid t \rangle$$

We remind that $\uplus$ is defined for disjoint stores only, so the above rule assumes that $l \notin \mathsf{dom}(h)$, which is always possible using α-conversion.

We define contextual equivalence on reference-closed terms as we expect programs to allocate their own store.

Definition 5. *Two reference-closed terms t and s are contextually equivalent, written $t \equiv s$, if for all reference-closed evaluation contexts E and closing substitutions $\int$, $\langle \emptyset \mid E[t] \rangle \int \Downarrow$ iff $\langle \emptyset \mid E[s] \rangle \int \Downarrow$.*

3.2 Bisimilarity

With local stores, an external observer no longer has direct access to the stored values. In presence of such information hiding, a sound bisimilarity relies on an *environment* to accumulate terms which should be tested in different stores [8].

Example 2. Let $f_1 \stackrel{\mathsf{def}}{=} \lambda x.\mathsf{if}\ !l = \mathsf{true}\ \mathsf{then}\ l := \mathsf{false}; \mathsf{true}\ \mathsf{else}\ \mathsf{false}$ and $f_2 \stackrel{\mathsf{def}}{=} \lambda x.\mathsf{true}$. If we compare $\mathsf{new}\ l := \mathsf{true}\ \mathsf{in}\ f_1$ and f_2 only once in the empty store, they would be seen as equivalent as they both return true, however f_1 modify its store, so running f_1 and f_2 a second time distinguishes them.

Environments generally contain only values [17], except in $\lambda\mu\rho$ [18], where plugged evaluation contexts are kept in the environment when comparing open-stuck configurations. In contrast with $\lambda\mu\rho$, our environment collects values, and we use a *stack* for registering contexts [7,10]. Unlike values, contexts are therefore tested only once, following a last-in first-out ordering. The next example shows that considering contexts repeatedly would lead to an overly-discriminating bisimilarity. For the stack discipline of testing contexts in action see Example 8 in Sect. 3.4.

Example 3. With the same f_1 and f_2 as in Example 2, the terms $t \stackrel{\text{def}}{=} \text{new } l := \text{true in } f_1\,(x\,\lambda y.y)$ and $s \stackrel{\text{def}}{=} f_2\,(x\,\lambda y.y)$ are contextually equivalent. Roughly, for all closing substitution $\int$, t and s either both diverge (if $\int(x)\,\lambda y.y$ diverges), or evaluate to true, since $\int(x)$ cannot modify the value in l. Testing $f_1\,\Box$ and $f_2\,\Box$ twice would discriminate them and wrongfully distinguish t and s.

Remark 1. The bisimilarity for $\lambda\mu\rho$ runs evaluation contexts several times and is still complete because of the μ operator, which, like call/cc, captures evaluation contexts, and may then execute them several times.

We let $\mathcal{E}$ range over sets of pairs of values, and ϵ over sets of values. Similarly, we write Σ for a stack of pairs of evaluation contexts and σ for a stack of evaluation contexts. We write $\odot$ for the empty stack, :: for the operator putting an element on top of a stack, and $+\!\!+$ for the concatenation of two stacks. The projection operator π_1 transforms a set or stack of pairs into respectively a set or stack of single elements by taking the first element of each pair. A candidate relation $\mathcal{R}$ can be composed of:

- quadruples $(\mathcal{E}, \Sigma, c, d)$, written $\mathcal{E}, \Sigma \vdash c \mathrel{\mathcal{R}} d$, meaning that c and d are related under $\mathcal{E}$ and Σ;
- quadruples $(\mathcal{E}, \Sigma, h, g)$, written $\mathcal{E}, \Sigma \vdash h \mathrel{\mathcal{R}} g$, meaning that the elements of $\mathcal{E}$ and the top of Σ should be related when run with the stores h and g;
- triples (ϵ, σ, c), written $\epsilon, \sigma \vdash c \in \mathcal{R}\!\uparrow$, meaning that either c is (deferred) diverging, or σ is non-empty and contains a (deferred) diverging context;
- triples (ϵ, σ, h), written $\epsilon, \sigma \vdash h \in \mathcal{R}\!\uparrow$, meaning that σ is non-empty and contains a (deferred) diverging context.

Definition 6. *A candidate relation $\mathcal{R}$ progresses to $\mathcal{S}, \mathcal{T}$ written $\mathcal{R} \rightarrowtail \mathcal{S}, \mathcal{T}$, if $\mathcal{R} \subseteq \mathcal{S}$, $\mathcal{S} \subseteq \mathcal{T}$, and*

1. $\mathcal{E}, \Sigma \vdash c \mathrel{\mathcal{R}} d$ *implies*
   - *if* $c \rightarrow c'$, *then* $d \rightarrow^* d'$ *and* $\mathcal{E}, \Sigma \vdash c' \mathrel{\mathcal{T}} d'$;
   - *if* $c = \langle h \mid v \rangle$, *then either*
     - $d \rightarrow^* \langle g \mid w \rangle$, *and* $\mathcal{E} \cup \{(v, w)\}, \Sigma \vdash h \mathrel{\mathcal{S}} g$, *or*
     - $\Sigma \neq \odot$ *and* $\pi_1(\mathcal{E}) \cup \{v\}, \pi_1(\Sigma) \vdash h \in \mathcal{S}\!\uparrow$;
   - *if* $c = \langle h \mid E[x\,v] \rangle$, *then either*
     - $d \rightarrow^* \langle g \mid F[x\,w] \rangle$, *and* $\mathcal{E} \cup \{(v, w)\}, (E, F) :: \Sigma \vdash h \mathrel{\mathcal{S}} g$, *or*
     - $\pi_1(\mathcal{E}) \cup \{v\}, E :: \pi_1(\Sigma) \vdash h \in \mathcal{S}\!\uparrow$.
2. $\mathcal{E}, \Sigma \vdash h \mathrel{\mathcal{R}} g$ *implies*
   - *if* $v \mathrel{\mathcal{E}} w$, *then* $\mathcal{E}, \Sigma \vdash \langle h \mid v\,x \rangle \mathrel{\mathcal{S}} \langle g \mid w\,x \rangle$ *for a fresh* x;
   - *if* $\Sigma = (E, F) :: \Sigma'$, *then* $\mathcal{E}, \Sigma' \vdash \langle h \mid E[x] \rangle \mathrel{\mathcal{S}} \langle g \mid F[x] \rangle$ *for a fresh* x.
3. $\epsilon, \sigma \vdash c \in \mathcal{R}\!\uparrow$ *implies*
   - *if* $c \rightarrow c'$, *then* $\epsilon, \sigma \vdash c' \in \mathcal{T}\!\uparrow$;
   - *if* $c = \langle h \mid v \rangle$, *then* $\sigma \neq \odot$ *and* $\epsilon \cup \{v\}, \sigma \vdash h \in \mathcal{S}\!\uparrow$;
   - *if* $c = \langle h \mid E[x\,v] \rangle$, *then* $\epsilon \cup \{v\}, E :: \sigma \vdash h \in \mathcal{S}\!\uparrow$.
4. $\epsilon, \sigma \vdash h \in \mathcal{R}\!\uparrow$ *implies that* $\sigma \neq \odot$ *and*
   - *if* $v \in \epsilon$, *then* $\epsilon, \sigma \vdash \langle h \mid v\,x \rangle \in \mathcal{S}\!\uparrow$ *for a fresh* x;
   - *if* $\sigma = E :: \sigma'$, *then* $\epsilon, \sigma' \vdash \langle h \mid E[x] \rangle \in \mathcal{S}\!\uparrow$ *for a fresh* x.

A normal-form simulation is a candidate relation $\mathcal{R}$ such that $\mathcal{R} \rightarrowtail \mathcal{R}, \mathcal{R}$, and a bisimulation is a candidate relation $\mathcal{R}$ such that $\mathcal{R}$ and $\mathcal{R}^{-1}$ are simulations. Normal-form bisimilarity $\approx$ is the union of all normal-form bisimulations.

When $\mathcal{E}, \Sigma \vdash c \mathrel{\mathcal{R}} d$, we reduce c until we get a value v or a stuck term $E[x\,v]$. At that point, either d also reduces to a normal form of the same kind, or we test (the first projection of) the stack Σ for divergence, assuming it is not empty. In the former case, we add the values to $\mathcal{E}$ and the evaluation contexts at the top of Σ, getting a judgment of the form $\mathcal{E}', \Sigma' \vdash h \mathrel{\mathcal{R}} g$, which then tests the environment and the stack by running either terms in $\mathcal{E}'$ or at the top of Σ'.

Example 4. We sketch the bisimulation proof for the terms t and s of Example 3. Because $\langle \emptyset \mid t \rangle \rightarrow^* \langle l := \mathsf{true} \mid f_1\,(x\,\lambda y.y)\rangle$ and $\langle \emptyset \mid s \rangle = \langle \emptyset \mid f_2\,(x\,\lambda y.y)\rangle$, we need to define $\mathcal{R}$ such that $\{(\lambda y.y, \lambda y.y)\}, (f_1\,\square, f_2\,\square) :: \odot \vdash l := \mathsf{true} \mathrel{\mathcal{R}} \emptyset$. Testing the equal values in the environment is easy with up-to techniques. For the contexts on the stack, we need $\{(\lambda y.y, \lambda y.y)\}, \odot \vdash \langle l := \mathsf{true} \mid f_1\,z\rangle \mathrel{\mathcal{R}} \langle \emptyset \mid f_2\,z\rangle$ for a fresh z. Since $\langle l := \mathsf{true} \mid f_1\,z\rangle \rightarrow^* \langle l := \mathsf{false} \mid \mathsf{true}\rangle$ and $\langle \emptyset \mid f_2\,z\rangle \rightarrow^* \langle \emptyset \mid \mathsf{true}\rangle$, we need $\{(\lambda y.y, \lambda y.y), (\mathsf{true}, \mathsf{true})\}, \odot \vdash l := \mathsf{false} \mathrel{\mathcal{R}} \emptyset$, which is simple to check.

Example 5. In contrast, we show that $t' \stackrel{\mathsf{def}}{=} \mathsf{new}\ l := \mathsf{true}\ \mathsf{in}\ f_1\ (x\ \lambda y.l := y; y)$ and $s' \stackrel{\mathsf{def}}{=} f_2\ (x\ \lambda y.y)$ are not bisimilar. We would need to build $\mathcal{R}$ such that $\{(\lambda y.l := y; y, \lambda y.y)\}, (f_1\,\square, f_2\,\square) :: \odot \vdash l := \mathsf{true} \mathrel{\mathcal{R}} \emptyset$. Testing the values in the environment, we want $\{(\lambda y.l := y; y, \lambda y.y), (z, z)\}, (f_1\,\square, f_2\,\square) :: \odot \vdash l := z \mathrel{\mathcal{R}} \emptyset$ for a fresh z. Executing the contexts on the stack, we get a stuck term of the form $\mathsf{if}\ z\ \mathsf{then}\ l := \mathsf{false}; \mathsf{true}\ \mathsf{else}\ \mathsf{false}$ and a value true, which cannot be related, because the former is not deferred diverging.

The terms t' and s' are therefore not bisimilar, and they are indeed not contextually equivalent, since t' gives access to its private reference by passing $\lambda y.l := y; y$ to x. The function represented by x can then change the value of l to false and break the equivalence.

The last two cases of the bisimulation definition aim at detecting a deferred diverging context. The judgment $\epsilon, \sigma \vdash h \in \mathcal{R}\!\uparrow$ roughly means that if $\sigma = E_n :: \ldots E_1 :: \odot$, then the configuration $\langle h' \mid E_1[\ldots E_n[x]]\rangle$ diverges for all fresh x and all h' obtained by running a term from $\mathcal{E}$ with the store h. As a result, when $\epsilon, \sigma \vdash h \in \mathcal{R}\!\uparrow$, we have two possibilities: either we run a term from $\mathcal{E}$ in h to potentially change h, or we run the context at the top of σ (which cannot be empty in that case) to check if it is diverging. In both cases, we get a judgment of the form $\epsilon, \sigma' \vdash c \in \mathcal{R}\!\uparrow$. In that case, either c diverges and we are done, or it terminates, meaning that we have to look for divergence in σ'.

Example 6. We prove that $\langle \emptyset \mid x\ v\ \Omega\rangle$ and $\langle \emptyset \mid \Omega\rangle$ are bisimilar. We define $\mathcal{R}$ such that $\emptyset, \odot \vdash \langle \emptyset \mid x\ v\ \Omega\rangle \mathrel{\mathcal{R}} \langle \emptyset \mid \Omega\rangle$, for which we need $\{v\}, \square\,\Omega :: \odot \vdash \emptyset \in \mathcal{R}\!\uparrow$, which itself holds if $\{v\}, \odot \vdash \langle \emptyset \mid y\ \Omega\rangle \in \mathcal{R}\!\uparrow$.

Finally, only the two clauses where a reduction step takes place are active; all the others are passive, because they are simply switching from one judgment to

$$\frac{\mathcal{E}, \Sigma \vdash c \mathrel{\mathcal{R}} d \qquad v \mathrel{\mathcal{E}} w \qquad x \notin \mathsf{fv}(v) \cup \mathsf{fv}(w)}{\mathcal{E}\{(v,w)/x\}, \Sigma\{(v,w)/x\} \vdash c\{v/x\} \mathrel{\mathsf{subst_c}(\mathcal{R})} d\{w/x\}}$$

$$\frac{\mathcal{E}, \Sigma_1 +\!\!+ (E_1, F_1) :: (E_2, F_2) :: \Sigma_2 \vdash \langle h \mid t \rangle \mathrel{\mathcal{R}} \langle g \mid s \rangle}{\mathcal{E}, \Sigma_1 +\!\!+ (E_2[E_1], F_2[F_1]) :: \Sigma_2 \vdash \langle h \mid t \rangle \mathrel{\mathsf{ccomp}(\mathcal{R})} \langle g \mid s \rangle}$$

$$\frac{\mathcal{E}, (E, F) :: \Sigma \vdash \langle h \mid t \rangle \mathrel{\mathcal{R}} \langle g \mid s \rangle}{\mathcal{E}, \Sigma \vdash \langle h \mid E[t] \rangle \mathrel{\mathsf{plug}(\mathcal{R})} \langle g \mid F[s] \rangle} \qquad \frac{c \rightarrow^* c' \qquad d \rightarrow^* d' \qquad \mathcal{E}, \Sigma \vdash c' \mathrel{\mathcal{R}} d'}{\mathcal{E}, \Sigma \vdash c \mathrel{\mathsf{red}(\mathcal{R})} d}$$

$$\frac{\epsilon, \sigma \vdash \langle h \mid t \rangle \in \mathcal{R}\!\uparrow \qquad \pi_1(\mathcal{E}) = \epsilon \qquad \pi_1(\Sigma) = \sigma}{\mathcal{E}, \Sigma \vdash \langle h \mid t \rangle \mathrel{\mathsf{div}(\mathcal{R})} \langle g \mid s \rangle} \qquad \frac{\mathcal{E}, \Sigma \vdash c \mathrel{\mathcal{R}} d \qquad \mathcal{E}' \subseteq \mathcal{E}}{\mathcal{E}', \Sigma \vdash c \mathrel{\mathsf{weak}(\mathcal{R})} d}$$

$$\frac{\mathcal{E}, \Sigma_1 +\!\!+ \Sigma_2 \vdash \langle h \mid t \rangle \mathrel{\mathcal{R}} \langle g \mid s \rangle \qquad \mathsf{fr}(E) \subseteq \mathsf{dom}(h')}{\mathcal{E}, \Sigma_1 +\!\!+ (E, E) :: \Sigma_2 \vdash \langle h \uplus h' \mid t \rangle \mathrel{\mathsf{refl}(\mathcal{R})} \langle g \uplus h' \mid s \rangle}$$

Fig. 2. Selected up-to techniques for the calculus with local store

the other without any real progress taking place. For example, when comparing value configurations, we go from a configuration judgment $\mathcal{E}, \Sigma \vdash c \mathrel{\mathcal{R}} d$ to a store judgment $\mathcal{E}, \Sigma \vdash h \mathrel{\mathcal{R}} g$ or a diverging store judgment $\mathcal{E}, \Sigma \vdash h \in \mathcal{R}\!\uparrow$. In a (diverging) store judgment, we simply decide whether we reduce a term from the store of from the stack, going back to a (diverging) configuration judgment. Actual progress is made only when we start reducing the chosen configuration.

3.3 Soundness and Completeness

We briefly discuss the up-to techniques we need to prove soundness. We write $\mathcal{E}\{(v,w)/x\}$ for the environment $\{(v'\{v/x\}, w'\{w/x\}) \mid v' \mathrel{\mathcal{E}} w'\}$, and we also define $\Sigma\{(x,w)/x\}$, $\epsilon\{v/x\}$, and $\sigma\{v/x\}$ as expected. To save space, Fig. 2 presents the up-to techniques for the configuration judgment only; see the report [4] for the other judgments.

As in Sect. 2.3, the techniques subst and plug allow to reason up to substitution and plugging into an evaluation context, except that the substituted values and plugged contexts must be taken from respectively the environment and the top of the stack. The technique div relates a diverging configuration to any configuration, like in the calculus with global store. The technique ccomp allows to merge successive contexts in the stack into one. The weakening technique weak, originally known as bisimulation up to environment [17], is an usual technique for environmental bisimulations. Making the environment smaller creates a weaker judgment, as having less testing terms means a less discriminating candidate relation. Bisimulation up to reduction red is also standard and allows for a big-step reasoning by ignoring reduction steps. Finally, the technique refl allows to introduce identical contexts in the stack, but also values in the environment or terms in configurations (see the report [4]).

We denote by $\mathsf{subst_c}$ the up to substitution technique restricted to the configuration and diverging configuration judgments, and by $\mathsf{subst_s}$ the restriction to the store and diverging store judgments.

Lemma 3. *The set* $\mathfrak{F} \stackrel{\mathsf{def}}{=} \{\mathsf{subst}_m, \mathsf{plug}, \mathsf{ccomp}, \mathsf{div}, \mathsf{weak}, \mathsf{red}, \mathsf{refl} \mid m \in \{\mathsf{c}, \mathsf{s}\}\}$ *is respectful, with* $\mathsf{strong}(\mathfrak{F}) = \{\mathsf{subst_s}, \mathsf{ccomp}, \mathsf{div}, \mathsf{weak}, \mathsf{red}, \mathsf{refl}\}$.

In contrast with Sect. 2.3 and our previous work [3], $\mathsf{subst_c}$ is *not* strong, because values are taken from the environment. Indeed, with $\mathsf{subst_c}$ strong, from $\{(v, w)\}, \odot \vdash \emptyset \mathrel{\mathcal{R}} \emptyset$, we could derive $\{(v, w)\}, \odot \vdash \langle \emptyset \mid x\, y\rangle \mathrel{\mathsf{refl}(\mathcal{R})} \langle \emptyset \mid x\, y\rangle$ and then $\{(v, w)\}, \odot \vdash \langle \emptyset \mid v\, x\rangle \mathrel{\mathsf{subst_c}(\mathsf{refl}(\mathcal{R}))} \langle \emptyset \mid w\, x\rangle$ for any v and w, which would be unsound.

The respectfulness proofs are in the report [4]. Using refl, plug, $\mathsf{subst_c}$, and Lemma 1 we prove that $\approx$ is preserved by evaluation contexts and substitution, from which we deduce it is sound w.r.t. contextual equivalence.

Theorem 3. *For all* t *and* s, *if* $\emptyset, \odot \vdash \langle \emptyset \mid t\rangle \approx \langle \emptyset \mid s\rangle$, *then* $t \equiv s$.

To establish completeness, we follow the proof of Theorem 2, i.e., we construct a candidate relation $\mathcal{R}$ that contains $\equiv$ and prove it is a simulation by case analysis on the behavior of the related terms.

Theorem 4. *For all* t *and* s, *if* $t \equiv s$, *then* $\emptyset, \odot \vdash \langle \emptyset \mid t\rangle \approx \langle \emptyset \mid s\rangle$.

The main difference is that the contexts and closing substitutions are built from the environment using compatible closures [17], to take into account the private resources of the related terms. We discuss the proof in the report [4].

3.4 Examples

Example 7. We start by the so-called awkward example [5,6,15]. Let

$$v \stackrel{\mathsf{def}}{=} \lambda f. l := 0; f\,(); l := 1; f\,(); !l \qquad w \stackrel{\mathsf{def}}{=} \lambda f. f\,(); f\,(); 1.$$

We equate $\mathsf{new}\ l := 0\ \mathsf{in}\ v$ and w, building the candidate $\mathcal{R}$ incrementally, starting from $\{(v, w)\}, \odot \vdash l := 0 \mathrel{\mathcal{R}} \emptyset$.

Running v and w with a fresh variable f, we obtain $\langle l := 0 \mid E_1[f\,()]\rangle$ and $\langle \emptyset \mid E_2[f\,()]\rangle$ with $E_1 \stackrel{\mathsf{def}}{=} \square; l := 1; f\,(); !l$ and $F_1 \stackrel{\mathsf{def}}{=} \square; f\,(); 1$. Ignoring the identical unit arguments (using refl), we need $\{(v, w)\}, (E_1, F_1) :: \odot \vdash l := 0 \mathrel{\mathcal{R}} \emptyset$; from that point, we can either test v and w again, resulting into an extra pair (E_1, F_1) on the stack, or run $\langle l := 0 \mid E_1[g]\rangle$ and $\langle \emptyset \mid F_1[g]\rangle$ for a fresh g instead.

In the latter case, we get $\langle l := 1 \mid E_2[g\,()]\rangle$ and $\langle \emptyset \mid F_2[g\,()]\rangle$, with $E_2 \stackrel{\mathsf{def}}{=} \square; !l$ and $F_2 \stackrel{\mathsf{def}}{=} \square; 1$, so we want $\{(v, w)\}, (E_2, F_2) :: \odot \vdash l := 1 \mathrel{\mathcal{R}} \emptyset$ (ignoring again the units). From there, testing v and w produces $\{(v, w)\}, (E_1, F_1) :: (E_2, F_2) :: \odot \vdash l := 0 \mathrel{\mathcal{R}} \emptyset$, while executing $\langle l := 1 \mid E_2[x]\rangle$ and $\langle \emptyset \mid F_2[x]\rangle$ for a fresh x gives us $\langle l := 1 \mid 1\rangle$ and $\langle \emptyset \mid 1\rangle$. This analysis suggests that $\mathcal{R}$ should be composed only of judgments of the form $\{(v, w)\}, \Sigma \vdash l := n \mathrel{\mathcal{R}} \emptyset$ such that $n \in \{0, 1\}$ and

- Σ is an arbitrary stack composed only of pairs (E_1, F_1) or (E_2, F_2);
- if $\Sigma = (E_2, F_2) :: \Sigma'$, then $n = 1$.

We can check that such a candidate is a bisimulation, and it ensures that when l is read (when E_2 is executed), it contains the value 1.

Example 8. As a variation on the awkward example, let

$$v \stackrel{\mathsf{def}}{=} \lambda f.l := !l + 1; f\ (); l := !l - 1; !l > 0 \qquad w \stackrel{\mathsf{def}}{=} \lambda f.f\ (); \mathsf{true}.$$

We show that $\langle \emptyset \mid \mathsf{new}\ l := 1\ \mathsf{in}\ v\rangle$ and $\langle \emptyset \mid w\rangle$ are bisimilar. Let $E \stackrel{\mathsf{def}}{=} \Box; l := !l - 1; !l > 0$ and $F \stackrel{\mathsf{def}}{=} \Box; \mathsf{true}$. We write $(E, F)^n$ for the stack $\odot$ if $n = 0$ and $(E, F) :: (E, F)^{n-1}$ otherwise. Then the candidate $\mathcal{R}$ verifying $\{(v, w)\}, (E, F)^n \vdash l := n + 1\ \mathcal{R}\ \emptyset$ for any n is a bisimulation. Indeed, running v and w increases the value stored in l and adds a pair (E, F) on the stack. If $n > 0$, we can run a copy of E and F, thus decreasing the value in l by 1, and then returning true in both cases.

Example 9. This deferred divergence example comes from Dreyer et al. [5]. Let

$$v_1 \stackrel{\mathsf{def}}{=} \lambda x.\mathsf{if}\ !l\ \mathsf{then}\ \Omega\ \mathsf{else}\ k := \mathsf{true}; \lambda y.y \qquad w_1 \stackrel{\mathsf{def}}{=} \lambda x.\Omega$$

$$v_2 \stackrel{\mathsf{def}}{=} \lambda f.f\ v_1; \mathsf{if}\ !k\ \mathsf{then}\ \Omega\ \mathsf{else}\ l := \mathsf{true}; \lambda y.y \qquad w_2 \stackrel{\mathsf{def}}{=} \lambda f.f\ w_1; \lambda y.y$$

We prove that $\mathsf{new}\ l := \mathsf{false}\ \mathsf{in}\ \mathsf{new}\ k := \mathsf{false}\ \mathsf{in}\ v_2$ is equivalent to w_2. Informally, if f in w_2 applies its argument w_1, the term diverges. Divergence also happens in v_2 but in a delayed fashion, as v_1 first sets k to true, and the continuation $t \stackrel{\mathsf{def}}{=} \mathsf{if}\ !k\ \mathsf{then}\ \Omega\ \mathsf{else}\ l := \mathsf{true}; \lambda y.y$ then diverges. Similarly, if f stores w_1 or v_1 to later apply it, then divergence also occurs in both cases: in that case t sets l to true, and when v_1 is later applied, it diverges.

To build a candidate $\mathcal{R}$, we execute $\langle l := \mathsf{false}; k := \mathsf{false} \mid v_2\ f\rangle$ and $\langle \emptyset \mid w_2\ f\rangle$ for a fresh f, which gives us $\langle l := \mathsf{false}; k := \mathsf{false} \mid E[f\ v_1]\rangle$ and $\langle \emptyset \mid F[f\ w_1]\rangle$ with $E \stackrel{\mathsf{def}}{=} \Box; t$ and $F \stackrel{\mathsf{def}}{=} \Box; \lambda y.y$. We consider $\{(v_2, w_2), (v_1, w_1)\}, (E, F) :: \emptyset \vdash l := \mathsf{false}; k := \mathsf{false}\ \mathcal{R}\ \emptyset$, for which we have several checks to do. The interesting one is running $\langle l := \mathsf{false}; k := \mathsf{false} \mid v_1\ x\rangle$ and $\langle \emptyset \mid w_1\ x\rangle$, as we get $\langle l := \mathsf{false}; k := \mathsf{true} \mid \lambda y.y\rangle$ and $\langle \emptyset \mid \Omega\rangle$. In that case, we are showing that the stack contains divergence, by establishing that $\{v_2, v_1, \lambda y.y\}, E :: \emptyset \vdash l := \mathsf{false}; k := \mathsf{true} \in \mathcal{R}{\uparrow}$, and indeed, we have $\langle l := \mathsf{false}; k := \mathsf{true} \mid E[x]\rangle \rightarrow^* \langle l := \mathsf{false}; k := \mathsf{true} \mid \Omega\rangle$ for a fresh x. In the end, the relation $\mathcal{R}$ verifying

$$\{(v_2, w_2), (v_1, w_1)\}, (E, F)^n \vdash l := \mathsf{false}; k := \mathsf{false}\ \mathcal{R}\ \emptyset$$
$$\{(v_2, w_2), (v_1, w_1)\}, (E, F)^n \vdash \langle l := \mathsf{false}; k := \mathsf{true} \mid \lambda y.y\rangle\ \mathcal{R}\ \langle \emptyset \mid \Omega\rangle$$
$$\{v_2, v_1, \lambda y.y\}, E^n \vdash l := \mathsf{false}; k := \mathsf{true} \in \mathcal{R}{\uparrow}$$
$$\{v_2, v_1, \lambda y.y\}, E^n \vdash \langle l := \mathsf{false}; k := \mathsf{true} \mid \Omega\rangle \in \mathcal{R}{\uparrow}$$
$$\{(v_2, w_2), (v_1, w_1)\}, (E, F)^n \vdash l := \mathsf{true}; k := \mathsf{false}\ \mathcal{R}\ \emptyset$$
$$\{(v_2, w_2), (v_1, w_1)\}, (E, F)^n \vdash \langle l := \mathsf{true}; k := \mathsf{false} \mid \Omega\rangle\ \mathcal{R}\ \langle \emptyset \mid \Omega\rangle$$

for all n is a bisimulation up to refl and red.

4 Related Work and Conclusion

Related Work. As pointed out in Sect. 1, the other bisimilarities defined for state either feature universal quantification over testing arguments [9,12,17,19], or are complete only for a more expressive language [18]. Kripke logical relations [1,5] also involve quantification over arguments when testing terms of a functional type. Finally, denotational models [10,13] can also be used to prove program equivalence, by showing that the denotations of two terms are equal. However, computing such denotations is difficult in general, and the automation of this task is so far restricted to a language with first-order references [14].

The work most closely related to ours is Jaber and Tabareau's Kripke Open Bisimulation (KOB) [6]. A KOB tests functional terms with fresh variables and not with related values like a regular logical relation would do. To relate two given configurations, one has to provide a World Transition System (WTS) which states the invariants the heaps of the configurations should satisfy and how to go from one invariant to the other during the evaluation. Similarly, the bisimulations for the examples of Sect. 3.4 state properties which could be seen as invariants about the stores at different points of the evaluation.

The difficulty for KOB as well as with our bisimilarity is to come up with the right invariants about the heaps, expressed either as a WTS or as a bisimulation. We believe that choosing a technique over the other is just a matter of preference, depending on whether one is more comfortable with game semantics or with coinduction. It would be interesting to see if there is a formal correspondence between KOB and our bisimilarity; we leave this question as a future work.

Conclusion. We define a sound and complete normal-form bisimilarity for higher-order local state, with an environment to be able to run terms in different stores. We distinguish in the environment values which should be tested several times from the contexts which should be executed only once. The other difficulty is to relate deferred and regular diverging terms, which is taken care of by the specific judgments about divergence. The lack of quantification over arguments make the bisimulation proofs quite simple.

A future work would be to make these proofs even simpler by defining appropriate up-to techniques. The techniques we use in Sect. 3.3 to prove soundness turn out to be not that useful when establishing the equivalences of Sect. 3.4, except for trivial ones such as up to reduction or reflexivity. The difficulty in defining the candidate relations for the examples of Sect. 3.4 is in finding the right property relating the stack Σ to the store, so maybe an up-to technique could make this task easier.

As pointed out in Sect. 1, our results can be seen as an indication of what kind of additional infrastructure in a complete normal-form bisimilarity is required when the considered syntactic theory becomes less discriminative—in our case, when control operators vanish from the picture, and mutable state is the only extension of the λ-calculus. A question one could then ask is whether we can find a less expressive calculus—maybe the plain λ-calculus itself—for which a suitably enhanced normal-form bisimilarity is still complete.

Acknowledgements. We thank Guilhem Jaber and the anonymous reviewers for their comments. This work was supported by the National Science Centre, Poland, grant no. 2014/15/B/ST6/00619 and by COST Action EUTypes CA15123.

References

1. Ahmed, A., Dreyer, D., Rossberg, A.: State-dependent representation independence. In: Pierce, B.C. (ed.) Proceedings of the Thirty-Fifth Annual ACM Symposium on Principles of Programming Languages, pp. 340–353. ACM Press, January 2009
2. Aristizábal, A., Biernacki, D., Lenglet, S., Polesiuk, P.: Environmental bisimulations for delimited-control operators with dynamic prompt generation. Logical Methods Comput. Sci. **13**(3) 2017
3. Biernacki, D., Lenglet, S., Polesiuk, P.: Proving soundness of extensional normal-form bisimilarities. In: Silva, A. (ed.) Proceedings of the 33rd Annual Conference on Mathematical Foundations of Programming Semantics (MFPS XXXIII), Ljubljana, Slovenia. Electronic Notes in Theoretical Computer Science, vol. 336, pp. 41–56, June 2017
4. Biernacki, D., Lenglet, S., Polesiuk, P.: A complete normal-form bisimilarity for state. Research report RR-9251, Inria, Nancy, France, January 2019
5. Dreyer, D., Neis, G., Birkedal, L.: The impact of higher-order state and control effects on local relational reasoning. J. Funct. Program. **22**(4–5), 477–528 (2012)
6. Jaber, G., Tabareau, N.: Kripke open bisimulation - a marriage of game semantics and operational techniques. In: Feng, X., Park, S. (eds.) APLAS 2015. LNCS, vol. 9458, pp. 271–291. Springer, Cham (2015)
7. Jagadeesan, R., Pitcher, C., Riely, J.: Open bisimulation for aspects. Trans. Aspect-Oriented Softw. Dev. **5**, 72–132 (2009)
8. Koutavas, V., Levy, P.B., Sumii, E.: From applicative to environmental bisimulation. In: Mislove, M., Ouaknine, J. (eds.) Proceedings of the 27th Annual Conference on Mathematical Foundations of Programming Semantics (MFPS XXVII), Pittsburgh, PA, USA. ENTCS, vol. 276, pp. 215–235, May 2011
9. Koutavas, V., Wand, M.: Small bisimulations for reasoning about higher-order imperative programs. In: Morrisett, J.G., Jones, S.L.P. (eds.) POPL 2006, Charleston, SC, USA, pp. 141–152. ACM Press (2006)
10. Laird, J.: A fully abstract trace semantics for general references. In: Arge, L., Cachin, C., Jurdziński, T., Tarlecki, A. (eds.) ICALP 2007. LNCS, vol. 4596, pp. 667–679. Springer, Heidelberg (2007)
11. Lassen, S.B.: Eager normal form bisimulation. In: Panangaden, P. (ed.) LICS 2005, Chicago, IL, pp. 345–354. IEEE Computer Society Press (2005)
12. Madiot, J.-M., Pous, D., Sangiorgi, D.: Bisimulations up-to: beyond first-order transition systems. In: Baldan, P., Gorla, D. (eds.) CONCUR 2014. LNCS, vol. 8704, pp. 93–108. Springer, Heidelberg (2014)
13. Murawski, A.S., Tzevelekos, N.: Game semantics for good general references. In: Proceedings of the 26th Annual IEEE Symposium on Logic in Computer Science, LICS 2011, pp. 75–84. IEEE Computer Society, June 2011
14. Murawski, A.S., Tzevelekos, N.: Algorithmic games for full ground references. Formal Methods Syst. Des. **52**(3), 277–314 (2018)
15. Pitts, A., Stark, I.: Operational reasoning for functions with local state. In: Gordon, A., Pitts, A. (eds.) Higher Order Operational Techniques in Semantics, pp. 227–273. Publications of the Newton Institute, Cambridge University Press (1998)

16. Sangiorgi, D.: The lazy lambda calculus in a concurrency scenario. In: Scedrov, A. (ed.) LICS 1992, Santa Cruz, California, pp. 102–109. IEEE Computer Society (1992)
17. Sangiorgi, D., Kobayashi, N., Sumii, E.: Environmental bisimulations for higher-order languages. ACM Trans. Program. Lang. Syst. **33**(1), 1–69 (2011)
18. Støvring, K., Lassen, S.B.: A complete, co-inductive syntactic theory of sequential control and state. In: Felleisen, M. (ed.) SIGPLAN Notices, POPL 2007, Nice, France, vol. 42, no. 1, pp. 161–172. ACM Press (2007)
19. Sumii, E.: A complete characterization of observational equivalence in polymorphic λ-calculus with general references. In: Grädel, E., Kahle, R. (eds.) CSL 2009. LNCS, vol. 5771, pp. 455–469. Springer, Heidelberg (2009)

Identifiers in Registers
Describing Network Algorithms with Logic

Benedikt Bollig, Patricia Bouyer, and Fabian Reiter(✉)

LSV, CNRS, ENS Paris-Saclay, Université Paris-Saclay, Cachan, France
{bollig,bouyer}@lsv.fr, fabian.reiter@gmail.com

Abstract. We propose a formal model of distributed computing based on register automata that captures a broad class of synchronous network algorithms. The local memory of each process is represented by a finite-state controller and a fixed number of registers, each of which can store the unique identifier of some process in the network. To underline the naturalness of our model, we show that it has the same expressive power as a certain extension of first-order logic on graphs whose nodes are equipped with a total order. Said extension lets us define new functions on the set of nodes by means of a so-called partial fixpoint operator. In spirit, our result bears close resemblance to a classical theorem of descriptive complexity theory that characterizes the complexity class PSPACE in terms of partial fixpoint logic (a proper superclass of the logic we consider here).

1 Introduction

This paper is part of an ongoing research project aiming to develop a *descriptive complexity* theory for *distributed computing.*

In classical sequential computing, descriptive complexity is a well-established field that connects computational complexity classes to equi-expressive classes of logical formulas. It began in the 1970s, when Fagin showed in [6] that the graph properties decidable by nondeterministic Turing machines in polynomial time are exactly those definable in existential second-order logic. This provided a logical—and thus machine-independent—characterization of the complexity class NP. Subsequently, many other popular classes, such as P, PSPACE, and EXPTIME were characterized in a similar manner (see for instance the textbooks [8,12,15]).

Of particular interest to us is a result due to Abiteboul, Vianu [1], and Vardi [19], which states that on structures equipped with a total order relation, the properties decidable in PSPACE coincide with those definable in *partial fixpoint logic.* The latter is an extension of first-order logic with an operator that allows us to inductively define new relations of arbitrary arity. Basically, this means that new relations can occur as free (second-order) variables in the logical formulas that define them. Those variables are initially interpreted as empty relations and then iteratively updated, using the defining formulas as update

M. Bojańczyk and A. Simpson (Eds.): FOSSACS 2019, LNCS 11425, pp. 115–132, 2019.
https://doi.org/10.1007/978-3-030-17127-8_7

rules. If the sequence of updates converges to a fixpoint, then the ultimate interpretations are the relations reached in the limit. Otherwise, the variables are simply interpreted as empty relations. Hence the term "partial fixpoint".

While well-developed in the classical case, descriptive complexity has so far not received much attention in the setting of distributed network computing. As far as the authors are aware, the first step in this direction was taken by Hella et al. in [10,11], where they showed that basic *modal logic* evaluated on finite graphs has the same expressive power as a particular class of *distributed automata* operating in constant time. Those automata constitute a weak model of distributed computing in arbitrary network topologies, where all nodes synchronously execute the same finite-state machine and communicate with each other by broadcasting messages to their neighbors. Motivated by this result, several variants of distributed automata were investigated by Kuusisto and Reiter in [14,18] and [17] to establish similar connections with standard logics such as the *modal μ-calculus* and *monadic second-order logic*. However, since the models of computation investigated in those works are based on anonymous finite-state machines, they are much too weak to solve many of the problems typically considered in distributed computing, such as leader election or constructing a spanning tree. It would thus be desirable to also characterize stronger models.

A common assumption underlying many distributed algorithms is that each node of the considered network is given a unique identifier. This allows us, for instance, to elect a leader by making the nodes broadcast their identifiers and then choose the one with the smallest identifier as the leader. To formalize such algorithms, we need to go beyond finite-state machines because the number of bits required to encode a unique identifier grows logarithmically with the number of nodes in the network. Recently, in [2,3], Aiswarya, Bollig and Gastin introduced a synchronous model where, in addition to a finite-state controller, nodes also have a fixed number of registers in which they can store the identifiers of other nodes. Access to those registers is rather limited in the sense that their contents can be compared with respect to a total order, but their numeric values are unknown to the nodes. (This restriction corresponds precisely to the notion of *order-invariant* distributed algorithms, which was introduced by Naor and Stockmeyer in [16].) Similarly, register contents can be copied, but no new values can be generated. Since the original motivation for the model was to automatically verify certain distributed algorithms running on ring networks, its formal definition is tailored to that particular setting. However, the underlying principle can be generalized to arbitrary networks of unbounded maximum degree, which was the starting point for the present work.

Contributions. While on an intuitive level, the idea of finite-state machines equipped with additional registers might seem very natural, it does not immediately yield a formal model for distributed algorithms in arbitrary networks. In particular, it is not clear what would be the canonical way for nodes to communicate with a non-constant number of peers, if we require that they all follow the same, finitely representable set of rules.

The model we propose here, dubbed *distributed register automata*, is an attempt at a solution. As in [2,3], nodes proceed in synchronous rounds and have a fixed number of registers, which they can compare and update without having access to numeric values. The new key ingredient that allows us to formalize communication between nodes of unbounded degree is a local computing device we call *transition maker*. This is a special kind of register machine that the nodes can use to scan the states and register values of their entire neighborhood in a sequential manner. In every round, each node runs the transition maker to update its own local configuration (i.e., its state and register valuation) based on a snapshot of the local configurations of its neighbors in the previous round. A way of interpreting this is that the nodes communicate by broadcasting their local configurations as messages to their neighbors. Although the resulting model of computation is by no means universal, it allows formalizing algorithms for a wide range of problems, such as constructing a spanning tree (see Example 5) or testing whether a graph is Hamiltonian (see Example 6).

Nevertheless, our model is somewhat arbitrary, since it could be just one particular choice among many other similar definitions capturing different classes of distributed algorithms. What justifies our choice? This is where descriptive complexity comes into play. By identifying a logical formalism that has the same expressive power as distributed register automata, we provide substantial evidence for the naturalness of that model. Our formalism, referred to as *functional fixpoint logic*, is a fragment of the above-mentioned partial fixpoint logic. Like the latter, it also extends first-order logic with a partial fixpoint operator, but a weaker one that can only define unary functions instead of arbitrary relations. We show that on totally ordered graphs, this logic allows one to express precisely the properties that can be decided by distributed register automata. The connection is strongly reminiscent of Abiteboul, Vianu and Vardi's characterization of PSPACE, and thus contributes to the broader objective of extending classical descriptive complexity to the setting of distributed computing. Moreover, given that logical formulas are often more compact and easier to understand than abstract machines (compare Examples 6 and 8), logic could also become a useful tool in the formal specification of distributed algorithms.

The remainder of this paper is structured around our main result:

Theorem 1. *When restricted to finite graphs whose nodes are equipped with a total order, distributed register automata are effectively equivalent to functional fixpoint logic.*

After giving some preliminary definitions in Sect. 2, we formally introduce distributed register automata in Sect. 3 and functional fixpoint logic in Sect. 4. We then sketch the proof of Theorem 1 in Sect. 5, and conclude in Sect. 6.

2 Preliminaries

We denote the empty set by $\emptyset$, the set of nonnegative integers by $\mathbb{N} = \{0, 1, 2, \dots\}$, and the set of integers by $\mathbb{Z} = \{\dots, -1, 0, 1, \dots\}$. The cardinality of any set S is written as $|S|$ and the power set as 2^S.

In analogy to the commonly used notation for real intervals, we define the notation $[m:n] := \{i \in \mathbb{Z} \mid m \leq i \leq n\}$ for any $m, n \in \mathbb{Z}$ such that $m \leq n$. To indicate that an endpoint is excluded, we replace the corresponding square bracket with a parenthesis, e.g., $(m:n] := [m:n] \setminus \{m\}$. Furthermore, if we omit the first endpoint, it defaults to 0. This gives us shorthand notations such as $[n] := [0:n]$ and $[n) := [0:n) = [0:n-1]$.

All graphs we consider are finite, simple, undirected, and connected. For notational convenience, we identify their nodes with nonnegative integers, which also serve as unique identifiers. That is, when we talk about the *identifier* of a node, we mean its numerical representation. A *graph* is formally represented as a pair $G = (V, E)$, where the set V of *nodes* is equal to $[n)$, for some integer $n \geq 2$, and the set E consists of undirected *edges* of the form $e = \{u, v\} \subseteq V$ such that $u \neq v$. Additionally, E must satisfy that every pair of nodes is connected by a sequence of edges. The restriction to graphs of size at least two is for technical reasons; it ensures that we can always encode Boolean values as nodes.

We refer the reader to [5] for standard graph theoretic terms such as *neighbor*, *degree*, *maximum degree*, *distance*, and *spanning tree*.

Graphs are used to model computer networks, where nodes correspond to processes and edges to communication links. To represent the current configuration of a system as a graph, we equip each node with some additional information: the current state of the corresponding process, taken from a nonempty finite set Q, and some pointers to other processes, modeled by a finite set R of registers.

We call $\Sigma = (Q, R)$ a *signature* and define a *Σ-configuration* as a tuple $C = (G, \mathfrak{q}, \mathfrak{r})$, where $G = (V, E)$ is a graph, called the *underlying* graph of C, $\mathfrak{q}\colon V \to Q$ is a *state function* that assigns to each node a state $q \in Q$, and $\mathfrak{r}\colon V \to V^R$ is a *register valuation function* that associates with each node a *register valuation* $\rho \in V^R$. The set of all Σ-configurations is denoted by $\mathbb{C}(\Sigma)$. Figure 1 on page 6 illustrates part of a $(\{q_1, q_2, q_3\}, \{r_1, r_2, r_3\})$-configuration.

If $R = \emptyset$, then we are actually dealing with a tuple $(G, \mathfrak{q})$, which we call a *Q-labeled graph*. Accordingly, the elements of Q may also be called *labels*. A set P of labeled graphs will be referred to as a *graph property*. Moreover, if the labels are irrelevant, we set Q equal to the singleton $\mathbb{1} := \{\varepsilon\}$, where ε is our dummy label. In this case, we identify $(G, \mathfrak{q})$ with G and call it an *unlabeled* graph.

3 Distributed Register Automata

Many distributed algorithms can be seen as *transducers*. A leader-election algorithm, for instance, takes as input a network and outputs the same network, but with every process storing the identifier of the unique leader in some dedicated register r. Thus, the algorithm transforms a $(\mathbb{1}, \emptyset)$-configuration into a $(\mathbb{1}, \{r\})$-configuration. We say that it defines a $(\mathbb{1}, \emptyset)$-$(\mathbb{1}, \{r\})$-transduction. By the same token, if we consider distributed algorithms that *decide* graph properties (e.g., whether a graph is Hamiltonian), then we are dealing with a $(I, \emptyset)$-$(\{\text{YES}, \text{NO}\}, \emptyset)$-transduction, where I is some set of labels. The idea is that a graph will be accepted if and only if every process eventually outputs YES.

Let us now formalize the notion of transduction. For any two signatures $\Sigma^{in} = (I, R^{in})$ and $\Sigma^{out} = (O, R^{out})$, a Σ^{in}-Σ^{out}-*transduction* is a *partial* mapping $T\colon \mathbb{C}(\Sigma^{in}) \to \mathbb{C}(\Sigma^{out})$ such that, if defined, $T(G, \mathfrak{q}, \mathfrak{r}) = (G, \mathfrak{q}', \mathfrak{r}')$ for some $\mathfrak{q}'$ and $\mathfrak{r}'$. That is, a transduction does not modify the underlying graph but only the states and register valuations. We denote the set of all Σ^{in}-Σ^{out}-transductions by $\mathbb{T}(\Sigma^{in}, \Sigma^{out})$ and refer to Σ^{in} and Σ^{out} as the *input* and *output signatures* of T. By extension, I and O are called the sets of *input* and *output labels*, and R^{in} and R^{out} the sets of *input* and *output registers*. Similarly, any Σ^{in}-configuration C can be referred to as an *input configuration* of T and $T(C)$ as an *output configuration*.

Next, we introduce our formal model of distributed algorithms.

Definition 2 (Distributed register automaton). *Let $\Sigma^{in} = (I, R^{in})$ and $\Sigma^{out} = (O, R^{out})$ be two signatures. A* distributed register automaton *(or simply* automaton*) with input signature Σ^{in} and output signature Σ^{out} is a tuple $A = (Q, R, \iota, \Delta, H, o)$ consisting of a nonempty finite set Q of* states, *a finite set R of* registers *that includes both R^{in} and R^{out}, an* input function *$\iota\colon I \to Q$, a transition maker Δ whose specification will be given in Definition 3 below, a set $H \subseteq Q$ of* halting states, *and an* output function *$o\colon H \to O$. The registers in $R \setminus (R^{in} \cup R^{out})$ are called* auxiliary registers.

Automaton A computes a transduction $T_A \in \mathbb{T}(\Sigma^{in}, \Sigma^{out})$. To do so, it runs in a sequence of synchronous rounds on the input configuration's underlying graph $G = (V, E)$. After each round, the automaton's global configuration is a (Q, R)-configuration $C = (G, \mathfrak{q}, \mathfrak{r})$, i.e., the underlying graph is always G. As mentioned before, for a node $v \in V$, we interpret $\mathfrak{q}(v) \in Q$ as the current state of v and $\mathfrak{r}(v) \in V^R$ as the current register valuation of v. Abusing notation, we let $C(v) := (\mathfrak{q}(v), \mathfrak{r}(v))$ and say that $C(v)$ is the *local configuration* of v. In Fig. 1, the local configuration node 17 is $(q_1, \{r_1, r_2, r_3 \mapsto 17, 34, 98\})$.

For a given input configuration $C = (G, \mathfrak{q}, \mathfrak{r}) \in \mathbb{C}(\Sigma^{in})$, the automaton's *initial configuration* is $C' = (G, \iota \circ \mathfrak{q}, \mathfrak{r}')$, where for all $v \in V$, we have $\mathfrak{r}'(v)(r) = \mathfrak{r}(v)(r)$ if $r \in R^{in}$, and $\mathfrak{r}'(v)(r) = v$ if $r \in R \setminus R^{in}$. This means that every node v is initialized to state $\iota(\mathfrak{q}(v))$, and v's initial register valuation $\mathfrak{r}'(v)$ assigns v's own identifier (provided by G) to all non-input registers while keeping the given values assigned by $\mathfrak{r}(v)$ to the input registers.

Each subsequent configuration is obtained by running the transition maker Δ synchronously on all nodes. As we will see, Δ computes a function

$$[\![\Delta]\!]\colon (Q \times V^R)^+ \to Q \times V^R$$

that maps from nonempty sequences of local configurations to local configurations. This allows the automaton A to transition from a given configuration C to the next configuration C' as follows. For every node $u \in V$ of degree d, we consider the list $v_1, \ldots v_d$ of u's neighbors sorted in ascending (identifier) order, i.e., $v_i < v_{i+1}$ for $i \in [1:d)$. (See Fig. 1 for an example, where u corresponds to node 17.) If u is already in a halting state, i.e., if $C(u) = (q, \rho) \in H \times V^R$,

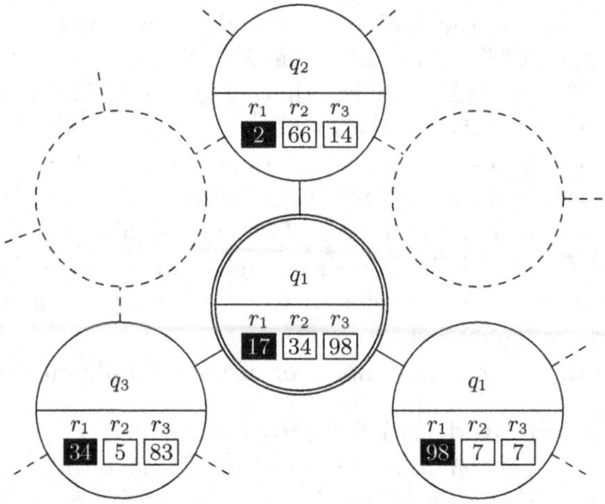

Fig. 1. Part of a configuration, as seen by a single node. Assuming the identifiers of the nodes are the values represented in black boxes (i.e., those stored in register r_1), the automaton at node 17 will update its own local configuration (q_1, $\{r_1, r_2, r_3 \mapsto$ 17, 34, 98$\}$) by running the transition maker on the sequence consisting of the local configurations of nodes 17, 2, 34, and 98 (in that exact order).

then its local configuration does not change anymore, i.e., $C'(u) = C(u)$. Otherwise, we define $C'(u) = [\![\Delta]\!](C(u), C(v_1), \ldots, C(v_d))$, which we may write more suggestively as

$$[\![\Delta]\!]\colon C(u) \xmapsto{C(v_1),\ldots,C(v_d)} C'(u).$$

Intuitively, node u updates its own local configuration by using Δ to scan a snapshot of its neighbors' local configurations. As the system is synchronous, this update procedure is performed simultaneously by all nodes.

A configuration $C = (G, \mathfrak{q}, \mathfrak{r})$ is called a *halting configuration* if all nodes are in a halting state, i.e., if $\mathfrak{q}(v) \in H$ for all $v \in V$. We say that A *halts* if it reaches a halting configuration.

The output configuration produced by a halting configuration $C = (G, \mathfrak{q}, \mathfrak{r})$ is the Σ^{out}-configuration $C' = (G, o \circ \mathfrak{q}, \mathfrak{r}')$, where for all $v \in V$ and $r \in R^{out}$, we have $\mathfrak{r}'(v)(r) = \mathfrak{r}(v)(r)$. In other words, each node v outputs the state $o(\mathfrak{q}(v))$ and keeps in its output registers the values assigned by $\mathfrak{r}(v)$.

It is now obvious that A defines a transduction $T_A\colon \mathbb{C}(\Sigma^{in}) \to \mathbb{C}(\Sigma^{out})$. If A receives the input configuration $C \in \mathbb{C}(\Sigma^{in})$ and eventually halts and produces the output configuration $C' \in \mathbb{C}(\Sigma^{out})$, then $T_A(C) = C'$. Otherwise (if A does not halt), $T_A(C)$ is undefined.

Deciding graph properties. Our primary objective is to use distributed register automata as decision procedures for graph properties. Therefore, we will focus on automata A that halt in a finite number of rounds on *every* input configuration, and we often restrict to input signatures of the form $(I, \emptyset)$ and the output signature $(\{\text{YES}, \text{NO}\}, \emptyset)$. For example, for $I = \{a, b\}$, we may be interested in the set of I-labeled graphs that have exactly one a-labeled node v (the "leader"). We stipulate that A *accepts* an input configuration C with underlying graph $G = (V, E)$ if $T_A(C) = (G, \mathfrak{q}, \mathfrak{r})$ such that $\mathfrak{q}(v) = \text{YES}$ for *all* $v \in V$. Conversely, A *rejects* C if $T_A(C) = (G, \mathfrak{q}, \mathfrak{r})$ such that $\mathfrak{q}(v) = \text{NO}$ for *some* $v \in V$. This corresponds to the usual definition chosen in the emerging field of *distributed decision* [7]. Accordingly, a graph property P is *decided* by A if the automaton accepts all input configurations that satisfy P and rejects all the others.

It remains to explain how the transition maker Δ works internally.

Definition 3 (Transition maker). *Suppose that $A = (Q, R, \iota, \Delta, H, o)$ is a distributed register automaton. Then its* transition maker $\Delta = (\tilde{Q}, \tilde{R}, \tilde{\iota}, \tilde{\delta}, \tilde{o})$ *consists of a nonempty finite set $\tilde{Q}$ of* inner states, *a finite set $\tilde{R}$ of* inner registers *that is disjoint from R, an* inner initial state $\tilde{\iota} \in \tilde{Q}$, *an* inner transition function $\tilde{\delta}\colon \tilde{Q} \times Q \times 2^{(\tilde{R} \cup R)^2} \to \tilde{Q} \times (\tilde{R} \cup R)^{\tilde{R}}$, *and an* inner output function $\tilde{o}\colon \tilde{Q} \to Q \times \tilde{R}^R$.

Basically, a transition maker $\Delta = (\tilde{Q}, \tilde{R}, \tilde{\iota}, \tilde{\delta}, \tilde{o})$ is a sequential register automaton (in the spirit of [13]) that reads a nonempty sequence $(q_0, \rho_0), \ldots, (q_d, \rho_d) \in (Q \times V^R)^+$ of local configurations of A in order to produce a new local configuration (q', ρ'). While reading this sequence, it traverses itself a sequence $(\tilde{q}_0, \tilde{\rho}_0), \ldots, (\tilde{q}_{d+1}, \tilde{\rho}_{d+1})$ of *inner configurations*, which each consist of an inner state $\tilde{q}_i \in \tilde{Q}$ and an *inner register valuation* $\tilde{\rho}_i \in (V \cup \{\bot\})^{\tilde{R}}$, where the symbol $\bot$ represents an undefined value. For the initial inner configuration, we set $\tilde{q}_0 = \tilde{\iota}$ and $\tilde{\rho}_0(\tilde{r}) = \bot$ for all $\tilde{r} \in \tilde{R}$. Now for $i \in [d]$, when Δ is in the inner configuration $(\tilde{q}_i, \tilde{\rho}_i)$ and reads the local configuration (q_i, ρ_i), it can compare all values assigned to the inner registers and registers by $\tilde{\rho}_i$ and ρ_i (with respect to the order relation on V). In other words, it has access to the binary relation $\prec_i \subseteq (\tilde{R} \cup R)^2$ such that for $\tilde{r}, \tilde{s} \in \tilde{R}$ and $r, s \in R$, we have $\tilde{r} \prec_i r$ if and only if $\tilde{\rho}_i(\tilde{r}) < \rho_i(r)$, and analogously for $r \prec_i \tilde{r}$, $\tilde{r} \prec_i \tilde{s}$, and $r \prec_i s$. In particular, if $\tilde{\rho}_i(\tilde{r}) = \bot$, then $\tilde{r}$ is incomparable with respect to $\prec_i$. Equipped with this relation, Δ transitions to $(\tilde{q}_{i+1}, \tilde{\rho}_{i+1})$ by evaluating $\tilde{\delta}(\tilde{q}_i, q_i, \prec_i) = (\tilde{q}_{i+1}, \tilde{\alpha})$ and computing $\tilde{\rho}_{i+1}$ such that $\tilde{\rho}_{i+1}(\tilde{r}) = \tilde{\rho}_i(\tilde{s})$ if $\tilde{\alpha}(\tilde{r}) = \tilde{s}$, and $\tilde{\rho}_{i+1}(\tilde{r}) = \rho_i(s)$ if $\tilde{\alpha}(\tilde{r}) = s$, where $\tilde{r}, \tilde{s} \in \tilde{R}$ and $s \in R$. Finally, after having read the entire input sequence and reached the inner configuration $(\tilde{q}_{d+1}, \tilde{\rho}_{d+1})$, the transition maker outputs the local configuration (q', ρ') such that $\tilde{o}(\tilde{q}_{d+1}) = (q', \tilde{\beta})$ and $\tilde{\beta}(r) = \tilde{r}$ implies $\rho'(r) = \tilde{\rho}_{d+1}(\tilde{r})$. Here we assume without loss of generality that Δ guarantees that $\rho'(r) \neq \bot$ for all $r \in R$.

Remark 4. Recall that $V = [n]$ for any graph $G = (V, E)$ with n nodes. However, as registers cannot be compared with constants, this actually represents an arbitrary assignment of unique, totally ordered identifiers. To determine the smallest identifier (i.e., 0), the nodes can run an algorithm such as the following.

Example 5 (Spanning tree). We present a simple automaton $A = (Q, R, \iota, \Delta, H, o)$ with input signature $\Sigma^{in} = (\mathbb{1}, \emptyset)$ and output signature $\Sigma^{out} = (\mathbb{1}, \{\mathit{parent}, \mathit{root}\})$ that computes a (breadth-first) spanning tree of its input

Algorithm 1. Transition maker of the automaton from Example 5

```
if ∃ neighbor nb (nb.root < my.root):                                   } Rule 1
    my.state ← 1;   my.parent ← nb.self;   my.root ← nb.root

else if my.state = 1
        ∧ ∀ neighbor nb [ nb.root = my.root ∧                            } Rule 2
                          (nb.parent ≠ my.self ∨ nb.state = 2) ] :
    my.state ← 2

else if (my.state = 2 ∧ my.root = my.self) ∨ (my.parent.state = 3):    } Rule 3
    my.state ← 3

else do nothing
```

graph $G = (V, E)$, rooted at the node with the smallest identifier. More precisely, in the computed output configuration $C = (G, \mathfrak{q}, \mathfrak{r})$, every node will store the identifier of its tree parent in register *parent* and the identifier of the root (i.e., the smallest identifier) in register *root*. Thus, as a side effect, A also solves the leader election problem by electing the root as the leader.

The automaton operates in three phases, which are represented by the set of states $Q = \{1, 2, 3\}$. A node terminates as soon as it reaches the third phase, i.e., we set $H = \{3\}$. Accordingly, the (trivial) input and output functions are $\iota\colon \varepsilon \mapsto 1$ and $o\colon 3 \mapsto \varepsilon$. In addition to the output registers, each node has an auxiliary register *self* that will always store its own identifier. Thus, we choose $R = \{self, parent, root\}$. For the sake of simplicity, we describe the transition maker Δ in Algorithm 1 using pseudocode rules. However, it should be clear that these rules could be relatively easily implemented according to Definition 3.

All nodes start in state 1, which represents the tree-construction phase. By Rule 1, whenever an active node (i.e., a node in state 1 or 2) sees a neighbor whose *root* register contains a smaller identifier than the node's own *root* register, it updates its *parent* and *root* registers accordingly and switches to state 1. To resolve the nondeterminism in Rule 1, we stipulate that *nb* is chosen to be the neighbor with the smallest identifier among those whose *root* register contains the smallest value seen so far.

As can be easily shown by induction on the number of communication rounds, the nodes have to apply Rule 1 no more than diameter(G) times in order for the pointers in register *parent* to represent a valid spanning tree (where the root points to itself). However, since the nodes do not know when diameter(G) rounds have elapsed, they must also check that the current configuration does indeed represent a single tree, as opposed to a forest. They do so by propagating a signal, in form of state 2, from the leaves up to the root.

By Rule 2, if an active node whose neighbors all agree on the same root realizes that it is a leaf or that all of its children are in state 2, then it switches to state 2 itself. Assuming the *parent* pointers in the current configuration already represent a single tree, Rule 2 ensures that the root will eventually be notified of this fact (when all of its children are in state 2). Otherwise, the *parent* pointers represent a forest, and every tree contains at least one node that has a neighbor outside of the tree (as we assume the underlying graph is connected).

Depending on the input graph, a node can switch arbitrarily often between states 1 and 2. Once the spanning tree has been constructed and every node is in state 2, the only node that knows this is the root. In order for the algorithm to terminate, Rule 3 then makes the root broadcast an acknowledgment message down the tree, which causes all nodes to switch to the halting state 3. □

Building on the automaton from Example 5, we now give an example of a graph property that can be decided in our model of distributed computing. The following automaton should be compared to the logical formula presented later in Example 8, which is much more compact and much easier to specify.

Example 6 (Hamiltonian cycle). We describe an automaton with input signature $\Sigma^{in} = (\mathbb{1}, \{parent, root\})$ and output signature $\Sigma^{out} = (\{\text{YES}, \text{NO}\}, \emptyset)$ that decides if the underlying graph $G = (V, E)$ of its input configuration $C = (G, \mathfrak{q}, \mathfrak{r})$ is Hamiltonian, i.e., whether G contains a cycle that goes through each node exactly once. The automaton works under the assumption that $\mathfrak{r}$ encodes a valid spanning tree of G in the registers *parent* and *root*, as constructed by the automaton from Example 5. Hence, by combining the two automata, we could easily construct a third one that decides the graph property of Hamiltonicity.

The automaton $A = (Q, R, \iota, \Delta, H, o)$ presented here implements a simple backtracking algorithm that tries to traverse G along a Hamiltonian cycle. Its set of states is $Q = \big(\{unvisited, visited, backtrack\} \times \{idle, request, good, bad\}\big) \cup H$, with the set of halting states $H = \{\text{YES}, \text{NO}\}$. Each non-halting state consists of two components, the first one serving for the backtracking procedure and the second one for communicating in the spanning tree. The input function ι initializes every node to the state $(unvisited, idle)$, while the output function simply returns the answers chosen by the nodes, i.e., $o\colon \text{YES} \mapsto \text{YES},\ \text{NO} \mapsto \text{NO}$. In addition to the input registers, each node has a register *self* storing its own identifier and a register *successor* to point to its successor in a (partially constructed) Hamiltonian path. That is, $R = \{self, parent, root, successor\}$. We now describe the algorithm in an informal way. It is, in principle, easy to implement in the transition maker Δ, but a thorough formalization would be rather cumbersome.

In the first round, the root marks itself as *visited* and updates its *successor* register to point towards its smallest neighbor (the one with the smallest identifier). Similarly, in each subsequent round, any *unvisited* node that is pointed to by one of its neighbors marks itself as *visited* and points towards its smallest *unvisited* neighbor. However, if all neighbors are already *visited*, the node instead sends the *backtrack* signal to its predecessor and switches back to *unvisited* (in the following round). Whenever a *visited* node receives the *backtrack* signal from its *successor*, it tries to update its *successor* to the next-smallest *unvisited* neighbor. If no such neighbor exists, it resets its *successor* pointer to itself, propagates the *backtrack* signal to its predecessor, and becomes *unvisited* in the following round.

There is only one exception to the above rules: if a node that is adjacent to the root cannot find any *unvisited* neighbor, it chooses the root as its *successor*.

This way, the constructed path becomes a cycle. In order to check whether that cycle is Hamiltonian, the root now broadcast a *request* down the spanning tree. If the *request* reaches an *unvisited* node, that node replies by sending the message *bad* towards the root. On the other hand, every *visited* leaf replies with the message *good*. While *bad* is always forwarded up to the root, *good* is only forwarded by nodes that receive this message from all of their children. If the root receives only *good*, then it knows that the current cycle is Hamiltonian and it switches to the halting state YES. The information is then broadcast through the entire graph, so that all nodes eventually accept. Otherwise, the root sends the *backtrack* signal to its predecessor, and the search for a Hamiltonian cycle continues. In case there is none (in particular, if there is not even an arbitrary cycle), the root will eventually receive the *backtrack* signal from its greatest neighbor, which indicates that all possibilities have been exhausted. If this happens, the root switches to the halting state NO, and all other nodes eventually do the same. □

4 Functional Fixpoint Logic

In order to introduce functional fixpoint logic, we first give a definition of first-order logic that suits our needs. Formulas will always be evaluated on *ordered*, undirected, connected, I-labeled graphs, where I is a fixed finite set of labels.

Throughout this paper, let $\mathcal{N}$ be an infinite supply of *node variables* and $\mathcal{F}$ be an infinite supply of *function variables*; we refer to them collectively as *variables*. The corresponding set of *terms* is generated by the grammar $t ::= x \mid f(t)$, where $x \in \mathcal{N}$ and $f \in \mathcal{F}$. With this, the set of *formulas* of *first-order logic* over I is given by the grammar

$$\varphi ::= \langle a \rangle\, t \mid s < t \mid s \leftrightarrow t \mid \neg\varphi \mid \varphi \vee \varphi \mid \exists x\, \varphi,$$

where s and t are terms, $a \in I$, and $x \in \mathcal{N}$. As usual, we may also use the additional operators $\wedge$, $\Rightarrow$, $\Leftrightarrow$, $\forall$ to make our formulas more readable, and we define the notations $s \leq t$, $s = t$, and $s \neq t$ as abbreviations for $\neg(t < s)$, $(s \leq t) \wedge (t \leq s)$, and $\neg(s = t)$, respectively.

The sets of *free variables* of a term t and a formula φ are denoted by free(t) and free(φ), respectively. While node variables can be bound by the usual quantifiers $\exists$ and $\forall$, function variables can be bound by a partial fixpoint operator that we will introduce below.

To interpret a formula φ on an I-labeled graph $(G, \mathfrak{q})$ with $G = (V, E)$, we are given a *variable assignment* σ for the variables that occur freely in φ. This is a partial function $\sigma\colon \mathcal{N} \cup \mathcal{F} \to V \cup V^V$ such that $\sigma(x) \in V$ if x is a free node variable and $\sigma(f) \in V^V$ if f is a free function variable. We call $\sigma(x)$ and $\sigma(f)$ the *interpretations* of x and f under σ, and denote them by x^σ and f^σ, respectively. For a composite term t, the corresponding interpretation t^σ under σ is defined in the obvious way.

We write $(G, \mathfrak{q}), \sigma \models \varphi$ to denote that $(G, \mathfrak{q})$ *satisfies* φ under assignment σ. If φ does not contain any free variables, we simply write $(G, \mathfrak{q}) \models \varphi$ and refer to the set P of I-labeled graphs that satisfy φ as the graph property *defined* by φ. Naturally enough, we say that two devices (i.e., automata or formulas) are *equivalent* if they specify (i.e., decide or define) the same graph property and that two classes of devices are equivalent if their members specify the same class of graph properties.

As we assume that the reader is familiar with first-order logic, we only define the semantics of the atomic formulas (whose syntax is not completely standard):

$$\begin{array}{llll} (G,\mathfrak{q}),\sigma \models \langle a \rangle\, t & \text{iff} & \mathfrak{q}(t^\sigma) = a & (\text{“}t \text{ has label } a\text{”}), \\ (G,\mathfrak{q}),\sigma \models s < t & \text{iff} & s^\sigma < t^\sigma & (\text{“}s \text{ is smaller than } t\text{”}), \\ (G,\mathfrak{q}),\sigma \models s \leftrightarrow t & \text{iff} & \{s^\sigma, t^\sigma\} \in E & (\text{“}s \text{ and } t \text{ are adjacent”}). \end{array}$$

We now turn to *functional fixpoint logic.* Syntactically, it is defined as the extension of first-order logic that allows us to write formulas of the form

$$\mathrm{pfp}\begin{bmatrix} f_1\colon\ \varphi_1(f_1,\dots,f_\ell,\textsc{in},\textsc{out}) \\ \vdots \\ f_\ell\colon\ \varphi_\ell(f_1,\dots,f_\ell,\textsc{in},\textsc{out}) \end{bmatrix} \psi, \qquad (*)$$

where $f_1,\dots,f_\ell \in \mathcal{F}$, $\textsc{in},\textsc{out} \in \mathcal{N}$, and $\varphi_1,\dots,\varphi_\ell,\psi$ are formulas. We use the notation “$\varphi_i(f_1,\dots,f_\ell,\textsc{in},\textsc{out})$” to emphasize that $f_1,\dots,f_\ell,\textsc{in},\textsc{out}$ may occur freely in φ_i (possibly among other variables). The free variables of formula $(*)$ are given by $\bigcup_{i\in(\ell]}\big[\mathrm{free}(\varphi_i) \setminus \{f_1,\dots,f_\ell,\textsc{in},\textsc{out}\}\big] \cup \big[\mathrm{free}(\psi) \setminus \{f_1,\dots,f_\ell\}\big]$.

The idea is that the *partial fixpoint operator* pfp binds the function variables $f_1,\dots,f_\ell$. The ℓ lines in square brackets constitute a system of function definitions that provide an interpretation of $f_1,\dots,f_\ell$, using the special node variables IN and OUT as helpers to represent input and output values. This is why pfp also binds any free occurrences of IN and OUT in $\varphi_1,\dots,\varphi_\ell$, but not in ψ.

To specify the semantics of $(*)$, we first need to make some preliminary observations. As before, we consider a fixed I-labeled graph $(G,\mathfrak{q})$ with $G = (V,E)$ and assume that we are given a variable assignment σ for the free variables of $(*)$. With respect to $(G,\mathfrak{q})$ and σ, each formula φ_i induces an operator $F_{\varphi_i}\colon (V^V)^\ell \to V^V$ that takes some interpretation of the function variables $f_1,\dots,f_\ell$ and outputs a new interpretation of f_i, corresponding to the function graph defined by φ_i via the node variables IN and OUT. For inputs on which φ_i does not define a functional relationship, the new interpretation of f_i behaves like the identity function. More formally, given a variable assignment $\hat\sigma$ that extends σ with interpretations of $f_1,\dots,f_\ell$, the operator F_{φ_i} maps $f_1^{\hat\sigma},\dots,f_\ell^{\hat\sigma}$ to the function f_i^{new} such that for all $u \in V$,

$$f_i^{\mathrm{new}}(u) = \begin{cases} v & \text{if } v \text{ is the unique node in } V \text{ } s.t.\ (G,\mathfrak{q}),\hat\sigma[\textsc{in},\textsc{out} \mapsto u,v] \models \varphi_i, \\ u & \text{otherwise.} \end{cases}$$

Here, $\hat{\sigma}[\text{IN}, \text{OUT} \mapsto u, v]$ is the extension of $\hat{\sigma}$ interpreting IN as u and OUT as v.

In this way, the operators $F_{\varphi_1}, \ldots, F_{\varphi_\ell}$ give rise to an infinite sequence $(f_1^k, \ldots, f_\ell^k)_{k \geq 0}$ of tuples of functions, called *stages*, where the initial stage contains solely the identity function id_V and each subsequent stage is obtained from its predecessor by componentwise application of the operators. More formally,

$$f_i^0 = \text{id}_V = \{u \mapsto u \mid u \in V\} \qquad \text{and} \qquad f_i^{k+1} = F_{\varphi_i}(f_1^k, \ldots, f_\ell^k),$$

for $i \in (\ell]$ and $k \geq 0$. Now, since we have not imposed any restrictions on the formulas φ_i, this sequence might never stabilize, i.e, it is possible that $(f_1^k, \ldots, f_\ell^k) \neq (f_1^{k+1}, \ldots, f_\ell^{k+1})$ for all $k \geq 0$. Otherwise, the sequence reaches a (simultaneous) fixpoint at some position k no greater than $|V|^{|V| \cdot \ell}$ (the number of ℓ-tuples of functions on V).

We define the *partial fixpoint* $(f_1^\infty, \ldots, f_\ell^\infty)$ of the operators $F_{\varphi_1}, \ldots, F_{\varphi_\ell}$ to be the reached fixpoint if it exists, and the tuple of identity functions otherwise. That is, for $i \in (\ell]$,

$$f_i^\infty = \begin{cases} f_i^k & \text{if there exists } k \geq 0 \text{ such that } f_j^k = f_j^{k+1} \text{ for all } j \in (\ell], \\ \text{id}_V & \text{otherwise.} \end{cases}$$

Having introduced the necessary background, we can finally provide the semantics of the formula $\text{pfp}[f_i \colon \varphi_i]_{i \in (\ell]}\, \psi$ presented in (*):

$$(G, \mathfrak{q}), \sigma \models \text{pfp}[f_i \colon \varphi_i]_{i \in (\ell]}\, \psi \qquad \text{iff} \qquad (G, \mathfrak{q}), \sigma[f_i \mapsto f_i^\infty]_{i \in (\ell]} \models \psi,$$

where $\sigma[f_i \mapsto f_i^\infty]_{i \in (\ell]}$ is the extension of σ that interprets f_i as f_i^∞, for $i \in (\ell]$. In other words, the formula $\text{pfp}[f_i \colon \varphi_i]_{i \in (\ell]}\, \psi$ can intuitively be read as

"if $f_1, \ldots, f_\ell$ are interpreted as the partial fixpoint of $\varphi_1, \ldots, \varphi_\ell$, then ψ holds".

Syntactic Sugar

Before we consider a concrete formula (in Example 8), we first introduce some "syntactic sugar" to make using functional fixpoint logic more pleasant.

Set variables. According to our definition of functional fixpoint logic, the operator pfp can bind only function variables. However, functions can be used to encode sets of nodes in a straightforward manner: any set U may be represented by a function that maps nodes outside of U to themselves and nodes inside U to nodes distinct from themselves. Therefore, we may fix an infinite supply $\mathcal{S}$ of *set variables*, and extend the syntax of first-order logic to allow atomic formulas of the form $t \in X$, where t is a term and X is a set variable in $\mathcal{S}$. Naturally, the semantics is that "t is an element of X". To bind set variables, we can then write partial fixpoint formulas of the form $\text{pfp}\big[(f_i \colon \varphi_i)_{i \in (\ell]}, (X_i \colon \vartheta_i)_{i \in (m]}\big]\, \psi$, where $f_1, \ldots, f_\ell \in \mathcal{F}$, $X_1, \ldots, X_m \in \mathcal{S}$, and $\varphi_1, \ldots, \varphi_\ell, \vartheta_1, \ldots, \vartheta_m, \psi$ are formulas. The stages of the partial fixpoint induction are computed as before, but each set variable X_i is initialized to $\emptyset$, and falls back to $\emptyset$ in case the sequence of stages does not converge to a fixpoint.

Quantifiers over functions and sets. Partial fixpoint inductions allow us to iterate over various interpretations of function and set variables and thus provide a way of expressing (second-order) quantification over functions and sets. Since we restrict ourselves to graphs whose nodes are totally ordered, we can easily define a suitable order of iteration and a corresponding partial fixpoint induction that traverses all possible interpretations of a given function or set variable. To make this more convenient, we enrich the language of functional fixpoint logic with second-order quantifiers, allowing us to write formulas of the form $\exists f\, \varphi$ and $\exists X\, \varphi$, where $f \in \mathcal{F}$, $X \in \mathcal{S}$, and φ is a formula. Obviously, the semantics is that "there exists a function f, or a set X, respectively, such that φ holds".

As a consequence, it is possible to express any graph property definable in *monadic second-order logic*, the extension of first-order logic with set quantifiers.

Corollary 7. *When restricted to finite graphs equipped with a total order, functional fixpoint logic is strictly more expressive than monadic second-order logic.*

The strictness of the inclusion in the above corollary follows from the fact that even on totally ordered graphs, Hamiltonicity cannot be defined in monadic second-order logic (see, e.g., the proof in [4, Prp. 5.13]). As the following example shows, this property is easy to express in functional fixpoint logic.

Example 8 (Hamiltonian cycle). The following formula of functional fixpoint logic defines the graph property of Hamiltonicity. That is, an unlabeled graph G satisfies this formula if and only if there exists a cycle in G that goes through each node exactly once.

$$\exists f \begin{bmatrix} \forall x\big(f(x) \leftrightarrow x\big) \;\wedge\; \forall x\, \exists y\big[f(y) = x \;\wedge\; \forall z\big(f(z) = x \;\Rightarrow\; z = y\big)\big] \;\wedge \\ \forall X\Big(\big[\exists x(x \in X) \wedge \forall y(y \in X \Rightarrow f(y) \in X)\big] \;\Rightarrow\; \forall y(y \in X)\Big) \end{bmatrix}$$

Here, $x, y, z \in \mathcal{N}$, $X \in \mathcal{S}$, and $f \in \mathcal{F}$. Intuitively, we represent a given Hamiltonian cycle by a function f that tells us for each node x, which of x's neighbors we should visit next in order to traverse the entire cycle. Thus, f actually represents a directed version of the cycle.

To ensure the existence of a Hamiltonian cycle, our formula states that there is a function f satisfying the following two conditions. By the first line, each node x must have exactly one f-predecessor and one f-successor, both of which must be neighbors of x. By the second line, if we start at any node x and collect into a set X all the nodes reachable from x (by following the path specified by f), then X must contain all nodes. □

5 Translating Between Automata and Logic

Having introduced both automata and logic, we can proceed to explain the first part of Theorem 1 (stated in Sect. 1), i.e., how distributed register automata can be translated into functional fixpoint logic.

Proposition 9. *For every distributed register automaton that decides a graph property, we can construct an equivalent formula of functional fixpoint logic.*

Proof (sketch). Given a distributed register automaton $A = (Q, R, \iota, \Delta, H, o)$ deciding a graph property P over label set I, we can construct a formula φ_A of functional fixpoint logic that defines P. For each state $q \in Q$, our formula uses a set variable X_q to represent the set of nodes of the input graph that are in state q. Also, for each register $r \in R$, it uses a function variable f_r to represent the function that maps each node u to the node v whose identifier is stored in u's register r. By means of a partial fixpoint operator, we enforce that on any I-labeled graph $(G, \mathfrak{q})$, the final interpretations of $(X_q)_{q \in Q}$ and $(f_r)_{r \in R}$ represent the halting configuration reached by A on $(G, \mathfrak{q})$. The main formula is simply

$$\varphi_A := \mathrm{pfp}\begin{bmatrix}(X_q\colon \varphi_q)_{q\in Q}\\ (f_r\colon \varphi_r)_{r\in R}\end{bmatrix} \forall x \Big(\bigvee_{p \in H:\, o(p)=\textsc{yes}} x \in X_p \Big),$$

which states that all nodes end up in a halting state that outputs YES.

Basically, the subformulas $(\varphi_q)_{q \in Q}$ and $(\varphi_r)_{r \in R}$ can be constructed in such a way that for all $i \in \mathbb{N}$, the $(i+1)$-th stage of the partial fixpoint induction represents the configuration reached by A in the i-th round. To achieve this, each of the subformulas contains a nested partial fixpoint formula describing the result computed by the transition maker Δ between two consecutive synchronous rounds, using additional set and function variables to encode the inner configurations of Δ at each node. Thus, each stage of the nested partial fixpoint induction corresponds to a single step in the transition maker's sequential scanning process. □

Let us now consider the opposite direction and sketch how to go from functional fixpoint logic to distributed register automata.

Proposition 10. *For every formula of functional fixpoint logic that defines a graph property, we can construct an equivalent distributed register automaton.*

Proof (sketch). We proceed by structural induction: each subformula φ will be evaluated by a dedicated automaton A_φ, and several such automata can then be combined to build an automaton for a composite formula. For this purpose, it is convenient to design *centralized* automata, which operate on a givenspanning tree (as computed in Example 5) and are coordinated by the root in a fairly sequential manner. In A_φ, each free node variable x of φ is represented by a corresponding input register x whose value at the root is the current interpretation x^σ of x. Similarly, to represent a function variable f, every node v has a register f storing $f^\sigma(v)$. The nodes also possess some auxiliary registers whose purpose will be explained below. In the end, for any formula φ (potentially with free variables), we will have an automaton A_φ computing a transduction $T_{A_\varphi}\colon \mathbb{C}(I, \{\mathit{parent}, \mathit{root}\} \cup \mathrm{free}(\varphi)) \to \mathbb{C}(\{\textsc{yes}, \textsc{no}\}, \emptyset)$, where *parent* and *root* are supposed to constitute a spanning tree. The computation is triggered by the root, which means that the other nodes are waiting for a signal to wake up.

Algorithm 2. A_φ for $\varphi = \text{pfp}[f_i : \varphi_i]_{i \in [1:\ell]}\, \psi$, as controlled by the root

```
1 init(A_inc)
2 repeat
3     @every node do for i ∈ [1:ℓ] do f_i ← f_i^new
4     for i ∈ [1:ℓ] do update(f_i^new)
5     if @every node (∀i ∈ [1:ℓ] : f_i^new = f_i) then goto 8
6 until execute(A_inc) returns NO      /* until global counter at maximum */
7 @every node do for i ∈ [1:ℓ] do f_i ← self
8 execute(A_ψ)
```

Essentially, the nodes involved in the evaluation of φ collect some information, send it towards the root, and go back to sleep. The root then returns YES or NO, depending on whether or not φ holds in the input graph under the variable assignment provided by the input registers. Centralizing A_φ in that way makes it very convenient (albeit not efficient) to evaluate composite formulas. For example, in $A_{\varphi \vee \psi}$, the root will first run A_φ, and then A_ψ in case A_φ returns NO.

The evaluation of atomic formulas is straightforward. So let us focus on the most interesting case, namely when $\varphi = \text{pfp}[f_i : \varphi_i]_{i \in (\ell]}\, \psi$. The root's program is outlined in Algorithm 2. Line 1 initializes a counter that ranges from 0 to $n^{\ell n} - 1$, where n is the number of nodes in the input graph. This counter is distributed in the sense that every node has some dedicated registers that together store the current counter value. Every execution of A_{inc} will increment the counter by 1, or return NO if its maximum value has been exceeded. Now, in each iteration of the loop starting at Line 2, all registers f_i and f_i^{new} are updated in such a way that they represent the current and next stage, respectively, of the partial fixpoint induction. For the former, it suffices that every node copies, for all i, the contents of f_i^{new} to f_i (Line 3). To update f_i^{new}, Line 4 calls a subroutine *update*(f_i^{new}) whose effect is that $f_i^{\text{new}} = F_{\varphi_i}((f_i)_{i \in (\ell]})$ for all i, where $F_{\varphi_i} : (V^V)^\ell \to V^V$ is the operator defined in Sect. 4. Line 5 checks whether we have reached a fixpoint: The root asks every node to compare, for all i, its registers f_i^{new} and f_i. The corresponding truth value is propagated back to the root, where *false* is given preference over *true*. If the result is *true*, we exit the loop and proceed with calling A_ψ to evaluate ψ (Line 8). Otherwise, we try to increment the global counter by executing A_{inc} (Line 6). If the latter returns NO, the fixpoint computation is aborted because we know that it has reached a cycle. In accordance with the partial fixpoint semantics, all nodes then write their own identifier to every register f_i (Line 7) before ψ is evaluated (Line 8). □

6 Conclusion

This paper makes some progress in the development of a descriptive distributed complexity theory by establishing a logical characterization of a wide class of network algorithms, modeled as distributed register automata.

In our translation from logic to automata, we did not pay much attention to algorithmic efficiency. In particular, we made extensive use of centralized subroutines that are triggered and controlled by a leader process. A natural question for future research is to identify cases where we can understand a distributed architecture as an opportunity that allows us to evaluate formulas faster. In other words, is there an expressive fragment of functional fixpoint logic that gives rise to efficient distributed algorithms in terms of running time? What about the required number of messages? We are then entering the field of automatic *synthesis of practical distributed algorithms* from logical specifications. This is a worthwhile task, as it is often much easier to declare what should be done than how it should be done (cf. Examples 6 and 8).

As far as the authors are aware, this area is still relatively unexplored. However, one noteworthy advance was made by Grumbach and Wu in [9], where they investigated distributed evaluation of first-order formulas on bounded-degree graphs and planar graphs. We hope to follow up on this in future work.

Acknowledgments. We thank Matthias Függer for helpful discussions. Work supported by ERC *EQualIS* (FP7-308087) (http://www.lsv.fr/~bouyer/equalis) and ANR *FREDDA* (17-CE40-0013) (https://www.irif.fr/anr/fredda/index).

References

1. Abiteboul, S., Vianu, V.: Fixpoint extensions of first-order logic and datalog-like languages. In: Proceedings of the Fourth Annual Symposium on Logic in Computer Science (LICS 1989), Pacific Grove, California, USA, 5–8 June 1989, pp. 71–79. IEEE Computer Society (1989). https://doi.org/10.1109/LICS.1989.39160
2. Aiswarya, C., Bollig, B., Gastin, P.: An automata-theoretic approach to the verification of distributed algorithms. In: Aceto, L., de Frutos-Escrig, D. (eds.) 26th International Conference on Concurrency Theory, CONCUR 2015, Madrid, Spain, 14 September 2015. LIPIcs, vol. 42, pp. 340–353. Schloss Dagstuhl - Leibniz-Zentrum fuer Informatik (2015). https://doi.org/10.4230/LIPIcs.CONCUR.2015.340
3. Aiswarya, C., Bollig, B., Gastin, P.: An automata-theoretic approach to the verification of distributed algorithms. Inf. Comput. **259**(Part 3), 305–327 (2018). https://doi.org/10.1016/j.ic.2017.05.006
4. Courcelle, B., Engelfriet, J.: Graph Structure and Monadic Second-Order Logic: A Language-Theoretic Approach. Encyclopedia of Mathematics and Its Applications, vol. 138. Cambridge University Press, Cambridge (2012). https://hal.archives-ouvertes.fr/hal-00646514. https://doi.org/10.1017/CBO9780511977619
5. Diestel, R.: Graph Theory. GTM, vol. 173. Springer, Heidelberg (2017). https://doi.org/10.1007/978-3-662-53622-3
6. Fagin, R.: Generalized first-order spectra and polynomial-time recognizable sets. In: Karp, R.M. (ed.) Complexity of Computation. SIAM-AMS Proceedings, vol. 7, pp. 43–73 (1974). http://www.almaden.ibm.com/cs/people/fagin/genspec.pdf
7. Feuilloley, L., Fraigniaud, P.: Survey of distributed decision. Bull. EATCS **119** (2016). http://eatcs.org/beatcs/index.php/beatcs/article/view/411
8. Grädel, E., et al.: Finite Model Theory and Its Applications. Texts in Theoretical Computer Science. An EATCS Series, 1st edn. Springer, Heidelberg (2007). https://doi.org/10.1007/3-540-68804-8

9. Grumbach, S., Wu, Z.: Logical locality entails frugal distributed computation over graphs (extended abstract). In: Paul, C., Habib, M. (eds.) WG 2009. LNCS, vol. 5911, pp. 154–165. Springer, Heidelberg (2010). https://doi.org/10.1007/978-3-642-11409-0_14
10. Hella, L., et al.: Weak models of distributed computing, with connections to modal logic. In: Kowalski, D., Panconesi, A. (eds.) ACM Symposium on Principles of Distributed Computing, PODC 2012, Funchal, Madeira, Portugal, 16–18 July 2012, pp. 185–194. ACM (2012). https://doi.org/10.1145/2332432.2332466
11. Hella, L., et al.: Weak models of distributed computing, with connections to modallogic. Distrib. Comput. **28**(1), 31–53 (2015). https://arxiv.org/abs/1205.2051. http://dx.doi.org/10.1007/s00446-013-0202-3
12. Immerman, N.: Descriptive Complexity. Texts in Computer Science. Springer, New York (1999). https://doi.org/10.1007/978-1-4612-0539-5
13. Kaminski, M., Francez, N.: Finite-memory automata. Theor. Comput. Sci. **134**(2), 329–363 (1994). https://doi.org/10.1016/0304-3975(94)90242-9
14. Kuusisto, A.: Modal logic and distributed message passing automata. In: Rocca, S.R.D. (eds.) Computer Science Logic 2013 (CSL 2013), Torino, Italy, 2–5 September 2013, LIPIcs, vol. 23, pp. 452–468. Schloss Dagstuhl - Leibniz-Zentrum fuer Informatik (2013). https://doi.org/10.4230/LIPIcs.CSL.2013.452
15. Libkin, L., et al.: Elements of Finite Model Theory. Texts in Theoretical Computer Science. An EATCS Series, 1st edn. Springer, Heidelberg (2004). https://doi.org/10.1007/978-3-662-07003-1
16. Naor, M., Stockmeyer, L.J.: What can be computed locally? SIAM J. Comput. **24**(6), 1259–1277 (1995). https://doi.org/10.1137/S0097539793254571
17. Reiter, F.: Distributed graph automata. In: 30th Annual ACM/IEEE Symposium on Logic in Computer Science, LICS 2015, Kyoto, Japan, 6–10 July 2015, pp. 192–201. IEEE Computer Society (2015). https://arxiv.org/abs/1408.3030. https://doi.org/10.1109/LICS.2015.27
18. Reiter, F.: Asynchronous distributed automata: a characterization of the modal MU-fragment. In: Chatzigiannakis, I., Indyk, P., Kuhn, F., Muscholl, A. (eds.) 44th International Colloquium on Automata, Languages, and Programming, ICALP 2017, Warsaw, Poland, 10–14 July 2017. LIPIcs, vol. 80, pp. 100:1–100:14. Schloss Dagstuhl - Leibniz-Zentrum fuer Informatik (2017). http://arxiv.org/abs/1611.08554. https://doi.org/10.4230/LIPIcs.ICALP.2017.100
19. Vardi, M.Y.: The complexity of relational query languages (extended abstract). In: Lewis, H.R., Simons, B.B., Burkhard, W.A., Landweber, L.H. (eds.) Proceedings of the 14th Annual ACM Symposium on Theory of Computing, San Francisco, California, USA, 5–7 May 1982, pp. 137–146. ACM (1982). https://doi.org/10.1145/800070.802186

The Impatient May Use Limited Optimism to Minimize Regret

Michaël Cadilhac[1], Guillermo A. Pérez[2(✉)], and Marie van den Bogaard[3]

[1] University of Oxford, Oxford, UK
michael@cadilhac.name
[2] University of Antwerp, Antwerp, Belgium
guillermoalberto.perez@uantwerpen.be
[3] Université libre de Bruxelles, Brussels, Belgium
marie.van.den.bogaard@ulb.ac.be

Abstract. Discounted-sum games provide a formal model for the study of reinforcement learning, where the agent is enticed to get rewards early since later rewards are discounted. When the agent interacts with the environment, she may realize that, with hindsight, she could have increased her reward by playing differently: this difference in outcomes constitutes her *regret value*. The agent may thus elect to follow a *regret-minimal* strategy. In this paper, it is shown that (1) there always exist regret-minimal strategies that are admissible—a strategy being inadmissible if there is another strategy that always performs better; (2) computing the minimum possible regret or checking that a strategy is regret-minimal can be done in $\mathsf{coNP}^{\mathsf{NP}}$, disregarding the computational cost of numerical analysis (otherwise, this bound becomes PSpace).

Keywords: Admissibility · Discounted-sum games · Regret minimization

1 Introduction

A pervasive model used to study the strategies of an agent in an unknown environment is *two-player infinite horizon games played on finite weighted graphs.* Therein, the set of vertices of a graph is split between two players, Adam and Eve, playing the roles of the environment and the agent, respectively. The play starts in a given vertex, and each player decides where to go next when the play reaches one of their vertices. Questions asked about these games are usually of the form: *Does there exist a strategy of Eve such that... ?* For such a question to be well-formed, one should provide:

1. A valuation function: given an infinite play, what is Eve's reward?
2. Assumptions about the environment: is Adam trying to help or hinder Eve?

The valuation function can be Boolean, in which case one says that Eve *wins* or *loses* (one very classical example has Eve winning if the maximum value

M. Bojańczyk and A. Simpson (Eds.): FOSSACS 2019, LNCS 11425, pp. 133–149, 2019.
https://doi.org/10.1007/978-3-030-17127-8_8

appearing infinitely often along the edges is even). In this setting, it is often assumed that Adam is adversarial, and the question then becomes: *Can Eve always win?* (The names of the players stem from this view: *is there* a strategy of ∃ve that *always* beats ∀dam?) The literature on that subject spans more than 35 years, with newly found applications to this day (see [4] for comprehensive lecture notes, and [7] for an example of recent use in the analysis of attacks in cryptocurrencies).

The valuation function can also aggregate the numerical values along the edges into a reward value. We focus in this paper on *discounted sum*: if w is the weight of the edge taken at the n-th step, Eve's reward grows by $\lambda^n \cdot w$, where $\lambda \in (0, 1)$ is a prescribed discount factor. Discounting future rewards is a classical notion used in economics [18], Markov decision processes [9,16], systems theory [1], and is at the heart of Q-learning, a reinforcement learning technique widely used in machine learning [19]. In this setting, we consider three attitudes towards the environment:

- The adversarial environment hypothesis translates to Adam trying to minimize Eve's reward, and the question becomes: *Can Eve always achieve a reward of x?* This problem is in NP ∩ coNP [20] and showing a P upper-bound would constitute a major breakthrough (namely, it would imply the same for so-called parity games [15]). A strategy of Eve that maximizes her rewards against an adversarial environment is called *worst-case optimal.* Conversely, a strategy that maximizes her rewards assuming a *collaborative* environment is called *best-case optimal.*
- Assuming that the environment is adversarial is drastic, if not pessimistic. Eve could rather be interested in settling for a strategy σ which is not *consistently* bad: if another strategy σ' gives a better reward in one environment, there should be another environment for which σ is better than σ'. Such strategies, called *admissible* [5], can be seen as an *a priori* rational choice.
- Finally, Eve could put no assumption on the environment, but regret not having done so. Formally, the *regret value* of Eve's strategy is defined as the maximal difference, for all environments, between the best value Eve *could* have obtained and the value she actually obtained. Eve can thus be interested in following a strategy that achieves the minimal regret value, aptly called a *regret-minimal* strategy [10]. This constitutes an *a posteriori* rational choice [12]. Regret-minimal strategies were explored in several contexts, with applications including competitive online algorithm synthesis [3,11] and robot-motion planning [13,14].

In this paper, we single out a class of strategies for Eve that first follow a best-case optimal strategy, then switch to a worst-case optimal strategy after some precise time; we call these strategies *optipess.* Our main contributions are then:

1. Optipess strategies are not only regret-minimal (a fact established in [13]) but also admissible—note that there are regret-minimal strategies that are not admissible and *vice versa.* On the way, we show that for any strategy of

Eve there is an admissible strategy that performs at least as well; this is a peculiarity of discounted-sum games.
2. The regret value of a given time-switching strategy can be computed with an NP algorithm (disregarding the cost of numerical analysis). The main technical hurdle is showing that exponentially long paths can be represented succinctly, a result of independent interest.
3. The question *Can Eve's regret be bounded by x?* is decidable in $\mathsf{NP}^{\mathsf{coNP}}$ (again disregarding the cost of numerical analysis, PSpace otherwise), improving on the implicit NExp algorithm of [13]. The algorithm consists in guessing a time-switching strategy and computing its regret value; since optipess strategies are time-switching strategies that are regret-minimal, the algorithm will eventually find the minimal regret value of the input game.

Structure of the Paper. Notations and definitions are introduced in Sect. 2. The study of admissibility appears in Sect. 3, and is independent from the complexity analysis of regret. The main algorithm devised in this paper (point 2 above) is presented in Theorem 5, Sect. 6; it relies on technical lemmas that are the focus of Sects. 4 and 5. We encourage the reader to go through the statements of the lemma sections, then through the proof of Theorem 5, to get a good sense of the role each lemma plays.

In more details, in Sect. 4 we provide a crucial lemma that allows to represent long paths succinctly, and in Sect. 5, we argue that the important values of a game (regret, best-case, worst-case) have short witnesses. In Sect. 6, we use these lemmas to devise our algorithms.

2 Preliminaries

We assume familiarity with basic graph and complexity theory. Some more specific definitions and known results are recalled here.

Game, Play, History. A *(discounted-sum) game* $\mathcal{G}$ is a tuple $(V, v_0, V_\exists, E, w, \lambda)$ where V is a finite set of vertices, v_0 is the starting vertex, $V_\exists \subseteq V$ is the subset of vertices that belong to Eve, $E \subseteq V \times V$ is a set of directed edges, $w\colon E \to \mathbb{Z}$ is an (edge-)weight function, and $0 < \lambda < 1$ is a rational *discount factor*. The vertices in $V \setminus V_\exists$ are said to belong to Adam. Since we consider games played for an infinite number of turns, we will always assume that every vertex has at least one outgoing edge.

A *play* is an infinite path $v_1 v_2 \cdots \in V^\omega$ in the digraph (V, E). A *history* $h = v_1 \cdots v_n$ is a finite path. The *length of h*, written $|h|$, is the number of *edges* it contains: $|h| \stackrel{\text{def}}{=} n - 1$. The set **Hist** consists of all histories that start in v_0 and end in a vertex from $V_\exists$.

Strategies. A *strategy of Eve* in $\mathcal{G}$ is a function σ that maps histories ending in some vertex $v \in V_\exists$ to a neighbouring vertex v' (i.e., $(v, v') \in E$). The strategy

σ is *positional* if for all histories h, h' ending in the same vertex, $\sigma(h) = \sigma(h')$. *Strategies of Adam* are defined similarly.

A history $h = v_1 \cdots v_n$ is said to be *consistent with a strategy* σ of Eve if for all $i \geq 2$ such that $v_i \in V_\exists$, we have that $\sigma(v_1 \cdots v_{i-1}) = v_i$. Consistency with strategies of Adam is defined similarly. We write $\mathbf{Hist}(\sigma)$ for the set of histories in **Hist** that are consistent with σ. A play is consistent with a strategy (of either player) if all its prefixes are consistent with it.

Given a vertex v and both Adam and Eve's strategies, τ and σ respectively, there is a unique play starting in v that is consistent with both, called the *outcome* of τ and σ on v. This play is denoted $\mathbf{out}^v(\sigma, \tau)$.

For a strategy σ of Eve and a history $h \in \mathbf{Hist}(\sigma)$, we let σ_h be the strategy of Eve that assumes h has already been played. Formally, $\sigma_h(h') = \sigma(h \cdot h')$ for any history h' (we will use this notation only on histories h' that start with the ending vertex of h).

Values. The *value of a history* $h = v_1 \cdots v_n$ is the discounted sum of the weights on the edges:

$$\mathbf{Val}(h) \stackrel{\text{def}}{=} \sum_{i=0}^{|h|-1} \lambda^i w(v_i, v_{i+1}) \ .$$

The *value of a play* is simply the limit of the values of its prefixes.

The *antagonistic value* of a strategy σ of Eve with history $h = v_1 \cdots v_n$ is the value Eve achieves when Adam tries to hinder her, after h:

$$\mathbf{aVal}^h(\sigma) \stackrel{\text{def}}{=} \mathbf{Val}(h) + \lambda^{|h|} \cdot \inf_{\tau} \mathbf{Val}(\mathbf{out}^{v_n}(\sigma_h, \tau)) \ ,$$

where τ ranges over all strategies of Adam. The *collaborative value* $\mathbf{cVal}^h(\sigma)$ is defined in a similar way, by substituting "sup" for "inf." We write $\mathbf{aVal}^h$ (resp. $\mathbf{cVal}^h$) for the best antagonistic (resp. collaborative) value achievable by Eve with any strategy.

Types of Strategies. A strategy σ of Eve is *strongly worst-case optimal* (SWO) if for every history h we have $\mathbf{aVal}^h(\sigma) = \mathbf{aVal}^h$; it is *strongly best-case optimal* (SBO) if for every history h we have $\mathbf{cVal}^h(\sigma) = \mathbf{cVal}^h$.

We single out a class of SWO strategies that perform well if Adam turns out to be helping. A SWO strategy σ of Eve is *strongly best worst-case optimal* (SBWO) if for every history h we have $\mathbf{cVal}^h(\sigma) = \mathbf{acVal}^h$, where:

$$\mathbf{acVal}^h \stackrel{\text{def}}{=} \sup\{\mathbf{cVal}^h(\sigma') \mid \sigma' \text{ is a SWO strategy of Eve}\} \ .$$

In the context of discounted-sum games, strategies that are positional and strongly optimal always exist. Furthermore, the set of all such strategies can be characterized by local conditions.

Lemma 1 (Follows from [20, Theorem 5.1]). *There exist positional SWO, SBO, and SBWO strategies in every game. For any positional strategy σ of Eve:*

- $(\forall v \in V)\,[\mathbf{aVal}^v(\sigma) = \mathbf{aVal}^v]$ *iff σ is SWO;*
- $(\forall v \in V)\,[\mathbf{cVal}^v(\sigma) = \mathbf{cVal}^v]$ *iff σ is SBO;*
- $(\forall v \in V)\,[\mathbf{aVal}^v(\sigma) = \mathbf{aVal}^v \wedge \mathbf{cVal}^v(\sigma) = \mathbf{acVal}^v]$ *iff σ is SBWO.*

Regret. The *regret* of a strategy σ of Eve is the maximal difference between the value obtained by using σ and the value obtained by using an alternative strategy:

$$\mathbf{Reg}\,(\sigma) \stackrel{\text{def}}{=} \sup_{\tau} \left(\left(\sup_{\sigma'} \mathbf{Val}(\mathbf{out}^{v_0}(\sigma', \tau)) \right) - \mathbf{Val}(\mathbf{out}^{v_0}(\sigma, \tau)) \right) ,$$

where τ and σ' range over all strategies of Adam and Eve, respectively. The *(minimal) regret of $\mathcal{G}$* is then $\mathbf{Reg} \stackrel{\text{def}}{=} \inf_\sigma \mathbf{Reg}\,(\sigma)$.

Regret can also be characterized by considering the point in history when Eve should have done things differently. Formally, for any vertices u and v let $\mathbf{cVal}^u_{\neg v}$ be the maximal $\mathbf{cVal}^u(\sigma)$ for strategies σ verifying $\sigma(u) \neq v$. Then:

Lemma 2 ([13, Lemma 13]). *For all strategies σ of Eve:*

$$\mathbf{Reg}\,(\sigma) = \sup \left\{ \lambda^n \left(\mathbf{cVal}^{v_n}_{\neg\sigma(h)} - \mathbf{aVal}^{v_n}(\sigma_h) \right) \,\middle|\, h = v_0 \cdots v_n \in \mathbf{Hist}(\sigma) \right\} .$$

Switching and Optipess Strategies. Given strategies σ_1, σ_2 of Eve and a *threshold function* $t\colon V_\exists \to \mathbb{N} \cup \{\infty\}$, we define the *switching strategy* $\sigma_1 \xrightarrow{t} \sigma_2$ for any history $h = v_1 \cdots v_n$ ending in $V_\exists$ as:

$$\sigma_1 \xrightarrow{t} \sigma_2(h) = \begin{cases} \sigma_2(h) & \text{if } (\exists i)[i \geq t(v_i)], \\ \sigma_1(h) & \text{otherwise.} \end{cases}$$

We refer to histories for which the first condition above holds as *switched histories*, to all others as *unswitched histories.* The strategy $\sigma = \sigma_1 \xrightarrow{t} \sigma_2$ is said to be *bipositional* if both σ_1 and σ_2 are positional. Note that in that case, for all histories h, if h is switched then $\sigma_h = \sigma_2$, and otherwise σ_h is the same as σ but with $t(v)$ changed to $\max\{0, t(v) - |h|\}$ for all $v \in V_\exists$. In particular, if $|h|$ is greater than $\max\{t(v) < \infty\}$, then σ_h is nearly positional: it switches to σ_2 as soon as it sees a vertex with $t(v) \neq \infty$.

A strategy σ is *perfectly optimistic-then-pessimistic* (optipess, for short) if there are positional SBO and SBWO strategies σ^{sbo} and σ^{sbwo} such that $\sigma = \sigma^{\text{sbo}} \xrightarrow{t} \sigma^{\text{sbwo}}$ where $t(v) = \inf \left\{ i \in \mathbb{N} \,\middle|\, \lambda^i \left(\mathbf{cVal}^v - \mathbf{aVal}^v \right) \leq \mathbf{Reg} \right\}$.

Theorem 1 (**[13]**)**.** *For all optipess strategies σ of Eve,* $\mathbf{Reg}\,(\sigma) = \mathbf{Reg}$.

Conventions. *As we have done so far, we will assume throughout the paper that a game $\mathcal{G}$ is fixed—with the notable exception of the results on complexity, in which we assume that the game is given with all numbers in binary. Regarding strategies, we assume that bipositional strategies are given as two positional strategies and a threshold function encoded as a table with binary-encoded entries.*

⋆

⋆ ⋆

Example 1. Consider the following game, where round vertices are owned by Eve, and square ones by Adam. The double edges represent Eve's positional strategy σ:

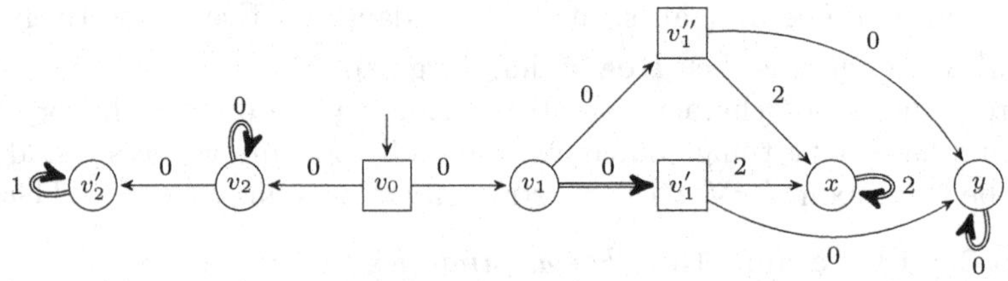

Eve's strategy has a regret value of $2\lambda^2/(1-\lambda)$. This is realized when Adam plays from v_0 to v_1, from v_1'' to x, and from v_1' to y. Against that strategy, Eve ensures a discounted-sum value of 0 by playing according to σ while regretting not having played to v_1'' to obtain $2\lambda^2/(1-\lambda)$. ∎

3 Admissible Strategies and Regret

There is no reason for Eve to choose a strategy that is consistently worse than another one. This classical idea is formalized using the notions of *strategy domination* and *admissible strategies.* In this section, which is independent from the rest of the paper, we study the relation between admissible and regret-minimal strategies. Let us start by formally introducing the relevant notions:

Definition 1. *Let σ_1, σ_2 be two strategies of Eve. We say that σ_1 is* weakly dominated *by σ_2 if* $\mathbf{Val}(\mathbf{out}^{v_0}(\sigma_1, \tau)) \leq \mathbf{Val}(\mathbf{out}^{v_0}(\sigma_2, \tau))$ *for every strategy τ of Adam. We say that σ_1 is* dominated *by σ_2 if σ_1 is* weakly dominated *by σ_2 but not conversely. A strategy σ of Eve is* admissible *if it is not dominated by any other strategy.*

In other words, admissible strategies are maximal elements for the weak-domination pre-order.

Example 2. Consider the following game, where the strategy σ of Eve is shown by the double edges:

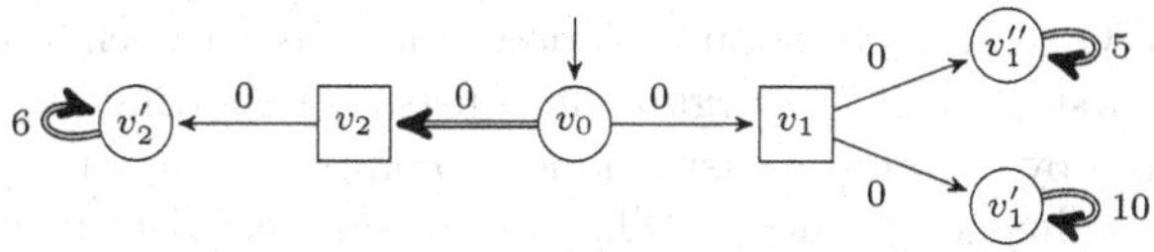

This strategy guarantees a discounted-sum value of $6\lambda^2(1-\lambda)$ against any strategy of Adam. Furthermore, it is worst-case optimal since playing to v_1 instead of v_2 would allow Adam the opportunity to ensure a strictly smaller value by playing to v_1''. The latter also implies that σ is admissible. Interestingly, playing to v_1 is also an admissible behavior of Eve since, against a strategy of Adam that plays from v_1 to v_1', it obtains $10\lambda^2(1-\lambda) > 6\lambda^2(1-\lambda)$. ■

The two examples above can be used to argue that the sets of strategies that are regret minimal and admissible, respectively, are in fact incomparable.

Proposition 1. *There are regret-optimal strategies that are not admissible and admissible strategies that have suboptimal regret.*

Proof (Sketch). Consider once more the game depicted in Example 1 and recall that the strategy σ of Eve corresponding to the double edges has minimal regret. This strategy is *not* admissible: it is dominated by the alternative strategy σ' of Eve that behaves like σ from v_1 but plays to v_2' from v_2. Indeed, if Adam plays to v_1 from v_0 then the outcomes of σ and σ' are the same. However, if Adam plays to v_2 then the value of the outcome of σ is 0 while the value of the outcome of σ' is strictly greater than 0.

Similarly, the strategy σ depicted by double edges in the game from Example 2 is admissible but *not* regret-minimizing. In fact, her strategy σ' that consists in playing v_1 from v_0 has a smaller regret. □

In the rest of this section, we show that (1) any strategy is weakly dominated by an admissible strategy; (2) being dominated entails more regret; (3) optipess strategies are both regret-minimal and admissible. We will need the following:

Lemma 3 ([6]). *A strategy σ of Eve is admissible if and only if for every history $h \in$ **Hist**(σ) the following holds: either $\mathbf{cVal}^h(\sigma) > \mathbf{aVal}^h$ or $\mathbf{aVal}^h(\sigma) = \mathbf{cVal}^h(\sigma) = \mathbf{aVal}^h = \mathbf{acVal}^h$.*

The above characterization of admissible strategies in so-called *well-formed games* was proved in [6, Theorem 11]. Lemma 3 follows from the fact that discounted-sum games are well-formed.

3.1 Any Strategy Is Weakly Dominated by an Admissible Strategy

We show that discounted-sum games have the distinctive property that every strategy is weakly dominated by an admissible strategy. This is in stark contrast with most cases where admissibility has been studied previously [6].

Theorem 2. *Any strategy of Eve is weakly dominated by an admissible strategy.*

Proof (Sketch). The main idea is to construct, based on σ, a strategy σ' that will switch to a SBWO strategy as soon as σ does not satisfy the characterization of Lemma 3. The first part of the argument consists in showing that σ is indeed weakly dominated by σ'. This is easily done by comparing, against each strategy τ of Adam, the values of σ and σ'. The second part consists in verifying that σ' is indeed admissible. This is done by checking that each history h consistent with σ' satisfies the characterization of Lemma 3, that is $\mathbf{cVal}^h(\sigma') > \mathbf{aVal}^h$ or $\mathbf{aVal}^h(\sigma') = \mathbf{cVal}^h(\sigma') = \mathbf{aVal}^h = \mathbf{acVal}^h$. □

3.2 Being Dominated Is Regretful

Theorem 3. *For all strategies σ, σ' of Eve such that σ is weakly dominated by σ', it holds that* $\mathbf{Reg}(\sigma') \leq \mathbf{Reg}(\sigma)$.

Proof. Let σ, σ' be such that σ is weakly dominated by σ'. This means that for every strategy τ of Adam, we have that $\mathbf{Val}(\pi) \leq \mathbf{Val}(\pi')$ where $\pi = \mathbf{out}^{v_0}(\sigma, \tau)$ and $\pi' = \mathbf{out}^{v_0}(\sigma', \tau)$. Consequently: we obtain

$$\left(\sup_{\sigma''} \mathbf{Val}(\mathbf{out}^{v_0}(\sigma'', \tau))\right) - \mathbf{Val}(\pi') \leq \left(\sup_{\sigma''} \mathbf{Val}(\mathbf{out}^{v_0}(\sigma'', \tau))\right) - \mathbf{Val}(\pi) .$$

As this holds for any τ, we can conclude that $\sup_\tau \sup_{\sigma''}(\mathbf{Val}(\mathbf{out}^{v_0}(\sigma'', \tau)) - \mathbf{Val}(\mathbf{out}^{v_0}(\sigma', \tau))) \leq \sup_\tau \sup_{\sigma''}(\mathbf{Val}(\mathbf{out}^{v_0}(\sigma'', \tau)) - \mathbf{Val}(\mathbf{out}^{v_0}(\sigma, \tau)))$, that is $\mathbf{Reg}(\sigma') \leq \mathbf{Reg}(\sigma)$. □

It follows from Proposition 1, however, that the converse of the theorem is false.

3.3 Optipess Strategies Are both Regret-Minimal and Admissible

Recall that there are admissible strategies that are not regret-minimal and *vice versa* (Proposition 1). However, as a direct consequence of Theorems 2 and 3, there always exist regret-minimal admissible strategies. It turns out that optipess strategies, which are regret-minimal (Theorem 1), are also admissible:

Theorem 4. *All optipess strategies of Eve are admissible.*

Proof. Let $\sigma = \sigma^{\text{sbo}} \xrightarrow{t} \sigma^{\text{sbwo}}$ be an optipess strategy; we show it is admissible. To this end, let $h = v_0 \ldots v_n \in \mathbf{Hist}(\sigma)$; we show that one of the properties of Lemma 3 holds. There are two cases:

(h is switched.) In that case, $\sigma_h = \sigma^{\text{sbwo}}$. Since σ^{sbwo} is an SBWO strategy, $\mathbf{cVal}^h(\sigma^{\text{sbwo}}) = \mathbf{acVal}^h$. Now if $\mathbf{acVal}^h > \mathbf{aVal}^h$, then:

$$\mathbf{cVal}^h(\sigma) = \mathbf{cVal}^h(\sigma^{\text{sbwo}}) = \mathbf{acVal}^h > \mathbf{aVal}^h ,$$

and σ satisfies the first property of Lemma 3. Otherwise $\mathbf{acVal}^h = \mathbf{aVal}^h$ and the second property holds: we have that $\mathbf{cVal}^h(\sigma) = \mathbf{acVal}^h$, and as σ^{sbwo} is an SWO and $\mathbf{aVal}^h(\sigma) = \mathbf{aVal}^h(\sigma^{\text{sbwo}})$, we also have that $\mathbf{aVal}^h(\sigma) = \mathbf{aVal}^h$.

(h is unswitched.) We show that $\mathbf{cVal}^h(\sigma) > \mathbf{aVal}^h$. Since h is unswitched, we have in particular that:

$$\mathbf{Reg}\,(\sigma) = \mathbf{Reg} < \lambda^n\left(\mathbf{cVal}^{v_n} - \mathbf{aVal}^{v_n}\right)\ . \tag{1}$$

Furthermore:

$$\begin{aligned}\lambda^n\left(\mathbf{cVal}^{v_n} - \mathbf{aVal}^{v_n}\right) &= \left(\mathbf{Val}(h) + \lambda^n \mathbf{cVal}^{v_n}\right) - \left(\mathbf{Val}(h) + \lambda^n \mathbf{aVal}^{v_n}\right)\\ &= \mathbf{cVal}^h - \mathbf{aVal}^h\ ,\end{aligned}$$

and combining the previous equation with Eq. 1, we obtain:

$$\mathbf{cVal}^h - \mathbf{Reg}\,(\sigma) > \mathbf{aVal}^h\ .$$

To conclude, we show that $\mathbf{Reg}\,(\sigma) \geq \mathbf{cVal}^h - \mathbf{cVal}^h(\sigma)$. Consider a strategy τ of Adam such that h is consistent with both σ^{sbo} and τ and satisfying $\mathbf{Val}(\mathbf{out}^{v_0}(\sigma^{\text{sbo}}, \tau)) = \mathbf{cVal}^h$. (That such a τ exists is intuitively clear since σ has been following the SBO strategy σ^{sbo} along h.) It holds immediately that $\mathbf{cVal}^h(\sigma) \geq \mathbf{Val}(\mathbf{out}^{v_0}(\sigma, \tau))$. Now by definition of the regret:

$$\begin{aligned}\mathbf{Reg}\,(\sigma) &\geq \mathbf{Val}(\mathbf{out}^{v_0}(\sigma^{\text{sbo}}, \tau)) - \mathbf{Val}(\mathbf{out}^{v_0}(\sigma, \tau))\\ &\geq \mathbf{cVal}^h - \mathbf{cVal}^h(\sigma)\ .\end{aligned}$$

□

4 Minimal Values Are Witnessed by a Single Iterated Cycle

We start our technical work towards a better algorithm to compute the regret value of a game. Here, we show that there are succinctly presentable histories that witness small values in the game. Our intention is to later use this result to apply a modified version of Lemma 2 to bipositional strategies to argue there are small witnesses of a strategy having too much regret.

More specifically, we show that for any history h, there is another history h' of the same length that has smaller value and such that $h' = \alpha \cdot \beta^k \cdot \gamma$ where $|\alpha\beta\gamma|$ is small. This will allow us to find the smallest possible value among exponentially long histories by guessing α, β, γ, and k, which will all be small. This property holds for a wealth of different valuation functions, hinting at possible further applications. For discounted-sum games, the following suffices to prove the desired property holds.

Lemma 4. *For any history $h = \alpha \cdot \beta \cdot \gamma$ with α and γ same-length cycles:*

$$\min\{\mathbf{Val}(\alpha^2 \cdot \beta), \mathbf{Val}(\beta \cdot \gamma^2)\} \leq \mathbf{Val}(h)\ .$$

Within the proof of the key lemma of this section, and later on when we use it (Lemma 9), we will rely on the following notion of cycle decomposition:

Definition 2. *A* simple-cycle decomposition *(SCD) is a pair consisting of paths and iterated simple cycles. Formally, an SCD is a pair $D = \langle(\alpha_i)_{i=0}^n, (\beta_j, k_j)_{j=1}^n\rangle$, where each α_i is a path, each β_j is a simple cycle, and each k_j is a positive integer. We write $D(j) = \beta_j^{k_j} \cdot \alpha_j$ and $D(\star) = \alpha_0 \cdot D(1)D(2)\cdots D(n)$.*

By carefully iterating Lemma 4, we have:

Lemma 5. *For any history h there exists an history $h' = \alpha \cdot \beta^k \cdot \gamma$ with:*

- *h and h' have the same starting and ending vertices, and the same length;*
- *$\mathbf{Val}(h') \leq \mathbf{Val}(h)$;*
- *$|\alpha\beta\gamma| \leq 4|V|^3$ and β is a simple cycle.*

Proof. In this proof, we focus on SCDs for which each path α_i is simple; we call them ßCDs. We define a wellfounded partial order on ßCDs. Let $D = \langle(\alpha_i)_{i=0}^n, (\beta_j, k_j)_{j=1}^n\rangle$ and $D' = \langle(\alpha'_i)_{i=0}^{n'}, (\beta'_j, k'_j)_{j=1}^{n'}\rangle$ be two ßCDs; we write $D' < D$ iff all the following holds:

- $D(\star)$ and $D'(\star)$ have the same starting and ending vertices, the same length, and satisfy $\mathbf{Val}(D'(\star)) \leq \mathbf{Val}(D(\star))$ and $n' \leq n$;
- Either $n' < n$, or $|\alpha'_0 \cdots \alpha'_{n'}| < |\alpha_0 \cdots \alpha_n|$, or $|\{k'_i \geq |V|\}| < |\{k_i \geq |V|\}|$.

That this order has no infinite descending chain is clear. We show two claims:

1. Any ßCD with n greater than $|V|$ has a smaller ßCD;
2. Any ßCD with two $k_j, k_{j'} > |V|$ has a smaller ßCD.

Together they imply that for a smallest ßCD D, $D(\star)$ is of the required form. Indeed let j be the unique value for which $k_j > |V|$, then the statement of the Lemma is satisfied by letting $\alpha = \alpha_0 \cdot D(1) \cdots D(j-1)$, $\beta = \beta_j$, $k = k_j$, and $\gamma = \alpha_j \cdot D(j+1) \cdots D(n)$.

Claim 1. Suppose D has $n > |V|$. Since all cycles are simple, there are two cycles $\beta_j, \beta_{j'}$, $j < j'$, of same length. We can apply Lemma 4 on the path $\beta_j \cdot (\alpha_j D(j+1) \cdots D(j'-1)) \cdot \beta_{j'}$, and remove one of the two cycles while duplicating the other; we thus obtain a similar path of smaller value. This can be done repeatedly until we obtain a path with only one of the two cycles, say $\beta_{j'}$, the other case being similar. Substituting this path in $D(\star)$ results in:

$$\alpha_0 \cdot D(1) \cdots D(j) \cdot \left(\alpha_j \cdot D(j+1) \cdots D(j'-1) \cdot \beta_{j'}^{k_j + k_{j'}}\right) \cdot \alpha_{j'} \cdot D(j'+1) \cdots D(n) \ .$$

This gives rise to a smaller ßCD as follows. If $\alpha_{j-1}\alpha_j$ is still a simple path, then the above history is expressible as an ßCD with a smaller number of cycles. Otherwise, we rewrite $\alpha_{j-1}\alpha_j = \alpha'_{j-1}\beta'_j\alpha'_j$ where α'_{j-1} and α'_j are simple paths and β'_j is a simple cycle; since $|\alpha'_{j-1}\alpha'_j| < |\alpha_{j-1}\alpha_j|$, the resulting ßCD is smaller.

Claim 2. Suppose D has two $k_j, k_{j'} > |V|$, $j < j'$. Since each cycle in the ßCD is simple, k_j and $k_{j'}$ are greater than both $|\beta_j|$ and $|\beta_{j'}|$; let us write $k_j = b|\beta_{j'}| + r$ with $0 \leq r < |\beta_{j'}|$, and similarly, $k_{j'} = b'|\beta_j| + r'$. We have:

$$D(j) \cdots D(j') = \beta_j^r \cdot \left((\beta_j^{|\beta_{j'}|})^b \cdot \alpha_j \cdot D(j+1) \cdots D(j'-1) \cdot (\beta_{j'}^{|\beta_j|})^{b'}\right) \cdot \beta_{j'}^{r'} \cdot \alpha_{j'} \ .$$

Noting that $\beta_{j'}^{|\beta_j|}$ and $\beta_j^{|\beta_{j'}|}$ are cycles of the same length, we can transfer all the occurrences of one to the other, as in Claim 1. Similarly, if two simple paths get merged and give rise to a cycle, a smaller ßCD can be constructed; if not, then there are now at most $r < |V|$ occurrences of $\beta_{j'}$ (or conversely, r' of β_j), again resulting in a smaller ßCD. □

5 Short Witnesses for Regret, Antagonistic, and Collaborative Values

We continue our technical work towards our algorithm for computing the regret value. In this section, the overarching theme is that of *short witnesses*. We show that (1) the regret value of a strategy is witnessed by histories of bounded length; (2) the collaborative value of a game is witnessed by a simple path and an iterated cycle; (3) the antagonistic value of a strategy is witnessed by an SCD and an iterated cycle.

5.1 Regret Is Witnessed by Histories of Bounded Length

Lemma 6. *Let* $\sigma = \sigma_1 \xrightarrow{t} \sigma_2$ *be an arbitrary bipositional switching strategy of Eve and let* $C = 2|V| + \max\{t(v) < \infty\}$*. We have that:*

$$\mathbf{Reg}(\sigma) = \max\left\{\lambda^n \left(\mathbf{cVal}^{v_n}_{\neg\sigma(h)} - \mathbf{aVal}^{v_n}(\sigma_h)\right) \,\middle|\, h = v_0 \dots v_n \in \mathbf{Hist}(\sigma), n \leq C\right\} \ .$$

Proof. Consider a history h of length greater than C, and write $h = h_1 \cdot h_2$ with $|h_1| = \max\{t(v) < \infty\}$. Let $h_2 = p \cdot p'$ where p is the maximal prefix of h_2 such that $h_1 \cdot p$ is unswitched—we set $p = \epsilon$ if h is switched. Note that one of p or p' is longer than $|V|$—say p, the other case being similar. This implies that there is a cycle in p, i.e., $p = \alpha \cdot \beta \cdot \gamma$ with β a cycle. Let $h' = h_1 \cdot \alpha \cdot \gamma \cdot p'$; this history has the same starting and ending vertex as h. Moreover, since $|h_1|$ is larger than any value of the threshold function, $\sigma_h = \sigma_{h'}$. Lastly, h' is still in $\mathbf{Hist}(\sigma)$, since the removed cycle did not play a role in switching strategy. This shows:

$$\mathbf{cVal}^{v_n}_{\neg\sigma(h)} - \mathbf{aVal}^{v_n}(\sigma_h) = \mathbf{cVal}^{v_n}_{\neg\sigma(h')} - \mathbf{aVal}^{v_n}(\sigma_{h'}) \ .$$

Since the length of h is greater than the length of h', the discounted value for h' will be greater than that of h, resulting in a higher regret value. There is thus no need to consider histories of size greater than C. □

It may seem from this lemma and the fact that $t(v)$ may be very large that we will need to guess histories of important length. However, since we will be considering bipositional switching strategies, we will only be interested in guessing *some* properties of the histories that are not hard to verify:

Lemma 7. *The following problem is decidable in* NP*:*

Given:	*A game, a bipositional switching strategy* σ*, a number* n *in binary, a Boolean* b*, and two vertices* v, v'
Question:	*Is there a* $h \in \mathbf{Hist}(\sigma)$ *of length* n*, switched if* b*, ending in* v*, with* $\sigma(h) = v'$*?*

Proof. This is done by guessing multiple flows within the graph (V, E). Here, we call *flow* a valuation of the edges E by integers, that describes the number of times a path crosses each edge. Given a vector in $\mathbb{N}^E$, it is not hard to check whether there is a path that it represents, and to extract the initial and final vertices of that path [17].

We first order the different thresholds from the strategy $\sigma = \sigma_1 \xrightarrow{t} \sigma_2$: let $V_\exists = \{v_1, v_2, \ldots, v_k\}$ with $t(v_i) \leq t(v_{i+1})$ for all i. We analyze the structure of histories consistent with σ. Let $h \in \mathbf{Hist}(\sigma)$, and write $h = h' \cdot h''$ where h' is the maximal unswitched prefix of h. Naturally, h' is consistent with σ_1 and h'' is consistent with σ_2. Then $h' = h_0 h_1 \cdots h_i$, for some $i < |V_\exists|$, with:

- $|h_0| = t(v_1)$ and for all $1 \leq j < i$, $|h_j| = t(v_{j+1}) - t(v_j)$;
- For all $0 \leq j \leq i$, h_j does not contain a vertex v_k with $k \leq j$.

To confirm the existence of a history with the given parameters, it is thus sufficient to guess the value $i \leq |V_\exists|$, and to guess i connected flows (rather than paths) with the above properties that are consistent with σ_1. Finally, we guess a flow for h'' consistent with σ_2 if we need a switched history, and verify that it is starting at a switching vertex. The flows must sum to $n + 1$, with the last vertex being v', and the previous v. □

5.2 Short Witnesses for the Collaborative and Antagonistic Values

Lemma 8. *There is a set* P *of pairs* (α, β) *with* α *a simple path and* β *a simple cycle such that:*

- $\mathbf{cVal}^{v_0} = \max\{\mathbf{Val}(\alpha \cdot \beta^\omega) \mid (\alpha, \beta) \in P\}$ *and*
- *membership in* P *is decidable in polynomial time w.r.t. the game.*

Proof. We argue that the set P of all pairs (α, β) with α a simple path, β a simple cycle, and such that $\alpha \cdot \beta$ is a path, gives us the result.

The first part of the claim is a consequence of Lemma 1: Consider positional SBO strategies τ and σ of Adam and Eve, respectively. Since they are positional, the path $\mathbf{out}^{v_0}(\sigma, \tau)$ is of the form $\alpha \cdot \beta^\omega$, as required, and its value is $\mathbf{cVal}^{v_0}$. We can thus let P be the set of all pairs obtained from such SBO strategies.

Moreover, it can be easily checked that for all pairs (α, β) such that $\alpha \cdot \beta$ is a path in the game there exists a pair of strategies with outcome $\alpha \cdot \beta^\omega$. (Note that verifying whether $\alpha \cdot \beta$ is a path can indeed be done in polynomial time given α and β.) Finally, the value $\mathbf{Val}(\alpha \cdot \beta^\omega)$ will, by definition, be at most $\mathbf{cVal}^{v_0}$. □

Lemma 9. *Let σ be a bipositional switching strategy of Eve. There is a set K of pairs (D, β) with D an SCD and β a simple cycle such that:*

- $\mathbf{aVal}^{v_0}(\sigma) = \min\{\mathbf{Val}(D(\star) \cdot \beta^\omega) \mid (D, \beta) \in K\}$ *and*
- *the size of each pair is polynomially bounded, and membership in K is decidable in polynomial time w.r.t. σ and the game.*

Proof. We will prove that the set K of all pairs (D, β) with D an SCD of polynomial length (which will be specified below), β a simple cycle, and such that $D(\star) \cdot \beta$ is a path, satisfies our claims.

Let $C = \max\{t(v) < \infty\}$, and consider a play π consistent with σ that achieves the value $\mathbf{aVal}^{v_0}(\sigma)$. Write $\pi = h \cdot \pi'$ with $|h| = C$, and let v be the final vertex of h. Naturally:

$$\mathbf{aVal}^{v_0}(\sigma) = \mathbf{Val}(\pi) = \mathbf{Val}(h) + \lambda^{|h|}\mathbf{Val}(\pi') \ .$$

We first show how to replace π' by some $\alpha \cdot \beta^\omega$, with α a simple path and β a simple cycle. First, since π witnesses $\mathbf{aVal}^{v_0}(\sigma)$, we have that $\mathbf{Val}(\pi') = \mathbf{aVal}^{v}(\sigma_h)$. Now σ_h is positional, because $|h| \geq C$.[1] It is known that there are optimal positional antagonistic strategies τ for Adam, that is, that satisfy $\mathbf{aVal}^{v}(\sigma_h) = \mathbf{out}^{v}(\sigma_h, \tau)$. As in the proof of Lemma 8, this implies that $\mathbf{aVal}^{v}(\sigma_h) = \mathbf{Val}(\alpha \cdot \beta^\omega) = \mathbf{Val}(\pi')$ for some α and β; additionally, any (α, β) that are consistent with σ_h and a potential strategy for Adam will give rise to a larger value.

We now argue that $\mathbf{Val}(h)$ is witnessed by an SCD of polynomial size. This bears similarity to the proof of Lemma 7. Specifically, we will reuse the fact that histories consistent with σ can be split into histories played "between thresholds."

Let us write $\sigma = \sigma_1 \xrightarrow{t} \sigma_2$. Again, we let $V_\exists = \{v_1, v_2, \ldots, v_k\}$ with $t(v_i) \leq t(v_{i+1})$ for all i and write $h = h' \cdot h''$ where h' is the maximal unswitched prefix of h. We note that h' is consistent with σ_1 and h'' is consistent with σ_2. Then $h' = h_0 h_1 \cdots h_i$, for some $i < |V_\exists|$, with:

- $|h_0| = t(v_1)$ and for all $1 \leq j < i$, $|h_j| = t(v_{j+1}) - t(v_j)$;
- For all $0 \leq j \leq i$, h_j does not contain a vertex v_k with $k \leq j$.

We now diverge from the proof of Lemma 7. We apply Lemma 5 on each h_j in the game where the strategy σ_1 is hardcoded (that is, we first remove every edge $(u, v) \in V_\exists \times V$ that does not satisfy $\sigma_1(u) = v$). We obtain a history $h'_0 h'_1 \cdots h'_i$ that is still in $\mathbf{Hist}(\sigma)$, thanks to the previous splitting of h. We also apply Lemma 5 to h', this time in the game where σ_2 is hardcoded, obtaining h''. Since each h'_j and h'' are expressed as $\alpha \cdot \beta^k \cdot \gamma$, there is an SCD D with no more

[1] Technically, σ_h is positional in the game that records whether the switch was made.

than $|V_\exists|$ elements that satisfies $\mathbf{Val}(D(\star)) \leq \mathbf{Val}(h)$—naturally, since $\mathbf{Val}(h)$ is minimal and $D(\star) \in \mathbf{Hist}(\sigma)$, this means that the two values are equal. Note that it is not hard, given an SCD D, to check whether $D(\star) \in \mathbf{Hist}(\sigma)$, and that SCDs that are not valued $\mathbf{Val}(h)$ have a larger value. □

6 The Complexity of Regret

We are finally equipped to present our algorithms. To account for the cost of numerical analysis, we rely on the problem PosSLP [2]. This problem consists in determining whether an arithmetic circuit with addition, subtraction, and multiplication gates, together with input values, evaluates to a positive integer. PosSLP is known to be decidable in the so-called counting hierarchy, itself contained in the set of problems decidable using polynomial space.

Theorem 5. *The following problem is decidable in* $\mathsf{NP}^{\mathsf{PosSLP}}$*:*

Given: *A game, a bipositional switching strategy* σ*, a value* $r \in \mathbb{Q}$ *in binary*

Question: *Is* $\mathbf{Reg}(\sigma) > r$*?*

Proof. Let us write $\sigma = \sigma_1 \xrightarrow{t} \sigma_2$. Lemma 6 indicates that $\mathbf{Reg}(\sigma) > r$ holds if there is a history h of some length $n \leq C = 2|V| + \max\{t(v) < \infty\}$, ending in some v_n such that:

$$\lambda^n \left(\mathbf{cVal}^{v_n}_{\neg\sigma(h)} - \mathbf{aVal}^{v_n}(\sigma_h)\right) > r \ . \tag{2}$$

Note that since σ is bipositional, we do not need to know everything about h. Indeed, the following properties suffice: its length n, final vertex v_n, $v' = \sigma(h)$, and whether it is switched. Rather than guessing h, we can thus rely on Lemma 7 to get the required information. We start by simulating the NP machine that this lemma provides, and verify that n, v_n, and v are consistent with a potential history.

Let us now concentrate on the collaborative value that we need to evaluate in Eq. 2. To compute $\mathbf{cVal}$, we rely on Lemma 8, which we apply in the game where v_n is set initial, and its successor forced not to be v. We guess a pair $(\alpha_c, \beta_c) \in P$; we thus have $\mathbf{Val}(\alpha_c \cdot \beta_c^\omega) \leq \mathbf{cVal}^{v_n}_{\neg\sigma(h)}$, with at least one guessed pair (α_c, β_c) reaching that latter value.

Let us now focus on computing $\mathbf{aVal}^{v_n}(\sigma_h)$. Since σ is a bipositional switching strategy, σ_h is simply σ where $t(v)$ is changed to $\max\{0, t(v) - n\}$. Lemma 9 can thus be used to compute our value. To do so, we guess a pair $(D, \beta_a) \in K$; we thus have $\mathbf{Val}(D(\star) \cdot \beta_a^\omega) \geq \mathbf{aVal}^{v_n}(\sigma_h)$, and at least one pair (D, β_a) reaches that latter value.

Our guesses satisfy:

$$\mathbf{cVal}^{v_n}_{\neg\sigma(h)} - \mathbf{aVal}^{v_n}(\sigma_h) \geq \mathbf{Val}(\alpha_c \cdot \beta_c^\omega) - \mathbf{Val}(D(\star) \cdot \beta_a^\omega) \ ,$$

and there is a choice of our guessed paths and SCD that gives exactly the left-hand side. Comparing the left-hand side with r can be done using an oracle to PosSLP, concluding the proof. □

Theorem 6. *The following problem is decidable in* $\mathsf{coNP}^{\mathsf{NP}^{\mathsf{PosSLP}}}$*:*

Given: *A game, a value* $r \in \mathbb{Q}$ *in binary*

Question: *Is* **Reg** $> r$*?*

Proof. To decide the problem at hand, we ought to check that *every* strategy has a regret value greater than r. However, optipess strategies being regret-minimal, we need only check this for a class of strategies that contains optipess strategies: bipositional switching strategies form one such class.

What is left to show is that optipess strategies can be encoded in *polynomial space*. Naturally, the two positional strategies contained in an optipess strategy can be encoded succinctly. We thus only need to show that, with t as in the definition of optipess strategies (page 5), $t(v)$ is at most exponential for every $v \in V_\exists$ with $t(v) \in \mathbb{N}$. This is shown in the long version of this paper. □

Theorem 7. *The following problem is decidable in* $\mathsf{coNP}^{\mathsf{NP}^{\mathsf{PosSLP}}}$*:*

Given: *A game, a bipositional switching strategy* σ

Question: *Is* σ *regret optimal?*

Proof. A consequence of the proof of Theorem 5 and the existence of optipess strategies is that the value **Reg** of a game can be computed by a polynomial size arithmetic circuit. Moreover, our reliance on PosSLP allows the input r in Theorem 5 to be represented as an arithmetic circuit without impacting the complexity. We can thus verify that for all bipositional switching strategies σ' (with sufficiently large threshold functions) and all possible polynomial size arithmetic circuits, $\mathbf{Reg}(\sigma) > r$ implies that $\mathbf{Reg}(\sigma') > r$. The latter holds if and only if σ is regret optimal since, as we have argued in the proof of Theorem 6, such strategies σ' include optipess strategies and thus regret-minimal strategies. □

7 Conclusion

We studied *regret*, a notion of interest for an agent that does not want to assume that the environment she plays in is simply adversarial. We showed that there are strategies that both minimize regret, and are not consistently worse than any other strategies. The problem of computing the minimum regret value of a game was then explored, and a better algorithm was provided for it.

The exact complexity of this problem remains however open. The only known lower bound, a straightforward adaptation of [14, Lemma 3] for discounted-sum games, shows that it is at least as hard as solving parity games [15].

Our upper bound could be significantly improved if we could efficiently solve the following problem:

PosRatBase

Given: $(a_i)_{i=1}^n \in \mathbb{Z}^n$, $(b_i)_{i=1}^n \in \mathbb{N}^n$, and $r \in \mathbb{Q}$ all in binary,

Question: Is $\sum_{i=1}^n a_i \cdot r^{b_i} > 0$?

This can be seen as the problem of comparing succinctly represented numbers in a rational base. The PosSLP oracle in Theorem 5 can be replaced by an oracle for this seemingly simpler arithmetic problem. The variant of PosRatBase in which r is an integer was shown to be in P by Cucker, Koiran, and Smale [8], and they mention that the complexity is open for rational values. To the best of our knowledge, the exact complexity of PosRatBase is open even for $n = 3$.

Acknowledgements. We thank Raphaël Berthon and Ismaël Jecker for helpful conversations on the length of maximal (and minimal) histories in discounted-sum games, James Worrell and Joël Ouaknine for pointers on the complexity of comparing succinctly represented integers, and George Kenison for his writing help.

References

1. de Alfaro, L., Henzinger, T.A., Majumdar, R.: Discounting the future in systems theory. In: Baeten, J.C.M., Lenstra, J.K., Parrow, J., Woeginger, G.J. (eds.) ICALP 2003. LNCS, vol. 2719, pp. 1022–1037. Springer, Heidelberg (2003). https://doi.org/10.1007/3-540-45061-0_79
2. Allender, E., Bürgisser, P., Kjeldgaard-Pedersen, J., Miltersen, P.B.: On the complexity of numerical analysis. SIAM J. Comput. **38**(5), 1987–2006 (2009). https://doi.org/10.1137/070697926
3. Aminof, B., Kupferman, O., Lampert, R.: Reasoning about online algorithms with weighted automata. ACM Trans. Algorithms **6**(2), 28:1–28:36 (2010). https://doi.org/10.1145/1721837.1721844
4. Apt, K.R., Grädel, E.: Lectures in Game Theory for Computer Scientists. Cambridge University Press, New York (2011)
5. Brenguier, R., et al.: Non-zero sum games for reactive synthesis. In: Dediu, A.-H., Janoušek, J., Martín-Vide, C., Truthe, B. (eds.) LATA 2016. LNCS, vol. 9618, pp. 3–23. Springer, Cham (2016). https://doi.org/10.1007/978-3-319-30000-9_1
6. Brenguier, R., Pérez, G.A., Raskin, J.F., Sankur, O.: Admissibility in quantitative graph games. In: Lal, A., Akshay, S., Saurabh, S., Sen, S. (eds.) 36th IARCS Annual Conference on Foundations of Software Technology and Theoretical Computer Science, FSTTCS 2016. LIPIcs, Chennai, India, 13–15 December 2016, vol. 65, pp. 42:1–42:14. Schloss Dagstuhl - Leibniz-Zentrum fuer Informatik (2016). https://doi.org/10.4230/LIPIcs.FSTTCS.2016.42
7. Chatterjee, K., Goharshady, A.K., Ibsen-Jensen, R., Velner, Y.: Ergodic mean-payoff games for the analysis of attacks in crypto-currencies. In: Schewe, S., Zhang, L. (eds.) 29th International Conference on Concurrency Theory, CONCUR 2018. LIPIcs, Beijing, China, 4–7 September 2018, vol. 118, pp. 11:1–11:17. Schloss Dagstuhl - Leibniz-Zentrum fuer Informatik (2018). https://doi.org/10.4230/LIPIcs.CONCUR.2018.11

8. Cucker, F., Koiran, P., Smale, S.: A polynomial time algorithm for diophantine equations in one variable. J. Symb. Comput. **27**(1), 21–29 (1999). https://doi.org/10.1006/jsco.1998.0242
9. Filar, J., Vrieze, K.: Competitive Markov Decision Processes. Springer, Heidelberg (2012). https://doi.org/10.1007/978-1-4612-4054-9
10. Filiot, E., Le Gall, T., Raskin, J.-F.: Iterated regret minimization in game graphs. In: Hliněný, P., Kučera, A. (eds.) MFCS 2010. LNCS, vol. 6281, pp. 342–354. Springer, Heidelberg (2010). https://doi.org/10.1007/978-3-642-15155-2_31
11. Filiot, E., Jecker, I., Lhote, N., Pérez, G.A., Raskin, J.F.: On delay and regret determinization of max-plus automata. In: 32nd Annual ACM/IEEE Symposium on Logic in Computer Science, LICS 2017, Reykjavik, Iceland, 20–23 June 2017, pp. 1–12. IEEE Computer Society (2017). https://doi.org/10.1109/LICS.2017.8005096
12. Halpern, J.Y., Pass, R.: Iterated regret minimization: a new solution concept. Games Econ. Behav. **74**(1), 184–207 (2012). https://doi.org/10.1016/j.geb.2011.05.012
13. Hunter, P., Pérez, G.A., Raskin, J.F.: Minimizing regret in discounted-sum games. In: Talbot, J.M., Regnier, L. (eds.) 25th EACSL Annual Conference on Computer Science Logic, CSL 2016. LIPIcs, Marseille, France, 29 August–1 September 2016, vol. 62, pp. 30:1–30:17. Schloss Dagstuhl - Leibniz-Zentrum fuer Informatik (2016). https://doi.org/10.4230/LIPIcs.CSL.2016.30
14. Hunter, P., Pérez, G.A., Raskin, J.F.: Reactive synthesis without regret. Acta Inf. **54**(1), 3–39 (2017). https://doi.org/10.1007/s00236-016-0268-z
15. Jurdzinski, M.: Deciding the winner in parity games is in UP ∩ co-UP. Inf. Process. Lett. **68**(3), 119–124 (1998). https://doi.org/10.1016/S0020-0190(98)00150-1
16. Puterman, M.L.: Markov Decision Processes. Wiley-Interscience, New York (2005)
17. Reutenauer, C.: The Mathematics of Petri Nets. Prentice-Hall Inc., Upper Saddle River (1990)
18. Shapley, L.S.: Stochastic games. Proc. Natl. Acad. Sci. **39**(10), 1095–1100 (1953)
19. Watkins, C.J.C.H., Dayan, P.: Technical note Q-learning. Mach. Learn. **8**, 279–292 (1992). https://doi.org/10.1007/BF00992698
20. Zwick, U., Paterson, M.: The complexity of mean payoff games on graphs. Theor. Comput. Sci. **158**(1&2), 343–359 (1996). https://doi.org/10.1016/0304-3975(95)00188-3

Causality in Linear Logic
Full Completeness and Injectivity (Unit-Free Multiplicative-Additive Fragment)

Simon Castellan(✉) and Nobuko Yoshida

Imperial College London, London, UK
simon@phis.me

Abstract. Commuting conversions of Linear Logic induce a notion of dependency between rules inside a proof derivation: a rule depends on a previous rule when they cannot be permuted using the conversions. We propose a new interpretation of proofs of Linear Logic as *causal invariants* which captures *exactly* this dependency. We represent causal invariants using game semantics based on general event structures, carving out, inside the model of [6], a submodel of causal invariants. This submodel supports an interpretation of unit-free Multiplicative Additive Linear Logic with MIX (MALL^-) which is (1) *fully complete*: every element of the model is the denotation of a proof and (2) *injective*: equality in the model characterises exactly commuting conversions of MALL^-. This improves over the standard fully complete game semantics model of MALL^-.

Keywords: Event structures · Linear Logic · Proof nets · Game semantics

1 Introduction

Proofs up to commuting conversions. In the sequent calculus of Linear Logic, the order between rules need not always matter: allowed reorderings are expressed by *commuting conversions.* These conversions are necessary for confluence of cut-elimination by mitigating the sequentiality of the sequent calculus. The real proof object is often seen as an equivalence class of proofs modulo commuting conversions. The problem of providing a canonical representation of proofs up to those commuting conversions is as old as Linear Logic itself, and proves to be a challenging problem. The traditional solution interprets a proof by a graphical representation called *proof net* and dates back to Girard [17]. Girard's solution is only satisfactory in the multiplicative-exponential fragment of Linear Logic. For additives, a well-known solution is due to Hughes and van Glabbeck [22], where proofs are reduced to their set of axiom linkings. However, the correctness criterion relies on the difficult *toggling* condition.

Proof nets tend to be based on specific representations such as graphs or sets of linkings. Denotational semantics has not managed to provide a semantic counterpart to proof nets, which would be a model where every element is

M. Bojańczyk and A. Simpson (Eds.): FOSSACS 2019, LNCS 11425, pp. 150–168, 2019.
https://doi.org/10.1007/978-3-030-17127-8_9

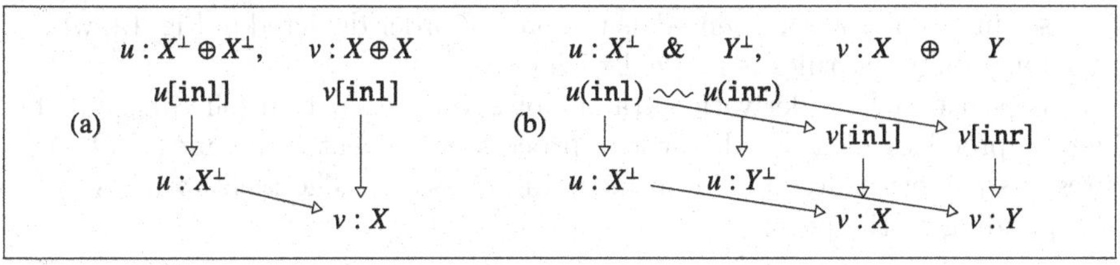

Fig. 1. Examples of causal invariants

the interpretation of a proof (*full completeness*) and whose equational theory coincides with commuting conversions (*injectivity*). We believe this is because denotational semantics views conversions as extensional principles, hence models proofs with extensional objects (relations, functions) too far from the syntax.

Conversions essentially state that the order between rules applied to different premises does not matter, as evidenced in the two equivalent proofs of the sequent $\vdash X^{\perp} \oplus X^{\perp}, X \oplus X$ depicted on the right. These two proofs are equal in extensional models of Linear Logic because *they have the same extensional behaviour.* Unfortunately, characterising the image of the interpretation proved to be a difficult task in extensional models. The first fully complete models used game semantics, and are due to Abramsky and Melliès (MALL) [1] and Melliès (Full LL) [24]. However, their models use an *extensional* quotient on strategies to satisfy the conversions, blurring the concrete nature of strategies.

$$\dfrac{\dfrac{\dfrac{}{\vdash X^{\perp}, X}\,Ax}{\vdash X^{\perp}, X \oplus X}\,\oplus_1}{\vdash X^{\perp} \oplus X^{\perp}, X \oplus X}\,\oplus_1 \qquad \dfrac{\dfrac{\dfrac{}{\vdash X^{\perp}, X}\,Ax}{\vdash X^{\perp} \oplus X^{\perp}, X}\,\oplus_1}{\vdash X^{\perp} \oplus X^{\perp}, X \oplus X}\,\oplus_1$$

The true concurrency of conversions. Recent work [5] highlights an interpretation of Linear Logic as communicating processes. Rules become actions whose polarity (input or output) is tied to the polarity of the connective (negative or positive), and cut-elimination becomes communication. In this interpretation, each assumption in the context is assigned a channel on which the proof communicates. Interestingly, commuting conversions can be read as asynchronous permutations. For instance, the conversion mentioned above becomes the equation in the syntax of Wadler [27]:

$$(1) \quad u[\mathtt{inl}].\, v[\mathtt{inl}].\, [u \leftrightarrow v] \equiv v[\mathtt{inl}].\, u[\mathtt{inl}].\, [u \leftrightarrow v] \;\triangleright\; u : X^{\perp} \oplus X^{\perp}, v : X \oplus X,$$

where $u[\mathtt{inl}]$ corresponds to a $\oplus_1$-introduction rule on (the assumption corresponding to) u, and $[u \leftrightarrow v]$ is the counterpart to an axiom between the hypothesis corresponding to u and v. It becomes then natural to consider that the canonical object representing these two proofs should be a concurrent process issuing the two outputs in parallel. A notion of causality emerges from this interpretation, where a rule *depends on* a previous rule below in the tree when these two rules cannot be permuted using the commuting conversions. This leads us to causal models to make this dependency explicit. For instance, the two processes in (1) can be represented as the partial order depicted in Fig. 1a, where dependency between rules is marked with $\twoheadrightarrow$.

In presence of &, a derivation stands for several execution (slices), given by different premises of a &-rule (whose process equivalent is $u.\texttt{case}\,(P, Q)$ and represents pattern matching on an incoming message). The identity on $X \oplus Y$, corresponding to the proof

$$u.\texttt{case}\,(v[\texttt{inl}].\,[u \leftrightarrow v],\;\; v[\texttt{inr}].\,[u \leftrightarrow v]) \;\triangleright\; u : X^\perp \,\&\, Y^\perp, v : X \oplus Y,$$

is interpreted by the *event structure* depicted in Fig. 1b. Event structures [28] combine a partial order, representing causality, with a conflict relation representing when two events cannot belong to the same execution (here, same slice). Conflict here is indicating with $\sim$ and separates the slices. The &-introduction becomes two conflicting events.

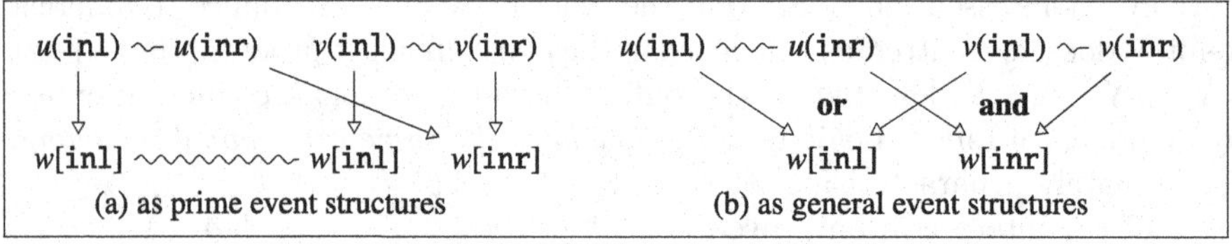

Fig. 2. Representations of or

Conjunctive and disjunctive causalities. Consider the process on the context $u : (X \oplus X)^\perp, v : (Y \oplus Y)^\perp, w : (X \otimes Y) \oplus (X \otimes Y)$ implementing disjunction:

$$\texttt{or} = u.\texttt{case}\begin{pmatrix} v.\texttt{case}\,(w[\texttt{inl}].\,P, w[\texttt{inl}].\,P), \\ v.\texttt{case}\,(w[\texttt{inl}].\,P, w[\texttt{inr}].\,P) \end{pmatrix} \text{ where } P = w[x].\,([u \leftrightarrow w] \mid [v \leftrightarrow x]).$$

Cuts of or against a proof starting with $u[\texttt{inl}]$ or $v[\texttt{inl}]$ answer on w after reduction:

$$(\nu u)(\texttt{or} \mid u[\texttt{inl}]) \to^* w[\texttt{inl}].v.\texttt{case}\,(P, P) \quad (\nu v)(\texttt{or} \mid v[\texttt{inl}]) \to^* w[\texttt{inl}].u.\texttt{case}\,(P, P)$$

where $(\nu u)(P \mid Q)$ is the process counterpart to logical cuts. This operational behaviour is related to *parallel or*, evaluating its arguments in parallel and returning true as soon as one returns true. Due to this intentional behaviour, the interpretation of or in prime event structures is nondeterministic (Fig. 2a), as causality in event structures is *conjunctive* (an event may only occur after all its predecessors have occurred). By moving to *general* event structures, however, we can make the disjunctive causality explicit and recover determinism (Fig. 2b).

Contributions and outline. Drawing inspiration from the interpretation of proofs in terms of processes, we build a fully complete and injective model of unit-free Multiplicative Additive Linear Logic with MIX (MALL^-), interpreting proofs as general event structures living in a submodel of the model introduced by [6]. Moreover, our model captures the dependency between rules, which makes sequentialisation a local operation, unlike in proof nets, and has a more uniform acyclicity condition than [22].

We first recall the syntax of MALL$^-$ and its reading in terms of processes in Sect. 2. Then, in Sect. 3, we present a slight variation on the model of [6], where we call the (pre)strategies *causal structures*, by analogy with proof structures. Each proof tree can be seen as a (sequential) causal structure. However, the space of causal structures is too broad and there are many causal structures which do not correspond to any proofs. A major obstacle to sequentialisation is the presence of *deadlocks*. In Sect. 4, we introduce a condition on causal structures, ensuring deadlock-free composition, inspired by the interaction between $\parr$ and $\otimes$ in Linear Logic. Acyclic causal structures are still allowed to only explore partially the game, contrary to proofs which must explore it exhaustively, hence in Sect. 5, we introduce further conditions on causal structures, ensuring a *strong* sequentialisation theorem (Theorem 2): we call them *causal nets*. In Sect. 6, we define causal invariants as maximal causal nets. Every causal net embeds in a *unique* causal invariant; and a particular proof P embeds inside a unique causal invariant which forms its denotation $[\![P]\!]$. Moreover, two proofs embed in the same causal invariant if and only if they are convertible (Theorem 4). Finally, we show how to equip causal invariants with the structure of $*$-autonomous category with products and deduce that they form a fully complete model of MALL$^-$ (Theorem 6) for which the interpretation is injective.

The proofs are available in the technical report [7].

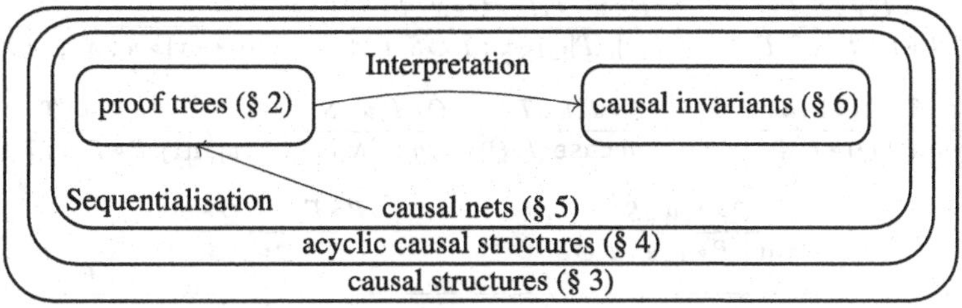

2 MALL$^-$ and Its Commuting Conversions

In this section, we introduce MALL$^-$ formulas and proofs as well as the standard commuting conversions and cut elimination for this logic. As mentioned in the introduction, we use a process-like presentation of proofs following [27]. This highlights the communicating aspect of proofs which is an essential intuition for the model; and it offers a concise visualisation of proofs and conversions.

Formulas. We define the formulas of MALL$^-$: $T, S ::= X \mid X^\perp \mid T \otimes S \mid T \parr S \mid T \oplus S \mid T \,\&\, S$, where X and $X^\perp$ are *atomic formulas* (or *ltterals*) belonging to a set $\mathbb{A}$. Formulas come with the standard notion of duality $(\cdot)^\perp$ given by the De Morgan rules: $\otimes$ is dual to $\parr$, and $\oplus$ to $\&$. An *environment* is a partial mapping of *names* to formulas, instead of a multiset of formulas – names disambiguate which assumption a rule acts on.

Proofs as processes. We see proofs of MALL⁻ (with MIX) as typing derivations for a variant of the π-calculus [27]. The (untyped) syntax for the processes is as follows:

$$\begin{array}{rll} P, Q ::= & u(v).\, P \mid u[v].\,(P \mid Q) & \text{(multiplicatives)} \\ \mid & u.\mathtt{case}\,(P, Q) \mid u[\mathtt{inl}].\, P \mid u[\mathtt{inr}].\, P & \text{(additives)} \\ \mid & [u \leftrightarrow v] \mid (\nu u)(P \mid Q) \mid \ (P \mid Q) & \text{(logical and mix)} \end{array}$$

$u(v).P$ denotes an input of v on channel u (used in $⅋$-introduction) while $u[v].(P \mid Q)$ denotes output of a fresh channel v along channel u (used in $\otimes$-introduction); The term $[u \leftrightarrow v]$ is a *link*, forwarding messages received on u to v, corresponds to axioms, and conversely; and $(\nu u)(P \mid Q)$ represents a restriction of u in P and Q and corresponds to cuts; $u.\mathtt{case}\,(P, Q)$ is an input branching representing &-introductions, which interacts with selection, either $u[\mathtt{inl}].\, R$ or $u[\mathtt{inr}].\, R$; in $(\nu u)(P \mid Q)$, u is bound in both P and Q, in $u(v).\, P$, v is bound in P, and in $u[v].\,(P \mid Q)$, v is only bound in Q.

We now define MALL⁻ proofs as typing derivations for processes. The inference rules, recalled in Fig. 3, are from [27]. The links (axioms) are restricted to literals – for composite types, one can use the usual η-expansion laws. There is a straightforward bijection between standard (η-expanded) proofs of MALL⁻ and typing derivations.

$$\frac{P \triangleright u : T, v : S, \Gamma}{u(v).\, P \triangleright u : T ⅋ S, \Gamma} \qquad \frac{P \triangleright u : T, \Gamma \quad Q \triangleright v : S, \Delta}{u[v].\,(P|Q) \triangleright u : T \otimes S, \Gamma, \Delta} \qquad \frac{}{[u \leftrightarrow v] \triangleright u : X^\perp, v : X}$$

$$\frac{P \triangleright \Gamma, u : T \quad Q \triangleright \Delta, u : T^\perp}{(\nu u)(P \mid Q) \triangleright \Gamma, \Delta} \qquad \frac{P \triangleright \Gamma, u : T \qquad Q \triangleright \Gamma, u : S}{u.\mathtt{case}\,(P, Q) \triangleright \Gamma, u : T \& S} \qquad \frac{P \triangleright \Gamma, u : T}{u[\mathtt{inl}].\, P \triangleright \Gamma, u : T \oplus S}$$

$$\frac{P \triangleright \Gamma, u : S}{u[\mathtt{inr}].\, P \triangleright \Gamma, u : T \oplus S} \qquad \frac{P \triangleright \Gamma \qquad Q \triangleright \Delta}{P \mid Q \triangleright \Gamma, \Delta}$$

$$\frac{}{\vdash u[\mathtt{inl}].\, [] : \Gamma, u : T \Rightarrow \Gamma, u : T \oplus S} \qquad \frac{Q \triangleright \Delta, v : S}{\vdash u[v].\,([] \mid Q) : \Gamma, u : T \Rightarrow u : T \otimes S, \Gamma, \Delta}$$

$$\frac{}{\vdash u[\mathtt{inr}].\, [] : \Gamma, u : S \Rightarrow \Gamma, u : T \oplus S} \qquad \frac{P \triangleright \Gamma, u : T}{\vdash u[v].\,(P \mid []) : \Delta, v : S \Rightarrow u : T \otimes S, \Gamma, \Delta}$$

$$\frac{}{\vdash u.\mathtt{case}\,([]_1, []_2) : (\Gamma, u : T) \times (\Gamma, u : S) \Rightarrow \Gamma, u : T \,\&\, S}$$

$$\frac{}{\vdash u(v).\, [] : \Gamma, u : T, v : S \Rightarrow \Gamma, u : T ⅋ S} \qquad \frac{P \triangleright \Delta}{\vdash ([] \mid P) : \Gamma \Rightarrow \Gamma, \Delta} \qquad \frac{P \triangleright \Gamma}{\vdash (P \mid []) : \Delta \Rightarrow \Gamma, \Delta}$$

Fig. 3. Typing rules for MALL⁻ (above) and contexts (below)

Commutation rules and cut elimination. We now explain the valid commutations rules in our calculus. We consider contexts $C\,[[]_1, \ldots, []_n]$ with several holes to accomodate & which has two branches. Contexts are defined in Fig. 3, and are assigned a type $\Gamma_1 \times \ldots \times \Gamma_n \Rightarrow \Delta$. It intuitively means that if we plug proofs of Γ_i in the holes, we get back a proof of Δ. We use the notation $C[P_i]_i$ for $C[P_1, \ldots, P_n]$ when (P_i) is a family of processes. Commuting conversion is the smallest congruence $\equiv$ satisfying all well-typed instances of the rule $C[D[P_{i,j}]_j]_i \equiv D[C[P_{i,j}]_i]_j$ for C and D two contexts. For instance $a[\texttt{inl}].\, b.\texttt{case}\,(P, Q) \equiv b.\texttt{case}\,(a[\texttt{inl}].\, P, a[\texttt{inl}].\, Q)$. Figure 4 gives reduction rules $P \to Q$. The first four rules are the *principal* cut rules and describe the interaction of two dual terms, while the last one allows cuts to move inside contexts.

3 Concurrent Games Based on General Event Structures

This section introduces a slight variation on the model of [6]. In Sect. 3.1, we define *games* as prime event structures with polarities, which are used to interpret formulas. We then introduce general event structures in Sect. 3.2, which are used to define causal structures.

$$(\nu u)([u \leftrightarrow v] \mid P) \to P[v/u] \qquad (\nu u)(u[x].\,(P \mid Q) \mid u(x).\,R) \to (\nu u)(P \mid (\nu x)(Q \mid R))$$

$$(\nu u)(u[\texttt{inl}].\,R \mid u.\texttt{case}\,(P, Q)) \to (\nu u)(R \mid P) \qquad (\nu u)(u[\texttt{inr}].\,R \mid u.\texttt{case}\,(P, Q)) \to (\nu u)(R \mid Q)$$

$$(\nu u)(C[P_i]_i \mid Q) \to C[(\nu u)(P_i \mid Q)]_i \quad (u \notin C)$$

Fig. 4. Cut elimination in MALL$^-$

3.1 Games as Prime Event Structures with Polarities

Definition of games. Prime event structures [28] (simply event structures in the rest of the paper) are a causal model of nondeterministic and concurrent computation. We use here prime event structures *with binary conflict.* An **event structure** is a triple $(E, \leq_E, \#_E)$ where $(E, \leq_E)$ is a partial order and $\#_E$ is an irreflexive symmetric relation (representing **conflict**) satisfying: (1) if $e \in E$, then $[e] := \{e' \in E \mid e' \leq_E e\}$ is finite; and (2) if $e \#_E e'$ and $e \leq_E e''$ then $e'' \#_E e'$. We often omit the E subscripts when clear from the context.

A **configuration** of E is a downclosed subset of E which does not contain two conflicting events. We write $\mathscr{C}(E)$ for the set of *finite* configurations of E. For any $e \in E$, $[e]$ is a configuration, and so is $[e) := [e] \setminus \{e\}$. We write $e \rightarrowtriangle e'$ for the immediate causal relation of E defined as $e < e'$ with no event between. Similarly, a conflict $e \# e'$ is **minimal**, denoted $e \sim e'$, when the $[e] \cup [e')$ and $[e) \cup [e']$ are configurations. When drawing event structures, only $\rightarrowtriangle$ and $\sim$ are represented. We write $\max(E)$ for the set of maximal events of E for $\leq_E$. An event e is maximal in x when it has no successor for $\leq_E$ in x. We write $\max_E x$ for the maximal events of a configuration $x \in \mathscr{C}(E)$.

An event structure E is **confusion-free** when (1) for all $e \sim_E e'$ then $[e) = [e')$ and (2) if $e \sim_E e'$ and $e' \sim_E e''$ then $e = e''$ or $e \sim_E e''$. As a result, the relation "$e \sim e'$ or $e = e'$" is an equivalence relation whose equivalent classes $\mathfrak{a}$ are called **cells**.

Definition 1. *A* ***game*** *is a confusion-free event structure A along with an assignment pol : $A \to \{-, +\}$ such that cells contain events of the same polarity, and a function atom:* $\max(A) \to \mathbb{A}$ *mapping every maximal event of A to an atom. Events with polarity $-$ (resp. $+$) are* ***negative*** *(resp.* ***positive****).*

Events of a game are usually called *moves*. The restriction imposes branching to be polarised (*i.e.* belonging to a player). A game is **rooted** when two minimal events are in conflict. Single types are interpreted by rooted games, while contexts are interpreted by arbitrary games. When introducing moves of a game, we will indicate their polarity in exponent, *e.g.* "let $a^+ \in A$" stands for assuming a positive move of A.

Interpretation of formulas. To interpret formulas, we make use of standard constructions on prime event structures. The event structure $a \cdot E$ is E prefixed with a, *i.e.* $E \cup \{a\}$ where *all* events of E depends on a. The parallel composition of E and E' represents parallel executions of E and E' without interference:

Definition 2. *The parallel composition of event structures A_0 and A_1 is the event structure $A_0 \parallel A_1 = (\{0\} \times A_0 \cup \{1\} \times A_1, \leq_{A_0 \parallel A_1}, \#_{A_0 \parallel A_1})$ with $(i, a) \leq_{A_0 \parallel A_1} (j, a')$ iff $i = j$ and $a \leq_{A_i} a'$; and $(i, a) \#_{A_0 \parallel A_1} (j, a')$ when $i = j$ and $a \#_{A_j} a'$.*

The sum of event structure $E + F$ is the nondeterministic analogue of parallel composition.

Definition 3. *The sum $A_0 + A_1$ of the two event structures A_0 and A_1 has the same partial order as $A_0 \parallel A_1$, and conflict relation $(i, a) \#_{A_0 + A_1} (j, a')$ iff $i \neq j$ or $i = j$ and $a \#_{A_j} a'$.*

Prefixing, parallel composition and sum of event structures extend to games. The dual of a game A, obtained by reversing the polarity labelling, is written $A^\perp$. Given $x \in \mathscr{C}(A)$, we define A/x ("A after x") as the subgame of A comprising the events $a \in A \setminus x$ not in conflict with events in x.

Interpretation of formulas. The interpretation of the atom X is the game with a single positive event simply written X with $atom(X) = X$, and the interpretation of $X^\perp$ is $[\![X]\!]^\perp$, written simply $X^\perp$ in diagrams. For composite formulas, we let (where `send`, `inl` and `inr` are simply labels):

$$[\![S \otimes T]\!] = \mathtt{send}^+ \cdot ([\![S]\!] \parallel [\![T]\!]) \qquad [\![S \parr T]\!] = \mathtt{send}^- \cdot ([\![S]\!] \parallel [\![T]\!])$$
$$[\![S \oplus T]\!] = (\mathtt{inl}^+ \cdot [\![S]\!]) + (\mathtt{inr}^+ \cdot [\![T]\!]) \qquad [\![S \,\&\, T]\!] = (\mathtt{inl}^- \cdot [\![S]\!]) + (\mathtt{inr}^- \cdot [\![T]\!])$$

Parallel composition is used to interpret contexts: $[\![u_1 : T_1, \ldots, u_n : T_n]\!] = [\![T_1]\!] \parallel \ldots \parallel [\![T_n]\!]$. The interpretation commutes with duality: $[\![T]\!]^\perp = [\![T^\perp]\!]$.

In diagrams, we write moves of a context following the syntactic convention: for instance $u[\mathtt{inl}]$ denotes the minimal `inl` move of the u component. For tensors and pars, we use the notation $u[v]$ and $u(v)$ to make explicit the variables we use in the rest of the diagram, instead of $\mathtt{send}^+$ and $\mathtt{send}^-$ respectively. For atoms, we use $u : X$ and $u : X^\perp$.

3.2 Causal Structures as Deterministic General Event Structures

As we discussed in Sect. 1, prime event structures cannot express disjunctive causalities deterministically, hence fail to account for the determinism of LL. Our notion of causal structure is based on *general event structures*, which allow more complex causal patterns. We use a slight variation on the definition of deterministic general event structures given by [6], to ensure that composition is well-defined without further assumptions.

Instead of using the more concrete representation of general event structures in terms of a set of events and an enabling relation, we use the following formulation in terms of set of configurations, more adequate for mathematical reasoning. Being only sets of configurations, they can be reasoned on with very simple set-theoretic arguments.

Definition 4. *A* ***causal structure*** *(abbreviated as causal struct) on a game A is a subset $\sigma \subseteq \mathscr{C}(A)$ containing $\emptyset$ and satisfying the following conditions:*

Coincidence-freeness *If $e, e' \in x \in \sigma$ then there exists $y \in \sigma$ with $y \subseteq x$ and $y \cap \{e, e'\}$ is a singleton.*
Determinism *for $x, y \in \sigma$ such that $x \cup y$ does not contain any minimal negative conflict, then $x \cup y \in \sigma$.*

Configurations of prime event structures satisfy a further axiom, *stability*, which ensures the absence of disjunctive causalities. When σ is a causal struct on A, we write $\sigma : A$. We draw as regular event structures, using $\rightarrowtriangle$ and $\rightsquigarrow$. To indicate disjunctive causalities, we annotate joins with **or**. This convention is not powerful enough to draw *all* causal structs, but enough for the examples in this paper. As an example, on $A = a \parallel b \parallel c$ the diagram on the right denotes the following causal struct $\sigma = \{x \in \mathscr{C}(A) \mid c \in x \Rightarrow x \cap \{a, b\} \neq \emptyset\}$.

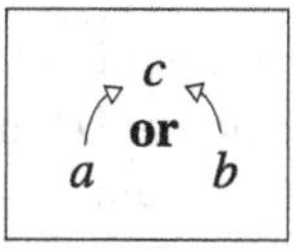

A **minimal event** of $\sigma : A$ is an event $a \in A$ with $\{a\} \in \sigma$. An event $a \in x \in \sigma$ is **maximal** in x when $x \setminus \{a\} \in \sigma$. A **prime configuration** of $a \in A$ is a configuration $x \in \sigma$ such that a is its unique maximal event. Because of disjunctive causalities, an event $a \in A$ can have several distinct prime configurations in σ (unlike in event structures). In the previous example, since c can be caused by either a or b, it has two prime configurations: $\{a, c\}$ and $\{b, c\}$. We write $\max \sigma$ for the set of **maximal configurations** of σ, ie. those configurations that cannot be further extended.

Even though causality is less clear in general event structures than in prime event structures, we give here a notion of immediate causal dependence that will be central to define acyclic causal structs. Given a causal struct $\sigma : A$ and $x \in \sigma$, we define a relation $\rightarrowtriangle_{x,\sigma}$ on x as follows: $a \rightarrowtriangle_{x,\sigma} a'$ when there exists a prime configuration y of a' such that $x \cup y \in \sigma$, and that a is maximal in $y \setminus \{a'\}$. This notion is compatible with the drawing above: we have $a \rightarrowtriangle_{\emptyset} c$ and $b \rightarrowtriangle_{\emptyset} c$ as c has two prime configurations: $\{a, c\}$ and $\{b, c\}$. Causality needs to be contextual, since different slices can implement different causal patterns. Parallel composition and prefixing structures extend to causal structs:

$$\sigma \parallel \tau = \{x \parallel y \in \mathscr{C}(A \parallel B) \mid (x, y) \in \sigma \times \tau\} \qquad a \cdot \sigma = \{x \in \mathscr{C}(a \cdot A) \mid x \cap A \in \sigma\}.$$

Categorical setting. Causal structs can be composed using the definitions of [6]. Consider $\sigma : A^\perp \parallel B$ and $\tau : B^\perp \parallel C$. A **synchronised configuration** is a configuration $x \in \mathscr{C}(A \parallel B \parallel C)$ such that $x \cap (A \parallel B) \in \sigma$ and $x \cap (B \parallel C) \in \tau$. A synchronised configuration x is **reachable** when there exists a sequence (**covering chain**) of synchronised configurations $x_0 = \emptyset \subseteq x_1 \subseteq \ldots \subseteq x_n = x$ such that $x_{i+1} \setminus x_i$ is a singleton. The reachable configurations are used to define the interaction $\tau \circledast \sigma$, and then after hiding, the composition $\tau \odot \sigma$:

$$\tau \circledast \sigma = \{x \text{ is a reachable synchronised configuration}\} \quad \tau \odot \sigma = \{x \cap (A \parallel C) \mid x \in \tau \circledast \sigma\}.$$

Unlike in [6], our determinism is strong enough for $\tau \odot \sigma$ to be a causal struct.

Lemma 1. *If $\sigma : A^\perp \parallel B$ and $\tau : B^\perp \parallel C$ are causal structs then $\tau \odot \sigma$ is a causal struct.*

Composition of causal structs will be used to interpret cuts between proofs of Linear Logic. In concurrent game semantics, composition has a natural identity, asynchronous copycat [25], playing on the game $A^\perp \parallel A$, forwarding negative moves on one side to the positive occurrence on the other side. Following [6], we define $\text{æ}_A = \{x \parallel y \in \mathscr{C}(A^\perp \parallel A) \mid y \supseteq^-_A x \cap y \subseteq^+_A x\}$ where $x \subseteq^p y$ means $x \subseteq y$ and $pol(y \setminus x) \subseteq \{p\}$.

However, in general copycat is not an identity on all causal structs, only $\sigma \subseteq \text{æ}_A \odot \sigma$ holds. Indeed, copycat represents an asynchronous buffer, and causal structs which expects messages to be transmitted synchronously may be affected by composition with copycat. We call causal structs that satisfy the equality **asynchronous**. From [6], we know that asynchronous causal structs form a compact-closed category.

The syntactic tree. The syntactic tree of a derivation $P \triangleright \Delta$ can be read as a causal struct $Tr(P)$ on $[\![\Delta]\!]$, which will be the basis for our interpretation. It is defined by induction:

$$Tr(u(v).\,P) = u(v) \cdot Tr(P) \qquad Tr(u[v].\,(P \mid Q)) = u[v] \cdot (Tr(P) \parallel Tr(Q))$$
$$Tr(a.\mathtt{case}\,(P, Q)) = (a(\mathtt{inl}) \cdot Tr(P)) \cup (a(\mathtt{inr}) \cdot Tr(Q))$$
$$Tr(a[\mathtt{inl}].\,P) = a[\mathtt{inl}] \cdot Tr(P) \qquad Tr(a[\mathtt{inr}].\,P) = a[\mathtt{inr}] \cdot Tr(P)$$
$$Tr([a \leftrightarrow b]) = \text{æ}_{[\![X]\!]} \text{ where } \Delta = a : X^\perp, b : X \qquad Tr(P \mid Q) = Tr(P) \parallel Tr(Q)$$
$$Tr((\nu a)(P \mid Q)) = Tr(P) \odot Tr(Q)$$

We use the convention in the diagram, for instance $u[v]$ means the initial `send` move of the u component. An example of this construction is given in Fig. 5a. Note that it is not asynchronous.

4 Acyclicity of Causal Structures

The space of causal structs is unfortunately too broad to provide a notion of causal nets, due in particular to the presence of deadlocks during composition.

As a first step towards defining causal nets, we introduce in this section a condition on causal structs inspired by the tensor rule in Linear Logic. In Sect. 4.1, we propose a notion of communication between actions, based on causality. In Sect. 4.2, we introduce a notion of acyclicity which is shown to be stable under composition and ensure deadlock-free composition.

4.1 Communication in Causal Structures

The tensor rule of Linear Logic says that after a tensor $u[v]$, the proof splits into two independent subproofs, one handling u and the other v. This syntactic condition is there to ensure that there are no communications between u and v. More precisely, we want to prevent any dependence between subsequent actions on u and an action v. Indeed such a causal dependence could create a deadlock when facing a par rule $u(v)$, which is allowed to put arbitrary dependence between such subsequent actions.

Communication in MLL. Let us start by the case of MLL, which corresponds to the case where games do not have conflicts. Consider the following three causal structs:

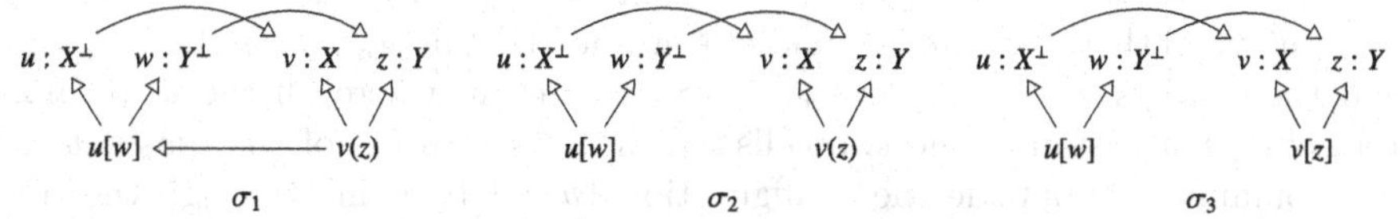

The causal structs σ_1 and σ_2 play on the game $[\![u : X^\perp \otimes Y^\perp, v : X \parr Y]\!]$, while σ_3 plays on the game $[\![u : X^\perp \otimes Y^\perp, v : X \otimes Y]\!]$. The causal structs σ_2 and σ_3 are very close to proof nets, and it is easy to see that σ_2 represents a correct proof net while σ_3 does not. In particular, there exists a proof P such that $Tr(P) \subseteq \sigma_2$ but there are no such proof Q for σ_3. Clearly, σ_3 should not be acyclic. But should σ_2? After all it is sequentialisable. But, in all sequentialisations of σ_2, the par rule $v(z)$ is applied *before* the tensor $u[w]$, and this dependency is not reflected by σ_2. Since our goal is exactly to compute these implicit dependencies, we will only consider σ_1 to be acyclic, by using a stronger sequentialisation criterion:

Definition 5. *A causal struct $\sigma : [\![\Gamma]\!]$ is* ***strongly sequentialisable*** *when for all $x \in \sigma$, there exists $P \triangleright \Gamma$ with $x \in Tr(P)$ and $Tr(P) \subseteq \sigma$.*

To understand the difference between σ_1 and σ_2, we need to look at causal chains. In both σ_1 and σ_2, we can go from $u : X^\perp$ to $w : Y^\perp$ by following immediate causal links $\rightarrowtriangle$ in any direction, but observe that in σ_1 they must all cross an event below $u[w]$ (namely $v(z)$ or $u[w]$). This prompts us to define a notion of communication *outside a configuration* x:

Definition 6. *Given $\sigma : A$ and $x \in \sigma$ we say that $a, a' \in A \setminus x$ communicate* outside x *(written $a \leftrightsquigarrow_{x,\sigma} a'$) when there exists a chain $a \leftrightarrow_{x,\sigma} a_0 \leftrightarrow_\sigma \cdots \leftrightarrow_{x,\sigma} a_n \leftrightarrow_\sigma a'$ where all the $a_i \in A \setminus x$, and $\leftrightarrow_{x,\sigma}$ denotes the symmetric closure of $\rightarrowtriangle_{x,\sigma}$.*

Communication in MALL. In presence of additives, immediate causality is not the only vector of communication. Consider the following causal struct σ_4, playing on the context $u : (A \,\&\, A) \otimes (A \,\&\, A), v : (A \oplus A) \,\&\, (A \oplus A)$ where A is irrelevant:

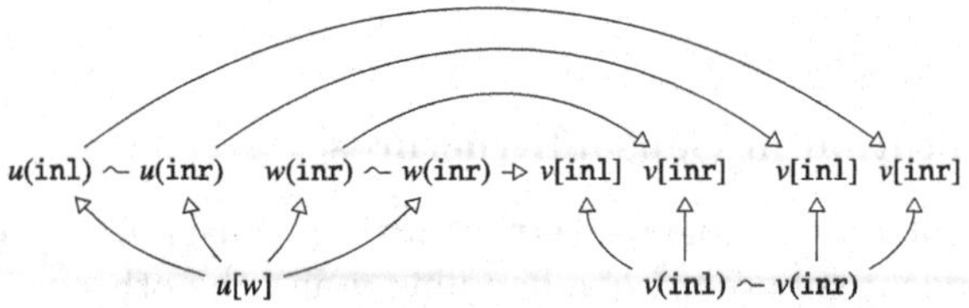

This pattern is not strongly sequentialisable: the tensor $u[w]$ must always go after the &-introduction on v, since we need this information to know how whether v should go with u or w when splitting the context. Yet, it is not possible to find a communication path from one side to the other by following purely causal links without crossing $u[w]$. There is however a path that uses both immediate causality and *minimal conflict.* This means that we should identify events in minimal conflict, since they represent the same (&-introduction rule). Concretely, this means lifting the previous definition at the level of cells. Given an causal struct $\sigma : A$ and $x \in \sigma$, along with two cells $\mathfrak{a}, \mathfrak{a}'$ of A/x, we define the relation $\mathfrak{a} \leftrightarrow_{x,\sigma} \mathfrak{a}'$ when there exists $a \in \mathfrak{a}$ and $a' \in \mathfrak{a}'$ such that $a \leftrightarrow_{x,\sigma} a'$; and $\mathfrak{a} \leftrightsquigarrow_{x,\sigma} \mathfrak{a}'$ when there exists $\mathfrak{a} \leftrightarrow_{x,\sigma} \mathfrak{a}_0 \leftrightarrow_{x,\sigma} \cdots \leftrightarrow_{\sigma} \mathfrak{a}_n \leftrightarrow_{\sigma} \mathfrak{a}'$ where all the $\mathfrak{a}_i$ do not intersect x. For instance, the two cells which are successors of the tensor $u[w]$ in σ_4 communicate outside the configuration $\{u[w]\}$ by going through the cell $\{v(\texttt{inl}), v(\texttt{inr})\}$.

4.2 Definition of Acyclicity on Casual Structures

Since games are trees, two events a, a' are either incomparable or have a meet $a \wedge a'$. If $a \wedge a'$ is defined and positive, we say that a and a' **have positive meet**, and means that they are on two distinct branches of a tensor. If $a \wedge a'$ is undefined, or defined and negative, we say that $a \wedge a'$ has a **negative meet**. When the meet is undefined, it means that a and a' are events of different components of the context. We consider the meet to be negative in this case, since components of a context are related by an implicit par.

These definitions are easily extended to cells. The meet $\mathfrak{a} \wedge \mathfrak{a}'$ of two cells $\mathfrak{a}$ and $\mathfrak{a}'$ of A is the meet $a \wedge a'$ for $a \in \mathfrak{a}$ and $a' \in \mathfrak{a}'$: by confusion-freeness, it does not matter which ones are chosen. Similarly, we say that $\mathfrak{a}$ and $\mathfrak{a}'$ have positive meet if $\mathfrak{a} \wedge \mathfrak{a}'$ is defined and positive; and have negative meet otherwise. These definitions formalise the idea of "the two sides of a tensor", and allow us to define acyclicity.

Definition 7. *A causal struct $\sigma : A$ is* **acyclic** *when for all $x \in \sigma$, for any cells $\mathfrak{a}, \mathfrak{a}'$ not intersecting x and with positive meet, if $\mathfrak{a} \leftrightsquigarrow_{x,\sigma} \mathfrak{a}'$ then $\mathfrak{a} \wedge \mathfrak{a}' \notin x$.*

This captures the desired intuition: if $\mathfrak{a}$ and $\mathfrak{a}'$ are on two sides of a tensor a (ie. have positive meet), and there is a communication path outside x relating them, then a must also be outside x (and implicitly, the communication path must be going through a).

Reasoning on the interaction of acyclic strategies proved to be challenging. We prove that acyclic strategies compose, and their interaction are deadlock-free, when composition is on a rooted game B. This crucial assumption arises from the fact that in linear logic, cuts are on *formulas*. It entails that for any $b, b' \in B$, $b \wedge b'$ is defined, hence must be positive either from the point of view of σ or of τ.

Theorem 1. *For acyclic causal structs $\sigma : A^\perp \parallel B$ and $\tau : B^\perp \parallel C$, (1) their interaction is* deadlock-free*: $\tau \circledast \sigma = (\sigma \parallel C) \cap (A \parallel \tau)$; and (2) the causal struct $\tau \odot \sigma$ is acyclic.*

As a result, acyclic and asynchronous causal structs form a category. We believe this intermediate category is interesting in its own right since it generalises the deadlock-freeness argument of Linear Logic without having to assume other constraints coming from Linear Logic, such as linearity. In the next section, we study further restriction on acyclic causal structs which guarantee strong sequentialisability.

5 Causal Nets and Sequentialisation

We now ready to introduce causal nets. In Sect. 5.1, we give their definition by restricting acyclic causal structs and in Sect. 5.2 we prove that causal nets are strongly sequentialisable.

5.1 Causal Nets: Totality and Well-Linking Casual Structs

To ensure that our causal structs are strongly sequentialisable, acyclicity is not enough. First, we need to require causal structs to respect the linearity discipline of Linear Logic:

Definition 8. *A causal struct $\sigma : A$ is* **total** *when* (1) *for $x \in \sigma$, if x is maximal in σ, then it is maximal in $\mathscr{C}(A)$; and* (2) *for $x \in \sigma$ and $a^- \in A \setminus x$ such that $x \cup \{a\} \in \sigma$, then whenever $\boldsymbol{a} \sim_A \boldsymbol{a}'$, we also have $x \cup \{a'\} \in \sigma$ as well.*

The first condition forces a causal struct to play until there are no moves to play, and the second forces an causal struct to be receptive to all Opponent choices, not a subset.

Our last condition constrains axiom links. A **linking** of a game A is a pair (x, ℓ) of a $x \in \max \mathscr{C}(A)$, and a bijection $\ell : (\max_A x)^- \simeq (\max_A x)^+$ preserving the *atom* labelling.

Definition 9. *A total causal struct $\sigma : A$ is* **well-linking** *when for each $x \in$ $\max(\sigma)$, there exists a linking ℓ_x of x, such that if y is a prime configuration of $\ell_x(e)$ in x, then $\max(y \setminus \{\ell_x(e)\}) = \{e\}$.*

This ensures that every positive atom has a unique predecessor which is a negative atom.

Definition 10. *A* ***causal net*** *is an acyclic, total and well-linking causal struct.*

A causal net $\sigma : A$ induces a set of linkings A, $\mathsf{link}(\sigma) := \{\ell_x \mid x \in \max \sigma\}$. The mapping $\mathsf{link}(\cdot)$ maps causal nets to the proof nets of [22].

5.2 Strong Sequentialisation of Causal Nets

Our proof of sequentialisation relies on an induction on causal nets. To this end, we provide an inductive deconstruction of parallel proofs. Consider $\sigma : A$ a causal net and a minimal event $a \in \sigma$ not an atom. We write A/a for $A/\{a\}$. Observe that if $A = [\![\Delta]\!]$, it is easy to see that there exists a context Δ/a such that $[\![\Delta/a]\!] \cong A/a$. Given a causal struct $\sigma : A$, we define the causal struct $\sigma/a = \{x \in \mathscr{C}(A/a) \mid x \cup \{a\} \in \sigma\} : A/a$.

Lemma 2. *σ/a is a causal net on A/a.*

When a is positive, we can further decompose σ/a in disjoint parts thanks to acyclicity. Write $\mathfrak{a}_1, \ldots, \mathfrak{a}_n$ for the minimal cells of A/a and consider for $n \geq k > 0$, $A_k = \{a' \in A/a \mid \mathrm{cell}(a') \leftrightsquigarrow_{\{a\},\sigma} \mathfrak{a}_k\}$. A_k contains the events of A/a which σ connects to the k-th successor of a. We also define the set $A_0 = A/a \setminus \bigcup_{1 \leq k \leq n} A_k$, of events not connected to any successor of a (this can happen with MIX). It inherits a game structure from A.

Each subset inherits a game structure from A/a. By acyclicity of σ, the A_k are pairwise disjoint, so $A/a \cong A_0 \parallel \ldots \parallel A_n$. For $0 \leq k \leq n$, define $\sigma_k = \mathscr{C}(A_k) \cap \sigma/a$.

Lemma 3. *σ_k is a causal net on A_k and we have $\sigma/a = \sigma_0 \parallel \ldots \parallel \sigma_n$.*

This formalises the intuition that after a tensor, an acyclic causal net must be a parallel composition of proofs (following the syntactic shape of the tensor rule of Linear Logic). From this result, we show by induction that any causal net is strongly sequentialisable.

Theorem 2. *If $\sigma : A$ is a causal net, then σ is strongly sequentialisable.*

We believe sequentialisation without MIX requires causal nets to be *connected*: two cells with negative meets always communicate outside any configuration they are absent from. We leave this lead for future work.

6 Causal Invariants and Completeness

Causal nets are naturally ordered by inclusion. When $\sigma \subseteq \tau$, we can regard τ as a less sequential implementation of σ. Two causal nets which are upper bounded by a causal net should represent the same proof, but with varying degrees of sequentiality. Causal nets which are maximal for inclusion (among causal nets) are hence most parallel implementations of a certain behaviour and capture our intuition of causal invariants.

Definition 11. *A* ***causal invariant*** *is a causal net $\sigma : A$ maximal for inclusion.*

6.1 Causal Invariants as Maximal Causal Nets

We start by characterising when two causal nets are upper-bounded for inclusion:

Proposition 1. *Given two causal nets $\sigma, \tau : A$, the following are equivalent:*

1. *there exists a causal net $\upsilon : A$ such that $\sigma \subseteq \upsilon$ and $\tau \subseteq \upsilon$,*
2. *the set $\sigma \vee \tau = \{x \cup y \mid x \in \sigma, y \in \tau, x \cup y \in \mathscr{C}(A)\}$ is a causal net on A,*
3. *$\mathsf{link}(\sigma) = \mathsf{link}(\tau)$.*

In this case we write $\sigma \uparrow \tau$ and $\sigma \vee \tau$ is the least upper bound of σ and τ for $\subseteq$.

It is a direct consequence of Proposition 1 that any causal net σ is included in a unique causal invariant $\sigma^\uparrow : A$, defined as: $\sigma^\uparrow = \bigvee_{\sigma \subseteq \tau} \tau$, where τ ranges over causal nets.

Lemma 4. *For $\sigma, \tau : A$ causal nets, $\sigma \uparrow \tau$ iff $\sigma^\uparrow = \tau^\uparrow$. Moreover, if σ and τ are causal invariants, $\sigma \uparrow \tau$ if and only if $\sigma = \tau$.*

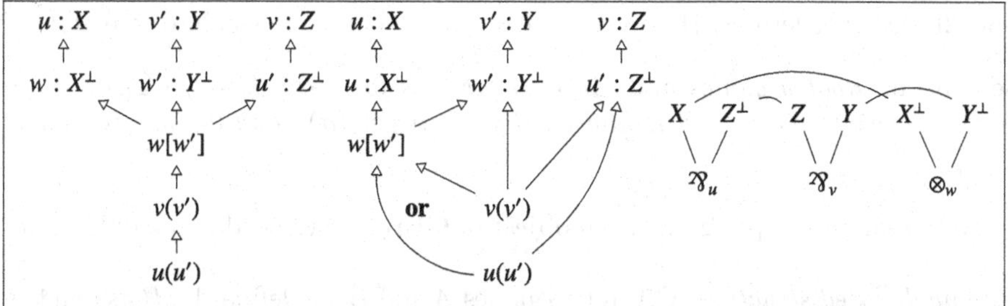

(a) *Tr(P)* (b) ⟦*P*⟧, its interpretation (c) The proof net for *P*

Fig. 5. Interpreting $P = u(u').\, v(v').\, w[w'].\, ([u \leftrightarrow w] \mid ([w' \leftrightarrow v'] \mid [u' \leftrightarrow v]))$ in the context $u : X \parr Z^\perp, v : Z \parr Y, w : X^\perp \otimes Y^\perp$

The interpretation of a proof $P \triangleright \Delta$ is simply defined as $[\![P]\!] = Tr(P)^\uparrow$. Figure 5c illustrates the construction on a proof of MLL+mix. The interpretation features a disjunctive causality, as the tensor can be introduced as soon as *one* of the two pars has been.

Defining $\mathsf{link}(P) = \mathsf{link}(Tr(P))$, we have from Lemma 4: $\mathsf{link}(P) = \mathsf{link}(Q)$ if and only if $[\![P]\!] = [\![Q]\!]$. This implies that our model has the same equational theory than the proof nets of [22]. Such proof nets are already complete:

Theorem 3 ([22]). *For P, Q two proofs of Γ, we have $P \equiv Q$ iff $\mathsf{link}(P) = \mathsf{link}(Q)$.*

As a corollary, we get:

Theorem 4. *For cut-free proofs P, Q we have $P \equiv Q$ iff $[\![P]\!] = [\![Q]\!]$.*

The technical report [7] also provides an inductive proof not using the result of [22]. A consequence of this result, along with *strong* sequentialisation is: $[\![P]\!] = \bigcup_{Q \equiv P} Tr(Q)$. This equality justifies our terminology of "causal completeness", as for instance it implies that the minimal events of $[\![P]\!]$ correspond exactly the possible rules in P that can be pushed to the front using the commuting conversions.

6.2 The Category of Causal Invariants

So far we have focused on the static. Can we integrate the dynamic aspect of proofs as well? In this section, we show that causal invariants organise themselves in a category. First, we show that causal nets are stable under composition:

Lemma 5. *If $\sigma : A^\perp \parallel B$ and $\tau : B^\perp \parallel C$ are causal nets, then so is $\tau \odot \sigma$.*

Note that totality requires acyclicity (and deadlock-freedom) to be stable under composition. However, causal invariants are not stable under composition: $\tau \odot \sigma$ might not be maximal, even if τ and σ are. Indeed, during the interaction, some branches of τ will not be explored by σ and vice-versa which can lead to new allowed reorderings. However, we can always embed $\tau \odot \sigma$ into $(\tau \odot \sigma)^\uparrow$:

Lemma 6. *Rooted games and causal invariants form a category **CInv**, where the composition of $\sigma : A^\perp \parallel B$ and $\tau : B^\perp \parallel C$ is $(\tau \odot \sigma)^\uparrow$ and the identity on A is $\text{æ}_A^\uparrow$.*

Note that the empty game is an object of **CInv**, as we need a monoidal unit.

Monoidal-closed structure. Given two games A and B we define $A \otimes B$ as $\mathtt{send}^+ \cdot (A \parallel B)$, and 1 as the empty game. There is an obvious isomorphism $A \otimes 1 \cong A$ and $A \otimes (B \otimes C) \cong (A \otimes B) \otimes C$ in **CInv**. We now show how to compute directly the functorial action of $\otimes$, without resorting to $^\uparrow$. Consider $\sigma \in \mathbf{CInv}(A, B)$ and $\tau \in \mathbf{CInv}(C, D)$. Given $x \in \mathscr{C}((A \otimes C)^\perp \parallel (B \otimes D))$, we define $x\langle\sigma\rangle = x \cap (A^\perp \parallel B)$ and $x\langle\tau\rangle = x \cap (C^\perp \parallel D)$. If $x\langle\sigma\rangle \in \sigma$ and $x\langle\tau\rangle \in \tau$, we say that x is connected when there exists cells $\mathfrak{a}, \mathfrak{b}, \mathfrak{c}$ and $\mathfrak{d}$ of A, B, C and D respectively such that $\mathfrak{a} \leftrightsquigarrow_{x\langle\sigma\rangle,\sigma} \mathfrak{c}$ and $\mathfrak{b} \leftrightsquigarrow_{x\langle\tau\rangle,\tau} \mathfrak{d}$. We define:

$$\sigma \otimes \tau = \left\{ \begin{array}{l} x \in \mathscr{C}((A \otimes C)^\perp \parallel (B \otimes D)) \text{ such that :} \\ (1)\ x\langle\sigma\rangle \in \sigma \text{ and } x\langle\tau\rangle \in \tau \\ (2) \text{ if } x \text{ is connected and contains } \mathtt{send}^+, \text{ then } \mathtt{send}^- \in x \end{array} \right\}$$

In (2), $\mathtt{send}^-$ refers to the minimal move of $(A \otimes C)^\perp$ and $\mathtt{send}^+$ to the one of $B \otimes D$. (2) ensures that $\sigma \otimes \tau$ is acyclic.

Lemma 7. *The tensor product defines a symmetric monoidal structure on **CInv**.*

Define $A \parr B = (A^\perp \otimes B^\perp)^\perp$, $\perp = 1 = \emptyset$ and $A \multimap B = A^\perp \parr B$.

Lemma 8. *We have a bijection $\parr_{B,C}$ between causal invariants on $A \parallel B \parallel C$ and on $A \parallel (B \parr C)$. As a result, there is an adjunction $A \otimes \_ \dashv A \multimap \_$.*

Lemma 8 implies that $\mathbf{CInv}((A \multimap \bot) \multimap \bot) \simeq \mathbf{CInv}(A)$, and $\mathbf{CInv}$ is $*$-autonomous.

Cartesian products. Given two games A, B in $\mathbf{CInv}$, we define their product $A \,\&\, B = \mathtt{inl}^- \cdot A + \mathtt{inr}^- \cdot B$. We show how to construct the pairing of two causal invariants concretely. Given $\sigma \in \mathbf{CInv}(A, B)$ and $\tau \in \mathbf{CInv}(A, C)$, we define the common behaviour of σ and τ on A to be those $x \in \mathscr{C}(A^\perp) \cap \sigma \cap \tau$ such that for all $\mathfrak{a}, \mathfrak{a}'$ outside of x with positive meet, $\mathfrak{a} \leftrightsquigarrow_{x,\sigma} \mathfrak{a}'$ iff $\mathfrak{a} \leftrightsquigarrow_{x,\tau} \mathfrak{a}'$. We write $\sigma \cap_A \tau$ for the set of common behaviours of σ and τ and define: $\langle \sigma, \tau \rangle = (L^- \cdot \sigma) \cup (R^- \cdot \tau) \cup \sigma \cap_A \tau$. The projections are defined using copycat: $\pi_1 = \{x \in \mathscr{C}((A \,\&\, B)^\perp \parallel A) \mid x \cap (A^\perp \parallel A) \in \mathfrak{c}_A^\uparrow\}$ (and similarly for π_2).

Theorem 5. *CInv has products. As it is also $*$-autonomous, it is a model of MALL.*

It is easy to see that the interpretation of MALL$^-$ in $\mathbf{CInv}$ following the structure is the same as $[\![\cdot]\!]$, however it is computed compositionally without resorting to the $^\uparrow$ operator. We deduce that our interpretation is invariant by cut-elimination: if $P \to Q$, then $[\![P]\!] = [\![Q]\!]$. Putting the pieces together, we get the final result.

Theorem 6. *CInv is an injective and fully complete model of MALL$^-$.*

7 Extensions and Related Work

The model provides a representation of proofs which retains only the necessary sequentiality. We study the phenomenon in Linear Logic, but commuting conversions of additives arise in other languages, eg. in functional languages with sums and products, where proof nets do not necessarily exist. Having an abstract representation of which reorderings are allowed could prove useful (reasoning on the possible commuting conversions in a language with sum types is notoriously difficult).

Extensions. Exponentials are difficult to add, as their conversions are not as canonical as those of MALL. Cyclic proofs [2] could be accomodated via recursive event structures.

Adding multiplicative units while keep determinism is difficult, as their commuting conversion is subtle (*e.g.* conversion for MLL is PSPACE-complete [18]), and exhibit apparent nondeterminism. For instance the following proofs are convertible in MLL:

$$a().\,b[] \mid c[] \equiv a().\,(b[] \mid c[]) \equiv b[] \mid a().\,c[] \rhd a : \bot, b : 1, c : 1$$

where $a().\,P$ is the process counterpart to introduction of $\bot$ and $a[]$ of 1. Intuitively, $b[]$ and $c[]$ can be performed at the start, but as soon as one is performed, the other has to wait for the input on a. This cannot be modelled inside deterministic general event structures, as it is only deterministic against an environment that will emit on b. In contrast, proofs of MALL$^-$ remain deterministic even if their environment is not total.

We would also be interested in recast multifocusing [9] in our setting by defining a class of focussed causal nets, where there are no concurrency between positive and negative events, and show that sequentialisation always give a focused proof.

Related work. The first fully complete model of MALL$^-$ is based on closure operators [1], later extended to full Linear Logic [24]. True concurrency is used to define innocence, on which the full completeness result rests. However their model does not take advantage of concurrency to account for permutations, as strategies are sequential. This investigation has been extended to concurrent strategies by Mimram and Melliès [25,26]. De Carvalho showed that the relational model is injective for MELL [11]. In another direction, [4] provides a fully complete model for MALL without game semantics, by using a glueing construction on the model of hypercoherences. [21] explores proof nets a weaker theory of commuting conversions for MALL.

The idea of having intermediate representations between proof nets and proofs has been studied by Faggian and coauthors using l-nets [10,13–16], leading to a similar analysis to ours: they define a space of causal nets as partial orders and compare different versions of proofs with varying degree of parallelism. Our work recasts this idea using event structures and adds the notion of causal completeness: keeping jumps that cannot be undone by a permutation, which leads naturally to step outside partial orders, as well as full completeness: which causal nets can be strongly sequentialised?

The notion of dependency between logical rules has also been studied in [3] in the case of MLL. From a proof net R, they build a partial order $D_{\parr,\otimes}(R)$ which we believe is very related to $[\![P]\!]$ where P is a sequentialisation of R. Indeed, in the case of MLL *without MIX* a partial order is enough to capture the dependency between rules. The work [12] shows that permutation rules of Linear Logic, understood as asynchronous optimisations on processes, are included in the observational equivalence. [19] studies mutual embedding between polarised proof nets [23] and the control π-calculus [20]. In another direction, we have recently built a fully-abstract, concurrent game semantics model of the synchronous session π-calculus [8]. The difficulty there was to understand name passing and the synchrony of the π-calculus, which is the dual of our objective here: trying to understand the asynchrony behind the conversions of MALL$^-$.

Acknowledgements. We would like to thank Willem Heijltjes, Domenico Ruoppolo, and Olivier Laurent for helpful discussions, and the anonymous referees for their insightful comments. This work has been partially sponsored by: EPSRC EP/K034413/1, EP/K011715/1, EP/L00058X/1, EP/N027833/1, and EP/N028201/1.

References

1. Abramsky, S., Melliès, P.-A.: Concurrent games and full completeness. In: 14th Annual IEEE Symposium on Logic in Computer Science, Trento, Italy, 2–5 July 1999, pp. 431–442 (1999). http://dx.doi.org/10.1109/LICS.1999.782638
2. Baelde, D., Doumane, A., Saurin, A.: Infinitary proof theory: the multiplicative additive case. In: CSL. Leibniz International Proceedings in Informatics (LIPIcs), vol. 62, pp. 42:1–42:17. Schloss Dagstuhl-Leibniz-Zentrum fuer Informatik (2016)
3. Bagnol, M., Doumane, A., Saurin, A.: On the dependencies of logical rules. In: Pitts, A. (ed.) FoSSaCS 2015. LNCS, vol. 9034, pp. 436–450. Springer, Heidelberg (2015). https://doi.org/10.1007/978-3-662-46678-0_28
4. Blute, R., Hamano, M., Scott, P.J.: Softness of hypercoherences and MALL full completeness. Ann. Pure Appl. Logic **131**(1–3), 1–63 (2005). https://doi.org/10.1016/j.apal.2004.05.002
5. Caires, L., Pfenning, F.: Session types as intuitionistic linear propositions. In: Gastin, P., Laroussinie, F. (eds.) CONCUR 2010. LNCS, vol. 6269, pp. 222–236. Springer, Heidelberg (2010). https://doi.org/10.1007/978-3-642-15375-4_16
6. Castellan, S., Clairambault, P., Winskel, G.: Observably deterministic concurrent strategies and intensional full abstraction for parallel-or. In: 2nd International Conference on Formal Structures for Computation and Deduction, FSCD 2017, Oxford, UK, 3–9 September 2017, pp. 12:1–12:16 (2017). https://doi.org/10.4230/LIPIcs.FSCD.2017.12
7. Castellan, S., Yoshida, N.: Causality in linear logic: full completeness and injectivity (unit-free multiplicative-additive fragment). Technical report (2019). http://iso.mor.phis.me/publis/Causality_in_Linear_Logic_FOSSACS19.pdf
8. Castellan, S., Yoshida, N.: Two sides of the same coin: session types and game semantics. Accepted for publication at POPL 2019 (2019)
9. Chaudhuri, K., Miller, D., Saurin, A.: Canonical sequent proofs via multi-focusing. In: Ausiello, G., Karhumäki, J., Mauri, G., Ong, L. (eds.) TCS 2008. IIFIP, vol. 273, pp. 383–396. Springer, Boston, MA (2008). https://doi.org/10.1007/978-0-387-09680-3_26
10. Curien, P.-L., Faggian, C.: L-nets, strategies and proof-nets. In: Ong, L. (ed.) CSL 2005. LNCS, vol. 3634, pp. 167–183. Springer, Heidelberg (2005). https://doi.org/10.1007/11538363_13
11. de Carvalho, D.: The relational model is injective for multiplicative exponential linear logic. In: 25th EACSL Annual Conference on Computer Science Logic, CSL 2016, Marseille, France, 29 August–1 September 2016, pp. 41:1–41:19 (2016). https://doi.org/10.4230/LIPIcs.CSL.2016.41
12. DeYoung, H., Caires, L., Pfenning, F., Toninho, B.: Cut reduction in linear logic as asynchronous session-typed communication. In: CSL, pp. 228–242 (2012)
13. Giamberardino, P.D.: Jump from parallel to sequential proofs: additives. Technical report (2011). https://hal.archives-ouvertes.fr/hal-00616386
14. Faggian, C., Maurel, F.: Ludics nets, a game model of concurrent interaction. In: 20th IEEE Symposium on Logic in Computer Science (LICS 2005), Chicago, IL, USA, 26–29 June 2005, Proceedings, pp. 376–385. IEEE Computer Society (2005). http://dx.doi.org/10.1109/LICS.2005.25
15. Faggian, C., Piccolo, M.: A graph abstract machine describing event structure composition. Electr. Notes Theor. Comput. Sci. **175**(4), 21–36 (2007). https://doi.org/10.1016/j.entcs.2007.04.014

16. Di Giamberardino, P., Faggian, C.: Jump from parallel to sequential proofs: multiplicatives. In: Ésik, Z. (ed.) CSL 2006. LNCS, vol. 4207, pp. 319–333. Springer, Heidelberg (2006). https://doi.org/10.1007/11874683_21
17. Girard, J.Y.: Linear logic. Theor. Comput. Sci. **50**(1), 1–101 (1987)
18. Heijltjes, W., Houston, R.: No proof nets for MLL with units: proof equivalence in MLL is PSPACE-complete. In: CSL-LICS 2014, pp. 50:1–50:10. ACM (2014)
19. Honda, K., Laurent, O.: An exact correspondence between a typed pi-calculus and polarised proof-nets. Theor. Comput. Sci. **411**(22–24), 2223–2238 (2010). https://doi.org/10.1016/j.tcs.2010.01.028
20. Honda, K., Yoshida, N., Berger, M.: Process types as a descriptive tool for interaction. In: Dowek, G. (ed.) RTA 2014. LNCS, vol. 8560, pp. 1–20. Springer, Cham (2014). https://doi.org/10.1007/978-3-319-08918-8_1
21. Hughes, D.J.D., Heijltjes, W.: Conflict nets: efficient locally canonical MALL proof nets. In: Proceedings of the 31st Annual ACM/IEEE Symposium on Logic in Computer Science, LICS 2016, New York, NY, USA, 5–8 July 2016, pp. 437–446 (2016). http://doi.acm.org/10.1145/2933575.2934559
22. Hughes, D.J.D., van Glabbeek, R.J.: Proof nets for unit-free multiplicative-additive linear logic. ACM Trans. Comput. Logic **6**(4), 784–842 (2005)
23. Laurent, O.: Polarized proof-nets and $\lambda\mu$-calculus. Theor. Comput. Sci. **290**(1), 161–188 (2003). https://doi.org/10.1016/S0304-3975(01)00297-3
24. Melliès, P.-A.: Asynchronous games 4: a fully complete model of propositional linear logic. In: 20th IEEE Symposium on Logic in Computer Science (LICS 2005), Chicago, IL, USA, 26–29 June 2005, Proceedings, pp. 386–395 (2005). http://dx.doi.org/10.1109/LICS.2005.6
25. Melliès, P.-A., Mimram, S.: Asynchronous games: innocence without alternation. In: Caires, L., Vasconcelos, V.T. (eds.) CONCUR 2007. LNCS, vol. 4703, pp. 395–411. Springer, Heidelberg (2007). https://doi.org/10.1007/978-3-540-74407-8_27
26. Mimram, S.: Sémantique des jeux asynchrones et réécriture 2-dimensionnelle (asynchronous game semantics and 2-dimensional rewriting systems). Ph.D. thesis, Paris Diderot University, France (2008). https://tel.archives-ouvertes.fr/tel-00338643
27. Wadler, P.: Propositions as sessions. J. Funct. Program. **24**(2–3), 384–418 (2014)
28. Winskel, G.: Event structures. In: Brauer, W., Reisig, W., Rozenberg, G. (eds.) ACPN 1986. LNCS, vol. 255, pp. 325–392. Springer, Heidelberg (1987). https://doi.org/10.1007/3-540-17906-2_31

Rewriting Abstract Structures: Materialization Explained Categorically

Andrea Corradini[1], Tobias Heindel[2], Barbara König[3], Dennis Nolte[3(✉)], and Arend Rensink[4]

[1] Università di Pisa, Pisa, Italy
andrea@di.unipi.it
[2] University of Hawaii, Honolulu, USA
heindel@hawaii.edu
[3] Universität Duisburg-Essen, Duisburg, Germany
{barbara_koenig,dennis.nolte}@uni-due.de
[4] University of Twente, Enschede, Netherlands
arend.rensink@utwente.nl

Abstract. The paper develops an abstract (over-approximating) semantics for double-pushout rewriting of graphs and graph-like objects. The focus is on the so-called materialization of left-hand sides from abstract graphs, a central concept in previous work. The first contribution is an accessible, general explanation of how materializations arise from universal properties and categorical constructions, in particular partial map classifiers, in a topos. Second, we introduce an extension by enriching objects with annotations and give a precise characterization of strongest post-conditions, which are effectively computable under certain assumptions.

1 Introduction

Abstract interpretation [12] is a fundamental static analysis technique that applies not only to conventional programs but also to general infinite-state systems. Shape analysis [30], a specific instance of abstract interpretation, pioneered an approach for analyzing pointer structures that keeps track of information about the "heap topology", e.g., out-degrees or existence of certain paths. One central idea of shape analysis is *materialization*, which arises as companion operation to summarizing distinct objects that share relevant properties. Materialization, a.k.a. partial concretization, is also fundamental in verification approaches based on separation logic [5,6,24], where it is also known as rearrangement [26], a special case of frame inference. Shape analysis—construed in a wide sense—has been adapted to graph transformation [29], a general purpose modelling language for systems with dynamically evolving topology, such as network protocols and cyber-physical systems. Motivated by earlier work of shape analysis for graph

T. Heindel—Partially supported by AFOSR.

M. Bojańczyk and A. Simpson (Eds.): FOSSACS 2019, LNCS 11425, pp. 169–188, 2019.
https://doi.org/10.1007/978-3-030-17127-8_10

transformation [1,2,4,27,28,31], we want to put the materialization operation on a new footing, widening the scope of shape analysis.

A natural abstraction mechanism for transition systems with graphs as states "summarizes" all graphs over a specific *shape graph*. Thus a single graph is used as abstraction for all graphs that can be mapped homomorphically into it. Further annotations on shape graphs, such as cardinalities of preimages of its nodes and general first-order formulas, enable fine-tuning of the granularity of abstractions. While these natural abstraction principles have been successfully applied in previous work [1,2,4,27,28,31], their companion materialization constructions are notoriously difficult to develop, hard to understand, and are redrawn from scratch for every single setting. Thus, we set out to explain materializations based on mathematical principles, namely universal properties (in the sense of category theory). In particular, partial map classifiers in the topos of graphs (and its slice categories) cover the purely structural aspects of materializations; this is related to final pullback complements [13], a fundamental construction of graph rewriting [7,25]. Annotations of shape graphs are treated orthogonally via op-fibrations.

The first milestones of a general framework for shape analysis of graph transformation and more generally rewriting of objects in a topos are the following:

▷ A rewriting formalism for graph abstractions that lifts the rule-based rewriting from single graphs to *abstract graphs*; it is developed for (abstract) objects in a topos.

▷ We characterize the materialization operation for abstract objects in a topos in terms of partial map classifiers, giving a sound and complete description of all occurrences of right-hand sides of rules obtained by rewriting an abstract object. → Sect. 3

▷ We decorate abstract objects with annotations from an ordered monoid and extend abstract rewriting to abstract objects with annotations. For the specific case of graphs, we consider global annotations (counting the nodes and edges in a graph), local annotations (constraining the degree of a node), and path annotations (constraining the existence of paths between certain nodes). → Sect. 4

▷ We show that abstract rewriting with annotations is sound and, with additional assumptions, complete. Finally, we derive strongest post-conditions for the case of graph rewriting with annotations. → Sect. 5

Related work: The idea of shape graphs together with shape constraints was pioneered in [30] where the constraints are specified in a three-valued logic. A similar approach was proposed in [31], using first-order formulas as constraints. In partner abstraction [3,4], cluster abstraction [1,2], and neighbourhood abstraction [28] nodes are clustered according to local criteria, such as their neighbourhood and the resulting graph structures are enriched with counting constraints, similar to our constraints. The idea of counting multiplicities of nodes and edges is also found in canonical graph shapes [27]. The uniform treatment of monoid annotations was introduced in previous work [9,10,20], in the context of type systems and with the aim of studying decidability and closure properties, but not for abstract rewriting.

a collection of structures—the *language* of the abstract object. The simplest example of an abstract structure is a plain object of a category to which we associate the language of objects that can be mapped to it; the formal definition is as follows (see also [10]).

Definition 5 (Language of an object). *Let A be an object of a category $\mathbf{C}$. Given another object X, we write $X \dashrightarrow A$ whenever there exists an arrow from X to A. We define the* language[1] *of A, denoted by $\mathcal{L}(A)$, as $\mathcal{L}(A) = \{X \in \mathbf{C} \mid X \dashrightarrow A\}$.*

Whenever $X \in \mathcal{L}(A)$ holds, we will say that X is *abstracted by* A, and A is called the *abstract object.* In the following we will also need to characterize a class of (co-)matches which are represented by a given (co-)match (which is a mono).

Definition 6 (Language of a mono). *Let $\varphi\colon L \rightarrowtail A$ be a mono in $\mathbf{C}$. The* language *of φ is the set of monos m with source L that factor φ such that the square on the right is a pullback:*

$$\mathcal{L}(\varphi) = \{m\colon L \rightarrowtail X \mid \exists(\psi\colon X \to A) \text{ such that square (1) is a pullback}\}. \qquad \begin{array}{ccc} L & \overset{m}{\rightarrowtail} & X \\ {\scriptstyle id_L}\downarrow & \text{(PB)} & \downarrow{\scriptstyle \psi} \\ L & \underset{\varphi}{\rightarrowtail} & A \end{array} \qquad (1)$$

Intuitively, for any arrow $(L \overset{m}{\to} X) \in \mathcal{L}(\varphi)$ we have $X \in \mathcal{L}(A)$ and X has a distinguished subobject L which corresponds precisely to the subobject $L \rightarrowtail A$. In fact ψ restricts and co-restricts to an isomorphism between the images of L in X and A. For graphs, no nodes or edges in X outside of L are mapped by ψ into the image of L in A.

3 Materialization

Given a production $p : L \leftarrowtail I \rightarrowtail R$, an abstract object A, and a (possibly non-monic) arrow $\varphi\colon L \to A$, we want to transform the abstract object A in order to characterize all successors of objects in $\mathcal{L}(A)$, i.e., those obtained by rewriting via p at a match compatible with φ. (Note that φ is not required to be monic, because a monic image of the left-hand side of p in an object of $\mathcal{L}(A)$ could be mapped non-injectively to A.) Roughly, we want to lift DPO rewriting to the level of abstract objects.

For this, it is necessary to use the materialization construction, defined categorically in Sect. 3.1, that enables us to concretize an instance of a left-hand side in a given abstract object. This construction is refined in Sect. 3.2 where we restrict to materializations that satisfy the gluing condition and can thus be rewritten via p. Finally in Sect. 3.3 we present the main result about materializations showing that we can fully characterize the co-matches obtained by rewriting.

[1] Here we assume that $\mathbf{C}$ is essentially small, so that a language can be seen as a set instead of a proper class of objects.

3.1 Materialization Category and Existence of Materialization

From now on we assume $\mathbf{C}$ to be an elementary topos. We will now define the materialization, which, given an arrow $\varphi \colon L \to A$, characterizes all objects X, abstracted over A, which contain a (monic) occurrence of the left-hand side compatible with φ.

Definition 7 (Materialization). *Let* $\varphi \colon L \to A$ *be an arrow in* $\mathbf{C}$. *The* materialization category *for* φ, *denoted* $\mathbf{Mat}_\varphi$, *has as*

objects *all factorizations* $L \rightarrowtail X \to A$ *of* φ *whose first factor* $L \rightarrowtail X$ *is a mono, and as*

arrows *from a factorization* $L \rightarrowtail X \to A$ *to another one* $L \rightarrowtail Y \to A$, *all arrows* $f \colon X \to Y$ *in* $\mathbf{C}$ *such that the diagram to the right is made of a commutative triangle and a pullback square.*

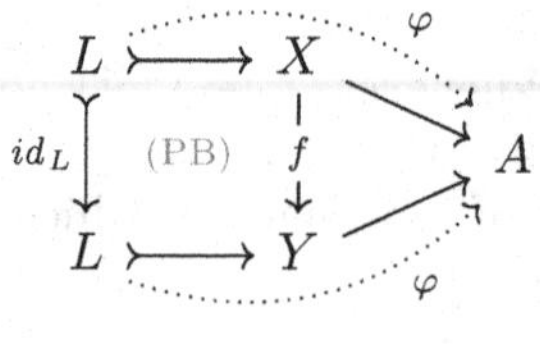

If $\mathbf{Mat}_\varphi$ *has a terminal object it is denoted by* $L \rightarrowtail \langle\varphi\rangle \to A$ *and is called the* materialization *of* φ.

Sometimes we will also call the object $\langle\varphi\rangle$ the materialization of φ, omitting the arrows.

Since we are working in a topos by assumption, the slice category over A provides us with a convenient setting to construct materializations. Note in particular that in the diagram in Definition 7 above, the span $X \leftarrowtail L \rightarrowtail L$ is a partial map from X to L in the slice category over A. Hence the materialization $\langle\varphi\rangle$ corresponds to the partial map classifier for L in this slice category.

Proposition 8 (Existence of materialization). *Let* $\varphi \colon L \to A$ *be an arrow in* $\mathbf{C}$, *and let* $\eta_\varphi \colon \varphi \to F(\varphi)$, *with* $F(\varphi) \colon \bar{A} \to A$, *be the partial map classifier of* φ *in the slice category* $\mathbf{C} \downarrow A$ *(which also is a topos).*[2] *Then* $L \overset{\eta_\varphi}{\to} \bar{A} \overset{F(\varphi)}{\to} A$ *is the materialization of* φ, *hence* $\langle\varphi\rangle = \bar{A}$.

As a direct consequence of Propositions 4 and 8 (and the fact that final pullback complements in the slice category correspond to those in the base category [25]), the terminal object of the materialization category can be constructed for each arrow of a topos by taking final pullback complements.

Corollary 9 (Construction of the materialization). *Let* $\varphi \colon L \to A$ *be an arrow of* $\mathbf{C}$ *and let* $\mathtt{true}_A \colon A \rightarrowtail A \times \Omega$ *be the subobject classifier (in the slice category* $\mathbf{C} \downarrow A$*) from* $id_A \colon A \to A$ *to the projection* $\pi_1 \colon A \times \Omega \to A$. *Then the terminal object* $L \overset{\eta_\varphi}{\rightarrowtail} \langle\varphi\rangle \overset{\psi}{\to} A$ *in the materialization category consists of the arrows* η_φ *and* $\psi = \pi_1 \circ \chi_{\eta_\varphi}$, *where* $L \overset{\eta_\varphi}{\rightarrowtail} \langle\varphi\rangle \overset{\chi_{\eta_\varphi}}{\to} A \times \Omega$ *is the final pullback complement of* $L \overset{\varphi}{\to} A \overset{\mathtt{true}_A}{\rightarrowtail} A \times \Omega$.

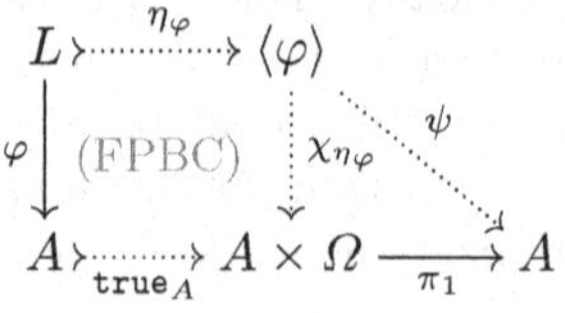

[2] This is by the Fundamental Theorem of topos theory [17, Theorem 2.31].

Example 10. We construct the materialization $L \overset{\eta_\varphi}{\rightarrowtail} \langle\varphi\rangle \overset{\psi}{\rightarrow} A$ for the following morphism $\varphi\colon L \to A$ of graphs with a single (omitted) label:

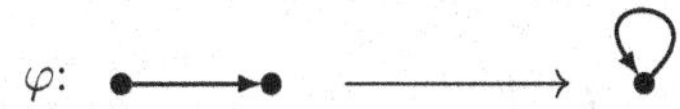

In particular, the materialization is obtained as a final pullback complement as depicted to the right (compare with the corresponding diagram in Corollary 9). Note that edges which are not in the image of η_φ resp. $\mathtt{true}_A$ are dashed.

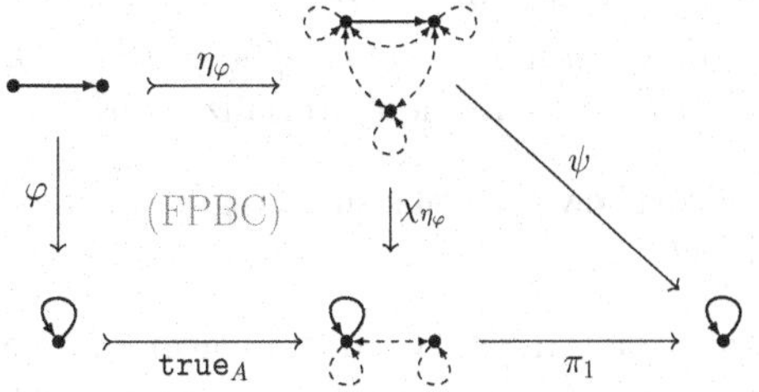

This construction corresponds to the usual intuition behind materialization: the left-hand side and the edges that are attached to it are "pulled out" of the given abstract graph.

We can summarize the result of our constructions in the following proposition:

Proposition 11 (Language of the materialization). *Let $\varphi\colon L \to A$ be an arrow in $\mathbf{C}$ and let $L \overset{\eta_\varphi}{\rightarrowtail} \langle\varphi\rangle \to A$ be the corresponding materialization. Then we have*

$$\mathcal{L}(L \overset{\eta_\varphi}{\rightarrowtail} \langle\varphi\rangle) = \{L \overset{m_L}{\rightarrowtail} X \mid \exists\psi\colon (X \to A).\ (\varphi = \psi \circ m_L)\}.$$

3.2 Characterizing the Language of Rewritable Objects

A match obtained through the materialization of the left-hand side of a production from a given object may not allow a DPO rewriting step because of the gluing condition. We illustrate this problem with an example.

Example 12. Consider the materialization $L \rightarrowtail \langle\varphi\rangle \to A$ from Example 10 and the production $L \leftarrowtail I \rightarrowtail R$ shown in the diagram to the right. It is easy to see that the pushout complement of morphisms $I \rightarrowtail L \rightarrowtail \langle\varphi\rangle$ does not exist.

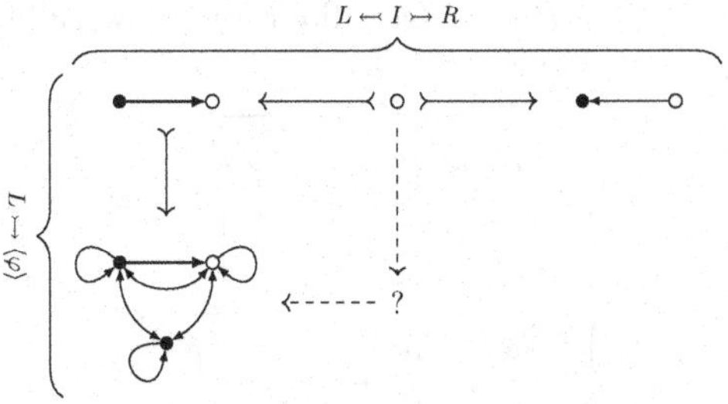

Nevertheless there exist factorizations $L \rightarrowtail X \to A$ abstracted by $\langle\varphi\rangle$ that could be rewritten using the production.

In order to take the existence of pushout complements into account, we consider a subcategory of the materialization category.

Definition 13 (Materialization subcategory of rewritable objects). *Let $\varphi\colon L \to A$ be an arrow of $\mathbf{C}$ and let $\varphi_L\colon I \rightarrowtail L$ be a mono (corresponding to the left leg of a production). The* materialization subcategory of rewritable objects *for φ and φ_L, denoted* $\mathbf{Mat}_{\varphi}^{\varphi_L}$*, is the full subcategory of* $\mathbf{Mat}_{\varphi}$ *containing as objects all factorizations $L \overset{m}{\rightarrowtail} X \to A$ of φ, where m is a mono and $I \overset{\varphi_L}{\rightarrowtail} L \overset{m}{\rightarrowtail} X$ has a pushout complement.*

Its terminal element, if it exists, is denoted by $L \overset{n_L}{\rightarrowtail} \langle\!\langle \varphi, \varphi_L \rangle\!\rangle \to A$ and is called the rewritable materialization.

We show that this subcategory of the materialization category has a terminal object.

Proposition 14 (Construction of the rewritable materialization). *Let $\varphi\colon L \to A$ be an arrow and let $\varphi_L\colon I \rightarrowtail L$ be a mono of* $\mathbf{C}$. *Then the* rewritable materialization of φ w.r.t. φ_L *exists and can be constructed as the following factorization $L \overset{n_L}{\rightarrowtail} \langle\!\langle \varphi, \varphi_L \rangle\!\rangle \overset{\psi\circ\alpha}{\longrightarrow} A$ of φ. In the left diagram, F is obtained as the final pullback complement of $I \overset{\varphi_L}{\rightarrowtail} L \rightarrowtail \langle\varphi\rangle$, where $L \rightarrowtail \langle\varphi\rangle \overset{\psi}{\to} A$ is the materialization of φ (Definition 7). Next in the right diagram $L \overset{n_L}{\rightarrowtail} \langle\!\langle \varphi, \varphi_L \rangle\!\rangle \overset{\beta}{\leftarrowtail} F$ is the pushout of the span $L \overset{\varphi_L}{\leftarrowtail} I \rightarrowtail F$ and α is the resulting mediating arrow.*

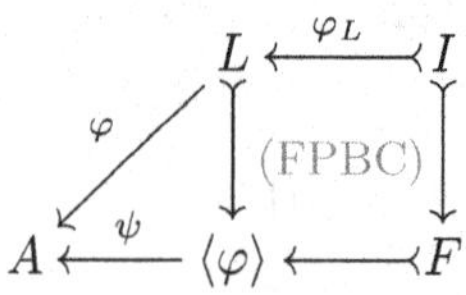

(2)

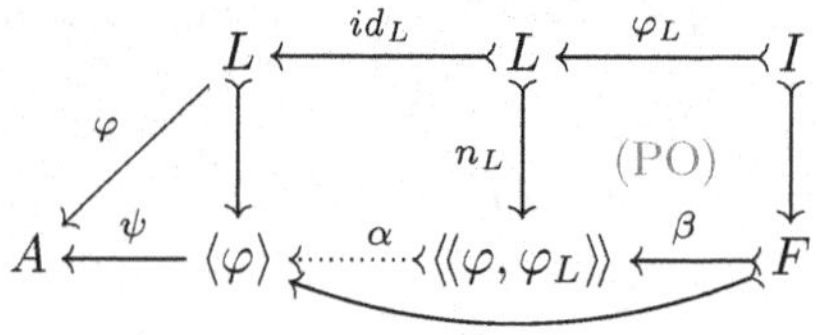

(3)

Example 15. We come back to the running example (Example 12) and, as in Proposition 14, determine the final pullback complement $I \rightarrowtail F \rightarrowtail \langle\varphi\rangle$ of $I \overset{\varphi_L}{\rightarrowtail} L \rightarrowtail \langle\varphi\rangle$ (see diagram below left) and obtain $\langle\!\langle \varphi, \varphi_L \rangle\!\rangle$ by taking the pushout over $L \leftarrowtail I \rightarrowtail F$ (see diagram below right).

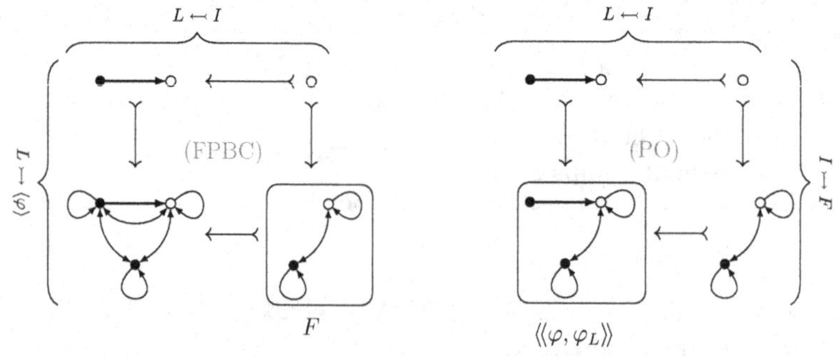

It remains to be shown that $L \rightarrowtail \langle\!\langle \varphi, \varphi_L \rangle\!\rangle \to A$ represents every factorization which can be rewritten. As before we obtain a characterization of the rewritable objects, including the match, as the language of an arrow.

Proposition 16 (Language of the rewritable materialization). *Assume there is a production* $p\colon L \overset{\varphi_L}{\leftarrowtail} I \overset{\varphi_R}{\rightarrowtail} R$ *and let* $L \overset{n_L}{\rightarrowtail} \langle\!\langle \varphi, \varphi_L \rangle\!\rangle$ *be the match for the rewritable materialization for* φ *and* φ_L. *Then we have*

$$\mathcal{L}(L \overset{n_L}{\rightarrowtail} \langle\!\langle \varphi, \varphi_L \rangle\!\rangle) = \{L \overset{m_L}{\rightarrowtail} X \mid \exists \psi\colon (X \to A).\ (\varphi = \psi \circ m_L \wedge X \overset{p,m_L}{\Longrightarrow})\}.$$

3.3 Rewriting Materializations

In the next step we will now rewrite the rewritable materialization $\langle\!\langle \varphi, \varphi_L \rangle\!\rangle$ with the match $L \overset{n_L}{\rightarrowtail} \langle\!\langle \varphi, \varphi_L \rangle\!\rangle$, resulting in a co-match $R \rightarrowtail B$. In particular, we will show that this co-match represents all co-matches that can be obtained by rewriting an object X of $\mathcal{L}(A)$ at a match compatible with φ. We first start with an example.

Example 17. We can rewrite the materialization $L \rightarrowtail \langle\!\langle \varphi, \varphi_L \rangle\!\rangle \to A$ as follows:

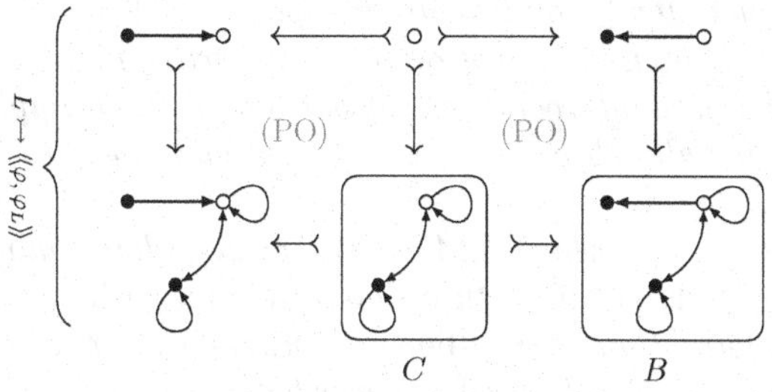

Proposition 18 (Rewriting abstract matches). *Let a match* $n_L\colon L \rightarrowtail \tilde{A}$ *and a production* $p\colon L \leftarrowtail I \rightarrowtail R$ *be given. Assume that* $\tilde{A}$ *is rewritten along the match* n_L, *i.e.,* $(L \overset{n_L}{\rightarrowtail} \tilde{A}) \overset{p}{\Rightarrow} (R \overset{n_R}{\rightarrowtail} B)$. *Then*

$$\mathcal{L}(R \overset{n_R}{\rightarrowtail} B) = \{R \overset{m_R}{\rightarrowtail} Y \mid \exists (L \overset{m_L}{\rightarrowtail} X) \in \mathcal{L}(L \overset{n_L}{\rightarrowtail} \tilde{A}).\ ((L \overset{m_L}{\rightarrowtail} X) \overset{p}{\Rightarrow} (R \overset{m_R}{\rightarrowtail} Y))\}$$

If we combine Propositions 16 and 18, we obtain the following corollary that characterizes the co-matches obtained from rewriting a match compatible with $\varphi\colon L \to A$.

Corollary 19 (Co-match language of the rewritable materialization). *Let* $\varphi\colon L \to A$ *and a production* $p\colon L \overset{\varphi_L}{\leftarrowtail} I \overset{\varphi_R}{\rightarrowtail} R$ *be given. Assume that* $\langle\!\langle \varphi, \varphi_L \rangle\!\rangle$ *is obtained as the rewritable materialization of* φ *and* φ_L *with match* $L \overset{n_L}{\rightarrowtail} \langle\!\langle \varphi, \varphi_L \rangle\!\rangle$ *(see Proposition 14). Furthermore let* $(L \overset{n_L}{\rightarrowtail} \langle\!\langle \varphi, \varphi_L \rangle\!\rangle) \overset{p}{\Rightarrow} (R \overset{n_R}{\rightarrowtail} B)$. *Then*

$$\mathcal{L}(R \overset{n_R}{\rightarrowtail} B) = \{R \overset{m_R}{\rightarrowtail} Y \mid \exists (L \overset{m_L}{\rightarrowtail} X), (X \overset{\psi}{\to} A).\ (\varphi = \psi \circ m_L \wedge$$
$$(L \overset{m_L}{\rightarrowtail} X) \overset{p}{\Rightarrow} (R \overset{m_R}{\rightarrowtail} Y))\}$$

This result does not yet enable us to construct post-conditions for languages of objects. The set of co-matches can be fully characterized as the language of a mono, which can only be achieved by fixing the right-hand side R and thus ensuring that exactly one occurrence of R is represented. However, as soon as we forget about the co-match, this effect is gone and can only be retrieved by adding annotations, which will be introduced next.

4 Annotated Objects

We now endow objects with annotations, thus making object languages more expressive. In particular we will use ordered monoids in order to annotate objects. Similar annotations have already been studied in [20] in the context of type systems and in [10] with the aim of studying decidability and closure properties, but not for abstract rewriting.

Definition 20 (Ordered monoid). *An* ordered monoid $(\mathcal{M}, +, \leq)$ *consists of a set* $\mathcal{M}$*, a partial order* $\leq$ *and a binary operation* $+$ *such that* $(\mathcal{M}, +)$ *is a monoid with unit* 0 *(which is the bottom element wrt.* $\leq$*) and the partial order is compatible with the monoid operation. In particular* $a \leq b$ *implies* $a + c \leq b + c$ *and* $c + a \leq c + b$ *for all* $a, b, c \in \mathcal{M}$*. An ordered monoid is commutative if* $+$ *is commutative.*

A tuple $(\mathcal{M}, +, -, \leq)$*, where* $(\mathcal{M}, +, \leq)$ *is an ordered monoid and* $-$ *is a binary operation on* $\mathcal{M}$*, is called an* ordered monoid with subtraction.

We say that subtraction is well-behaved *whenever for all* $a, b \in \mathcal{M}$ *it holds that* $a - a = 0$ *and* $(a - b) + b = a$ *whenever* $b \leq a$*.*

For now subtraction is just any operation, without specific requirements. Later we will concentrate on specific subtraction operations and demand that they are well-behaved.

In the following we will consider only commutative monoids.

Definition 21 (Monotone maps and homomorphisms). *Let* $\mathcal{M}_1$*,* $\mathcal{M}_2$ *be two ordered monoids. A map* $h\colon \mathcal{M}_1 \to \mathcal{M}_2$ *is called* monotone *if* $a \leq b$ *implies* $h(a) \leq h(b)$ *for all* $a, b \in \mathcal{M}_1$*. The category of ordered monoids with subtraction and monotone maps is called* **Mon**.

A monotone map h *is called a* homomorphism *if* $h(0) = 0$ *and* $h(a + b) = h(a) + h(b)$*. If* $\mathcal{M}_1, \mathcal{M}_2$ *are ordered monoids with subtraction, we say that* h *preserves subtraction if* $h(a - b) = h(a) - h(b)$*.*

Example 22. Let $n \in \mathbb{N}\backslash\{0\}$ and take $\mathcal{M}_n = \{0, 1, \ldots, n, *\}$ (zero, one, ..., n, many) with $0 \leq 1 \leq \cdots \leq n \leq *$ and addition as (commutative) monoid operation with the proviso that $a + b = *$ if the sum is larger than n. In addition $a + * = *$ for all $a \in \mathcal{M}_n$. Subtraction is truncated subtraction where $a - b = 0$ if $a \leq b$. Furthermore $* - a = *$ for all $a \in \mathbb{N}$. It is easy to see that subtraction is well-behaved.

Given a set S and an ordered monoid (with subtraction) $\mathcal{M}$, it is easy to check that also $\mathcal{M}^S$ is an ordered monoid (with subtraction), where the elements are functions from S to $\mathcal{M}$ and the partial order, the monoidal operation and the subtraction are taken pointwise.

The following path monoid is useful if we want to annotate a graph with information over which paths are present. Note that due to the possible fusion of nodes and edges caused by the abstraction, a path in the abstract graph does not necessarily imply the existence of a corresponding path in a concrete graph. Hence annotations based on such a monoid, which provide information about the existence of paths, can yield useful additional information.

Example 23. Given a graph G, we denote by $E_G^+ \subseteq V_G \times V_G$ the transitive closure of the edge relation $E_G^{\rightarrow} = \{(src_G(e), tgt_G(e)) \mid e \in E_G\}$. The *path monoid* $\mathcal{P}_G$ of G has the carrier set $\mathcal{P}(E_G^+)$. The partial order is simply inclusion and the monoid operation is defined as follows: given $P_0, P_1 \in \mathcal{P}_G$, we have

$$P_0 + P_1 = \{(v_0, v_n) \mid \exists v_1, \ldots, v_{n-1} \colon (v_i, v_{i+1}) \in P_{j_i},\\ j_0 \in \{0,1\}, j_{i+1} = 1 - j_i, i \in \{0, \ldots, n-1\} \text{ and } n \in \mathbb{N}\}.$$

That is, new paths can be formed by concatenating alternating path fragments from P_0, P_1. It is obvious to see that $+$ is commutative and one can also show associativity. $P = \emptyset$ is the unit. Subtraction simply returns the first parameter: $P_0 - P_1 = P_0$.

We will now formally define annotations for objects via a functor from a given category to **Mon**.

Definition 24 (Annotations for objects). *Given a category* **C** *and a functor* $\mathcal{A}\colon \mathbf{C} \to \mathbf{Mon}$, *an* annotation based on $\mathcal{A}$ *for an object* $X \in \mathbf{C}$ *is an element* $a \in \mathcal{A}(X)$. *We write* $\mathcal{A}_\varphi$, *instead of* $\mathcal{A}(\varphi)$, *for the action of functor* $\mathcal{A}$ *on a* **C***-arrow* φ. *We assume that for each object* X *there is a* standard annotation *based on* $\mathcal{A}$ *that we denote by* s_X, *thus* $s_X \in \mathcal{A}(X)$.

It can be shown quite straightforwardly that the forgetful functor mapping an annotated object $X[a]$, with $a \in \mathcal{A}(X)$, to X is an op-fibration (or co-fibration [19]), arising via the Grothendieck construction.

Our first example is an annotation of graphs with global multiplicities, counting nodes and edges, where the action of the functor is to sum up those multiplicities.

Example 25. Given $n \in \mathbb{N}\backslash\{0\}$, we define the functor $\mathcal{B}^n$: **Graph** $\to$ **Mon**: For every graph G, $\mathcal{B}^n(G) = \mathcal{M}_n^{V_G \cup E_G}$. For every graph morphism $\varphi\colon G \to H$ and $a \in \mathcal{B}^n(G)$, we have $\mathcal{B}^n_\varphi(a) \in \mathcal{M}_n^{V_H \cup E_H}$ with:

$$\mathcal{B}^n_\varphi(a)(y) = \sum_{\varphi(x)=y} a(x), \quad \textit{where } x \in (V_G \cup E_G) \textit{ and } y \in (V_H \cup E_H).$$

Therefore an annotation based on a functor $\mathcal{B}^n$ associates every item of a graph with a number (or the top value $*$). We will call such annotations *multiplicities.* Furthermore the action of the functor on a morphism transforms a multiplicity by summing up (in $\mathcal{M}_n$) the values of all items of the source graph that are mapped to the same item of the target graph.

For a graph G, its *standard multiplicity* $s_G \in \mathcal{B}^n(G)$ is defined as the function which maps every node and edge of G to 1.

As another example we consider local annotations which record the out-degree of a node and where the action of the functor is to take the supremum instead of the sum.

Example 26. Given $n \in \mathbb{N}\backslash\{0\}$, we define the functor $\mathcal{S}^n$: **Graph** $\to$ **Mon** as follows: For every graph G, $\mathcal{S}^n(G) = \mathcal{M}_n^{V_G}$. For every graph morphism $\varphi \colon G \to H$ and $a \in \mathcal{S}^n(G)$, we have $\mathcal{S}^n_\varphi(a) \in \mathcal{M}_n^{V_H}$ with:

$$\mathcal{S}^n_\varphi(a)(w) = \bigvee_{\varphi(v)=w} a(v), \quad \textit{where } v \in V_G \textit{ and } w \in V_H.$$

For a graph G, its *standard annotation* $s_G \in \mathcal{S}^n(G)$ is defined as the function which maps every node of G to its out-degree (or $*$ if the out-degree is larger than n).

Finally, we consider annotations based on the path monoid (see Example 23).

Example 27. We define the functor $\mathcal{T}$: **Graph** $\to$ **Mon** as follows: For every graph G, $\mathcal{T}(G) = \mathcal{P}_G$. For every graph morphism $\varphi \colon G \to H$ and $P \in \mathcal{T}(G)$, we have $\mathcal{T}_\varphi(P) \in \mathcal{P}_H$ with:

$$\mathcal{T}_\varphi(P) = \{(\varphi(v), \varphi(w)) \mid (v, w) \in P\}.$$

For a graph G, its *standard annotation* $s_G \in \mathcal{T}(G)$ is the transitive closure of the edge relation, i.e., $s_G = E_G^+$.

In the following we will consider only annotations satisfying certain properties in order to achieve soundness and completeness.

Definition 28 (Properties of annotations). *Let $\mathcal{A}$: $\mathbf{C} \to$ **Mon** be an annotation functor, together with standard annotations. In this setting we say that*

- *the* homomorphism property *holds if whenever φ is a mono, then $\mathcal{A}_\varphi$ is a monoid homomorphism, preserving also subtraction.*
- *the* adjunction property *holds if whenever $\varphi \colon A \rightarrowtail B$ is a mono, then*
  - *$\mathcal{A}_\varphi \colon \mathcal{A}(A) \to \mathcal{A}(B)$ has a right adjoint $red_\varphi \colon \mathcal{A}(B) \to \mathcal{A}(A)$, i.e., red_φ is monotone and satisfies $a \le red_\varphi(\mathcal{A}_\varphi(a))$ for $a \in \mathcal{A}(A)$ and $\mathcal{A}_\varphi(red_\varphi(b)) \le b$ for $b \in \mathcal{A}(B)$.*[3]

[3] This amounts to saying that the forgetful functor is a bifibration when we restrict to monos, see [19, Lem. 9.1.2].

- *red_φ is a monoid homomorphism that preserves subtraction.*
- *it holds that $red_\varphi(s_B) = s_A$, where s_A, s_B are standard annotations.*

Furthermore, assuming that $\mathcal{A}_\varphi$ has a right adjoint red_φ, we say that

– *the* pushout property *holds, whenever for each pushout as shown in the diagram to the right, with all arrows monos where $\eta = \psi_1 \circ \varphi_1 = \psi_2 \circ \varphi_2$, it holds that for every $d \in \mathcal{A}(D)$:*

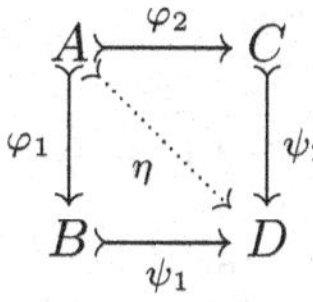

$$d = \mathcal{A}_{\psi_1}(red_{\psi_1}(d)) + (\mathcal{A}_{\psi_2}(red_{\psi_2}(d)) - \mathcal{A}_\eta(red_\eta(d))).$$

We say that the pushout property for standard annotations *holds if we replace d by s_D, $red_\eta(d)$ by s_A, $red_{\psi_1}(d)$ by s_B and $red_{\psi_2}(d)$ by s_C.*

– *the* Beck-Chevalley property *holds if whenever the square shown to the right is a pullback with φ_1, ψ_2 mono, then it holds for every $b \in \mathcal{A}(B)$ that*

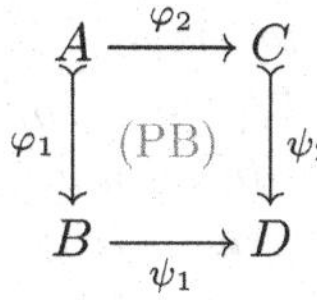

$$\mathcal{A}_{\varphi_2}(red_{\varphi_1}(b)) = red_{\psi_2}(\mathcal{A}_{\psi_1}(b)).$$

Note that the annotation functor from Example 25 satisfies all properties above, whereas the functors from Examples 26 and 27 satisfy both the homomorphism property and the pushout property for standard annotations, but do not satisfy all the remaining requirements [8].

We will now introduce a more flexible notion of language, by equipping the abstract objects with two annotations, establishing lower and upper bounds.

Definition 29 (Doubly annotated object). *Given a topos* **C** *and a functor $\mathcal{A}$:* **C** $\to$ **Mon**, *a* doubly annotated object $A[a_1, a_2]$ *is an object A of* **C** *with two annotations $a_1, a_2 \in \mathcal{A}(A)$.An arrow $\varphi\colon A[a_1, a_2] \to B[b_1, b_2]$, also called a* legal arrow, *is a* **C***-arrow $\varphi\colon A \to B$ such that $\mathcal{A}_\varphi(a_1) \geq b_1$ and $\mathcal{A}_\varphi(a_2) \leq b_2$.*

The language of a doubly annotated object $A[a_1, a_2]$ *(also called the language of objects which are abstracted by $A[a_1, a_2]$) is defined as follows:*

$$\mathcal{L}(A[a_1, a_2]) = \{X \in \mathbf{C} \mid \textit{there exists a legal arrow } \varphi\colon X[s_X, s_X] \to A[a_1, a_2]\}$$

Note that legal arrows are closed under composition [9]. Examples of doubly annotated objects are given in Example 36 for global annotations from Example 25 (providing upper and lower bounds for the number of nodes resp. edges in the preimage of a given element). Graph elements without annotation are annotated by $[0, *]$ by default.

Definition 30 (Isomorphism property). *An annotation functor $\mathcal{A}$:* **C** $\to$ **Mon**, *together with standard annotations, satisfies the* isomorphism property *if the following holds: whenever $\varphi\colon X[s_X, s_X] \to Y[s_Y, s_Y]$ is legal, then φ is an isomorphism, i.e., $\mathcal{L}(Y[s_Y, s_Y])$ contains only Y itself (and objects isomorphic to Y).*

5 Abstract Rewriting of Annotated Objects

We will now show how to actually rewrite annotated objects. The challenge is both to find suitable annotations for the materialization and to "rewrite" the annotations.

5.1 Abstract Rewriting and Soundness

We first describe how the annotated rewritable materialization is constructed and then we investigate its properties.

Definition 31 (Construction of annotated rewritable materialization). *Let $p\colon L \overset{\varphi_L}{\leftarrowtail} I \overset{\varphi_R}{\rightarrowtail} R$ be a production and let $A[a_1, a_2]$ be a doubly annotated object. Furthermore let $\varphi\colon L \to A$ be an arrow.*

We first construct the factorization $L \overset{n_L}{\rightarrowtail} \langle\!\langle \varphi, \varphi_L \rangle\!\rangle \overset{\psi}{\to} A$, obtaining the rewritable materialization $\langle\!\langle \varphi, \varphi_L \rangle\!\rangle$ from Definition 13. Next, let M contain all maximal[4] elements of the set

$$\{(a_1', a_2') \in \mathcal{A}(\langle\!\langle \varphi, \varphi_L \rangle\!\rangle)^2 \mid \mathcal{A}_{n_L}(s_L) \le a_2', a_1 \le \mathcal{A}_\psi(a_1'), \mathcal{A}_\psi(a_2') \le a_2\}.$$

Then the doubly annotated objects $\langle\!\langle \varphi, \varphi_L \rangle\!\rangle[a_1', a_2']$ with $(a_1', a_2') \in M$ are the annotated rewritable materializations for $A[a_1, a_2]$, φ and φ_L.

Note that in general there can be several such materializations, differing by the annotations only, or possibly none. The definition of M ensures that the upper bound a_2' of the materialization covers the annotations arising from the left-hand side. We cannot use a corresponding condition for the lower bound, since the materialization might contain additional structures, hence the arrow n_L is only "semi-legal". A more symmetric condition will be studied in Sect. 5.2.

Proposition 32 (Annotated rewritable materialization is terminal). *Given a production $p\colon L \overset{\varphi_L}{\leftarrowtail} I \overset{\varphi_R}{\rightarrowtail} R$, let $L \overset{m_L}{\rightarrowtail} X$ be the match of L in an object X such that $X \overset{p,m_L}{\Longrightarrow}$, i.e., X can be rewritten. Assume that X is abstracted by $A[a_1, a_2]$, witnessed by ψ. Let $\varphi = \psi \circ m_L$ and let $L \overset{n_L}{\rightarrowtail} \langle\!\langle \varphi, \varphi_L \rangle\!\rangle \overset{\psi'}{\to} A$ the the corresponding rewritable materialization. Then there exists an arrow ζ_A and a pair of annotations $(a_1', a_2') \in M$ for $\langle\!\langle \varphi, \varphi_L \rangle\!\rangle$ (as described in Definition 31) such that the diagram below commutes and the square is a pullback in the underlying category. Furthermore the triangle consists of legal arrows. This means in particular that ζ_A is legal.*

$$\begin{array}{ccccc}
L[s_L, s_L] & \overset{m_L}{\rightarrowtail} & X[s_X, s_X] & \overset{\psi}{\longrightarrow} & A[a_1, a_2] \\
{\scriptstyle id_L}\downarrow & \text{(PB)} & \downarrow{\scriptstyle \zeta_A} & \nearrow{\scriptstyle \psi'} & \\
L[s_L, s_L] & \underset{n_L}{\rightarrowtail} & \langle\!\langle \varphi, \varphi_L \rangle\!\rangle[a_1', a_2'] & &
\end{array}$$

4 "Maximal" means maximality with respect to the interval order $(a_1, a_2) \sqsubseteq (a_1', a_2') \iff a_1' \le a_1, a_2 \le a_2'$.

Having performed the materialization, we will now show how to rewrite annotated objects. Note that we cannot simply take pushouts in the category of annotated objects and legal arrows, since this would result in taking the supremum of annotations, when instead we need the sum (subtracting the annotation of the interface I, analogous to the inclusion-exclusion principle).

Definition 33 (Abstract rewriting step$\rightsquigarrow$). *Let $p\colon L \overset{\varphi_L}{\hookleftarrow} I \overset{\varphi_R}{\rightarrowtail} R$ be a production and let $A[a_1,a_2]$ be an annotated abstract object. Furthermore let $\varphi\colon L \to A$ be a match of a left-hand side, let $n_L\colon L \rightarrowtail \langle\langle \varphi,\varphi_L\rangle\rangle$ be the match obtained via materialization and let $(a_1',a_2') \in M$ (as in Definition 31).*

Then $A[a_1,a_2]$ can be transformed to $B[b_1,b_2]$ via p if there are arrows such that the two squares below are pushouts in the base category and b_1,b_2 are defined as:

$$b_i = \mathcal{A}_{\varphi_B}(c_i) + (\mathcal{A}_{n_R}(s_R) - \mathcal{A}_{n_R\circ\varphi_R}(s_I)) \qquad \text{for } i \in \{1,2\}$$

where c_1,c_2 are maximal annotations such that:

$$a_1' \le \mathcal{A}_{\varphi_A}(c_1) + (\mathcal{A}_{n_L}(s_L) - \mathcal{A}_{n_L\circ\varphi_L}(s_I)) \quad \mathcal{A}_{\varphi_A}(c_2) + (\mathcal{A}_{n_L}(s_L) - \mathcal{A}_{n_L\circ\varphi_L}(s_I)) \le a_2'$$

$$\begin{array}{ccccc} L[s_L,s_L] & \xleftarrow{\varphi_L} & I[s_I,s_I] & \xrightarrow{\varphi_R} & R[s_R,s_R] \\ \downarrow n_L & & \downarrow n_I & & \downarrow n_R \\ \langle\langle \varphi,\varphi_L\rangle\rangle[a_1',a_2'] & \xleftarrow{\varphi_A} & C[c_1,c_2] & \xrightarrow{\varphi_B} & B[b_1,b_2] \end{array}$$

In this case we write $A[a_1,a_2] \overset{p,\varphi}{\rightsquigarrow} B[b_1,b_2]$ and say that $A[a_1,a_2]$ makes an abstract rewriting step *to $B[b_1,b_2]$.*

We will now show soundness of abstract rewriting, i.e., whenever an object X is abstracted by $A[a_1,a_2]$ and X is rewritten to Y, then there exists an abstract rewriting step from $A[a_1,a_2]$ to $B[b_1,b_2]$ such that Y is abstracted by $B[b_1,b_2]$.

Assumption: In the following we will require that the homomorphism property as well as the pushout property for standard annotations hold (cf. Definition 28).

Proposition 34 (Soundness for $\rightsquigarrow$). *Relation $\rightsquigarrow$ is sound in the following sense: Let $X \in \mathcal{L}(A[a_1,a_2])$ (witnessed via a legal arrow $\psi\colon X[s_X,s_X] \to A[a_1,a_2]$) where $X \overset{p,m_L}{\Longrightarrow} Y$. Then there exists an abstract rewriting step $A[a_1,a_2] \overset{p,\psi\circ m_L}{\rightsquigarrow} B[b_1,b_2]$ such that $Y \in \mathcal{L}(B[b_1,b_2])$.*

5.2 Completeness

The conditions on the annotations that we imposed so far are too weak to guarantee completeness, that is the fact that every object represented by $B[b_1,b_2]$ can be obtained by rewriting an object represented by $A[a_1,a_2]$. This can be clearly seen by the fact that the requirements hold also for the singleton monoid and, as discussed before, the graph structure of B is insufficient to characterize the successor objects or graphs.

Hence we will now strengthen our requirements in order to obtain completeness.

Assumption: In addition to the assumptions of Sect. 5.1, we will need that subtraction is well-behaved and that the adjunction property, the pushout property, the Beck-Chevalley property (Definition 28) and the isomorphism property (Definition 30) hold.

The global annotations from Example 25 satisfy all these properties. In particular, given an injective graph morphism $\varphi\colon G \rightarrowtail H$ the right adjoint $red_\varphi : \mathcal{M}_n^{V_H \cup E_H} \to \mathcal{M}_n^{V_G \cup E_G}$ to $\mathcal{B}_\varphi^n$ is defined as follows: given an annotation $b \in \mathcal{M}_n^{V_H \cup E_H}$, $red_\varphi(b)(x) = b(\varphi(x))$, i.e., red_φ simply provides a form of reindexing.

We will now modify the abstract rewriting relation and allow only those abstract annotations for the materialization that reduce to the standard annotation of the left-hand side.

Definition 35 (Abstract rewriting step $\hookrightarrow$). *Given $\varphi\colon L \to A$, assume that $B[b_1, b_2]$ is constructed from $A[a_1, a_2]$ via the construction described in Definitions 31 and 33, with the modification that the set of annotations from which the set of maximal annotations M of the materialization $\langle\!\langle \varphi, \varphi_L \rangle\!\rangle$ are taken, is replaced by:*

$$\{(a_1', a_2') \in \mathcal{A}(\langle\!\langle \varphi, \varphi_L \rangle\!\rangle)^2 \mid red_{n_L}(a_i') = s_L, i \in \{1,2\}, a_1 \le \mathcal{A}_\psi(a_1'), \mathcal{A}_\psi(a_2') \le a_2\}.$$

In this case we write $A[a_1, a_2] \overset{p,\varphi}{\hookrightarrow} B[b_1, b_2]$.

Due to the adjunction property we have $\mathcal{A}_{n_L}(s_L) = \mathcal{A}_{n_L}(red_{n_L}(a_2')) \le a_2'$ and hence the set M of annotations of Definition 35 is a subset of the corresponding set of Definition 33.

Example 36. We give a small example of an abstract rewriting step (a more extensive, worked example can be found in the full version [8]). Elements without annotation are annotated by $[0,*]$ by default and those with annotation $[0,0]$ are omitted. Furthermore elements in the image of the match and co-match are annotated by the standard annotation $[1,1]$ to specify the concrete occurrence of the left-hand and right-hand side.

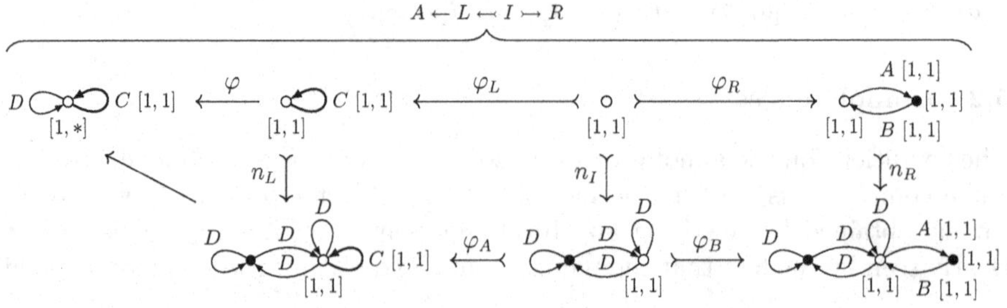

The variant of abstract rewriting introduced in Definition 35 can still be proven to be sound, assuming the extra requirements stated above.

Proposition 37 (Soundness for $\hookrightarrow$). *Relation $\hookrightarrow$ is sound in the sense of Proposition 34.*

Using the assumptions we can now show completeness.

Proposition 38 (Completeness for $\hookrightarrow$). *If $A[a_1, a_2] \overset{p,\varphi}{\hookrightarrow} B[b_1, b_2]$ and $Y \in \mathcal{L}(B[b_1, b_2])$, then there exists $X \in \mathcal{L}(A[a_1, a_2])$ (witnessed via a legal arrow $\psi\colon X[s_X, s_X] \to A[a_1, a_2]$) such that $X \overset{p,m_L}{\Longrightarrow} Y$ and $\varphi = \psi \circ m_L$.*

Finally, we can show that annotated graphs of this kind are expressive enough to construct a strongest post-condition. If we would allow several annotations for objects, as in [9], we could represent the language with a single (multiply) annotated object.

Corollary 39 (Strongest post-condition). *Let $A[a_1, a_2]$ be an annotated object and let $\varphi\colon L \to A$. We obtain (several) abstract rewriting steps $A[a_1, a_2] \overset{p,\varphi}{\hookrightarrow} B[b_1, b_2]$, where we always obtain the same object B. (B is dependent on φ, but not on the annotation.) Now let $N = \{(b_1, b_2) \mid A[a_1, a_2] \overset{p,\varphi}{\hookrightarrow} B[b_1, b_2]\}$. Then*

$$\bigcup_{(b_1,b_2)\in N} \mathcal{L}(B[b_1, b_2]) = \{Y \mid \exists(X \in \mathcal{L}(A[a_1, a_2]), \textit{witnessed by } \psi), (L \overset{m_L}{\rightarrowtail} X). \\ (\varphi = \psi \circ m_L \wedge X \overset{p,m_L}{\Longrightarrow} Y)\}$$

6 Conclusion

We have described a rewriting framework for abstract graphs that also applies to objects in any topos, based on existing work for graphs [1,2,4,27,28,31]. In particular, we have given a blueprint for materialization in terms of the universal property of partial map classifiers. This is a first theoretical milestone towards shape analysis as a general static analysis method for rule-based systems with graph-like objects as states. Soundness and completeness results for the rewriting of abstract objects with annotations in an ordered monoid provide an effective verification method for the special case of graphs We plan to implement the materialization construction and the computation of rewriting steps of abstract graphs in a prototype tool.

The extension of annotations with logical formulas is the natural next step, which will lead to a more flexible and versatile specification language, as described in previous work [30,31]. The logic can possibly be developed in full generality using the framework of nested application conditions [18,23] that applies to objects in adhesive categories. This logical approach might even reduce the proof obligations for annotation functors. Another topic for future work is the integration of widening or similar approximation techniques, which collapse abstract objects and ideally lead to finite abstract transition systems that (over-)approximate the typically infinite transitions systems of graph transformation systems.

References

1. Backes, P.: Cluster abstraction of graph transformation systems. Ph.D. thesis, Saarland University (2015)
2. Backes, P., Reineke, J.: Analysis of infinite-state graph transformation systems by cluster abstraction. In: D'Souza, D., Lal, A., Larsen, K.G. (eds.) VMCAI 2015. LNCS, vol. 8931, pp. 135–152. Springer, Heidelberg (2015). https://doi.org/10.1007/978-3-662-46081-8_8
3. Bauer, J.: Analysis of communication topologies by partner abstraction. Ph.D. thesis, Saarland University (2006)
4. Bauer, J., Wilhelm, R.: Static analysis of dynamic communication systems by partner abstraction. In: Nielson, H.R., Filé, G. (eds.) SAS 2007. LNCS, vol. 4634, pp. 249–264. Springer, Heidelberg (2007). https://doi.org/10.1007/978-3-540-74061-2_16
5. Calcagno, C., Distefano, D., O'Hearn, P.W., Yang, H.: Compositional shape analysis by means of bi-abduction. J. ACM **58**(6), 26:1–26:66 (2011)
6. Chang, B.-Y.E., Rival, X.: Relational inductive shape analysis. In: Proceedings of POPL 2008, pp. 247–260. ACM (2008)
7. Corradini, A., Heindel, T., Hermann, F., König, B.: Sesqui-pushout rewriting. In: Corradini, A., Ehrig, H., Montanari, U., Ribeiro, L., Rozenberg, G. (eds.) ICGT 2006. LNCS, vol. 4178, pp. 30–45. Springer, Heidelberg (2006). https://doi.org/10.1007/11841883_4
8. Corradini, A., Heindel, T., König, B., Nolte, D., Rensink, A.: Rewriting abstract structures: materialization explained categorically (2019). arXiv:1902.04809
9. Corradini, A., König, B., Nolte, D.: Specifying graph languages with type graphs. In: de Lara, J., Plump, D. (eds.) ICGT 2017. LNCS, vol. 10373, pp. 73–89. Springer, Cham (2017). https://doi.org/10.1007/978-3-319-61470-0_5
10. Corradini, A., König, B., Nolte, D.: Specifying graph languages with type graphs. J. Log. Algebraic Methods Program. (to appear)
11. Corradini, A., Montanari, U., Rossi, F., Ehrig, H., Heckel, R., Löwe, M.: Algebraic approaches to graph transformation–part I: basic concepts and double pushout approach, Chap. 3. In: Rozenberg, G. (ed.) Handbook of Graph Grammars and Computing by Graph Transformation: Foundations, vol. 1. World Scientific (1997)
12. Cousot, P.: Abstract interpretation. ACM Comput. Surv. **28**(2), 324–328 (1996). https://dl.acm.org/citation.cfm?id=234740
13. Dyckhoff, R., Tholen, W.: Exponentiable morphisms, partial products and pullback complements. J. Pure Appl. Algebra **49**(1–2), 103–116 (1987)
14. Ehrig, H., Golas, U., Hermann, F., et al.: Categorical frameworks for graph transformation and HLR systems based on the DPO approach. Bull. EATCS **3**(102), 111–121 (2013)
15. Ehrig, H., Habel, A., Padberg, J., Prange, U.: Adhesive high-level replacement categories and systems. In: Ehrig, H., Engels, G., Parisi-Presicce, F., Rozenberg, G. (eds.) ICGT 2004. LNCS, vol. 3256, pp. 144–160. Springer, Heidelberg (2004). https://doi.org/10.1007/978-3-540-30203-2_12
16. Ehrig, H., Pfender, M., Schneider, H.J.: Graph-grammars: an algebraic approach. In: 14th Annual Symposium on Switching and Automata Theory, Iowa City, Iowa, USA, 15–17 October 1973, pp. 167–180 (1973)
17. Freyd, P.: Aspects of topoi. Bull. Aust. Math. Soc. **7**(1), 1–76 (1972)
18. Habel, A., Pennemann, K.-H.: Nested constraints and application conditions for high-level structures. In: Kreowski, H.-J., Montanari, U., Orejas, F., Rozenberg,

G., Taentzer, G. (eds.) Formal Methods in Software and Systems Modeling. LNCS, vol. 3393, pp. 293–308. Springer, Heidelberg (2005). https://doi.org/10.1007/978-3-540-31847-7_17
19. Jacobs, B.: Categorical Logic and Type Theory. Studies in Logic and the Foundation of Mathematics, vol. 141. Elsevier, Amsterdam (1999)
20. König, B.: Description and verification of mobile processes with graph rewriting techniques. Ph.D. thesis, Technische Universität München (1999)
21. Lack, S., Sobociński, P.: Adhesive and quasiadhesive categories. RAIRO - Theor. Inform. Appl. **39**(3), 511–545 (2005)
22. Lack, S., Sobociński, P.: Toposes are adhesive. In: Corradini, A., Ehrig, H., Montanari, U., Ribeiro, L., Rozenberg, G. (eds.) ICGT 2006. LNCS, vol. 4178, pp. 184–198. Springer, Heidelberg (2006). https://doi.org/10.1007/11841883_14
23. Lambers, L., Orejas, F.: Tableau-based reasoning for graph properties. In: Giese, H., König, B. (eds.) ICGT 2014. LNCS, vol. 8571, pp. 17–32. Springer, Cham (2014). https://doi.org/10.1007/978-3-319-09108-2_2
24. Li, H., Rival, X., Chang, B.-Y.E.: Shape analysis for unstructured sharing. In: Blazy, S., Jensen, T. (eds.) SAS 2015. LNCS, vol. 9291, pp. 90–108. Springer, Heidelberg (2015). https://doi.org/10.1007/978-3-662-48288-9_6
25. Löwe, M.: Graph rewriting in span-categories. In: Ehrig, H., Rensink, A., Rozenberg, G., Schürr, A. (eds.) ICGT 2010. LNCS, vol. 6372, pp. 218–233. Springer, Heidelberg (2010). https://doi.org/10.1007/978-3-642-15928-2_15
26. O'Hearn, P.W.: A primer on separation logic (and automatic program verification and analysis). In: Software Safety and Security: Tools for Analysis and Verification. NATO Science for Peace and Security Series, vol. 33, pp. 286–318 (2012)
27. Rensink, A.: Canonical graph shapes. In: Schmidt, D. (ed.) ESOP 2004. LNCS, vol. 2986, pp. 401–415. Springer, Heidelberg (2004). https://doi.org/10.1007/978-3-540-24725-8_28
28. Rensink, A., Zambon, E.: Neighbourhood abstraction in GROOVE. In: Proceedings of GraBaTs 2010 (Workshop on Graph-Based Tools). Electronic Communications of the EASST, vol. 32 (2010)
29. Rozenberg, G. (ed.): Handbook of Graph Grammars and Computing by Graph Transformation: Foundations, vol. 1. World Scientific, Singapore (1997)
30. Sagiv, M., Reps, T., Wilhelm, R.: Parametric shape analysis via 3-valued logic. TOPLAS (ACM Trans. Program. Lang. Syst.) **24**(3), 217–298 (2002)
31. Steenken, D., Wehrheim, H., Wonisch, D.: Sound and complete abstract graph transformation. In: Simao, A., Morgan, C. (eds.) SBMF 2011. LNCS, vol. 7021, pp. 92–107. Springer, Heidelberg (2011). https://doi.org/10.1007/978-3-642-25032-3_7

Two-Way Parikh Automata with a Visibly Pushdown Stack

Luc Dartois[1(✉)], Emmanuel Filiot[2], and Jean-Marc Talbot[3]

[1] LACL-Université Paris-Est Créteil, Créteil, France
ldartois@lacl.fr
[2] Université Libre de Bruxelles, Brussels, Belgium
[3] LIM-Aix-Marseille Université, Marseille, France

Abstract. In this paper, we investigate the complexity of the emptiness problem for Parikh automata equipped with a pushdown stack. Pushdown Parikh automata extend pushdown automata with counters which can only be incremented and an acceptance condition given as a semi-linear set, which we represent as an existential Presburger formula over the final values of the counters. We show that the non-emptiness problem both in the deterministic and non-deterministic cases is NP-c. If the input head can move in a two-way fashion, emptiness gets undecidable, even if the pushdown stack is visibly and the automaton deterministic. We define a restriction, called the single-use restriction, to recover decidability in the presence of two-wayness, when the stack is visibly. This syntactic restriction enforces that any transition which increments at least one dimension is triggered only a bounded number of times per input position. Our main contribution is to show that non-emptiness of two-way visibly Parikh automata which are single-use is NExpTime-c. We finally give applications to decision problems for expressive transducer models from nested words to words, including the equivalence problem.

1 Introduction

Parikh automata. Since the classical automata-based approach to model-checking [28], finite automata have been extended in many ways to tackle the automatic verification of more realistic and powerful systems against more expressive specifications. For instance, they have been extended to pushdown systems [3,26,30], concurrent systems [5], and systems with counters or specifications with arithmetic constraints have been the focus of many works in verification [7,11,15–18,23].

Along this line of work, Parikh automata (or PA), introduced in [22], are an important instance of automata extension with arithmetic constraints. They are automata on finite words whose transitions are equipped with counter operations. The counters can only be incremented, and do not influence the run (enabling a transition requires no test on counter values), but the acceptance of a run is defined by the membership of the final counter valuations to some semi-linear set S. Expressivity of PAs goes beyond regularity, as the language

M. Bojańczyk and A. Simpson (Eds.): FOSSACS 2019, LNCS 11425, pp. 189–206, 2019.
https://doi.org/10.1007/978-3-030-17127-8_11

$L = \{w \mid |w|_a = |w|_b\}$ of words having the same numbers of as and bs is realised by a simple automaton counting the numbers of as and bs in counters x_1 and x_2 respectively, and the accepting condition is given by the linear-set $\{(i, i) \mid i \in \mathbb{N}\}$. Semi-linear sets can be defined by formulas in existential Presburger arithmetic, ie first-order formulas with equality and sum predicates over integers, whose free variables are evaluated by the counter values calculated by the run.

A central problem in automata theory is the non-emptiness problem: does the automaton accepts at least one input. Although PAs go beyond regular languages, they retain relatively good algorithmic properties. The emptiness problem is decidable, and it is NP-c [12]. The hardness holds even if the semi-linear set is represented as a set of generator vectors. Motivated by applications in transducer theory for well-nested words, we investigate in this article extensions of Parikh automata with a pushdown stack.

First contribution: pushdown Parikh automata. As a first contribution, we study the complexity of the emptiness problem for Parikh automata with a pushdown store. Parikh automata extend finite automata with counter operations and an acceptance condition given as a semi-linear set, *pushdown Parikh automata* extend pushdown automata in the same way. We show that adding a stack can be done for free with respect to the emptiness problem, which remains, as for stack-free Parikh automata, NP-c. However in this case, we are able to strengthen the lower bound: it remains NP-hard even if there are only two counters, the automaton is deterministic, and the Presburger formula only tests for equality of these two counters. In the stack-free setting, it is necessary to have an unfixed number of counters to get such a lower bound.

Contribution 1. The emptiness problem for pushdown Parikh automata (PPA) is NP-c. The lower bound holds even if the automaton is deterministic, has only two counters whose operations are encoded in unary, and they are eventually tested for equality.

Second contribution: adding two-wayness. We investigate the complexity of pushdown Parikh automata when the input head is allowed to move in two directions. It is not difficult to see that in that case emptiness gets undecidable, since, already without counters, one can simulate the intersection of two deterministic pushdown automata, by performing two passes over the input (visiting each input position at most three times). We consider a first restriction on the stack behaviour, which is required to be *visibly*.

A pushdown stack is called visibly if it is driven by the type of letters it reads, which can be either call symbols, return symbols or internal symbols. Words formed over such a structured alphabet are called nested words, and well-nested words if additionally the call/return structure of the word is well-balanced, such as in the following example:

$$c\ \ c\ r\ \ r\ \ c\ r$$

Automata for nested words, called *visibly pushdown automata* (or VPA), have been introduced in [2]. They are pushdown automata whose stack behaviour is constrained by the input in the following way. Upon reading a call symbol, exactly one symbol is pushed onto the stack. Upon reading a return symbol, exactly one symbol is popped from it. Upon reading an internal symbol, the stack is left unchanged. Hence, the symbol that is pushed while reading a given call symbol is popped while reading its matching return symbol. Consequently, visibly pushdown automata enjoy nice properties, such as closure under Boolean operations and determinisation.

VPA have been extended to two-way VPA (2VPA) [8] with the following stack constraints: in a backward reading mode, the role of the return and call symbols regarding the stack are inverted: when reading a call, exactly one symbol is popped from the stack and when reading a return, one symbol is pushed. It was shown in [8] that adding this visibly condition to two-way pushdown automata allows one to recover decidability for the emptiness problem. However, for Parikh acceptance, this restriction is not sufficient. Indeed, by encoding diophantine equations, we show the following undecidability result:

Contribution 2. The emptiness problem for two-way visibly pushdown Parikh automata (2VPPA) is undecidable.

Single-use property. The problem is that by using the combination of two-wayness and a pushdown stack, it is possible to encode polynomially, and even exponentially large counter values, with respect to the length of the input word. We consider therefore the single-use restriction, which appears in several transducer models [6,8,10], by which it is possible to keep a linear behaviour for the counters. Informally, a *single-use* two-way machine bounds the size of the production per input positions. It is syntactically enforced by asking that transitions which strictly increment at least one counter are triggered at most once per input position. Our main result is the decidability of 2VPPA emptiness under the single-use restriction, with tight complexity.

Contribution 3 (Main). The emptiness problem for two-way single-use visibly pushdown Parikh automata ($2VPPA_{su}$) is NExpTime-c. The hardness holds even if the automaton is deterministic, has only two counters whose operations are encoded in unary, and they are eventually tested for equality.

To prove the upper-bound, we show that two-wayness can be removed from single-use 2VPPA, at the price of one exponential. In other words, single-use 2VPPA and VPPA have the same expressive power, although it can be shown that the former model is exponentially more succinct. The lower bound is obtained by encoding the succinct variant of the subset sum problem, based on a reduction which uses the fact that, by combining the pushdown and two-way features, single-use 2VPPA can encode doubly-exponential values 2^{2^n} with a polynomial number of states (in n).

	Visibly Pushdown	Pushdown
one-way	NP-complete	NP-complete
2-way Single-use	NExptime-complete	Undecidable
2-way	Undecidable	Undecidable

Fig. 1. Complexity of the emptiness of different Pushdown Parikh Automata. All results hold for deterministic and non-deterministic machines.

Contribution 4 (Applications). As an application, we give an elementary upper-bound (NExpTime) for the equivalence problem of functional single-use two-way visibly pushdown transducers [8], while an ExpTime lower bound was known. This transducer model defines transductions from well-nested words to words and, as shown in [8], they are well-suited to define XML transformations, have the same expressive power as Courcelle's MSO-transducers [6] (casted to well-nested words), and admit a memory-efficient evaluation algorithm. We also provide two other new results on single-use 2VPT (not necessarily functional). First, we show that given a positive integer k, it is decidable whether a single-use 2VPT produces at most k different output words per input (k-valuedness problem). Then, we show the decidability of a typechecking problem: given a single-use 2VPT T and a finite (stack-free) Parikh automaton P, it is decidable whether the codomain of T has a non-empty intersection with P. This allows for instance to decide whether a single-use 2VPT produces only well-nested words and thus describes a well-nested words to well-nested words transformation, since the property of a word to be non well-nested is definable, as we show, by a Parikh automaton.

Finite-visit vs single-useness. The single-use property is more general than the more classical *finite-visit* restriction, used for instance in [9,19]: it requires to visit any input position a (machine-dependent) constant number of times, while single-useness only bounds the number of visits by producing transitions. Although, consequently to our results, 2VPPA single-use and finite-visit have the same expressive power, this extra modelling feature is desirable, for instance when using 2VPPA to test properties of 2VPT: single-use 2VPT are strictly more expressive than finite-visit ones, and this relaxation is crucial to capture MSO transductions [8]. Moreover, we somehow get it for free: we show that the NExpTime lower bound also holds for finite-visit 2VPPA. Finally, we note that as we deal with single-use machines rather than finite-visit ones, the usual ingredient for going from two-way to one-way consisting of memorizing simply crossing sections of states, is not sufficient to get the result here, since we cannot bound the size of these crossing sections.

Related work. Parikh automata are closely related to reversal-bounded counter machines [18]. In fact, both models have equivalent expressiveness in the non-deterministic case [22]. The difference of expressive power in the deterministic case is due to the fact that counter machines can perform tests on its counters that can influence the run, while counters in Parikh automata only matter at the end of the run. Several extensions of reversal-bounded counter machines were studied, whether they are two-way or equipped with a (visibly) pushdown stack. However, to the best of our knowledge, the combination of the two features has never been studied (see [19] for a survey). It is possible to define a model of single-use reversal-bounded two-way visibly pushdown counter machines, where the single-useness is put on transitions that modify the counters. This model is expressively equivalent to $\mathsf{2VPPA_{su}}$ in the non-determinstic case, and thanks to our result, has a decidable emptiness problem. The non-emptiness problem for reversal-bounded (one-way) pushdown counter machines for fixed numbers of counters and reversals is known to be in NP [13] and NP-hard [16]. Converting PPA into reversal-bounded counter machines would yield an unfixed number of counters. Our NP lower-bound for PPA however follows ideas of [16] about encoding, using the stack, integers n with $O(log(n))$ states and stack symbols.

Two-way (stack-free) reversal-bounded counter machines, even deterministic, are known to have undecidable emptiness problem [19]. Decidability is recovered by taking the finite-visit restriction [19]. Our result on $\mathsf{2VPPA_{su}}$ entails the decidability of emptiness of two-way reversal-bounded counter machines which are single-use.

Finally, all the decidability results we prove on two-way visibly pushdown transducers were already known in the one-way case [13]. Two-way visibly pushdown transducers, which are strictly more expressive, can also be seen as a model of unranked tree-to-word transducers, modulo tree linearisation. To the best of our knowledge, this is the first model of unranked tree-to-word transducers for which k-valuedness and codomain well-nestedness is shown to be decidable. Another model, introduced in [1], is known to be expressively equivalent to $\mathsf{2VPT_{su}}$ [8], and in the functional case, has decidable equivalence problem in NExpTime. However, translating $\mathsf{2VPT_{su}}$ to this model requires an exponential blow-up, yielding a worst complexity for equivalence testing.

Structure. Section 2 introduces the computing models used, the proof of the lower bound for $\mathsf{2VPPA_{su}}$ is given in Sect. 3 and the upper bound in Sect. 4. Finally, some applications to the main theorem to transducers are given in Sect. 5.

2 Two-Way Visibly Pushdown (Parikh) Automata

In this section, we first recall the definition of two-way visibly pushdown automata and later on extend them to two-way visibly pushdown Parikh automata.

We consider a structured alphabet Σ defined as the disjoint union of call symbols Σ_c, return symbols Σ_r and internal symbols Σ_i. The set of words over Σ is Σ^*. As usual, ϵ denotes the empty word. Amongst nested words, the set of well-nested words Σ^*_{wn} is defined as the least set such that $\Sigma_i \cup \{\epsilon\}$ is included into Σ^*_{wn} and if $w_1, w_2 \in \Sigma^*_{\mathsf{wn}}$ then both $w_1 w_2$ and $c w_1 r$ (for all $c \in \Sigma_c$ and $r \in \Sigma_r$) belong to Σ^*_{wn}.

When dealing with two-way machines, we assume the structured alphabet Σ to be extended to $\overline{\Sigma}$ by adding a left and right marker symbols $\triangleright, \triangleleft$ in $\overline{\Sigma}_c$ and $\overline{\Sigma}_r$ respectively, and we consider words in the language $\triangleright\Sigma^*\triangleleft$.

Definition 1. *A two way visibly pushdown automaton (*2VPA *for short) A over $\overline{\Sigma}$ is given by $(Q, q_I, F, \Gamma, \delta)$ where Q is a finite set of states, $q_I \in Q$ is the initial state, $F \subseteq Q$ is a set of final states and Γ is a finite stack alphabet. Given the set $\mathbb{D} = \{\leftarrow, \rightarrow\}$ of directions, the transition relation δ is defined by $\delta^{push} \cup \delta^{pop} \cup \delta^{int}$ where*

- $\delta^{push} \subseteq ((Q \times \{\rightarrow\} \times \Sigma_c) \cup (Q \times \{\leftarrow\} \times \Sigma_r)) \times ((Q \times \mathbb{D}) \times \Gamma)$
- $\delta^{pop} \subseteq ((Q \times \{\leftarrow\} \times \Sigma_c \times \Gamma) \cup (Q \times \{\rightarrow\} \times \Sigma_r \times \Gamma)) \times (Q \times \mathbb{D})$
- $\delta^{int} \subseteq ((Q \times \mathbb{D} \times \Sigma_i) \times (Q \times \mathbb{D})$

Additionally, we require that for any states q, q' and any stack symbol γ, if $(q, \leftarrow, \triangleright, \gamma, q', d) \in \delta^{pop}$ then $d = \rightarrow$ and if $(q, \rightarrow, \triangleleft, \gamma, q', d) \in \delta^{pop}$ then $d = \leftarrow$ ensuring that the reading head stays within the bounds of the input word.

Informally, a 2VPA has a reading head pointing between symbols (and possibly on the left of $\triangleright$ or the right of $\triangleleft$). A configuration of the machine is given by a state, a direction d and a stack content. The next symbol to be read is on the right of the head if $d = \rightarrow$ and on the left if $d = \leftarrow$. Note that when reading the left marker from right to left $\leftarrow$ (resp. the right marker from left to right $\rightarrow$), the next direction can only be $\rightarrow$ (resp. $\leftarrow$). The structure of the alphabet induces the behavior of the machine regarding the stack when reading the input word: when reading on the right, a call symbol leads to push one symbol onto the stack while a return symbol pops one symbol from the stack. When reading on the left, a dual behaviour holds. In any direction internal transitions from δ^{int} read internal symbols and do not affect the stack; hence, at a given position in the input word, the height of the stack is always constant at each visit of that position in the run of the machine. The triggering of a transition leads to the update of the state of the machine, the future direction as well as the stack content. For a direction d, a natural i ($0 \leq i \leq |w|$) and a word w, we denote by

- $\text{move}(d, i)$ the integer $i - 1$ if $d = \leftarrow$ and $i + 1$ if $d = \rightarrow$.
- $\text{read}(w, d, i)$ the symbol $w(i)$ if $d = \leftarrow$ and $w(i + 1)$ if $d = \rightarrow$.

Note that when switching directions (i.e. when the direction of the first part of the transition is different from the second part), we read twice the same letter. This ensures the good behavior of the stack, as reading a call letter from left to right pushes a stack symbol, we need to pop it if we start moving from right to left.

Formally, a stack σ is a finite word over Γ. The empty stack/word over Γ is denoted $\perp$. For a word w from $\overline{\Sigma}$ and a 2VPA $A = (Q, q_I, F, \Gamma, \delta)$, a *configuration* κ of A is a tuple (q, i, d, σ) where $q \in Q$, $0 \leq i \leq |w|$, $d \in \mathbb{D}$ and σ is a stack. A *run* of A on a word w is a finite sequence ρ from $K(\delta K)^*$, where K is the set of all configurations κ (that is a sequence starting and ending with a configuration and alternating between configurations and transitions); a run ρ is of the form

$(q_0, i_0, d_0, \sigma_0)\tau_1(q_1, i_1, d_1, \sigma_1)\tau_2 \dots \tau_\ell(q_\ell, i_\ell, d_\ell, \sigma_\ell)$ where for all $0 \leq j < \ell$, we have:

- either $d_j =\rightarrow$ and $\text{read}(w, d_j, i_j) \in \Sigma_c$ or $d_j =\leftarrow$ and $\text{read}(w, d_j, i_j) \in \Sigma_r$, $\tau_{j+1} = (q_j, d_j, \text{read}(w, d_j, i_j), q_{j+1}, d_{j+1}, \gamma) \in \delta^{\text{push}}$, $i_{j+1} = \text{move}(i_j, d_j)$ and $\sigma_{j+1} = \sigma_j\gamma$
- either $d_j =\leftarrow$ and $\text{read}(w, d_j, i_j) \in \Sigma_c$ or $d_j =\rightarrow$ and $\text{read}(w, d_j, i_j) \in \Sigma_r$, $\tau_{j+1} = (q_j, d_j, \text{read}(w, d_j, i_j), \gamma, q_{j+1}, d_{j+1}) \in \delta^{\text{pop}}$, $i_{j+1} = \text{move}(i_j, d_j)$ and $\sigma_{j+1}\gamma = \sigma_j$
- $\text{read}(w, d_j, i_j) \in \Sigma_i$, $\tau_{j+1} = (q_j, d_j, \text{read}(w, d_j, i_j), q_{j+1}, d_{j+1}) \in \delta^{\text{int}}$, $i_{j+1} = i_j$ and $\sigma_{j+1} = \sigma_j$.

Note that any configuration is actually a run on the empty word ϵ. The initial configuration is $(q_I, 0, \rightarrow, \perp)$. A configuration $(q, i, d, \perp)$ is *final* if $q \in F$ and i is the last position. A run for the word w is accepting if its first configuration is initial and its last configuration is final. A two-way visibly pushdown automaton A is:

- *deterministic* (denoted D2VPA) if δ^{push} (resp. δ^{pop}, δ^{int}) is a function from $Q \times \mathbb{D} \times \Sigma$ (resp. $Q \times \mathbb{D} \times \Sigma \times \Gamma$, $Q \times \mathbb{D} \times \Sigma$) to $Q \times \mathbb{D} \times \Gamma$ (resp. $Q \times \mathbb{D}$, $Q \times \mathbb{D}$).
- *one-way* (denoted VPA) if all transitions in A have $\rightarrow$ for direction.
- *finite-visit* if for some $k \geq 0$, any run visits at most k times the same input position.

The size of a 2VPA is the number of states times the size of the stack alphabet. For A an automaton, we denote by $L(A)$ the language recognized by A.

Lemma 1 ([8]). *Given a* 2VPA *A, deciding if $L(A)$ is empty is* ExpTime-complete.

Parikh automata. Parikh automata were introduced in [22]. Informally, they are automata with counters that can only be incremented, and do not act on the transition relation. Acceptance of runs is done by evaluating a Presburger formula whose free variables are set to the counter values. In our setting, a *Presburger formula* is a positive formula $\psi(x_1, \dots, x_n) = \exists y_1 \dots y_m \varphi(x_1, \dots, x_n, y_1, \dots, y_m)$ such that φ is a boolean combination of atoms $s + s' \leq t + t'$, for $s, s', t, t' \in \{0, 1, x_1, \dots, x_n, y_1, \dots, y_m\}$. For a set S and some positive number m, we denote by S^m the set of all mappings from $[1 \dots m]$ to S. If $(s_1, \dots, s_m)$ and $(t_1, \dots, t_m)$ are two tuples of S^m and $+$ is an binary operation on S, we extend $+$ to S^m by considering the operation element-wise, i.e. $(s_1, \dots, s_m) + (t_1, \dots, t_m) = (s_1 + t_1, \dots, s_m + t_m)$.

Definition 2. *A two-way visibly pushdown Parikh automaton (*2VPPA *for short) is a tuple $P = (A, \lambda, \phi)$ where A is a* 2VPA *and for some natural dim, λ is a mapping from δ to $\mathbb{N}^{dim}$, the set of vectors of length dim of naturals and $\phi(x_1, \dots, x_{dim})$ is a Presburger formula with dim free variables.*

When clear from context, we may omit the free variables from the Presburger formula, and simply note ϕ. A run of a 2VPPA is a run of its underlying 2VPA. We extend canonically the mapping λ to runs. For a run ρ of the form $(q_0, i_0, d_0, \sigma_0)\tau_1(q_1, i_1, d_1, \sigma_1)\tau_2 \ldots \tau_\ell(q_\ell, i_\ell, d_\ell, \sigma_\ell)$, we set

$$\lambda(\rho) = \lambda(\tau_1) + \lambda(\tau_2) + \ldots + \lambda(\tau_\ell)$$

We recall that a single configuration c is a run over the empty word ϵ. For such a run c, we set $\lambda(c) = 0^{dim}$. A run $(q_0, i_0, d_0, \sigma_0)\tau_1(q_1, i_1, d_1, \sigma_1)$ $\tau_2 \ldots \tau_\ell(q_\ell, i_\ell, d_\ell, \sigma_\ell)$ is accepted if $(q_0, i_0, d_0, \sigma_0)$, $(q_\ell, i_\ell, d_\ell, \sigma_\ell)$ are respectively an initial and a final configuration of the underlying automaton and for $\lambda(\rho) = (n_1, \ldots, n_{dim})$, $[x_1 \leftarrow n_1, \ldots, x_\ell \leftarrow n_{dim}] \models \phi(x_1, \ldots, x_{dim})$. The language $L(P)$ is the set of words which admit an accepting run. We define the set of values computed by P as $Val(P) = \{\lambda(\rho) \mid \rho$ a valid run of the underlying automaton of $P\}$. We define the size of P as the size of A plus the number of symbols in ϕ and $|\delta| \cdot dim \cdot log(W)$ where W is the maximal value occurring in the codomain of λ.

It is deterministic (resp. one-way), denoted D2VPPA (resp. VPPA) if its underlying automaton is deterministic (resp. one-way). It is known from [4] that DPA (i.e. deterministic one-way and stack-free Parikh automata in our setting) are strictly less expressive than their nondeterministic counterpart. As a counter example, they exhibit the language $L = \{w \mid w_{\#_a(w)} = b\}$, ie all words w such that if n is the number of a in w, the letter at the nth position is a b. Note that even in the two-way case, a deterministic machine recognizing L needs to either have access, during the computation, to the number of a's, or be able to store, in counters, the position of each b. As the first solution cannot be done since Parikh automata only access their counters at the end of the run, and the second is also impossible since there are only a finite number of counters, this language is also non definable by a D2VPPA, furthering the separation between deterministic and nondeterministic Parikh automata.

Example 1. As an example, we give a deterministic 2VPPA P that, given an input $i^n c^k i^\ell r^k$ with c, i, r in Σ_c, Σ_i and Σ_r respectively, accepts if $k = \ell$ and $n = k^2$. The 2VPPA P uses 4 variables x_n, x_k, x_ℓ and y. The first 3 variables are used to count the number of the first block of is, the number of calls and the second block of is respectively. The handling of these 3 variables is straightforward and can be done in a single pass over the input. The fourth variables y counts the multiplication $k \cdot \ell$ and doing so is more involved. The part of the underlying 2VPA of P handling y is given in Fig. 2. On this part, the mapping λ simply increments the counter on transitions going to state 2 (i.e. on reading the letters i from left to right). It makes as many passes on the set of internal symbols in state 2 as there are call symbols, and the state of the stack upon reading i^ℓ for the jth time is $1^j 0^{k-j}$. Finally, the accepting formula ϕ of P is defined by $x_n = y \wedge x_k = x_\ell$. Note that this widget allows us to compute the set $\{(k^2, k, k, k^2) \mid k \in \mathbb{N}\}$ which is not semilinear.

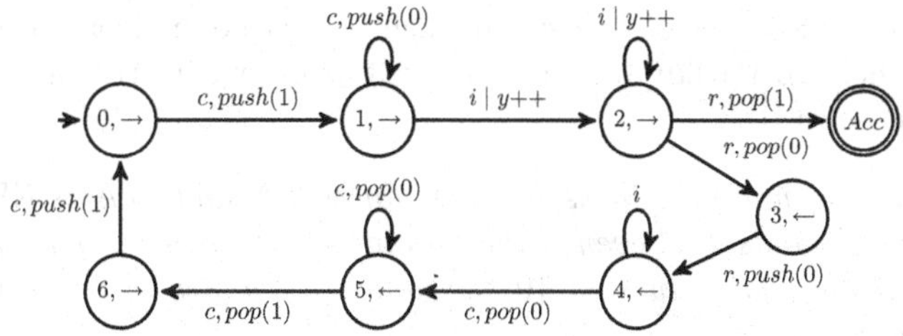

Fig. 2. A 2VPPA reading words $c^k i^\ell r^k$ and making k passes on i^ℓ, adding $k \cdot \ell$ to the variable y. The transitions have two components, the first being the letter read, and the second being the stack operation. There is no stack operation upon reading internal symbols. The variable y is incremented in transitions going to state 2 only.

As we have seen in the previous example, the set $Val(P)$ is not necessarily semi-linear, even with P a D2VPPA. We use this fact to encode diophantine equations, and get the following undecidability result:

Theorem 1. *The emptiness problem of* D2VPPA *is undecidable.*

Single-useness. In order to recover decidability, we adapt to Parikh Automata the notion of single-useness introduced in [8]. Simply put, a 2VPPA is *single-use* (denoted $2VPPA_{su}$) if the transitions that affect the variables can only be taken once on any given input position, thus effectively bounding the size of variables linearly with respect to the size of the input. Formally, a state p of a 2VPPA P is *producing* if there exists a transition t from p on some symbol and $\lambda(t) \neq 0^{dim}$. A 2VPPA is single-use if for every input w and every accepting run ρ over w, there do not exist two different configurations (p, i, d, σ) and (p, i, d, σ') with p a producing state, meaning that ρ does not reach any position in the same direction twice in any given state of P. This property is a syntaxic restriction of the model. However, since this property is regular, it can equivalently be seen as a semantic one. Moreover, deciding the single-useness of a 2VPPA is ExpTime-c (see [8] for the same result but on transducers). Note that the Parikh automaton given in Example 1 is not single-use, since it passes over the second subword of internal letters i in state 2 as many times as there are call symbols. In the following, we prove that $2VPPA_{su}$ have the same expressiveness as VPPA, while being exponentially more succinct. In particular, this equivalence implies by Parikh's Theorem [24], semi-linearity of $Val(P)$ for any $2VPPA_{su}$ P.

3 Emptiness Complexity

We show that the non-emptiness problem for VPPA is NP-complete. We actually show the upper-bound for the strictly more expressive *Pushdown Parikh Automata* (PPA), i.e. VPPA without the visibly restriction. While decidability was known [20,21], the precise complexity was, to the best of our knowledge, unknown. Let us also remark that the model and the proof are similar to the proof of NP-completeness of k-reversal pushdown systems from [16]. However, it is adapted here to Parikh automata as well as deterministic machines, which was not the case in [16].

Theorem 2. *The non-emptiness problem for* VPPA *and* PPA *is* NP-complete. *The complexity bounds hold even if the automata are deterministic, with a fixed dimension* 2*, tuples of values in* $\{0,1\}^2$ *and with a fixed Presburger formula* $\phi(x_1,x_2) \equiv x_1 = x_2$.

From $\mathsf{2VPPA_{su}}$ *to* VPPA From a two-way visibly pushdown Parikh automaton satisfying the single-useness restriction, one can build an equivalent one-way visibly pushdown Parikh automaton. The construction induces an exponential blow-up, which cannot be avoided, as with most constructions from two-way to one-way machines.

Theorem 3. *For any* $\mathsf{2VPPA_{su}}$ A*, one can construct a* VPPA B *whose size is at most exponential in the size of* A *and such that* $L(A)=L(B)$*. Moreover, the procedure can be done in exponential time.*

Proof (Sketch). The goal is to be able to correctly guess all the transitions exactly taken by a run of the two-way machine at once. More precisely, the one-way machine guesses the behavior of the two-way machine on each well-nested subword of the input, i.e. a set of partial runs over a subword. A partial run is a pair from $Q \times \{\leftarrow, \rightarrow\}$. Informally, they describe a maximal subrun over a subword of the input. We call these sets of partial runs *profiles*, and we define relations C and $N_{c,r}$ to describe compatible profiles. Formally, the relation $C \subseteq \mathcal{P}^3$ is the *concatenation* relation, defined as set of triples (P, P', P'') such that there exists a word $u = u_1vv'u_2$ where v and v' are well-nested subwords of u, and a run r on u such that P (resp. P') is the profile of v in r (resp. of v') and P'' is the profile of vv' in r. Similarly, the relation $N_{c,r} \subseteq \mathcal{P}^2$ for c, r call and return letters respectively, is the *cr-nesting* relation, and defined as the set of pairs (P, P') such that there exists a word $u = u_1cvru_2$ where v is well-nested, and a run r of A on u such that P is the profile of v in r and P' is the profile of cvr in r. We prove that these relations are computable in exponential time.

Given these relations, we can compute a VPPA B whose runs are bijective to the runs of A. Moreover, we can recover from a run of B which transitions are effectively taken at each positions by its bijective run of A. Then, the increment function simply does all the increments done by the run at a given position at once. Since the operation is the addition on integers, it is commutative and the variables are updated in the same way they were by the run of A. Note that we only recover which transitions are taken, and not how many times they are taken, which can depend on the size of the input. However, since A is single-use, we only have to add each non zero transition once, which gives the result.

As a direct corollary of Theorems 3 and 2, we get the following.

Corollary 1. *The emptiness of* $\mathsf{2VPPA_{su}}$ *can be decided in* NExpTime.

4 NExpTime-Hardness

In this section, we show that the problem of deciding whether the language of a $\mathsf{2VPPA_{su}}$ is non-empty is hard for NExpTime. Moreover, we show that this hardness does not depend on the fact that we have taken existential Presburger formulas, nor on the vector dimensions, and nor on the fact that the values in the tuples are encoded in binary.

Theorem 4. *The non-emptiness problem for* $\mathsf{2VPPA_{su}}$ *is NExpTime-hard. The result holds even if the automaton is deterministic, of dimension* 2, *with counter updates in* $\{0,1\}$, *the Presburger formula is* $\phi(x_1,x_2) \equiv x_1 = x_2$, *and it is finite-visit.*

Succinct Subset Sum Problem. We reduce to the succinct subset sum problem (SSSP), which is NExpTime-hard [16]. Let us define SSSP. Let $m,k \geq 1$, $X = \{x_1,\ldots,x_k\}$ and $Y = \{y_1,\ldots,y_m\}$ be sets of Boolean variables. Let θ be a Boolean formula over $X \cup Y$. Any word $v \in \{0,1\}^{k+m}$ naturally defines a valuation of $X \cup Y$ (the first bit of v is the value of x_1, etc.). We denote by $\theta[v] \in \{0,1\}$ the truth value of θ under the valuation v. The formula θ defines 2^k non-negative integers $a_1,\ldots,a_{2^k}$ each with 2^m bits, as follows:

$$a_i = \theta[b_i d_1].2^{2^m-1} + \theta[b_i d_2].2^{2^m-2} + \cdots + \theta[b_i d_{2^m}].2^0$$

where b_i is the binary encoding over k bits of i, and $d_1,\ldots,d_{2^m}$ is the lexicographic enumeration of $\{0,1\}^m$, starting from 0^m. Note that for all $i \in \{1,\ldots,2^k\}$, $a_i \in \{0,\ldots,2^{2^m}-1\}$. The *Succinct Subset Sum Problem* asks, given X, Y and θ, whether there exists $J \subseteq \{1,\ldots,2^k-1\}$ such that $\sum_{j\in J} a_j = a_{2^k}$.

Overview of the construction and encoding the values a_i. Given an instance of SSSP $\mathcal{I}$, our goal is to construct a $\mathsf{D2VPPA_{su}}$ $\mathcal{P} = (C,\rho,\phi)$ of dimension 2 such that $|\mathcal{P}|$ is polynomial in $|\theta| + k + m$ and $\mathcal{L}(\mathcal{P}) \neq \varnothing$ iff $\mathcal{I}$ has a solution.

The main idea is to ensure that $\mathcal{L}(C) = \{X_1 e_1 \ldots X_{2^k-1} e_{2^k-1} \# e_{2^k} \mid X_i \in \{0,1\}\}$ where the X_i are internal symbols which are used to encode a subset $J \subseteq \{1,\ldots,2^k-1\}$, and each e_i is an encoding of a_i, defined later, over some alphabet containing the symbol $\mathbb{1}$, and such that the number of occurrences of $\mathbb{1}$ in e_i is a_i. In other words, e_i somehow encodes a_i in unary. For the vector part, the machine $\mathcal{P}$, when running over $X_i e_i$, updates its dimensions depending on two cases: (1) if $X_i = 1$ ("put value a_i in J"), then any transition reading $\mathbb{1}$ has weight $(1,0)$ and any other transition has weight $(0,0)$, (2) if $X_i = 0$, then every transition has weight $(0,0)$. So, if $X_i = 1$, the value in the first dimension after processing $X_i e_i$ has been incremented by a_i. Similarly, when processing $\# e_{2^k}$, any transition reading $\mathbb{1}$ increments the 2nd dimension by 1, so that after processing $\# e_{2^k}$, this dimension has value a_{2^k}. The formula $\phi(x_1,x_2)$ then only requires equality of x_1 and x_2, i.e. $\phi(x_1,x_2) \equiv x_1 = x_2$.

We now explain how to encode a_i by a well-nested word e_i. Due to the finite-visit restriction, every incremental transition can be triggered at most once for each input position. Since the value a_i is possibly doubly exponential in m and

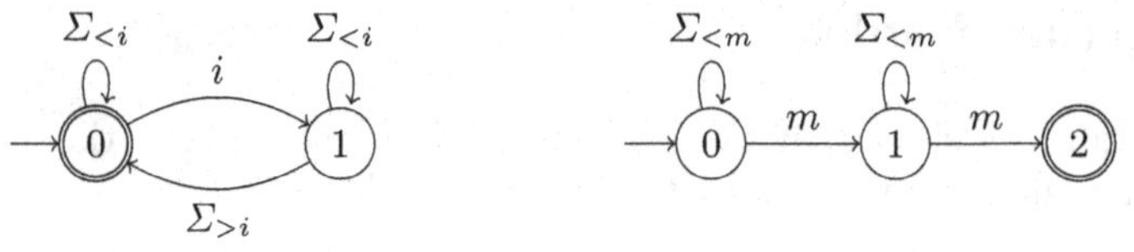

Fig. 3. On the left, the automaton A_i, for $i < m$. On the right, the automaton A_m.

we are allowed to have a polynomial number of transitions (in $|\theta| + k + m$), necessarily e_i must be of doubly exponential length. The main idea is to use the stack and the two-wayness to recognise with a polynomial number of states well-nested words which are of doubly exponential length. We need a series of intermediate lemmas to achieve this idea. We start with a useful result about intersection of finite automata, here *reversible* finite automata (deterministic and backward deterministic). Let $\Sigma = \{1, \ldots, m\}$ and let us define recursively the sequence of words $(u_i)_{0 \leq i \leq m} \in \Sigma^*$ as follows: $u_0 = 1$, $u_i = u_{i-1} i u_{i-1}$ for $1 \leq i < m$ and $u_m = u_{m-1} m u_{m-1} m$.

Lemma 2. *The word u_m has length 2^m, and there exist m reversible finite automata $A_0, \ldots, A_m$ (Fig. 3) such that (i) each A_i has $O(1)$ states, and (ii) $\bigcap_{i=1}^{m} L(A_i) = \{u_m\}$.*

Encoding of the values a_i. The idea is to define a well-nested word e_i over an alphabet of call symbols $\Sigma_c = \{c_1, \ldots, c_m\}$, an alphabet of return symbols $\Sigma_r = \{r_1, \ldots, r_m\}$ and an alphabet of internal symbols $\Sigma_\iota = \{0, 1, \mathbb{1}, \mathbb{0}\}$. The number of occurrences of $\mathbb{1}$ in e_i will be exactly a_i, i.e. $\#_{\mathbb{1}}(e_i) = a_i$ and hence, the Parikh automaton will just have to count the number of $\mathbb{1}$ occurrences. Let us remind the reader that a_i is actually given by θ, and therefore, the automaton $\mathcal{P}$ will somehow have to evaluate θ for valuations of its variables that will be contained in e_i. Let us now define the words e_i. For that, we call a *binary tree* either an internal symbol $\mathbb{1}, \mathbb{0}$, or a well-nested word of the form $c_j t_1 t_2 r_j$ where t_1, t_2 are themselves binary trees. For a well-nested word of the form cwr, a root-to-leaf branch π is a sequence of calls $x_1 \ldots x_n$ such that $cwr = x_1 w_1 x_2 w_2 \ldots x_n w_n r_n w'_n r_{n-1} w'_{n-1} \ldots r_2 w'_2 r_1$ where $x_1 = c$, $r_1 = r$ and for some w_i, w'_i well-nested words such that w_n contains only internal symbols. The *height* of a binary tree t is the maximal length of a root-to-leaf branch, and it is *complete* if all root-to-leaf branches have the same length. Note that the number of internal symbols of a complete binary tree of height n is 2^n.

Then, e_i is the well-nested word defined by $e_i = c_{j_1} b_i d_1 t_1 c_{j_2} b_i d_2 t_2 \ldots c_{j_{2^m}} b_i d_{2^m} t_{2^m} r_{j_{2^m}} \ldots r_{j_1}$ where

1. the words t_i are binary trees
2. every root-to-leaf branch $\pi = c_{i_1} \ldots c_{i_\ell}$ of e_i satisfies $i_1 \ldots i_\ell = u_m$
3. $b_i \in \{0,1\}^k$ and $d_1, \ldots, d_{2^m}$ is a lexicographic enumeration of $\{0,1\}^m$ (starting from 0^m)
4. for all j, all internal symbols occurring in t_j are $\mathbb{1}$ if $\theta[b_i d_j] = 1$, $\mathbb{0}$ otherwise.

Our goal is now to prove that e_i is a correct encoding of a_i.

Lemma 3. *For all $i \in \{1, \ldots, 2^k\}$, $\#_{\mathbb{1}}(e_i) = a_i$, where $\#_{\mathbb{1}}(e_i)$ denotes the number of occurrences of $\mathbb{1}$ in e_i.*

Proof. By Condition 2, every root-to-leaf branch of e_i has length 2^m. Therefore, for all $j \in \{1, \ldots, 2^m\}$, every root-to-leaf branch in t_j has length $2^m - j$. In particular, t_{2^m} does not contain any call symbol. Hence all the trees t_j are complete binary trees of height $2^m - j$. So, every t_j has $2^{2^m - j}$ internal symbols and by Condition 4, we get $\#_{\mathbb{1}}(t_j) = \theta[b_i d_j].2^{2^m - j}$. Therefore, $\#_{\mathbb{1}}(e_i) = \sum_{j=1}^{2^m} \#_{\mathbb{1}}(t_j) = \sum_{j=1}^{2^m} \theta[b_i d_j].2^{2^m - j} = a_i$.

Note that Condition 3 was not used in the previous proof, but it will be useful to define a succinct D2VPA recognising e_i. The key result is the following. It states the existence of a succinct D2VPA which recognises exactly the candidate solutions to SSSP.

Lemma 4. *One can construct a* D2VPA $\mathcal{B}$ *such that* $\mathcal{B}$ *has polynomially many states in* $|\theta| + k + m$ *and* $L(\mathcal{B}) = \{X_1 e_1 \ldots X_{2^k - 1} e_{2^k - 1} \# e_{2^k} \mid X_i \in \{0, 1\}\}$.

Proof (Sketch). First, we show the existence of a D2VPA $\mathcal{A}$ with polynomially many states in $|\theta| + k + m$ such that $L(\mathcal{A}) = \{e_i \mid i \in \{1, \ldots, 2^k\}\}$ (Proposition ?? in Appendix). The main idea is to construct succinct D2VPA which check each of the conditions 1 to 4 of the definition of the encoding independently, and then to take their intersection (by running the first, then the second, etc.). Condition 1 is easy to check. For condition 2, we rely on Lemma 2, and run sequentially the automata A_i (in m passes) to check independently that for all i, each root-to-leaf branch has a sequence of indices that belongs to A_i. Thanks to the reversibility of A_i, it is possible when going upward in the tree, to recover the previous state of A_i. For condition 3, we rely on the two-wayness to check that a sequence of m bits is a successor of another sequence succinctly, by doing $O(m)$ passes over the two successor vectors. The stack is not necessary there. For condition 4, we rely on the existence of a succinct 2DFA which accepts all the valuations that satisfy a given Boolean formula.

We can finally construct the D2VPPA$_{\mathsf{su}}$ $\mathcal{P} = (\mathcal{C}, \rho, \phi)$ of dimension 2 whose language is non-empty iff the SSSP instance $\mathcal{I}$ has a solution. The automaton $\mathcal{C}$ performs a first pass on the whole word by running the automaton $\mathcal{B}$ of Lemma 4, to check that the input is of the form $X_1 e_1 \ldots X_{2^k - 1} e_{2^k - 1} \# e_{2^k}$. During this pass, no vector dimension is incremented. During a second pass, $\mathcal{C}$, when reading some $X_i = 1$, it goes to some state q_1 from which it increments the 1st dimension whenever $\mathbb{1}$ is read (all other transitions have value $(0, 0)$). When reading some X_{i+1}, it stays in q_1 if $X_{i+1} = 1$ or to q_0 otherwise, from which no transition touches the counters. When reading $\#$, it goes to a state from which it increments only the 2nd dimension on reading $\mathbb{1}$. Note that this automaton is *single-use*: any symbol $\mathbb{1}$ occurring in the whole input word is counted at most once. It is even finite-visit (each position is visited $O(m + k + |\theta|)$ times). Finally, one only needs to check whether the first dimension equals the second one, using a formula $\phi(x_1, x_2) \equiv x_1 = x_2$. Note that the following lemma proves Theorem 4, since SSSP is NExpTime-c.

Lemma 5. *Given an instance X, Y, θ of SSSP, one can construct a* D2VPPA$_{su}$ *$\mathcal{P}$ of polynomial size in $|\theta| + |X| + |Y|$ such that $L(\mathcal{P}) \neq \varnothing$ iff SSSP has a solution.*

5 Applications to Decision Problems for Nested Word Transducers

In this section, we give two applications of 2VPPA, namely on decision problems for two-way visibly pushdown transducers (2VPT). 2VPT were introduced in [8] as a model to define transductions from well-nested words to words, or, modulo tree linearisation, from tree to words. It was shown that they can express, even in their deterministic and single-use version, all functions from well-nested words to words definable in MSOT, in the sense of Courcelle [6], while having decidable equivalence problem. No upper bound was provided however. Using 2VPPA, we show that the equivalence of 2VPT$_{su}$ defining functions can be tested in NExpTime. We also consider other standard problems from transducer theory and show, again using 2VPPA, their decidability. First, let us define formally 2VPT.

A *two-way visibly pushdown transducer* (2VPT for short) is a pair (A, μ) where A is a 2VPA and μ is a morphism from the sequences of transitions δ^* to some output alphabet Γ^*. A run of a 2VPT is a run of its underlying 2VPA. The *output* of a run ρ of the form $(q_0, i_0, d_0, \sigma_0)\tau_1(q_1, i_1, d_1, \sigma_1)\tau_2 \dots \tau_\ell(q_\ell, i_\ell, d_\ell, \sigma_\ell)$ is $\mu(\tau_1 ... \tau_\ell)$. A run is accepted if it is accepted by its underlying automaton. The transduction defined by a 2VPT is the set of pairs (u, v) such that v is the output of some accepting run on u. A state p of a 2VPT is *producing* if there exists a transition τ such that p is the first component of τ and $\mu(\tau) \neq \epsilon$. Similarly to Parikh automata, a 2VPT T is single-use (denoted 2VPT$_{su}$) if for any valid run of T, we do not reach the same position twice in the same producing state. It is deterministic, denoted D2VPT, if its underlying automaton is deterministic.

Deciding the k-valuedness and equivalence problems. For any positive integer k, we say that a transducer is *k-valued* if all input word have at most k different outputs. In particular, it is 1-valued if it defines a (partial) function, and also called *functional* in that case.

Theorem 5. *Let T be a* 2VPT$_{su}$*, and k an integer. Then the k-valuedness of T can be decided in* NExpTime. *It is also* ExpTime-hard.

The theorem is proved by reducing the k-valuedness of T to the emptiness of a 2VPPA$_{su}$ $\mathcal{P}$ that guesses $k+1$ runs of T that produce $k+1$ different outputs. To ensure that the output are different, during each run $\mathcal{P}$ guesses, and stores in counters, k output positions and the letters produced at these positions. The formula of $\mathcal{P}$ at the end simply checks, for each pairs of runs, that the same positions were guessed by both runs, and that the letters were different, ensuring that the guessed runs have different output pairwise. As two functional transducers are equivalent if they have the same domain and their union is 1-valued, we get the following corollary.

Corollary 2. *The equivalence of two functional* $\mathsf{2VPT_{su}}$ T *and* T' *can be decided in* NExpTime. *It is also* ExpTime-hard.

The NexpTime complexity of equivalence of tree to string transducers was already established for *Streaming Tree to string transducers* (STST), introduced in [1]. However, the conversion between the $\mathsf{2VPT_{su}}$ and STST yields an exponential blow-up.

We can generalize Corollary 2 to *strictly* k-valued transducers. We say that a transducer T is strictly k-valued if each input word in the domain of T has *exactly* k different images. Then similarly to the previous corollary, two strictly k-valued transducers are equivalent if, and only if, they have same domain and their union is k-valued.

Corollary 3. *The equivalence of two strictly* k*-valued* $\mathsf{2VPT_{su}}$ T *and* T' *can be decided in* NExpTime. *It is also* ExpTime-hard.

Strict k-valuedness is however an undecidable property (this can be shown by using the Post correspondence problem), even for $k = 2$. Deciding the equivalence problem for k-valued $\mathsf{2VPT_{su}}$ (which are not necessarily strictly k-valued) is open already in the stack-less case, and a (very) particular case has been solved in [14].

Type-checking against Parikh properties. Given a 2VPT T, it might be desirable to check some properties of the output words it produces, i.e., for a language L, whether the codomain of T is included in L. Formally, the *type-checking problem* asks, given a transducer T and a language L, whether $T(\Sigma^*) \subseteq L$. Unfortunately, this problem is undecidable when L is given by a visibly pushdown automaton (and T is a VPT) [13]. Nevertheless, we show that the type-checking problem is decidable when T is a $\mathsf{2VPT_{su}}$ and L is the complement of the language given by a (stack-less) Parikh Automaton. As a consequence, we are able to decide whether a $\mathsf{2VPT_{su}}$ T produces only well-nested words, i.e. if the output alphabet of T is structured and for every input word u and any $v \in T(u)$, v is a well-nested word.

Theorem 6. *Let* T *be a* $\mathsf{2VPT_{su}}$ *and* P *be a (stack-free) Parikh Automaton over the output alphabet of* T*. Then we can decide whether* $T(\Sigma^*) \cap L(P) = \emptyset$ *in* NExpTime. *It is also* ExpTime-hard.

This is done by constructing a $\mathsf{2VPPA_{su}}$ P' which simulates T, and instead of producing letters, simulates P on the output of T. A word w on a structured alphabet Σ is not well-nested if either $|w|_c \neq |w|_r$, i.e. the number of call letters is not equal to the number of return letters, or if there exists a prefix u of w such that $|u|_c < |u|_r$. As this can be checked by a (non-deterministic) Parikh automata, we get the following corollary.

Corollary 4. *Let T be a* $2\mathsf{VPT}_{\mathsf{su}}$ *whose output alphabet is structured. It can be decided in* CoNExpTime *whether T only produces well-nested words.*

Acknowledgements. This work was supported by the Belgian FNRS CDR project Flare (J013116), the ARC project Transform (Fédération Wallonie Bruxelles) and by the ANR Project *DELTA*, ANR-16-CE40-0007. Emmanuel Filiot is an FNRS research associate (Chercheur Qualifié).

References

1. Alur, R., D'Antoni, L.: Streaming tree transducers. In: Czumaj, A., Mehlhorn, K., Pitts, A., Wattenhofer, R. (eds.) ICALP 2012. LNCS, vol. 7392, pp. 42–53. Springer, Heidelberg (2012). https://doi.org/10.1007/978-3-642-31585-5_8
2. Alur, R., Madhusudan, P.: Adding nesting structure to words. J. ACM **56**(3), 16:1–16:43 (2009)
3. Burkart, O., Steffen, B.: Model checking the full modal mu-calculus for infinite sequential processes. In: Degano, P., Gorrieri, R., Marchetti-Spaccamela, A. (eds.) ICALP 1997. LNCS, vol. 1256, pp. 419–429. Springer, Heidelberg (1997). https://doi.org/10.1007/3-540-63165-8_198
4. Cadilhac, M., Finkel, A., McKenzie, P.: On the expressiveness of Parikh automata and related models. In: Proceedings of the Third Workshop on Non-Classical Models for Automata and Applications - NCMA 2011, Milan, Italy, 18 July–19 July 2011, pp. 103–119 (2011)
5. Clarke, E.M., Emerson, E.A., Sistla, A.P.: Automatic verification of finite state concurrent systems using temporal logic specifications. In: Conference Record of the Tenth Annual ACM Symposium on Principles of Programming Languages, pp. 117–126. ACM, January 1983
6. Courcelle, B., Engelfriet, J.: Graph Structure and Monadic Second-Order Logic - A Language-Theoretic Approach. Encyclopedia of Mathematics and its Applications, vol. 138. Cambridge University Press (2012). http://www.cambridge.org/fr/knowledge/isbn/item5758776/?site_locale=fr_FR
7. Dang, Z., Ibarra, O.H., Bultan, T., Kemmerer, R.A., Su, J.: Binary reachability analysis of discrete pushdown timed automata. In: Emerson, E.A., Sistla, A.P. (eds.) CAV 2000. LNCS, vol. 1855, pp. 69–84. Springer, Heidelberg (2000). https://doi.org/10.1007/10722167_9
8. Dartois, L., Filiot, E., Reynier, P.-A., Talbot, J.-M.: Two-way visibly pushdown automata and transducers. In: Proceedings of the 31st Annual ACM/IEEE Symposium on Logic in Computer Science, LICS 2016, New York, NY, USA, 5–8 July 2016, pp. 217–226 (2016). https://doi.org/10.1145/2933575.2935315
9. Engelfriet, J., Hoogeboom, H.J.: MSO definable string transductions and two-way finite-state transducers. ACM Trans. Comput. Logic **2**(2), 216–254 (2001)
10. Engelfriet, J., Maneth, S.: Macro tree transducers, attribute grammars, and MSO definable tree translations. Inf. Comput. **154**(1), 34–91 (1999). https://doi.org/10.1006/inco.1999.2807. http://www.sciencedirect.com/science/article/pii/S0890540199928079
11. Esparza, J., Ganty, P.: Complexity of pattern-based verification for multithreaded programs. ACM SIGPLAN Not. - POPL 2011 **46**(1), 499–510 (2011). https://doi.org/10.1145/1925844.1926443

12. Figueira, D., Libkin, L.: Path logics for querying graphs: combining expressiveness and efficiency. In: 30th Annual ACM/IEEE Symposium on Logic in Computer Science, LICS 2015, Kyoto, Japan, 6–10 July 2015, pp. 329–340 (2015). https://doi.org/10.1109/LICS.2015.39
13. Filiot, E., Raskin, J.-F., Reynier, P.-A., Servais, F., Talbot, J.-M.: Properties of visibly pushdown transducers. In: Hliněný, P., Kučera, A. (eds.) MFCS 2010. LNCS, vol. 6281, pp. 355–367. Springer, Heidelberg (2010). https://doi.org/10.1007/978-3-642-15155-2_32
14. Gallot, P., Muscholl, A., Puppis, G., Salvati, S.: On the decomposition of finite-valued streaming string transducers. In: STACS 2017. LIPIcs, vol. 66, pp. 34:1–34:14 (2017). https://doi.org/10.4230/LIPIcs.STACS.2017.34. http://drops.dagstuhl.de/opus/volltexte/2017/6999
15. Haase, C.: On the complexity of model checking counter automata. Ph.D. thesis, University of Oxford, UK (2012). http://ora.ox.ac.uk/objects/uuid:f43bf043-de93-4b5c-826f-88f1bd4c191d
16. Hague, M., Lin, A.W.: Model checking recursive programs with numeric data types. In: Gopalakrishnan, G., Qadeer, S. (eds.) CAV 2011. LNCS, vol. 6806, pp. 743–759. Springer, Heidelberg (2011). https://doi.org/10.1007/978-3-642-22110-1_60
17. Hague, M., Lin, A.W.: Synchronisation- and reversal-bounded analysis of multithreaded programs with counters. In: Madhusudan, P., Seshia, S.A. (eds.) CAV 2012. LNCS, vol. 7358, pp. 260–276. Springer, Heidelberg (2012). https://doi.org/10.1007/978-3-642-31424-7_22
18. Ibarra, O.H.: Reversal-bounded multicounter machines and their decision problems. J. ACM **25**(1), 116–133 (1978). http://doi.acm.org/10.1145/322047.322058
19. Ibarra, O.H.: Automata with reversal-bounded counters: a survey. In: Jürgensen, H., Karhumäki, J., Okhotin, A. (eds.) DCFS 2014. LNCS, vol. 8614, pp. 5–22. Springer, Cham (2014). https://doi.org/10.1007/978-3-319-09704-6_2
20. Karianto, W.: Parikh automata with pushdown stack. Technical report (2004)
21. Klaedtke, F.: Parikh automata and monadic second-order logics with linear cardinality constraints. Technical report, 30 July 2002
22. Klaedtke, F., Rueß, H.: Monadic second-order logics with cardinalities. In: Baeten, J.C.M., Lenstra, J.K., Parrow, J., Woeginger, G.J. (eds.) ICALP 2003. LNCS, vol. 2719, pp. 681–696. Springer, Heidelberg (2003). https://doi.org/10.1007/3-540-45061-0_54. http://dl.acm.org/citation.cfm?id=1759210.1759277
23. König, B., Esparza, J.: Verification of graph transformation systems with context-free specifications. In: Ehrig, H., Rensink, A., Rozenberg, G., Schürr, A. (eds.) ICGT 2010. LNCS, vol. 6372, pp. 107–122. Springer, Heidelberg (2010). https://doi.org/10.1007/978-3-642-15928-2_8
24. Parikh, R.J.: On context-free languages. J. ACM **13**(4), 570–581 (1966). https://doi.org/10.1145/321356.321364
25. Scarpellini, B.: Complexity of subcases of Presburger arithmetic. Trans. Am. Math. Soc. **284**(1), 203–218 (1984)
26. Schwoon, S.: Model checking pushdown systems. Ph.D. thesis, Technical University Munich, Germany (2002). http://tumb1.biblio.tu-muenchen.de/publ/diss/in/2002/schwoon.html
27. Shepherdson, J.C.: The reduction of two-way automata to one-way automata. IBM J. Res. Dev. **3**(2), 198–200 (1959)
28. Vardi, M.Y., Wolper, P.: An automata-theoretic approach to automatic program verification (preliminary report). In: LICS, pp. 332–344. IEEE Computer Society (1986)

29. Verma, K.N., Seidl, H., Schwentick, T.: On the complexity of equational horn clauses. In: Nieuwenhuis, R. (ed.) CADE 2005. LNCS (LNAI), vol. 3632, pp. 337–352. Springer, Heidelberg (2005). https://doi.org/10.1007/11532231_25
30. Walukiewicz, I.: Pushdown processes: games and model-checking. Inf. Comput. **164**(2), 234–263 (2001). https://doi.org/10.1006/inco.2000.2894. http://www.sciencedirect.com/science/article/pii/S0890540100928943

Kleene Algebra with Hypotheses

Amina Doumane[1,2], Denis Kuperberg[1(✉)], Damien Pous[1], and Pierre Pradic[1,2]

[1] Univ Lyon, EnsL, UCBL, CNRS, LIP, 69342 Lyon Cedex 07, France
denis.kuperberg@ens-lyon.fr
[2] Warsaw University, MIMUW, Warsaw, Poland

Abstract. We study the Horn theories of Kleene algebras and star continuous Kleene algebras, from the complexity point of view. While their equational theories coincide and are PSPACE-complete, their Horn theories differ and are undecidable. We characterise the Horn theory of star continuous Kleene algebras in terms of downward closed languages and we show that when restricting the shape of allowed hypotheses, the problems lie in various levels of the arithmetical or analytical hierarchy. We also answer a question posed by Cohen about hypotheses of the form $1 = S$ where S is a sum of letters: we show that it is decidable.

Keywords: Kleene algebra · Hypotheses · Horn theory · Complexity

1 Introduction

Kleene algebras [6,10] are idempotent semirings equipped with a unary operation *star* such that x^* intuitively corresponds to the sum of all powers of x. They admit several models which are important in practice: formal languages, where L^* is the Kleene star of a language L; binary relations, where R^* is the reflexive transitive closure of a relation R; matrices over various semirings, where M^* can be used to perform flow analysis.

A fundamental result is that their equational theory is decidable, and actually PSPACE-complete. This follows from a completeness result which was proved independently by Kozen [11] and Krob [17] and Boffa [3], and the fact that checking language equivalence of two regular expressions is PSPACE-complete: given two regular expressions, we have

$$\mathsf{KA} \vdash e \leq f \quad \text{iff} \quad [e] \subseteq [f]$$

(where $\mathsf{KA} \vdash e \leq f$ denotes provability from Kleene algebra axioms, and $[e]$ is the language of a regular expression e).

This work has been supported by the European Research Council (ERC) under the European Union's Horizon 2020 programme (CoVeCe, grant agreement No 678157) and by the LABEX MILYON (ANR-10-LABX-0070) of Université de Lyon, within the program "Investissements d'Avenir" (ANR-11-IDEX-0007) operated by the French National Research Agency (ANR).

M. Bojańczyk and A. Simpson (Eds.): FOSSACS 2019, LNCS 11425, pp. 207–223, 2019.
https://doi.org/10.1007/978-3-030-17127-8_12

Because of their interpretation in the algebra of binary relations, Kleene algebras and their extensions have been used to reason abstractly about program correctness [1,2,9,12,15]. For instance, if two programs can be abstracted into two relational expressions $(R^*; S)^*$ and $((R \cup S)^*; S)^=$, then we can deduce that these programs are equivalent by checking that the regular expression $(a^*b)^*$ and $(a+b)^*b+1$ denote the same language. This technique made it possible to automate reasoning steps in proof assistants [4,16,19].

In such a scenario, one often has to reason under assumptions. For instance, if we can abstract our programs into relational expressions $(R+S)^*$ and $S^*; R^*$, then we can deduce algebraically that the starting programs are equal if we know that $R; S = R$ (i.e., that S is a no-op when executed after R). When doing so, we move from the equational theory of Kleene algebras to their Horn theory: we want to know whether a given set of equations, the *hypotheses*, entails another equation in all Kleene algebras. Unfortunately, this theory is undecidable in general [13]. In this paper, we continue the work initiated by Cohen [5] and pursued by Kozen [13], by characterising the precise complexity of new subclasses of this general problem.

A few cases have been shown to be decidable in the literature, when we restrict the form of the hypotheses:

- when they are of the form $e = 0$ [5],
- when they are of the form $a \leq 1$ for a a letter [5],
- when they are of the form $1 = w$ or $a = w$ for a a letter and w a word, provided that those equations seen as a word rewriting system satisfy certain properties [14,18]; this includes equations like idempotency ($x = xx$) or self-invertibility ($1 = xx$).

(In the first two cases, the complexity can be shown to remain in PSPACE.) We add one positive case, which was listed as open by Cohen [5], and which is typically useful to express that a certain number of predicates cover all cases:

- when hypotheses are of the form $S = 1$ for S a sum of letters.

Conversely, Kozen also studied the precise complexity of various undecidable sub-classes of the problem [13]. For those, one has to be careful about the precise definition of Kleene algebras. Indeed, these only form a quasi-variety (their definition involves two implications), and one often consider $*$-*continuous* Kleene algebras [6], which additionally satisfy an infinitary implication (We define these formally in Sect. 2). While the equational theory of Kleene algebras coincides with that of $*$-continuous Kleene algebras, this is not the case for their Horn theories: there exist Horn sentences which are valid in all $*$-continuous Kleene algebras but not in all Kleene algebras.

Kozen [13] showed for instance that when hypotheses are of the form $pq = qp$ for pairs of letters (p, q), then validity of an implication in all $*$-continuous Kleene algebras is Π^0_1-complete, while it is only known to be EXPSPACE-hard for plain Kleene algebras. In fact, for plain Kleene algebras, the only known negative result is that the problem is undecidable for hypotheses of the form $u = v$ for

	$1 = \sum a$	$a \leq \sum b$	$a \leq \sum w$	$a \leq g$
$\mathsf{KA}_H \vdash u \leq f$	Decidable	EXPTIME – complete	Σ_1^0–complete	Σ_1^0–complete
$\mathsf{KA}_H \vdash e \leq f$	Decidable	Undecidable	Σ_1^0–complete	Σ_1^0–complete
$\mathsf{KA}_H^* \vdash u \leq f$	Decidable	EXPTIME – complete	Σ_1^0–complete	Π_1^1–complete
$\mathsf{KA}_H^* \vdash e \leq f$	Decidable	Π_1^0–complete	Π_2^0–complete	Π_1^1–complete

Fig. 1. Summary of the main results.

pairs (u, v) of words (Kleene star plays no role in this undecidability result: this is just the word problem). We show that it is already undecidable, and in fact Σ_1^0-complete when hypotheses are of the form $a \leq S$ where a is a letter and S is a sum of letters. We use a similar encoding as in [13] to relate the Horn theories of KA and KA* to runs of Turing Machines and alternating linearly bounded automata. This allows us to show that deciding whether an inequality $w \leq f$ holds where w is a word, in presence of sum-of-letters hypotheses, is EXPTIME-complete. We also refine the Π_1^1-completeness result obtained in [13] for general hypotheses, by showing that hypotheses of the form $a \leq g$ where a is a letter already make the problem Π_1^1-complete.

The key notion we define and exploit in this paper is the following: given a set H of equations, and given a language L, write $\mathrm{cl}_H(L)$ for the smallest language containing L such that for all hypotheses $(e \leq f) \in H$ and all words u, v,

$$\text{if} \quad u[f]v \subseteq \mathrm{cl}_H(L) \quad \text{then} \quad u[e]v \subseteq \mathrm{cl}_H(L) \ .$$

This notion makes it possible to characterise the Horn theory of $*$-continuous Kleene algebras, and to approximate that of Kleene algebras: we have

$$\mathsf{KA}_H \vdash e \leq f \quad \Rightarrow \quad \mathsf{KA}_H^* \vdash e \leq f \quad \Leftrightarrow \quad [e] \subseteq \mathrm{cl}_H([f])$$

where $\mathsf{KA}_H \vdash e \leq f$ (resp. $\mathsf{KA}_H^* \vdash e \leq f$) denotes provability in Kleene algebra (resp. $*$-continuous Kleene algebra). We study downward closed languages and prove the above characterisation in Sect. 3.

The first implication can be strengthened into an equivalence in a few cases, for instance when the regular expression e and the right-hand sides of all hypotheses denote finite languages, or when hypotheses have the form $1 = S$ for S a sum of letters. We obtain decidability in those cases (Sect. 4).

Then we focus on cases where hypotheses are of the form $a \leq e$ for a a letter, and we show that most problems are already undecidable there. We do so by exploiting the characterisation in terms of downward closed languages to provide encodings of various undecidable problems on Turing machines, total Turing machines, and linearly bounded automata (Sect. 5).

We summarise our results in Fig. 1. The top of each column restricts the type of allowed hypotheses. Variables e, f stand for general expressions, u, w for words, and a, b for letters. Grayed statements are implied by non-grayed ones.

Notations. We let a, b range over the letters of a finite alphabet Σ. We let u, v, w range over the words over Σ, whose set is written Σ^*. We write ϵ for the empty word; uv for the concatenation of two words u, v; $|w|$ for the length of a word w. We write Σ^+ for the set of non-empty words. We let e, f, g range over the regular expressions over Σ, whose set is written Exp_Σ. We write $[e]$ for the language of such a an expression e: $[e] \subseteq \Sigma^*$. We sometimes implicitly regard a word as a regular expression. If X is a set, $\mathcal{P}(X)$ (resp. $\mathcal{P}_{\mathsf{fin}}(X)$) is the set of its subsets (resp. finite subsets) and $|X|$ for its cardinality.

A long version of this extended abstract is available on HAL [8], with most proofs in appendix.

2 The Systems KA and KA*

Definition 1 ($\mathsf{KA}, \mathsf{KA}^*$). *A* Kleene algebra *is a tuple* $(M, 0, 1, +, \cdot, *)$ *where* $(M, 0, 1, +, \cdot)$ *is an idempotent semiring and the following axioms and implications, where the partial order* $\leq$ *is defined by* $x \leq y$ *if* $x + y = y$*, hold for all* $x, y \in M$.

$$1 + xx^* \leq x^* \qquad\qquad xy \leq y \ \Rightarrow\ x^*y \leq y$$

$$1 + x^*x \leq x^* \qquad\qquad yx \leq y \ \Rightarrow\ yx^* \leq y$$

A Kleene algebra is $*$*-continuous if it satisfies the following implication:*

$$(\forall i \in \mathbb{N},\ xy^iz \leq t) \ \Rightarrow\ xy^*z \leq t$$

A hypothesis *is an inequation of the form* $e \leq f$*, where* e *and* f *are regular expressions. If* H *is a set of hypotheses, and* e, f *are regular expressions, we write* $\mathsf{KA}_H \vdash e \leq f$ *(resp.* $\mathsf{KA}^*_H \vdash e \leq f$*) if* $e \leq f$ *is derivable from the axioms and implications of* KA *(resp.* KA***) *as well as the hypotheses from* H*. We omit the subscript when* H *is empty.*

Note that the letters appearing in the hypotheses are constants: they are not universally quantified. In particular if $H = \{aa \leq a\}$, we may deduce $\mathsf{KA}_H \vdash a^* \leq a$ but not $\mathsf{KA}_H \vdash b^* \leq b$.

Languages over the alphabet Σ form a $*$-continuous Kleene algebra, as well as binary relations over an arbitrary set.

In absence of hypotheses, provability in KA is coincides with provability in KA* and with language inclusion:

Theorem 1 (Kozen [11]).

$$\mathsf{KA} \vdash e \leq f \quad \Leftrightarrow \quad \mathsf{KA}^* \vdash e \leq f \quad \Leftrightarrow \quad [e] \subseteq [f]$$

We will classify the theories based on the shape of hypotheses we allow; we list them below (I is a finite non-empty set):

Name of the hypothesis	Its shape		
$(1 = \sum x)$ – hypothesis	$1 = \sum_{i \in I} a_i$	where	$a_i \in \Sigma$
$(w \leq \sum w)$ – hypothesis	$v \leq \sum_{i \in I} v_i$	where	$v, v_i \in \Sigma^*$
$(x \leq \sum w)$ – hypothesis	$a \leq \sum_{i \in I} v_i$	where	$a \in \Sigma, v_i \in \Sigma^*$
$(x \leq \sum x)$ – hypothesis	$a \leq \sum_{i \in I} a_i$	where	$a, a_i \in \Sigma$
$(1 \leq \sum x)$ – hypothesis	$1 \leq \sum_{i \in I} a_i$	where	$a_i \in \Sigma$
$(x \leq 1)$ – hypothesis	$a \leq 1$	where	$a \in \Sigma$

We call *letter hypotheses* any class of hypotheses where the left-hand side is a letter (the last four ones). In the rest of the paper, we study the following problem from a complexity point of view: given a set of C-hypotheses H, where C is one of the classes listed above, and two expressions $e, f \in \mathsf{Exp}_\Sigma$, can we decide whether $\mathsf{KA}_H \vdash e \leq f$ (resp. $\mathsf{KA}^*_H \vdash e \leq f$) holds? We call it the problem of **deciding** KA **(resp.** KA^***) under** C**-hypotheses.**

3 Closure of Regular Languages

It is known that provability in KA and KA^* can be characterised by language inclusions (Theorem 1). In the presence of hypotheses, this is not the case anymore: we need to take the hypotheses into account in the semantics. We do so by using the following notion of *downward closure* of a language.

3.1 Definition of the Closure

Definition 2 (H**-closure).** *Let H be a set of hypotheses and $L \subseteq \Sigma^*$ be a language. The* H-closure *of L, denoted $\mathrm{cl}_H(L)$, is the smallest language K such that $L \subseteq K$ and for all hypotheses $e \leq f \in H$ and all words $u, v \in \Sigma^*$, we have*

$$u[f]v \subseteq C \qquad \Rightarrow \qquad u[e]v \subseteq K$$

Alternatively, $\mathrm{cl}_H(L)$ can be defined as the least fixed point of the function $\phi_L : \mathcal{P}(\Sigma^*) \to \mathcal{P}(\Sigma^*)$ defined by $\phi_L(X) = L \cup \psi_H(X)$, where

$$\psi_H(X) = \bigcup_{(e \leq f) \in H} \{u[e]v \mid u, v \in \Sigma^*, u[f]v \subseteq X\}.$$

Example 1. If $H = \{ab \leq ba\}$ then $\mathrm{cl}_H([b^*a^*]) = [(a+b)^*]$, while $\mathrm{cl}_H([a^*b^*]) = [a^*b^*]$.

In order to manipulate closures more conveniently, we introduce a syntactic object witnessing membership in a closure: derivation trees.

Definition 3. *Let H be a set of hypotheses and L a regular language. We define an infinitely branching proof system related to $\mathrm{cl}_H(L)$, where statements are regular expressions, and rules are the following, called respectively* axiom, extension, *and* hypothesis*:*

$$\frac{}{u}\ u \in L \qquad \frac{(u)_{u\in[e]}}{e} \qquad \frac{ufv}{uwv}\ w \in [e],\ e \leq f \in H$$

We write $\vdash_{H,L} e$ if e is derivable in this proof system, i.e. if there is a well-founded tree using these rules, with root e and all leaves labelled by words in L. Such a tree will be called a derivation tree *for $[e] \subseteq \mathrm{cl}_H(L)$ (or $e \in \mathrm{cl}_H(L)$ if e is a word).*

Example 2. The following derivation is a derivation tree for $bababa \in \mathrm{cl}_H([b^*a^*])$, where $H = \{ab \leq ba\}$.

$$\cfrac{\cfrac{\cfrac{\cfrac{}{bbbaaa}}{bb\mathbf{ab}aa}}{bba\mathbf{ab}a}}{b\mathbf{ab}aba}$$

Derivation trees witness membership to the closure as shown by the following proposition.

Proposition 1. $[e] \subseteq \mathrm{cl}_H(L)$ *iff* $\vdash_{H,L} e$.

(See [8, App. A] for a proof.)

3.2 Properties of the Closure Operator

We summarise in this section some useful properties of the closure. Lemma 1 shows in particular that the closure is idempotent, monotonic (both for the set of hypotheses and its language argument) and invariant by context application. Lemma 2 shows that internal closure operators can be removed in the evaluation of regular expressions. Those two lemmas are proved in [8, App. A].

Lemma 1. *Let $A, B, U, V \subseteq \Sigma^*$. We have*

1. $A \subseteq \mathrm{cl}_H(A)$
2. $\mathrm{cl}_H(\mathrm{cl}_H(A)) = \mathrm{cl}_H(A)$
3. $A \subseteq B$ *implies* $\mathrm{cl}_H(A) \subseteq \mathrm{cl}_H(B)$
4. $H \subseteq H'$ *implies* $\mathrm{cl}_H(A) \subseteq \mathrm{cl}_{H'}(A)$
5. $\mathrm{cl}_H(A) \subseteq \mathrm{cl}_H(B)$ *if and only if* $A \subseteq \mathrm{cl}_H(B)$.
6. $A \subseteq \mathrm{cl}_H(B)$ *implies* $UAV \subseteq \mathrm{cl}_H(UBV)$.

Lemma 2. *Let $A, B \subseteq \Sigma^*$, then*

1. $\mathrm{cl}_H(A + B) = \mathrm{cl}_H(\mathrm{cl}_H(A) + \mathrm{cl}_H(B))$,
2. $\mathrm{cl}_H(AB) = \mathrm{cl}_H(\mathrm{cl}_H(A)\mathrm{cl}_H(B))$,
3. $\mathrm{cl}_H(A^*) = \mathrm{cl}_H(\mathrm{cl}_H(A)^*)$

3.3 Relating Closure and Provability in KA_H and KA^*_H

We show that provability in KA^* can be characterized by closure inclusions. In KA, provability implies closure inclusions but the converse is not true in general.

Theorem 2. *Let H be a set of hypotheses and e, f be two regular expressions.*

$$\mathsf{KA}_H \vdash e \leq f \quad \Rightarrow \quad \mathsf{KA}^*_H \vdash e \leq f \quad \Leftrightarrow \quad [e] \subseteq \mathrm{cl}_H([f])$$

Proof. Let $\mathsf{CReg}_{H,\Sigma} = \{\mathrm{cl}_H(L) \mid L \in \mathsf{Reg}_\Sigma\}$, on which we define the following operations:

$$X \oplus Y = \mathrm{cl}_H(X + Y) \qquad X \odot Y = \mathrm{cl}_H(X \cdot Y) \qquad X^{\circledast} = \mathrm{cl}_H(X^*).$$

We define the *closure model* $F_{H,\Sigma} = (\mathsf{CReg}_{H,\Sigma}, \emptyset, \{\epsilon\}, \oplus, \odot, \circledast)$.

We write $\leq$ for the inequality induced by $\oplus$ in $F_{H,\Sigma}$: $X \leq Y$ if $X \oplus Y = Y$.

Lemma 3. *$F_{H,\Sigma} = (\mathsf{CReg}_{H,\Sigma}, \emptyset, \{\epsilon\}, \oplus, \odot, \circledast)$ is a $*$-continuous Kleene algebra. The inequality $\leq$ of $F_{H,\Sigma}$ coincides with inclusion of languages.*

Proof. By Lemma 2, the function $\mathrm{cl}_H : (\mathcal{P}(\Sigma^*), +, \cdot, *) \to (\mathsf{CReg}_{H,\Sigma}, \oplus, \odot, \circledast)$ is a homomorphism. We show that $F_{H,\Sigma}$ is a $*$-continuous Kleene algebra. First, identities of $\mathsf{Lang}_\Sigma = (\mathcal{P}(\Sigma^*), +, \cdot, *)$ are propagated through the morphism cl_H, so only Horn formulas defining $*$-continuous Kleene algebras remain to be verified. It suffices to prove that $F_{H,\Sigma}$ satisfies the $*$-continuity implication, because the implication $xy \leq y \rightarrow x^*y \leq y$ and its dual can be deduced from it. Let $A, B, C \in F_{H,\Sigma}$ such that for all $i \in \mathbb{N}$, $A \odot B^{ⓘ} \odot C \leq D$, where $B^{ⓘ} = B \odot \cdots \odot B$. By Lemma 2, $A \odot B^{ⓘ} \odot C = \mathrm{cl}_H(AB^iC)$, so we have $\mathrm{cl}_H(AB^iC) \leq D$, and in particular $AB^iC \leq D$ for all i. By $*$-continuity of Lang_Σ, we obtain $AB^*C \leq D$. By Lemma 1 and using $D = \mathrm{cl}_H(D)$, we obtain $\mathrm{cl}_H(AB^*C) \leq D$ and finally by Lemma 2, $A \odot B^{\circledast} \odot C \leq D$. This achieves the proof that $F_{H,\Sigma}$ is a $*$-continuous Kleene algebra.

Let $A, B \in \mathsf{CReg}_{H,\Sigma}$. We have $A \leq B \Leftrightarrow A \oplus B = B \Leftrightarrow \mathrm{cl}_H(A + B) = B \Leftrightarrow A \subseteq B$. Finally, if $e \leq f$ is a hypothesis from H, then we have $\mathrm{cl}_H[e] \subseteq \mathrm{cl}_H([f])$, so the hypothesis is verified in $F_{H,\Sigma}$. □

The implications $\mathsf{KA}^{(*)}_H \vdash e \leq f \Rightarrow [e] \subseteq \mathrm{cl}_H(f)$ follow from the fact that if an inequation $e \leq f$ is derivable in KA_H (resp. KA^*_H) then it is true in every model, in particular in the model $F_{H,\Sigma}$, thus $\mathrm{cl}_H([e]) \subseteq \mathrm{cl}_H([f])$ or, equivalently. $[e] \subseteq \mathrm{cl}_H([f])$.

Let us prove that for any regular expressions e, f, if $[e] \subseteq \mathrm{cl}_H([f])$ then $\mathsf{KA}^*_H \vdash e \leq f$. Let e, f be two such expressions and let T be a derivation tree for $[e] \subseteq \mathrm{cl}_H([f])$, i.e. witnessing $\vdash_{H,L} e \leq f$. We show that we can transform this tree T into a proof tree in KA^*_H. The extension rule is an occurrence of [8, App. A, Lem. 12]. Finally, the hypothesis rule is also provable in KA^*_H, using the hypothesis $e \leq f$ together with compatibility of $\leq$ with concatenation, and completeness of KA^* for membership of $u \in [e]$. We can therefore build from the tree T a proof in KA^*_H witnessing $\mathsf{KA}^*_H \vdash e \leq f$. □

When we restrict the shape of the expression e to words, and hypotheses to $(w \leq \sum w)$-hypotheses, we get the implication missing from Theorem 2.

Proposition 2. *Let H be a set of $(w \leq \sum w)$-hypotheses, $w \in \Sigma^*$ and $f \in \mathsf{Exp}_\Sigma$.*

$$\mathsf{KA}_H \vdash w \leq f \qquad \Leftrightarrow \qquad w \in \mathrm{cl}_H([f])$$

Proof. Let us show that $w \in \mathrm{cl}_H([f])$ implies $\mathsf{KA}_H \vdash w \leq f$. We proceed by induction on the height of a derivation tree for $w \in \mathrm{cl}_H([f])$. If this tree is just a leaf, then $w \in [f]$ and by Theorem 1 $\mathsf{KA} \vdash w \leq f$. Otherwise, this derivation starts with the following steps:

$$\dfrac{\dfrac{\left(\dfrac{\cdots}{uw_iv}\right)_i}{u(\sum_i w_i)v}}{uwv}\; w \leq \textstyle\sum_i w_i \in H$$

Our inductive assumption is that $\mathsf{KA}_H \vdash uw_iv \leq f$ for all i, thus $\mathsf{KA}_H \vdash \sum_i uw_iv \leq f$. We also have $\mathsf{KA}_H \vdash w \leq (\sum_i w_i)$ hence $\mathsf{KA} \vdash w \leq f$ by distributivity. □

4 Decidability of KA and KA* with $(1 = \sum x)$-Hypotheses

In this section, we answer positively the decidability problem of KA_H, where H is a set of $(1 = \sum x)$-hypotheses, posed by Cohen [5]:

Theorem 3. *If H is a set of $(1 = \sum x)$-hypotheses, then KA_H is decidable.*

To prove this theorem we show that in the case of $(1 = \sum x)$-hypotheses:

(P1) $\mathsf{KA}_H \vdash e \leq f$ if and only if $[e] \subseteq \mathrm{cl}_H([f])$.
(P2) $\mathrm{cl}_H([f])$ is regular and we can compute effectively an expression for it.

Decidability of KA_H follows immediately from (P1) and (P2), since it amounts to checking language inclusion for two regular expressions.

To show $(P1)$ and $(P2)$, it is enough to prove the following result:

Theorem 4. *Let H be a set of $(1 = \sum x)$-hypotheses and let f be a regular expression. The language $\mathrm{cl}_H([f])$ is regular and we can compute effectively an expression c such that $[c] = \mathrm{cl}_H([f])$ and $\mathsf{KA}_H \vdash c \leq f$.*

(P2) follows immediately from Theorem 4. To show (P1), it is enough to prove that $[e] \subseteq \mathrm{cl}_H([f])$ implies $\mathsf{KA}_H \vdash e \leq f$, since the other implication is always true (Theorem 2). Let e, f such that $[e] \subseteq \mathrm{cl}_H([f])$. If c is the expression given by Theorem 4, we have $\mathsf{KA}_H \vdash c \leq f$ and $[e] \subseteq [c]$ so by Theorem 1 $\mathsf{KA} \vdash e \leq c$, and this concludes the proof.

To prove Theorem 4, we first show that the closure of $(1 = \sum x)$-hypotheses can be decomposed into the closure of $(x \leq 1)$-hypotheses followed by the closure of $(1 \leq \sum x)$-hypotheses:

Proposition 3 (Decomposition result). *Let $H = \{1 = S_j \mid j \in J\}$ be a set of $(1 = \sum x)$-hypotheses.*

We set $H_{sum} = \{1 \leq S_j \mid j \in J\}$ and $H_{id} = \{a \leq 1 \mid a \in [S_j], j \in J\}$. For every language $L \subseteq \Sigma^*$, we have $\mathrm{cl}_H(L) = \mathrm{cl}_{H_{sum}}(\mathrm{cl}_{H_{id}}(L))$.

Sketch. We show that rules from H_{id} can be locally permuted with rules of H_{sum} in a derivation tree. This allows to compute a derivation tree where all rules from H_{id} occur after (i.e. closer to leaves than) rules from H_{sum}. □

Now, we will show results similar to Theorem 4, but which apply to $(x \leq 1)$-hypotheses and $(1 \leq \sum x)$-hypotheses (Propositions 5 and 6 below). To prove Theorem 4, the idea is to decompose H into H_{id} and H_{sum} using the decomposition property Proposition 3, then applying Propositions 5 and 6 to H_{id} and H_{sum} respectively.

To show these two propositions, we make use of a result from [7]:

Definition 4. *Let $\mathcal{A} = (Q, \Delta, \iota, F)$ be an NFA, H be a set of hypotheses and $\varphi : Q \to \mathsf{Exp}_\Sigma$ a function from states to expressions. We say that φ is H-*compatible *with $\mathcal{A}$ if:*

- $\mathsf{KA}_H \vdash 1 \leq \varphi(q)$ *whenever* $q \in F$,
- $\mathsf{KA}_H \vdash a\varphi(r) \leq \varphi(q)$ *for all transitions* $(q, a, r) \in \Delta$.

We set $\varphi^{\mathcal{A}} = \varphi(\iota)$.

Proposition 4 ([7])**.** *Let $\mathcal{A}$ be a NFA, H be a set of hypothesis and φ be a function H-compatible with $\mathcal{A}$. We can construct a regular expression $f_{\mathcal{A}}$ such that:*

$$[f_{\mathcal{A}}] = [\mathcal{A}] \quad \text{and} \quad \mathsf{KA}_H \vdash f_{\mathcal{A}} \leq \varphi^{\mathcal{A}}$$

Proposition 5. *Let H be a set of $(x \leq 1)$-hypotheses and let f be a regular expression. The language $\mathrm{cl}_H([f])$ is regular and we can compute effectively an expression c such that $[c] = \mathrm{cl}_H([f])$ and $\mathsf{KA}_H \vdash c \leq f$.*

Proof. Let $K = \mathrm{cl}_H([f])$ and $\Gamma = \{a \mid (a \leq 1) \in H\}$, we show that K is regular. If $\mathcal{A}$ is a NFA for f, a NFA $\mathcal{A}_{id}$ recognizing K can be built from $\mathcal{A}$ by adding a Γ-labelled loop on every state. It is straightforward to verify that the resulting NFA recognizes K, by allowing to ignore any letter from Γ.

For every $q \in Q$, let f_q be a regular expression such that $[f_q] = [q]_{\mathcal{A}}$, where $[q]_{\mathcal{A}}$ denotes the language accepted from q in $\mathcal{A}$. Let $\varphi : Q \to \mathsf{Exp}_\Sigma$ which maps each state q of $\mathcal{A}_{id}$ (which is also a state of $\mathcal{A}$) to $\varphi(q) = f_q$. Let us show that φ is H-compatible with $\mathcal{A}$. If $q \in F$, then $1 \in [f_q]$, so by completeness of KA, we have $\mathsf{KA} \vdash 1 \leq f_q$. Let (p, a, q) be a transition of $\mathcal{A}_{id}$. Either $(p, a, q) \in \Delta$, in which case we have $a[f_q] \subseteq [f_p]$, and so by Theorem 1 $\mathsf{KA} \vdash af_q \leq f_p$. Or $p = q$ (this transition is a loop that we added). Then $\mathsf{KA}_H \vdash a \leq 1$, so $\mathsf{KA}_H \vdash af_p \leq f_p$, and this concludes the proof.

By Proposition 4, we can now construct a regular expression c which satisfies the desired properties. □

Definition 5. *Let Γ be a set of letters. A language L is said to be Γ-*closed *if:*

$$\forall u, v \in \Sigma^*, \forall a \in \Gamma \qquad uv \in L \qquad \Rightarrow \qquad uav \in L$$

If $H = \{1 \leq S_i \mid i \in I\}$ is a set of $(1 \leq \sum x)$-hypotheses, we say that a language L is H-closed if if it is Γ-closed where $\Gamma = \cup_{i \in I}[S_i]$.

Remark 1. If H is a set of $(x \leq 1)$-hypothesis, and $\Gamma = \{a \mid (a \leq 1) \in H\}$, then $\mathrm{cl}_H(L)$ is Γ-closed for every language L.

Proposition 6. *Let H be a set of $(1 \leq \sum x)$-hypotheses and let f be a regular expression whose language is H-closed. The language $\mathrm{cl}_H([f])$ is regular and we can compute effectively an expression c such that $[c] = \mathrm{cl}_H([f])$ and $\mathsf{KA}_H \vdash c \leq f$.*

Proof. We set $L = [f]$, $H = \{1 \leq S_j \mid j \in J\}$ and $\Gamma = \{a \mid a \in [S_j], j \in J\}$.

Let us show that $\mathrm{cl}_H(L)$ is regular. The idea is to construct a set of words $L_\sharp$, where each word $u_\sharp$ is obtained from a word u of $\mathrm{cl}_H(L)$, by adding at the position where a rule $(1 \leq S_j)$ is applied in the derivation tree for $\mathrm{cl}_H(L) \vdash u$, a new symbol $\sharp_j$. We will show that this set satisfies the two following properties:

- $\mathrm{cl}_H(L)$ is obtained from $L_\sharp$ by erasing the symbols $\sharp_j$.
- $L_\sharp$ is regular.

Since the operation that erases letters preserves regularity, we obtain as a corollary that $\mathrm{cl}_H(L)$ is regular.

Let us now introduce more precisely the language $L_\sharp$ and show the properties that it satisfies. Let $\Theta_\sharp = \{\sharp_j \mid j \in J\}$ be a set of new letters and $\Sigma_\sharp = \Sigma \cup \Theta_\sharp$ be the alphabet Σ enriched with these new letters.

We define the function $exp : \Sigma_\sharp \to \mathcal{P}(\Sigma)$ that expands every letter $\sharp_j$ into the sum of the letters corresponding to its rule in H as follows:

$$\begin{aligned} exp(a) &= a && \text{if } a \in \Sigma \\ exp(\sharp_j) &= \{a \mid a \in [S_j]\} && \forall j \in J \end{aligned}$$

This function can naturally be extended to $exp : (\Sigma_\sharp)^* \to \mathcal{P}(\Sigma^*)$.
If $L \subseteq \Sigma^*$, we define $L_\sharp \subseteq (\Sigma_\sharp)^*$ as follows:

$$L_\sharp = exp^{-1}(\mathcal{P}(L)) = \{u \in (\Sigma_\sharp)^* \mid exp(u) \subseteq L\}$$

We define the morphism $\pi : (\Sigma_\sharp)^* \to \Sigma^*$ that erases the letters from $\Theta_\sharp$ as follows: $\pi(a) = a$ if $a \in \Sigma$ and $\pi(\sharp_j) = \epsilon$ for all $j \in J$. Our goal is to prove that $\mathrm{cl}_H(L) = \pi(L_\sharp)$ and that $L_\sharp$ is regular. To prove the first part, we need an alternative presentation of $L_\sharp$ as the closure of a new set of hypotheses $H_\sharp$ which we define as follows:

$$H_\sharp = \{\sharp_j \leq S_j \mid j \in J\} \cup \{\sharp_j \leq 1 \mid j \in J\}$$

Lemma 4. *We have $L_\sharp = \mathrm{cl}_{H_\sharp}(L)$. In particular $L_\sharp$ is $\Theta_\sharp$-closed.*

See App. B for a detailed proof of Lemma 4.

Lemma 5. $\mathrm{cl}_H(L) = \pi(L_\sharp)$.

Proof. If $u \in \pi(L_\sharp)$, let $v \in L_\sharp$ such that $u = \pi(v)$. By Lemma 4, there is a derivation tree T_v for $v \in \mathrm{cl}_{H_\sharp}(L)$. Erasing all occurrences of $\sharp_j$ in T_v yields a derivation tree for $u \in \mathrm{cl}_H(L)$.

Conversely, if $u \in \mathrm{cl}_H(L)$ is witnessed by some derivation tree T_u, we show by induction on T_u that there exists $v \in L_\sharp \cap \pi^{-1}(u)$. If T_u is a single leaf, we have $u \in L$, and therefore it suffices to take $v = u$.

Otherwise, the rule applied at the root of T_u partitions u into $u = wz$, and has premises $\{wbz \mid b \in [S_j]\}$ for some $j \in J$ and $w, z \in \Sigma^*$. By induction hypothesis, for all $b \in [S_j]$, there is $v_b \in L_\sharp \cap \pi^{-1}(wbz)$. Let $w = w_1 \dots w_n$ and $z = z_1 \dots z_m$ be the decompositions of w, z into letters of Σ. By definition of π, for all $b \in [S_j]$, v_b can be written $v_b = \alpha_{b,1} w_1 \alpha_{b,2} w_2 \dots w_n \alpha_{b,n} b \alpha_{b,n+1} z_1 \alpha_{b,n+2} \dots z_m \alpha_{b,n+m+3}$, with $\alpha_{b,0} \dots \alpha_{b,n+m+3} \in (\Theta_\sharp)^*$. For each $k \in [0, n+m+3]$, let $\alpha_k = \Pi_{b \in [S_j]} \alpha_{b,k}$. Let $w' = \alpha_0 w_1 \alpha_1 \dots w_n \alpha_{n+1}$ and $z' = \alpha_{n+2} z_1 \alpha_{n+3} \dots z_m \alpha_{n+m+3}$. By Lemma 4, $L_\sharp$ is $\Theta_\sharp$-closed, so for each $b \in [S_j]$ the word $v'_b = w' b z'$ is in $L_\sharp$, since v'_b is obtained from v_b by adding letters from $\Theta_\sharp$. We can finally build $v = w' \sharp_j z'$. We have $exp(v) = \bigcup_{b \in [S_j]} exp(v'_b) \subseteq L$, and $\pi(v) = \pi(w')\pi(z') = wz = u$. □

Lemma 6. *$L_\sharp$ is a regular language, computable effectively.*

Sketch. From a DFA $\mathcal{A} = (\Sigma, Q, q_0, F, \delta)$ for for L, we first build a DFA $\mathcal{A}_\wedge = (\Sigma, \mathcal{P}(Q), q_0, \mathcal{P}(F), \delta_\wedge)$, which corresponds to a powerset construction, except that accepting states are $\mathcal{P}(F)$. This means that the semantic of a state P is the conjunction of its members. We then build $\mathcal{A}_\sharp = (\Sigma, \mathcal{P}(Q), q_0, \mathcal{P}(F), \delta_\sharp)$ based on $\mathcal{A}_\wedge$, which can additionally read letters of the form $\sharp_j$, by expanding them using the powerset structure of $\mathcal{A}_\wedge$. □

Lemma 7. *We can construct a regular expression c such that $[c] = \mathrm{cl}_H(L)$ and $\mathsf{KA}_H \vdash c \leq f$.*

Proof. Let $\mathcal{A}_\sharp$ be the DFA constructed for $L_\sharp$ in the proof of Lemma 6. We will use the notations of this proof in the following.

Let $\pi(\mathcal{A}_\sharp) = (\Sigma, \mathcal{P}(Q), q_0, \mathcal{P}(F), \pi(\delta_\sharp))$ be the NFA obtained from $\mathcal{A}_\sharp$ by replacing every transition $\delta_\sharp(P, \sharp_j) = R$, where $j \in J$, by a transition $\pi(\delta_\sharp)(P, \epsilon) = R$. By Lemma 5, the automaton $\pi(\mathcal{A}_\sharp)$ recognizes the language $\mathrm{cl}_H(L)$. Let us construct a regular expression c for this automaton such that $\mathsf{KA}_H \vdash c \leq f$.

For every $P \in \mathcal{P}(Q)$, let f_P be a regular expression such that $[f_P] = [P]_{\mathcal{A}_\wedge}$.

Let $\varphi : \mathcal{P}(Q) \to \mathsf{Exp}_\Sigma$ be the function which maps each state P of $\pi(\mathcal{A}_\sharp)$ to $\varphi(P) = f_P$. Let us show that φ is H-compatible.

If $P \in \mathcal{P}(F)$, then P is a final state of $\mathcal{A}_\wedge$, so $1 \in [f_P]$, and by completeness of KA, $\mathsf{KA} \vdash 1 \leq f_P$. Let $(P, a, R) \in \pi(\Delta_\sharp)$. Either $a \in \Sigma$, so $(P, a, R) \in \Delta_\wedge$ and $a[f_R] \subseteq [f_P]$, so by Theorem 1 $\mathsf{KA} \vdash a f_R \leq f_P$. Or $a = \epsilon$ so there is $j \in J$ such that $(P, \sharp_j, R) \in \Delta_\sharp$. This means that $R = \cup_{b \in [S_j]} R_b$ where $\delta_\wedge(P, b) = R_b, \forall b \in [S_j]$. We have then that $b[f_{R_b}] \subseteq [f_P]$ for all $b \in [S_j]$. Note that for all $b \in [S_j]$, $R_b \subseteq R$, so $[f_R] \subseteq [f_{R_b}]$ and then $S_j[f_R] \subseteq [f_P]$. By Theorem 1 $\mathsf{KA} \vdash S_j f_R \leq f_P$. We have also that $\mathsf{KA}_H \vdash \sharp_j \leq S_j$, so $\mathsf{KA}_H \vdash \sharp_j f_R \leq f_P$.

By Proposition 4, we can construct the desired regular expression c. □

5 Complexity Results for Letter Hypotheses

In this section, we give a recursion-theoretic characterization of KA_H and KA_H^* where H is a set of letter hypotheses or $(w \leq \sum w)$-hypotheses. In all the section, by "deciding $\mathsf{KA}_H^{(*)}$" we mean deciding whether $\mathsf{KA}_H^{(*)} \vdash e \leq f$, given e, f, H as input.

Theses various complexity classes will be obtained by reduction from some known problems concerning Turing Machines (TM) and alternating linearly bounded automata (LBA), such as halting problem and universality.

To obtain these reductions, we build on a result which bridges TMs and LBAs on one hand and closures on the other: the set of co-reachable configurations of a TM (resp. LBA) can be seen as the closure of a well-chosen set of hypotheses.

We present this result in Sect. 5.1, and show in Sect. 5.2 how to instantiate it to get our complexity classes.

5.1 Closure and Co-reachable States of TMs and LBAs

Definition 6. *An* alternating Turing Machine *over Σ is a tuple $\mathcal{M} = (Q, Q_F, \Gamma, \iota, B, \Delta)$ consisting of a finite set of states Q and final states $Q_F \subseteq Q$, a finite set of states Q, a finite working alphabet $\Gamma \supseteq \Sigma$, an initial state $\iota \in Q$, $B \in \Gamma$ the blank symbol and a transition function $\Delta : (Q \setminus Q_F) \times \Gamma \to \mathcal{P}(\mathcal{P}(\{L, R\} \times \Gamma \times Q))$. Let $\#_L, \#_R \notin \Gamma$ be fresh symbols to mark the ends of the tape, and $\Gamma_\# = \Gamma \cup \{\#_L, \#_R\}$.*

A configuration *is a word $uqav = \#_L\Gamma^*Q\Gamma^+\#_R$, where $\#_L$ and $\#_R$ are special symbols not in Γ, meaning that the head of the TM points to the letter a. We denote by C the set of configurations of $\mathcal{M}$. A configuration is* final *if it is of the form $\#_L\Gamma^*Q_F\Gamma^+\#_L$.*

The execution of the TM $\mathcal{M}$ over input $w \in \Sigma$ may be seen as a game-like scenario between two players ∃loise *and* ∀belard *over a graph $C \sqcup (C \times \mathcal{P}(\{L, R\} \times \Gamma \times Q))$, with initial position ιw which proceeds as follows.*

- *over a configuration $uqav$ with $a \in \Gamma$, $u, v \in \Gamma_\#^*$,* ∃loise *picks a transition $X \in \Delta(q, a)$ to move to position $(uqav, X)$*
- *over a position $(uqav, X)$ with $a \in \Gamma$, $u, v \in \Gamma^*$,* ∀belard *picks a triple $(d, c, r) \in X$ to move in configuration*
  - *$ucrB\#_R$ if $v = \#_R$ and $d = R$*
  - *$ucrv$ if $v \neq \#_R$ and $d = R$*
  - *$\#_LrBcv$ if $u = \#_L$ and $d = L$*
  - *$u'rbcv$ if $u = \#_Ru'b$ and $d = L$*

Given a subset of configurations $D \subseteq C$, we define $\mathrm{Attr}^{\exists\mathrm{loise}}(D)$ the ∃loise *attractor for D as the set of configurations from which* ∃loise *may force the execution to go through D.*

A deterministic *TM $\mathcal{M}$ is one where every $\Delta(q, a) \subseteq \{\{(d, c, r)\}\}$ for some $(d, c, r) \in \{L, R\} \times \Gamma \times Q$ In such a case, we may identify $\mathcal{M}$ with the underlying partial function $[\mathcal{M}] : \Sigma^* \rightharpoonup Q_F$.*

An alternating linearly bounded automaton *over the alphabet* Σ *is a tuple* $\mathcal{A} = (Q, Q_F, \Gamma, \iota, \Delta)$ *where* $(Q, Q_F, \Gamma \sqcup \{B\}, \iota, B, \Delta)$ *is a TM that does not insert* B *symbols. This means that the head can point to* $\sharp_d$*, and for every* $X \in \Delta(q, \#_d)$ *and* $(d', a, r) \in X$*, we have* $d \neq d'$ *and* $a = \#_d$.

An LBA is deterministic if its underlying TM is.

Definition 7. *A set of* $(w \leq \sum w)$*-hypotheses is said to be* length-preserving *if for every* $(v \leq \sum_{i \in I} v_i) \in H$*, we have that* $|v| = |v_i|$ *for all* $i \in I$.

The following lemma generalizes a similar construction from [13].

Lemma 8. *For every TM* $\mathcal{M}$ *of working alphabet* Γ*, there exists a set of* $(w \leq \sum w)$*-hypotheses* $H_{\mathcal{M}}$ *over the alphabet* $\Theta = Q \cup \Gamma$ *such that, for any set of configurations* $D \subseteq C$ *we have that:* $\mathrm{cl}_{H_{\mathcal{A}}}(D) = \mathrm{Attr}^{\exists \mathrm{loise}}(D)$*. Furthermore, this reduction is polytime computable, and* $H_{\mathcal{A}}$ *is length-preserving if* $\mathcal{M}$ *is an LBA.*

A configuration c is *co-reachable* if $\exists$loise has a strategy to reach a final configuration from c. Lemma 8 shows that the set of co-reachable configurations can be seen as the closure by $(w \leq \sum w)$-hypotheses. Since we are also interested in $(x \leq \sum x)$-hypotheses, we will show that $(w \leq \sum w)$ hypotheses can be transformed into letter hypotheses. Moreover, this transformation preserves the length-preserving property.

Theorem 5. *Let* Σ *be an alphabet,* H *be a set of* $(w \leq \sum w)$*-hypotheses over* Σ*. There exists an extended alphabet* $\Sigma' \supseteq \Sigma$*, a set of* $(x \leq \sum w)$*-hypotheses* H' *over* Σ' *and a regular expression* $h \in \mathsf{Exp}_{\Sigma'}$ *such that the following holds for every* $f \in \mathsf{Exp}_{\Sigma}$ *and* $w \in \Sigma^*$.

$$w \in \mathrm{cl}_H([f]) \qquad \textit{if and only if} \qquad w \in \mathrm{cl}_{H'}([f + h])$$

Furthermore, we guarantee the following:

- (Σ', H', h) *can be computed in polynomial time from* (Σ, H).
- H' *is length-preserving whenever* H *is.*

5.2 Complexity Results

Lemma 9. *If* H *is a set of length-preserving* $(w \leq \sum w)$*-hypotheses (resp. a set of* $(x \leq \sum x)$*-hypotheses),* $w \in \Sigma^*$ *and* $f \in \mathsf{Exp}_{\Sigma}$*, deciding* $\mathsf{KA}_H \vdash w \leq f$ *is* EXPTIME – complete.

Proof. We actually show that our problem is complete in alternating-PSPACE (APSPACE), which enables us to conclude as EXPTIME and APSPACE coincide. First, notice that by completeness of KA_H over this fragment (Proposition 2), we have $\mathsf{KA}_H \vdash w \leq f \Leftrightarrow w \in \mathrm{cl}_H([f])$. Hence, we work directly with the latter notion. It suffices to show hardness for the $(x \leq \sum x)$ case and membership for the $(w \leq \sum w)$ case.

Given an arbitrary alternating Turing Machine $\mathcal{M}$ in APSPACE there exists a polynomial $p \in \mathbb{N}[X]$ such that executions of $\mathcal{M}$ over words w are bisimilar to executions of the LBA($\mathcal{M}$) over $wB^{p(|w|)}$. Hence, by Lemma 8 and Theorem 5, the problem with ($x \leq \sum x$)-hypotheses is APSPACE-hard. Conversely, we may show that our problem with ($w \leq \sum w$)-hypotheses falls into APSPACE. On input w, the alternating algorithm first checks whether $w \in [f]$ in linear time. If it is the case, it returns "yes". Otherwise, it non-deterministically picks a factorization $w = uxv$ with $x \in \Sigma^*$ and a hypothesis $x \leq \sum_i y_i$. It then universally picks $y_i \in \Sigma^{|x|}$, and replaces x by y_i on the tape, so that the new tape content is $w' = uy_iv$. Then the algorithm loops back to its first step. In parallel, we keep track of the number of steps and halt by returning "no" as soon as we reach $|\Sigma|^{|w|}$ steps. This is correct because, if there is a derivation tree witnessing $w \in \text{cl}_H([f])$, there is one where on every path, all nodes have distinct labels, so the nondeterministic player can play according to this tree, while the universal player selects a branch. □

Theorem 6. *Deciding* KA^*_H *is* Π^0_1–complete *for* ($x \leq \sum x$)*-hypotheses.*

Proof. By Lemma 9 and the fact that regular expressions are in recursive bijection with natural numbers, our set is clearly Π^0_1. To show completeness, we effectively reduce the set of universal LBAs, which is known to be Π^0_1–complete, to our set of triples. Indeed, by Lemma 8, an LBA $\mathcal{A}$ is universal if and only if $\#_L\{\iota\}\Sigma^*\#_R \subseteq \text{cl}_H(C_F)$ where C_F is the set of final configurations. □

Theorem 7. *If H is a set of ($x \leq \sum w$)-hypotheses, $w \in \Sigma^*$ and $f \in \mathsf{Exp}_\Sigma$, deciding* $\mathsf{KA}^{(*)}_H \vdash w \leq f$ *is* Σ^0_1–complete.

Proof. As KA_H is a recursively enumerable theory, our set is Σ^0_1. By the completeness theorem (Proposition 2), we have $\mathsf{KA}_H \vdash w \leq f \Leftrightarrow \mathsf{KA}^*_H \vdash w \leq f \Leftrightarrow w \in \text{cl}_H([f])$, so we may work directly with closure. In order to show completeness, we reduce the halting problem for Turing machines (on empty input) to this problem. Let $\mathcal{M}$ be a Turing machine with alphabet Σ and final state q_f, and $H_\mathcal{M}$ be the set of ($w \leq \sum w$)-hypotheses given effectively by Lemma 8. Let $f = \Sigma^* q_f \Sigma^*$, by Lemma 8 we have $\mathcal{M}$ halts on empty input if and only if $q_0 \in \text{cl}_{H_\mathcal{M}}(f)$. Notice that hypotheses of H' are of the form $u \leq V$ where $u \in \Theta^3$ and $V \subseteq \Theta^3$. By Theorem 5, we can compute a set H' of ($x \leq \sum x$)-hypotheses, and an expression h on an extended alphabet such that $q_0 \in \text{cl}_{H_\mathcal{M}}([f]) \Leftrightarrow q_0 \in \text{cl}_{H'}([f + h])$. □

Theorem 8. *Deciding* KA^*_H *is* Π^0_2–complete *for* ($x \leq \sum w$)*-hypotheses.*

Proof. This set is Π^0_2 by Theorem 7. It is complete by reduction from the set of Turing Machines accepting all inputs, which is known to be Π^0_2. Indeed, let $\mathcal{M}$ be a Turing Machine on alphabet Σ with final state q_f, by Lemma 8, we can compute a set of ($w \leq \sum w$)-hypotheses $H_\mathcal{M}$ with finite language in second components such that $c \in \text{cl}_{H_\mathcal{M}}(c')$ if and only if configuration c' is reachable from c. As before, by Theorem 5, we can compute a set of letter hypotheses H' with finite languages in second components, and a regular expression h on an extended alphabet, such that for any $\text{cl}_{H'}([f + h]) \cap \Theta^* = \text{cl}_H([f])$ for any $f \in \mathsf{Exp}_\Theta$. Let $C_f = \Sigma^* q_f \Sigma^*$, we obtain that $\mathcal{M}$ accepts all inputs if and only if $[q_0\Sigma^*] \subseteq \text{cl}_{H'}([C_f + h])$, which achieves the proof of Π^0_2-completeness. □

Theorem 9. *Deciding* KA_H^* *is* Π_1^1–complete *for* $(x \leq g)$*-hypotheses* $(g \in \mathsf{Exp}_\Sigma)$.

Sketch. It is shown in [13] that the problem is complete with hypotheses of the form $H = H_w \cup \{x \leq g\}$, where H_w is a set of length-preserving $(w \leq \sum w)$ hypotheses. A slight refinement of Theorem 5 allows us to reduce this problem to hypotheses of the form $x \leq g$. □

5.3 Undecidability of KA_H for Sums of Letters

Fix an alphabet Σ, a well-behaved coding function $\lceil\cdot\rceil$ of Turing machines with final states $\{0, 1\}$ into Σ^* and a recursive pairing function $\langle\cdot,\cdot\rangle : \Sigma^* \times \Sigma^* \to \Sigma^*$. A *universal total* $F : \Sigma^* \to \{0, 1\}$ is a function such that, for every total Turing machine $\mathcal{M}$ and input $w \in \Sigma^*$ we have $F(\langle\lceil\mathcal{M}\rceil, w\rangle) = [M](w)$. In particular, F should be total and is not uniquely determined over codes of partial Turing machines. The next folklore lemma follows from an easy diagonal argument.

Lemma 10. *There is no universal total Turing machine.*

Our strategy is to show that decidability of KA_H with $(x \leq \sum x)$ hypotheses would imply the existence of a universal total TM. To do so, we need one additional lemma.

Lemma 11. *Suppose that* $\mathcal{M} = (Q, Q_F, \Gamma, \iota, B, \Delta)$ *is a total Turing machine with final states* $\{0, 1\}$ *and initial state* ι. *Let* $w \in \Sigma^*$ *be an input word for* $\mathcal{M}$.

Then there is effectively a set of length-preserving $(w \leq \sum w)$*-hypotheses* H *and expressions* e_w, h *such that* $[\mathcal{M}](w) = 1$ *if and only if* $\mathsf{KA}_H \vdash e_w \leq h$.

Theorem 10. KA_H *is undecidable for* $(x \leq \sum x)$*-hypotheses.*

Proof. Assume that KA_H is decidable. This means that we have an algorithm $\mathcal{A}$ taking tuples (Σ, w, f, H), with H consisting only of sum-of-letters hypotheses and returning true when $\mathsf{KA}_H \vdash w \leq f$ and false otherwise. Without loss of generality, we can assume that $\mathcal{A}$ is total. By Theorem 5, we may even provide an algorithm $\mathcal{A}'$ taking as input tuples (w, f, H) where H is a set of length-preserving $(w \leq \sum w)$-hypotheses with a similar behaviour: $\mathcal{A}'$ returns true when $\mathsf{KA}_H \vdash w \leq f$ and false otherwise.

Given $\mathcal{A}'$, consider $\mathcal{M}$ defined so that $[\mathcal{M}](\lceil\mathcal{N}\rceil, w) = [\mathcal{A}'](e_w, h, H)$, where the last tuple is given by Lemma 11. We show that $\mathcal{M}$ is a total universal Turing machine. Since such a machine cannot exist by Lemma 10, this is enough to conclude. Since $\mathcal{A}'$ is total, so is $\mathcal{M}$. For total Turing Machines $\mathcal{N}$, Lemma 11 guarantees that $[\mathcal{N}](w) = 1$ if and only if $[\mathcal{A}'](e_w, h, H) = [\mathcal{M}](\lceil\mathcal{N}\rceil, w) = 1$. Since both $[\mathcal{A}']$ and $[\mathcal{M}]$ are total with codomain $\{0, 1\}$, we really have $[\mathcal{M}](\lceil\mathcal{N}\rceil, w) = [\mathcal{N}](w)$. □

References

1. Anderson, C.J., et al.: NetKAT: semantic foundations for networks. In: Proceedings of the POPL, pp. 113–126. ACM (2014). https://doi.org/10.1145/2535838.2535862
2. Angus, A., Kozen, D.: Kleene algebra with tests and program schematology. Technical report TR2001-1844, CS Dpt., Cornell University, July 2001. http://hdl.handle.net/1813/5831
3. Boffa, M.: Une remarque sur les systèmes complets d'identités rationnelles. Informatique Théorique et Applications **24**, 419–428 (1990). http://archive.numdam.org/article/ITA19902444190.pdf
4. Braibant, T., Pous, D.: An efficient Coq tactic for deciding Kleene algebras. In: Kaufmann, M., Paulson, L.C. (eds.) ITP 2010. LNCS, vol. 6172, pp. 163–178. Springer, Heidelberg (2010). https://doi.org/10.1007/978-3-642-14052-5_13
5. Cohen, E.: Hypotheses in Kleene algebra. Technical report, Bellcore, Morristown, N.J. (1994). http://www.researchgate.net/publication/2648968_Hypotheses_in_Kleene_Algebra
6. Conway, J.H.: Regular Algebra and Finite Machines. Chapman and Hall, London (1971)
7. Das, A., Doumane, A., Pous, D.: Left-handed completeness for Kleene algebra, via cyclic proofs. In: Proceedings of the LPAR. EPiC Series in Computing, vol. 57, pp. 271–289. EasyChair (2018). https://doi.org/10.29007/hzq3
8. Doumane, A., Kuperberg, D., Pous, D., Pradic, P.: Kleene algebra with hypotheses. Full version of this extended abstract (2019). https://hal.archives-ouvertes.fr/hal-02021315
9. Hoare, C.A.R.T., Möller, B., Struth, G., Wehrman, I.: Concurrent Kleene algebra. In: Bravetti, M., Zavattaro, G. (eds.) CONCUR 2009. LNCS, vol. 5710, pp. 399–414. Springer, Heidelberg (2009). https://doi.org/10.1007/978-3-642-04081-8_27
10. Kleene, S.C.: Representation of events in nerve nets and finite automata. In: Automata Studies, pp. 3–41. Princeton University Press (1956). http://www.rand.org/pubs/research_memoranda/2008/RM704.pdf
11. Kozen, D.: A completeness theorem for Kleene algebras and the algebra of regular events. Inform. Comput. **110**(2), 366–390 (1994). https://doi.org/10.1006/inco.1994.1037
12. Kozen, D.: On Hoare logic and Kleene algebra with tests. ACM Trans. Comput. Log. **1**(1), 60–76 (2000). https://doi.org/10.1145/343369.343378
13. Kozen, D.: On the complexity of reasoning in Kleene algebra. Inform. Comput. **179**, 152–162 (2002). https://doi.org/10.1006/inco.2001.2960
14. Kozen, D., Mamouras, K.: Kleene algebra with equations. In: Esparza, J., Fraigniaud, P., Husfeldt, T., Koutsoupias, E. (eds.) ICALP 2014. LNCS, vol. 8573, pp. 280–292. Springer, Heidelberg (2014). https://doi.org/10.1007/978-3-662-43951-7_24
15. Kozen, D., Patron, M.-C.: Certification of compiler optimizations using Kleene algebra with tests. In: Lloyd, J., et al. (eds.) CL 2000. LNCS (LNAI), vol. 1861, pp. 568–582. Springer, Heidelberg (2000). https://doi.org/10.1007/3-540-44957-4_38
16. Krauss, A., Nipkow, T.: Proof pearl: regular expression equivalence and relation algebra. JAR **49**(1), 95–106 (2012). https://doi.org/10.1007/s10817-011-9223-4
17. Krob, D.: Complete systems of B-rational identities. TCS **89**(2), 207–343 (1991). https://doi.org/10.1016/0304-3975(91)90395-I

18. Mamouras, K.: Extensions of Kleene algebra for program verification. Ph.D. thesis, Cornell University, Ithaca, NY (2015). https://ecommons.cornell.edu/handle/1813/40960
19. Pous, D.: Kleene algebra with tests and Coq tools for while programs. In: Blazy, S., Paulin-Mohring, C., Pichardie, D. (eds.) ITP 2013. LNCS, vol. 7998, pp. 180–196. Springer, Heidelberg (2013). https://doi.org/10.1007/978-3-642-39634-2_15

Trees in Partial Higher Dimensional Automata

Jérémy Dubut[1,2](✉)

[1] National Institute of Informatics, Tokyo, Japan
dubut@nii.ac.jp
[2] Japanese-French Laboratory for Informatics, Tokyo, Japan

Abstract. In this paper, we give a new definition of partial Higher Dimension Automata using lax functors. This definition is simpler and more natural from a categorical point of view, but also matches more clearly the intuition that pHDA are Higher Dimensional Automata with some missing faces. We then focus on trees. Originally, for example in transition systems, trees are defined as those systems that have a unique path property. To understand what kind of unique property is needed in pHDA, we start by looking at trees as colimits of paths. This definition tells us that trees are exactly the pHDA with the unique path property modulo a notion of homotopy, and without any shortcuts. This property allows us to prove two interesting characterisations of trees: trees are exactly those pHDA that are an unfolding of another pHDA; and trees are exactly the cofibrant objects, much as in the language of Quillen's model structure. In particular, this last characterisation gives the premisses of a new understanding of concurrency theory using homotopy theory.

Keywords: Higher Dimensional Automata · Trees · Homotopy theories

1 Introduction

Higher Dimensional Automata (HDA, for short), introduced by Pratt in [23], are a geometric model of true concurrency. Geometric, because they are defined very similarly to simplicial sets, and can be interpreted as glueings of geometric objects, here hypercubes of any dimension. Similarly to other models of concurrency much as event structures [21], asynchronous systems [1,25], or transition systems with independence [22], they model true concurrency, in the sense that they distinguish interleaving behaviours from simultaneous behaviours. In [12], van Glabbeek proved that they form the most powerful models of a hierarchy of concurrent models. In [6], Fahrenberg described a notion of bisimilarity of HDA using the general framework of open maps from [17]. If this work is very natural,

The author was supported by ERATO HASUO Metamathematics for Systems Design Project (No. JPMJER1603), JST.

M. Bojańczyk and A. Simpson (Eds.): FOSSACS 2019, LNCS 11425, pp. 224–241, 2019.
https://doi.org/10.1007/978-3-030-17127-8_13

it is confronted with a design problem: paths (or executions) cannot be nicely encoded as HDA. Indeed, in a HDA, it is impossible to model the fact that two actions *must* be executed at the same time, or that two actions are executed at the same time but one *must* start before the other. From a geometric point of view, those impossibilities are expressed by the fact that we deal with closed cubes, that is, cubes that must contain all of their faces. Motivated by those examples, Fahrenberg, in [7], extended HDA to partial HDA, intuitively, HDA with cubes with some missing faces. If the intuition is clear, the formalisation is still complicated to achieve: the definition from [7] misses the point that faces can be not uniquely defined. This comes from the fact that Fahrenberg wanted to stick to the 'local' definition of precubical sets, that is, that cubes must satisfy some local conditions about faces. As we will show, those local equations are not enough in the partial case. Another missed point is the notion of morphisms of partial HDA: as defined in [7], the natural property that morphisms map executions to executions is not satisfied. In Sect. 2, we address those issues by giving a new definition of partial HDA in terms of lax functors. This definition, similar to the presheaf theoretic definition of HDA, avoid the issues discussed above by considering global inclusions, instead of local equations. This illustrates more clearly the intuition of partial HDA being HDA with missing faces: we coherently replace sets and total functions by sets and partial functions. From this similarity with the original definition of HDA, we can prove that it is possible to complete a partial HDA to turn it into a HDA, by adding the missing faces, and from this completion, it is possible to define a geometric realisation of pHDA (which was impossible with Fahrenberg's definition).

The geometry of Higher Dimensional Automata, and more generally, of true concurrency, has been studied since Goubault's PhD thesis [13]. Since then, numerous pieces of work relating algebraic topology and true concurrency have been achieved (for example, see the textbooks [9,14]). In particular, some attempts of defining nice homotopy theories for true concurrency (or directed topology), through the language of model structures of Quillen [24], have been made by Gaucher [10], and the author [3]. In the second part of this paper (Sects. 3, 4 and 5), we consider another point of view of this relationship between HDA and model structures. The goal is not to understand the true concurrency of HDA, that is, understanding the homotopy theory of HDA as an abstract homotopy theory, but to understand the concurrency theory of HDA. By this we mean to understand how paths (or executions) and extensions of paths can be understood using (co)fibrations (in Quillen's sense). Also, the goal is not to construct a model structure, as Quillen's axioms would fail, but to give intuitions and some preliminary formal statements toward the understanding of concurrency using homotopy theory. Using this point of view, many constructions in concurrency can be understood using the language of model structures:

- Open maps from [17] can be understood as trivial fibrations, namely weak equivalences (here, bisimulations) that have the right lifting properties with respect to some morphisms.

- Those morphisms are precisely extensions of executions, which means that they can be seen as cofibration generators (in the language of cofibrantly generated model structures [15]).
- Cofibrations are then morphisms that have the left lifting property with respect to open maps. In particular, this allows us to define cofibrant objects as those objects whose unique morphisms from the initial object is a cofibration. In a way, cofibrant objects are those objects that are constructed by just using extensions of paths, and should correspond to trees.
- The cofibrant replacement is then given by canonically constructing a cofibrant object, which is weakly equivalent (here, bisimilar) to a given object. That should correspond to the unfolding.

The main ingredient is to understand what trees are in this context. In the case of transition systems for semantics of CCS [19], synchronisation trees are those systems with exactly one path from the initial state to any state. Those trees are then much simpler to reason on, but they are still powerful enough to capture any bisimulation type: by unfolding, it is possible to canonically construct a tree from a system. The goal of Sects. 3 and 4 will be to understand how to generalise this to pHDA. In this context, it is not clear what kind of unique path property should be considered as, in general, in truly concurrent systems, we have to deal with homotopies, namely, equivalences of paths modulo permutation of independent actions. Following [4], we will first consider trees as colimits of paths. This will guide us to determine what kind of unique path property is needed: a tree is a pHDA with exactly one class of paths modulo a notion of homotopy, from the initial state to any state, and without any shortcuts. This will be proved by defining a suitable notion of unfolding of pHDA. Finally, in Sect. 5, we prove that those trees coincide exactly with the cofibrant objects, illustrating the first steps of this new understanding of concurrency, using homotopy theory.

2 Fixing the Definition of pHDA

In this Section, we review the definitions of HDA (Sect. 2.1), the first one using face maps, and the second one using presheaves. In Sect. 2.2, we describe the definition of partial HDA from [7] and explain why it does not give us what we are expecting. We tackle those issues by introducing a new definition in Sect. 2.3, extending the presheaf theoretic definition, using lax functors instead of strict functors. Finally, in Sect. 2.4, we prove that HDA form a reflective subcategory of partial HDA, by constructing a completion of a partial HDA.

2.1 Higher Dimensional Automata

Higher Dimensional Automata are an extension of transition systems: they are labeled graphs, except that, in addition to vertices and edges, the graph structure also has higher dimensional data, expressing the fact that several actions can be made at the same time. Those additional data are intuitively cubes filling up interleaving: if a and b can be made at the same time, instead of having an empty square as on the left figure, with $a.b$ and $b.a$ as only behaviours, we have a full square as on the right figure, with any possible behaviours in-between. This requires to extend the notion of graph to add those higher dimensional cubical data: that is the notion of *precubical sets*.

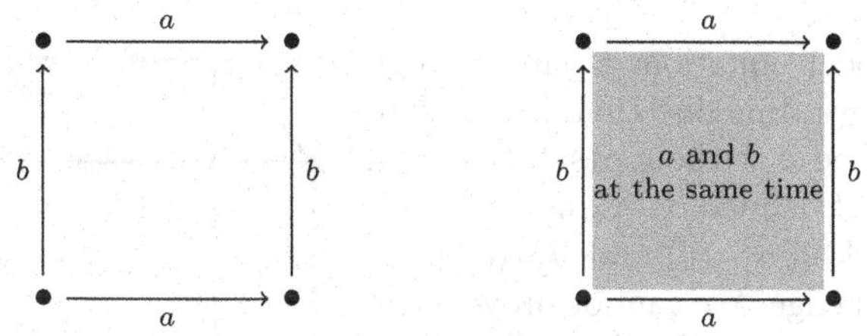

Concrete Definition of Precubical Sets. A **precubical set** X is a collection of sets $(X_n)_{n\in\mathbb{N}}$ together with a collection of functions $(\partial^\alpha_{i,n} : X_n \longrightarrow X_{n-1})_{n>0, 1\leq i\leq n, \alpha\in\{0,1\}}$ satisfying the local equations $\partial^\alpha_{i,n} \circ \partial^\beta_{j,n+1} = \partial^\beta_{j,n} \circ \partial^\alpha_{i+1,n+1}$ for every $\alpha, \beta \in \{0,1\}$, $n > 0$ and $1 \leq j \leq i \leq n$. A **morphism of precubical sets** from X to Y is a collection of functions $(f_n : X_n \longrightarrow Y_n)_{n\in\mathbb{N}}$ satisfying the equations $f_n \circ \partial^\alpha_{i,n} = \partial^\alpha_{i,n} \circ f_{n+1}$ for every $n \in \mathbb{N}$, $1 \leq i \leq n$ and $\alpha \in \{0,1\}$. The elements of X_0 are called **points**, X_1 **segments**, X_2 **squares**, X_n **n-cubes**. In the following, we will call **past** (resp. **future**) **i-face maps** the $\partial^0_{i,n}$ (resp. $\partial^1_{i,n}$). We denote this category of precubical sets by **pCub**.

Precubical Sets as Presheaves. Equivalently, **pCub** is the category of preshea-ves over the cubical category $\square$. $\square$ is the subcategory of **Set** whose objects are the sets $\{0,1\}^n$ for $n \in \mathbb{N}$ and whose morphisms are generated by the so-called **coface maps**:

$$d^\alpha_{i,n} : \{0,1\}^{n-1} \longrightarrow \{0,1\}^n \quad (\beta_1, \ldots, \beta_{n-1}) \longmapsto (\beta_1, \ldots, \beta_{i-1}, \alpha, \beta_i, \ldots, \beta_{n-1})$$

A precubical set is a functor $X : \square^{op} \longrightarrow \mathbf{Set}$, that is, a presheaf over $\square$, and a morphism of precubical sets is a natural transformation.

Higher Dimensional Automata [11]. From now on, fix a set L, called the **alphabet**. We can form a precubical set also noted L such that $L_n = L^n$ and the i-face maps are given by $\delta^\alpha_i(a_1 \ldots a_n) = a_1 \ldots a_{i-1}.a_{i+1} \ldots a_n$. We can also form the following precubical set $*$ such that $*_0 = \{*\}$ and $*_n = \varnothing$ for $n > 0$. A **HDA** X on L is a bialgebra $* \to X \to L$ in **pCub**. In other words, a HDA X is a precubical set, also noted X, together with a specified point, the **initial state**, $i \in X_0$ and a **labelling function** $\lambda : X_1 \longrightarrow L$ satisfying the equations $\lambda \circ \partial^0_{i,2} = \lambda \circ \partial^1_{i,2}$ for $i \in \{1,2\}$ (see previous figure, right). A **morphism of HDA** from X to Y is a morphism f of precubical sets from X to Y such that $f_0(i_X) = i_Y$ and $\lambda_X = \lambda_Y \circ f_1$. HDA on L and morphisms of HDA form a category that we denote by $\mathbf{HDA_L}$. This category can also be defined as a the double slice category $*/\mathbf{pCub}/L$. Remark that we are only concerned with labelling-preserving morphisms, not general morphisms as described in [5].

2.2 Original Definition of Partial Higher Dimensional Automata

Originally [7], partial HDA are defined similarly to the concrete definition of HDA, except that the face maps can be partial functions and the local equations hold only when *both* sides are well defined. There are two reasons why it fails to give the good intuition:

– first the 'local' equations are not enough in the partial case. Imagine that we want to model a full cube c without its lower face, that is, $\partial^0_{3,3}$ is not defined on c, and such that $\partial^1_{1,2}$ is undefined on $\partial^1_{1,3}(c)$ and $\partial^1_{2,3}(c)$, that is, we remove an edge. We cannot prove using the local equations that $\partial^1_1 \circ \partial^0_2 \circ \partial^1_1(c) = \partial^1_1 \circ \partial^0_2 \circ \partial^1_2(c)$, that is, that the vertices of the cube are uniquely defined. Indeed, to prove this equality using the local equations, you can only permute two consecutive ∂. From $\partial^1_1 \circ \partial^0_2 \circ \partial^1_1(c)$, you can:

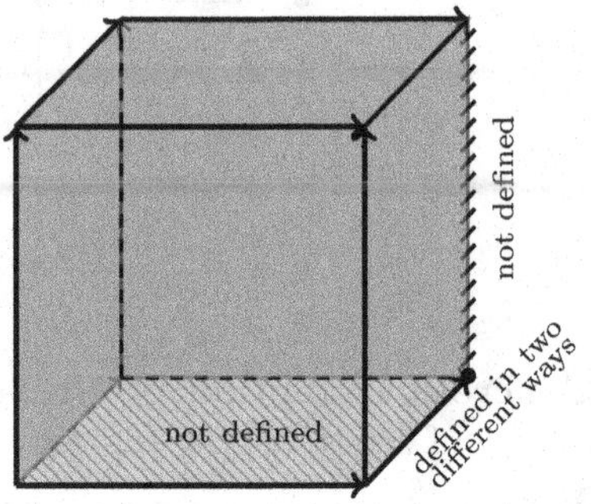

- either permute the first two and you obtain $\partial^1_1 \circ \partial^1_1 \circ \partial^0_3(c)$,
- or permute the last two and you obtain $\partial^0_1 \circ \partial^1_1 \circ \partial^1_1(c)$.

and both faces are not defined. On the other hand, those two should be equal because the comaps $d^1_1 \circ d^0_2 \circ d^1_1$ and $d^1_2 \circ d^0_2 \circ d^1_1$ are equal in $\square$, and $\partial^1_1 \circ \partial^0_2 \circ \partial^1_1$ and $\partial^1_1 \circ \partial^0_2 \circ \partial^1_2$ are both defined on c.

– secondly, the notion of morphism is not good (or at least, ambiguous). The equations $f_n \circ \partial^\alpha_{i,n,X} = \partial^\alpha_{i,n,Y} \circ f_{n+1}$ hold in [7] only when *both* face maps are defined, which authorises many morphisms. For example, consider the segment I, and the 'split' segment I' which is defined as I, except that no face maps are defined (geometrically, this corresponds to two points and an open segment). The identity map from I to I' is a morphism of partial precubical sets in the sense of [7], which is unexpected. A bad consequence of that is that the notion of paths in a partial HDA does not correspond to morphisms from some particular partial HDA, and paths are not preserved by morphisms, as we will see later.

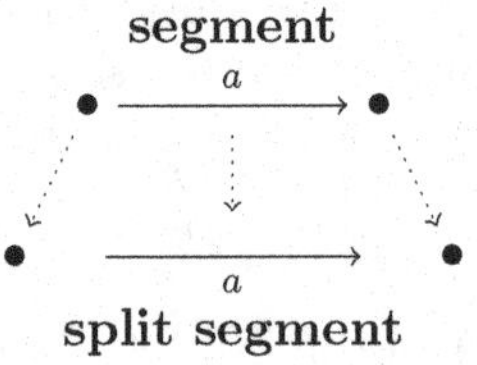

2.3 Partial Higher Dimensional Automata as Lax Functors

The idea is to generalise the 'presheaf' definition of precubical sets. The problem is to deal with partial functions and when two of them should coincide. Let **pSet** be the category of sets and partial functions. A partial function $f : X \longrightarrow Y$ can be either seen as a pair (A, f) of a subset $A \subseteq X$ and a total function $f : A \longrightarrow Y$, or as a functional relation $f \subseteq X \times Y$, that is, a relation such that for every $x \in X$, there is at most one $y \in Y$ with $(x, y) \in f$. We will freely use both views in the following. For two partial maps $f, g : X \longrightarrow Y$, we denote by $f \equiv g$ if and only if for every $x \in X$ such that $f(x)$ and $g(x)$ are defined, then

$f(x) = g(x)$. Note that this is not equality, but equality on the intersection of the domains. We also write $f \subseteq g$ if and only if f is include in g as a relation, that is, if and only if, for every $x \in X$ such that $f(x)$ is defined, then $g(x)$ is defined and $f(x) = g(x)$. By a **lax functor** $F : \mathcal{C} \rightharpoonup \mathbf{pSet}$, we mean the following data [20]:

- for every object c of $\mathcal{C}$, a set Fc,
- for every morphism $i : c \longrightarrow c'$, a partial function $Fi : Fc \longrightarrow Fc'$

satisfying that $F\mathrm{id}_c = \mathrm{id}_{Fc}$ and $Fj \circ Fi \subseteq F(j \circ i)$.

The point is that partial precubical sets as defined in [7] do not satisfy the second condition, while they should. In addition, this definition will authorise a square to have vertices, that is, that some $\partial\partial$ are defined, while having no edge, that is, no ∂ defined. This may be useful to define paths as discrete traces in [8] (that we will call *shortcuts* later), that is, paths that can go directly from a point to a square for example. Observe also that if $j \circ i = j' \circ i'$ then $Fj \circ Fi \equiv Fj' \circ Fi'$, which gives us the local equations from [7]. A **partial precubical set** X is then a lax functor $F : \square^{op} \rightharpoonup \mathbf{pSet}$. It becomes harder to describe explicitly what a partial precubical set is, since we cannot restrict to the ∂_i^α anymore. It is a collection of sets $(X_n)_{n\in\mathbb{N}}$ together with a collection of *partial* functions $(\partial_{i_1<\ldots<i_k}^{\alpha_1,\ldots,\alpha_k} : X_{n+k} \longrightarrow X_n)$ satisfying the inclusions $\partial_{j_1<\ldots<j_m}^{\beta_1,\ldots,\beta_m} \circ \partial_{i_1<\ldots<i_n}^{\alpha_1,\ldots,\alpha_n} \subseteq \partial_{k_1<\ldots<k_{n+m}}^{\gamma_1,\ldots,\gamma_{n+m}}$ where the k_s and γ_s are defined as follows. $(k_1 < \ldots < k_{n+m}; \gamma_1, \ldots, \gamma_{n+m}) = (i_1 < \ldots < i_n; \alpha_1, \ldots, \alpha_n) \star (j_1 < \ldots < j_m; \beta_1, \ldots, \beta_m)$ where $\star$ is defined by induction on $n + m$:

- if $n = 0$, $\epsilon \star (j_1 < \ldots < j_m; \beta_1, \ldots, \beta_m) = (j_1 < \ldots < j_m; \beta_1, \ldots, \beta_m)$,
- if $m = 0$, $(i_1 < \ldots < i_n; \alpha_1, \ldots, \alpha_n) \star \epsilon = (i_1 < \ldots < i_n; \alpha_1, \ldots, \alpha_n)$,
- if $i_1 \leq j_1$, $(i_1 < \ldots < i_n; \alpha_1, \ldots, \alpha_n) \star (j_1 < \ldots < j_m; \beta_1, \ldots, \beta_m) = (i_1; \alpha_1).((i_2 < \ldots < i_n; \alpha_2, \ldots, \alpha_n) \star (j_1 + 1 < \ldots < j_m + 1; \beta_1, \ldots, \beta_m))$,
- if $i_1 > j_1$, $(i_1 < \ldots < i_n; \alpha_1, \ldots, \alpha_n) \star (j_1 < \ldots < j_m; \beta_1, \ldots, \beta_m) = (j_1; \beta_1).((i_1 < \ldots < i_n; \alpha_1, \ldots, \alpha_n) \star (j_2 < \ldots < j_m; \beta_2, \ldots, \beta_m))$.

A **function-valued op-lax transformation** [20] from $F : \mathcal{C} \rightharpoonup \mathbf{pSet}$ to $G : \mathcal{C} \rightharpoonup \mathbf{pSet}$ is a collection $(f_c)_{c\in Ob(\mathcal{C})}$ of *total* functions such that for every $i : c \longrightarrow c'$, $f_{c'} \circ F(i) \subseteq G(i) \circ f_c$. A **morphism of partial precubical sets** from X to Y is then a function-valued op-lax transformation. In other words, this is a collection of *total* functions $(f_n : X_n \longrightarrow Y_n)_{n\in\mathbb{N}}$ satisfying the equations $f_n \circ \partial_{i_1<\ldots<i_k}^{\alpha_1,\ldots,\alpha_k} \subseteq \partial_{i_1<\ldots<i_k}^{\alpha_1,\ldots,\alpha_k} \circ f_{n+k}$. Partial precubical sets and morphisms of partial precubical sets form a category that we denote by **ppCub**. **pCub** is a full subcategory of **ppCub**. In particular, the precubical sets $*$ and L are partial precubical sets. A **partial HDA** X on L is a partial precubical set, also noted X, together with a specified point, the **initial state** $i \in X_0$ and a morphism of ppCub, the **labelling functions**, $(\lambda_n : X_n \longrightarrow L^n)_{n\in\mathbb{N}}$. A **morphism of pHDA** from X to Y is a morphism f of partial precubical sets from X to Y such that $f_0(i_X) = i_Y$ and $\lambda_X = \lambda_Y \circ f$. Partial HDA on L and morphisms of partial HDA form a category that we note $\mathbf{pHDA_L}$. In other words, this is the double slice category $*/\mathbf{ppCub}/L$.

2.4 Completion of a pHDA

Let us describe how it is possible to construct a HDA from a pHDA X, by 'completing' X, that is, by adding the faces that are missing, and by connecting the faces that are not. Let

$$Y_n = \{((i_1 < \ldots < i_k; \alpha_1, \ldots, \alpha_k), x) \mid x \in X_{n+k} \wedge i_k \leq n + k\}$$

$Y = (Y_n)_{n \in \mathbb{N}}$ is intuitively the collection of all abstract faces of all cubes of X, that is, pairs of a cube and all possible ways to define a face from it. Of course, some of those are the same, since there are several ways to describe a cube as the face of some other cube. Define $\sim$ as the smallest equivalence relation such that:

- if $\partial^{\alpha_1,\ldots,\alpha_k}_{i_1<\ldots<i_k}(x)$ is defined, then

  $$((i_1 < \ldots < i_k; \alpha_1, \ldots, \alpha_k), x) \sim (\epsilon, \partial^{\alpha_1,\ldots,\alpha_k}_{i_1<\ldots<i_k}(x)).$$

  This means that, if a face of a cube exists in X, this face is identified with both abstract faces $(\epsilon, \partial^{\alpha_1,\ldots,\alpha_k}_{i_1<\ldots<i_k}(x))$ (i.e., the cube $\partial^{\alpha_1,\ldots,\alpha_k}_{i_1<\ldots<i_k}(x)$ itself) and $((i_1 < \ldots < i_k; \alpha_1, \ldots, \alpha_k), x)$ (i.e., the face of x, which consists of taking the (i_k, α_k) face, then the (i_{k-1}, α_{k-1}) face, and so on).
- if $((i_1 < \ldots < i_k; \alpha_1, \ldots, \alpha_k), x) \sim ((j_1 < \ldots < j_l; \beta_1, \ldots, \beta_l), y)$, then $((i_1 < \ldots < i_k; \alpha_1, \ldots, \alpha_k) \star (i, \alpha), x) \sim ((j_1 < \ldots < j_l; \beta_1, \ldots, \beta_l) \star (i, \alpha), y)$. This means that if two abstract faces coincide, then taking both their (i, α) face gives two abstract faces that also coincide.
  Let $\chi(X)_n = Y_n/\sim$ and we denote by $\ll (i_1 < \ldots < i_k; \alpha_1, \ldots, \alpha_k), x \gg$ the equivalence class of $((i_1 < \ldots < i_k; \alpha_1, \ldots, \alpha_k), x)$ modulo $\sim$. We define the i-face map as $\partial_i^\alpha(\ll (i_1 < \ldots < i_k; \alpha_1, \ldots, \alpha_k), x \gg) = \ll (i_1 < \ldots < i_k; \alpha_1, \ldots, \alpha_k) \star (i, \alpha), x \gg$, the initial state as $\ll \epsilon, i \gg$ and the labelling function as $\lambda(\ll (i_1 < \ldots < i_k; \alpha_1, \ldots, \alpha_k), x \gg) = \delta^{\alpha_1}_{i_1} \circ \ldots \circ \delta^{\alpha_k}_{i_k}(\lambda(x))$.

Theorem 1. *χ is a well-defined functor and is the left adjoint of τ, the injection of **HDA**$_L$ into **pHDA**$_L$. Furthermore, **HDA**$_L$ is a reflective subcategory of **pHDA**$_L$.*

Now, we can define the **geometric realisation** of a pHDA X as the subspace of the realisation of $\chi(X)$ consisting of points whose carrier is of the form $\ll \epsilon, x \gg$ for some $x \in X$. This really corresponds to the drawings we have been using to depict pHDA until now.

3 Paths in Partial Higher Dimensional Automata

Executions of HDA are defined using the notion of paths. Those paths describe the succession of starting and finishing of actions in a HDA. For example, a HDA can start an action then start another at the same time, and finish the two actions. This sequence is then not just a sequence of 1-dimensional transitions, since some actions can be made at the same time, but a sequence of hypercubes corresponding to the evolution of the state of the system. We will formalise this idea in Sect. 3.2, and we will see in particular that those paths can be encoded in the category $\mathbf{pHDA_L}$ (while it is not possible in the category $\mathbf{HDA_L}$) as morphisms from particular pHDA, called path shapes. In Sect. 3.1, let us first start by recalling the general framework of [17].

3.1 Path Category, Open Maps, Coverings

In the general framework of [17], we start with a category $\mathcal{M}$ of systems, together with a subcategory $\mathcal{P}$ of execution shapes. For example, keep in mind the case where $\mathcal{M}$ is the category of transition systems and $\mathcal{P}$ is the full subcategory of finite linear systems. One interesting remark about this case is that executions of a given systems are in bijective correspondance with morphisms from a finite linear system to this given system. This means that to reason about behaviours of such systems, it is enough to reason about morphisms and execution shapes.

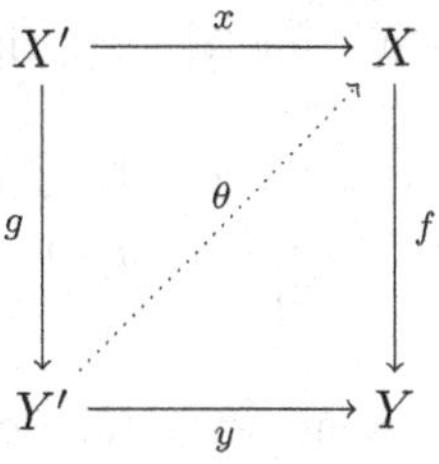

This idea was formalised by describing precisely which morphisms are witnesses for the existence of a bisimulation between systems. This description uses right lifting properties: we say that a morphism $f : X \longrightarrow Y$ has the **right lifting property with respect to** $g : X' \longrightarrow Y'$ if for every $x : X' \longrightarrow X$ and $y : Y' \longrightarrow Y$ such that $f \circ x = y \circ g$, there exists $\theta : Y' \longrightarrow X$ such that $x = \theta \circ g$ and $f \circ \theta = y$. For example, let us assume that f is a morphism of transition systems and that X' and Y' are finite linear systems. Then x (resp. y) is the same as an execution in X (resp. Y), and $f \circ x = y \circ g$ means that the execution y is a extension of the image of the execution x by f. The right lifting property means that the longer execution y of Y can be lifted to a longer execution θ of X, that is, θ is an extension of x and the image of θ by f is y. This property of lifting longer executions is precisely the property needed on a morphism to make its graph relation a bisimulation. They are also very similar to morphisms of coalgebras [16]. We call **$\mathcal{P}$-open** (or simply open when $\mathcal{P}$ is clear), a morphism that has the right lifting property with respect to every morphism in $\mathcal{P}$. From open maps, it is possible to describe similarity and bismilarity as the existence of a span of morphisms/open maps, and many kinds of bisimilarities can be captured in this way [17]. An open map is said to be a **$\mathcal{P}$-covering** (or simply covering) if furthermore the lifts in the right lifting properties are unique. Being a covering is a very strong requirement, as they correspond to partial unfolding of a system.

3.2 Encoding Paths in pHDA

In this section, we describe the classical notion of execution of HDA from [12], extended to partial HDA in [7], defined using the notion of path. We then show that those executions can be encoded as an execution shapes subcategory, as in the general framework of [17], proving in particular that paths are in bijective correspondance with a class of morphisms. A **path** π of a HDA X is a sequence $i = x_0 \xrightarrow{j_1,\alpha_1} x_1 \xrightarrow{j_2,\alpha_2} \dots \xrightarrow{j_n,\alpha_n} x_n$ where $x_k \in X$, $j_k > 0$ and $\alpha_k \in \{0,1\}$ are such that for every k:

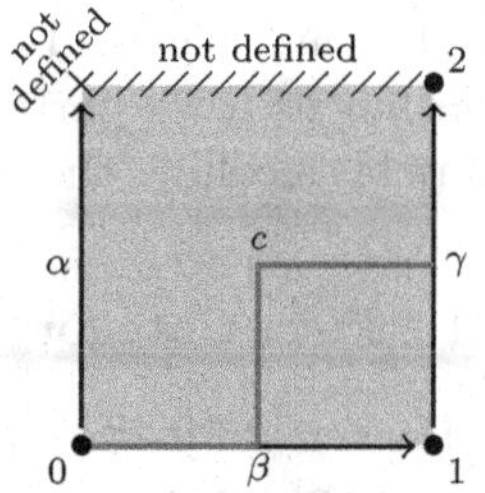

In red: path $\pi = 0 \xrightarrow{1,0} \beta \xrightarrow{2,0} c \xrightarrow{1,1} \gamma$ in the pHDA X

- if $\alpha_k = 0$, then $x_{k-1} = \partial^0_{j_k}(x_k)$,
- if $\alpha_k = 1$, then $x_k = \partial^1_{j_k}(x_{k-1})$.

This definition can easily be extended to pHDA, by requiring that the j_k-face maps are defined on x_k or x_{k-1}. A natural property of executions and morphisms is that morphisms map executions to executions. This is the case here (while it is not for [7], e.g., the split segment):

Proposition 1. *If $f : X \longrightarrow Y$ is a map of pHDA and if $\pi = x_0 \xrightarrow{j_1,\alpha_1} x_1 \xrightarrow{j_2,\alpha_2} \dots \xrightarrow{j_n,\alpha_n} x_n$ is a path in X, then $\pi' = f(x_0) \xrightarrow{j_1,\alpha_1} f(x_1) \xrightarrow{j_2,\alpha_2} \dots \xrightarrow{j_n,\alpha_n} f(x_n)$ is a path in Y.*

One advantage of considering pHDA instead of HDA is that paths can be encoded in pHDA, which is not really possible in HDA. It is done as follows. A **spine** σ is a sequence $(0,\epsilon) = (d_0, w_0) \xrightarrow{j_1,\alpha_1} (d_1, w_1) \xrightarrow{j_2,\alpha_2} \dots \xrightarrow{j_n,\alpha_n} (d_n, w_n)$ where $j_k > 0$, $d_k \in \mathbb{N}$, $w_k \in L^{d_k}$ and $\alpha_k \in \{0,1\}$ are such that:

- if $\alpha_k = 0$, then $d_{k-1} = d_k - 1$, $\delta_{j_k}(w_k) = w_{k-1}$ and $j_k \leq d_k$,
- if $\alpha_k = 1$, then $d_k = d_{k-1} - 1$, $\delta_{j_k}(w_{k-1}) = w_k$ and $j_k \leq d_{k-1}$.

A path π has a underlying spine σ_π by mapping x_k to the pair of its dimension and its label. A spine σ induces a pHDA $B\sigma$ as follows:

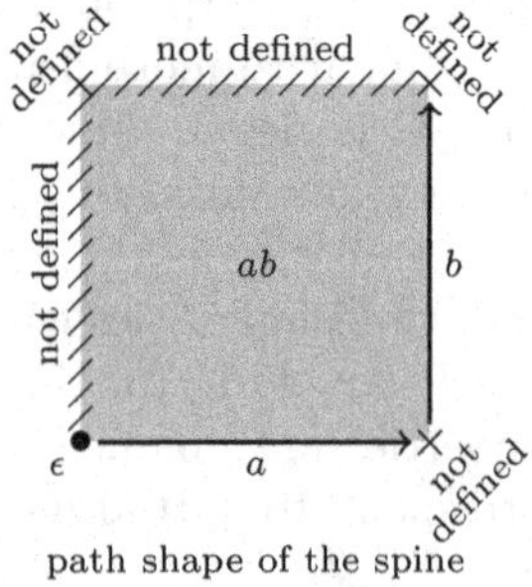

path shape of the spine $\sigma = (0,\epsilon) \xrightarrow{1,0} (1,a) \xrightarrow{2,0} (2,ab) \xrightarrow{1,1} (1,b)$

- $B\sigma_p = \{k \in \{0,\dots,n\} \mid d_k = p\}$,
- the partial face maps $\partial^{\alpha_1,\dots,\alpha_n}_{i_1<\dots<i_n}$ are the smallest (as relations ordered by inclusion) partial functions such that:
  - if $\alpha_k = 0$, then $\partial^0_{j_k}(k) = k-1$,
  - if $\alpha_k = 1$, then $\partial^1_{j_k}(k-1) = k$,
  - $\partial^{\beta_1,\dots,\beta_m}_{j_1<\dots<j_m} \circ \partial^{\alpha_1,\dots,\alpha_n}_{i_1<\dots<i_n} \subseteq \partial^{\gamma_1,\dots,\gamma_{n+m}}_{k_1<\dots<k_{n+m}}$, for $(k_1,\dots,k_{n+m};\gamma_1,\dots,\gamma_{n+m}) = (i_1,\dots,i_n;\alpha_1,\dots,\alpha_n) \star (j_1,\dots,j_m;\beta_1,\dots,\beta_m)$.

- the initial state is 0,
- the labelling functions λ_n map k to w_k.

By a **path shape**, we mean such a pHDA $B\sigma$. The set $\mathbf{Spine_L}$ of spines can be partially ordered by prefix. B can then be extended to an embedding from $\mathbf{Spine_L}$ to $\mathbf{pHDA_L}$. We note $\mathbf{PS_L}$ the image of this embedding, i.e., the full sub-category of path shapes.

Proposition 2. *There is a bijection between paths in a pHDA X and morphisms of pHDA from a path shape to X.*

Again, this is not true with the definition of morphisms from [7] (e.g., the split segment). As an example, the red path π above corresponds to a morphism from the path shape $B\sigma$ to X.

4 Trees and Unfolding in pHDA

In this section, we introduce our notion of trees. Following [4], we consider trees as colimits (or glueings of paths). Section 4.1 is dedicated to proving that those colimits actually exist, by giving an explicit construction of those. From this explicit construction, we will describe the kind of unique path properties that are satisfied by those trees in Sect. 4.2. Starting by showing, that the strict unicity of path fails, we then describe a notion of homotopy, the confluent homotopy, which is weaker than the one from [12], for which every tree has the property that there is exactly one homotopy class of paths form the initial state to any state. We will also see that, because the face maps of trees are defined in a local way, they do not have any shortcuts, that is, paths that 'skip' dimensions, for example, going from a point to a square without going through a segment. Finally, in Sect. 4.3, we will prove that those two properties – the unicity of paths modulo confluent homotopy, and the non-existence of shortcuts – completely characterise trees. This proof will use a suitable notion of unfolding of pHDA, showing furthermore that trees form a coreflective subcategory of pHDA.

4.1 Trees, as Colimits of Paths in pHDA

In this section, we give an explicit construction of colimits of diagrams with values in path shapes. Those will be our first definition of trees in pHDA, following [4]. Let $D : \mathcal{C} \longrightarrow \mathbf{PS_L}$ be a small diagram with values in $\mathbf{PS_L}$, that is, a functor from $\mathcal{C}$ to $\mathbf{PS_L}$. Let us use some notations: for every object u of $\mathcal{C}$, $Du = B\sigma_u$ with $\sigma_u = (d_0^u, w_0^u) \xrightarrow{j_1^u, \alpha_1^u} (d_1^u, w_1^u) \xrightarrow{j_2^u, \alpha_2^u} \dots \xrightarrow{j_{l_u}^u, \alpha_{l_u}^u} (d_{l_u}^u, w_{l_u}^u)$. The definition of the colimit $\mathrm{col}\, D$ will be in two steps. The first step consists in putting all the paths Du side-by-side, and in glueing them together, along the morphisms Df, for every morphism f of $\mathcal{C}$. This is done as follows. Define $(X_n)_{n\in\mathbb{N}}$ to be:

- $X_0 = \{(u, k) \mid u \in \mathcal{C}, k \leq l_u \wedge d_k^u = 0\} \sqcup \{\epsilon\}$,
- $X_n = \{(u, k) \mid u \in \mathcal{C}, k \leq l_u \wedge d_k^u = n\}$.

We quotient X_n by the smallest equivalence relation $\sim$ (for inclusion) such that:

- for every u, $(u, 0) \sim \epsilon$,
- if $i : u \longrightarrow v \in \mathcal{C}$, and if $k \leq l_u, l_v$, then $(u, k) \sim (v, k)$.

We denote by Y_n the quotient $X_n/\sim$, and by $[u, k]$ the equivalence class of (u, k) modulo $\sim$.

At this stage, we still do not have the colimit because it is not possible to define the face maps. Let us consider the following example.

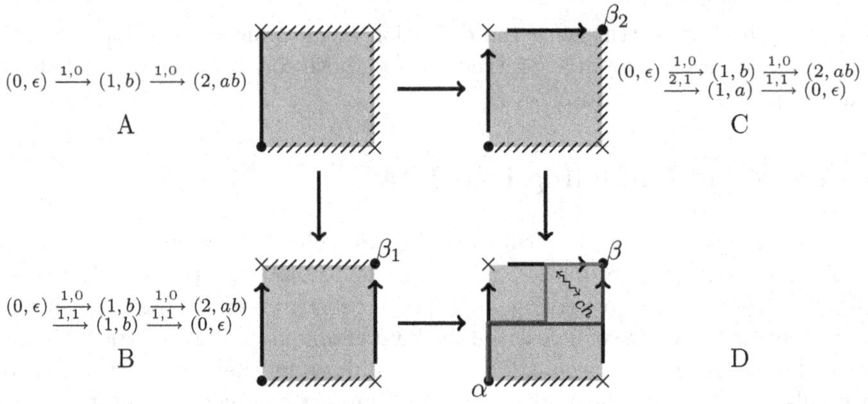

A, B and C are path shapes, and we would like to compute their pushout. The expected outcome is D, since we must identify the three squares by the previous construction. The problem is that the previous construction does not identify β_1 and β_2. Those two must be identified because they are both the top right corner of the same square (after identification). We hence need to quotient a little more to be able to define the face maps, as follows. Define Z_n to be the quotient of Y_n by the smallest equivalence relation $\approx$ such that if there are two sequences $u_0, \ldots, u_l$ and $v_0, \ldots, v_l$ such that:

- $[u_0, k] \approx [v_0, k]$,
- for every $0 \leq s \leq l$, $\alpha^{u_s}_{k+1+s} = \alpha^{v_s}_{k+1+s} = 1$,
- for every $0 \leq s < l$, $[u_s, k + s + 1] \approx [u_{s+1}, k + s + 1]$ and $[v_s, k + s + 1] \approx [v_{s+1}, k + s + 1]$,
- $(j^{u_0}_{k+1}; 1) \star \ldots \star (j^{u_l}_{k+l+1}; 1) = (j^{v_0}_{k+1}; 1) \star \ldots \star (j^{v_l}_{k+l+1}; 1)$,
  then, $[u_l, k+l+1] \approx [v_l, k+l+1]$. col D is the pHDA Z_N with the face maps being the smallest relations for inclusion such that:
- if $\alpha^u_k = 0$, then $\partial^0_{j^u_k}(\langle u, k\rangle)$ is defined and is equal to $\langle u, k-1\rangle$,
- if $\alpha^u_{k+1} = 1$ then $\partial^1_{j^u_k}(\langle u, k\rangle)$ is defined and is equal to $\langle u, k+1\rangle$,
- $\partial^{\beta_1,\ldots,\beta_m}_{j_1<\ldots<j_m} \circ \partial^{\alpha_1,\ldots,\alpha_n}_{i_1<\ldots<i_n} \subseteq \partial^{\gamma_1,\ldots,\gamma_{n+m}}_{k_1<\ldots<k_{n+m}}$, for $(k_1, \ldots, k_{n+m}; \gamma_1, \ldots, \gamma_{n+m}) = (i_1, \ldots, i_n; \alpha_1, \ldots, \alpha_n) \star (j_1, \ldots, j_m; \beta_1, \ldots, \beta_m)$.

The initial state is $\langle\epsilon\rangle$ and the labelling $\lambda : \text{col}\, D \longrightarrow L$ maps $\langle u, k\rangle$ to w^u_k.

Proposition 3. *$colD$ is the colimit of D in* $\mathbf{pHDA}_L$

By **tree** we mean any pHDA that is the colimit of a diagram with values in path shapes. We denote by $\mathbf{Tr_L}$ the full subcategory of trees.

4.2 The Unique Path Properties of Trees

Failure of the Unicity of Paths. Let us consider the pushout square above again. In particular, the pHDA on the bottom-right corner is a tree, by definition. However, there are two paths from α to β (in red and blue). This actually comes from the fact that we needed to identify β_1 and β_2 to be able to define the face maps. This means that trees do not have the unique path property.

Confluent Homotopy. A careful reader may have observed that the only difference between the two previous paths is that some future faces are swapped. Actually, this is the only obstacle for the unicity of paths for trees: there is a unique path modulo equivalence of paths that permutes arrows of the form $\xrightarrow{-,1}$. That is what we call **confluent homotopy**. This confluent homotopy will be defined by restricting the elementary homotopies of [12] to be of only one type out of the four possible, which means our notion of homotopy makes fewer paths equivalent than the one from [12].

We say that a path $\pi = x_0 \xrightarrow{j_1,\alpha_1} x_1 \xrightarrow{j_2,\alpha_2} \dots \xrightarrow{j_n,\alpha_n} x_n$ is **elementary confluently homotopic** to a path $\pi' = x'_0 \xrightarrow{j'_1,\alpha'_1} x'_1 \xrightarrow{j'_2,\alpha'_2} \dots \xrightarrow{j'_n,\alpha'_n} x'_n$, and denote by $\pi \leftrightsquigarrow_{ch} \pi'$, if and only if there are $0 < s < t \leq n$ such that:

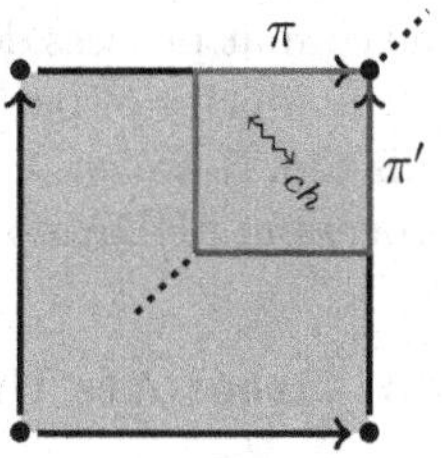

- for all $k < s$ or $k \geq t$, $x_k = x'_k$,
- for all $k < s$ or $k > t$, $j_k = j'_k$ and $\alpha_k = \alpha'_k$,
- for all $s \leq k \leq t$, $\alpha_k = \alpha'_k = 1$,
- $(j_s, \alpha_s) \star \dots \star (j_t, \alpha_t) = (j'_s, \alpha'_s) \star \dots \star (j'_t, \alpha'_t)$.

We denote by $\sim_{ch}$, and call **confluent homotopy**, the reflexive transitive closure of $\leftrightsquigarrow_{ch}$.

Lemma 1. *If X is a tree, then for every element (of any dimension) x of X, there is exactly one path modulo confluent homotopy from the initial state to x.*

Shortcuts. The face maps of path shapes and of the colimits we computed in Sect. 4.1 are of a very particular form: we start by defining the ∂_j^α and we extend this definition to general $\partial_{j_1<\dots<j_n}^{\alpha_1,\dots,\alpha_n}$. In a way, they are locally defined, and then extended to higher face maps. This means in particular that, in addition to having unique paths modulo confluent homotopy, they also do not have any 'shortcut'. A possible shortcut can be defined as a generalisation of paths, in which we allow to make transitions that go, for example, from a point to a square or to a cube, not only to segments, a shortcut being such a possible shortcut which is not confluently homotopic to a path. Those shortcuts may occur in a

pHDA, even if it has the unique path property. Concretely, by **shortcut** we mean the following situation: the face $\partial_{i_1<\ldots<i_n}^{\alpha_1,\ldots,\alpha_n}(x)$ is defined, but there is no sequence $(j_1;\beta_1)\star\ldots\star(j_n;\beta_n) = (i_1<\ldots<i_n;\alpha_1,\ldots,\alpha_n)$ such that $\partial_{j_n}^{\alpha_n}\circ\ldots\circ\partial_{j_1}^{\alpha_1}(x)$ is defined. By local-definedness of the face maps:

Lemma 2. *Trees do not have any shortcuts.*

Trees. We say that a pHDA **has the unique path property modulo confluent homotopy** if it has no shortcut, and there is exactly one class of paths modulo confluent homotopy from the initial state to any state. Given such a pHDA X and an element x of X, by **depth of** x we mean the length of a path from the initial state to x in X. Since homotopic paths have the same length, this is uniquely defined. We deduce from the previous discussions that:

Proposition 4. *Trees have unique path property modulo confluent homotopy.*

In the following, we will prove the converse: trees, defined as colimits of path shapes are exactly those pHDA that have the unique path property modulo confluent homotopy. This will be done by proving that such a pHDA X is isomorphic to its unfolding. A question that occurs now is the following. Much as the general framework of [4], trees are colimits of paths. Everything tends to work well when those trees have a nice property, which we called **accessibility**, intuitively, that the colimit process do not 'create' paths. This property is actually deeply related to the unicity of paths. Since this unicity fails in the case of pHDA, accessibility fails too. However, an accessibility modulo confluent homotopy holds: the colimit process in pHDA do not create confluent homotopy classes of paths.

4.3 Trees Are Unfoldings

We are now constructing our unfolding $U(X)$ of a pHDA X by giving an explicit definition, similar to [6,11], and proving that this is a tree. We will prove that there is a covering $\mathrm{unf}_X : U(X) \longrightarrow X$, which in particular means that the unfolding $U(X)$ is $\mathbf{PS_L}$-bisimilar (in the general sense of [17]) to X, and that this covering is actually an isomorphism when X has the unique path property modulo confluent homotopy.

Unfolding of a pHDA. Let us start with a few notations. Given a path $\pi = x_0 \xrightarrow{j_1,\alpha_1} x_1 \xrightarrow{j_2,\alpha_2} \ldots \xrightarrow{j_n,\alpha_n} x_n$ we note $e(\pi) = x_n$, $l(\pi) = n$ and $\pi_{-k} = x_0 \xrightarrow{j_1,\alpha_1} x_1 \xrightarrow{j_2,\alpha_2} \ldots \xrightarrow{j_{n-k},\alpha_{n-k}} x_{n-k}$. Given a pHDA X, its **unfolding** is the following pHDA:

- $U(X)_n$ is the set of equivalence classes $[\pi]$ of paths modulo confluent homotopy, such that $e(\pi)$ is of dimension n,
- the face maps are the smallest relations for inclusion such that:
  - $\partial_i^1(\alpha) = [\pi \xrightarrow{i,1} \partial_i^1(e(\pi))]$, for any $\pi \in \alpha$ such that $\partial_i^1(e(\pi))$ is defined,
  - $\partial_i^0(\alpha) = [\pi_{-1}]$ for any $\pi \in \alpha$ such that $\pi = \pi_{-1} \xrightarrow{i,0} e(\pi)$,

- $\partial^{\beta_1,\ldots,\beta_m}_{j_1<\ldots<j_m} \circ \partial^{\alpha_1,\ldots,\alpha_n}_{i_1<\ldots<i_n} \subseteq \partial^{\gamma_1,\ldots,\gamma_{n+m}}_{k_1<\ldots<k_{n+m}}$, for $(k_1, \ldots, k_{n+m}; \gamma_1, \ldots, \gamma_{n+m}) = (i_1, \ldots, i_n; \alpha_1, \ldots, \alpha_n) \star (j_1, \ldots, j_m; \beta_1, \ldots, \beta_m)$.

- the initial state is $[i]$,
- the labelling is given by $\lambda(\alpha) = \lambda(e(\pi))$ for $\pi \in \alpha$.

Following ideas from [4] again, the unfolding can be seen as the glueing of all possible executions of a system, but with care needed to handle confluent homotopy. Concretely:

Proposition 5. *The unfolding of a pHDA is a tree.*

We can also define $\mathrm{unf}_X : U(X) \longrightarrow X$ as the function that maps $[\pi]$ to $e(\pi)$.

Proposition 6. *unf_X is a covering, and so, $U(X)$ is $\boldsymbol{PS_L}$-bisimilar to X.*

The Unique Path Property Characterises Trees. When X has exactly one class of paths modulo confluent homotopy from the initial state to any state, it is possible to define a function $\eta_X : X \longrightarrow U(X)$ that maps any element x of X to the unique confluent homotopy class to x. When furthermore X does not have shortcuts, then η is actually a morphism of pHDA.

Proposition 7. *When X has the unique path property modulo confluent homotopy, then η_X is the inverse of unf_X. In particular, X is a tree.*

Together with Proposition 4, this implies the following:

Theorem 2. *Trees are exactly the pHDA that have the unique path property modulo confluent homotopy.*

Another consequence is that this isomorphism η_X is actually natural (in the categorical sense) and is part of an adjunction, which implies that trees form a coreflective subcategory of pHDA:

Corollary 1. *U extends to a functor, which is the right adjoint of the embedding $\iota : \boldsymbol{Tr_L} \longrightarrow \boldsymbol{pHDA_L}$. Furthermore, this is a coreflection.*

5 Cofibrant Objects

Cofibrant objects are another type of 'simple objects', coming from homotopy theory, more particularly the language of model categories from [24]. Those cofibrant objects are those whose unique morphism from the initial object is a cofibration. Intuitively (intuition which holds at least in cofibrantly generated model structures [15]), this means that cofibrant objects are those objects constructed from 'nothing', using only very basic constructions (generators of cofibrations). In the case of the classical model structure on topological spaces (Kan-Quillen), those spaces are those constructed from the empty space by adding 'cells', which produces what is called CW-complexes. In this section, we want to mimic this idea with trees: trees are those pHDA constructed from an initial state by only extending paths. We also want to emphasize that much as CW-complexes gives a kind of homotopy type of a space, trees gives a concurrency type of a pHDA, in the sense that there is a canonical way to produce an equivalent cofibrant object out of any object, which is called the **cofibrant replacement** in homotopy theory. In concurrency theory, this is the unfolding.

5.1 Cofibrant Objects in $\mathbf{pHDA}_\mathbf{L}$

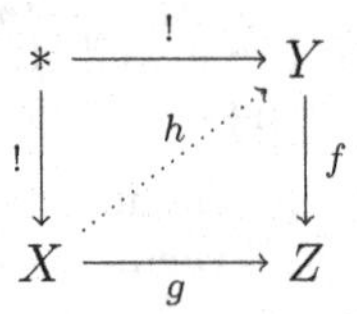

Following the language of model structures from [24], we say that a pHDA X is **cofibrant** if for every $\mathbf{PS}_\mathbf{L}$-open morphism $f : Y \longrightarrow Z$ and every morphism $g : X \longrightarrow Z$, there is a morphism $h : X \longrightarrow Y$, such that $f \circ h = g$. That is, a partial HDA X is cofibrant if and only if every $\mathbf{PS}_\mathbf{L}$-open morphism has the right lifting property with respect to the unique morphism from $*$ to X.

5.2 Cofibrant Objects Are Exactly Trees

In this section, we would like to prove the following:

Theorem 3. *The cofibrant objects are exactly trees.*

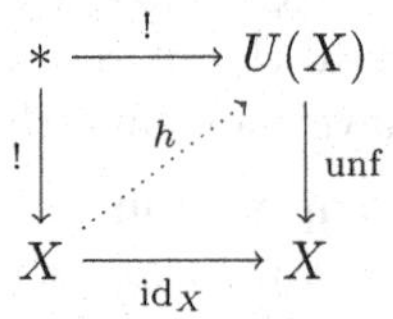

Let us start by giving the idea of the proof of the fact that cofibrant objects are trees. By Proposition 6, unf_X is a covering, so is open. This means that for every cofibrant object X, there is a morphism $h : X \longrightarrow U(X)$ such that $\mathrm{unf}_X \circ h = \mathrm{id}_X$, that is, X is a retract of its unfolding. Since we know that the unfolding is a tree by Proposition 5, it is enough to observe the following:

Lemma 3. *A retract of a tree is a tree.*

Intuitively, a pHDA is the retract of a tree only when it is obtain by retracting branches. This can only produce a tree. For the converse:

Proposition 8. *A tree is a cofibrant object. Furthermore, if $f : Y \longrightarrow Z$ is a covering, then the lift $h : X \longrightarrow Y$ is unique.*

The lift h is constructed by induction as follows. We define X_n as the restriction of X to elements whose depth is smaller than n, and the face maps $\partial_{j_1<\ldots<j_m}^{\alpha_1,\ldots,\alpha_m}(x)$ are defined if and only if $\partial_{j_1<\ldots<j_m}^{\alpha_1,\ldots,\alpha_m}(x)$ is defined in X and belongs to X_n. We then construct $h_n : X_n \longrightarrow Y$ using the unique path property modulo confluent homotopy, in a natural way (in the categorical meaning), i.e., such that $h_n \circ \kappa_n = h_{n-1}$, where $\kappa_n : X_{n-1} \longrightarrow X_n$ is the inclusion. h is then the inductive limit of those h_n. This proof can be seen as a small object argument.

5.3 The Unfolding Is Universal

As an application of the previous theorem, we would like to prove that the unfolding is universal. As in the case of covering spaces in algebraic topology, a covering corresponds to a partial unrolling of a system, in the sense that we can unroll some loops or even partially unroll a loop (imagine for example executing a few steps of a while-loop). In this sense, we can describe the fact that a covering unrolls more than another one, and that, an unfolding is a complete unrolling: since the domain is a tree, it is impossible to unroll more. Actually, much as the topological and the groupoidal cases (see [18] for example), unfoldings are the only such maximal unrollings among coverings: they are initial among coverings, that is why we call them 'universal'. In a way, this says that our definition of unfolding is the only reasonable one. Concretely, we say that a $\mathbf{PS_L}$-covering is **universal** if its domain is a tree.

Corollary 2. *If $f : Y \longrightarrow X$ is a universal covering, then for every covering $g : Z \longrightarrow X$ there is a unique map $h : Y \longrightarrow X$ such that $f = g \circ h$. Furthermore, h is itself a covering. Consequently, the universal covering is unique up-to isomorphism, and is given by the unfolding.*

This whole story is similar to the universal covering of a topological space: just replace pHDA by spaces and trees by simply-connected spaces [2].

6 Conclusion and Future Work

In this paper, we have given a cleaner definition of partial precubical sets and partial Higher Dimensional Automata, as they really correspond to collections of cubes with missing faces. From this categorical definition, we derived that pHDA can be completed, giving rise to a geometric realisation. We also describe the first premisses of a homotopy theory of the concurrency of pHDA where the cofibrant objects are trees, and replacement is the unfolding. As a future work, we could look at wider class of paths, typically allowing shortcuts as paths, or introducing general homotopies in the path category, which is possible because we can encode those inside the category of pHDA. Another direction would be to continue the description of this homotopy theory, to see if it corresponds to some kind of Quillen's model structure, or at least to some weaker version (e.g., category of cofibrant objects).

References

1. Bednarczyk, M.A.: Categories of asynchronous systems. Ph.D. thesis, University of Sussex (1987)
2. tom Dieck, T.: Algebraic Topology. Textbooks in Mathematics. European Mathematical Society, Zürich (2008)
3. Dubut, J.: Directed homotopy and homology theories for geometric models of true concurrency. Ph.D. thesis, ENS Paris-Saclay (2017)
4. Dubut, J., Goubault, E., Goubault-Larrecq, J.: Bisimulations and unfolding in P-accessible categorical models. In: Proceedings of the 27th International Conference on Concurrency Theory (CONCUR 2016). Leibniz International Proceedings in Informatics (LIPIcs), vol. 59, pp. 1–14. Schloss Dagstuhl-Leibniz-Zentrum fuer Informatik (2016)
5. Fahrenberg, U.: A category of higher-dimensional automata. In: Sassone, V. (ed.) FoSSaCS 2005. LNCS, vol. 3441, pp. 187–201. Springer, Heidelberg (2005). https://doi.org/10.1007/978-3-540-31982-5_12
6. Fahrenberg, U., Legay, A.: History-preserving bisimilarity for higher-dimensional automata via open maps. Electron. Notes Theor. Comput. Sci. **298**, 165–178 (2013)

7. Fahrenberg, U., Legay, A.: Partial higher-dimensional automata. In: CALCO 2015, pp. 101–115 (2015)
8. Fajstrup, L.: Dipaths and dihomotopies in a cubical complex. Adv. Appl. Math. **35**(2), 188–206 (2005)
9. Fajstrup, L., Goubault, E., Haucourt, E., Mimram, S., Raussen, M.: Directed Algebraic Topology and Concurrency. Springer, Cham (2016). https://doi.org/10.1007/978-3-319-15398-8
10. Gaucher, P.: Towards a homotopy theory of higher dimensional transition systems. Theory Appl. Categ. **25**, 295–341 (2011)
11. van Glabbeek, R.J.: Bisimulations for higher dimensional automata, June 1991. http://theory.stanford.edu/~rvg/hda
12. van Glabbeek, R.J.: On the expresiveness of higher dimensional automata: (extended abstract). Electron. Notes Theor. Comput. Sci. **128**(2), 5–34 (2005)
13. Goubault, E.: Géométrie du parallélisme. Ph.D. thesis, Ecole Polytechnique (1995)
14. Grandis, M.: Directed Algebraic Topology: Models of Non-Reversible Worlds. New Mathematical Monographs, vol. 13. Cambridge University Press, Cambridge (2009)
15. Hirschhorn, P.S.: Model Categories and Their Localizations. Mathematical Surveys and Monographs, vol. 99. American Mathematical Society, Providence (2003)
16. Jacobs, B.: Introduction to Coalgebra: Towards Mathematics of States and Observation. Cambridge Tracts in Theoretical Computer Science. Cambridge University Press, New York (2016)
17. Joyal, A., Nielsen, M., Winskel, G.: Bisimulation from open maps. Inf. Comput. **127**(2), 164–185 (1996)
18. May, J.P.: A Concise Course in Algebraic Topology. Chicago Lectures in Mathematics. University of Chicago Press, Chicago (1999)
19. Milner, R. (ed.): A Calculus of Communicating Systems. LNCS, vol. 92. Springer, Heidelberg (1980). https://doi.org/10.1007/3-540-10235-3
20. Niefield, S.: Lax presheaves and exponentiability. Theory Appl. Categ. **24**(12), 288–301 (2010)
21. Nielsen, M., Plotkin, G., Winskel, G.: Petri nets, event structures and domains, part I. Theor. Comput. Sci. **13**(1), 85–108 (1981)
22. Nielsen, M., Sassone, V., Winskel, G.: Relationships between models of concurrency. In: de Bakker, J.W., de Roever, W.-P., Rozenberg, G. (eds.) REX 1993. LNCS, vol. 803, pp. 425–476. Springer, Heidelberg (1994). https://doi.org/10.1007/3-540-58043-3_25
23. Pratt, V.: Modeling concurrency with geometry. In: Proceedings of the 18th ACM SIGPLAN-SIGACT Symposium on Principles of Programming Languages (POPL), pp. 311–322, January 1991
24. Quillen, D.G.: Homotopical Algebra. LNM, vol. 43. Springer, Heidelberg (1967). https://doi.org/10.1007/BFb0097438
25. Shields, M.W.: Concurrent machines. Comput. J. **28**(5), 449–465 (1985)

The Bernays-Schönfinkel-Ramsey Class of Separation Logic on Arbitrary Domains

Mnacho Echenim[1], Radu Iosif[2], and Nicolas Peltier[1](✉)

[1] Univ. Grenoble Alpes, CNRS, LIG, 38000 Grenoble, France
Nicolas.peltier@imag.fr

[2] Univ. Grenoble Alpes, CNRS, VERIMAG, 38000 Grenoble, France

Abstract. This paper investigates the satisfiability problem for Separation Logic with k record fields, with unrestricted nesting of separating conjunctions and implications, for prenex formulæ with quantifier prefix $\exists^*\forall^*$. In analogy with first-order logic, we call this fragment Bernays-Schönfinkel-Ramsey Separation Logic [$\mathsf{BSR}(\mathsf{SL}^k)$]. In contrast to existing work in Separation Logic, in which the universe of possible locations is assumed to be infinite, both finite and infinite universes are considered. We show that, unlike in first-order logic, the (in)finite satisfiability problem is undecidable for $\mathsf{BSR}(\mathsf{SL}^k)$. Then we define two non-trivial subsets thereof, that are decidable for finite and infinite satisfiability respectively, by controlling the occurrences of universally quantified variables within the scope of separating implications, as well as the polarity of the occurrences of the latter. Beside the theoretical interest, our work has natural applications in program verification, for checking that constraints on the shape of a data-structure are preserved by a sequence of transformations.

1 Introduction

Separation Logic [10,14], or SL, is a logical framework used in program verification to describe properties of the dynamically allocated memory, such as topologies of data structures (lists, trees), (un)reachability between pointers, etc. In a nutshell, given an integer $k \geq 1$, the logic SL^k is obtained from the first-order theory of a finite partial function $h : U \rightharpoonup U^k$ called a *heap*, by adding two substructural connectives: (i) the *separating conjunction* $\phi_1 * \phi_2$, that asserts a split of the heap into disjoint heaps satisfying ϕ_1 and ϕ_2 respectively, and (ii) the *separating implication* or *magic wand* $\phi_1 \mathrel{-\!\!*} \phi_2$, stating that each extension of the heap by a heap satisfying ϕ_1 must satisfy ϕ_2. Intuitively, U is the universe of possible of memory locations (cells) and k is the number of record fields in each memory cell.

The separating connectives $*$ and $\mathrel{-\!\!*}$ allow concise definitions of program semantics, via weakest precondition calculi [10] and easy-to-write specifications of recursive linked data structures (e.g. singly- and doubly-linked lists, trees with linked leaves and parent pointers, etc.), when higher-order inductive definitions are added [14]. Investigating the decidability and complexity of the satisfiability

M. Bojańczyk and A. Simpson (Eds.): FOSSACS 2019, LNCS 11425, pp. 242–259, 2019.
https://doi.org/10.1007/978-3-030-17127-8_14

problem for fragments of SL is of theoretical and practical interest. In this paper, we consider prenex SL formulæ with prefix $\exists^*\forall^*$. In analogy with first-order logic with equality and uninterpreted predicates [12], we call this fragment Bernays-Schönfinkel-Ramsey SL [$\mathsf{BSR}(\mathsf{SL}^k)$].

As far as we are aware, all existing work on SL assumes that the universe (set of available locations) is countably infinite. This assumption is not necessarily realistic in practice since the available memory is usually finite, although the bound depends on the hardware and is not known in advance. The finite universe hypothesis is especially useful when dealing with bounded memory issues, for instance checking that the execution of a program satisfies its postcondition, provided that there are sufficiently many available memory cells. In this paper we consider both the finite and infinite satisfiability problems. We show that both problems are undecidable for $\mathsf{BSR}(\mathsf{SL}^k)$ (unlike in first-order logic) and that they become PSPACE-complete under some additional restrictions, related to the occurrences of the magic wand and universal variables:

1. The infinite satisfiability problem is PSPACE-complete if the positive occurrences of $\multimap\!\!*$ (i.e., the occurrences of $\multimap\!\!*$ that are in the scope of an even number of negations) contain no universal variables.
2. The finite satisfiability problem is PSPACE-complete if there is no positive occurrence of $\multimap\!\!*$ (i.e., $\multimap\!\!*$ only occurs in the scope of an odd number of negations).

Reasoning on finite domains is more difficult than on infinite ones, due to possibility of asserting cardinality constraints on unallocated cells, which explains that the latter condition is more restrictive than the former one. Actually, the finite satisfiability problem is undecidable even if there is only one positive occurrence of a $\multimap\!\!*$ with no variable within the scope of $\multimap\!\!*$. These results establish sharp decidability frontiers within $\mathsf{BSR}(\mathsf{SL}^k)$.

Undecidability is shown by reduction from BSR first-order formulæ with two monadic function symbols. To establish the decidability results, we first show that every quantifier-free SL formula can be transformed into an equivalent boolean combination of formulæ of some specific patterns, called *test formulæ*. This result is interesting in itself, since it provides a precise and intuitive characterization of the expressive power of SL: it shows that separating connectives can be confined to a small set of test formulæ. Afterward, we show that such test formulæ can be transformed into first-order formulæ. If the above conditions (1) or (2) are satisfied, then the obtained first-order formulæ are in the BSR class, which ensures decidability. The PSPACE upper-bound relies on a careful analysis of the maximal size of the test formulæ. The analysis reveals that, although the boolean combination of test formulæ is of exponential size, its components (e.g., the conjunctions in its dnf) are of polynomial size and can be enumerated in polynomial space. For space reasons, full details and proofs are given in a technical report [8].

Applications. Besides theoretical interest, our work has natural applications in program verification. Indeed, purely universal SL formulæ are useful to express pre- or post-conditions asserting "local" constraints on the shape of the data structures manipulated by a program. Consider the atomic proposition $x \mapsto (y_1, \ldots, y_k)$ which states that the value of the heap at x is the tuple $(y_1, \ldots, y_k)$ and there is no value, other than x, in the domain of h. With this in mind, the following formula describes a well-formed doubly-linked list:

$$\forall x_1, x_2, x_3, x_4, x_5 \;.\; x_1 \mapsto (x_2, x_3) * x_2 \mapsto (x_4, x_5) * \top \Rightarrow x_5 \approx x_1 \wedge \neg x_3 \approx x_4 \quad (1)$$

Such constraints could also be expressed by using inductively defined predicates, unfortunately checking satisfiability of SL formulæ, even of very simple fragments with no occurrence of $-\!\!*$ in the presence of user-defined inductive predicates is undecidable, unless some rather restrictive conditions are fulfilled [9]. In contrast, checking entailment between two universal formulæ boils down to checking the satisfiability of a $\mathsf{BSR}(\mathsf{SL}^k)$ formula, which can be done thanks to the decidability results in our paper.

The separating implication (magic wand) seldom occurs in such shape constraints. However, it is useful to describe the dynamic transformations of the data structures, as in the following Hoare-style axiom, giving the weakest precondition of $\forall \mathbf{u} \;.\; \psi$ with respect to redirecting the i-th record field of x to z [10]:

$$\{\mathsf{x} \mapsto (y_1, \ldots, y_k) * [\mathsf{x} \mapsto (y_1, \ldots, y_{i-1}, \mathsf{z}, y_{i+1}, \ldots, y_k) -\!\!* \forall \mathbf{u} \;.\; \psi]\} \;\mathsf{x.i := z}\; \{\forall \mathbf{u} \;.\; \psi\}$$

It is easy to check that the precondition is equivalent to the formula $\forall \mathbf{u} \;.\; \mathsf{x} \mapsto (y_1, \ldots, y_k) * [\mathsf{x} \mapsto (y_1, \ldots, y_{i-1}, \mathsf{z}, y_{i+1}, \ldots, y_k) -\!\!* \psi]$ because, although hoisting universal quantifiers outside of the separating conjunction is unsound in general, this is possible here due to the special form of the left-hand side $\mathsf{x} \mapsto (y_1, \ldots, y_{i-1}, \mathsf{z}, \ldots, y_k)$ which unambiguously defines a single heap cell. Therefore, checking that $\forall \mathbf{u} \;.\; \psi$ is an invariant of the program statement $\mathsf{x.i := z}$ amounts to checking that the formula $\forall u \;.\; \psi \wedge \exists \mathbf{u} \;.\; \neg[\mathsf{x} \mapsto (y_1, \ldots, y_k) * (\mathsf{x} \mapsto (y_1, \ldots, y_{i-1}, \mathsf{z}, \ldots, y_k) -\!\!* \psi)]$ is unsatisfiable. Because the magic wand occurs negated, this formula falls into a decidable class defined in the present paper, for both finite and infinite satisfiability. The complete formalization of this deductive program verification technique and the characterization of the class of programs for which it is applicable is outside the scope of the paper and is left for future work.

Related Work. In contrast to first-order logic for which the decision problem has been thoroughly investigated [1], only a few results are known for SL. For instance, the problem is undecidable in general and PSPACE-complete for quantifier-free formulæ [4]. For $k = 1$, the problem is also undecidable, but it is PSPACE-complete if in addition there is only one quantified variable [6] and decidable but nonelementary if there is no magic wand [2]. In particular, we have also studied the prenex form of SL^1 [7] and found out that it is decidable and nonelementary, whereas $\mathsf{BSR}(\mathsf{SL}^1)$ is PSPACE-complete. In contrast, in this paper we show that undecidability occurs for $\mathsf{BSR}(\mathsf{SL}^k)$, for $k \geq 2$.

Expressive completeness results exist for quantifier-free SL^1 [2,11] and for SL^1 with one and two quantified variables [5,6]. There, the existence of equivalent boolean combinations of test formulæ is shown implicitly, using a finite enumeration of equivalence classes of models, instead of an effective transformation. Instead, here we present an explicit equivalence-preserving transformation of quantifier-free SL^k into boolean combinations of test formulæ, and translate the latter into first-order logic. Further, we extend the expressive completeness result to finite universes, with additional test formulæ asserting cardinality constraints on unallocated cells.

Another translation of quantifier-free SL^k into first-order logic with equality has been described in [3]. There, the small model property of quantifier-free SL^k [4] is used to bound the number of first-order variables to be considered and the separating connectives are interpreted as first-order quantifiers. The result is an equisatisfiable first-order formula. This translation scheme cannot be, however, directly applied to $\mathsf{BSR}(\mathsf{SL}^k)$, which does not have a small model property, being moreover undecidable. Theory-parameterized versions of $\mathsf{BSR}(\mathsf{SL}^k)$ have been shown to be undecidable, e.g. when integer linear arithmetic is used to reason about locations, and claimed to be PSPACE-complete for countably infinite and finite unbounded location sorts, with no relation other than equality [13]. In the present paper, we show that this claim is wrong, and draw a precise chart of decidability for both infinite and finite satisfiability of $\mathsf{BSR}(\mathsf{SL}^k)$.

2 Preliminaries

Basic Definitions. Let $\mathbb{Z}_\infty = \mathbb{Z} \cup \{\infty\}$ and $\mathbb{N}_\infty = \mathbb{N} \cup \{\infty\}$, where for each $n \in \mathbb{Z}$ we have $n + \infty = \infty$ and $n < \infty$. For a countable set S we denote by $||S|| \in \mathbb{N}_\infty$ the cardinality of S. Let Var be a countable set of variables, denoted as x, y, z and U be a sort. Vectors of variables are denoted by $\mathbf{x}$, $\mathbf{y}$, etc. A *function symbol* f has $\#(f) \geq 0$ arguments of sort U and a sort $\sigma(f)$, which is either the boolean sort Bool or U. If $\#(f) = 0$, we call f a *constant*. We use $\bot$ and $\top$ for the boolean constants false and true, respectively. First-order (FO) terms t and formulæ φ are defined by the following grammar:

$$t := x \mid f(\underbrace{t, \ldots, t}_{\#(f)}) \qquad \varphi := \bot \mid \top \mid t \approx t \mid p(\underbrace{t, \ldots, t}_{\#(p)}) \mid \varphi \wedge \varphi \mid \neg\varphi \mid \exists x \,.\, \varphi$$

where $x \in \mathsf{Var}$, f and p are function symbols, $\sigma(f) = U$ and $\sigma(p) = \mathsf{Bool}$. We write $\varphi_1 \vee \varphi_2$ for $\neg(\neg\varphi_1 \wedge \neg\varphi_2)$, $\varphi_1 \rightarrow \varphi_2$ for $\neg\varphi_1 \vee \varphi_2$, $\varphi_1 \leftrightarrow \varphi_2$ for $\varphi_1 \rightarrow \varphi_2 \wedge \varphi_2 \rightarrow \varphi_1$ and $\forall x \,.\, \varphi$ for $\neg\exists x \,.\, \neg\varphi$. The size of a formula φ, denoted as $\mathsf{size}(\varphi)$, is the number of symbols needed to write it down. Let $\mathsf{var}(\varphi)$ be the set of variables that occur free in φ, i.e. not in the scope of a quantifier. A *sentence* φ is a formula where $\mathsf{var}(\varphi) = \emptyset$.

First-order formulæ are interpreted over FO-structures (called structures, when no confusion arises) $\mathcal{S} = (\mathfrak{U}, \mathfrak{s}, \mathfrak{i})$, where $\mathfrak{U}$ is a countable set, called the *universe*, the elements of which are called *locations*, $\mathfrak{s} : \mathsf{Var} \rightharpoonup \mathfrak{U}$ is a mapping of variables to locations, called a *store* and $\mathfrak{i}$ interprets each function symbol f by a function $f^\mathfrak{i} : \mathfrak{U}^{\#(f)} \rightarrow \mathfrak{U}$, if $\sigma(f) = U$ and $f^\mathfrak{i} : \mathfrak{U}^{\#(f)} \rightarrow \{\bot^\mathfrak{i}, \top^\mathfrak{i}\}$ if $\sigma(f) = \mathsf{Bool}$.

A structure $(\mathfrak{U}, \mathfrak{s}, \mathfrak{i})$ is *finite* when $||\mathfrak{U}|| \in \mathbb{N}$ and *infinite* otherwise. We write $\mathcal{S} \models \varphi$ iff φ is true when interpreted in $\mathcal{S}$. This relation is defined recursively on the structure of φ, as usual. When $\mathcal{S} \models \varphi$, we say that $\mathcal{S}$ is a *model* of φ. A formula is [finitely] *satisfiable* when it has a [finite] model. We write $\varphi_1 \equiv \varphi_2$ when $(\mathfrak{U}, \mathfrak{s}, \mathfrak{i}) \models \varphi_1 \Leftrightarrow (\mathfrak{U}, \mathfrak{s}, \mathfrak{i}) \models \varphi_2$, for every structure $(\mathfrak{U}, \mathfrak{s}, \mathfrak{i})$.

The Bernays-Schönfinkel-Ramsey fragment of FO, denoted by BSR(FO), is the set of sentences $\exists x_1 \ldots \exists x_n \forall y_1 \ldots \forall y_m \,.\, \varphi$, where φ is a quantifier-free formula in which all function symbols f of arity $\#(f) > 0$ have sort $\sigma(f) = \mathsf{Bool}$.

Separation Logic. Let k be a strictly positive integer. The logic SL^k is the set of formulæ generated by the grammar:

$$\varphi := \bot \mid \top \mid \mathsf{emp} \mid x \approx y \mid x \mapsto (y_1, \ldots, y_k) \mid \varphi \wedge \varphi \mid \neg\varphi \mid \varphi * \varphi \mid \varphi -\!\!* \varphi \mid \exists x \,.\, \varphi$$

where $x, y, y_1, \ldots, y_k \in \mathsf{Var}$. The connectives $*$ and $-\!\!*$ are respectively called the *separating conjunction* and *separating implication* (*magic wand*). We write $\varphi_1 -\!\!\circ\, \varphi_2$ for $\neg(\varphi_1 -\!\!* \neg\varphi_2)$ ($-\!\!\circ$ is called *septraction*). The size and set of free variables of an SL^k formula φ are defined as for first-order formulæ.

Given an SL^k formula ϕ and a subformula ψ of ϕ, we say that ψ *occurs at polarity* $p \in \{-1, 0, 1\}$ iff one of the following holds: (i) $\phi = \psi$ and $p = 1$, (ii) $\phi = \neg\phi_1$ and ψ occurs at polarity $-p$ in ϕ_1, (iii) $\phi = \phi_1 \wedge \phi_2$ or $\phi = \phi_1 * \phi_2$, and ψ occurs at polarity p in ϕ_i, for some $i = 1, 2$, or (iv) $\phi = \phi_1 -\!\!* \phi_2$ and either ψ is a subformula of ϕ_1 and $p = 0$, or ψ occurs at polarity p in ϕ_2. A polarity of $1, 0$ or -1 is also referred to as positive, neutral or negative, respectively. Note that our notion of polarity is slightly different than usual, because the antecedent of a separating implication is of neutral polarity while the antecedent of an implication is usually of negative polarity. This is meant to strengthen upcoming decidability results, see Remark 2.

SL^k formulæ are interpreted over SL-*structures* $\mathcal{I} = (\mathfrak{U}, \mathfrak{s}, \mathfrak{h})$, where $\mathfrak{U}$ and $\mathfrak{s}$ are as before and $\mathfrak{h} : \mathfrak{U} \rightharpoonup_{fin} \mathfrak{U}^k$ is a finite partial mapping of locations to k-tuples of locations, called a *heap*. As before, a structure $(\mathfrak{U}, \mathfrak{s}, \mathfrak{h})$ is finite when $||\mathfrak{U}|| \in \mathbb{N}$ and infinite otherwise. We denote by $\mathrm{dom}(\mathfrak{h})$ the domain of the heap $\mathfrak{h}$ and by $||\mathfrak{h}|| \in \mathbb{N}$ the cardinality of $\mathrm{dom}(\mathfrak{h})$. Two heaps $\mathfrak{h}_1$ and $\mathfrak{h}_2$ are *disjoint* iff $\mathrm{dom}(\mathfrak{h}_1) \cap \mathrm{dom}(\mathfrak{h}_2) = \emptyset$, in which case $\mathfrak{h}_1 \uplus \mathfrak{h}_2$ denotes their union. A heap $\mathfrak{h}'$ is an *extension* of $\mathfrak{h}$ by $\mathfrak{h}''$ iff $\mathfrak{h}' = \mathfrak{h} \uplus \mathfrak{h}''$. The relation $(\mathfrak{U}, \mathfrak{s}, \mathfrak{h}) \models \varphi$ is defined inductively, as follows:

$$
\begin{array}{lll}
(\mathfrak{U}, \mathfrak{s}, \mathfrak{h}) \models \mathsf{emp} & \Leftrightarrow & \mathfrak{h} = \emptyset \\
(\mathfrak{U}, \mathfrak{s}, \mathfrak{h}) \models x \mapsto (y_1, \ldots, y_k) & \Leftrightarrow & \mathfrak{h} = \{\langle \mathfrak{s}(x), (\mathfrak{s}(y_1), \ldots, \mathfrak{s}(y_k))\rangle\} \\
(\mathfrak{U}, \mathfrak{s}, \mathfrak{h}) \models \varphi_1 * \varphi_2 & \Leftrightarrow & \text{there exist disjoint heaps } h_1, h_2 \text{ such that } h = h_1 \uplus h_2 \\
 & & \text{and } (\mathfrak{U}, \mathfrak{s}, \mathfrak{h}_i) \models \varphi_i, \text{ for } i = 1, 2 \\
(\mathfrak{U}, \mathfrak{s}, \mathfrak{h}) \models \varphi_1 -\!\!* \varphi_2 & \Leftrightarrow & \text{for all heaps } \mathfrak{h}' \text{ disjoint from } \mathfrak{h} \text{ such that } (\mathfrak{U}, \mathfrak{s}, \mathfrak{h}') \models \varphi_1 \\
 & & \text{we have } (\mathfrak{U}, \mathfrak{s}, \mathfrak{h} \uplus \mathfrak{h}') \models \varphi_2
\end{array}
$$

The semantics of equality, boolean and first-order connectives is the usual one. Satisfiability, entailment and equivalence are defined for SL^k as for FO formulæ.

The Bernays-Schönfinkel-Ramsey fragment of SL^k, denoted by $\mathsf{BSR}(\mathsf{SL}^k)$, is the set of sentences $\exists x_1 \ldots \exists x_n \forall y_1 \ldots \forall y_m \,.\, \phi$, where ϕ is a quantifier-free SL^k formula. Since there is no function symbol of arity greater than zero in SL^k, there is no restriction, other than the form of the quantifier prefix defining $\mathsf{BSR}(\mathsf{SL}^k)$.

3 Test Formulæ for SL^k

We define a small set of SL^k patterns of formulæ, possibly parameterized by a positive integer, called *test formulæ*. These patterns capture properties related to allocation, points-to relations in the heap and cardinality constraints.

Definition 1. *The following patterns are called* test formulæ*:*

$$x \hookrightarrow \mathbf{y} \stackrel{\mathsf{def}}{=} x \mapsto \mathbf{y} * \top \qquad |U| \geq n \stackrel{\mathsf{def}}{=} \top \multimap |h| \geq n,\ n \in \mathbb{N}$$

$$\mathsf{alloc}(x) \stackrel{\mathsf{def}}{=} x \mapsto \underbrace{(x, \ldots, x)}_{k \text{ times}} \mathrel{-\!\!*} \bot \qquad |h| \geq |U| - n \stackrel{\mathsf{def}}{=} |h| \geq n+1 \mathrel{-\!\!*} \bot, n \in \mathbb{N}$$

$$x \approx y \qquad |h| \geq n \stackrel{\mathsf{def}}{=} \begin{cases} |h| \geq n-1 * \neg\mathsf{emp}, & \text{if } n > 0 \\ \top, & \text{if } n = 0 \\ \bot, & \text{if } n = \infty \end{cases}$$

where $x, y \in \mathsf{Var}$, $\mathbf{y} \in \mathsf{Var}^k$ *and* $n \in \mathbb{N}_\infty$ *is a positive integer or* ∞.

The semantics of test formulæ is very natural: $x \hookrightarrow \mathbf{y}$ means that x points to vector $\mathbf{y}$, $\mathsf{alloc}(x)$ means that x is allocated, and the arithmetic expressions are interpreted as usual, where $|h|$ and $|U|$ respectively denote the number of allocated cells and the number of locations (possibly ∞). Formally:

Proposition 1. *Given an* SL*-structure* $(\mathfrak{U}, \mathfrak{s}, \mathfrak{h})$, *the following equivalences hold, for all variables* $x, y_1, \ldots, y_k \in \mathsf{Var}$ *and integers* $n \in \mathbb{N}$*:*

$$(\mathfrak{U}, \mathfrak{s}, \mathfrak{h}) \models x \hookrightarrow \mathbf{y} \Leftrightarrow \mathfrak{h}(\mathfrak{s}(x)) = \mathfrak{s}(\mathbf{y}) \qquad (\mathfrak{U}, \mathfrak{s}, \mathfrak{h}) \models |h| \geq |U| - n \Leftrightarrow ||\mathfrak{h}|| \geq ||\mathfrak{U}|| - n$$

$$(\mathfrak{U}, \mathfrak{s}, \mathfrak{h}) \models |U| \geq n \Leftrightarrow ||\mathfrak{U}|| \geq n \qquad (\mathfrak{U}, \mathfrak{s}, \mathfrak{h}) \models |h| \geq n \Leftrightarrow ||\mathfrak{h}|| \geq n$$

$$(\mathfrak{U}, \mathfrak{s}, \mathfrak{h}) \models \mathsf{alloc}(x) \Leftrightarrow \mathfrak{s}(x) \in \mathrm{dom}(\mathfrak{h})$$

Not all atoms of SL^k are test formulæ, for instance $x \mapsto \mathbf{y}$ and emp are not test formulæ. However, by Proposition 1, we have the equivalences $x \mapsto \mathbf{y} \equiv x \hookrightarrow \mathbf{y} \wedge \neg|h| \geq 2$ and $\mathsf{emp} \equiv \neg|h| \geq 1$. Note that, for any $n \in \mathbb{N}$, the test formulæ $|U| \geq n$ and $|h| \geq |U| - n$ are trivially true and false respectively, if the universe is infinite. We write $t < u$ for $\neg(t \geq u)$.

We need to introduce a few notations useful to describe upcoming transformations in a concise and precise way. A *literal* is a test formula or its negation. Unless stated otherwise, we view a conjunction T of literals as a set[1] and we use the same symbol to denote both a set and the formula obtained by conjoining the elements of the set. The equivalence relation $x \approx_T y$ is defined as $T \models x \approx y$ and we write $x \not\approx_T y$ for $T \models \neg x \approx y$. Observe that $x \not\approx_T y$ is not the complement of $x \approx_T y$. For a set X of variables, $|X|_T$ is the number of equivalence classes of $\approx_T$ in X.

[1] The empty set is thus considered to be true.

Definition 2. *A variable x is* allocated *in an* SL*-structure* $\mathcal{I}$ *iff* $\mathcal{I} \models \mathsf{alloc}(x)$. *For a set of variables* $X \subseteq \mathsf{Var}$, *let* $\mathsf{alloc}(X) \stackrel{\text{def}}{=} \bigwedge_{x \in X} \mathsf{alloc}(x)$ *and* $\mathsf{nalloc}(X) \stackrel{\text{def}}{=} \bigwedge_{x \in X} \neg \mathsf{alloc}(x)$. *For a set T of literals, let:*

$$\begin{aligned}
\mathsf{av}(T) &\stackrel{\text{def}}{=} \{x \in \mathsf{Var} \mid x \approx_T x',\ T \cap \{\mathsf{alloc}(x'), x' \hookrightarrow \mathbf{y} \mid \mathbf{y} \in \mathsf{Var}^k\} \neq \emptyset\} \\
\mathsf{nv}(T) &\stackrel{\text{def}}{=} \{x \in \mathsf{Var} \mid x \approx_T x',\ \neg\mathsf{alloc}(x') \in T\} \\
\mathsf{fp}_X(T) &\stackrel{\text{def}}{=} T \cap \{\mathsf{alloc}(x), \neg\mathsf{alloc}(x), x \hookrightarrow \mathbf{y}, \neg x \hookrightarrow \mathbf{y} \mid x \in X, \mathbf{y} \in \mathsf{Var}^k\}
\end{aligned}$$

We let $\#_a(T) \stackrel{\text{def}}{=} |\mathsf{av}(T)|_T$ *be the number of equivalence classes of* $\approx_T$ *containing variables allocated in every model of T and* $\#_n(X,T) \stackrel{\text{def}}{=} |X \cap \mathsf{nv}(T)|_T$ *be the number of equivalence classes of* $\approx_T$ *containing variables from X that are not allocated in any model of T. We also let* $\mathsf{fp}_a(T) \stackrel{\text{def}}{=} \mathsf{fp}_{\mathsf{av}(T)}(T)$.

Intuitively, $\mathsf{av}(T)$ [$\mathsf{nv}(T)$] is the set of variables that must be [are never] allocated in every [any] model of T, and $\mathsf{fp}_X(T)$ is the *footprint* of T relative to the set $X \subseteq \mathsf{Var}$, i.e. the set of formulæ describing allocation and points-to relations over variables from X. For example, if $T = \{x \approx z, \mathsf{alloc}(x), \neg\mathsf{alloc}(y), \neg z \hookrightarrow \mathbf{y}\}$, then $\mathsf{av}(T) = \{x, z\}$, $\mathsf{nv}(T) = \{y\}$, $\mathsf{fp}_a(T) = \{\mathsf{alloc}(x), \neg z \hookrightarrow \mathbf{y}\}$ and $\mathsf{fp}_{\mathsf{nv}(T)}(T) = \{\neg\mathsf{alloc}(y)\}$.

3.1 From Test Formulæ to FO

The introduction of test formulæ (Definition 1) is motivated by the reduction of the (in)finite satisfiability problem for quantified boolean combinations thereof to the same problem for FO. The reduction is devised in such a way that the obtained formula is in the BSR class, if possible. Given a quantified boolean combination of test formulæ ϕ, the FO formula $\mathfrak{r}(\phi)$ is defined by induction on the structure of ϕ:

$$\begin{aligned}
\mathfrak{r}(|h| \geq n) &\stackrel{\text{def}}{=} \mathfrak{a}_n & \mathfrak{r}(|U| \geq n) &\stackrel{\text{def}}{=} \mathfrak{b}_n \\
\mathfrak{r}(|h| \geq |U| - n) &\stackrel{\text{def}}{=} \neg \mathfrak{c}_{n+1} & \mathfrak{r}(\neg \phi_1) &\stackrel{\text{def}}{=} \neg \mathfrak{r}(\phi_1) \\
\mathfrak{r}(x \hookrightarrow \mathbf{y}) &\stackrel{\text{def}}{=} \mathfrak{p}(x, y_1, \ldots, y_k) & \mathfrak{r}(\mathsf{alloc}(x)) &\stackrel{\text{def}}{=} \exists y_1 \ldots \exists y_k \,.\, \mathfrak{p}(x, y_1, \ldots, y_k) \\
\mathfrak{r}(\phi_1 \wedge \phi_2) &\stackrel{\text{def}}{=} \mathfrak{r}(\phi_1) \wedge \mathfrak{r}(\phi_2) & \mathfrak{r}(\exists x \,.\, \phi_1) &\stackrel{\text{def}}{=} \exists x \,.\, \mathfrak{r}(\phi_1) \\
\mathfrak{r}(x \approx y) &\stackrel{\text{def}}{=} x \approx y
\end{aligned}$$

where $\mathfrak{p}$ is a $(k+1)$-ary function symbol of sort Bool and $\mathfrak{a}_n, \mathfrak{b}_n$ and $\mathfrak{c}_n$ are constants of sort Bool, for all $n \in \mathbb{N}$. These function symbols are related by the following axioms, where $\mathfrak{u}_n, \mathfrak{v}_n$ and $\mathfrak{w}_n$ are constants of sort U, for all $n > 0$:

$$\begin{aligned}
& & P &: \forall x \forall \mathbf{y} \forall \mathbf{y}' \,.\, \mathfrak{p}(x, \mathbf{y}) \wedge \mathfrak{p}(x, \mathbf{y}') \rightarrow \textstyle\bigwedge_{i=1}^{k} y_i \approx y'_i \\
A_0 &: \mathfrak{a}_0 & A_n &: \left\{ \begin{array}{l} \exists \mathbf{y} \,.\, \mathfrak{a}_n \rightarrow \mathfrak{a}_{n-1} \wedge \mathfrak{p}(\mathfrak{u}_n, \mathbf{y}) \wedge \bigwedge_{i=1}^{n-1} \neg \mathfrak{u}_i \approx \mathfrak{u}_n \\ \wedge\ \forall x \forall \mathbf{y} \,.\, \neg \mathfrak{a}_n \wedge \mathfrak{p}(x, \mathbf{y}) \rightarrow \bigvee_{i=1}^{n-1} x \approx \mathfrak{u}_i \end{array} \right\} \\
B_0 &: \mathfrak{b}_0 & B_n &: \left\{ \begin{array}{l} \mathfrak{b}_n \rightarrow \mathfrak{b}_{n-1} \wedge \bigwedge_{i=1}^{n-1} \neg \mathfrak{v}_i \approx \mathfrak{v}_n \\ \wedge\ \forall x \,.\, \neg \mathfrak{b}_n \rightarrow \bigvee_{i=1}^{n-1} x \approx \mathfrak{v}_i \end{array} \right\} \\
C_0 &: \mathfrak{c}_0 & C_n &: \forall \mathbf{y} \,.\, \mathfrak{c}_n \rightarrow \mathfrak{c}_{n-1} \wedge \neg \mathfrak{p}(\mathfrak{w}_n, \mathbf{y}) \wedge \textstyle\bigwedge_{i=1}^{n-1} \neg \mathfrak{w}_n \approx \mathfrak{w}_i
\end{aligned}$$

Intuitively, $\mathfrak{p}$ encodes the heap and $\mathfrak{a}_n$ (resp. $\mathfrak{b}_n$) is true iff there are at least n cells in the domain of the heap (resp. in the universe), namely $\mathfrak{u}_1, \ldots, \mathfrak{u}_n$ (resp. $\mathfrak{v}_1, \ldots, \mathfrak{v}_n$). If $\mathfrak{c}_n$ is true, then there are at least n locations $\mathfrak{w}_1, \ldots, \mathfrak{w}_n$ outside of the domain of the heap (free), but the converse does not hold. The C_n axioms do not state the equivalence of $\mathfrak{c}_n$ with the existence of at least n free locations, because such an equivalence cannot be expressed in $\mathsf{BSR(FO)}$[2]. As a consequence, the transformation preserves sat-equivalence only if the formulæ $|h| \geq |U| - n$ occur only at negative polarity (see Lemma 1, Point 2). If the domain is infinite then this problem does not arise since the formulæ $|h| \geq |U| - n$ are always false.

Definition 3. *For a quantified boolean combination of test formulæ ϕ, we let $\mathcal{N}(\phi)$ be the maximum integer n occurring in a test formula θ of the form $|h| \geq n$, $|U| \geq n$, or $|h| \geq |U| - n$ from ϕ and define $\mathcal{A}(\phi) \stackrel{\text{def}}{=} \{P\} \cup \{A_i\}_{i=0}^{\mathcal{N}(\phi)} \cup \{B_i\}_{i=0}^{\mathcal{N}(\phi)} \cup \{C_i\}_{i=0}^{\mathcal{N}(\phi)+1}$ as the set of axioms related to ϕ.*

The relationship between ϕ and $\tau(\phi)$ is stated below.

Lemma 1. *Let ϕ be a quantified boolean combination of test formulæ. The following hold, for any universe $\mathfrak{U}$ and any store $\mathfrak{s}$:*

1. *if $(\mathfrak{U}, \mathfrak{s}, \mathfrak{h}) \models \phi$, for a heap $\mathfrak{h}$, then $(\mathfrak{U}, \mathfrak{s}, \mathfrak{i}) \models \tau(\phi) \wedge \mathcal{A}(\phi)$, for an interpretation $\mathfrak{i}$;*
2. *if each test formula $|h| \geq |U| - n$ in ϕ occurs at a negative polarity and $(\mathfrak{U}, \mathfrak{s}, \mathfrak{i}) \models \tau(\phi) \wedge \mathcal{A}(\phi)$ for an interpretation $\mathfrak{i}$ such that $||\mathfrak{p}^{\mathfrak{i}}|| \in \mathbb{N}$, then $(\mathfrak{U}, \mathfrak{s}, \mathfrak{h}) \models \phi$, for a heap $\mathfrak{h}$.*

The translation of $\mathsf{alloc}(x)$ introduces existential quantifiers depending on x. For instance, $\forall x \,.\, \mathsf{alloc}(x)$ is translated as $\forall x \exists y_1 \ldots \exists y_k \,.\, \mathfrak{p}(x, y_1, \ldots, y_k)$, which lies outside of the $\mathsf{BSR(FO)}$ fragment. Because upcoming decidability results (Theorem 2) require that $\tau(\phi)$ be in $\mathsf{BSR(FO)}$, we end this section by delimiting a fragment of SL^k whose translation falls into $\mathsf{BSR(FO)}$.

Lemma 2. *Given an SL^k formula $\varphi = \forall z_1 \ldots \forall z_m \,.\, \phi$, where ϕ is a boolean combination of test formulæ containing no positive occurrence of $\mathsf{alloc}(z_i)$ for any $i \in [1, m]$, $\tau(\varphi)$ is equivalent (up to transformation into prenex form) to a $\mathsf{BSR(FO)}$ formula with the same constants and free variables as $\tau(\varphi)$.*

Intuitively, if a formula $\mathsf{alloc}(x)$ occurs negatively then the quantifiers $\exists y_1 \ldots \exists y_k$ added when translating $\mathsf{alloc}(x)$ can be transformed into universal ones by transformation into nnf, and if x is not universal then they may be shifted at the root of the formula since $y_1, \ldots, y_k$ depend only on x. In both cases, the quantifier prefix $\exists^* \forall^*$ is preserved.

[2] The converse of C_n: $\forall x \,.\, (\neg \mathfrak{c}_n \wedge \forall \mathbf{y} \,.\, \neg \mathfrak{p}(x, \mathbf{y})) \rightarrow \bigvee_{i=1}^{n-1} x \approx \mathfrak{w}_i$ is not in $\mathsf{BSR(FO)}$.

4 From Quantifier-Free SL^k to Test formulæ

This section states the expressive completeness result of the paper, namely that any quantifier-free SL^k formula is equivalent, on both finite and infinite models, to a boolean combination of test formulæ. Starting from a quantifier-free SL^k formula φ, we define a set $\mu(\varphi)$ of conjunctions of test formulæ and their negations, called *minterms*, such that $\varphi \equiv \bigvee_{M \in \mu(\varphi)} M$. Although the number of minterms in $\mu(\varphi)$ is exponential in the size of φ, checking the membership of a given minterm M in $\mu(\varphi)$ can be done in PSPACE. Together with the translation of minterms into FO (Sect. 3.1), this fact is used to prove PSPACE membership of the two decidable fragments of $\mathsf{BSR}(\mathsf{SL}^k)$, defined next (Sect. 5.2).

4.1 Minterms

A *minterm* M is a set (conjunction) of literals containing: exactly one literal $|h| \geq \min_M$ and one literal $|h| < \max_M$, where $\min_M \in \mathbb{N} \cup \{|U| - n \mid n \in \mathbb{N}\}$ and $\max_M \in \mathbb{N}_\infty \cup \{|U| - n \mid n \in \mathbb{N}\}$, and at most one literal of the form $|U| \geq n$, respectively $|U| < n$.

A minterm may be viewed as an abstract description of a heap. The conditions are for technical convenience only and are not restrictive. For instance, tautological test formulæ of the form $|h| \geq 0$ and/or $|h| < \infty$ may be added if needed so that the first condition holds. If M contains two literals $t \geq n_1$ and $t \geq n_2$ with $n_1 < n_2$ and $t \in \{|h|, |U|\}$ then $t \geq n_1$ is redundant and can be removed – and similarly if M contains literals $|h| \geq |U| - n_1$ and $|h| \geq |U| - n_2$. Heterogeneous constraints are merged by performing a case split on the value of $|U|$. For example, if M contains both $|h| \geq |U| - 4$ and $|h| \geq 1$, then the first condition prevails if $|U| \geq 5$ yielding the equivalence disjunction: $|h| \geq 1 \wedge |U| < 5 \vee |h| \geq |U| - 4 \wedge |U| \geq 5$. Thus, in the following, we assume that any conjunction of literals can be transformed into a disjunction of minterms [8].

Definition 4. *Given a minterm M, we define the sets:*

$$M^e \stackrel{\mathsf{def}}{=} M \cap \{x \approx y, \neg x \approx y \mid x, y \in \mathsf{Var}\} \quad M^a \stackrel{\mathsf{def}}{=} M \cap \{\mathsf{alloc}(x), \neg \mathsf{alloc}(x) \mid x \in \mathsf{Var}\}$$
$$M^u \stackrel{\mathsf{def}}{=} M \cap \{|U| \geq n, |U| < n \mid n \in \mathbb{N}\} \quad M^p \stackrel{\mathsf{def}}{=} M \cap \{x \hookrightarrow \mathbf{y}, \neg x \hookrightarrow \mathbf{y} \mid x, \mathbf{y} \in \mathsf{Var}^{k+1}\}$$

Thus, $M = M^e \cup M^u \cup M^a \cup M^p \cup \{|h| \geq \min_M, |h| < \max_M\}$, for each minterm M. Given a set of variables $X \subseteq \mathsf{Var}$, a minterm M is (1) *E-complete* for X iff for all $x, y \in X$ exactly one of $x \approx y \in M$, $\neg x \approx y \in M$ holds, and (2) *A-complete* for X iff for each $x \in X$ exactly one of $\mathsf{alloc}(x) \in M$, $\neg\mathsf{alloc}(x) \in M$ holds.

For a literal ℓ, we denote by $\overline{\ell}$ its complement, i.e. $\overline{\theta} \stackrel{\mathsf{def}}{=} \neg\theta$ and $\overline{\neg\theta} \stackrel{\mathsf{def}}{=} \theta$, where θ is a test formula. Let $\overline{M}$ be the minterm obtained from M by replacing each literal with its complement. The *complement closure* of M is $\mathsf{cc}(M) \stackrel{\mathsf{def}}{=} M \cup \overline{M}$. Two tuples $\mathbf{y}, \mathbf{y}' \in \mathsf{Var}^k$ are *M-distinct* if $y_i \not\approx_M y'_i$, for some $i \in [1, k]$. Given a minterm M that is E-complete for $\mathsf{var}(M)$, its *points-to closure* is $\mathsf{pc}(M) \stackrel{\mathsf{def}}{=} \bot$ if there exist literals $x \hookrightarrow \mathbf{y}, x' \hookrightarrow \mathbf{y}' \in M$ such that $x \approx_M x'$ and $\mathbf{y}$, $\mathbf{y}'$ are M-distinct, and $\mathsf{pc}(M) \stackrel{\mathsf{def}}{=} M$, otherwise. Intuitively, $\mathsf{pc}(M)$ is $\bot$ iff M contradicts the fact that the heap is a partial function[3]. The *domain closure* of M is $\mathsf{dc}(M) \stackrel{\text{def}}{=} \bot$ if either $\min_M = n_1$ and $\max_M = n_2$ for some $n_1, n_2 \in \mathbb{Z}$ such that $n_1 \geq n_2$, or $\min_M = |U| - n_1$ and $\max_M = |U| - n_2$, where $n_2 \geq n_1$; and otherwise:

$$\begin{aligned}\mathsf{dc}(M) \stackrel{\text{def}}{=} M &\cup \{|U| \geq \lceil \sqrt[k]{\max_{x \in \mathsf{av}(M)}(\delta_x(M)+1)} \rceil\} \\ &\cup \{|U| \geq n_1 + n_2 + 1 \mid \min_M = n_1, \max_M = |U| - n_2, n_1, n_2 \in \mathbb{N}\} \\ &\cup \{|U| < n_1 + n_2 \mid \min_M = |U| - n_1, \max_M = n_2, n_1, n_2 \in \mathbb{N}\}\end{aligned}$$

where $\delta_x(M)$ is the number of pairwise M-distinct tuples $\mathbf{y}$ for which there exists $\neg x' \hookrightarrow \mathbf{y} \in M$ such that $x \approx_M x'$. Intuitively, $\mathsf{dc}(M)$ asserts that $\min_M < \max_M$ and that the domain contains enough elements to allocate all cells. Essentially, given a structure $(\mathfrak{U}, \mathfrak{s}, \mathfrak{h})$, if $\mathfrak{h}(x)$ is known to be defined and distinct from n pairwise distinct vectors of locations $\mathbf{v}_1, \ldots, \mathbf{v}_n$, then necessarily at least $n+1$ vectors must exist. Since there are $||\mathfrak{U}||^k$ vectors of length k, we must have $||\mathfrak{U}||^k \geq n+1$, hence $||\mathfrak{U}|| \geq \sqrt[k]{n+1}$. For instance, if $M = \{\neg x \hookrightarrow y_i, \mathsf{alloc}(x), y_i \not\approx y_j \mid i, j \in [1, n], i \neq j\}$, then it is clear that M is unsatisfiable if there are less than n locations, since x cannot be allocated in this case.

Definition 5. *A minterm M is* footprint-consistent[4] *if for all $x, x' \in \mathsf{Var}$ and $\mathbf{y}, \mathbf{y}' \in \mathsf{Var}^k$, such that $x \approx_M x'$ and $y_i \approx_M y'_i$ for all $i \in [1, k]$, we have (1) if $\mathsf{alloc}(x) \in M$ then $\neg\mathsf{alloc}(x') \notin M$, and (2) if $x \hookrightarrow \mathbf{y} \in M$ then $\neg\mathsf{alloc}(x'), \neg x' \hookrightarrow \mathbf{y}' \notin M$.*

We are now ready to define a boolean combination of test formulæ that is equivalent to $M_1 * M_2$, where M_1 and M_2 are minterms satisfying a number of additional conditions. Let $\mathsf{npto}(M_1, M_2) \stackrel{\text{def}}{=} (M_1 \cap M_2) \cap \{\neg x \hookrightarrow \mathbf{y} \mid x \notin \mathsf{av}(M_1 \cup M_2), \mathbf{y} \in \mathsf{Var}^k\}$ be the set of negative points-to literals common to M_1 and M_2, involving left-hand side variables not allocated in either M_1 or M_2.

Lemma 3. *Let M_1, M_2 be two footprint-consistent minterms that are and E-complete for $\mathsf{var}(M_1 \cup M_2)$, with $\mathsf{cc}(M_1^p) = \mathsf{cc}(M_2^p)$. Then $M_1 * M_2 \equiv \mathsf{elim}_*(M_1, M_2)$, where*

$$\mathsf{elim}_*(M_1, M_2) \stackrel{\text{def}}{=} M_1^e \wedge M_2^e \wedge \mathsf{dc}(M_1)^u \wedge \mathsf{dc}(M_2)^u \wedge \tag{2}$$

$$\bigwedge_{x \in \mathsf{av}(M_1),\ y \in \mathsf{av}(M_2)} \neg x \approx y \wedge \mathsf{fp}_a(M_1) \wedge \mathsf{fp}_a(M_2) \wedge \tag{3}$$

$$\mathsf{nalloc}(\mathsf{nv}(M_1) \cap \mathsf{nv}(M_2)) \wedge \mathsf{npto}(M_1, M_2) \wedge \tag{4}$$

$$|h| \geq \min_{M_1} + \min_{M_2} \ \wedge\ |h| < \max_{M_1} + \max_{M_2} - 1 \tag{5}$$

$$\wedge\ \eta_{12} \wedge \eta_{21} \tag{6}$$

and $\eta_{ij} \stackrel{\text{def}}{=} \bigwedge_{Y \subseteq \mathsf{nv}(M_j) \setminus \mathsf{av}(M_i)} \mathsf{alloc}(Y) \rightarrow \begin{pmatrix} |h| \geq \#_a(M_i) + |Y|_{M_i} + \min_{M_j} \\ \wedge\ \#_a(M_i) + |Y|_{M_i} < \max_{M_i} \end{pmatrix}.$

[3] Note that we do not assert the equality $\mathbf{y} \approx \mathbf{y}'$, instead we only check that it is not falsified. This is sufficient for our purpose because in the following we always assume that the considered minterms are E-complete.

[4] Footprint-consistency is a necessary, yet not sufficient, condition for satisfiability of minterms. For example, the minterm $M = \{x \hookrightarrow y, x' \hookrightarrow y', \neg y \approx y', |h| < 2\}$ is at the same time footprint-consistent and unsatisfiable.

Intuitively, if M_1 and M_2 hold separately, then all heap-independent literals from $M_1 \cup M_2$ must be satisfied (2), the variables allocated in M_1 and M_2 must be pairwise distinct and their footprints, relative to the allocated variables, jointly asserted (3). Moreover, unallocated variables on both sides must not be allocated and common negative points-to literals must be asserted (4). Since the heap satisfying $\mathsf{elim}_*(M_1, M_2)$ is the disjoint union of the heaps for M_1 and M_2, its bounds are the sum of the bounds on both sides (5) and, moreover, the variables that M_2 never allocates $[\mathsf{nv}(M_2)]$ may occur allocated in the heap of M_1 and viceversa, thus the constraints η_{12} and η_{21}, respectively (6).

Next, we show a similar result for the separating implication. For technical convenience, we translate the septraction $M_1 \multimap M_2$, instead of $M_1 \mathrel{-\!\!*} M_2$, as an equivalent boolean combination of test formulæ. This is without loss of generality, because $M_1 \mathrel{-\!\!*} M_2 \equiv \neg(M_1 \multimap \neg M_2)$. Unlike with the case of the separating conjunction (Lemma 3), here the definition of the boolean combination of test formulæ depends on whether the universe is finite or infinite.

If the complement of some literal $\ell \in \mathsf{fp}_a(M_1)$ belongs to M_2 then no extension by a heap that satisfies ℓ may satisfy $\overline{\ell}$. Therefore, as an additional simplifying assumption, we suppose that $\mathsf{fp}_a(M_1) \cap \overline{M_2} = \emptyset$, so that $M_1 \multimap M_2$ is not trivially unsatisfiable. We write $\phi \equiv^{fin} \psi$ $[\phi \equiv^{inf} \psi]$ if ϕ has the same truth value as ψ in all finite [infinite] structures.

Lemma 4. *Let M_1 and M_2 be footprint-consistent minterms that are E-complete for $\mathsf{var}(M_1 \cup M_2)$, such that: M_1 is A-complete for $\mathsf{var}(M_1 \cup M_2)$, $M_2^a \cup M_2^p \subseteq \mathsf{cc}(M_1^a \cup M_1^p)$ and $\mathsf{fp}_a(M_1) \cap \overline{M_2} = \emptyset$.*

Then, $M_1 \multimap M_2 \equiv^{fin} \mathsf{elim}_{\multimap}^{fin}(M_1, M_2)$ and $M_1 \multimap M_2 \equiv^{inf} \mathsf{elim}_{\multimap}^{inf}(M_1, M_2)$, where:

$$\begin{aligned}
\mathsf{elim}_{\multimap}^{\dagger}(M_1, M_2) \stackrel{\mathsf{def}}{=}\ & \mathsf{pc}(M_1)^e \wedge M_2^e \wedge \mathsf{dc}(M_1)^u \wedge \mathsf{dc}(M_2)^u \wedge & (7)\\
& \mathsf{nalloc}(\mathsf{av}(M_1)) \wedge \mathsf{fp}_{\mathsf{nv}(M_1)}(M_2) \wedge & (8)\\
& |h| \geq \min_{M_2} - \max_{M_1} + 1 \wedge |h| < \max_{M_2} - \min_{M_1} & (9)\\
& \wedge\ \lambda^{\dagger} & (10)
\end{aligned}$$

with

$$\lambda^{fin} \stackrel{\mathsf{def}}{=} \bigwedge_{Y \subseteq \mathsf{var}(M_1 \cup M_2)} \mathsf{nalloc}(Y) \rightarrow \begin{pmatrix} |h| < |U| - \min_{M_1} - \#_n(Y, M_1) + 1 \\ \wedge\ |U| \geq \min_{M_2} + \#_n(Y, M_1) \end{pmatrix},$$

$$\lambda^{inf} \stackrel{\mathsf{def}}{=} \top.$$

A heap satisfies $M_1 \multimap M_2$ iff it has an extension, by a disjoint heap satisfying M_1, that satisfies M_2. Thus, $\mathsf{elim}_{\multimap}^{\dagger}(M_1, M_2)$ must entail the heap-independent literals of both M_1 and M_2 (7). Next, no variable allocated by M_1 must be allocated by $\mathsf{elim}_{\multimap}^{\dagger}(M_1, M_2)$, otherwise no extension by a heap satisfying M_1 is possible and, moreover, the footprint of M_2 relative to the unallocated variables of M_1 must be asserted (8). The heap's cardinality constraints depend on the bounds of M_1 and M_2 (9) and, if Y is a set of variables not allocated in the heap, these variables can be allocated in the extension (10). Actually, this is where the finite universe assumption first comes into play. If the universe is infinite, then there are enough locations outside the heap to be assigned to Y. However, if the universe is finite, then it is necessary to ensure that there are at least $\#_n(Y, M_1)$ free locations to be assigned to Y (10).

4.2 Translating Quantifier-Free SL^k into Minterms

We prove next that each quantifier-free SL^k formula is equivalent to a finite disjunction of minterms:

Lemma 5. *Given a quantifier-free* SL^k *formula* ϕ*, there exist two sets of minterms* $\mu^{fin}(\phi)$ *and* $\mu^{inf}(\phi)$ *such that the following equivalences hold: (1)* $\phi \equiv^{fin} \bigvee_{M \in \mu^{fin}(\phi)} M$*, and (2)* $\phi \equiv^{inf} \bigvee_{M \in \mu^{inf}(\phi)} M$*.*

The formal definition of $\mu^{fin}(\phi)$ and $\mu^{inf}(\phi)$ is given in [8] and omitted for the sake of conciseness and readability. Intuitively, these sets are defined by induction on the structure of the formula. For base cases, the following equivalences are used:

$$x \mapsto \mathbf{y} \equiv x \hookrightarrow \mathbf{y} \wedge |h| \approx 1 \qquad \mathsf{emp} \equiv |h| \approx 0 \qquad x \approx y \equiv x \approx y \wedge |h| \geq 0 \wedge |h| < \infty$$

For formulæ $\neg\psi_1$ or $\psi_1 \wedge \psi_2$, the transformation is first applied recursively on ψ_1 and ψ_2, then the obtained formula is transformed into dnf. For formulæ $\psi_1 * \psi_2$ or $\psi_1 \multimap \psi_2$, the transformation is applied on ψ_1 and ψ_2, then the following equivalences are used to shift $*$ and $\multimap$ innermost in the formula:

$$(\phi_1 \vee \phi_2) * \phi \equiv (\phi_1 * \phi) \vee (\phi_2 * \phi) \qquad (\phi_1 \vee \phi_2) \multimap \phi \equiv (\phi_1 \multimap \phi) \vee (\phi_2 \multimap \phi)$$
$$\phi * (\phi_1 \vee \phi_2) \equiv (\phi * \phi_1) \vee (\phi * \phi_2) \qquad \phi \multimap (\phi_1 \vee \phi_2) \equiv (\phi \multimap \phi_1) \vee (\phi \multimap \phi_2)$$

Afterwards, the operands of $*$ and $\multimap$ are minterms, and the result is obtained using the equivalences in Lemmas 3 and 4, respectively (up to a transformation into dnf). The only difficulty is that these lemmas impose some additional conditions on the minterms (e.g., being E-complete, or A-complete). However, the conditions are easy to enforce by case splitting, as illustrated by the following example:

Example 1. Consider the formula $x \mapsto x \multimap y \mapsto y$. It is easy to check that $\mu^\dagger(x \mapsto x) = \{M_1\}$, for $\dagger \in \{fin, inf\}$, where $M_1 = x \hookrightarrow x \wedge |h| \geq 1 \wedge |h| < 2$ and $\mu^\dagger(y \mapsto y) = \{M_2\}$, where $M_2 = y \hookrightarrow y \wedge |h| \geq 1 \wedge |h| < 2$. To apply Lemma 4, we need to ensure that M_1 and M_2 are E-complete, which may be done by adding either $x \approx y$ or $x \not\approx y$ to each minterm. We also have to ensure that M_1 is A-complete, thus for $z \in \{x, y\}$, we add either $\mathsf{alloc}(z)$ or $\neg\mathsf{alloc}(z)$ to M_1. Finally, we must have $M_2^a \cup M_2^p \subseteq \mathsf{cc}(M_1^a \cup M_1^p)$, thus we add either $y \hookrightarrow y$ or $\neg y \hookrightarrow y$ to M_1. After removing redundancies, we get (among others) the minterms: $M_1' = x \hookrightarrow x \wedge |h| \geq 1 \wedge |h| < 2 \wedge x \approx y$ and $M_2' = y \hookrightarrow y \wedge |h| \geq 1 \wedge |h| < 2 \wedge x \approx y$. Afterwards we compute $\mathsf{elim}^{fin}_{\multimap}(M_1', M_2') = x \approx y \wedge \neg\mathsf{alloc}(x) \wedge |h| \geq 0 \wedge |h| < 1$. ■

As explained in Sect. 3.1, boolean combinations of minterms can only be transformed into sat-equivalent $\mathsf{BSR}(\mathsf{FO})$ formulæ if there is no positive occurrence of test formulæ $|h| \geq |U| - n$ or $\mathsf{alloc}(x)$ (see the conditions in Lemmas 1 (2) and 2). Consequently, we relate the polarity of these formulæ in some minterm $M \in \mu^{fin}(\phi) \cup \mu^{inf}(\phi)$ with that of a separating implication within ϕ. The analysis depends on whether the universe is finite or infinite.

Lemma 6. *For any quantifier-free* SL^k *formula* ϕ*, the following properties hold:*

1. *For all* $M \in \mu^{inf}(\phi)$*, we have* $M \cap \{|h| \geq |U| - n, |h| < |U| - n \mid n \in \mathbb{N}\} = \emptyset$.
2. *If* $|h| \geq |U| - n \in M$ *[*$|h| < |U| - n \in M$*] for some minterm* $M \in \mu^{fin}(\phi)$*, then a formula* $\psi_1 \multimap \psi_2$ *occurs at a positive [negative] polarity in* ϕ.
3. *If* $\mathsf{alloc}(x) \in M$ *[*$\neg\mathsf{alloc}(x) \in M$*] for some minterm* $M \in \mu^{inf}(\phi)$*, then a formula* $\psi_1 \multimap \psi_2$*, such that* $x \in \mathsf{var}(\psi_1) \cup \mathsf{var}(\psi_2)$*, occurs at a positive [negative] polarity in* ϕ.
4. *If* $M \cap \{\mathsf{alloc}(x), \neg\mathsf{alloc}(x) \mid x \in \mathsf{Var}\} \neq \emptyset$ *for some minterm* $M \in \mu^{fin}(\phi)$*, then a formula* $\psi_1 \multimap \psi_2$*, such that* $x \in \mathsf{var}(\psi_1) \cup \mathsf{var}(\psi_2)$*, occurs in* ϕ *at some polarity* $p \in \{-1, 1\}$*. Moreover,* $\mathsf{alloc}(x)$ *occurs at a polarity* $-p$*, only if* $\mathsf{alloc}(x)$ *is in the scope of a* λ^{fin} *subformula (10) of a formula* $\mathsf{elim}^{fin}_{\multimap}(M_1, M_2)$ *used to compute* $\bigvee_{M \in \mu^{fin}(\phi)} M$.

Given a quantifier-free SL^k formula ϕ, the number of minterms occurring in $\mu^{fin}(\phi)$ [$\mu^{inf}(\phi)$] is exponential in the size of ϕ, in the worst case. Therefore, an optimal decision procedure cannot generate and store these sets explicitly, but rather must enumerate minterms lazily. We show that (i) the size of the minterms in $\mu^{fin}(\phi) \cup \mu^{inf}(\phi)$ is bounded by a polynomial in the size of ϕ, and that (ii) the problem "*given a minterm M, does M occur in* $\mu^{fin}(\phi)$ *[resp. in* $\mu^{inf}(\phi)$*]?*" is in PSPACE. To this aim, we define a measure on a quantifier-free formula ϕ, which bounds the size of the minterms in the sets $\mu^{fin}(\phi)$ and $\mu^{inf}(\phi)$, inductively on the structure of the formulæ:

$$
\begin{aligned}
\mathcal{M}(x \approx y) &\stackrel{\text{def}}{=} 0 & \mathcal{M}(\bot) &\stackrel{\text{def}}{=} 0 \\
\mathcal{M}(\mathsf{emp}) &\stackrel{\text{def}}{=} 1 & \mathcal{M}(x \mapsto \mathbf{y}) &\stackrel{\text{def}}{=} 2 \\
\mathcal{M}(\neg\phi_1) &\stackrel{\text{def}}{=} \mathcal{M}(\phi_1) & \mathcal{M}(\phi_1 \wedge \phi_2) &\stackrel{\text{def}}{=} \max(\mathcal{M}(\phi_1), \mathcal{M}(\phi_2)) \\
\mathcal{M}(\phi_1 * \phi_2) &\stackrel{\text{def}}{=} \textstyle\sum_{i=1}^{2}(\mathcal{M}(\phi_i) + ||\mathsf{var}(\phi_i)||) & \mathcal{M}(\phi_1 \multimap \phi_2) &\stackrel{\text{def}}{=} \textstyle\sum_{i=1}^{2}(\mathcal{M}(\phi_i) + ||\mathsf{var}(\phi_i)||)
\end{aligned}
$$

Definition 6. *A minterm M is* $\mathcal{M}$-bounded *by a formula* ϕ*, if for each literal* $\ell \in M$*, the following hold: (i)* $\mathcal{M}(\ell) \leq \mathcal{M}(\phi)$ *if* $\ell \in \{|h| \geq \min_{M_i}, |h| < \max_{M_i}\}$ *(ii)* $\mathcal{M}(\ell) \leq 2\mathcal{M}(\phi) + 1$*, if* $\ell \in \{|U| \geq n, |U| < n \mid n \in \mathbb{N}\}$.

The following lemma provides the desired result:

Lemma 7. *Given a quantifier-free* SL^k *formula* ϕ*, each minterm* $M \in \mu^{fin}(\phi) \cup \mu^{inf}(\phi)$ *is* $\mathcal{M}$*-bounded by* ϕ.

The proof goes by a careful analysis of the test formulæ introduced in Lemmas 3 and 4 or created by minterm transformations (see [8] for details). Since $\mathcal{M}(\phi)$ is polynomially bounded by $\mathsf{size}(\phi)$, this entails that it is possible to check whether $M \in \mu^{fin}(\phi)$ [resp. $\mu^{inf}(\phi)$] using space bounded also by a polynomial in $\mathsf{size}(\phi)$.

Lemma 8. *Given a minterm M and an* SL^k *formula* ϕ*, the problems of checking whether* $M \in \mu^{fin}(\phi)$ *and* $M \in \mu^{inf}(\phi)$ *are in* PSPACE.

Remark 1. Observe that the formulæ $\mathsf{elim}_*(M_1, M_2)$ and $\mathsf{elim}^{fin}_{-\circ}(M_1, M_2)$ in Lemmas 3 and 4 are of exponential size, because Y ranges over sets of variables. However these formulæ do not need to be constructed explicitly. To check that $M \in \mu^{fin}(\phi)$ or $M \in \mu^{inf}(\phi)$, we only have to guess such sets Y. See [8] for details.

5 Bernays-Schönfinkel-Ramsey SL^k

This section gives the results concerning decidability of the (in)finite satisfiability problems within the $\mathsf{BSR}(\mathsf{SL}^k)$ fragment. $\mathsf{BSR}(\mathsf{SL}^k)$ is the set of sentences $\forall y_1 \ldots \forall y_m \,.\, \phi$, where ϕ is a quantifier-free SL^k formula, with $\mathsf{var}(\phi) = \{x_1, \ldots, x_n, y_1, \ldots, y_m\}$, where the existentially quantified variables $x_1, \ldots, x_n$ are left free. First, we show that, contrary to $\mathsf{BSR}(\mathsf{FO})$, the satisfiability of $\mathsf{BSR}(\mathsf{SL}^k)$ is undecidable for $k \geq 2$. Second, we carve two nontrivial fragments of $\mathsf{BSR}(\mathsf{SL}^k)$, for which the infinite and finite satisfiability problems are both PSPACE-complete. These fragments are defined based on restrictions of (i) polarities of the occurrences of the separating implication, and (ii) occurrences of universally quantified variables in the scope of separating implications. These results draw a rather precise chart of decidability within the $\mathsf{BSR}(\mathsf{SL}^k)$ fragment. For $k = 1$, the satisfiability problem of $\mathsf{BSR}(\mathsf{SL}^1)$ is in PSPACE [7] (it is undecidable for arbitrary SL^1 formulæ [2] and decidable but nonelementary for *prenex* formulæ [7]).

5.1 Undecidability of $\mathsf{BSR}(\mathsf{SL}^k)$

Theorem 1. *The finite and infinite satisfiability problems are both undecidable for* $\mathsf{BSR}(\mathsf{SL}^k)$.

We provide a brief sketch of the proof, see [8] for details. We consider the finite satisfiability problem of the $[\forall, (0), (2)]_=$ fragment of FO, which consists of sentences of the form $\forall x \,.\, \phi(x)$, where ϕ is a quantifier-free boolean combination of atomic propositions $t_1 \approx t_2$, and t_1, t_2 are terms built using two function symbols f and g, of arity one, the variable x and constant c. It is known (see e.g. [1, Theorem 4.1.8]) that finite satisfiability is undecidable for $[\forall, (0), (2)]_=$. We reduce this problem to $\mathsf{BSR}(\mathsf{SL}^k)$ satisfiability. The idea is to encode the value of f and g into the heap, in such a way that every element x points to $(f(x), g(x))$. Given a sentence $\varphi = \forall x \,.\, \phi(x)$ in $[\forall, (0), (2)]_=$, we proceed by first *flattening* each term in ϕ consisting of nested applications of f and g. The result is an equivalent sentence $\varphi_{flat} = \forall x_1 \ldots \forall x_n \,.\, \phi_{flat}$, in which the only terms are x_i, c, $f(x_i)$, $g(x_i)$, $f(c)$ and $g(c)$, for $i \in [1, n]$. For example, the formula $\forall x \,.\, f(g(x)) \approx c$ is flattened into $\forall x_1 \forall x_2 \,.\, g(x_1) \not\approx x_2 \vee f(x_2) \approx c$. We define the following $\mathsf{BSR}(\mathsf{SL}^2)$ sentences $\varphi^\dagger_{\mathsf{sl}}$, for $\dagger \in \{fin, inf\}$:

$$\alpha^\dagger \wedge x_c \hookrightarrow (y_c, z_c) \wedge \forall x_1 \ldots \forall x_n \forall y_1 \ldots \forall y_n \forall z_1 \ldots \forall z_n \,.\, \bigwedge_{i=1}^{n} (x_i \hookrightarrow (y_i, z_i) \rightarrow \phi_{\mathsf{sl}}) \tag{11}$$

with $\alpha^{fin} \stackrel{\mathsf{def}}{=} \forall x \,.\, \mathsf{alloc}(x)$ or $\alpha^{fin} \stackrel{\mathsf{def}}{=} |h| \geq |U| - 0$, $\alpha^{inf} \stackrel{\mathsf{def}}{=} \forall x \forall y \forall z \,.\, x \hookrightarrow (y, z) \rightarrow \mathsf{alloc}(y) \wedge \mathsf{alloc}(z)$ and ϕ_{sl} is obtained from ϕ_{flat} by replacing each occurrence of c by x_c, each term $f(c)$ $[g(c)]$ by y_c $[z_c]$ and each term $f(x_i)$ $[g(x_i)]$ by y_i $[z_i]$. Intuitively, α^{fin} asserts that the heap is a total function, and α^{inf} states that every referenced cell is allocated[5]. It is easy to check that φ and φ_{sl} are equisatisfiable. The undecidability result still holds for finite satisfiability if a single occurrence of $-\!\!*$ is allowed, in a (ground) formula $|h| \geq |U| - 0$ (see the definition of α^{fin} above).

5.2 Two Decidable Fragments of $\mathsf{BSR}(\mathsf{SL}^k)$

The reductions (11) use either positive occurences of $\mathsf{alloc}(x)$, where x is universally quantified, or test formulæ $|h| \geq |U| - n$. We obtain decidable subsets of $\mathsf{BSR}(\mathsf{SL}^k)$ by eliminating the positive occurrences of both (i) $\mathsf{alloc}(x)$, with x universally quantified, and (ii) $|h| \geq |U| - n$, from $\mu^\dagger(\phi)$, where $\dagger \in \{fin, inf\}$ and $\forall y_1 \ldots \forall y_m . \phi$ is any $\mathsf{BSR}(\mathsf{SL}^k)$ formula. Note that $\mu^{inf}(\phi)$ contains no formulæ of the form $|h| \geq |U| - n$, which explains why slightly less restrictive conditions are needed for infinite structures.

Definition 7. *Given an integer $k \geq 1$, we define:*

1. $\mathsf{BSR}^{inf}(\mathsf{SL}^k)$ *as the set of sentences $\forall y_1 \ldots \forall y_m . \phi$ such that for all $i \in [1, m]$ and all formulæ $\psi_1 -\!\!* \psi_2$ occurring at polarity 1 in ϕ, we have $y_i \notin \mathsf{var}(\psi_1) \cup \mathsf{var}(\psi_2)$,*
2. $\mathsf{BSR}^{fin}(\mathsf{SL}^k)$ *as the set of sentences $\forall y_1 \ldots \forall y_m . \phi$ such that no formula $\psi_1 -\!\!* \psi_2$ occurs at polarity 1 in ϕ.*

Note that $\mathsf{BSR}^{fin}(\mathsf{SL}^k) \subsetneq \mathsf{BSR}^{inf}(\mathsf{SL}^k) \subsetneq \mathsf{BSR}(\mathsf{SL}^k)$, for any $k \geq 1$.

Remark 2. Because the polarity of the antecedent of a $-\!\!*$ is neutral, Definition 7 imposes no constraint on the occurrences of separating implications at the *left* of a $-\!\!*$[6].

The decidability result of this paper is stated below:

Theorem 2. *For any integer $k \geq 1$ not depending on the input, the infinite satisfiability problem for $\mathsf{BSR}^{inf}(\mathsf{SL}^k)$ and the finite satisfiability problem for $\mathsf{BSR}^{fin}(\mathsf{SL}^k)$ are both PSPACE-complete.*

We provide a brief sketch of the proof (all details are available in [8]). In both cases, PSPACE-hardness is an immediate consequence of the fact that the quantifier-free fragment of SL^k, without the separating implication, but with the separating conjunction and negation, is PSPACE-hard [4]. For PSPACE-membership, consider a formula φ in $\mathsf{BSR}^{inf}(\mathsf{SL}^k)$, and its equivalent disjunction of minterms φ' (of exponential size). Lemma 8 gives us an upper bound on the size of test

[5] Note that the two definitions of α^{fin} are equivalent. The formula α^{fin} is unsatisfiable on infinite universes, which explains why the definitions of α^{fin} and α^{inf} differ.

[6] The idea is that if a formula $\mathsf{alloc}(x)$ or $|h| \geq |U| - n$ occurs in the antecedent of a $-\!\!*$, then it will be eliminated by the transformation in Lemma 4. In contrast, such test formulæ will not be eliminated if they occur in the subsequent of a $-\!\!*$.

formulæ in φ', hence on the number of constant symbols occurring in $\tau(\varphi')$. This, in turns, gives a bound on the cardinality of the model of $\tau(\varphi')$. We may thus guess such an interpretation, and check that it is indeed a model of $\tau(\varphi')$ by enumerating all the minterms in φ' (this is feasible in polynomial space thanks to Lemma 8) and translating them on-the-fly into first-order formulæ. The only subtle point is that the model obtained in this way is finite, whereas our aim is to test that the obtained formula has a *infinite* model. This difficulty can be overcome by adding an axiom ensuring that the domain contains more *unallocated* elements than the total number of constant symbols and variables in the formula. This is sufficient to prove that the obtained model – although finite – can be extended into an infinite model, obtained by creating infinitely many copies of these elements.

The proof for $\mathsf{BSR}^{fin}(\mathsf{SL}^k)$ is similar, but far more involved. The problem is that, if the universe is finite, then $\mathsf{alloc}(x)$ test formulæ may occur at a positive polarity, even if every $\phi_1 \multimap\!\!* \phi_2$ subformula occurs at a negative polarity, due to the positive occurrences of $\mathsf{alloc}(x)$ within λ^{fin} (10) in the definition of $\mathsf{elim}^{fin}_{\multimap}(M_1, M_2)$. As previously discussed, positive occurrences of $\mathsf{alloc}(x)$ hinder the translation into $\mathsf{BSR(FO)}$, because of the existential quantifiers that may occur in the scope of a universal quantifier. The solution is to distinguish a class of finite structures $(\mathfrak{U}, \mathfrak{s}, \mathfrak{h})$, the so-called *$\alpha$-controlled structures*, for some $\alpha \in \mathbb{N}$, for which there are locations $\ell_1, \ldots, \ell_\alpha$, such that every location $\ell \in \mathfrak{U}$ is either ℓ_i or points to a tuple from the set $\{\ell_1, \ldots, \ell_\alpha, \ell\}$. For such structures, the formulæ $\mathsf{alloc}(x)$ can be eliminated in a straightforward way because they are equivalent to $\bigwedge_{i=1}^{\alpha}(x \approx \ell_i \rightarrow \mathsf{alloc}(\ell_i))$. If the structure is not α-controlled, then we can show that there exist sufficiently many unallocated cells, so that all the cardinality constraints of the form $|h| \leq |U| - n$ or $|U| \geq n$ are always satisfied. This ensures that the truth value of the positive occurrences of $\mathsf{alloc}(x)$ are irrelevant, because they only occur in formulæ λ^{fin} that are always true if all test formulæ $|h| \leq |U| - n$ or $|U| \geq n$ are true (see the definition of λ^{fin} in Lemma 4).

6 Conclusions and Future Work

We have studied the decidability problem for SL formulæ with quantifier prefix in the language $\exists^*\forall^*$, denoted as $\mathsf{BSR}(\mathsf{SL}^k)$. Although the fragment was found to be undecidable, we identified two non-trivial subfragments for which the infinite and finite satisfiability are PSPACE-complete. These fragments are defined by restricting the use of universally quantified variables within the scope of separating implications at positive polarity. The universal quantifiers and separating conjunctions are useful to express local constraints on the shape of the data-structure, whereas the separating implications allow one to express dynamic transformations of these data-structures. As a consequence, separating implications usually occur negatively in the formulæ tested for satisfiability, and the decidable classes found in this work are of great practical interest. Future work involves formalizing and implementing an invariant checking algorithm based on the above ideas, and using the techniques for proving decidability (namely the translation of quantifier-free $\mathsf{SL}(k)$ formulæ into boolean combinations of test formulæ) to solve other logical problems, such as frame inference, abduction and possibly interpolation.

Acknowledgments. The authors wish to acknowledge the contributions of Stéphane Demri and Étienne Lozes to the insightful discussions during the early stages of this work.

References

1. Börger, E., Grädel, E., Gurevich, Y.: The Classical Decision Problem. Perspectives in Mathematical Logic. Springer, Heidelberg (1997)
2. Brochenin, R., Demri, S., Lozes, E.: On the almighty wand. Inf. Comput. **211**, 106–137 (2012)
3. Calcagno, C., Gardner, P., Hague, M.: From separation logic to first-order logic. In: Sassone, V. (ed.) FoSSaCS 2005. LNCS, vol. 3441, pp. 395–409. Springer, Berlin, Heidelberg (2005). https://doi.org/10.1007/978-3-540-31982-5_25
4. Calcagno, C., Yang, H., O'Hearn, P.W.: Computability and complexity results for a spatial assertion language for data structures. In: Hariharan, R., Vinay, V., Mukund, M. (eds.) FSTTCS 2001. LNCS, vol. 2245, pp. 108–119. Springer, Berlin, Heidelberg (2001). https://doi.org/10.1007/3-540-45294-X_10
5. Demri, S., Deters, M.: Expressive completeness of separation logic with two variables and no separating conjunction. In: Henzinger, T.A., Miller, D. (eds.), Joint Meeting of the Twenty-Third EACSL Annual Conference on Computer Science Logic (CSL) and the Twenty-Ninth Annual ACM/IEEE Symposium on Logic in Computer Science (LICS), CSL-LICS 2014, Vienna, Austria, 14–18 July 2014, pp. 37:1–37:10. ACM (2014)
6. Demri, S., Galmiche, D., Larchey-Wendling, D., Méry, D.: Separation logic with one quantified variable. Theory Comput. Syst. **61**(2), 371–461 (2017)
7. Echenim, M., Iosif, R., Peltier, N.: The complexity of prenex separation logic with one selector. CoRR, abs/1804.03556 (2018)
8. Echenim, M., Iosif, R., Peltier, N.: On the expressive completeness of Bernays-Schoenfinkel-Ramsey separation logic. ArXiv e-prints (2018)
9. Iosif, R., Rogalewicz, A., Simacek, J.: The tree width of separation logic with recursive definitions. In: Bonacina, M.P. (ed.) CADE 2013. LNCS (LNAI), vol. 7898, pp. 21–38. Springer, Heidelberg (2013). https://doi.org/10.1007/978-3-642-38574-2_2
10. Ishtiaq, S.S., O'Hearn, P.W.: BI as an assertion language for mutable data structures. ACM SIGPLAN Not. **36**, 14–26 (2001)
11. Lozes, É.: Expressivité des logiques spatiales. Thèse de doctorat, Laboratoire de l'Informatique du Parallélisme, ENS Lyon, France, November 2004
12. Ramsey, F.P.: On a problem of formal logic. In: Classic Papers in Combinatorics, pp. 1–24 (1987)

13. Reynolds, A., Iosif, R., Serban, C.: Reasoning in the Bernays-Schönfinkel-Ramsey fragment of separation logic. In: Bouajjani, A., Monniaux, D. (eds.) VMCAI 2017. LNCS, vol. 10145, pp. 462–482. Springer, Cham (2017). https://doi.org/10.1007/978-3-319-52234-0_25
14. Reynolds, J.C.: Separation logic: a logic for shared mutable data structures. In: Proceedings of the 17th Annual IEEE Symposium on Logic in Computer Science, LICS 2002, pp. 55–74. IEEE Computer Society (2002)

Continuous Reachability for Unordered Data Petri Nets is in PTime

Utkarsh Gupta[1], Preey Shah[1], S. Akshay[1(✉)], and Piotr Hofman[2]

[1] Department of CSE, IIT Bombay, Mumbai, India
akshayss@cse.iitb.ac.in
[2] University of Warsaw, Warsaw, Poland

Abstract. Unordered data Petri nets (UDPN) are an extension of classical Petri nets with tokens that carry data from an infinite domain and where transitions may check equality and disequality of tokens. UDPN are well-structured, so the coverability and termination problems are decidable, but with higher complexity than for Petri nets. On the other hand, the problem of reachability for UDPN is surprisingly complex, and its decidability status remains open. In this paper, we consider the continuous reachability problem for UDPN, which can be seen as an over-approximation of the reachability problem. Our main result is a characterization of continuous reachability for UDPN and polynomial time algorithm for solving it. This is a consequence of a combinatorial argument, which shows that if continuous reachability holds then there exists a run using only polynomially many data values.

Keywords: Petri nets · Continuous reachability · Unordered data · Polynomial time

1 Introduction

The theory of Petri nets has been developing since more than 50 years. On one hand, from a theory perspective, Petri nets are interesting due to their deep mathematical structure and despite exhibiting nice properties, like being a well structured transition system [1], we still don't understand them well. On the other hand, Petri nets are a useful pictorial formalism for modeling and thus found their way to the industry. To connect this theory and practice, it would be desirable to use the developed theory of Petri nets [2–4] for the symbolic analysis and verification of Petri nets models. However, we already know that this is difficult in its full generality. It suffices to recall two results that were proved more than 30 years apart. An old but classical result by Lipton [5] shows that even coverability is ExpSpace-hard, while the non-elementary hardness of the reachability relation has just been

Supported by Polish NCN grant UMO-2016/21/D/ST6/01368, DST Inspire faculty award IFA12-MA-17, and DST/CEFIPRA project EQuaVe.
U. Gupta and P. Shah—Contributed equally to this work.

M. Bojańczyk and A. Simpson (Eds.): FOSSACS 2019, LNCS 11425, pp. 260–276, 2019.
https://doi.org/10.1007/978-3-030-17127-8_15

established this year [6]. Moreover, when we look at Petri nets based formalisms that are needed to model various aspects of industrial systems, we see that they go beyond the expressivity of Petri nets. For instance, colored Petri nets, which are used in modeling workflows [7], allow the tokens to be colored with an infinite set of colors, and introduce a complex formalism to describe dependencies between colors. This makes all verification problems undecidable for this generic model. Given the basic nature and importance of the reachability problem in Petri nets (and its extensions), there have been several efforts to sidestep the complexity-theoretic hardness results. One common approach is to look for easy subclasses (such as bounded nets [8], free-choice nets [9] etc.). The other approach, which we adopt in this work, is to compute over-approximations of the reachability relation.

Continuous Reachability. A natural question regarding the dynamics of a Petri net is to ask what would happen if tokens instead of behaving like discrete units start to behave like a continuous fluid? This simple question led to an elegant theory of so-called continuous Petri nets [10–12]. Petri nets with continuous semantics allow markings to be functions from places to *nonnegative rational numbers* (i.e., in $\mathbb{Q}^+$) instead of natural numbers. Moreover, whenever a transition is fired a positive rational coefficient is chosen and both the number of consumed and produced tokens are multiplied with the coefficient. This allows to split tokens into arbitrarily small parts and process them independently. This may occur, e.g., in applications related to hybrid systems where the discrete part is used to control the continuous system [13,14]. Interestingly, this makes things simpler to analyze. For example reachability under the continuous semantics for Petri nets is *PTime*-complete [11]. However, when one wants to analyze extensions of Petri nets, e.g., reset Petri nets with continuous semantics, it turns out that reachability is as hard as reachability in reset Petri nets under the usual semantics i.e. it is undecidable[1]. In this paper we identify an extension of Petri nets with unordered data, for which this is not the case and continuous semantics leads to a substantial reduction in the complexity of the reachability problem.

Unordered Data Petri Nets. The possibility of equipping tokens with some additional information is one of the main lines of research regarding extensions of Petri nets, the best known being Colored Petri nets [15] and various types of timed Petri nets [16,17]. In [18] authors equipped tokens with data and restricted interactions between data in a way that allow to transfer techniques for well structured transition systems. They identified various classes of nets exhibiting interesting combinatorial properties which led to a number of results [19–23]. Unordered Data Petri Nets (UDPN), are simplest among them: every token carries a single datum like a barcode and transitions may check equality or disequality of data in consumed and produced tokens. UDPN are the only class identified in [18] for which the reachability is still unsolved, although in [20] authors show that the problem is at least Ackermannian-hard (for all other data extensions, reachability is undecidable). A recent attempt to over-approximate the reachability relation for UDPN in [22]

[1] This can be seen on the same lines as the proof of undecidability of continuous reachability for Petri nets with zero tests [12].

considers integer reachability i.e. number of tokens may get negative during the run (also called solution of the state equation). From the above perspective, this paper is an extension of the mentioned line of research.

Our Contribution. Our main contribution is a characterization of continuous reachability in UDPN and a polynomial time algorithm for solving it. Observe that if we find an upper bound on the minimal number of data required by a run between two configurations (if any run exists), then we can reduce continuous reachability in UDPN to continuous reachability in vanilla Petri nets with an exponential blowup and use the already developed characterization from [11]. In Sect. 5 we prove such a bound on the minimal number of required data. The bound is novel and exploits techniques that did not appear previously in the context of data nets. Further, the obtained bounds are lower than bounds on the number of data values required to solve the state equation [22], which is surprising considering that existence of a continuous run requires a solution of a sort of state equation. Precisely, the difference is that we are looking for solutions of the state equation over $\mathbb{Q}^+$ instead of $\mathbb{N}$ and in this case we prove better bounds for the number of data required. This also gives us an easy polytime algorithm for finding $\mathbb{Q}^+$-solutions of state equations of UDPN (we remark that for Petri nets without data, this appears among standard algebraic techniques [24]).

Finally, with the above bound, we solve continuous reachability in UDPN by adapting the techniques from the non-data setting of [12,25]. We adapt the characterization of continuous reachability to the data setting and next encode it as system of linear equations with implications. In doing so, however, we face the problem that a naive encoding (representing data explicitly) gives a system of equations of exponential size, giving only an ExpTime-algorithm. To improve the complexity, we use histograms, a combinatorial tool developed in [22], to compress the description of solutions of state equations in UDPNs. However, this may lead to spurious solutions for continuous reachability. To eliminate them, we show that it suffices to first transform the net and then apply the idea of histograms to characterize continuous runs in the modified net. The whole procedure is described in Sect. 7.3 and leads us to our *PTime* algorithm for continuous reachability in UDPN. Note that since we easily have *PTime* hardness for the problem (even without data), we obtain that the problem of continuous reachability in UDPN is *PTime*-complete.

Towards Verification. Over-approximations are useful in verification of Petri nets and their extensions: as explained in [24], for many practical problems, over-approximate solutions are already correct. Further, we can use them as a sub-routine to improve the practical performance of verification algorithms. A remarkable example is the recent work in [25], where the *PTime* continuous reachability algorithm for Petri nets from [11] is used as a subroutine to solve the *ExpSpace* hard coverability problem in Petri nets, outperforming the best known tools for this problem, such as Petrinizer [26]. Our results can be seen as a first step in the same spirit towards handling practical instances of coverability, but for the extended model of UDPN, where the coverability problem for UDPN is known to be Ackermannian-hard [20].

Omitted proofs and details can be found in the extended version at [27].

2 Preliminaries

We denote integers, non-negative integers, rationals, and reals as $\mathbb{Z}, \mathbb{N}, \mathbb{Q}$, and $\mathbb{R}$, respectively. For a set $\mathbb{X} \subseteq \mathbb{R}$ denote by $\mathbb{X}^+$, the set of all non-negative elements of $\mathbb{X}$. We denote by $\mathbf{0}$, a vector whose entries are all zero. We define in a standard point-wise way operations on vectors i.e. *scalar multiplication* $\cdot$, *addition* $+$, *subtraction* $-$, *and vector comparison* $\leq$. In this paper, we use functions of the type $X \to (Y \to Z)$, and instead of $(f(x))(y)$, we write $f(y, x)$. For functions f, g where the range of g is a subset of the domain of f, we denote their composition by $f \circ g$. If π is an injection then by π^{-1} we mean a partial function such that $\pi^{-1} \circ \pi$ is the identity function. Let $f : X_1 \to Y$, $g : X_2 \to Y$ be two functions with addition and scalar multiplication operations defined on Y. A *scalar multiplication* of a function is defined as follows $(a \cdot f)(x) = a \cdot f(x)$ for all $x \in X_1$. We lift *addition* operation to functions pointwise, i.e. $f + g : X_1 \cup X_2 \to Y$ such that

$$(f+g)(x) = \begin{cases} f(x) & \text{if } x \in X_1 \setminus X_2 \\ g(x) & \text{if } x \in X_2 \setminus X_1 \\ f(x) + g(x) & \text{if } x \in X_1 \cap X_2. \end{cases}$$

Similarly for *subtraction* $(f - g)(x) = f(x) + -1 \cdot g(x)$, and $f \leq g$ if for all $x \in X_1 \cup X_2, (g - f)(x) \leq 0$.

We use *matrices* with *rows and columns* indexed by sets $\mathbb{S}_1, \mathbb{S}_2$, possibly infinite. For a matrix M, let $M(r, c)$ denote the entry at column c and row r, and $M(r, \bullet)$, $M(\bullet, c)$ denote the row vector indexed by r and column vector indexed by c, respectively. Denote by $col(M)$, $row(M)$ the set of indices of nonzero columns and nonzero rows of the matrix M, respectively. Even if we have infinitely many rows or columns, our matrices will have only finitely many *nonzero* rows and columns, and only this nonzero part will be represented. Following our nonstandard matrix definition we precisely define operations on them, although they are natural. First, a *multiplication by a constant number* produces a new matrix with row and columns labelled with the same sets $\mathbb{S}_1, \mathbb{S}_2$ and defined as follows $(a \cdot M)(r, c) = a \cdot (M(r, c))$ for all $(r, c) \in \mathbb{S}_1 \times \mathbb{S}_2$. *Addition* of two matrices is only defined if the sets indexing rows $\mathbb{S}_1$ and columns $\mathbb{S}_2$ are the same for both summands M_1 and M_2, $\forall (r, c) \in \mathbb{S}_1 \times \mathbb{S}_2$ the sum $(M_1 + M_2)(r, c) = M_1(r, c) + M_2(r, c)$, the *subtraction* $M_1 - M_2$ is a shorthand for $M_1 + (-1) \cdot M_2$. Observe that all but finitely many entries in matrices are 0, and therefore when we do computation on matrices we can restrict to rows $row(M_1) \cup row(M_2)$ and columns $col(M_1) \cup col(M_2)$. Similarly the *comparison* for two matrices M_1, M_2 is defined as follows $M_1 \leq M_2$ if $\forall (r, c) \in (row(M_1) \cup row(M_2)) \times (col(M_1) \cup col(M_2))$ $M_1(r, c) \leq M_2(r, c)$; relations $>, \geq, \leq$ are defined analogically. The last operation which we need is matrix multiplication $M_1 \cdot M_2 = M_3$, it is only allowed if the set of columns of the first matrix M_1 is the same as the set of rows of the second matrix M_2, the sets of rows and columns of the resulting matrix M_3 are rows of the matrix M_1 and columns of M_2, respectively. $M_3(r, c) = \sum_k M_1(r, k) M_2(k, c)$

where k runs through columns of M_1. Again, observe that if the row or a column is equal to 0 for all entries then the effect of multiplication is 0, thus we may restrict to $row(M_1)$ and $col(M_2)$. Moreover in the sum it suffices to write $\sum_{k \in col(M_1)} M_1(r,k)M_2(k,c)$.

3 UDPN, Reachability and Its Variants: Our Main Results

Unordered data Petri nets extend the classical model of Petri nets by allowing each token to hold a data value from a countably-infinite domain $\mathbb{D}$. Our definition is closest to the definition of ν-Petri nets from [28]. For simplicity we choose this one instead of using the equivalent but complex one from [18].

Definition 1. *Let $\mathbb{D}$ be a countably infinite set. An unordered data Petri net (UDPN) over domain $\mathbb{D}$ is a tuple (P, T, F, Var) where P is a finite set of places, T is a finite set of transitions, Var is a finite set of variables, and $F : (P \times T) \cup (T \times P) \rightarrow (Var \rightarrow \mathbb{N})$ is a flow function that assigns each place $p \in P$ and transition $t \in T$ a function over variables in Var.*

For each transition $t \in T$ we define functions $F(\bullet, t)$ and $F(t, \bullet)$, $Var \rightarrow (P \rightarrow \mathbb{N})$ as $F(\bullet, t)(p, x) = F(p, t)(x)$ and analogously $F(t, \bullet)(p, x) = F(t, p)(x)$. *Displacement* of the transition t is a function $\Delta(t) : Var \rightarrow (P \rightarrow \mathbb{Z})$ defined as $\Delta(t) \stackrel{\text{def}}{=} F(t, \bullet) - F(\bullet, t)$.

For $\mathbb{X} \in \{\mathbb{N}, \mathbb{Z}, \mathbb{Q}, \mathbb{Q}^+\}$, we define an $\mathbb{X}$-*marking* as a function $M : \mathbb{D} \rightarrow (P \rightarrow \mathbb{X})$ that is constant 0 on all except finitely many values of $\mathbb{D}$. Intuitively, $M(p, \alpha)$ denotes the number of *tokens* with the data value α at place p. The fact that it is 0 at all but finitely many data means that the number of tokens in any $\mathbb{X}$-marking is finite. We denote the infinite set of all $\mathbb{X}$-markings by $\mathcal{M}_{\mathbb{X}}$.

We define an $\mathbb{X}$-*step* as a triple (c, t, π) for a transition $t \in T$, *mode* π being an injective map $\pi : Var \rightarrow \mathbb{D}$, and a scalar constant $c \in \mathbb{X}^+$. An $\mathbb{X}$-step (c, t, π) is *fireable* at a $\mathbb{X}$-marking $\boldsymbol{i}$ if $\boldsymbol{i} - c \cdot F(\bullet, t) \circ \pi^{-1} \in \mathcal{M}_{\mathbb{X}}$.

The $\mathbb{X}$-marking $\boldsymbol{f}$ reached after *firing* an $\mathbb{X}$-step (c, t, π) at $\boldsymbol{i}$ is given as $\boldsymbol{f} = \boldsymbol{i} + c \cdot \Delta(t) \circ \pi^{-1}$. We also say that an $\mathbb{X}$-step (c, t, π) when fired *consumes* tokens $c \cdot F(\bullet, t) \circ \pi^{-1}$ and *produces* tokens $c \cdot F(t, \bullet) \circ \pi^{-1}$. We define an $\mathbb{X}$-run as a sequence of $\mathbb{X}$-steps and we can represent it as $\{(c_i, t_i, \pi_i)\}_{|\rho|}$ where (c_i, t_i, π_i) is the ith $\mathbb{X}$-step and $|\rho|$ is the number of $\mathbb{X}$-steps. A run $\rho = \{(c_i, t_i, \pi_i)\}_{|\rho|}$ is fireable at a $\mathbb{X}$-marking $\boldsymbol{i}$ if, $\forall 1 \leq i \leq |\rho|$, the step (c_i, t_i, π_i) is fireable at $\boldsymbol{i} + \sum_{j=1}^{i-1} c_j \Delta(t_j) \circ \pi_j^{-1}$. By $\boldsymbol{i} \xrightarrow{\rho}_{\mathbb{X}} \boldsymbol{f}$ we denote that ρ is fireable at $\boldsymbol{i}$ and after firing ρ at $\boldsymbol{i}$ we reach $\mathbb{X}$-marking $\boldsymbol{f} = \boldsymbol{i} + \sum_{i=1}^{|\rho|} c_i \cdot \Delta(t_i) \circ \pi_i^{-1}$. We call (the function computed by) the mentioned sum $\sum_{i=1}^{|\rho|} c_i \Delta(t_i) \circ \pi_i^{-1}$ as the *effect* of the run and denote it by $\Delta(\rho)$.

We fix some notations for the rest of the paper. We use Greek letters α, β, γ to denote data values from data domain $\mathbb{D}$, ρ, σ to denote a run, π to denote a mode and x, y, z to denote the variables. When clear from the context, we may omit $\mathbb{X}$ from $\mathbb{X}$-marking, $\mathbb{X}$-run and just write marking, run, etc. Further, we will use letters in bold, e.g., $\boldsymbol{m}$ to denote markings, where $\boldsymbol{i}, \boldsymbol{f}$ will be used for initial and final markings respectively. Further, throughout the paper, unless stated explicitly otherwise, we will refer to a UDPN $\mathcal{N} = (P, T, F, Var)$, therefore P, T, F, Var will denote the places, transitions, flow, and variables of this UDPN.

Example 1. An example of a simple UDPN $\mathcal{N}_1$ is given in Fig. 1. For this example, we have $P = \{p_1, p_2, p_3, p_4\}$, $T = \{t\}$, $Var = \{x, y, z\}$, and the flow relation is given by $F(p_1, t) = \{y \mapsto 1\}$, $F(p_2, t) = \{x \mapsto 1\}$, $F(t, p_3) = \{y \mapsto 2\}$, $F(t, p_4) = \{x \mapsto 1, z \mapsto 1\}$, and an assignment of 0 to every variable for the remaining of the pairs. Thus, for enabling transition p_1 and p_2 must have one token each with a different data value (since $x \neq y$) and after firing two tokens are produced in p_3 with same data value as was consumed from p_1 and two tokens are produced in p_4, one of whom has same data as consumed from p_2.

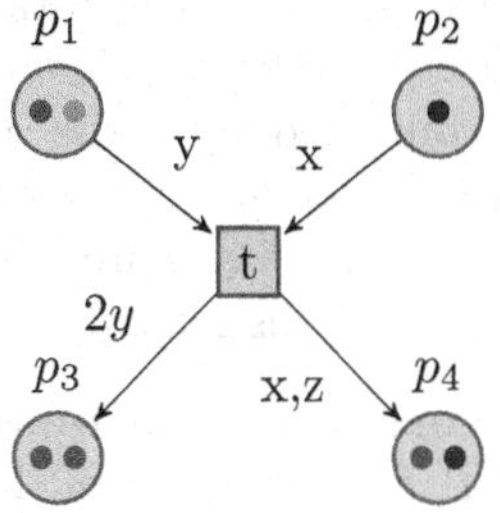

Fig. 1. A simple UDPN $\mathcal{N}_1$

Definition 2. *Given $\mathbb{X}$-markings $\boldsymbol{i}, \boldsymbol{f}$, we say $\boldsymbol{f}$ is $\mathbb{X}$-reachable from $\boldsymbol{i}$ if there exists an $\mathbb{X}$-run ρ s.t., $\boldsymbol{i} \xrightarrow{\rho}_{\mathbb{X}} \boldsymbol{f}$.*

When $\mathbb{X} = \mathbb{N}$, $\mathbb{X}$-reachability is the classical reachability problem, whose decidability is still unknown, while $\mathbb{Z}$-reachability for UDPN is in NP [22].

In this paper we tackle $\mathbb{Q}$ and $\mathbb{Q}^+$-reachability, also called *continuous* reachability in UDPN.

The first step towards the solution is showing that if a $\mathbb{Q}^+$-marking $\boldsymbol{f}$ is $\mathbb{Q}^+$-reachable from a $\mathbb{Q}^+$-marking $\boldsymbol{i}$, then there exists a $\mathbb{Q}^+$-run ρ which uses polynomially many data values and $\boldsymbol{i} \xrightarrow{\rho}_{\mathbb{Q}^+} \boldsymbol{f}$. We first formalize the set of distinct data values associated with $\mathbb{X}$-markings, data values *used* in $\mathbb{X}$-runs and variables associated with a transition.

Definition 3. *For $\mathcal{N} = (P, T, F, Var)$ a UDPN, $\mathbb{X}$-marking $\boldsymbol{m}$, $t \in T$, and $\mathbb{X}$-run $\rho = \{(c_i, t_i, \pi_i)\}_{|\rho|}$, we define*

1. $vars(t) = \{x \in Var \mid \exists p \in P : F(p, t)(x) \neq 0 \lor F(t, p)(x) \neq 0\}$.
2. $dval(\boldsymbol{m}) = \{\alpha \in \mathbb{D} \mid \exists p \in P : \boldsymbol{m}(p, \alpha) \neq 0\}$.
3. $dval(\rho) = \{\alpha \in \mathbb{D} \mid \exists i \leq |\rho| \; \exists x \in vars(t_i) : (\pi_i(x) = \alpha)\}$.

With this we state the first main result of this paper, which provides a bound on witnesses of $\mathbb{Q}, \mathbb{Q}^+$-reachability, and is proved in Sect. 5.

Theorem 1. *For $\mathbb{X} \in \{\mathbb{Q}, \mathbb{Q}^+\}$, if an $\mathbb{X}$-marking $\boldsymbol{f}$ is $\mathbb{X}$-reachable from an initial $\mathbb{X}$-marking $\boldsymbol{i}$, then there is an $\mathbb{X}$-run ρ such that $\boldsymbol{i} \xrightarrow{\rho}_{\mathbb{X}} \boldsymbol{f}$ and $|dval(\rho)| \leq |dval(\boldsymbol{i}) \cup dval(\boldsymbol{f})| + 1 + \max_{t \in T}(|vars(t)|)$.*

Using the above bound, we obtain a polynomial time algorithm for $\mathbb{Q}$-reachability, as detailed in Sect. 6.

Theorem 2. *Given $\mathcal{N} = (P, T, F, Var)$ a UDPN and two $\mathbb{Q}$-markings $\boldsymbol{i}$, $\boldsymbol{f}$, deciding if $\boldsymbol{f}$ is $\mathbb{Q}$-reachable from $\boldsymbol{i}$ in $\mathcal{N}$ is in polynomial time.*

Finally, we consider continuous, i.e., $\mathbb{Q}^+$-reachability for UDPN. We adapt the techniques used for $\mathbb{Q}^+$-reachability of Petri nets without data from [11,12] to the setting with data, and obtain a characterization of $\mathbb{Q}^+$-reachability for UDPN in Sect. 7.1. Finally, in Sect. 7.3, we show how the characterization can be combined with the above bound and compression techniques from [22] to obtain a polynomial sized system of linear equations with implications over $\mathbb{Q}^+$. To do so, we require a slight transformation of the net which is described in Sect. 7.2. This leads to our headline result, stated below.

Theorem 3 (Continuous reachability for UDPN). *Given a UDPN $\mathcal{N} = (P, T, F, Var)$ and two $\mathbb{Q}^+$-markings $\boldsymbol{i}$, $\boldsymbol{f}$, deciding if $\boldsymbol{f}$ is $\mathbb{Q}^+$-reachable from $\boldsymbol{i}$ in $\mathcal{N}$ is in polynomial time.*

The rest of this paper is dedicated to proving these theorems. First, we present an equivalent formulation via matrices, which simplifies the technical arguments.

4 Equivalent Formulation via Matrices

From now on, we restrict $\mathbb{X}$ to a symbol denoting $\mathbb{Q}$ or $\mathbb{Q}^+$. We formulate the definitions presented earlier in terms of matrices, since defining object such as $\mathbb{X}$-marking as functions is intuitive to define but difficult to operate upon.

In the following, we abuse the notation and use the same names for objects as well as matrices representing them. We remark that this is safe as all arithmetic operations on objects correspond to matching operations on matrices.

An $\mathbb{X}$-marking $\boldsymbol{m}$ is a $P \times \mathbb{D}$ matrix M, where $\forall p \in P, \forall \alpha \in \mathbb{D}, M(p, \alpha) = \boldsymbol{m}(p, \alpha)$. As a *finite representation*, we keep only a $P \times dval(\boldsymbol{m})$ matrix of non-zero columns. For a transition $t \in T$, we represent $F(t, \bullet), F(\bullet, t)$ as $P \times Var$ matrices. Note that $(t, \bullet)$ is not the position in the matrix, but is part of the name of the matrix; its entry at $(i, j) \in P \times Var$ is given by $F(t, \bullet)(i, j)$. For a place $p \in row(F(t, \bullet))$, the row $F(t, \bullet)(p, \bullet)$ is a vector in $\mathbb{N}^{Var}$, given by an equation $F(\bullet, t)(p, \bullet)(x) = F(p, t)(x)$ for $p \in P, t \in T, x \in Var$. Similarly, $\Delta(t)$ is a $P \times Var$ matrix with $\Delta(t)(p, x) = F(t, \bullet)(p, x) - F(\bullet, t)(p, x)$ for $t \in T, p \in P$, and $x \in Var$. Although, both $\Delta(t)$ and $F(\bullet, t)$ are defined as $P \times Var$ matrices, only the columns for variables in $vars(t)$ may be non-zero, so often we will iterate only over $vars(t)$ instead of Var.

Finally, we capture a mode $\pi : Var \to \mathbb{D}$ as a $Var \times \mathbb{D}$ permutation matrix $\mathcal{P}$. Although $\mathcal{P}$ may not be a square matrix, we abuse notation and call them permutation matrices. $\mathcal{P}$ basically represents assignment of variables in Var to data values just like π does. An entry of 1 represents that the corresponding variable is assigned corresponding data value in mode π. Thus, for each mode $\pi : Var \to \mathbb{D}$ there is a permutation matrix $\mathcal{P}_\pi$, such that for all $x \in Var$, $\alpha \in \mathbb{D}$, $\mathcal{P}_\pi(x, \alpha) = 1$ if $\pi(x) = \alpha$, and $\mathcal{P}_\pi(x, \alpha) = 0$ otherwise. Formulating a mode as a permutation matrix has the advantage that $\Delta(t) \circ \pi^{-1}$ is captured by $\Delta(t) \cdot \mathcal{P}_\pi$.

Example 2. In the UDPN $\mathcal{N}_1$ from Example 1, if $\mathbb{D} = \{red, blue, green, black\}$ then the initial marking $\boldsymbol{i}$ can be represented by the matrix $\boldsymbol{i}$ below and the function $\Delta(t)$ by the matrix $\Delta(t)$

$$\boldsymbol{i} = \begin{array}{c} \begin{array}{cccc} red & blue & green & black \end{array} \\ \begin{pmatrix} 1 & 0 & 1 & 0 \\ 0 & 1 & 0 & 0 \\ 2 & 0 & 0 & 0 \\ 1 & 1 & 0 & 0 \end{pmatrix} \begin{array}{c} p_1 \\ p_2 \\ p_3 \\ p_4 \end{array} \end{array} \qquad \Delta(t) = \begin{array}{c} \begin{array}{ccc} x & y & z \end{array} \\ \begin{pmatrix} 0 & -1 & 0 \\ -1 & 0 & 0 \\ 0 & 2 & 0 \\ 1 & 0 & 1 \end{pmatrix} \begin{array}{c} p_1 \\ p_2 \\ p_3 \\ p_4 \end{array} \end{array}$$

If we fire transition t with the assignment $x = blue, y = green, z = black$, we get the following net depicted below (left), with marking $\boldsymbol{f}$ (below center). The permutation matrix corresponding to the mode of fired transition is given by $\mathcal{P}$ matrix on the right. Note that the matrix $\boldsymbol{f} - \boldsymbol{i}$ is indeed the matrix $\Delta(t) \cdot \mathcal{P}$.

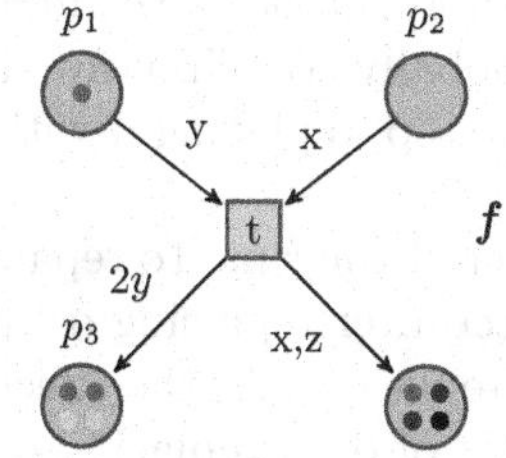

$$\boldsymbol{f} = \begin{array}{c} \begin{array}{cccc} red & blue & green & black \end{array} \\ \begin{pmatrix} 1 & 0 & 0 & 0 \\ 0 & 0 & 0 & 0 \\ 2 & 0 & 2 & 0 \\ 1 & 2 & 0 & 1 \end{pmatrix} \begin{array}{c} p_1 \\ p_2 \\ p_3 \\ p_4 \end{array} \end{array} \quad \mathcal{P} = \begin{array}{c} \begin{array}{cccc} red & blue & green & black \end{array} \\ \begin{array}{c} x \\ y \\ z \end{array} \begin{pmatrix} 0 & 1 & 0 & 0 \\ 0 & 0 & 1 & 0 \\ 0 & 0 & 0 & 1 \end{pmatrix} \end{array}$$

Using the representations developed so far we can represent an $\mathbb{X}$-run ρ as $\{(c_i, t_i, \mathcal{P}_i)\}_{|\rho|}$ where $(c_i, t_i, \mathcal{P}_i)$ denotes the ith $\mathbb{X}$-step fired with coefficient c_i using transition t_i with a mode corresponding to permutation matrix $\mathcal{P}_i$. The sum of the matrices $(\sum_{i=1}^{|\rho|} c_i \Delta(t_i) \cdot \mathcal{P}_i)$ gives us the effect of the run i.e. $\Delta(\rho) = \boldsymbol{f} - \boldsymbol{i}$ where $\boldsymbol{i} \xrightarrow{\rho}_{\mathbb{X}} \boldsymbol{f}$. Effect of an $\mathbb{X}$-run ρ on a data value α is $\Delta(\rho)(\bullet, \alpha)$. Also, for an $\mathbb{X}$-run $\rho = \{(c_i, t_i, \mathcal{P}_i)\}_{|\rho|}$, define $k\rho = \{(kc_i, t_i, \mathcal{P}_i)\}_{|\rho|}$ where $k \in \mathbb{X}^+$.

5 Bounding Number of Data Values Used in $\mathbb{Q}, \mathbb{Q}^+$-run

We now prove the first main result of the paper, namely, Theorem 1, which shows a linear upper bound on the number of data values required in a $\mathbb{Q}^+$-run and a $\mathbb{Q}$-run. Theorem 1 is an immediate consequence of the following lemma, which states that if more than a linearly bounded number of data values are used in a $\mathbb{Q}$ or $\mathbb{Q}^+$ run, then there is another such run in which we use at least one less data value.

Lemma 1. *Let $\mathbb{X} \in \{\mathbb{Q}, \mathbb{Q}^+\}$. If there exists an $\mathbb{X}$-run σ such that $\boldsymbol{i} \xrightarrow{\sigma}_{\mathbb{X}} \boldsymbol{f}$ and $|dval(\sigma)| > |dval(\boldsymbol{i}) \cup dval(\boldsymbol{f})| + 1 + \max_{t \in T}(|vars(t)|)$, then there exists an $\mathbb{X}$-run ρ such that $\boldsymbol{i} \xrightarrow{\rho}_{\mathbb{X}} \boldsymbol{f}$ and $|dval(\rho)| \leq |dval(\sigma)| - 1$.*

By repeatedly applying this lemma, Theorem 1 follows immediately. The rest of this section is devoted to proving this lemma. The central idea is to take any $\mathbb{Q}$ or $\mathbb{Q}^+$-run between $\boldsymbol{i}, \boldsymbol{f}$ and transform it to use at least one less data value.

5.1 Transformation of an $\mathbb{X}$-run

The transformation which we call *decrease* is defined as a combination of two separate operations on an $\mathbb{X}$-run; we name them *uniformize* and *replace* and denote them by $\mathcal{U}$ and $\mathcal{R}$ respectively.

- *uniformize* takes an $\mathbb{X}$-step and a non-empty set of data values $\mathbb{E}$ as input and produces an $\mathbb{X}$-run, such that in the resultant run, the effect of the run for each data value in $\mathbb{E}$ is equal.
- *replace* takes an $\mathbb{X}$-step, a single data value α, and a non-empty set of data values $\mathbb{E}$ as input and outputs an $\mathbb{X}$-step which doesn't use data value α.

The intuition behind the decrease operation is that we would like to take two data values α and β used in the run such that effect on both of them is $\mathbf{0}$ (they exists as the effect on every data value not present in the initial of final configuration is $\mathbf{0}$) and replace usage of α by β. However, such a replacement can only be done if both data are not used together in a single step (indeed, a mode π cannot assign the same data values to two variables). Unfortunately we cannot guarantee the existence of such a β that may replace α globally. We circumvent this by applying the *replace* operation separately for every step, replacing α with different data values in different steps.

But such a transformation would not preserve the effect of the run. To repair this aspect we uniformize i.e. guarantee that the final effect after replacing α by other data values is equal for every datum that is used to replace α. As the effect on α was $\mathbf{0}$ then if we split it uniformly it adds $\mathbf{0}$ to effects of data replacing α, which is exactly what we want. We now formalize this intuition below.

The Uniformize Operator. By © we denote an operator of concatenation of two sequences. Although the data set $\mathbb{D}$ is unordered, the following definitions require access to an arbitrary but fixed linear order on its elements. The definition of the *uniformize* operator needs another operator to act on an $\mathbb{X}$-step, which we call *rotate* and denote by *rot*.

Definition 4. *For a non-empty finite set of data values $\mathbb{E} \subset \mathbb{D}$ and an $\mathbb{X}$-step, $\omega = (c, t, \mathcal{P})$, define $rot(\mathbb{E}, \omega) = (c, t, \mathcal{P}')$ where $\mathcal{P}'$ is obtained from $\mathcal{P}$ as follows.*

- $\forall \alpha \in col(\mathcal{P}) \setminus \mathbb{E}$, $\mathcal{P}'(\bullet, \alpha) = \mathcal{P}(\bullet, \alpha)$.
- $\forall \alpha \in \mathbb{E}$, $\mathcal{P}'(\bullet, \alpha) = \mathcal{P}(\bullet, next_{\mathbb{E}}(\alpha))$, *where* $next_{\mathbb{E}}(\alpha) = \min(\{\beta \in \mathbb{E} \mid \beta > \alpha\})$ *if* $|\{\beta \in \mathbb{E} \mid \beta > \alpha\}| > 0$ *and* $\min(\mathbb{E})$ *otherwise.*

For a fixed set $\mathbb{E}$, we can repeatedly apply $rot(\mathbb{E}, \bullet)$ operation on an $\mathbb{X}$-step, which we denote by $rot^k(\mathbb{E}, \omega)$, where k is the number of times we applied the operation (for example: $rot^2(\mathbb{E}, \omega) = rot(\mathbb{E}, (rot(\mathbb{E}, \omega)))$.

Definition 5. *For a finite and non-empty set of data values* $\mathbb{E} \subset \mathbb{D}$ *and an* $\mathbb{X}$*-step* $\omega = (c, t, \mathcal{P})$*, we define uniformize as follows*

$$\mathcal{U}(\mathbb{E}, \omega) = rot^0(\mathbb{E}, \tfrac{\omega}{|\mathbb{E}|}) \circledcirc rot^1(\mathbb{E}, \tfrac{\omega}{|\mathbb{E}|}) \circledcirc rot^2(\mathbb{E}, \tfrac{\omega}{|\mathbb{E}|}) \circledcirc \ldots \circledcirc rot^{|\mathbb{E}|-1}(\mathbb{E}, \tfrac{\omega}{|\mathbb{E}|}).$$

An important property of uniformize is its effect on data values.

Lemma 2. *For a finite and non-empty set of data values* $\mathbb{E} \subset \mathbb{D}$ *and an* $\mathbb{X}$*-step* $\omega = (c, t, \mathcal{P})$*,* $\boldsymbol{i} \xrightarrow{\omega}_{\mathbb{Q}^+} \boldsymbol{f}$*, if* $\boldsymbol{i}' \xrightarrow{\mathcal{U}(\mathbb{E},\omega)} \boldsymbol{f}'$*, then*

1. $\forall \alpha \in dval(\omega) \backslash \mathbb{E},\ \boldsymbol{f}'(\bullet, \alpha) - \boldsymbol{i}'(\bullet, \alpha) = \boldsymbol{f}(\bullet, \alpha) - \boldsymbol{i}(\bullet, \alpha)$
2. $\forall \alpha \in \mathbb{E},\ ,\boldsymbol{f}'(\bullet, \alpha) - \boldsymbol{i}'(\bullet, \alpha) = \frac{\sum_{\beta \in \mathbb{E}}(\boldsymbol{f}(\bullet,\beta) - \boldsymbol{i}(\bullet,\beta))}{|\mathbb{E}|}$.

This lemma tells us the effect of the run on the initial marking is equalized for data values in $\mathbb{E}$ by the $\mathcal{U}$ operation, and is unchanged for the other data values.

The Replace Operator. To define the *replace* operator it is useful to introduce $swap_{\alpha,\beta}(\mathcal{P})$ which exchanges columns α and β in the matrix $\mathcal{P}$.

Definition 6. *For a finite set of data values* $\mathbb{E}$*, an* $\mathbb{X}$*-step* $\omega = (c, t, \mathcal{P})$*, and* $\alpha \notin \mathbb{E}$ *we define replace as follows*

$$\mathcal{R}(\alpha, \mathbb{E}, \omega) = \begin{cases} (c, t, \mathcal{P}) & \text{if } (F(t, \bullet) \cdot \mathcal{P})(\bullet, \alpha) = (F(\bullet, t) \cdot \mathcal{P})(\bullet, \alpha) = \mathit{0} \\ (c, t, swap_{\alpha,\beta}(\mathcal{P})) & \text{else, if } \beta \text{ is the smallest datum in } \mathbb{E} \text{ s.t.,} \\ & (F(t, \bullet) \cdot \mathcal{P})(\bullet, \beta) = (F(\bullet, t) \cdot \mathcal{P})(\bullet, \beta) = \mathit{0} \\ undefined & \text{otherwise.} \end{cases}$$

After applying the *replace* operation α is no longer used in the run, which reduces the number of data values used in the run. Observe that *replace* can not be always applied to an $\mathbb{X}$-step. It requires a zero column labelled with an element from $\mathbb{E}$ in the permutation matrix corresponding to the $\mathbb{X}$-step.

The Decrease Transformation. Finally, we define the transformation on an $\mathbb{X}$-run between two markings which we call *decrease* and denote by *dec*.

Definition 7. *For two* $\mathbb{X}$*-markings* $\boldsymbol{i}$*,* $\boldsymbol{f}$*, and an* $\mathbb{X}$*-run* σ *such that* $\boldsymbol{i} \xrightarrow{\sigma}_{\mathbb{X}} \boldsymbol{f}$ *and* $|dval(\sigma)| > |dval(\boldsymbol{i}) \cup dval(\boldsymbol{f})| + 1 + \max_{t \in T}(|vars(t)|)$*, let* $\{\alpha\} \cup \mathbb{E} = dval(\sigma) \setminus (dval(\boldsymbol{i}) \cup dval(\boldsymbol{f}))$ *and* $\alpha \notin \mathbb{E}$*. We define decrease by,* $dec(\mathbb{E}, \alpha, \sigma) =$

$$\mathcal{U}(\mathbb{E}, \mathcal{R}(\alpha, \mathbb{E}, \sigma(1))) \circledcirc \mathcal{U}(\mathbb{E}, \mathcal{R}(\alpha, \mathbb{E}, \sigma(2))) \circledcirc \ldots \circledcirc \mathcal{U}(\mathbb{E}, \mathcal{R}(\alpha, \mathbb{E}, \sigma(|\sigma|))).$$

where $\sigma(j)$ *denotes the* j^{th} $\mathbb{X}$*-step of* σ*.*

Observe that the required size of $dval(\sigma)$ guarantees existence of a $\beta \in \mathbb{E}$ which can be replaced with α, for every application of the $\mathcal{R}$ operation. Note that the exchanged data value β could be different for each step. Finally, we can analyze the *decrease* transformation and show that if the original run allows for the *decrease* transformation (as given in the above definition), then after the application of it, the resulting sequence of transitions is a valid run of the system.

Lemma 3. *Let σ be an $\mathbb{X}$-run such that $\boldsymbol{i} \xrightarrow{\sigma}_{\mathbb{X}} \boldsymbol{f}$ and $|dval(\sigma)| > |dval(\boldsymbol{i}) \cup dval(\boldsymbol{f})| + 1 + \max_{t \in T}(|dval(t)|)$. Let $\alpha \in dval(\sigma) \setminus (dval(\boldsymbol{i}) \cup dval(\boldsymbol{f}))$ and $\mathbb{E} = dval(\sigma) \setminus (dval(\boldsymbol{i}) \cup dval(\boldsymbol{f}) \cup \{\alpha\})$. Then for $\rho = dec(\mathbb{E}, \alpha, \sigma)$, we obtain $\boldsymbol{i} \xrightarrow{\rho}_{\mathbb{X}} \boldsymbol{f}$.*

Proof. Suppose $\sigma = \sigma_1\sigma_2\ldots\sigma_l$ where each $\sigma_j = (c_j, t_j, \mathcal{P}_j)$, for $1 \le j \le l$ is an $\mathbb{X}$-step. Then $\rho = \rho_1 © \ldots © \rho_l$, where each ρ_j is an $\mathbb{X}$-run defined by $\rho_j = \mathcal{U}(\mathbb{E}, \mathcal{R}(\alpha, \mathbb{E}, \sigma_j))$. It will be useful to identify intermediate $\mathbb{X}$-markings

$$\boldsymbol{i} = \boldsymbol{m}_0 \xrightarrow{\sigma_1}_{\mathbb{X}} \boldsymbol{m}_1 \xrightarrow{\sigma_2}_{\mathbb{X}} \boldsymbol{m}_2 \xrightarrow{\sigma_3}_{\mathbb{X}} \ldots \xrightarrow{\sigma_l}_{\mathbb{X}} \boldsymbol{m}_l = \boldsymbol{f} \tag{1}$$

$$\boldsymbol{i} = \boldsymbol{m}'_o \xrightarrow{\mathcal{U}(\mathbb{E},\mathcal{R}(\alpha,\mathbb{E},\sigma_1))}_{\mathbb{Q}} \boldsymbol{m}'_1 \xrightarrow{\mathcal{U}(\mathbb{E},\mathcal{R}(\alpha,\mathbb{E},\sigma_2))}_{\mathbb{Q}} \boldsymbol{m}'_2 \ldots \xrightarrow{\mathcal{U}(\mathbb{E},\mathcal{R}(\alpha,\mathbb{E},\sigma_l))}_{\mathbb{Q}} \boldsymbol{m}'_l = \boldsymbol{f}' \tag{2}$$

We split the proof: first we show that $\boldsymbol{f} = \boldsymbol{f}'$ and then ρ is $\mathbb{X}$-fireable from $\boldsymbol{i}$.

Step 1: Showing that the final markings reached are the same. We prove a stronger statement which implies that $\boldsymbol{f} = \boldsymbol{f}'$, namely:

Claim 1. *For all $0 \le j \le l$,*

1. $\boldsymbol{m}'_j(\bullet, \alpha) = \boldsymbol{0}$
2. $\forall \gamma \in dval(\boldsymbol{i}) \cup dval(\boldsymbol{f})$, $\boldsymbol{m}'_j(\bullet, \gamma) = \boldsymbol{m}_j(\bullet, \gamma)$
3. $\forall \gamma \in \mathbb{E}$ $\boldsymbol{m}'_j(\bullet, \gamma) = \frac{1}{|\mathbb{E}|}\left(\sum_{\delta \in \mathbb{E} \cup \{\alpha\}} \boldsymbol{m}_j(\bullet, \delta)\right)$.

The proof is obtained by induction on j. Intuitively, point 1 holds as we shift effects on α to β, point 2 holds as the transformation does not touch $\gamma \in dval(\boldsymbol{i}) \cup dval(\boldsymbol{f})$. The last and most complicated point follows from the fact that the number of tokens consumed and produced along each segment $\xrightarrow{\mathcal{U}(\mathbb{E},\mathcal{R}(\alpha,\mathbb{E},\sigma_j))}$ is the same as for σ_j, but uniformized over $\mathbb{E}$.

Step 2: Showing that ρ is an $\mathbb{X}$-run. If $\mathbb{X} = \mathbb{Q}$ then the run ρ is fireable, as any $\mathbb{Q}$-run is fireable, so in this case this step is trivial. The case when $\mathbb{X} = \mathbb{Q}^+$ is more involved. As we know from Claim 1, each $\boldsymbol{m}'_j$ is a $\mathbb{Q}^+$-marking, so it suffices to prove that for every j, $\boldsymbol{m}'_j \xrightarrow{\mathcal{U}(\mathbb{E},\mathcal{R}(\alpha,\mathbb{E},\sigma_j))}_{\mathbb{Q}^+} \boldsymbol{m}'_{j+1}$. Consider a data vector of tokens consumed along the $\mathbb{Q}^+$-run $\mathcal{U}(\mathbb{E}, \mathcal{R}(\alpha, \mathbb{E}, \sigma_j))$. If we show that it is smaller than or equal to $\boldsymbol{m}'_j$ (component-wise), then we can conclude that $\mathcal{U}(\mathbb{E}, \mathcal{R}(\alpha, \mathbb{E}, \sigma_j))$ is indeed $\mathbb{Q}^+$-fireable from $\boldsymbol{m}'_j$. To show this, we examine the consumed tokens for each datum γ separately. There are three cases:

(i) $\gamma = \alpha$. For this case, every step in $\mathcal{U}(\mathbb{E}, \mathcal{R}(\alpha, \mathbb{E}, \sigma_j))$ does not make any change on α so tokens with data value α are not consumed along the $\mathbb{Q}^+$-run $\mathcal{U}(\mathbb{E}, \mathcal{R}(\alpha, \mathbb{E}, \sigma_j))$.

(ii) $\gamma \in dval(\boldsymbol{i}) \cup dval(\boldsymbol{f})$. This is similar to the above case. Consider any data value $\gamma \in (dval(\sigma) \backslash \mathbb{E}) \setminus \{\alpha\}$. Since γ does not change on *rotate* operation, the $\mathcal{U}$ operation causes each $\mathbb{Q}$-step in $\mathcal{U}(\mathbb{E}, \mathcal{R}(\alpha, \mathbb{E}, \sigma_j))$ to consume $\frac{1}{|\mathbb{E}|}$ of the tokens with data value γ consumed when σ_j is fired. This is repeated $|\mathbb{E}|$ times and hence the vector of tokens with data value γ consumed along $\mathcal{U}(\mathbb{E}, \mathcal{R}(\alpha, \mathbb{E}, \sigma_j))$ is equal to the vector of tokens with value γ consumed by step σ_j. But we know that, it is smaller than $\boldsymbol{m}_j(\bullet,\gamma)$ and concluding smaller than $\boldsymbol{m}'_j(\bullet,\gamma)$. The last inequality is true as $\boldsymbol{m}_j(\bullet,\gamma) = \boldsymbol{m}'_j(\bullet,\gamma)$ according to Claim 1.

(iii) $\gamma \in \mathbb{E}$. Let ω be a triple $(c_j, F(\bullet,t_j), \mathcal{P}_j)$ where $(c_j, t_j, \mathcal{P}_j) = \sigma_j$. ω simply describes tokens consumed by σ_j. We slightly overload the notation and treat a triple ω like a step, where $F(\bullet,t_j)$ represents a transition "_" for which $F(\bullet,\_) = F(\bullet,t_j)$ and $F(\_,\bullet)$ is a zero matrix. We calculate the vector of consumed tokens with data value γ as follows: *consumed*$(\bullet,\gamma) =$

$$\frac{1}{|\mathbb{E}|}\sum_{k=0}^{|\mathbb{E}|-1} \Delta(rot^k(\mathbb{E},\mathcal{R}(\alpha,\mathbb{E},\omega)))(\bullet,\gamma) = \frac{1}{|\mathbb{E}|}\sum_{k=0}^{|\mathbb{E}|} \Delta(rot^k(\mathbb{E}\cup\{\alpha\},\omega))(\bullet,\gamma)$$

the first equality is from definition and the second by the *replace* operation,

$$= \frac{c_j}{|\mathbb{E}|}\sum_{k=0}^{|\mathbb{E}|}(rot^k(\mathbb{E}\cup\{\alpha\},(1,F(\bullet,t_j),\mathcal{P}_j)))(\bullet,\gamma) = \frac{c_j}{|\mathbb{E}|}\sum_{\delta\in\mathbb{E}\cup\{\alpha\}}(F(\bullet,t_j)\cdot\mathcal{P}_j)(\bullet,\delta)$$

Further, observe that as σ_j can fired in $\boldsymbol{m}_j$

$$c_j(F(\bullet,t_j)\cdot\mathcal{P}_j)(\bullet,\delta) \leq \boldsymbol{m}_j(\bullet,\delta) \text{ for all } \delta\in\mathbb{D},$$

summing up over $\delta \in \mathbb{E}\cup\{\alpha\}$ and multiplying with $\frac{1}{|\mathbb{E}|}$ we get

$$\frac{1}{|\mathbb{E}|}c_j\sum_{\delta\in\mathbb{E}\cup\{\alpha\}}(F(\bullet,t_j)\cdot\mathcal{P}_j)(\bullet,\delta) \leq \frac{1}{|\mathbb{E}|}\sum_{\delta\in\mathbb{E}\cup\{\alpha\}}\boldsymbol{m}_j(\bullet,\delta) = \boldsymbol{m}'_j(\delta,\gamma),$$

where the last equality comes from Claim 1 point 3. Combining inequalities we get *consumed*$(\bullet,\gamma) \leq \boldsymbol{m}'_i(\bullet,\gamma)$.

Proof (of Lemma 1). Now the proof of Lemma 1 (and hence Theorem 1) follow immediately, since we can use the *decrease* transformation, to decrease the number of data values required in an $\mathbb{X}$-run. We simply take $\alpha \in dval(\sigma) \setminus (dval(\boldsymbol{i}) \cup dval(\boldsymbol{f}))$ and $\mathbb{E} = dval(\sigma) \setminus (dval(\boldsymbol{i}) \cup dval(\boldsymbol{f})) \setminus \{\alpha\}$. Next, let $\rho = dec(\mathbb{E},\alpha,\sigma)$. Due to Lemma 3 we know that $\boldsymbol{i} \xrightarrow{\rho}_{\mathbb{X}} \boldsymbol{f}$. Moreover, observe that $dval(\rho) \subseteq dval(\sigma)$. But in addition, $\alpha \notin dval(\rho)$ as due to the one of properties of the *decrease* operation α does not participate in the run ρ. So $dval(\rho) \subset dval(\sigma)$. Therefore $|dval(\rho)| \leq |dval(\sigma)| - 1$.

6 $\mathbb{Q}$-reachability is in PTime

We recall the definition of histograms from [22].

Definition 8. *A histogram M of order $q \in \mathbb{Q}$ is a Var $\times\, \mathbb{D}$ matrix having non-negative rational entries such that,*

1. $\sum_{\alpha \in col(M)} M(x, \alpha) = q$ *for all* $x \in row(M)$.
2. $\sum_{x \in row(M)} M(x, \alpha) \le q$ *for all* $\alpha \in col(M)$.

A permutation matrix is a histogram of order 1.

In the following lemma, we state two properties of histograms. We say that a histogram of order a is an *[a]-histogram* if the histogram has only $\{0, a\}$ entries.

Lemma 4. *Let* $H, H_1, H_2, .., H_n$ *be histograms of order* $q, q_1, q_2, ..., q_n$ *respectively and of same row dimensions then (i)* $\sum_{i=1}^{n} H_i$ *is a histogram of order* $\sum_i^n q_i$*, (ii)* H *can be decomposed as a sum of* $[a_i]$*-histograms such that* $\sum_i a_i = q$.

Using histograms we define a representation $Hist(\rho)$ for an $\mathbb{X}$-run ρ, which captures $\Delta(\rho)$. From an $\mathbb{X}$-run $\rho = \{(c_j, t_j, \mathcal{P}_j)\}_{|\rho|}$ we obtain $Hist(\rho)$ as follows. For all transitions $t \in T$, define the set $I_t = \{j \in [1..|\rho|] | \ t_j = t\}$. Then calculate the matrix $H_t = \sum_{i \in I_t} c_i \mathcal{P}_i$. Observe that since permutation matrices are histograms and histograms are closed under scalar multiplication and addition, H_t is a histogram. If I_t is empty, then H_t is simply the null matrix. We define $Hist(\rho)$ as a mapping from T to histograms such that t is mapped to H_t.

Analogous to an $\mathbb{X}$-run we can represent $Hist(\rho)$ simply as $\{(t_j, H_{t_j})\}$, unlike an $\mathbb{X}$-run we don't indicate the length of the sequence since it is dependent on the net and not the individual run itself.

Proposition 1. *Let* $\mathcal{N} = (P, T, F, Var)$ *be a UDPN,* $\boldsymbol{i}, \boldsymbol{f}$ $\mathbb{X}$*-markings, and* σ *an* $\mathbb{X}$*-run such that* $\boldsymbol{i} \xrightarrow{\sigma}_{\mathbb{X}} \boldsymbol{f}$*. Then for each* $t \in T$ *there exists* H_t *such that:*

1. $\boldsymbol{f} - \boldsymbol{i} = \sum_{t \in T} \Delta(t) \cdot H_t$,
2. $col(H_t) \subseteq dval(\sigma)$ *for every* $t \in T$.

A PTime Procedure. We start by observing that from any $\mathbb{Q}$-marking $\boldsymbol{i}$, every $\mathbb{Q}$-step $(c, t, \mathcal{P})$ is fireable and every $\mathbb{Q}$ run is fireable. This follows from the fact that rationals are closed under addition, thus $\boldsymbol{i} + c \cdot F(\bullet, t) \cdot \mathcal{P}$ is a marking in $\mathcal{M}_{\mathbb{Q}}$. Thus if we have to find a $\mathbb{Q}$-run $\rho = \{(c_j, t_j, \mathcal{P}_j)\}_{|\rho|}$ between two $\mathbb{Q}$-markings, $\boldsymbol{i}, \boldsymbol{f}$ it is sufficient to ensure that $\boldsymbol{f} - \boldsymbol{i} = \sum_{j=1}^{|\rho|} c_j \Delta(t_j) \cdot \mathcal{P}_j$. Thus for a $\mathbb{Q}$-run all that matters is the difference in markings caused by the $\mathbb{Q}$-run which is captured succinctly by $Hist(\rho) = \{t_j, H_{t_j}\}$. This brings us to our characterization of $\mathbb{Q}$-run.

Lemma 5. *Let* $\mathcal{N} = (P, T, F, Var)$ *be a UDPN, a marking* $\boldsymbol{f}$ *is* $\mathbb{Q}$*-reachable from* $\boldsymbol{i}$ *iff there exists set* $\mathbb{E}$ *of size bounded by* $|\mathbb{E}| \le |dval(\boldsymbol{i}) \cup dval(\boldsymbol{f})| + 1 + \max_{t \in T}(|vars(t)|)$ *and a histogram* H_t *for each* $t \in T$ *such that* $\boldsymbol{f} - \boldsymbol{i} = \sum_{t \in T} \Delta(t) \cdot H_t$ *and* $\forall t \in T \ col(H_t) \subseteq \mathbb{E}$.

Using this characterization we can write a system of linear inequalities to encode the condition of Lemma 5. Thus, we obtain our second main result, namely, Theorem 2, with detailed proofs in [27].

7 $\mathbb{Q}^+$-reachability is in PTime

Finally, we turn to $\mathbb{Q}^+$-reachability for UDPNs and to the proof of Theorem 3. At a high level, the proof is in three steps. We start with a characterization of $\mathbb{Q}^+$-reachability in UDPNs. Then we present a polytime reduction of the continuous reachability problem to the same problem but for a special subclass of UDPN, called loop-less nets. Finally, we present how to encode the characterization for loop-less nets into a system of *linear equations with implications* to obtain a polytime algorithm for continuous reachability in UDPNs.

7.1 Characterizing $\mathbb{Q}^+$-reachability

We begin with a definition. For an $\mathbb{X}$-run we introduce the notion of the pre and post sets of $\mathbb{X}$–run. For an $\mathbb{X}$-run, $\rho = \{(c_i, t_i, \mathcal{P}_i)\}_{|\rho|}$ we define $Pre(\rho) = \{(p, \alpha) |\ \exists\ t_i, \exists\ x : F(p, t_i)(x) < 0 \wedge \mathcal{P}_i(x, \alpha) = 1\}$. We also define $Post(\rho) = \{(p, \alpha) |\ \exists\ t_i, \exists\ x : F(t_i, p)(x) > 0 \wedge \mathcal{P}_i(x, \alpha) = 1\}$. Intuitively, $Pre(\rho), Post(\rho)$ denote the set of (p, α) (place, data value) pairs describing tokens that are consumed, produced respectively by the run ρ.

Throughout this section, by a marking we denote a $\mathbb{Q}^+$-marking.

Lemma 6. *Let $\mathcal{N} = (P, T, F, Var)$ be an UDPN and $\boldsymbol{i}, \boldsymbol{f}$ are markings. For any $\mathbb{Q}^+$-run σ such that $\boldsymbol{i} \xrightarrow{\sigma}_{\mathbb{Q}^+} \boldsymbol{f}$ there exist markings $\boldsymbol{i}'$ and $\boldsymbol{f}'$ (possibly on a different run) such that*

1. *$\boldsymbol{i}'$ is $\mathbb{Q}^+$-reachable from $\boldsymbol{i}$ in at most $|P| \cdot |dval(\sigma)|$ $\mathbb{Q}^+$-steps*
2. *There is a run σ' such that $dval(\sigma') \subseteq dval(\sigma)$ and $\boldsymbol{i}' \xrightarrow{\sigma'}_{\mathbb{Q}} \boldsymbol{f}'$*
3. *$\boldsymbol{f}$ is $\mathbb{Q}^+$-reachable from $\boldsymbol{f}'$ in at most $|P| \cdot |dval(\sigma)|$ $\mathbb{Q}^+$-steps*
4. *$\forall (p, \alpha) \in Pre(\sigma'), \boldsymbol{i}'(p, \alpha) > 0$*
5. *$\forall (p, \alpha) \in Post(\sigma'), \boldsymbol{f}'(p, \alpha) > 0$*

Remark 1. If in conditions 1 and 3 we drop the requirement on the number of steps then the five conditions still imply continuous reachability.

Note that if there exist markings $\boldsymbol{i}'$ and $\boldsymbol{f}'$ and $\mathbb{Q}^+$ -runs ρ, ρ', ρ'' such that $\boldsymbol{i} \xrightarrow{\rho}_{\mathbb{Q}^+} \boldsymbol{i}', \boldsymbol{i}' \xrightarrow{\rho'}_{\mathbb{Q}^+} \boldsymbol{f}', \boldsymbol{f}' \xrightarrow{\rho''}_{\mathbb{Q}^+} \boldsymbol{f}$ then there is a $\mathbb{Q}^+$-run σ such that $\boldsymbol{i} \xrightarrow{\sigma}_{\mathbb{Q}^+} \boldsymbol{f}$. The above characterization and its proof are obtained by adapting to the data setting, the techniques developed for continuous reachability in Petri nets (without data) in [11] and [12].

7.2 Transforming UDPN to Loop-less UDPN

For a UDPN $\mathcal{N} = (P, T, F, Var)$, we construct a UDPN $\mathcal{N}'$ which is polynomial in the size of $\mathcal{N}$ and the $\mathbb{Q}^+$-reachability problem is equivalent. We define $PrePlace(t) = \{p \in P | \exists v \in Var\ s.t.\ F(p, t)(v) > 0\}$ and $PostPlace(t) = \{p \in P | \exists v \in Var\ s.t.\ F(t, p)(v) > 0\}$, where $t \in T$. The essential property of the transformed UDPN is that for every transition the sets of PrePlace and

PostPlace do not intersect. A UDPN $\mathcal{N} = (P, T, F, Var)$ is said to be *loop-less* if for all $t \in T$, $PrePlace(t) \cap PostPlace(t) = \emptyset$.

Any UDPN can easily be transformed in polynomial time into a loop-less UDPN such that $\mathbb{Q}^+$-reachability is preserved, by doubling the number of places and adding intermediate transitions. Formally, For every net $\mathcal{N}$ and two markings $\boldsymbol{i}, \boldsymbol{f}$ in polynomial time one can construct a loop-less net $\mathcal{N}'$ and two markings $\boldsymbol{i}'$, $\boldsymbol{f}'$ such that $\boldsymbol{i} \rightarrow_{\mathbb{Q}^+} \boldsymbol{f}$ in the net $\mathcal{N}$ iff $\boldsymbol{i}' \rightarrow_{\mathbb{Q}^+} \boldsymbol{f}'$ in $\mathcal{N}'$. Now, the following lemma which describes a property of loop-less nets will be crucial for our reachability algorithm:

Lemma 7. *In a loop-less net, for markings $\boldsymbol{i}$, $\boldsymbol{f}$, if there exist a histogram H, and a transition $t \in T$ such that $\boldsymbol{i} + \Delta(t) \cdot H = \boldsymbol{f}$, then there exist a $\mathbb{Q}^+$-run ρ such that $\boldsymbol{i} \xrightarrow{\rho}_{\mathbb{Q}^+} \boldsymbol{f}$.*

7.3 Encoding $\mathbb{Q}^+$-reachability as Linear Equations with Implications

Linear equations with implications, as we use them, are defined in [23], but were introduced in [12]. A system of linear equations with implications, also denoted a $\Longrightarrow$ system, is a finite set of linear inequalities over the same variables, plus a finite set of implications of the form $x > 0 \implies y > 0$, where x, y are variables appearing in the linear inequalities.

Lemma 8 *[12]. The $\mathbb{Q}^+$ solvability problem for a $\Longrightarrow$ system is in PTime.*

We then reduce the $\mathbb{Q}^+$-reachability problem to checking the solvability of a system of linear equations with implications, using the characterization established in Lemma 6 in the following lemma.

Lemma 9. *$\mathbb{Q}^+$-reachability in a UDPN $\mathcal{N} = (P, T, F, Var)$ between markings $\boldsymbol{i}, \boldsymbol{f}$ can be encoded as a set of linear equations with implications in P-time.*

Finally, we obtain Theorem 3 as a consequence of Lemmas 8 and 9.

8 Conclusion

In this paper, we provided a polynomial time algorithm for continuous reachability in UDPN, matching the complexity for Petri nets without data. This is in contrast to problems such as discrete coverability, termination, where Petri nets with and without data differ enormously in complexity, and to (discrete) reachability, where decidability is still open. As future work, we aim to implement the continuous reachability algorithm developed here, to build the first tool for discrete coverability in UDPN on the lines of what has been done for Petri nets without data. The main obstacle will be performance evaluation due to lack of benchmarks for UDPNs. Another interesting avenue for future work would be to tackle continuous reachability for Petri nets with ordered data, which would allow us to analyze continuous variants of Timed Petri nets.

Acknowledgments. We thank the anonymous reviewers for their careful reading and their helpful and insightful comments.

References

1. Finkel, A., Schnoebelen, P.: Well-structured transition systems everywhere! Theor. Comput. Sci. **256**(1–2), 63–92 (2001)
2. Rackoff, C.: The covering and boundedness problems for vector addition systems. Theor. Comput. Sci. **6**, 223–231 (1978)
3. Rao Kosaraju, S.: Decidability of reachability in vector addition systems (preliminary version). In: Proceedings of the 14th Annual ACM Symposium on Theory of Computing, San Francisco, California, USA, 5–7 May 1982, pp. 267–281 (1982)
4. Leroux, J., Schmitz, S.: Demystifying reachability in vector addition systems. In: 30th Annual ACM/IEEE Symposium on Logic in Computer Science, LICS 2015, Kyoto, Japan, 6–10 July 2015, pp. 56–67 (2015)
5. Cardoza, E., Lipton, R.J., Meyer, A.R.: Exponential space complete problems for Petri nets and commutative semigroups: preliminary report. In: Proceedings of the 8th Annual ACM Symposium on Theory of Computing, Hershey, Pennsylvania, USA, 3–5 May 1976, pp. 50–54 (1976)
6. Czerwinski, W., Lasota, S., Lazic, R., Leroux, J., Mazowiecki, F.: The reachability problem for Petri nets is not elementary (extended abstract). CoRR, abs/1809.07115 (2018)
7. van der Aalst, W.M.P.: The application of Petri nets to workflow management. J. Circ. Syst. Comput. **8**(1), 21–66 (1998)
8. Esparza, J.: Decidability and complexity of Petri net problems — an introduction. In: Reisig, W., Rozenberg, G. (eds.) ACPN 1996. LNCS, vol. 1491, pp. 374–428. Springer, Heidelberg (1998). https://doi.org/10.1007/3-540-65306-6_20
9. Desel, J., Esparza, J.: Free Choice Petri Nets. Cambridge University Press, New York (1995)
10. David, R., Alla, H.: Continuous Petri nets. In: Proceedings of the 8th European Workshop on Application and Theory of Petri Nets, Zaragoza, Spain, pp. 275–294 (1987)
11. Fraca, E., Haddad, S.: Complexity analysis of continuous Petri nets. Fundam. Inform. **137**(1), 1–28 (2015)
12. Blondin, M., Haase, C.: Logics for continuous reachability in Petri nets and vector addition systems with states. In: 32nd Annual ACM/IEEE Symposium on Logic in Computer Science, LICS 2017, Reykjavik, Iceland, 20–23 June 2017, pp. 1–12 (2017)
13. David, R., Alla, H.: Petri nets for modeling of dynamic systems: a survey. Automatica **30**(2), 175–202 (1994)
14. Alla, H., David, R.: Continuous and hybrid Petri nets. J. Circ. Syst. Comput. **8**, 159–188 (1998)
15. Jensen, K.: Coloured Petri nets - preface by the section editor. STTT **2**(2), 95–97 (1998)
16. Wang, J.: Timed Petri nets. Timed Petri Nets: Theory and Application. The Kluwer International Series on Discrete Event Dynamic Systems, vol. 9, pp. 63–123. Springer, Boston (1998). https://doi.org/10.1007/978-1-4615-5537-7_4
17. Abdulla, P.A., Nylén, A.: Timed Petri nets and BQOs. In: Colom, J.-M., Koutny, M. (eds.) ICATPN 2001. LNCS, vol. 2075, pp. 53–70. Springer, Heidelberg (2001). https://doi.org/10.1007/3-540-45740-2_5
18. Lazic, R., Newcomb, T.C., Ouaknine, J., Roscoe, A.W., Worrell, J.: Nets with tokens which carry data. Fundam. Inform. **88**(3), 251–274 (2008)

19. Rosa-Velardo, F., de Frutos-Escrig, D.: Forward analysis for Petri nets with name creation. In: Lilius, J., Penczek, W. (eds.) PETRI NETS 2010. LNCS, vol. 6128, pp. 185–205. Springer, Heidelberg (2010). https://doi.org/10.1007/978-3-642-13675-7_12
20. Lazić, R., Totzke, P.: What makes Petri nets harder to verify: stack or data? In: Gibson-Robinson, T., Hopcroft, P., Lazić, R. (eds.) Concurrency, Security, and Puzzles. LNCS, vol. 10160, pp. 144–161. Springer, Cham (2017). https://doi.org/10.1007/978-3-319-51046-0_8
21. Hofman, P., Lasota, S., Lazić, R., Leroux, J., Schmitz, S., Totzke, P.: Coverability trees for Petri nets with unordered data. In: Jacobs, B., Löding, C. (eds.) FOSSACS 2016. LNCS, vol. 9634, pp. 445–461. Springer, Heidelberg (2016). https://doi.org/10.1007/978-3-662-49630-5_26
22. Hofman, P., Leroux, J., Totzke, P.: Linear combinations of unordered data vectors. In: 32nd Annual ACM/IEEE Symposium on Logic in Computer Science, LICS 2017, Reykjavik, Iceland, 20–23 June 2017, pp. 1–11 (2017)
23. Hofman, P., Lasota, S.: Linear equations with ordered data. In: 29th International Conference on Concurrency Theory, CONCUR 2018, Beijing, China, 4–7 September 2018, pp. 24:1–24:17 (2018)
24. Silva, M., Terue, E., Colom, J.M.: Linear algebraic and linear programming techniques for the analysis of place/transition net systems. In: Reisig, W., Rozenberg, G. (eds.) ACPN 1996. LNCS, vol. 1491, pp. 309–373. Springer, Heidelberg (1998). https://doi.org/10.1007/3-540-65306-6_19
25. Blondin, M., Finkel, A., Haase, C., Haddad, S.: Approaching the coverability problem continuously. In: Chechik, M., Raskin, J.-F. (eds.) TACAS 2016. LNCS, vol. 9636, pp. 480–496. Springer, Heidelberg (2016). https://doi.org/10.1007/978-3-662-49674-9_28
26. Esparza, J., Ledesma-Garza, R., Majumdar, R., Meyer, P., Niksic, F.: An SMT-based approach to coverability analysis. In: Biere, A., Bloem, R. (eds.) CAV 2014. LNCS, vol. 8559, pp. 603–619. Springer, Cham (2014). https://doi.org/10.1007/978-3-319-08867-9_40
27. Gupta, U., Shah, P., Akshay, S., Hofman, P.: Continuous reachability for unordered data Petri nets is in PTime. CoRR abs/1902.05604 (2019). arxiv.org/abs/1902.05604
28. Rosa-Velardo, F., de Frutos-Escrig, D.: Decidability and complexity of Petri nets with unordered data. Theor. Comput. Sci. **412**(34), 4439–4451 (2011)

Optimal Satisfiability Checking for Arithmetic μ-Calculi

Daniel Hausmann(✉) and Lutz Schröder

Friedrich-Alexander-Universität Erlangen-Nürnberg, Erlangen, Germany
{daniel.hausmann,lutz.schroeder}@fau.de

Abstract. The coalgebraic μ-calculus provides a generic semantic framework for fixpoint logics with branching types beyond the standard relational setup, e.g. probabilistic, weighted, or game-based. Previous work on the coalgebraic μ-calculus includes an exponential time upper bound on satisfiability checking, which however requires a well-behaved set of tableau rules for the next-step modalities. Such rules are not available in all cases of interest, in particular ones involving either integer weights as in the graded μ-calculus, or real-valued weights in combination with non-linear arithmetic. In the present work, we prove the same upper complexity bound under more general assumptions, specifically regarding the complexity of the (much simpler) satisfiability problem for the underlying *one-step logic*, roughly described as the nesting-free next-step fragment of the logic. The bound is realized by a generic global caching algorithm that supports on-the-fly satisfiability checking. Example applications include new exponential-time upper bounds for satisfiability checking in an extension of the graded μ-calculus with polynomial inequalities (including positive Presburger arithmetic), as well as an extension of the (two-valued) probabilistic μ-calculus with polynomial inequalities.

1 Introduction

Modal fixpoint logics are a well-established tool in the temporal specification, verification, and analysis of concurrent systems. One of the most expressive logics of this type is the modal μ-calculus [2,3,20], which features explicit least and greatest fixpoint operators; roughly speaking, these serve to specify liveness properties (least fixpoints) and safety properties (greatest fixpoints), respectively. Like most modal logics, the modal μ-calculus is traditionally interpreted over relational models such as Kripke frames or labelled transition systems. The growing interest in more expressive models where transitions are governed, e.g., by probabilities, weights, or games has sparked a commensurate growth of temporal logics and fixpoint logics interpreted over such systems; prominent examples include probabilistic μ-calculi [5,17,24], the alternating-time μ-calculus [1], and the monotone μ-calculus, which contains Parikh's game logic [28]. The graded μ-calculus [21] features next-step modalities that count successors; it is standardly interpreted over Kripke frames but, as pointed out by D'Agostino and

M. Bojańczyk and A. Simpson (Eds.): FOSSACS 2019, LNCS 11425, pp. 277–294, 2019.
https://doi.org/10.1007/978-3-030-17127-8_16

Visser [6], graded modalities are more naturally interpreted over so-called multigraphs, where edges carry integer weights, and in fact this modification leads to better bounds on minimum model size for satisfiable formulas.

Coalgebraic logic [29,34] has emerged as a unifying framework for modal logics interpreted over such more general models. It is based on casting the transition type of the systems at hand as a set functor, and the systems in question as coalgebras for this type functor, following the paradigm of universal coalgebra [31]; additionally, modalities are interpreted as so-called *predicate liftings*. The *coalgebraic μ-calculus* [4] caters for fixpoint logics within this framework, and essentially covers all mentioned (two-valued) examples as instances. It has been shown that satisfiability checking in a coalgebraic μ-calculus is in ExpTime, *provided* that one exhibits a set of tableau rules for the modalities, so-called *one-step rules*, that is *tractable* in a suitable sense (an assumption made also in our own previous work on the flat [14] and alternation-free [16] fragments of the coalgebraic μ-calculus). Such rules are known for many important cases, notably including alternating-time logics, the probabilistic μ-calculus even when extended with linear inequalities, and game logic [4,22,36]. There are, however, important cases where such rule sets are currently missing, and where there is in fact little perspective for finding suitable rules. One prominent case of this kind is graded modal logic; further cases arise when logics over systems with non-negative real weights, such as probabilistic systems, are taken beyond linear arithmetic to include polynomial inequalities.

The object of the current paper is to fill this gap by proving a generic ExpTime upper bound for coalgebraic μ-calculi in the absence of tractable sets of modal tableau rules. The method we use instead is to analyse the so-called *one-step satisfiability* problem of the logic on a semantic level – this problem is essentially the satisfiability problem of a very small fragment of the logic, the *one-step logic*, which excludes not only fixpoints, but also nested next-step modalities, with a correspondingly simplified semantics that no longer involves actual transitions. E.g. the one-step logic of the relational μ-calculus is interpreted over models essentially consisting of a set with a distinguished subset, abstracting the successors of a single state that is not itself part of the model. We have applied this principle to satisfiability checking in coalgebraic (next-step) modal logics [35], coalgebraic hybrid logics [26], and reasoning with global assumptions in coalgebraic modal logics [23]. It also appears implicitly in work on automata for the coalgebraic μ-calculus [8], which however establishes only a doubly exponential upper bound in the case without tractable modal tableau rules.

Our main example applications are on the one hand the graded modal μ-calculus and its extension with (monotone) polynomial inequalities, including Presburger modalities, i.e. (monotone) linear inequalities, and on the other hand the extension of the (two-valued) probabilistic μ-calculus [4,24] with (monotone) polynomial inequalities. While the graded μ-calculus as such is known to be in ExpTime [21], the other mentioned instances of our result are, to our best knowledge, new. At the same time, our proofs are fairly simple, even compared to specific ones, e.g. for the graded μ-calculus.

Technically, we base our results on an automata-theoretic treatment by means of standard parity automata with singly exponential branching degree (in particular on modal steps), thus precisely enabling the singly exponential upper bound, in contrast to previous work in [8] where the introduced Λ-automata lead to doubly exponential branching on modal steps in the resulting satisfiability games. Our algorithm witnessing the singly exponential time bound is, in fact, a global caching algorithm [11,12], and is able to decide the satisfiability of nodes on-the-fly, that is, possibly before the tableau is fully expanded, thus offering a perspective for practically feasible reasoning. A side result of our approach is a criterion for a polynomial bound on branching in models, which holds in all our examples.

Organization. In Sect. 2, we recall the basics of the coalgebraic μ-calculus. We outline our automata-theoretic approach in Sect. 3, and present the global caching algorithm and its runtime analysis in Sect. 4. Soundness and completeness of the algorithm are proved in Sect. 5.

2 The Coalgebraic μ-Calculus

We recall basic definitions in coalgebraic logic [29,34] and the coalgebraic μ-calculus [4].

Syntax. We fix a *modal similarity type* Λ, that is, a set of modal operators with assigned finite arities, possibly including propositional atoms as nullary modalities. For readability, we restrict the technical development to unary modalities, noting that all proofs generalize to higher arities by just writing more indices; in fact, we will liberally use higher arities in examples. We assume that Λ is closed under duals, i.e., that for each modal operator $\heartsuit \in \Lambda$, there is a *dual* $\overline{\heartsuit} \in \Lambda$ such that $\overline{\overline{\heartsuit}} = \heartsuit$ for all $\heartsuit \in \Lambda$. Let $\mathbf{V}$ be an infinite set of *fixpoint variables*. Formulas of the *coalgebraic μ-calculus* (over Λ) are given by the grammar

$$\psi, \phi ::= \bot \mid \top \mid \psi \wedge \phi \mid \psi \vee \phi \mid \heartsuit\phi \mid X \mid \mu X.\, \psi \mid \nu X.\, \psi \qquad \heartsuit \in \Lambda, X \in \mathbf{V}.$$

As usual, μ and ν take least and greatest fixpoints, respectively. Negation is not included but can be defined as usual. Throughout, we use $\eta \in \{\mu, \nu\}$ as a placeholder for fixpoint operators; we briefly refer to formulas of the form $\eta X.\, \phi$ as *fixpoints*. Fixpoint operators *bind* their fixpoint variables, so that we have standard notions of bound and free fixpoint variables; a formula is closed if it contains no free fixpoint variables. We assume w.l.o.g. that all formulas are *clean*, i.e. each fixpoint variable appears in at most one fixpoint operator, and *irredundant*, i.e. each bound variable is used at least once. Moreover, we restrict to *guarded* formulas, in which all occurrences of fixpoint variables are separated by at least one modal operator from their binding fixpoint operator (this is standard although possibly not w.l.o.g. [9]). For $\heartsuit \in \Lambda$, we denote by $\mathsf{size}(\heartsuit)$ the length of a suitable representation of $\heartsuit$; for natural or rational numbers indexing $\heartsuit$, we assume binary representation. The *length* $|\psi|$ of a formula is its length over the alphabet $\{\bot, \top, \wedge, \vee\} \cup \Lambda \cup \mathbf{V} \cup \{\eta X. \mid X \in \mathbf{V}\}$, while the *size* $\mathsf{size}(\psi)$ of ψ is defined by counting $\mathsf{size}(\heartsuit)$ for each $\heartsuit \in \Lambda$ (and 1 for all other operators). The *alternation depth* $\mathsf{ad}(\eta X.\psi)$ of a fixpoint $\eta X.\psi$ is the maximal depth of nesting of such alternating least and greatest fixpoints in ψ that depend on X, tweaked to be *even* for least fixpoint formulas and *odd* for greatest fixpoint formulas (that is, starting with $\mathsf{ad}(\mu X.\psi) = 2$ and $\mathsf{ad}(\nu X.\psi) = 1$ for closed ψ). For a more detailed definition of various flavours of alternation depth, see e.g. [27].

Semantics. As indicated above, the branching type of the underlying systems is a parameter of the framework, given by fixing a **Set**-endofunctor T. Elements of TU should be thought of as structured collections over U that serve as collections of successors of states – e.g. in the most basic example, classical relational systems, T is powerset $\mathcal{P}$. Formulas are then interpreted over *T-coalgebras* (C, ξ) consisting of a set C of *states* and a *transition function* $\xi : C \to TC$ that assigns a structured collection $\xi(x) \in TC$ of successors (and observations) to $x \in C$; e.g. $\mathcal{P}$-coalgebras are just Kripke frames, as they assign a set of successors to each state. We interpret each modal operator $\heartsuit \in \Lambda$ as a *T-predicate lifting* $[\![\heartsuit]\!]$, that is, a natural transformation $[\![\heartsuit]\!] : \mathcal{Q} \to \mathcal{Q} \circ T^{op}$ where $\mathcal{Q} : \mathbf{Set}^{op} \to \mathbf{Set}$ denotes the contravariant powerset functor. Predicate liftings thus are families of functions $[\![\heartsuit]\!]_U : \mathcal{Q}(U) \to \mathcal{Q}(TU)$ satisfying *naturality*, i.e. $[\![\heartsuit]\!]_U(f^{-1}[A]) = (Tf)^{-1}[[\![\heartsuit]\!]_V(A)]$ for $f : U \to V$ and $A \subseteq V$, where f^{-1} denotes preimage. E.g. the standard relational box modality is interpreted by $[\![\Box]\!]_U(A) = \{B \in \mathcal{P}(U) \mid B \subseteq A\}$. For sets $U \subseteq V$, we write $\overline{U} = V \setminus U$ for the *complement* of U in V when V is understood from the context. We require that duality of modal operators is respected, i.e. $[\![\heartsuit]\!]_U(A) = \overline{[\![\overline{\heartsuit}]\!]_U \overline{A}}$ for $A \subseteq U$. To ensure existence of fixpoints, we require that all $[\![\heartsuit]\!]$ are *monotone*, i.e. $A \subseteq B \subseteq U$ implies $[\![\heartsuit]\!]_U(A) \subseteq [\![\heartsuit]\!]_U(B)$.

A *valuation* is a partial function $i : \mathbf{V} \rightharpoonup \mathcal{P}(C)$ that assigns sets $i(X)$ of states to fixpoint variables X. The *extension* $[\![\phi]\!]_i \subseteq C$ of a formula ϕ in a T-coalgebra (C, ξ) is defined by the expected clauses for propositional operators and

$$[\![\heartsuit\psi]\!]_i = \xi^{-1}[[\![\heartsuit]\!]_C([\![\psi]\!]_i)] \qquad [\![\mu X.\,\psi]\!]_i = \mathsf{LFP}([\![\psi]\!]_i^X)$$
$$[\![X]\!]_i = i(X) \qquad [\![\nu X.\,\psi]\!]_i = \mathsf{GFP}([\![\psi]\!]_i^X),$$

where LFP and GFP compute the least and greatest fixpoints of their argument functions, respectively, where $[\![\psi]\!]_i^X(A) = [\![\psi]\!]_{i[X \mapsto A]}$ for $A \subseteq C$, and where $(i[X \mapsto A])(X) = A$ and $(i[X \mapsto A])(Y) = i(Y)$ for $Y \neq X$. In particular, the extension is invariant under *unfolding* of fixpoints, i.e. $[\![\eta X.\,\psi]\!]_i = [\![\psi[X \mapsto \eta X.\,\psi]]\!]_i$. For closed formulas ψ, the valuation i is irrelevant, so we write $[\![\psi]\!]$ instead of $[\![\psi]\!]_i$. A state $x \in C$ *satisfies* a closed formula ψ (denoted $x \models \psi$) if $x \in [\![\psi]\!]$. Given a set Z, we define the set $\Lambda(Z) = \{\heartsuit z \mid \heartsuit \in \Lambda, z \in Z\}$ of *modal literals* (over Z). A closed formula χ is *satisfiable* if there is a coalgebra (C, ξ) and a state $x \in C$ such that $x \models \chi$.

Example 1. We now detail several instances of the coalgebraic μ-calculus; for further examples, e.g. the alternating-time μ-calculus, see [4].

1. To obtain the standard modal μ-calculus [19] (which contains CTL as a fragment), we take $\Lambda = \{\Diamond, \Box\} \cup P$ where P is a set of propositional atoms, seen as nullary modalities. The semantics is captured by $TU = \mathcal{P}(U) \times \mathcal{P}(P)$, so that T-coalgebras are Kripke models, as they assign to each state a set of successors and a set of atoms satisfied in the state. The relevant predicate liftings are

$$[\![\Diamond]\!]_U(A) = \{(B,Q) \in TU \mid A \cap B \neq \emptyset\} \quad [\![\Box]\!]_U(A) = \{(B,Q) \in TU \mid B \subseteq A\}$$

and $[\![p]\!]_U = \{(B,Q) \in TU \mid p \in Q\}$, a nullary predicate lifting. Standard example formulas include the CTL-formula $\mathsf{AF}\, p = \mu X.\,(p \vee \Box X)$, which states that on all paths, p eventually holds, and the fairness formula $\nu X.\,\mu Y.\,((p \wedge \Diamond X) \vee \Diamond Y)$, which asserts the existence of a path on which p holds infinitely often.
2. We interpret the *graded μ-calculus* [21] over multigraphs [6], i.e. T-coalgebras for the multiset functor $T = \mathcal{B}$, defined by

$$\mathcal{B}(U) = \{\theta : U \to \mathbb{N} \cup \{\infty\}\} \qquad \mathcal{B}(f)(\theta)(v) = \textstyle\sum_{u \in U \mid f(u)=v} \theta(u)$$

for sets U, V and functions $f : U \to V$, $\theta : U \to \mathbb{N} \cup \{\infty\}$. Thus $\mathcal{B}$-coalgebras (C, ξ) assign multisets $\xi(x)$ to states $x \in C$, with the intuition that x has $y \in C$ as successor with multiplicity m if $\xi(x)(y) = m$. We use the modal similarity type $\Lambda = \{\langle m \rangle, [m] \mid m \in \mathbb{N} \cup \{\infty\}\}$ and define the predicate liftings

$$[\![\langle m \rangle]\!]_U(A) = \{\theta \in \mathcal{B}(U) \mid \theta(A) > m\} \quad [\![[m]]\!]_U(A) = \{\theta \in \mathcal{B}(U) \mid \theta(\overline{A}) \leq m\}$$

for sets U and $A \subseteq U$, where $\theta(A) = \sum_{a \in A} \theta(a)$. E.g. a state satisfies $\nu X.\,(\psi \wedge \langle 1 \rangle X)$ if it is the root of an infinite binary tree in which ψ is satisfied globally.
3. Similarly, the two-valued *probabilistic μ-calculus* [4,24] is obtained by using the distribution functor $T = \mathcal{D}$ that maps sets U to probability distributions over U with countable support, defined by

$$\mathcal{D}(U) = \{d : U \to (\mathbb{Q} \cap [0,1]) \mid \textstyle\sum_{u \in U} d(u) = 1\}.$$

Then T-coalgebras are just Markov chains. We use the modal similarity type $\Lambda = \{\langle p \rangle, [p] \mid p \in \mathbb{Q} \cap [0,1]\}$ and define the predicate liftings

$$[\![\langle p \rangle]\!]_U(A) = \{d \in \mathcal{D}(U) \mid d(A) > p\} \quad [\![[p]]\!]_U(A) = \{d \in \mathcal{D}(U) \mid d(\overline{A}) \leq p\},$$

for sets U and $A \subseteq U$, where again $d(A) = \sum_{a \in A} d(a)$.
4. We interpret the *graded μ-calculus with polynomial inequalities* over the semantic domain from item 2 (i.e. multigraphs). We put $\Lambda = \{L_{p,b}, M_{p,b} \mid p \in \mathbb{N}_{>0}[X_1, \dots, X_n], b, n \in \mathbb{N}\}$ (that is, p ranges over multivariate polynomials with positive integer coefficients) and define the predicate liftings

$$[\![L_{p,b}]\!]_U(A_1, \dots, A_n) = \{\theta \in \mathcal{B}(U) \mid p(\theta(A_1), \dots, \theta(A_n)) > b)\}$$
$$[\![M_{p,b}]\!]_U(A_1, \dots, A_n) = \{\theta \in \mathcal{B}(U) \mid p(\theta(\overline{A_1}), \dots, \theta(\overline{A_n})) \leq b)\},$$

for sets U and $A_1, \dots, A_n \subseteq U$, where $\theta(A) = \sum_{a \in A} \theta(a)$. This logic subsumes the *Presburger μ-calculus*, that is, the extension of the graded μ-calculus with (monotone) linear inequalities, which may be seen as the fixpoint variant of *Presburger modal logic* [7]. E.g. the formula $\mu Y. (r \vee L_{2X_1+X_2^2,2}(p \wedge Y, q \wedge Y))$ says that the current state is the root of a finite tree all whose leaves satisfy r, and each of whose inner nodes has n_1 children satisfying p and n_2 children satisfying q where $2n_1 + n_2^2 > 2$. One sees an apparent coding of the logic into the graded μ-calculus, which however incurs exponential blowup.

5. Similarly, we use the semantic domain from item 3, Markov chains, to obtain the *probabilistic μ-calculus with polynomial inequalities* [23]: We put $\Lambda = \{L_{p,b}, M_{p,b} \mid p \in \mathbb{Q}_{>0}[X_1, \dots, X_n], b \in \mathbb{Q}_{\geq 0}, n \in \mathbb{N}\}$ (i.e. p ranges over polynomials) and

$$[\![L_{p,b}]\!]_U(A_1, \dots, A_n) = \{d \in \mathcal{D}(U) \mid p(d(A_1), \dots, d(A_n)) > b\}$$
$$[\![M_{p,b}]\!]_U(A_1, \dots, A_n) = \{d \in \mathcal{D}(U) \mid p(d(\overline{A_1}), \dots, d(\overline{A_n})) \leq b\}$$

for sets U and $A_1, \dots, A_n \subseteq U$. This logic presumably does not encode into the probabilistic μ-calculus as in 3 above, and can express constraints on independent products of events (see also [25]). E.g. the formula $\nu Y. L_{X_1 X_2, 0.9}(p \wedge Y, q \wedge Y)$ says roughly that two independently sampled successors of the current state will satisfy p and q, respectively, and then satisfy the same property again, with probability at least 0.9.

(The modalities in the last two items are inevitably less general than in the corresponding next-step logics [7,23], due to the need to ensure monotonicity.)

3 Tracking Automata

We use *parity automata* (e.g. [13]) that track single formulas along paths through potential models to decide whether it is possible to construct a model in which all least fixpoint formulas are eventually satisfied. Formally, (nondeterministic) parity automata are tuples $\mathsf{A} = (V, \Sigma, \Delta, q_0, \alpha)$ where V is a set of *nodes*; Σ is a finite set, the *alphabet*; $\Delta \subseteq V \times \Sigma \times V$ is the *transition relation* assigning a set $\Delta(v, a) = \{u \mid (v, a, u) \in \Delta\}$ of nodes to all $v \in V$ and $a \in \Sigma$; $q_0 \in V$ is the *initial node*; and $\alpha : \Delta \to \mathbb{N}$ is the *priority function*, assigning priorities $\alpha(v, a, u) \in \mathbb{N}$ to *transitions* $(v, a, u) \in \Delta$ (this is the standard in recent work since it yields slightly more succinct automata). If Δ is a (partial) functional relation, then A is said to be *deterministic*, and we denote the corresponding partial function by $\delta : V \times \Sigma \rightharpoonup V$. The automaton A *accepts* an infinite word $w = w_0, w_1, \dots \in \Sigma^\omega$ if there is a w-path through A on which the highest priority that is passed infinitely often is even; formally, the language that is accepted by A is defined by $L(\mathsf{A}) = \{w \in \Sigma^\omega \mid \exists \rho \in \mathsf{run}(\mathsf{A}, w).\ \max(\mathsf{Inf}(\alpha \circ \rho)) \text{ is even}\}$, where $\mathsf{run}(\mathsf{A}, w)$ denotes the set of infinite sequences $(\rho_0, w_0, \rho_1), (\rho_1, w_1, \rho_2), \dots \in \Delta^\omega$ such that $\rho_0 = q_0$ and where, given an infinite sequence S, $\mathsf{Inf}(S)$ denotes the elements that occur infinitely often in S. Here, we see infinite sequences $\rho \in U^\omega$ over some set U as functions $\mathbb{N} \to U$ and write ρ_i to denote the i-th element of ρ.

We now *fix a target formula* χ and put $n_0 = |\chi|$, $n_1 = \mathsf{size}(\chi)$. We let $\mathbf{F}$ denote the *Fischer-Ladner closure* [20] of χ; i.e. $\mathbf{F}$ contains all formulas that can arise as subformulas when unfolding each fixpoint in χ exactly once. We put $k = \max\{\mathsf{ad}(\psi) \mid \psi \in \mathbf{F}\}$ and $\mathsf{selections} = \mathcal{P}(\mathbf{F} \cap \Lambda(\mathbf{F}))$ ($\mathbf{F} \cap \Lambda(\mathbf{F})$ is the set of modal literals in $\mathbf{F}$). We have $|\mathbf{F}| \leq n$ and hence $|\mathsf{selections}| \leq 2^n$.

Definition 2 (Tracking automaton). The *tracking automaton* for χ is a nondeterministic parity automaton $\mathsf{A}_\chi = (\mathbf{F}, \Sigma, \Delta, q_0, \alpha)$, where $q_0 = \chi$,

$$\begin{aligned}\Sigma = &\{(\psi_0 \vee \psi_1, b) \in \mathbf{F} \times \{0,1\}\} \cup \{(\psi_0 \wedge \psi_1, 0) \in \mathbf{F} \times \{0\}\} \cup \\ &\{(\eta X.\, \psi_1, 0) \in \mathbf{F} \times \{0\}\} \cup \mathsf{selections}\ ,\end{aligned}$$

and for $\psi, \psi_0, \psi_1 \in \mathbf{F}$, $\kappa \in \mathsf{selections}$ and $b \in \{0,1\}$,

$$\begin{aligned}\Delta(\psi, \kappa) &= \{\psi_0 \in \mathbf{F} \mid \psi \in \kappa \cap \Lambda(\{\psi_0\})\} \\ \Delta(\psi, (\psi_0 \vee \psi_1, b)) &= \{\psi_b \mid \psi = \psi_0 \vee \psi_1\} \cup \{\psi \mid \psi \neq \psi_0 \vee \psi_1\} \\ \Delta(\psi, (\psi_0 \wedge \psi_1, 0)) &= \{\psi_0, \psi_1 \mid \psi = \psi_0 \wedge \psi_1\} \cup \{\psi \mid \psi \neq \psi_0 \wedge \psi_1\} \\ \Delta(\psi, (\eta X.\, \psi_1, 0)) &= \{\psi_1[X \mapsto \psi] \mid \psi = \eta X.\, \psi_1\} \cup \{\psi \mid \psi \neq \eta X.\, \psi_1\}\end{aligned}$$

E.g. the last clause means that when tracking the unfolding of a fixpoint $\eta X.\, \psi_1$ at ψ, we track ψ to the unfolding $\psi_1[X \mapsto \psi]$ if ψ equals the unfolded fixpoint, and to ψ otherwise; similarly for the other clauses, and in particular a modal literal $\psi = \heartsuit\psi_0$ is only tracked to ψ_0 through a selection κ if $\heartsuit\psi_0 \in \kappa$, i.e. if κ selects $\heartsuit\psi_0$ to be tracked. The priority function α is derived from the alternation depths of formulas, counting only unfoldings of fixpoints (i.e. all other transitions have priority 1). Formally, $\alpha(\psi, \sigma, \psi') = 1$ if $\psi = \psi'$ or ψ is not a fixpoint literal; if ψ is a fixpoint literal and $\psi \neq \psi'$, then we put $\alpha(\psi, \sigma, \psi') = \mathsf{ad}(\psi)$.

Intuitively, words from Σ^ω encode infinite paths through coalgebras (C, ξ) in which states $x \in C$ are labelled with sets $l(x)$ of formulas, where letters $\kappa \in \mathsf{selections}$ encode modal steps from states $x \in C$ with label $l(x)$ to states $y \in C$ with label $\{\psi \mid \heartsuit\psi \in \kappa \cap l(x)\}$. The automaton is built to accept $L(\mathsf{A}_\chi) = \mathsf{BadBranch}_\chi$ where $\mathsf{BadBranch}_\chi$ is the set of words that encode a path on which a least fixpoint formula ψ is unfolded infinitely often without being dominated by any outer fixpoint formula (i.e. one with alternation depth greater than $\mathsf{ad}(\psi)$). Letters $(\psi_0 \vee \psi_1, b)$ choose disjuncts according to b, while for letters $(\psi_0 \wedge \psi_1, 0)$, the tracking automaton is nondeterministic, reflecting the fact that bad fixpoints can reside in either ψ_0 or ψ_1. The automaton A_χ has size n_0 and priorities 1 to k. Using a standard construction (e.g. [18]), we transform A_χ into an equivalent Büchi automaton of size $n_0 k$. Then we determinize the Büchi automaton using, e.g., the Safra/Piterman-construction [30,32] and obtain an equivalent deterministic parity automaton with priorities 0 to $2n_0k - 1$ and size $\mathcal{O}(((n_0k)!)^2)$. Finally we complement this parity automaton by increasing every priority by 1, obtaining a deterministic parity automaton $\mathsf{B}_\chi = (D_\chi, \Sigma, \delta, v_0, \beta)$ of size $\mathcal{O}(((n_0k)!)^2)$, with priorities 1 to $2n_0k$ and with

$$L(\mathsf{B}_\chi) = \overline{L(\mathsf{A}_\chi)} = \overline{\mathsf{BadBranch}_\chi} =: \mathsf{GoodBranch}_\chi,$$

i.e. B_χ is a deterministic parity automaton that accepts the words that encode paths along which satisfaction of least fixpoints is never deferred indefinitely. We define a labelling function $l : D_\chi \to \mathcal{P}(\mathbf{F})$ mapping each state of B_χ (e.g. a Safra tree) to the set of formulas occurring in it.

Remark 3. It has been noted that the standard tracking automata for *alternation-free* formulas are, in fact, Co-Büchi automata [10,16] and that the tracking automata for *aconjunctive* formulas are *limit-deterministic* parity automata [15]. These considerably simpler automata can be determinized to deterministic Büchi automata of size 3^{n_0} and to deterministic parity automata of size $\mathcal{O}((n_0k)!)$ and with $2n_0k$ priorities, respectively. This observation also holds true for the tracking automata in this work so that for formulas of suitable syntactic shape, Lemma 11 below yields accordingly lower bounds on the runtime of our satisfiability checking algorithm.

4 Global Caching for the Coalgebraic μ-Calculus

We now introduce a generic global caching algorithm for satisfiability in the coalgebraic μ-calculus. Given an input formula χ, the algorithm expands the determinized and complemented tracking automaton B_χ step by step and propagates (un)satisfiability through this graph; the algorithm terminates as soon as the initial node v_0 is marked as (un)satisfiable. The algorithm bears similarity to standard game-based algorithms for μ-calculi [8,9,15]; however, it crucially deviates from these algorithms in the treatment of modal steps: Intuitively, our algorithm decides whether it is possible to remove some of the modal transitions as well as one of the transitions from each reachable pair $((\psi_0\vee\psi_1),0),((\psi_0\vee\psi_1),1)$ of disjunction transitions within the automaton B_χ in such a way that the resulting sub-automaton of B_χ is totally accepting, that is, accepts any word for which there is an infinite run. In doing so, it is crucial that the labels of state nodes v in the reduced automaton are *one-step satisfiable*, in a sense introduced next, in the set of states that are reachable from v by the remaining modal transitions. Propagating (un)satisfiability over modal transitions thus involves *one-step satisfiability checking*, a functor-specific problem that in many instances can be solved in time singly exponential in $\mathsf{size}(\chi)$. In previous work [8], a variant of one-step satisfiability has been used in satisfiability games for coalgebraic μ-calculi, which however leads to a doubly exponential number of modal moves for one of the players and hence does not yield a singly exponential upper bound on satisfiability checking (unless a suitable set of tableau rules is provided).

Definition 4 (One-step satisfiability problem [26,33,35]). Let V be a finite set, let $v \subseteq \Lambda(V)$ such that $a \neq b$ whenever $\heartsuit_1 a, \heartsuit_2 b \in v$, and let $U \subseteq \mathcal{P}(V)$. The *one-step satisfiability problem* for inputs v and U is to decide whether $TU \cap [\![v]\!]_1 \neq \emptyset$, where

$$[\![v]\!]_1 = \textstyle\bigcap_{\heartsuit a\in v}[\![\heartsuit]\!]\{u \in U \mid a \in u\}.$$

We put $\mathsf{size}(v) = \sum_{\heartsuit a\in v}\mathsf{size}(\heartsuit)$, and denote the time it takes to solve the problem on v, U with $\mathsf{size}(v) = a$ and $|V| = b$ (hence $|U| \leq 2^b$) by $t(a, b)$.

Remark 5. We keep the definition of the actual one-step logic as mentioned in the introduction somewhat implicit in the above definition of the one-step satisfiability problem. One can see that it contains two layers: a purely propositional layer embodied in U, which postulates which propositional formulas over V are satisfiable; and a modal layer with nesting depth of modalities uniformly equal to 1, embodied in the set v, which specifies constraints on an element of TU.

Example 6. For the standard modal μ-calculus (Example 1.1), the one-step satisfiability problem is to decide for given $v \subseteq \Lambda(V)$ and $U \subseteq \mathcal{P}(V)$ whether there is $A \in \mathcal{P}(U) \cap [\![v]\!]_1$, that is, a subset $A \subseteq U$ such that for each $\Diamond a \in v$, there is $u \in A$ such that $a \in u$, and for each $\Box a \in v$ and each $u \in A$, $a \in u$. Here we have $t(a,b) \leq a \cdot 2^b$ where $a = \mathsf{size}(v)$, $b = |V|$. For the graded μ-calculus (Example 1.2), the one-step satisfiability problem is to decide for v, U as above whether there is a multiset $\theta \in \mathcal{B}(U)$ such that $\sum_{u \in U | a \in u} \theta(u) > m$ for each $\langle m \rangle a \in v$ and $\sum_{u \in U | a \notin u} \theta(u) \leq m$ for each $[m]a \in v$.

Definition 7 (States and Prestates). A node v of B_χ is a *state* if its label contains only modal literals ($l(v) \subseteq \Lambda(\mathbf{F})$), and otherwise a *prestate*, in which case we fix $\psi_v \in l(v) \setminus \Lambda(\mathbf{F})$. We write $\mathsf{states}, \mathsf{prestates} \subseteq D_\chi$ for the sets of states and prestates, respectively.

We next define $2n_0k$-ary set functions f and g that compute one-step (un)satisfiability w.r.t. their argument sets.

Definition 8 (One-step propagation). For sets $G \subseteq D_\chi$ and $\mathbf{X} = X_1, \dots, X_{2n_0k} \in \mathcal{P}(G)^{2n_0k}$, we put

$$\begin{aligned} f(\mathbf{X}) =& \{v \in \mathsf{prestates} \mid \exists b \in \{0,1\}.\, \delta(v,(\psi_v,b)) \in X_{\beta(v,(\psi_v,b))}\} \cup \\ & \{v \in \mathsf{states} \mid T(\textstyle\bigcup_{1 \leq i \leq 2n_0k} X_i(v)) \cap [\![l(v)]\!]_1 \neq \emptyset\} \\ g(\mathbf{X}) =& \{v \in \mathsf{prestates} \mid \forall b \in \{0,1\}.\, \delta(v,(\psi_v,b)) \in X_{\beta(v,(\psi_v,b))}\} \cup \\ & \{v \in \mathsf{states} \mid T(\textstyle\bigcup_{1 \leq i \leq 2n_0k} \overline{X_i}(v)) \cap [\![l(v)]\!]_1 = \emptyset\}, \end{aligned}$$

where $\beta(v,(\psi_v,b))$ abbreviates $\beta(v,(\psi_v,b),\delta(v,(\psi_v,b)))$ and where

$$X_i(v) = \{l(u) \mid u \in X_i, \exists \kappa \in \mathsf{selections}.\, \delta(v,\kappa) = u, \beta(v,\kappa,u) = i\}.$$

Since for states v, $l(v) \subseteq \Lambda(\mathbf{F})$ and $X_i(v) \subseteq \mathcal{P}(\mathbf{F})$ for all i, one-step propagation steps for states are instances of the one-step satisfiability problem with $|V| = |\mathbf{F}|$, solvable in time $t(n_1, n_0)$ because $\mathsf{size}(l(v)) \leq n_1$ and $|\mathbf{F}| \leq n_0$.

Definition 9 (Propagation). Given a set G, we put

$$\begin{aligned} \mathbf{E}_G &= \eta_{2n_0k} X_{2n_0k}. \dots \eta_2 X_2. \eta_1 X_1. f(\mathbf{X}) \\ \mathbf{A}_G &= \overline{\eta_{2n_0k}} X_{2n_0k} \dots \overline{\eta_2} X_2. \overline{\eta_1} X_1. g(\mathbf{X}), \end{aligned}$$

where $\mathbf{X} = X_1, \dots, X_{2n_0k}$ for $X_i \subseteq G$, where $\eta_i = \mu$ for odd i, $\eta_i = \nu$ for even i and where $\overline{\nu} = \mu$ and $\overline{\mu} = \nu$.

The set $\mathbf{E}_G$ contains nodes $v \in G$ for which there are choices for all disjunctions and modal transitions that are reachable from v within G (as indicated at the beginning of the section) such that the labels of all reachable states in the chosen sub-automaton of B_χ are one-step satisfiable and such that on all paths through the chosen sub-automaton, the highest priority that is passed infinitely often is even, the intuition being that no least fixpoint is unfolded infinitely often without being dominated. Dually, the set $\mathbf{A}_G$ contains nodes for which there exist no such suitable choices.

We recall that $v_0 \in D_\chi$ is the initial state of the determinized and complemented tracking automaton B_χ. The algorithm expands B_χ step-by-step starting from v_0; for prestates u, the expansion step adds nodes according to the fixed non-modal formula ψ_u that is to be expanded next (Definition 7), and for states, the expansion follows all (matching) selections. The order of expansion can be chosen freely, e.g. by heuristic methods. Optional intermediate propagation steps can be used judiciously to realize on-the-fly solving.

Algorithm 10 (Global caching). To decide the satisfiability of the input formula χ, initialize the sets of *unexpanded* and *expanded* nodes, $U = \{v_0\}$ and $G = \emptyset$, respectively.

1. Expansion: Choose some unexpanded node $u \in U$, remove u from U, and add u to G. If u is a prestate, then add the set $\{\delta(u,\sigma) \mid \sigma \in \Sigma \cap (\psi_u \times \{0,1\})\}$ to U. If u is a state, then add the set $\{\delta(u,\kappa) \mid \kappa \in \mathsf{selections}\}$ to U.
2. Optional propagation: Compute $\mathbf{E}_G$ and/or $\mathbf{A}_G$. If $v_0 \in \mathbf{E}_G$, then return 'satisfiable', if $v_0 \in \mathbf{A}_G$, then return 'unsatisfiable'.
3. If $U \neq \emptyset$, then continue with step 1.
4. Final propagation: Compute $\mathbf{E}_G$. If $v_0 \in \mathbf{E}_G$, then return 'satisfiable', otherwise return 'unsatisfiable'.

Lemma 11. *Algorithm 10 runs in time* $\mathcal{O}(((n_0 k)!)^{4n_0 k} \cdot t(n_1, n_0))$.

Proof. The loop of the algorithm expands the determinized and complemented tracking automaton node by node and hence is executed at most $|D_\chi| \in \mathcal{O}(((n_0 k)!)^2) \subseteq 2^{\mathcal{O}(n_0 k \log n_0)}$ times. A single expansion step can be implemented in time $\mathcal{O}(2^{n_0})$ since propositional expansion is unproblematic and for the modal expansion of a state u, all (matching) selections, of which there are (at most) 2^{n_0}, have to be considered. A single propagation step consists in computing two fixpoints of nesting depth $2n_0 k$ of the functions f and g over $\mathcal{P}(D_\chi)^{2n_0 k}$ and can hence be implemented in time $2(|D_\chi|^{2n_0 k} \cdot t(n_1, n_0)) \in \mathcal{O}(((n_0 k!)^2)^{2n_0 k} \cdot t(n_1, n_0)) \subseteq 2^{\mathcal{O}(n_0^2 k^2 \log n_0 + \log(t(n_1, n_0)))}$, noting that a single computation of $f(\mathbf{X})$ and $g(\mathbf{X})$ for a tuple $\mathbf{X} \in \mathcal{P}(D_\chi)^{2n_0 k}$ can be implemented in time $\mathcal{O}(t(n_1, n_0))$ – this has been noted above for states, and prestates are unproblematic. Thus the complexity of the whole algorithm is dominated by the complexity of the propagation step. □

Corollary 12. *If the one-step satisfiability problem of a coalgebraic logic can be solved in time $t(a, b)$ exponential in $a + b$ on inputs $v \subseteq \Lambda(V)$, $U \subseteq \mathcal{P}(V)$*

with $\mathsf{size}(v) = a$, $|V| = b$, *then the satisfiability problem of the corresponding coalgebraic* μ*-calculus is in* ExpTime.

Since the existence of a tractable set of tableau rules implies the required time bound on one-step satisfiability, the above result subsumes earlier bounds obtained by tableau-based approaches in [4,15,16]; however, it covers additional example logics for which no suitable tableau rules are known. In particular we have

Proposition 13. *The satisfiability problems of the following logics are in* ExpTime*:*

1. *the standard* μ*-calculus,*
2. *the graded* μ*-calculus,*
3. *the (two-valued) probabilistic* μ*-calculus,*
4. *the graded* μ*-calculus with polynomial inequalities,*
5. *the (two-valued) probabilistic* μ*-calculus with polynomial inequalities.*

(Tractable sets of tableau rules have previously been claimed for the graded [36] and Presburger [22] μ-calculus but have since been discovered to be flawed [23].)

Proof. It suffices to show that the respective one-step satisfiability problems can be solved on inputs $v \subseteq \Lambda(V)$, $U \subseteq \mathcal{P}(V)$ with $\mathsf{size}(v) = a$ and $|V| = b$ in singly exponential time in $a + b$, i.e. in time $t(a, b) \in 2^{p(a+b)}$ for p at most polynomial. E.g. for standard relational modalities, we have $t(a, b) = a \cdot 2^b = 2^{b+\log a}$, see Example 6. While the bounds can be established by relatively easy arguments (e.g. using known bounds on sizes of solutions of systems of real or integer linear inequalities) for all of our remaining example logics, we import them from previous work for brevity. For the one-step satisfiability problem of graded modal logic, by [21, Lemma 1], we have $t(a, b) \leq (2 \cdot 2^a + 2)^b \leq 2^{ab+2b}$; the Lemma uses counters to check joint one-step satisfiability of constraints and directly extends to the one-step satisfiability problem of graded modal logic with monotone polynomial inequalities, in which case we require n counters for each n-ary polynomial. The bound for (two-valued) probabilistic modal logic (with polynomial inequalities) is shown in [23, Example 7]. □

Remark 14. We also obtain a polynomial bound on branching width in models for all our example logics simply by importing Lemma 6 and the observations in Example 7 from [23]. With the exception of the standard μ-calculus, this bound appears to be new in all our example logics. Of course, for graded and Presburger μ-calculi, polynomial branching holds only in their coalgebraic semantics, i.e. over multigraph models but not over Kripke models.

5 Soundness and Completeness

We now prove the central result, that is, the soundness and completeness of Algorithm 10. As the sets $\mathbf{E}_G$ and $\mathbf{A}_G$ grow monotonically with G, it suffices to prove equivalence of satisfiability and containment of the initial node v_0 in $\mathbf{E} := \mathbf{E}_{D_\chi}$. Our program is as follows: We show that $v_0 \in \mathbf{E}$ if and only if there is a *pre-semi-tableau* (Definition 15) for χ with *unfolding timeouts* (Definition 17), which in turn is the case if and only if χ is satisfiable. We establish the latter equivalence by constructing a model for χ from a given pre-semi-tableau with unfolding timeouts and, for the converse direction, extracting a pre-semi-tableau with unfolding timeouts from the model.

Definition 15 (Pre-semi-tableau). Given a ternary relation $R \subseteq A \times B \times A$ and $a \in A$, $b \in B$, we generally write $R(a) = \{a' \in A \mid \exists b \in B.\, (a, b, a') \in R\}$ and $R(a, b) = \{a' \in A \mid (a, b, a') \in R\}$. Let $W \subseteq D_\chi$ and put $U = W \cap \mathsf{prestates}$ and $V = W \cap \mathsf{states}$. Given a ternary relation $L \subseteq W \times \Sigma \times W$, the pair (W, L) is a *pre-semi-tableau* for χ if the following conditions hold: $L \subseteq \delta$; $T(L(v)) \cap [\![l(v)]\!]_1 \neq \emptyset$ for all $v \in V$; for each $u \in U$, there is exactly one $b \in \{0, 1\}$ such that $L(u, (\psi_u, b)) = \{\delta(u, (\psi_u, b))\}$, and for all other $\sigma \in \Sigma$, $L(u, \sigma) = \emptyset$; and there is no L-cycle that contains only elements from U. A *path* through a pre-semi-tableau is an infinite sequence $(v_0, \sigma_0), (v_1, \sigma_1), \ldots \in (W \times \Sigma)^\omega$ such that for all i, $v_{i+1} \in L(v_i, \sigma_i)$. We denote *the* first state that is reachable by zero or more L-steps from a node $v \in W$ by $\lceil v \rceil$ (since there is no L-cycle within U, such a state always exists).

Given a state v, the relation L of a pre-semi-tableau thus picks a set $L(v)$ of nodes in which $l(v)$ is one-step satisfiable; given a prestate u, L picks a single (pre)state that is obtained from u by transforming the formula ψ_u.

Definition 16 (Tracking timeouts). Given a path $\rho = (v_0, \sigma_0), (v_1, \sigma_1), \ldots$ through a pre-semi-tableau, we say that priority i *occurs* (at position j) in ρ if $\beta(v_j, \sigma_j, v_{j+1}) = i$, recalling that β is the priority function of the determinised and complemented tracking automaton B_χ. Then the path ρ has *tracking timeouts* $\overline{m} = (m_{2n_0k}, \ldots, m_1)$ if for each odd $1 \leq i < 2n_0k$, priority i occurs at most m_i times in ρ before some priority greater than i occurs in ρ. Nothing is said about the m_i for even i, which are in fact irrelevant and serve only to ease notation. A node $w \in W$ in a pre-semi-tableau (W, L) has *tracking timeouts* $\overline{m}$ if every path through (W, L) starting at w has tracking timeouts $\overline{m}$. A pre-semi-tableau (W, L) *has tracking timeouts* if each $w \in W$ has tracking timeouts $\overline{m}$ for some $\overline{m}$.

Intuitively, a pre-semi-tableau (W, L) has tracking timeouts if every word that encodes an infinite L-path through W is accepted by B_χ. The next definition is geared towards characterizing non-acceptance by A_χ:

Definition 17 (Traces and unfolding timeouts). Let (W, L) be a graph with $L \subseteq W \times \Sigma \times W$ and labeling function $l : W \to \mathcal{P}(\mathbf{F})$. Given an L-path $\rho = (v_0, \sigma_0), (v_1, \sigma_1), \ldots$ (with $(v_i, \sigma_i, v_{i+1}) \in L$ for $i \geq 0$) and a sequence of formulas $\Psi = \psi_0, \psi_1, \ldots$, we say that Ψ is a *trace* of ψ_0 along ρ (we also say that ρ *contains* the trace Ψ) if $\psi_i \in l(v_i)$ and $\psi_{i+1} \in \Delta(\psi_i, \sigma_i)$ for all i. For i with $\psi_i = \eta X.\psi$ for some fixpoint variable X and some formula ψ, we say that Ψ *unfolds at level* $\mathsf{ad}(\psi_i)$ at position i. Then the trace Ψ has *unfolding* *timeout* $m \in \mathbb{N}$ for ψ_0 at level j if Ψ unfolds at most m times at level j before Ψ unfolds at some level greater than j. The path ρ has *unfolding timeouts* for ψ_0 at level j if there is, for all its traces Ψ of ψ_0, some m such that Ψ has unfolding timeout m for ψ_0 at level j. A node $w \in W$ has *unfolding timeouts* at level j for some formula ψ if every path through (W, L) that starts at w and that contains infinitely many steps (v_i, σ_i) such that $\sigma_i \in \mathsf{selections}$ has unfolding timeouts for ψ at level i. (Since fixpoint variables are by assumption guarded by modal operators, it suffices to require timeouts just for such paths that contain infinitely many modal steps.) A node $w \in W$ has *unfolding timeouts* $\overline{m} = (m_k, \dots, m_1)$ for some formula ψ if every path through (W, L) that starts at w and that contains infinitely many steps (v_i, σ_i) such that $\sigma_i \in \mathsf{selections}$ has, for each odd $1 \leq i \leq k$, unfolding timeouts $\overline{m}$ for ψ at level i; again the unfolding timeouts for even i, that is, for greatest fixpoints, are irrelevant. The graph (W, L) has *unfolding timeouts* if for each element $w \in W$ and each formula $\psi \in l(v)$, there is some vector $\overline{m}$ such that w has unfolding timeouts $\overline{m}$ for ψ. We denote the set of nodes that have unfolding timeouts $\overline{m}$ for ψ by $\mathsf{uto}(\psi, \overline{m}) \subseteq W$.

A graph (W, L) has unfolding timeouts if for all words that encode an infinite L-path through (W, L), all runs of the nondeterministic tracking automaton A_χ on the word are *non*-accepting. We recall that a run of A_χ is accepting if it unfolds some least fixpoint infinitely often without having it dominated.

Lemma 18. *Let (W, L) be a pre-semi-tableau. Then (W, L) has tracking timeouts if and only if it has unfolding timeouts.*

Proof. We recall that B_χ is obtained from A_χ by determinization and subsequent complementation so that we have $L(\mathsf{B}_\chi) = \overline{L(\mathsf{A}_\chi)}$. The result thus follows directly from the fact that having tracking timeouts means that B_χ accepts all words that encode a path in (W, L) while having unfolding timeouts means that A_χ does not accept any word that encodes a path in (W, L). □

Lemma 19. *We have $v_0 \in \mathbf{E}$ if and only if there is a pre-semi-tableau for χ that has tracking timeouts.*

Combining Lemmas 19 and 18, we obtain

Corollary 20. *We have $v_0 \in \mathbf{E}$ if and only if there is a pre-semi-tableau for χ that has unfolding timeouts.*

We now show that satisfiability of χ and the existence of a semi-pre-tableau for χ with unfolding timeouts coincide.

Definition 21. Given a pre-semi-tableau (W, L) with set of states V, we put

$$\widehat{[\![\psi]\!]} = \{v \in V \mid l(v) \vdash_{\mathsf{PL}} \psi\} \qquad \widehat{[\![\psi]\!]}_{\overline{m}} = \widehat{[\![\psi]\!]} \cap \{\lceil u \rceil \in V \mid u \in \mathsf{uto}(\psi, \overline{m})\}$$

where $\psi \in \mathbf{F}$, where $\vdash_{\mathsf{PL}}$ denotes propositional entailment and where $\overline{m}$ is a vector of k natural numbers.

Thus we have $v \in [\![\psi]\!]_{\overline{m}}$ if there is a node $u \in W$ such that $\lceil u \rceil = v$ and u has timeouts $\overline{m}$ for ψ. This serves to ease the proofs of the upcoming existence and truth lemmas as it anchors the timeout vector $\overline{m}$ at the node u instead of anchoring it at the state v which may not have timeouts $\overline{m}$ for ψ (namely, if a greatest fixpoint is unfolded on the L-path from u to v).

Definition 22 (Strong coherence). Let (W, L) be a pre-semi-tableau with set V of states. A coalgebra $\mathcal{C} = (V, \xi)$ is *strongly coherent* if for all states $v \in V$, for all formulas $\heartsuit\psi \in \mathbf{F}$ and for all timeout-vectors $\overline{m}$,

$$v \in \widehat{[\![\heartsuit\psi]\!]_{\overline{m}}} \text{ implies } \xi(v) \in [\![\heartsuit]\!](\widehat{[\![\psi]\!]_{\overline{m}}}).$$

Strongly coherent coalgebras exist over pre-semi-tableaux:

Lemma 23 (Existence). *Let (W, L) be a pre-semi-tableau with set of states V. Then there is a strongly coherent coalgebra over V.*

Since all least fixpoint literals are satisfied after finitely many unfolding steps in strongly coherent coalgebras with unfolding timeouts, they are models, i.e. satisfy all the formulas in their labels:

Lemma 24 (Truth). *In strongly coherent coalgebras that have unfolding timeouts, we have that for all $\psi \in \mathbf{F}$,*

$$\widehat{[\![\psi]\!]} \subseteq [\![\psi]\!].$$

Definition 25 (Timed-out satisfaction). Given sets $U \subseteq W$, a function $f : \mathcal{P}(W) \to \mathcal{P}(W)$ and an ordinal number λ, we define $f^\lambda(U) = U$ if $\lambda = 0$, $f^\lambda(U) = f(f^{\lambda'}(U))$ if $\lambda = \lambda' + 1$ and $f^\lambda(U) = \bigcup_{k<\lambda} f^k(U)$ if λ is a limit-ordinal. The target formula χ is clean so that it contains, for each fixpoint variable $X \in \mathbf{V}$, at most a single fixpoint literal $\eta X.\psi_0$ as a subformula; we denote this formula by $\theta(X)$. Given a coalgebra (C, ξ), a formula ψ and a vector $\overline{\lambda} = (\lambda_k, \dots, \lambda_j)$ of ordinal numbers, we define $[\![\psi]\!]^{\overline{\lambda}} = [\![\psi]\!]_i$ where $i : \mathbf{V} \nrightarrow \mathcal{P}(C)$ is defined, for fixpoint variables X_j that occur freely in ψ and for which we have $\theta(X_j) = \eta X_j \psi_j$, by $i(X_j) = ([\![\psi_j]\!]_{i'}^{X_j})^{\lambda_j}(\emptyset)$ if $\eta = \mu$ and by $i(X_j) = [\![\nu X_j.\psi_j]\!]_{i'}$ if $\eta = \nu$, where $i'(X_{j'})$ is undefined for $j' \geq j$ and where $i'(X_{j'}) = i(X_{j'})$ for $j' < j$. Again the timeouts for greatest fixpoint variables are irrelevant and serve only to ease notation.

Definition 26 (Strongly supporting Kripke frame). Let (C, ξ) be a coalgebra. For states $x \in C$ and formulas ψ such that $x \in [\![\psi]\!]$, let $\overline{\lambda}_\psi$ denote the least vector of ordinal numbers such that $x \in [\![\psi]\!]^{\overline{\lambda}_\psi}$. Also let, for $\psi \in \mathbf{F}$, $\overline{\psi}$ be *the* subformula of χ such that ψ is obtained from $\overline{\psi}$ by repeatedly replacing free variables X by $\theta(X)$. A graph (C, L) with $L \subseteq C \times \Sigma \times C$ and with labeling function $l : C \to \mathcal{P}(\mathbf{F})$ such that $l(x) = \{\psi \in \mathbf{F} \mid x \in [\![\psi]\!]\}$ is a *strongly supporting Kripke frame* (for C, ξ) if

- for all $\psi \in \mathbf{F}$ and $x \in C$, if $x \notin [\![\psi]\!]$, then $L(x, (\psi, b)) = \emptyset$ for $b \in \{0,1\}$; if $x \in [\![\psi]\!]$, then we distinguish upon the shape of ψ: if $\psi = \psi_0 \vee \psi_1$, then we require $L(x, (\psi, b)) = \{x\}$ for exactly one $b \in \{0,1\}$ with $x \in [\![\overline{\psi_b}]\!]^{\overline{\lambda}_{\psi}}$ and $L(x, (\psi, \bar{b})) = \emptyset$, where $\bar{1} = 0$, $\bar{0} = 1$; if $\psi = \psi_0 \wedge \psi_1$ or $\psi = \eta X.\psi_0$, then we require $L(x, (\psi, 0)) = \{x\}$.
- for all $x \in C$ and $\kappa \in$ selections, we have $L(x, \kappa) = \{y\}$ for *some* $y \in A = \bigcap_{\heartsuit\psi \in \kappa} [\![\overline{\psi}^{\overline{\lambda}_{\heartsuit\psi}}]\!]$ if $A \neq \emptyset$, and $L(x, \kappa) = \emptyset$ otherwise.

Lemma 27. *Every coalgebra has a strongly supporting Kripke frame.*

Definition 28. Given a coalgebra (C, ξ) with strongly supporting Kripke frame (C, L), a formula ψ and a valuation $i : \mathbf{V} \nrightarrow \mathcal{P}(C)$, we define $[\![\psi]\!]_i^L$ by the same clauses as $[\![\psi]\!]_i$ in all cases except the following:

$$\begin{aligned}
[\![\psi_0 \vee \psi_1]\!]_i^L &= \{x \in C \mid x \in [\![\psi_b]\!]_i^L, b \in \{0,1\}, L(x, (\phi_0 \vee \phi_1, b)) = \{x\}\} \\
[\![\heartsuit\psi_0]\!]_i^L &= \{x \in C \mid (Tg_x)(\xi(x)) \in [\![\heartsuit]\!](g_x[[\![\psi_0]\!]_i^L])\} \\
[\![\mu X.\psi_0]\!]_i^L &= \{x \in C \mid x \text{ has unfolding timeouts at level } \mathsf{ad}(\mu X.\phi_0) \\
&\qquad\qquad \text{for } \mu X.\phi_0 \text{ in } (C, L)\},
\end{aligned}$$

where $\mu X.\psi_0 = \overline{\mu X.\phi_0}$ and $\psi_0 \vee \psi_1 = \overline{\phi_0 \vee \phi_1}$, and where $g_x : C \to \{y_\kappa \mid L(x, \kappa) = \{y_\kappa\}\}$ is defined by $g_x(c) = y_\kappa$ if and only if $\kappa = \{\heartsuit\psi \in \mathbf{F} \mid c \in [\![\psi]\!]\}$.

Strongly supporting Kripke frames have unfolding timeouts:

Lemma 29. *For all coalgebras (C, ξ) with strongly supporting Kripke frame (C, L), all formulas ψ and all valuations $i : \mathbf{V} \nrightarrow \mathcal{P}(C)$, we have $[\![\psi]\!]_i \subseteq [\![\psi]\!]_i^L$.*

Lemma 30 (Soundness). *Let χ be satisfiable. Then a pre-semi-tableau for χ with unfolding timeouts can be constructed over a subset of D_χ.*

Proof (Sketch). By Lemmas 27 and 29, any model of χ has a strongly supporting Kripke frame (C, L) with unfolding timeouts. We derive a pre-semi-tableau for χ from (C, L), inheriting unfolding timeouts. □

Corollary 31 (Soundness and completeness). *We have*

$$v_0 \in \mathbf{E} \textit{ if and only if } \chi \textit{ is satisfiable.}$$

Our model construction moreover yields the same bound on minimum model size as in earlier work on the coalgebraic μ-calculus [4]:

Corollary 32 (Small model property). *Let χ be a satisfiable coalgebraic μ-calculus formula. Then χ has a model of size $\mathcal{O}(((nk)!)^2) \in 2^{\mathcal{O}(nk \log n)}$.*

6 Conclusion

We have shown that the satisfiability problem of the coalgebraic μ-calculus is in ExpTime, subject to establishing a suitable time bound on the much simpler one-step satisfiability problem. Prominent examples include the graded μ-calculus, the (two-valued) probabilistic μ-calculus, and extensions of the probabilistic and the graded μ-calculus, respectively, with (monotone) polynomial inequalities; the ExpTime bound appears to be new for the last two logics. We have also presented a generic satisfiability algorithm that realizes the time bound and supports global caching and on-the-fly solving. Moreover, we have obtained a polynomial bound on minimum branching width in models for all example logics mentioned above.

References

1. Alur, R., Henzinger, T., Kupferman, O.: Alternating-time temporal logic. J. ACM **49**, 672–713 (2002)
2. Bradfield, J., Stirling, C.: Modal μ-calculi. In: Handbook of Modal Logic, pp. 721–756. Elsevier (2006)
3. Bradfield, J., Walukiewicz, I.: The mu-calculus and model checking. In: Clarke, E., Henzinger, T., Veith, H., Bloem, R. (eds.) Handbook of Model Checking, pp. 871–919. Springer, Cham (2018). https://doi.org/10.1007/978-3-319-10575-8_26
4. Cîrstea, C., Kupke, C., Pattinson, D.: EXPTIME tableaux for the coalgebraic μ-calculus. Log. Methods Comput. Sci. **7**, 1–33 (2011)
5. Cleaveland, R., Iyer, S., Narasimha, M.: Probabilistic temporal logics via the modal μ-calculus. Theor. Comput. Sci. **342**, 316–350 (2005)
6. D'Agostino, G., Visser, A.: Finality regained: a coalgebraic study of Scott-sets and multisets. Arch. Math. Logic **41**, 267–298 (2002)
7. Demri, S., Lugiez, D.: Presburger modal logic is PSPACE-complete. In: Furbach, U., Shankar, N. (eds.) IJCAR 2006. LNCS (LNAI), vol. 4130, pp. 541–556. Springer, Heidelberg (2006). https://doi.org/10.1007/11814771_44
8. Fontaine, G., Leal, R., Venema, Y.: Automata for coalgebras: an approach using predicate liftings. In: Abramsky, S., Gavoille, C., Kirchner, C., Meyer auf der Heide, F., Spirakis, P.G. (eds.) ICALP 2010, Part II. LNCS, vol. 6199, pp. 381–392. Springer, Heidelberg (2010). https://doi.org/10.1007/978-3-642-14162-1_32
9. Friedmann, O., Lange, M.: Deciding the unguarded modal μ-calculus. J. Appl. Non-Classical Log. **23**, 353–371 (2013)
10. Friedmann, O., Latte, M., Lange, M.: Satisfiability games for branching-time logics. Log. Methods Comput. Sci. **9**, 1–36 (2013)
11. Goré, R., Nguyen, L.A.: Exptime tableaux for $\mathcal{ALC}$ using sound global caching. J. Autom. Reason. **50**, 355–381 (2013)
12. Goré, R., Widmann, F.: Sound global state caching for $\mathcal{ALC}$ with inverse roles. In: Giese, M., Waaler, A. (eds.) TABLEAUX 2009. LNCS (LNAI), vol. 5607, pp. 205–219. Springer, Heidelberg (2009). https://doi.org/10.1007/978-3-642-02716-1_16
13. Grädel, E., Thomas, W., Wilke, T. (eds.): Automata Logics, and Infinite Games: A Guide to Current Research. LNCS, vol. 2500. Springer, Heidelberg (2002). https://doi.org/10.1007/3-540-36387-4

14. Hausmann, D., Schröder, L.: Global caching for the flat coalgebraic μ-calculus. In: Grandi, F., Lange, M., Lomuscio, A. (eds.) Temporal Representation and Reasoning, TIME 2015, pp. 121–143. IEEE (2015)
15. Hausmann, D., Schröder, L., Deifel, H.-P.: Permutation games for the weakly aconjunctive μ-calculus. In: Beyer, D., Huisman, M. (eds.) TACAS 2018, Part II. LNCS, vol. 10806, pp. 361–378. Springer, Cham (2018). https://doi.org/10.1007/978-3-319-89963-3_21
16. Hausmann, D., Schröder, L., Egger, C.: Global caching for the alternation-free coalgebraic μ-calculus. In: Desharnais, J., Jagadeesan, R. (eds.) Concurrency Theory, CONCUR 2016. LIPIcs, vol. 59, pp. 34:1–34:15. Schloss Dagstuhl - Leibniz-Zentrum für Informatik (2016)
17. Huth, M., Kwiatkowska, M.: Quantitative analysis and model checking. In: Logic in Computer Science, LICS 1997, pp. 111–122. IEEE (1997)
18. King, V., Kupferman, O., Vardi, M.Y.: On the complexity of parity word automata. In: Honsell, F., Miculan, M. (eds.) FOSSACS 2001. LNCS, vol. 2030, pp. 276–286. Springer, Heidelberg (2001). https://doi.org/10.1007/3-540-45315-6_18
19. Kozen, D.: Results on the propositional μ-calculus. Theor. Comput. Sci. **27**, 333–354 (1983)
20. Kozen, D.: A finite model theorem for the propositional μ-calculus. Stud. Log. **47**, 233–241 (1988)
21. Kupferman, O., Sattler, U., Vardi, M.Y.: The complexity of the graded μ-calculus. In: Voronkov, A. (ed.) CADE 2002. LNCS (LNAI), vol. 2392, pp. 423–437. Springer, Heidelberg (2002). https://doi.org/10.1007/3-540-45620-1_34
22. Kupke, C., Pattinson, D.: On modal logics of linear inequalities. In: Beklemishev, L., Goranko, V., Shehtman, V. (eds.) Advances in Modal Logic, AiML 2010, pp. 235–255. College Publications (2010)
23. Kupke, C., Pattinson, D., Schröder, L.: Reasoning with global assumptions in arithmetic modal logics. In: Kosowski, A., Walukiewicz, I. (eds.) FCT 2015. LNCS, vol. 9210, pp. 367–380. Springer, Cham (2015). https://doi.org/10.1007/978-3-319-22177-9_28
24. Liu, W., Song, L., Wang, J., Zhang, L.: A simple probabilistic extension of modal mu-calculus. In: Yang, Q., Wooldridge, M. (eds.) International Joint Conference on Artificial Intelligence, IJCAI 2015, pp. 882–888. AAAI Press (2015)
25. Mio, M.: Probabilistic modal μ-calculus with independent product. In: Hofmann, M. (ed.) FOSSACS 2011. LNCS, vol. 6604, pp. 290–304. Springer, Heidelberg (2011). https://doi.org/10.1007/978-3-642-19805-2_20
26. Myers, R., Pattinson, D., Schröder, L.: Coalgebraic hybrid logic. In: de Alfaro, L. (ed.) FOSSACS 2009. LNCS, vol. 5504, pp. 137–151. Springer, Heidelberg (2009). https://doi.org/10.1007/978-3-642-00596-1_11
27. Niwinski, D., Walukiewicz, I.: Games for the μ-calculus. Theor. Comput. Sci. **163**, 99–116 (1996)
28. Parikh, R.: The logic of games and its applications. Ann. Discret. Math. **24**, 111–140 (1985)
29. Pattinson, D.: Coalgebraic modal logic: soundness, completeness and decidability of local consequence. Theor. Comput. Sci. **309**, 177–193 (2003)
30. Piterman, N.: From nondeterministic Büchi and Streett automata to deterministic parity automata. Log. Methods Comput. Sci. **3**(3:5), 1–21 (2007)
31. Rutten, J.: Universal coalgebra: a theory of systems. Theor. Comput. Sci. **249**, 3–80 (2000)
32. Safra, S.: On the complexity of omega-automata. In: Foundations of Computer Science, FOCS 1988, pp. 319–327. IEEE Computer Society (1988)

33. Schröder, L.: A finite model construction for coalgebraic modal logic. J. Log. Algebr. Prog. **73**, 97–110 (2007)
34. Schröder, L.: Expressivity of coalgebraic modal logic: the limits and beyond. Theor. Comput. Sci. **390**(2–3), 230–247 (2008)
35. Schröder, L., Pattinson, D.: Shallow models for non-iterative modal logics. In: Dengel, A.R., Berns, K., Breuel, T.M., Bomarius, F., Roth-Berghofer, T.R. (eds.) KI 2008. LNCS (LNAI), vol. 5243, pp. 324–331. Springer, Heidelberg (2008). https://doi.org/10.1007/978-3-540-85845-4_40
36. Schröder, L., Pattinson, D.: PSPACE bounds for rank-1 modal logics. ACM Trans. Comput. Log. **10**(2), 13:1–13:33 (2009)

Constructing Inductive-Inductive Types in Cubical Type Theory

Jasper Hugunin(✉)

University of Washington, Seattle, WA, USA
jasper@hugunin.net

Abstract. Inductive-inductive types are a joint generalization of mutual inductive types and indexed inductive types. In extensional type theory, inductive-inductive types can be constructed from inductive types, and this construction has been conjectured to work in intensional type theory as well. In this paper, we show that the existing construction requires Uniqueness of Identity Proofs, and present a new construction (which we conjecture generalizes) of one particular inductive-inductive type in cubical type theory, which is compatible with homotopy type theory.

1 Introduction

Inductive-inductive types allow for the mutual inductive definition of a type and a family over that type. As an example, we can simultaneously define contexts and types defined in a context, with dependently typed context extension:

$$\begin{aligned} \mathrm{Ctx} &: \mathrm{Type}, & \mathrm{Ty} &: \mathrm{Ctx} \to \mathrm{Type}, \\ \epsilon &: \mathrm{Ctx}, & \mathrm{U} &: (\varGamma : \mathrm{Ctx}) \to \mathrm{Ty}\,\varGamma, \\ \mathtt{ext} &: (\varGamma : \mathrm{Ctx}) \to \mathrm{Ty}\,\varGamma \to \mathrm{Ctx}, & \mathrm{El} &: (\varGamma : \mathrm{Ctx}) \to \mathrm{Ty}\,(\mathtt{ext}\,\varGamma\,(\mathrm{U}\,\varGamma)). \end{aligned}$$

Such definitions have been used for example by Danielsson [9] and Chapman [5] to define intrinsically typed syntax of a dependent type theory, and Agda supports such definitions natively.

These types have been studied extensively in Nordvall Forsberg [15]. There, in §5.3, inductive-inductive types with simple elimination rules (defined in op. cit. §3.2.5) are constructed from indexed inductive types in extensional type theory, and in §5.4 this is conjectured to work in intensional type theory as well.

In this paper, we first show that this construction does not work in intensional type theory without assuming Uniqueness of Identity Proofs (UIP), which is incompatible with the Univalence axiom of Homotopy Type Theory [18]. We then give an alternate construction in cubical type theory [6], which is compatible with Univalence. Specifically, this paper makes the following contributions:[1]

[1] The formalization can be found at https://github.com/jashug/ConstructingII.

M. Bojańczyk and A. Simpson (Eds.): FOSSACS 2019, LNCS 11425, pp. 295–312, 2019.
https://doi.org/10.1007/978-3-030-17127-8_17

1. In Sect. 2, we show that, in intensional type theory, if the types constructed by Nordvall Forsberg satisfy the simple elimination rules, then UIP holds (formalized in both Coq and Agda).
2. In Sect. 3, we give the construction of a particular inductive-inductive type with simple elimination rules in cubical type theory (formalized in cubical Agda).

1.1 Syntax and Conventions

We mostly mimic Agda syntax. The double bar symbol $=$ is used for definitions directly and by pattern matching, and for equality of terms up to conversion. We write $(a : A) \to B$ for the dependent product type, and $A \to B$ for the non-dependent version. Functions are given by pattern matching $f\ x = y$ or by lambda expressions $f = \lambda x.y$. Similarly $(a : A) \times B$ is the dependent pair type, and $A \times B$ the non-dependent version. Pairs are (a, b), and projections are $p.1$ and $p.2$. The unit type is $\top$, with unique inhabitant $\star$. Identity types are $x \equiv_X y$ for the type of identifications of x with y in type X, and we write `refl` for a proof of reflexivity. We do not assume that axiom K holds for identity types. We write Type for a universe of types (where Agda uses Set). In Sect. 3 we work in cubical type theory, which will be explained there.

1.2 Running Example of an Inductive-Inductive Definition

For the purposes of this paper, we will focus on one relatively simple inductive-inductive definition (with only 5 clauses), parametrized by a type X, which is given in Fig. 1. We will use this definition to prove that Nordvall Forsberg's construction implies UIP in Sect. 2 and as a running example to demonstrate our construction in cubical type theory in Sect. 3.

Our example starts with the simplest inductive-inductive sorts, taking A : Type and $B : A \to$ Type, and then populates A and B with simple constructors which suffice for our proof of UIP. We have `inj`, which is supposed to give exactly one element of each $B\ a$, while `ext` lets us mix Bs back into the As (mirroring the type of context extension), and η gives us something to start with: one element of A for each element of X (following the use of η in [15, Example 3.3]). The proof of UIP in Sect. 2 proceeds by considering the type B (`ext` $(\eta\ x)$ (`inj` $(\eta\ x)$)) for some $x : X$, and noticing that, while the simple elimination rules tell us that there should only be one element of this type (given by `inj`), in Nordvall Forsberg's construction there are actually as many as there are proofs of $x \equiv_X x$.

Our goal in this paper is to construct $(A, B, \eta,$ `ext`, `inj`$)$ of the types given in Fig. 1 such that the simple elimination rules hold without using UIP. But first, we will show why Nordvall Forsberg's approach is not sufficient.

2 Deriving UIP

Uniqueness of Identity proofs (UIP) for a type X is the principle that, for all $x : X$, $y : X$, $p : x \equiv_X y$, $q : x \equiv_X y$, the type $p \equiv_{x \equiv_X y} q$ is inhabited.

Given X : Type, we consider the inductive-inductive definition

$$\begin{aligned} A &: \text{Type}, \\ B &: A \to \text{Type}, \\ \eta &: X \to A, \\ \texttt{ext} &: (a : A) \to B\ a \to A, \\ \texttt{inj} &: (a : A) \to B\ a. \end{aligned}$$

This has simple elimination rules stating that for all motives (PA, PB) and methods $(P\eta, P\texttt{ext}, P\texttt{inj})$

$$\begin{aligned} PA &: A \to \text{Type}, \\ PB &: (a : A) \to B\ a \to \text{Type}, \\ P\eta &: (x : X) \to PA\ (\eta\ x), \\ P\texttt{ext} &: (a : A) \to PA\ a \to (b : B\ a) \to PB\ a\ b \to PA\ (\texttt{ext}\ a\ b), \\ P\texttt{inj} &: (a : A) \to PA\ a \to PB\ a\ (\texttt{inj}\ a), \end{aligned}$$

we have eliminators (EA, EB) satisfying equalities $(E\eta, E\texttt{ext}, E\texttt{inj})$.

$$\begin{aligned} EA &: (a : A) \to PA\ a, \\ EB &: (a : A) \to (b : B\ a) \to PB\ a\ b, \\ E\eta &: (x : X) \to EA\ (\eta\ x) \equiv_{PA\ (\eta\ x)} P\eta\ x, \\ E\texttt{ext} &: (a : A) \to (b : B\ a) \to \\ &\qquad EA\ (\texttt{ext}\ a\ b) \equiv_{PA\ (\texttt{ext}\ a\ b)} P\texttt{ext}\ a\ (EA\ a)\ b\ (EB\ a\ b), \\ E\texttt{inj} &: (a : A) \to EB\ a\ (\texttt{inj}\ a) \equiv_{PB\ a\ (\texttt{inj}\ a)} P\texttt{inj}\ a\ (EA\ a). \end{aligned}$$

Fig. 1. Running example

Equivalently, for all $x : X$, $p : x \equiv_X x$, the type $p \equiv_{x \equiv_X x} \texttt{refl}$ is inhabited. It expresses that there is at most one proof of any equality. UIP is independent of standard intensional type theory [13], and is inconsistent with Homotopy Type Theory [18].

Nordvall Forsberg's construction of inductive-inductive types is described in [15, §5.3]. In this section, we show that if the simple elimination rules hold for this construction of the inductive-inductive type in Fig. 1, then UIP holds for the type X (Theorem 1). This argument has been formalized in both Coq version 8.8.0 [8] (see `UIP_from_Forsberg_II.v`) and Agda using the `--without-K` flag (see `UIP_from_Forsberg_II.agda`).

To recap, Nordvall Forsberg [15, §5.3] constructs an inductive-inductive type by first defining an approximation (the *pre-syntax*) which drops the A index from B leaving a mutual inductive definition. Concretely, we have A_{pre} and B_{pre} defined as in Fig. 2. Then a mutual indexed inductive definition is used to define the index relationship between A_{pre} and B_{pre}; these are the goodness predicates A_{good} and B_{good}. Finally, the inductive object $(A, B, \eta, \texttt{ext}, \texttt{inj})$ is defined by pairing the pre-syntax with goodness proofs (see Fig. 3).

In extensional type theory, Nordvall Forsberg proved that $A_{\text{good}}\ a$ is a mere proposition (all inhabitants are equal) [15, Lemma 5.37(ii)]. In intensional type theory as well, if function extensionality and UIP hold then A_{good} is a mere

Dropping the inductive index from B leaves a mutual inductive definition.

$$\begin{aligned} A_{\text{pre}} &: \text{Type}, \\ B_{\text{pre}} &: \text{Type}, \\ \eta_{\text{pre}} &: X \to A_{\text{pre}}, \\ \mathtt{ext}_{\text{pre}} &: A_{\text{pre}} \to B_{\text{pre}} \to A_{\text{pre}}, \\ \mathtt{inj}_{\text{pre}} &: A_{\text{pre}} \to B_{\text{pre}}. \end{aligned}$$

Fig. 2. Pre-syntax for the running example

A mutual indexed inductive definition is used to define the index relationship between A_{pre} and B_{pre}:

$$\begin{aligned} A_{\text{good}} &: A_{\text{pre}} \to \text{Type}, \\ B_{\text{good}} &: A_{\text{pre}} \to B_{\text{pre}} \to \text{Type}, \\ \eta_{\text{good}} &: (x : X) \to A_{\text{good}}\ (\eta_{\text{pre}}\ x), \\ \mathtt{ext}_{\text{good}} &: (a_{\text{pre}} : A_{\text{pre}}) \to A_{\text{good}}\ a_{\text{pre}} \to \\ &\qquad (b_{\text{pre}} : B_{\text{pre}}) \to B_{\text{good}}\ a_{\text{pre}}\ b_{\text{pre}} \to A_{\text{good}}\ (\mathtt{ext}_{\text{pre}}\ a_{\text{pre}}\ b_{\text{pre}}), \\ \mathtt{inj}_{\text{good}} &: (a_{\text{pre}} : A_{\text{pre}}) \to A_{\text{good}}\ a_{\text{pre}} \to B_{\text{good}}\ a_{\text{pre}}\ (\mathtt{inj}_{\text{pre}}\ a_{\text{pre}}). \end{aligned}$$

The inductive-inductive object is defined as

$$\begin{aligned} A &= (a_{\text{pre}} : A_{\text{pre}}) \times A_{\text{good}}\ a_{\text{pre}}, \\ B\ (a_{\text{pre}}, a_{\text{good}}) &= (b_{\text{pre}} : B_{\text{pre}}) \times B_{\text{good}}\ a_{\text{pre}}\ b_{\text{pre}}, \\ \eta\ x &= \eta_{\text{pre}}\ x, \eta_{\text{good}}\ x, \\ \mathtt{ext}\ (a_{\text{pre}}, a_{\text{good}})\ (b_{\text{pre}}, b_{\text{good}}) &= \mathtt{ext}_{\text{pre}}\ a_{\text{pre}}\ b_{\text{pre}}, \mathtt{ext}_{\text{good}}\ a_{\text{pre}}\ a_{\text{good}}\ b_{\text{pre}}\ b_{\text{good}}, \\ \mathtt{inj}\ (a_{\text{pre}}, a_{\text{good}}) &= \mathtt{inj}_{\text{pre}}\ a_{\text{pre}}, \mathtt{inj}_{\text{good}}\ a_{\text{pre}}\ a_{\text{good}}. \end{aligned}$$

Here, the sorts A and B are defined as pairs of the pre-syntax with a goodness proof, and operations are performed component-wise on both the pre-syntax and the goodness proof (using sort and operation in their algebraic sense).

Fig. 3. Construction given by Nordvall Forsberg

proposition. This uniqueness of goodness proofs justifies having the definition of B ignore the goodness proof a_{good}, since a_{good} can have at most one value.

In the next two subsections, we prove that:

1. If $A_{\text{good}}\ a$ is a mere proposition then UIP holds for the type X (Lemma 2).
2. If the simple elimination rules from Fig. 1 hold for the $(A, B, \eta, \mathtt{inj}, \mathtt{ext})$ constructed above then $A_{\text{good}}\ a$ is a mere proposition (Lemma 5).

Combining these results, we conclude that Nordvall Forsberg's construction satisfies the simple elimination rules in intensional type theory only if UIP holds (Theorem 1).

2.1 Unique Goodness Implies UIP

We define notation $(x == y)$ to mean the term

$$\texttt{ext}_{\text{pre}}\ (\eta_{\text{pre}}\ x)\ (\texttt{inj}_{\text{pre}}\ (\eta_{\text{pre}}\ y)) : A_{\text{pre}}.$$

We first prove that there are at least as many proofs of $A_{\text{good}}\ (x == y)$ as there are of $x \equiv_X y$.

Lemma 1 ($x \equiv_X y$ is a retract of A_{good}). *For all $x : X$ and $y : X$, there are functions*

$$f : x \equiv_X y \to A_{good}\ (x == y), \qquad g : A_{good}\ (x == y) \to x \equiv_X y,$$

such that for all $e : x \equiv_X y$, $g\ (f\ e) \equiv e$.

Proof. To define f, we let $f\ \texttt{refl} =$

$$\texttt{ext}_{\text{good}}\ (\eta_{\text{pre}}\ x)\ (\eta_{\text{good}}\ x)\ (\texttt{inj}_{\text{pre}}\ (\eta_{\text{pre}}\ x))\ (\texttt{inj}_{\text{good}}\ (\eta_{\text{pre}}\ x)\ (\eta_{\text{good}}\ x)).$$

To define g, pattern matching on a_{good} has only one possibility: $a_{\text{good}} =$

$$\texttt{ext}_{\text{good}}\ (\eta_{\text{pre}}\ x)\ (\eta_{\text{good}}\ x)\ (\texttt{inj}_{\text{pre}}\ (\eta_{\text{pre}}\ x))\ (\texttt{inj}_{\text{good}}\ (\eta_{\text{pre}}\ x)\ (\eta_{\text{good}}\ x)),$$

forcing y to be x, and in this case $x \equiv_X y$ holds by reflexivity. Then when $e = \texttt{refl}$, $f\ e$ returns a proof in the format matched by g, so $g\ (f\ \texttt{refl}) \equiv \texttt{refl}$, and thus $g\ (f\ e) \equiv e$.

Lemma 2 (Unique goodness implies UIP). *If $A_{good}\ t$ is a mere proposition for all $t : A_{pre}$, then UIP holds for the type X.*

Proof. Assume goodness proofs are unique, and take $x : X$, $y : X$, with $p : x \equiv y$, $q : x \equiv y$. We want to show that $p \equiv q$. Using the f and g from Lemma 1,

$$\begin{aligned} p &\equiv g\ (f\ p) && \text{by Lemma 1} \\ &\equiv g\ (f\ q) && \text{by uniqueness in } A_{\text{good}}\ (x == y), f\ p \equiv f\ q \\ &\equiv q && \text{by Lemma 1.} \end{aligned}$$

2.2 Simple Elimination Rules Imply Unique Goodness

Now we prove that there are at least as many proofs of $B\ (t_{\text{pre}}, t_{\text{good}})$ as there are of $A_{\text{good}}\ t_{\text{pre}}$.

Lemma 3 (A_{good} is a retract of B). *For all $t_{pre} : A_{pre}$ and $t_{good} : A_{good}\ t_{pre}$, there are functions*

$$f : A_{good}\ t_{pre} \to B\ (t_{pre}, t_{good}), \qquad g : B\ (t_{pre}, t_{good}) \to A_{good}\ t_{pre}$$

such that for all $a_{good} : A_{good}\ t_{pre}$, $g\ (f\ a_{good}) \equiv a_{good}$.

Proof. We define $f\ a_{\text{good}} = \texttt{inj}_{\text{pre}}\ t_{\text{pre}}, \texttt{inj}_{\text{good}}\ t_{\text{pre}}\ a_{\text{good}}$. By induction on B_{good}, we define a function

$$g' : (a_{\text{pre}} : A_{\text{pre}}) \to (b_{\text{pre}} : B_{\text{pre}}) \to B_{\text{good}}\ a_{\text{pre}}\ b_{\text{pre}} \to A_{\text{good}}\ a_{\text{pre}}$$

taking

$$g'\ a_{\text{pre}}\ (\texttt{inj}_{\text{pre}}\ a_{\text{pre}})\ (\texttt{inj}_{\text{good}}\ a_{\text{pre}}\ a_{\text{good}}) = a_{\text{good}}.$$

Then we can define $g\ (b_{\text{pre}}, b_{\text{good}}) = g'\ t_{\text{pre}}\ b_{\text{pre}}\ b_{\text{good}}$. Then $g\ (f\ a_{\text{good}}) \equiv a_{\text{good}}$ holds by reflexivity.

Lemma 4 ($B\ a$ is contractible). *Assuming the simple elimination rules from Fig. 1 hold for the* $(A, B, \eta, \mathit{inj}, \mathit{ext})$ *constructed above, for all* $a : A$ *and* $b : B\ a$, $\mathit{inj}\ a \equiv_{B\ a} b$.

Proof. Referring to the simple elimination rules given in Fig. 1, we pattern match on B by giving motives (PA, PB) and methods $(P\eta, P\texttt{ext}, P\texttt{inj})$, and then using the resulting EB.

We set $PA\ a = \top$, and take $PB\ a\ b = \texttt{inj}\ a \equiv_{B\ a} b$. Then we have $P\eta\ x = \star$, and $P\texttt{ext}\ a\ \star\ b\ H = \star$, and we take $P\texttt{inj}\ a\ \star = \texttt{refl} : \texttt{inj}\ a \equiv_{B\ a} \texttt{inj}\ a$. The conclusion follows by $EB : (a : A) \to (b : B\ a) \to \texttt{inj}\ a \equiv_{B\ a} b$.

Lemma 5 (Simple elimination rules imply unique goodness). *If the simple eliminators hold for the* $(A, B, \eta, \mathit{inj}, \mathit{ext})$ *constructed above, then for all* $t : A_{pre}$, $A_{good}\ t$ *is a mere proposition.*

Proof. Assume that the simple elimination rules hold, and take $t : A_{\text{pre}}$, and a_1 and a_2 in $A_{\text{good}}\ t$. We use the definition of f and g from Lemma 3 with $t_{\text{pre}} = t$ and $t_{\text{good}} = a_1$.

By Lemma 4, we know that

$$\texttt{inj}\ (t, a_1) \equiv_{B\ (t, a_1)} f\ a_2.$$

Applying g to both sides, and recognizing that $g\ (\texttt{inj}\ (t, a_1))$ computes to a_1, while $g\ (f\ a_2)$ computes to a_2 we find that

$$a_1 = g\ (\texttt{inj}\ (t, a_1)) \equiv_{A_{\text{good}}\ t} g\ (f\ a_2) = a_2.$$

2.3 Simple Elimination Rules for Nordvall Forsberg's Construction only if UIP

Theorem 1. *If the simple elimination rules hold for Nordvall Forsberg's construction, then UIP holds for the type* X.

Proof. Compose the results of Lemmas 2 and 5.

Therefore Nordvall Forsberg's approach to constructing inductive-inductive types requires UIP. Since UIP is inconsistent with the Univalence axiom at the center of Homotopy Type Theory (HoTT) [18], we have an incentive to come up with a different construction which is consistent with HoTT.

3 Constructing an Inductive-Inductive Type in Cubical Type Theory

Cubical type theory [6] is a recently developed type theory which gives a constructive interpretation of the Univalence axiom of Homotopy Type Theory. It has an implementation as a mode for Agda [19], which we use to formalize the construction given in this section of the running example from Fig. 1.

The most important difference between cubical type theory and standard intensional type theory as implemented by Coq or vanilla Agda is that the identity type $x \equiv_X y$ is represented (loosely speaking) by the type of functions p from an interval type $\mathbb{I}$ with two endpoints i_0 and i_1 to X such that $p\ i_0$ reduces to x and $p\ i_1$ reduces to y. This allows, for example, a simple proof of function extensionality: if we have A : Type, $B : A \to$ Type, f and g functions of type $(a : A) \to B\ a$, and $h : (a : A) \to f\ a \equiv g\ a$, then we have $(\lambda i.\lambda a.h\ a\ i) : f \equiv g$. Taking cong $f = \lambda p.\lambda i.f\ (p\ i) : x \equiv y \to f\ x \equiv f\ y$ and $\circ$ for function composition, we also have nice properties such as (cong f) $\circ$ (cong g) $=$ cong $(f \circ g)$.

In this section, we construct the running example from Fig. 1, along with the simple elimination rules, in cubical type theory. Our construction proceeds in several steps:

- In Sect. 3.1, we approximate by dropping the indices, leaving a standard mutual inductive definition called the *pre-syntax*. This is the same as the pre-syntax given in Fig. 2.
- In Sect. 3.2, we define *goodness algebras*, collections of predicates over the pre-syntax which define the index relationship (analogously to A_{good} and B_{good} from Sect. 2). We also show that a goodness algebra exists, and call it $\mathbb{O}$.
- In Sect. 3.3, we define a predicate *nice* on goodness algebras, such that if we have a nice goodness algebra, then we can construct the simple elimination rules. Being nice is similar to having proofs of goodness be unique as in Sect. 2.
- In Sect. 3.4, we use pattern matching over the pre-syntax to define a function S from goodness algebras to goodness algebras.
- In Sect. 3.5, we define the limit of the sequence

$$\mathbb{O}, S\ \mathbb{O}, S\ (S\ \mathbb{O}), \ldots, S^n\ \mathbb{O}, \ldots$$

  and show that it is nice. This is the only section that utilizes the differences between cubical type theory and standard intentional type theory.

Given the nice goodness algebra in Sect. 3.5 we can then construct the simple elimination rules by Sect. 3.3. This construction has been formalized in Agda[2] using the `--cubical` flag which implies `--without-K` (see `Running Example.agda`).

The intuition for our construction is that the Nordvall Forsberg's approach of pairing an approximation with goodness predicates can be repeated, and each time the approximation gets better. Using HoTT terminology, we showed in

[2] Agda version 2.6.0 commit `bd338484d`.

Sect. 2 that one iteration suffices only if X has homotopy level 0 (is a homotopy set, satisfies UIP). In general, $n+1$ iterations are sufficient if only if X has homotopy level n. The successor goodness algebra defined in Sect. 3.4 is a slightly simplified version of Nordvall Forsberg's construction, and taking the limit (in Sect. 3.5) gives a construction which works for arbitrary homotopy levels.

3.1 Pre-syntax

The pre-syntax is the same as that used in Sect. 2, defined as a mutually inductive type in Fig. 2. The constructors of the pre-syntax have the same types as the constructors of the full inductive-inductive definition (given in Fig. 1), except we replace $B\ a$ with B_{pre} everywhere, ignoring the dependence of B on A.

Consider this as the closest approximation of the target inductive-inductive type by a standard inductive type; the dependence of B on A is the only new element that inductive-inductive definitions add. Of course, this is only an approximation. We can form elements of the pre-syntax, such as

$$\mathtt{ext}_{\text{pre}}\ (\eta_{\text{pre}}\ x)\ (\mathtt{inj}_{\text{pre}}\ (\eta_{\text{pre}}\ y))$$

for $x \neq y$ that should be excluded from the inductive-inductive formulation, since $\mathtt{inj}\ (\eta\ y) : B\ (\eta\ y)$ while $\mathtt{ext}\ (\eta\ x) : B\ (\eta\ x) \to A$.

We will use definitions by induction and by pattern-matching on the pre-syntax in sections Sects. 3.3 and 3.4 respectively.

3.2 Goodness Algebras

As we saw in Sect. 3.1, the pre-syntax is too lenient, and contains terms we want to exclude from the inductive-inductive object. In this section, we define a notion of sub-algebra of the pre-syntax, which we will call a *goodness algebra*, and explain how to combine a goodness algebra with the pre-syntax to get an inductive-inductive object $(A, B, \eta, \mathtt{ext}, \mathtt{inj})$. We also define a goodness algebra $\mathbb{O}$.

In Fig. 4, for each clause of the inductive-inductive specification, we define 3 things:

1. For each sort X a type $\operatorname{Ix} X$ giving the data X depends on, and for each operation F constructing an element of sort X, a family $\operatorname{Arg} F : Y \to \operatorname{Ix} X \to$ Type where Y collects the arguments of the operation in the pre-syntax, where $\operatorname{Arg} F\ y\ \phi$ gives the data needed to justify that pre-syntax constructed by F_{pre} from y has index ϕ. In later sections we will also write $\operatorname{Ix} X\ \delta^G$ and $\operatorname{Arg} F\ \delta^G$ to specify which goodness algebra we are working in.
2. The type of the corresponding component in the goodness algebra. For sorts, this is a predicate relating Ix and the pre-syntax, while for the operations, this is a function witnessing that each element of Arg gives a goodness proof relating the index ϕ to the pre-syntax.

For our running example from Figure 1, a goodness algebra is the type of tuples of

$$\delta^G = (\delta^G.A, \delta^G.B, \delta^G.\eta, \delta^G.\mathtt{ext}, \delta^G.\mathtt{inj})$$

with the types defined below. Simultaneously, we define how to combine a goodness algebra δ^G with the pre-syntax to construct an inductive-inductive object $(A, B, \eta, \mathtt{ext}, \mathtt{inj})$.

$$\begin{aligned}
&\mathrm{Ix}\, A = \top, \\
&\delta^G.A : \mathrm{Ix}\, A \to A_{\mathrm{pre}} \to \mathrm{Type}, \\
&\quad A = (a : A_{\mathrm{pre}}) \times \delta^G.A \star a
\end{aligned}$$

$$\begin{aligned}
&\mathrm{Ix}\, B = A, \\
&\delta^G.B : \mathrm{Ix}\, B \to B_{\mathrm{pre}} \to \mathrm{Type}, \\
&B\, \phi = (b : B_{\mathrm{pre}}) \times \delta^G.B\, \phi\, b,
\end{aligned}$$

$$\begin{aligned}
&\mathrm{Arg}\, \eta\, x\, \phi = \top \times \star \equiv_{\mathrm{Ix}\, A} \phi, \\
&\quad \delta^G.\eta : (x : X) \to (\phi : \mathrm{Ix}\, A) \to \mathrm{Arg}\, \eta\, x\, \phi \to \delta^G.A\, \phi\, (\eta_{\mathrm{pre}}\, x), \\
&\quad \eta\, x = \eta_{\mathrm{pre}}\, x,\ \delta^G.\eta\, x \star (\star, \mathtt{refl}),
\end{aligned}$$

$$\begin{aligned}
&\mathrm{Arg}\, \mathtt{ext}\, (a, b)\, \phi = ((a^G : \delta^G.A \star a) \times \delta^G.B\, (a, a^G)\, b) \times \star \equiv_{\mathrm{Ix}\, A} \phi, \\
&\quad \delta^G.\mathtt{ext} : (p : A_{\mathrm{pre}} \times B_{\mathrm{pre}}) \to (\phi : \mathrm{Ix}\, A) \to \\
&\qquad\qquad \mathrm{Arg}\, \mathtt{ext}\, p\, \phi \to \delta^G.A\, \phi\, (\mathtt{ext}_{\mathrm{pre}}\, p), \\
&\quad \mathtt{ext}\, ((a_{\mathrm{pre}}, a_{\mathrm{good}}), (b_{\mathrm{pre}}, b_{\mathrm{good}})) = \\
&\qquad\qquad \mathtt{ext}_{\mathrm{pre}}\, a_{\mathrm{pre}}\, b_{\mathrm{pre}},\ \delta^G.\mathtt{ext}\, (a_{\mathrm{pre}}, b_{\mathrm{pre}}) \star ((a_{\mathrm{good}}, b_{\mathrm{good}}), \mathtt{refl}),
\end{aligned}$$

$$\begin{aligned}
&\mathrm{Arg}\, \mathtt{inj}\, a\, \phi = (a^G : \delta^G.A \star a) \times (a, a^G) \equiv_{\mathrm{Ix}\, B} \phi, \\
&\quad \delta^G.\mathtt{inj} : (a : A_{\mathrm{pre}}) \to (\phi : \mathrm{Ix}\, B) \to \mathrm{Arg}\, \mathtt{inj}\, a\, \phi \to \delta^G.B\, \phi\, (\mathtt{inj}_{\mathrm{pre}}\, a), \\
&\mathtt{inj}\, (a_{\mathrm{pre}}, a_{\mathrm{good}}) = \mathtt{inj}_{\mathrm{pre}}\, a_{\mathrm{pre}},\ \delta^G.\mathtt{inj}\, a_{\mathrm{pre}}\, (a_{\mathrm{pre}}, a_{\mathrm{good}})\, (a_{\mathrm{good}}, \mathtt{refl}).
\end{aligned}$$

We also define the goodness algebra $\mathbb{O}$ by

$$\mathbb{O}.A\, \phi\, a = \top, \qquad \mathbb{O}.B\, \phi\, b = \top,$$
$$\mathbb{O}.\eta\, x\, \phi\, t = \star, \qquad \mathbb{O}.\mathtt{ext}\, (a, b)\, \phi\, t = \star, \qquad \mathbb{O}.\mathtt{inj}\, a\, \phi\, t = \star.$$

Fig. 4. Goodness algebras

3. A way to combine the goodness algebra with the pre-syntax to form an inductive-inductive object. For sorts, we pair the pre-syntax with a goodness proof, while for operations we apply the operation given by the goodness algebra, mimicking the construction in Fig. 3.

Comparing this definition to the construction in Sect. 2, the mutual inductive definition of A_{good} and B_{good} (in Fig. 3) has types equivalent to the result of dropping the dependence of $\delta^G.B$ on $\delta^G.A$ (defined in Fig. 4), going from

$$\delta^G.B : (a : A_{\mathrm{pre}}) \times \delta^G.A \star a \to B_{\mathrm{pre}} \to \mathrm{Type} \text{ to } B_{\mathrm{good}} : A_{\mathrm{pre}} \to B_{\mathrm{pre}} \to \mathrm{Type}.$$

The other difference is that we replace the inductive index (call it s) in the conclusion by a fresh variable ϕ, with the condition $s = \phi$ included in Arg.

3.3 Niceness

In this section, we identify a property *niceness* that is sufficient for a goodness algebra to produce an inductive-inductive object $(A, B, \eta, \mathtt{ext}, \mathtt{inj})$ which satisfies the simple elimination rules, as given in Fig. 1.

To define niceness, we use the concept of equivalence, as defined in Univalent Foundations Program [18] (§4.4 Contractible fibers). Given a function $f : A \to B$, we write $\mathrm{isEquiv}\, f$ (leaving A and B implicit) to denote that f is an equivalence between A and B. We will also write $A \simeq B$ for the type of pairs of a function f with a proof that f is an equivalence.

We will say that a goodness algebra is *nice* if we have equivalence proofs $(\delta^N.\eta, \delta^N.\mathtt{ext}, \delta^N.\mathtt{inj})$, with types

$$\begin{aligned} \delta^N.\eta\ x\ \phi &: \mathrm{isEquiv}\ (\delta^G.\eta\ x\ \phi), \\ \delta^N.\mathtt{ext}\ (a,b)\ \phi &: \mathrm{isEquiv}\ (\delta^G.\mathtt{ext}\ (a,b)\ \phi), \\ \delta^N.\mathtt{inj}\ a\ \phi &: \mathrm{isEquiv}\ (\delta^G.\mathtt{inj}\ a\ \phi). \end{aligned}$$

Equivalences between types are very close to equalities between types (the Univalence axiom makes this precise). If we have a *nice* goodness algebra, the combined data looks similar to a recursive definition:

$$\begin{aligned} \delta^G.A &: \top \to A_{\mathrm{pre}} \to \mathrm{Type}, \\ \delta^G.B &: ((a : A_{\mathrm{pre}}) \times \delta^G.A \star a) \to B_{\mathrm{pre}} \to \mathrm{Type}, \\ \delta^G.A\ \phi\ (\eta_{\mathrm{pre}}\ x) &\simeq \mathrm{Arg}\, \eta\ x\ \phi, \\ \delta^G.A\ \phi\ (\mathtt{ext}_{\mathrm{pre}}\ a\ b) &\simeq \mathrm{Arg}\, \mathtt{ext}\ (a,b)\ \phi, \\ \delta^G.B\ \phi\ (\mathtt{inj}_{\mathrm{pre}}\ a) &\simeq \mathrm{Arg}\, \mathtt{inj}\ a\ \phi. \end{aligned}$$

However, the dependence of $\delta^G.B$ on $\delta^G.A$ makes this what Nordvall Forsberg calls a "recursive-recursive" definition, and so we cannot use the standard eliminator of the pre-syntax. In Sect. 3.5, we will expend much effort to construct a solution to this system. Once we have done so, the inductive-inductive object produced by the goodness algebra will satisfy the simple elimination rules, as we show in the following lemma.

Lemma 6 (Nice goodness algebras give simple elimination rules). *Given a goodness algebra δ^G with proof of niceness δ^N, the inductive-inductive object $(A, B, \eta, \mathtt{ext}, \mathtt{inj})$ produced from δ^G as specified in Sect. 3.2 satisfies the simple induction rules given in Fig. 1.*

Proof. The proof is formalized in `RunningExample.agda`. The main idea of the proof is to induct on the pre-syntax, and exploit the equivalences provided by niceness δ^N. In the `inj` case for example, we have a proof of $\delta^G.B\ \phi\ (\mathtt{inj}_{\text{pre}}\ a)$. But without loss of generality, we can replace that goodness proof with $\delta^G.\mathtt{inj}$ applied to an element of $\text{Arg}\,\mathtt{inj}\ a\ \phi$, which contains both a proof $a_{\text{good}} : \delta^G.A \star a$ and a proof that $(a, a_{\text{good}}) \equiv \phi$. Using J to eliminate that equality leaves a goal to which the provided simple induction step for `inj` applies. This proof does not use cubical type theory in any essential way.

3.4 Successor Goodness Algebra

We are trying to create a nice goodness algebra by taking the limit of successive approximations, so we need a step function, which we will call S, that takes a goodness algebra δ^G and returns a new goodness algebra $S\ \delta^G$, which is closer in some sense to being nice. We do so by pattern matching on the pre-syntax to unroll one level of the recurrence equations niceness encodes.

We define by pattern matching

$$(E\ \delta^G).A : (a : A_{\text{pre}}) \to (\phi : \text{Ix}\,A\ \delta^G) \to (Y : \text{Type}) \times (Y \to \delta^G.A\ \phi\ a),$$
$$(E\ \delta^G).B : (b : B_{\text{pre}}) \to (\phi : \text{Ix}\,B\ \delta^G) \to (Y : \text{Type}) \times (Y \to \delta^G.B\ \phi\ b),$$
$$(E\ \delta^G).A\ (\eta_{\text{pre}}\ x) = \lambda\phi.\ \text{Arg}\ \eta\ \delta^G\ x\ \phi, \delta^G.\eta\ x\ \phi,$$
$$(E\ \delta^G).A\ (\mathtt{ext}_{\text{pre}}\ a\ b) = \lambda\phi.\ \text{Arg}\ \mathtt{ext}\ \delta^G\ (a, b)\ \phi, \delta^G.\mathtt{ext}\ (a, b)\ \phi,$$
$$(E\ \delta^G).B\ (\mathtt{inj}_{\text{pre}}\ a) = \lambda\phi.\ \text{Arg}\ \mathtt{inj}\ \delta^G\ a\ \phi, \delta^G.\mathtt{inj}\ a\ \phi,$$

which gives a new property Y which maps back to $\delta^G.B\ \phi\ b$ for each b and ϕ, and similarly for A.

Then, in Fig. 5, we define the new goodness algebra $(S\ \delta^G)$ along with projection functions $(\delta^\pi\ \delta^G)$ which take Ix and Arg from $(S\ \delta^G)$ to δ^G.

The projection functions $(\delta^\pi\ \delta^G)$ consist of applying the map given by the second component of $(E\ \delta^G)$ everywhere in sight. The sorts are then defined by the first component of $(E\ \delta^G)$, while the operations can be defined to be the corresponding projection function itself.

Concretely, for the sort B, we define $(\delta^\pi\ \delta^G).B$ to map between $\text{Ix}\,B\ (S\delta^G)$ and $\text{Ix}\,B\ \delta^G$. This consists of applying the function $((E\ \delta^G).A\ a_{\text{pre}} \star .2)$ which we defined by pattern matching above to a_{good}. Then, since $(S\ \delta^G).B$ gets an inductive index ϕ in $(S\ \delta^G)$ but $((E\ \delta^G)\ b\ \phi\ .1)$ is expecting an inductive index in δ^G, we span the gap with the projection function $(\delta^\pi\ \delta^G).B$ just defined. The definition of A follows the same pattern, but $(\delta^\pi\ \delta^G).A$ is even simpler because $\text{Ix}\,A\ \delta^G = \top$ regardless of what goodness algebra we are working in.

For the operations, consider `inj`. Like with the sorts, we first define a projection function $(\delta^\pi\ \delta^G).\mathtt{inj}\ a\ \phi$, which maps from $\text{Arg}\,\mathtt{inj}\ (S\ \delta^G)$ to $\text{Arg}\,\mathtt{inj}\ \delta^G$, and we fix up the inductive index ϕ with $(\delta^\pi\ \delta^G).B$. For the first component of Arg, we use the function given by the second component of $(E\ \delta^G).A$ to fix up a_{good}. For the second component, applying the projection $(\delta^\pi\ \delta^G).B$ to the equality proof works out on the left hand side because all these projection functions

We define the successor algebra $(S\,\delta^G)$ along with projection functions $(\delta^\pi\,\delta^G)$ by:

$$(\delta^\pi\,\delta^G).A : \mathrm{Ix}\,A\,(S\,\delta^G) \to \mathrm{Ix}\,A\,\delta^G,$$
$$(\delta^\pi\,\delta^G).A = \lambda\star.\star,$$
$$(S\,\delta^G).A\,\phi\,a = (E\,\delta^G).A\,a\,((\delta^\pi\,\delta^G).A\,\phi)\,.1,$$
$$(\delta^\pi\,\delta^G).B : \mathrm{Ix}\,B\,(S\,\delta^G) \to \mathrm{Ix}\,B\,\delta^G,$$
$$(\delta^\pi\,\delta^G).B = \lambda(a_{\mathrm{pre}}, a_{\mathrm{good}}).(a_{\mathrm{pre}}, (E\,\delta^G).A\,a_{\mathrm{pre}}\,\star\,.2\,a_{\mathrm{good}}),$$
$$(S\,\delta^G).B\,\phi\,b = (E\,\delta^G).B\,b\,((\delta^\pi\,\delta^G).B\,\phi)\,.1,$$
$$(\delta^\pi\,\delta^G).\eta\,x\,\phi : \mathrm{Arg}\,\eta\,(S\,\delta^G)\,x\,\phi \to \mathrm{Arg}\,\eta\,\delta^G\,x\,((\delta^\pi\,\delta^G).A\,\phi),$$
$$(\delta^\pi\,\delta^G).\eta\,x\,\phi = \lambda(\star, p).(\star, \mathrm{cong}\,((\delta^\pi\,\delta^G).A)\,p),$$
$$(S\,\delta^G).\eta\,x\,\phi = (\delta^\pi\,\delta^G).\eta\,x\,\phi,$$
$$(\delta^\pi\,\delta^G).\mathtt{ext}\,(a,b)\,\phi : \mathrm{Arg}\,\mathtt{ext}\,(S\,\delta^G)\,(a,b)\,\phi \to \mathrm{Arg}\,\mathtt{ext}\,\delta^G\,(a,b)\,((\delta^\pi\,\delta^G).A\,\phi),$$
$$(\delta^\pi\,\delta^G).\mathtt{ext}\,(a,b)\,\phi = \lambda((a_{\mathrm{good}}, b_{\mathrm{good}}), p).\ \mathrm{let}\ a^G := (E\,\delta^G).A\,a\,\star\,.2\,a_{\mathrm{good}}\ \mathrm{in}$$
$$((a^G, (E\,\delta^G).B\,b\,(a, a^G)\,.2\,b_{\mathrm{good}}), \mathrm{cong}\,((\delta^\pi\,\delta^G).A)\,p),$$
$$(S\,\delta^G).\mathtt{ext}\,(a,b)\,\phi = (\delta^\pi\,\delta^G).\mathtt{ext}\,(a,b)\,\phi,$$
$$(\delta^\pi\,\delta^G).\mathtt{inj}\,a\,\phi : \mathrm{Arg}\,\mathtt{inj}\,(S\,\delta^G)\,a\,\phi \to \mathrm{Arg}\,\mathtt{inj}\,\delta^G\,a\,((\delta^\pi\,\delta^G).B\,\phi),$$
$$(\delta^\pi\,\delta^G).\mathtt{inj}\,a\,\phi = \lambda(a_{\mathrm{good}}, p).((E\,\delta^G).A\,a\,\star\,.2\,a_{\mathrm{good}}, \mathrm{cong}\,((\delta^\pi\,\delta^G).B)\,p),$$
$$(S\,\delta^G).\mathtt{inj}\,a\,\phi = (\delta^\pi\,\delta^G).\mathtt{inj}\,a\,\phi.$$

Fig. 5. Successor goodness algebra

are doing the same thing: applying the function given by the second component of $(E\,\delta^G)$ everywhere. Finally, we can define $(S\,\delta^G).\mathtt{inj} = (\delta^\pi\,\delta^G).\mathtt{inj}$, because $(S\,\delta^G).\mathtt{inj}\,a\,\phi$ is supposed to have codomain

$$(S\,\delta^G).B\,\phi\,(\mathtt{inj}_{\mathrm{pre}}\,a),$$

which is defined to be

$$(E\,\delta^G).B\,(\mathtt{inj}_{\mathrm{pre}}\,a)\,((\delta^\pi\,\delta^G).B\,\phi)\,.1,$$

which reduces on $(\mathtt{inj}_{\mathrm{pre}}\,a)$ to

$$\mathrm{Arg}\,\mathtt{inj}\,\delta^G\,a\,((\delta^\pi\,\delta^G).B\,\phi),$$

which is exactly the codomain of $(\delta^\pi\,\delta^G).\mathtt{inj}\,a\,\phi$.

3.5 Limit of Goodness Algebras

We will now construct a nice goodness algebra by taking the limit of the sequence $S^n\,\mathbb{O}$ and showing that it is nice, where $S^n\,\mathbb{O}$ is defined by recursion on n with $S^0\mathbb{O} = \mathbb{O}, S^{1+n}\mathbb{O} = S(S^n\,\mathbb{O})$. But first, we consider the limit of a chain of types.

Limit of Types. This subsection *Limit of Types* is formalized in `Chain.agda`.

In order to take the limit of successive goodness algebras, we need to know how to work with *chains* of types. Specifically, given $(X : \mathbb{N} \to \mathbb{I} \to \text{Type})$ and $\pi : (n : \mathbb{N}) \to X\ (n+1)\ i_0 \to X\ n\ i_1$, we consider the limit given by the type

$$\text{chain.t}\ X\ \pi = (f : (n : \mathbb{N}) \to X\ n\ i_0) \times ((n : \mathbb{N}) \to f\ n \equiv_{X\ n} \pi\ n\ (f\ (n+1))).$$

If we have $x : \text{chain.t}\ X\ \pi$, then let $x.p$ denote the second projection.

This definition is designed to work well in cubical type theory, and uses the interval $\mathbb{I}$ and native heterogeneous equality $x \equiv_X y$ where $X : \mathbb{I} \to \text{Type}$ (where we can form $p = \lambda i.w : x \equiv_X y$ when $p\ i_0 = x$, $p\ i_1 = y$, and $p\ i : X\ i$). In particular, this definition allows for dependent chains without transporting over the base equality, which is problematic in cubical type theory because transport gets stuck on neutral types; instead given

$$A : \mathbb{N} \to \text{Type} \quad \text{with} \quad f_A : (n : \mathbb{N}) \to A\ (1+n) \to A\ n \quad \text{and}$$
$$B : (n : \mathbb{N}) \to A\ n \to \text{Type} \quad \text{with}$$
$$f_B : (n : \mathbb{N}) \to (a : A\ (1+n)) \to B\ (1+n)\ a \to B\ n\ (f_A\ n\ a),$$

we can form

$$\begin{aligned} LA &= \text{chain.t}\ (\lambda n.\lambda i.A\ n)\ f_A && : \text{Type}, \\ LB &= \lambda a.\text{chain.t}(\lambda n.\text{cong}(B\ n)(a.p\ n))(\lambda n.f_B\ n\ (a.p\ (1+n)\ i_0)) && : LA \to \text{Type} \end{aligned}$$

using $\text{cong}(B\ n)(a.p\ n)$ which is particularly well behaved in cubical type theory.

This construction commutes with most type formers: dependent function types, dependent pair types, identity types, and constants. We also note a dependent version of the fact that the limit of a chain is equivalent to the limit of a shifted chain to substitute for Ahrens et al. [1, Lemma 12].

Lemma 7 (Dependent chain equivalent to shifted chain). *Given*

$$\begin{aligned} &X : \mathbb{N} \to \mathit{Type}, \qquad \pi_X : (n : \mathbb{N}) \to X\ (1+n) \to X\ n, \\ &Y_0 : (n : \mathbb{N}) \to X\ n \to \mathit{Type}, \qquad Y_1 : (n : \mathbb{N}) \to X\ n \to \mathit{Type}, \\ &f : (n : \mathbb{N}) \to (x : X\ n) \to Y_1\ n\ x \to Y_0\ n\ x, \\ &g : (n : \mathbb{N}) \to (x : X\ (1+n)) \to Y_0\ (1+n)\ x \to Y_1\ n\ (\pi_X\ n\ x), \\ &x : \mathit{chain.t}\ (\lambda n.\lambda i.X\ n)\ \pi_X, \end{aligned}$$

and letting the X arguments to f and g be implicit, we can define the types

$$\begin{aligned} t &= \mathit{chain.t}\ (\lambda n.\mathit{cong}\ (Y_0\ n)\ (x.p\ n))\ (\lambda n.\lambda y.f\ n\ (g\ n\ y)), \\ t^+ &= \mathit{chain.t}\ (\lambda n.\mathit{cong}\ (Y_1\ n)\ (x.p\ n))\ (\lambda n.\lambda y.g\ n\ (f\ (1+n)\ y)). \end{aligned}$$

Applying f component-wise gives a function from t^+ to t. This function is an equivalence.

We only use Lemma 7 when $Y_1\ n\ (\pi_X\ n\ x) = Y_0\ (1+n)\ x$, so we may take g to be the identity, leaving t^+ the shifted chain of t up to X arguments.

Limit of Goodness Algebras. Now we use the lemmas about chains to construct a nice goodness algebra, and then conclude by constructing an inductive-inductive object $(A, B, \eta, \mathtt{ext}, \mathtt{inj})$ that satisfies the simple elimination rules.

Lemma 8. *A nice goodness algebra exists.*

Proof. The sorts of the limit goodness algebra are defined as a chain, and operations act pointwise on each component of the chain. To prove that the operations are equivalences, we compose a proof that Arg commutes with chains (given by combining the lemmas about chains commuting with type formers) with a proof that for each sort, the chain given by the $(E\ (S^n\ \mathbb{O}))$ is equivalent to the chain given by $(S^n\ \mathbb{O})$ (given by Lemma 7). Since $(E\ (S^n\ \mathbb{O}))$ is defined by pattern matching to reduce to Arg, the right and left sides of these equivalences agree, and we find that the operations are indeed nice. See the formalization for details.

Theorem 2. *There exists an inductive-inductive object* $(A, B, \eta, \mathit{ext}, \mathit{inj})$ *that satisfies the simple elimination rules as defined in Fig. 1.*

Proof. A nice goodness algebra exists by Lemma 8, therefore we can construct $(A, B, \eta, \mathtt{ext}, \mathtt{inj})$ satisfying the simple elimination rules by Lemma 6.

We have therefore succeeded. In cubical type theory, the inductive-inductive definition from Fig. 1 is constructible.

4 Related Work

The principle of simultaneously defining a type and a family over that type has been used many times before. Danielsson [9] used an inductive-inductive-recursive definition to define the syntax of dependent type theory, and Chapman [5] used an inductive-inductive definition for the same purpose. Conway's surreal numbers [7] are given (up to a defined equivalence relation) by the inductive-inductive definition of number and less than, where less than is a relation indexed by two numbers [15, §7.1]. The HoTT book §11.3 gives a definition of the Cauchy reals as a higher inductive-inductive definition [18].

In his thesis and previous papers [15–17], Nordvall Forsberg studies the general theory of inductive-inductive types, axiomatizing a limited class of such definitions, and giving a set theoretic model showing that they are consistent. He also considers various extensions such as allowing a third type indexed by the first two, allowing the second type to be indexed by two elements of the first, or combining inductive-inductive definitions with inductive-recursive definitions from Dybjer and Setzer [10].

There have been several attempts to define a general class of inductive-inductive types larger than that in Nordvall Forsberg's thesis. Kaposi and Kovács [14] gives an external syntactic description of a class which includes higher inductive-inductive types, and Altenkirch et al. [2] gives a semantic description of a class including quotient inductive-inductive types, but neither gives a type of codes that can be reasoned about internally. Working with UIP, Altenkirch et al. [4] propose a class of quotient inductive-inductive types.

Nordvall Forsberg's thesis [15] appears to give the best previously known reduction of inductive-inductive types to regular inductive types known. As we have shown, Nordvall Forsberg's approach can only be applied to intensional type theory if UIP holds. Furthermore, the equations for both Nordvall Forsberg's approach and our approach only hold propositionally.

Many other structures have been reduced to plain inductive types. Our construction of inductive-inductive types can be seen as an adaptation of the technique in Ahrens et al. [1], where coinductive types are constructed from $\mathbb{N}$ by taking a limit. Indexed inductive types (which are used in Nordvall Forsberg's construction) are constructed from plain inductive types in Altenkirch et al. [3], with good computational properties (provided an identity type that satisfies J strictly). And small induction-recursion is reduced to plain indexed inductive types in Hancock et al. [11].

5 Conclusions and Future Work

In this paper, we have:

1. Shown that the construction of inductive-inductive types given by Nordvall Forsberg implies UIP.
2. Given an alternative construction of one particular inductive-inductive type in cubical type theory, which is compatible with Homotopy Type Theory.

We claim that the construction of our specific running example is straightforwardly generalizable to other inductive-inductive types, and have formalized the construction of a number of other examples including types with non-finitary constructors and indices to support this claim (see the GitHub repository referenced in the introduction).

Going forward, we would like to investigate

- An internal definition of inductive-inductive specifications in HoTT. Early experiments suggest that this requires surmounting difficulties related to increasingly complex coherence conditions similar to those encountered when defining semi-simplicial sets, c.f. Herbelin [12].
- Extending the proof given here to construct the general elimination rules. The general elimination rules were defined in Nordvall Forsberg [15], but that formulation they relies on either strict computation rules or extensional type theory to be well typed. Kaposi and Kovács [14] give equivalent rules which are well typed in intensional type theory.
- Identifying what needs to be added for the simple elimination rules to have the expected computational behavior. Given the similar construction method, this hopefully also allows the construction of coinductive types with nice computational behavior, c.f. Ahrens et al. [1].

- In the opposite direction from the previous point, rewriting the construction given here in Coq + Function Extensionality. While the elimination rules will have poor computational behavior, this would make using inductive-inductive types in Coq possible without requiring any change to Coq itself, while being compatible with HoTT. In particular, using cubical type theory makes the proofs in Sect. 3.5 simpler, but we speculate that axiomatic function extensionality is sufficient.

Acknowledgements. I would like to thank Talia Ringer and Dan Grossman from the UW PLSE lab, for their invaluable feedback throughout the revision process. I also thank Pavel Panchekha, John Leo, Remy Wang, and Fredrik Nordvall Forsberg for their comments.

Some of this work was completed while studying at Tokyo Institute of Technology under Professor Ryo Kashima. I would like to thank Professor Kashima, as well as my fellow lab members and mentors Asami and Maniwa for making my stay both productive and enjoyable.

References

1. Ahrens, B., Capriotti, P., Spadotti, R.: Non-wellfounded trees in homotopy type theory. In: TLCA (2015)
2. Altenkirch, T., Capriotti, P., Dijkstra, G., Kraus, N., Nordvall Forsberg, F.: Quotient inductive-inductive types. In: Baier, C., Dal Lago, U. (eds.) FoSSaCS 2018. LNCS, vol. 10803, pp. 293–310. Springer, Cham (2018). https://doi.org/10.1007/978-3-319-89366-2_16
3. Altenkirch, T., Ghani, N., Hancock, P., McBride, C., Morris, P.: Indexed containers. J. Funct. Program. **25**, e5 (2015)
4. Altenkirch, T., Kaposi, A., Kovács, A.: Constructing quotient inductive-inductive types. Proc. ACM Program. Lang. **3**(POPL), 2:1–2:24 (2019). http://doi.acm.org/10.1145/3290315
5. Chapman, J.: Type theory should eat itself. Electron. Notes Theoret. Comput. Sci. **228**, 21–36 (2009). http://www.sciencedirect.com/science/article/pii/S157106610800577X. Proceedings of the International Workshop on Logical Frameworks and Metalanguages: Theory and Practice (LFMTP 2008)
6. Cohen, C., Coquand, T., Huber, S., Mörtberg, A.: Cubical Type Theory: a constructive interpretation of the univalence axiom. IfCoLog J. Logics Appl. **4**(10), 3127–3169 (2017). http://www.collegepublications.co.uk/journals/ifcolog/?00019
7. Conway, J.: On Numbers and Games. AK Peters Series. Taylor & Francis, Milton Park (2000). https://books.google.com/books?id=tXiVo8qA5PQC
8. The Coq Development Team: The Coq proof assistant, version 8.8.0, April 2018. https://doi.org/10.5281/zenodo.1219885
9. Danielsson, N.A.: A formalisation of a dependently typed language as an inductive-recursive family. In: Altenkirch, T., McBride, C. (eds.) TYPES 2006. LNCS, vol. 4502, pp. 93–109. Springer, Heidelberg (2007). https://doi.org/10.1007/978-3-540-74464-1_7
10. Dybjer, P., Setzer, A.: A finite axiomatization of inductive-recursive definitions. In: Girard, J.Y. (ed.) TLCA 1999. LNCS, vol. 1581, pp. 129–146. Springer, Heidelberg (1999). https://doi.org/10.1007/3-540-48959-2_11

11. Hancock, P., McBride, C., Ghani, N., Malatesta, L., Altenkirch, T.: Small induction recursion. In: Hasegawa, M. (ed.) TLCA 2013. LNCS, vol. 7941, pp. 156–172. Springer, Heidelberg (2013). https://doi.org/10.1007/978-3-642-38946-7_13
12. Herbelin, H.: A dependently-typed construction of semi-simplicial types. Math. Struct. Comput. Sci. **25**(5), 1116–1131 (2015)
13. Hofmann, M., Streicher, T.: The groupoid interpretation of type theory. In: Twenty-Five Years of Constructive Type Theory (Venice, 1995), Oxford Logic Guides, vol. 36, pp. 83–111. Oxford University Press, New York (1998)
14. Kaposi, A., Kovács, A.: A syntax for higher inductive-inductive types. In: Kirchner, H. (ed.) 3rd International Conference on Formal Structures for Computation and Deduction, FSCD 2018. Leibniz International Proceedings in Informatics, LIPIcs, vol. 108, pp. 20:1–20:18. Schloss Dagstuhl–Leibniz-Zentrum fuer Informatik, Dagstuhl, Germany (2018). http://drops.dagstuhl.de/opus/volltexte/2018/9190
15. Nordvall Forsberg, F.: Inductive-inductive definitions. Ph.D. thesis. Swansea University (2013)
16. Nordvall Forsberg, F., Setzer, A.: Inductive-inductive definitions. In: Dawar, A., Veith, H. (eds.) CSL 2010. LNCS, vol. 6247, pp. 454–468. Springer, Heidelberg (2010). https://doi.org/10.1007/978-3-642-15205-4_35
17. Nordvall Forsberg, F., Setzer, A.: A finite axiomatisation of inductive-inductive definitions. In: Berger, U., Hannes, D., Schuster, P., Seisenberger, M. (eds.) Logic, Construction, Computation, Ontos Mathematical Logic, vol. 3, pp. 259–287. Ontos Verlag (2012)
18. The Univalent Foundations Program: Homotopy Type Theory: Univalent Foundations of Mathematics, Institute for Advanced Study (2013). https://homotopytypetheory.org/book
19. Vezzosi, A.: Adding cubes to agda (2017). https://hott-uf.github.io/2017/abstracts/cubicalagda.pdf

Causal Inference by String Diagram Surgery

Bart Jacobs[1], Aleks Kissinger[1], and Fabio Zanasi[2](✉)

[1] Radboud University, Nijmegen, The Netherlands
[2] University College London, London, UK
f.zanasi@ucl.ac.uk

Abstract. Extracting causal relationships from observed correlations is a growing area in probabilistic reasoning, originating with the seminal work of Pearl and others from the early 1990s. This paper develops a new, categorically oriented view based on a clear distinction between syntax (string diagrams) and semantics (stochastic matrices), connected via interpretations as structure-preserving functors.

A key notion in the identification of causal effects is that of an intervention, whereby a variable is forcefully set to a particular value independent of any prior dependencies. We represent the effect of such an intervention as an endofunctor which performs 'string diagram surgery' within the syntactic category of string diagrams. This diagram surgery in turn yields a new, interventional distribution via the interpretation functor. While in general there is no way to compute interventional distributions purely from observed data, we show that this is possible in certain special cases using a calculational tool called comb disintegration.

We showcase this technique on a well-known example, predicting the causal effect of smoking on cancer in the presence of a confounding common cause. We then conclude by showing that this technique provides simple sufficient conditions for computing interventions which apply to a wide variety of situations considered in the causal inference literature.

Keywords: Causality · String diagrams · Probabilistic reasoning

1 Introduction

An important conceptual tool for distinguishing correlation from causation is the possibility of *intervention*. For example, a randomised drug trial attempts to destroy any confounding 'common cause' explanation for correlations between drug use and recovery by randomly assigning a patient to the control or treatment group, independent of any background factors. In an ideal setting, the observed correlations of such a trial will reflect genuine causal influence. Unfortunately, it is not always possible (or ethical) to ascertain causal effects by means of actual interventions. For instance, one is unlikely to get approval to run a clinical trial on whether smoking causes cancer by randomly assigning 50% of the

M. Bojańczyk and A. Simpson (Eds.): FOSSACS 2019, LNCS 11425, pp. 313–329, 2019.
https://doi.org/10.1007/978-3-030-17127-8_18

patients to smoke, and waiting a bit to see who gets cancer. However, in certain situations it is possible to predict the effect of such a hypothetical intervention from purely observational data.

In this paper, we will focus on the problem of *causal identifiability*. For this problem, we are given observational data as a joint distribution on a set of variables and we are furthermore provided with a *causal structure* associated with those variables. This structure, which typically takes the form of a directed acyclic graph or some variation thereof, tells us which variables can in principle have a causal influence on others. The problem then becomes whether we can measure how strong those causal influences are, by means of computing an *interventional* distribution. That is, can we ascertain what would have happened if a (hypothetical) intervention had occurred?

Over the past 3 decades, a great deal of work has been done in identifying necessary and sufficient conditions for causal identifiability in various special cases, starting with very specific notions such as the *back-door* and *front-door* criteria [20] and progressing to more general necessary and sufficient conditions for causal identifiability based on the **do**-calculus [11], or combinatoric concepts such as confounded components in semi-Makovian models [25,26].

This style of causal reasoning relies crucially on a delicate interplay between syntax and semantics, which is often not made explicit in the literature. The syntactic object of interest is the causal structure (e.g. a causal graph), which captures something about our understanding of the world, and the mechanisms which gave rise to some observed phenomena. The semantic object of interest is the data: joint and conditional probability distributions on some variables. Fixing a causal structure entails certain constraints on which probability distributions can arise, hence it is natural to see distributions satisfying those constraints as models of the syntax.

In this paper, we make this interplay precise using functorial semantics in the spirit of Lawvere [17], and develop basic syntactic and semantic tools for causal reasoning in this setting. We take as our starting point a functorial presentation of Bayesian networks similar to the one appearing in [7]. The syntactic role is played by string diagrams, which give an intuitive way to represent morphisms of a monoidal category as boxes plugged together by wires. Given a directed acyclic graph (dag) G, we can form a free category Syn_G whose arrows are (formal) string diagrams which represent the causal structure syntactically. Structure-preserving functors from Syn_G to Stoch, the category of stochastic matrices, then correspond exactly to Bayesian networks based on the dag G.

Within this framework, we develop the notion of intervention as an operation of 'string diagram surgery'. Intuitively, this cuts a string diagram at a certain variable, severing its link to the past. Formally, this is represented as an endofunctor on the syntactic category $\mathsf{cut}_X : \mathsf{Syn}_G \to \mathsf{Syn}_G$, which propagates through a model $\mathcal{F} \colon \mathsf{Syn}_G \to \mathsf{Stoch}$ to send observational probabilities $\mathcal{F}(\omega)$ to interventional probabilities $\mathcal{F}(\mathsf{cut}_X(\omega))$.

The cut_X endofunctor gives us a diagrammatic means of computing interventional distributions given complete knowledge of $\mathcal{F}$. However, more interestingly, we can sometimes compute interventionals given only partial knowledge of $\mathcal{F}$, namely some observational data. We show that this can also be done via a technique we call *comb disintegration*, which is a string diagrammatic version of a technique called *c-factorisation* introduced by Tian and Pearl [26]. Our approach generalises disintegration, a calculational tool whereby a joint state on two variables is factored into a single-variable state and a channel, representing the marginal and conditional parts of the distribution, respectively. Disintegration has recently been formulated categorically in [5] and using string diagrams in [4]. We take the latter as a starting point, but instead consider a factorisation of a three-variable state into a channel and a *comb*. The latter is a special kind of map which allows inputs and outputs to be interleaved. They were originally studied in the context of quantum communication protocols, seen as games [8], but have recently been used extensively in the study of causally-ordered quantum [3,21] and generalised [15] processes. While originally imagined for quantum processes, the categorical formulation given in [15] makes sense in both the classical case (Stoch) and the quantum. Much like Tian and Pearl's technique, comb factorisation allows one to characterise when the confounding parts of a causal structure are suitably isolated from each other, then exploit that isolation to perform the concrete calculation of interventional distributions.

However, unlike in the traditional formulation, the syntactic and semantic aspects of causal identifiability within our framework exactly mirror one-another. Namely, we can give conditions for causal identifiability in terms of factorisation a morphism in Syn_G, whereas the actual concrete computation of the interventional distribution involves factorisation of its interpretation in Stoch. Thanks to the functorial semantics, the former immediately implies the latter.

To introduce the framework, we make use of a running example taken from Pearl's book [20]: identifying the causal effect of smoking on cancer with the help of an auxiliary variable (the presence of tar in the lungs). After providing some preliminaries on stochastic matrices and the functorial presentation of Bayesian networks in Sects. 2 and 3, we introduce the smoking example in Sect. 4. In Sect. 5 we formalise the notion of intervention as string diagram surgery, and in Sect. 6 we introduce the combs and prove our main calculational result: the existence and uniqueness of comb factorisations. In Sect. 7, we show how to apply this theorem in computing the interventional distribution in the smoking example, and in 8, we show how this theorem can be applied in a more general case which captures (and slightly generalises) the conditions given in [26]. In Sect. 9, we conclude and describe several avenues of future work.

2 Stochastic Matrices and Conditional Probabilities

Symmetric monoidal categories (SMCs) give a very general setting for studying processes which can be composed in sequence (via the usual categorical composition $\circ$) and in parallel (via the monoidal composition $\otimes$). Throughout this paper, we will use *string diagram* notation [24] for depicting composition of morphisms in an SMC. In this notation, morphisms are depicted as boxes with labelled input and output wires, composition $\circ$ as 'plugging' boxes together, and the monoidal product $\otimes$ as placing boxes side-by-side. Identity morphisms are depicted simply as a wire and the unit I of $\otimes$ as the empty diagram. The 'symmetric' part of the structure consists of symmetry morphisms, which enable us to permute inputs and outputs arbitrarily. We depict these as wire-crossings: . Morphisms whose domain is I are called *states*, and they will play a special role throughout this paper.

A monoidal category of prime interest in this paper is Stoch, whose objects are finite sets and morphisms $\boldsymbol{f} : A \to B$ are $|B| \times |A|$ dimensional stochastic matrices. That is, they are matrices of positive numbers (including 0) whose columns each sum to 1:

$$\boldsymbol{f} = \{\boldsymbol{f}_i^j \in \mathbb{R}^+ \mid i \in A, j \in B\} \qquad \text{with} \qquad \textstyle\sum_j \boldsymbol{f}_i^j = 1, \text{ for all } i.$$

Note we adopt the physicists convention of writing row indices as superscripts and column indices as subscripts. Stochastic matrices are of interest for probabilistic reasoning, because they exactly capture the data of a conditional probability distribution. That is, if we take $A := \{1, \ldots, m\}$ and $B := \{1, \ldots, n\}$, conditional probabilities naturally arrange themselves into a stochastic matrix:

$$\boldsymbol{f}_i^j := P(B = j | A = i) \quad \rightsquigarrow \quad \boldsymbol{f} = \begin{pmatrix} P(B = 1 | A = 1) & \cdots & P(B = 1 | A = m) \\ \vdots & \ddots & \vdots \\ P(B = n | A = 1) & \cdots & P(B = n | A = m) \end{pmatrix}$$

States, i.e. stochastic matrices from a trivial input $I := \{*\}$, are (non-conditional) probability distributions, represented as column vectors. There is only one stochastic matrix with trivial output: the row vector consisting only of 1's. The latter, with notation  as on the right, will play a special role in this paper (see (1) below).

Composition of stochastic matrices is matrix multiplication. In terms of conditional probabilities, that is multiplication followed by marginalization over the shared variable: $\sum_B P(C|B)P(B|A)$. Identities are thus given by identity matrices, which we will often express in terms of the Kronecker delta function $\boldsymbol{\delta}_i^j$.

The monoidal product $\otimes$ in Stoch is the cartesian product on objects, and Kronecker product of matrices: $(\boldsymbol{f} \otimes \boldsymbol{g})_{(i,j)}^{(k,l)} := \boldsymbol{f}_i^k \boldsymbol{g}_j^l$. We will typically omit parentheses and commas in the indices, writing e.g. $\boldsymbol{h}_{ij}^{kl}$ instead of $\boldsymbol{h}_{(i,j)}^{(k,l)}$ for an arbitrary matrix entry of $\boldsymbol{h} \colon A \otimes B \to C \otimes D$. In terms of conditional probabilities, the Kronecker product corresponds to taking product distributions. That is, if $\boldsymbol{f}$ represents the conditional probabilities $P(B|A)$ and $\boldsymbol{g}$ the probabilities $P(D|C)$, then $\boldsymbol{f} \otimes \boldsymbol{g}$ represents $P(B|A)P(D|C)$. Stoch also comes with a natural choice of 'swap' matrices $\boldsymbol{\sigma} : A \otimes B \to B \otimes A$ given by $\boldsymbol{\sigma}_{ij}^{kl} := \boldsymbol{\delta}_i^l \boldsymbol{\delta}_j^k$, making it into a symmetric monoidal category. Every object A in Stoch has three other pieces of structure which will play a key role in our formulation of Bayesian networks and interventions: the *copy* map, the *discarding* map, and the *uniform state*:

$\Big($  $\Big)_i^{jk} := \boldsymbol{\delta}_i^j \boldsymbol{\delta}_i^k$ $\qquad$ $\Big($  $\Big)_i := 1$ $\qquad$ $\Big($  $\Big)^i := \frac{1}{|A|}$ $\qquad$ (1)

Abstractly, this provides Stoch with the structure of a *CDU category.*

Definition 2.1. *A* CDU category *(for* copy, discard, uniform*) is a symmetric monoidal category* $(\mathsf{C}, \otimes, I)$ *where each object* A *has a copy map* $: A \to A \otimes A$, *a discarding map* $: A \to I$, *and a uniform state* $: I \to A$ *satisfying the following equations:*

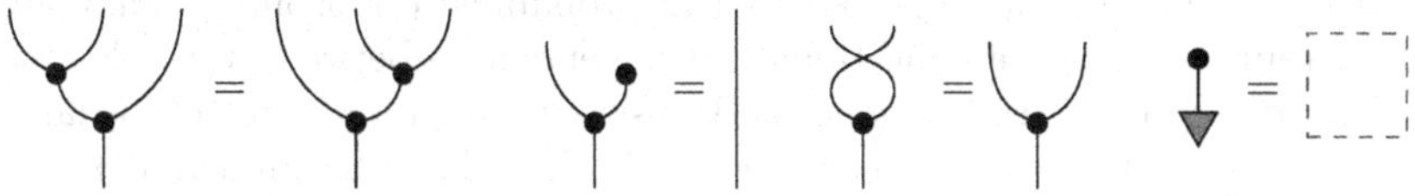

(2)

CDU functors *are symmetric monoidal functors between CDU categories preserving copy maps, discard maps and uniform states.*

We assume that the CDU structure on I is trivial and the CDU structure on $A \otimes B$ is constructed in the obvious way from the structure on A and B. We also use the first equation in (2) to justify writing 'copy' maps with arbitrarily many output wires: .

Similar to [2], we can form the free CDU category $\mathsf{FreeCDU}(X, \Sigma)$ over a pair (X, Σ) of a generating set of objects X and a generating set Σ of typed morphisms $f : u \to w$, with $u, w \in X^\star$ as follows. The category $\mathsf{FreeCDU}(X, \Sigma)$ has $X^\star$ as set of objects, and morphisms the string diagrams constructed from the elements of Σ and maps $: x \to x \otimes x$, $: x \to I$ and $: I \to x$ for each $x \in X$, taken modulo the equations (2).

Lemma 2.2. Stoch *is a CDU category, with CDU structure defined as in* (1).

An important feature of Stoch is that $I = \{\star\}$ is the final object, with $: B \to I$ the map provided by the universal property, for any set B. This yields Eq. (3) on the right, for any $\boldsymbol{f} : A \to B$, justifying the name "discarding map" for .

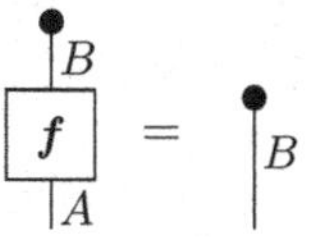

(3)

We conclude by recording another significant feature of Stoch: *disintegration* [4,5]. In probability theory, this is the mechanism of factoring a joint probability distribution $P(AB)$ as a product of the first marginal $P(A)$ and a conditional distribution $P(B|A)$. We recall from [4] the string diagrammatic rendition of this process. We say that a morphism $\boldsymbol{f} : X \to Y$ in Stoch has *full support* if, as a stochastic matrix, it has no zero entries. When $\boldsymbol{f}$ is a state, it is a standard result that full support ensures uniqueness of disintegrations of $\boldsymbol{f}$.

Proposition 2.3 (Disintegration). *For any state* $\boldsymbol{\omega} : I \to A \otimes B$ *with full support, there exists unique morphisms* $\boldsymbol{a} : I \to A$, $\boldsymbol{b} : A \to B$ *such that:*

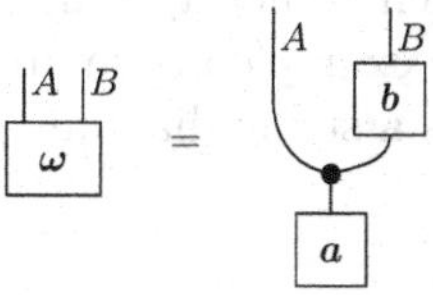

(4)

Note that Eq. (3) and the CDU rules immediately imply that the unique $\boldsymbol{a}\colon I \to A$ in Proposition 2.3 is the marginal of ω onto A: 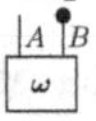 .

3 Bayesian Networks as String Diagrams

Bayesian networks are a widely-used tool in probabilistic reasoning. They give a succinct representation of conditional (in)dependences between variables as a directed acyclic graph. Traditionally, a Bayesian network on a set of variables $A, B, C, \ldots$ is defined as a directed acyclic graph (dag) G, an assignment of sets to each of the nodes $V_G := \{A, B, C, \ldots\}$ of G and a joint probability distribution over those variables which factorises as $P(V_G) = \prod_{A \in V_G} P(A \,|\, \mathrm{Pa}(A))$ where $\mathrm{Pa}(A)$ is the set of parents of A in G. Any joint distribution that factorises this way is said to satisfy the *global Markov property* with respect to the dag G. Alternatively, a Bayesian network can be seen as a dag equipped with a set of conditional probabilities $\{P(A \,|\, \mathrm{Pa}(A)) \mid A \in V_G\}$ which can be combined to form the joint state. Thanks to disintegration, these two perspectives are equivalent.

Much like in the case of disintegration in the previous section, Bayesian networks have a neat categorical description as string diagrams in the category Stoch [7, 13, 14]. For example, here is a Bayesian network in its traditional depiction as a dag with an associated joint distribution over its vertices, and as a string diagram in Stoch:

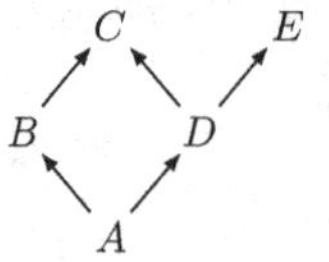

$$P(ABCDE) = P(A)P(B|A)P(D|A)P(C|BD)P(E|D)$$

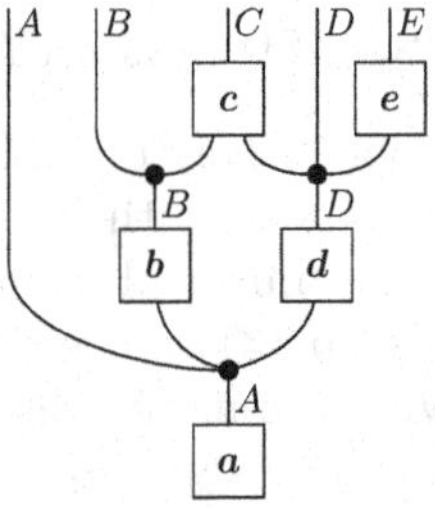

In the string diagram above, the stochastic matrix $\boldsymbol{a}\colon I \to A$ contains the probabilities $P(A)$, $\boldsymbol{b}\colon B \to A$ contains the conditional probabilities $P(B|A)$, $\boldsymbol{c}\colon B \otimes D \to C$ contains $P(C|BD)$, and so on. The entire diagram is then equal to a state $\omega\colon I \to A \otimes B \otimes C \otimes D \otimes E$ which represents $P(ABCDE)$.

Note the dag and the diagram above look similar in structure. The main difference is the use of copy maps to make each variable (even those that are not leaves of the dag, A, B and D) an output of the overall diagram. This corresponds to a variable being *observed*. We can also consider Bayesian networks with *latent* variables, which do not appear in the joint distribution due to marginalisation. Continuing the example above, making A into a latent variable yields the following depiction as a string diagram:

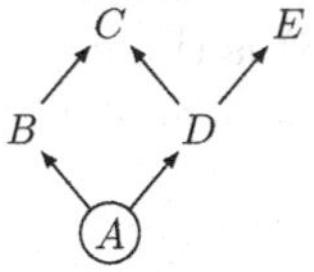

$$P(BCDE) = \sum_A P(A)P(B|A)P(D|A)P(C|BD)P(E|D)$$

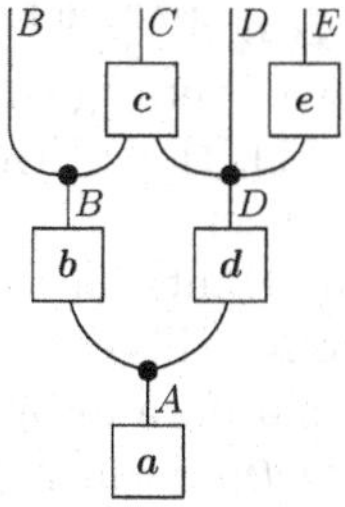

In general, a Bayesian network (with possible latent variables), is a string diagram in Stoch that (1) only has outputs and (2) consists only of copy maps and boxes which each have exactly one output.

By 'a string diagram in Stoch', we mean not only the stochastic matrix itself, but also its decomposition into components. We can formalise exactly what we mean by taking a perspective on Bayesian networks which draws inspiration from Lawvere's functorial semantics of algebraic theories [16]. In this perspective, which elaborates on [7, Ch. 4], we maintain a conceptual distinction between the purely syntactic object (the diagram) and its probabilistic interpretation.

Starting from a dag $G = (V_G, E_G)$, we construct a free CDU category Syn_G which provides the syntax of causal structures labelled by G. The objects of Syn_G are generated by the vertices of G, whereas the morphisms are generated by the following signature:

$$\Sigma_G = \left\{ \quad \right.$$ 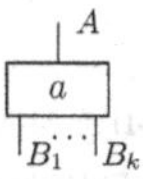

$$\left. \quad \middle| \quad A \in V_G \text{ with parents } B_1, \ldots, B_k \in V_G \right\}$$

Then $\mathsf{Syn}_G := \mathsf{FreeCDU}(V_G, \Sigma_G)$.[1] The following result establishes that models (*à la* Lawvere) of Syn_G coincide with G-based Bayesian networks.

Proposition 3.1. *There is a 1-1 correspondence between Bayesian networks based on the dag G and CDU functors of type* $\mathsf{Syn}_G \to \mathsf{Stoch}$.

We refer to [12] for a proof. This proposition justifies the following definition of a category BN_G of G-based Bayesian networks: objects are CDU functors $\mathsf{Syn}_G \to \mathsf{Stoch}$ and arrows are monoidal natural transformations between them.

4 Towards Causal Inference: The Smoking Scenario

We will motivate our approach to causal inference via a classic example, inspired by the one given in the Pearl's book [20]. Imagine a dispute between a scientist and a tobacco company. The scientist claims that smoking causes cancer. As a source of evidence, the scientist cites a joint probability distribution ω over variables S for smoking and C for cancer, which disintegrates as in (5) below,

[1] Note that E_G is implicitly used in the construction of Syn_G: the edges of G determine the parents of a vertex, and hence the input types of the symbols in Σ_G.

with matrix $\boldsymbol{c} = \left(\begin{smallmatrix} 0.9 & 0.7 \\ 0.1 & 0.3 \end{smallmatrix}\right)$. Inspecting this $\boldsymbol{c} : S \to C$, the scientist notes that the probability of getting cancer for smokers (0.3) is three times as high as for non-smokers (0.1). Hence, the scientist claims that smoking has a significant causal effect on cancer.

An important thing to stress here is that the scientist draws this conclusion using not only the observational data $\boldsymbol{\omega}$ but also from an assumed *causal structure* which gave rise to that data, as captured in the diagram in Eq. (5). That is, rather than treating diagram (5) simply as a calculation on the observational data, it can also be treated as an assumption about the actual, physical mechanism that gave rise to that data. Namely, this diagram encompasses the assumption that there is some prior propensity for people to smoke captured by $\boldsymbol{s} : I \to S$, which is both observed and fed into some other process $\boldsymbol{c} : S \to C$ whereby an individuals choice to smoke determines whether or not they get cancer.

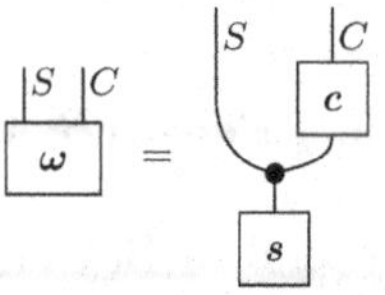

(5)

The tobacco company, in turn, says that the scientists' assumptions about the provenance of this data are too strong. While they concede that *in principle* it is possible for smoking to have some influence on cancer, the scientist should allow for the possibility that there is some latent common cause (e.g. genetic conditions, stressful work environment, etc.) which leads people both to smoke and get cancer. Hence, says the tobacco company, a 'more honest' causal structure to ascribe to the data ω is (6). This structure then allows for either party to be correct. If the scientist is right, the output of $\boldsymbol{c} : S \otimes H \to C$ depends mostly on its first input, i.e. the causal path from smoking to cancer. If the tabacco company is right, then $\boldsymbol{c}$ depends very little on its first input, and the correlation between S and C can be explained almost entirely from the hidden common cause.

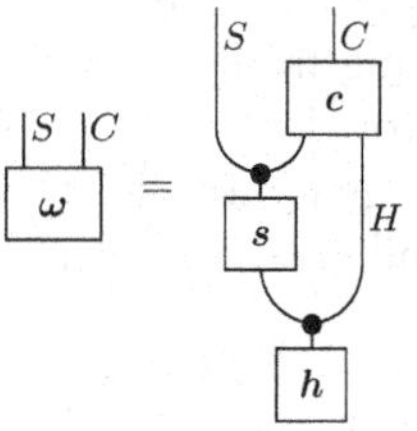

(6)

So, who is right after all? Just from the observed distribution $\boldsymbol{\omega}$, it is impossible to tell. So, the scientist proposes a clinical trial, in which patients are randomly required to smoke or not to smoke. We can model this situation by replacing $\boldsymbol{s}$ in (6) with a process that ignores its inputs and outputs the uniform state. Graphically, this looks like 'cutting' the link $\boldsymbol{s}$ between H and S:

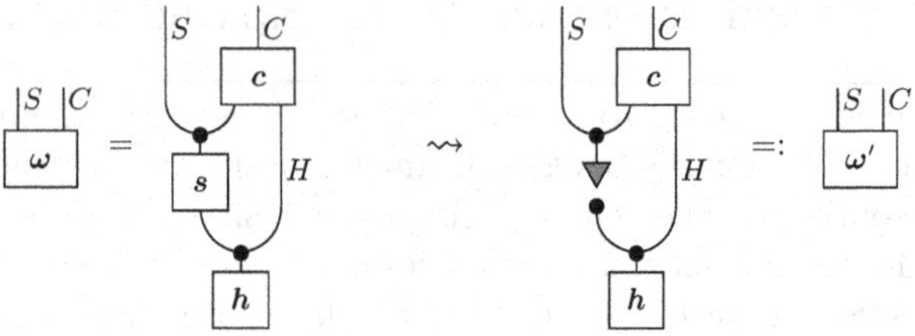

(7)

This captures the fact that variable S is now randomised and no longer dependent on any background factors. This new distribution $\boldsymbol{\omega}'$ represents the data the scientist would have obtained had they run the trial. That is, it gives the results of an *intervention* at s. If this ω' *still* shows a strong correlation between smoking and cancer, one can conclude that smoking indeed causes cancer even when we assume the weaker causal structure (6).

Unsurprisingly, the scientist fails to get ethical approval to run the trial, and hence has only the observational data ω to work with. Given that the scientist only knows ω (and not c and h), there is no way to compute ω' in this case. However, a key insight of statistical causal inference is that sometimes it *is* possible to compute interventional distributions from observational ones. Continuing the smoking example, suppose the scientist proposes the following revision to the causal structure: they posit a structure (8) that includes a third observed variable (the presence of T of tar in the lungs), which completely mediates the causal effect of smoking on cancer.

As with our simpler structure, the diagram (8) contains some assumptions about the provenance of the data ω. In particular, by omitting wires, we are asserting there is no *direct* causal link between certain variables. The absence of an H-labelled input to t says there is no direct causal link from H to T (only mediated by S), and the absence of an S-labelled input wire into c captures that there is no direct causal link from S to C (only mediated by T). In the traditional approach to causal inference, such relationships are typically captured by a graph-theoretic property called *d-separation* on the dag associated with the causal structure.

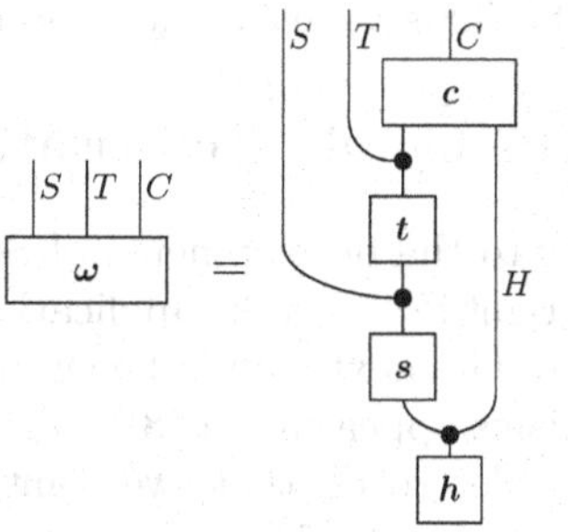

(8)

We can again imagine intervening at S by replacing $s : H \to S$ by 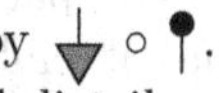 Again, this 'cutting' of the diagram will result in a new interventional distribution ω'. However, unike before, it *is* possible to compute this distribution from the observational distribution ω.

However, in order to do that, we first need to develop the appropriate categorical framework. In Sect. 5, we will model 'cutting' as a functor. In 6, we will introduce a generalisation of disintegration, which we call *comb disintegration*. These tools will enable us to compute ω' for ω, in Sect. 7.

5 Interventional Distributions as Diagram Surgery

The goal of this section is to define the 'cut' operation in (7) as an endofunctor on the category of Bayesian networks. First, we observe that such an operation exclusively concerns the string diagram part of a Bayesian network: following the functorial semantics given in Sect. 3, it is thus appropriate to define cut as an endofunctor on Syn_G, for a given dag G.

Definition 5.1. *For a fixed node $A \in V_G$ in a graph G, let $\mathsf{cut}_A \colon \mathsf{Syn}_G \to \mathsf{Syn}_G$ be the CDU functor freely obtained by the following action on the generators (V_G, Σ_G) of Syn_G:*

- *For each object $B \in V_G$, $\mathsf{cut}_A(B) = B$.*

- $\mathsf{cut}_A\big(\begin{smallmatrix} A \\ \boxed{a} \\ B_i \cdots B_k \end{smallmatrix}\big) = \begin{smallmatrix} A \\ \blacktriangledown \\ B_i \cdots B_k \end{smallmatrix}$ *and* $\mathsf{cut}_A\big(\begin{smallmatrix} B \\ \boxed{b} \\ C_i \cdots C_j \end{smallmatrix}\big) = \begin{smallmatrix} B \\ \boxed{b} \\ C_i \cdots C_j \end{smallmatrix}$ *for any other* $\begin{smallmatrix} B \\ \boxed{b} \\ C_i \cdots C_j \end{smallmatrix} \in \Sigma_G$.

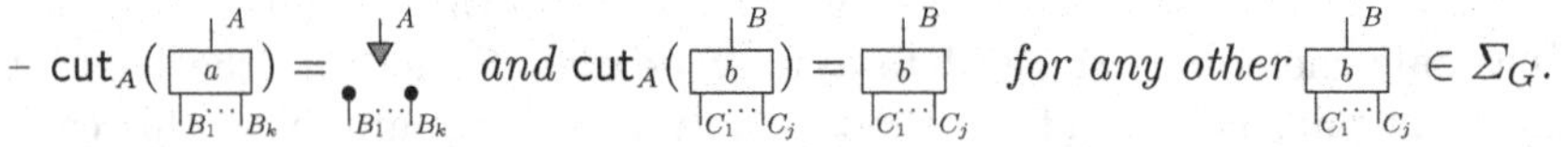

Intuitively, cut_A applied to a string diagram f of Syn_G removes from f each occurrence of a box with output wire of type A.

Proposition 3.1 allows us to "transport" the cutting operation over to Bayesian networks. Given any Bayesian network based on G, let $\mathcal{F} \colon \mathsf{Syn}_G \to \mathsf{Stoch}$ be the corresponding CDU functor given by Proposition 3.1. Then, we can define its A-cutting as the Bayesian network identified by the CDU functor $\mathcal{F} \circ \mathsf{cut}_A$. This yields an (idempotent) endofunctor $\mathsf{Cut}_A \colon \mathsf{BN}_G \to \mathsf{BN}_G$.

6   The Comb Factorisation

Thanks to the developments of Sect. 5, we can understand the transition from left to right in (7) as the application of the functor Cut_S applied to the 'Smoking' node S. The next step is being able to actually compute the individual Stoch-morphisms appearing in (8), to give an answer to the causality question.

In order to do that, we want to work in a setting where $t \colon S \to T$ can be isolated and 'extracted' from (8). What is left behind is a stochastic matrix with a 'hole' where t has been extracted. To define 'morphisms with holes', it is convenient to pass from SMCs to compact closed categories (see e.g. [24]). Stoch is not itself compact closed, but it embeds into $\mathsf{Mat}(\mathbb{R}^+)$, whose morphisms are *all* matrices over positive numbers. $\mathsf{Mat}(\mathbb{R}^+)$ has a (self-dual) compact closed structure; that means, for any set A there is a 'cap' $\cap \colon A \otimes A \to I$ and a 'cup' $\cup \colon I \to A \otimes A$, which satisfy the 'yanking' equations on the right. As matrices, caps and cups are defined by $\cap_{ij} = \cup^{ij} = \delta_i^j$. Intuitively, they amount to 'bent' identity wires. Another aspect of $\mathsf{Mat}(\mathbb{R}^+)$ that is useful to recall is the following handy characterisation of the subcategory Stoch.

Lemma 6.1. *A morphism $f \colon A \to B$ in $\mathsf{Mat}(\mathbb{R}^+)$ is a stochastic matrix (thus a morphism of Stoch) if and only if (3) holds.*

A suitable notion of 'stochastic map with a hole' is provided by a *comb*. These structures originate in the study of certain kinds of quantum channels [3].

Definition 6.2. *A 2-comb in Stoch is a morphism $f \colon A_1 \otimes A_2 \to B_1 \otimes B_2$ satisfying, for some other morphism $f' \colon A_1 \to B_1$,*

$$
\begin{array}{c}
B_1 \quad B_2 \\
\boxed{f} \\
A_1 \quad A_2
\end{array}
=
\begin{array}{c}
B_1 \\
\boxed{f'} \quad \bullet \\
A_1 \quad A_2
\end{array}
$$

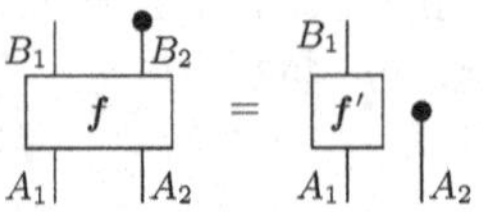

This definition extends inductively to *n-combs*, where we require that discarding the rightmost output yields $\boldsymbol{f}' \otimes \top$, for some $(n-1)$-comb $\boldsymbol{f}'$. However, for our purposes, restricting to 2-combs will suffice.

The intuition behind condition (9) is that the contribution from input A_2 is only visible via output B_2. Thus, if we discard B_2 we may as well discard A_2. In other words, the input/output pair A_2, B_2 happen 'after' the pair A_1, B_1. Hence, it is typical to depict 2-combs in the shape of a (hair) comb, with 2 'teeth', as in (10) below:

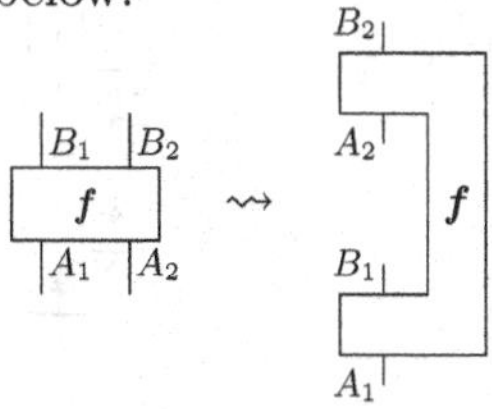

(10)

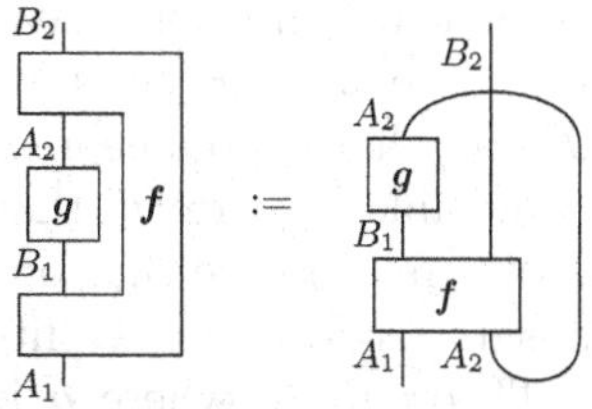

(11)

While combs themselves live in Stoch, $\mathsf{Mat}(\mathbb{R}^+)$ accommodates a second-order reading of the transition $\rightsquigarrow$ in (10): we can treat $\boldsymbol{f}$ as a map which expects as input a map $\boldsymbol{g}\colon B_1 \to A_2$ and produces as output a map of type $A_1 \to B_2$. Plugging $\boldsymbol{g}\colon B_1 \to A_2$ into the 2-comb can be formally defined in $\mathsf{Mat}(\mathbb{R}^+)$ by composing $\boldsymbol{f}$ and $\boldsymbol{g}$ in the usual way, then feeding the output of $\boldsymbol{g}$ into the second input of $\boldsymbol{f}$, using caps and cups, as in (11).

Importantly, for generic $\boldsymbol{f}$ and $\boldsymbol{g}$ of Stoch, there is no guarantee that forming the composite (11) in $\mathsf{Mat}(\mathbb{R}^+)$ yields a valid Stoch-morphism, i.e. a morphism satisfying the finality Eq. (3). However, if $\boldsymbol{f}$ is a 2-comb and $\boldsymbol{g}$ is a Stoch-morphism, Eq. (9) enables a discarding map plugged into the output B_2 in (11) to 'fall through' the right side of $\boldsymbol{f}$, which guarantees that the composed map satisfies the finality equation for discarding. See [12, § ??] for the explicit diagram calculation.

With the concept of 2-combs in hand, we can state our factorisation result.

Theorem 6.3. *For any state* $\boldsymbol{\omega}\colon I \to A \otimes B \otimes C$ *of* Stoch *with full support, there exists a unique 2-comb* $\boldsymbol{f} : B \to A \otimes C$ *and stochastic matrix* $\boldsymbol{g}\colon A \to B$ *such that, in* $\mathsf{Mat}(\mathbb{R}^+)$*:*

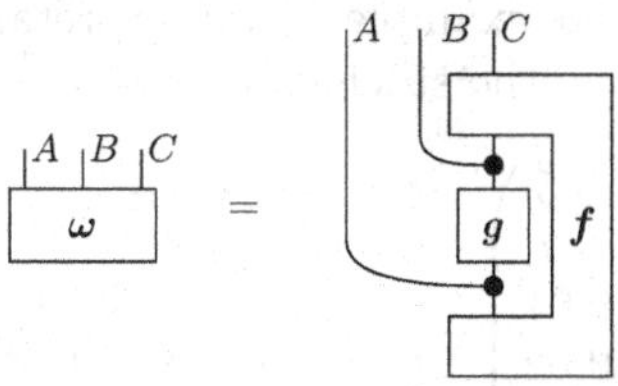

(12)

Proof. The construction of $\boldsymbol{f}$ and $\boldsymbol{g}$ mimics the one of c-factors in [26], using string diagrams and (diagrammatic) disintegration. We first use $\boldsymbol{\omega}$ to construct maps $\boldsymbol{a} : I \to A, \boldsymbol{b} : A \to B, \boldsymbol{c} : A \otimes B \to C$, then construct $\boldsymbol{f}$ using $\boldsymbol{a}$ and $\boldsymbol{c}$ and construct $\boldsymbol{g}$ using $\boldsymbol{b}$. For the full proof, including uniqueness, see [12]. □

Note that Theorem 6.3 generalises the normal disintegration property given in Proposition 2.3. The latter is recovered by taking $A := I$ (or $C := I$) above.

7 Returning to the Smoking Scenario

We now return to the smoking scenario of Sect. 4. There, we concluded by claiming that the introduction of an intermediate variable T to the observational distribution $\boldsymbol{\omega} : I \to S \otimes T \otimes C$ would enable us to calculate the interventional distribution. That is, we can calculate $\boldsymbol{\omega}' = \mathcal{F}(\mathsf{cut}_S(\omega))$ from $\boldsymbol{\omega} := \mathcal{F}(\omega)$. Thanks to Theorem 6.3, we are now able to perform that calculation. We first observe that our assumed causal structure for $\boldsymbol{\omega}$ fits the form of Theorem 6.3, where $\boldsymbol{g}$ is $\boldsymbol{t}$ and $\boldsymbol{f}$ is a 2-comb containing everything else, as in the diagram on the side.

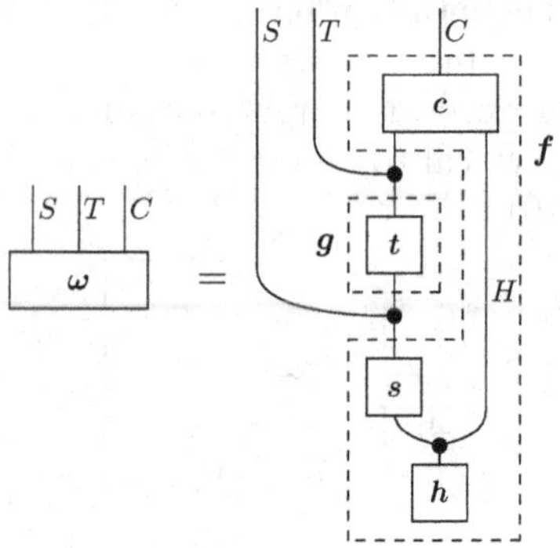

Hence, $\boldsymbol{f}$ and $\boldsymbol{g}$ are computable from $\boldsymbol{\omega}$. If we plug them back together as in (12), we will get $\boldsymbol{\omega}$ back. However, if we insert a 'cut' between $\boldsymbol{f}$ and $\boldsymbol{g}$:

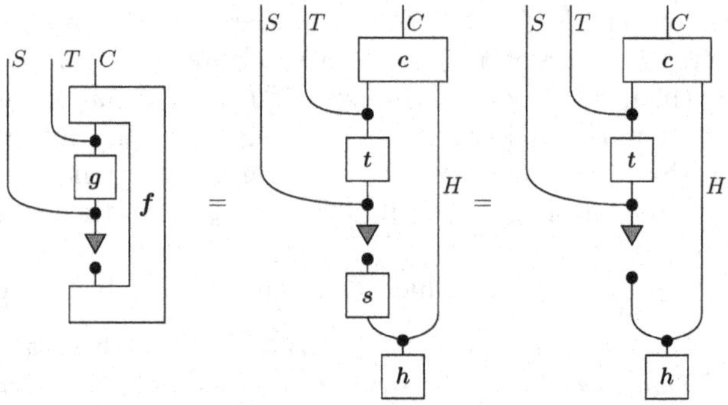

(13)

we obtain $\boldsymbol{\omega}' = \mathcal{F}(\mathsf{cut}_S(\omega))$.

We now consider a concrete example. Fix interpretations $S = T = C = \{0, 1\}$ and let $\boldsymbol{\omega} : I \to S \otimes T \otimes C$ be the stochastic matrix:

$$\boldsymbol{\omega} := \begin{pmatrix} 0.5 \\ 0.1 \\ 0.01 \\ 0.02 \\ 0.1 \\ 0.05 \\ 0.02 \\ 0.2 \end{pmatrix} \begin{matrix} \leftarrow P(S=0, T=0, C=0) \\ \leftarrow P(S=0, T=0, C=1) \\ \leftarrow P(S=0, T=1, C=0) \\ \leftarrow P(S=0, T=1, C=1) \\ \leftarrow P(S=1, T=0, C=0) \\ \leftarrow P(S=1, T=0, C=1) \\ \leftarrow P(S=1, T=1, C=0) \\ \leftarrow P(S=1, T=1, C=1) \end{matrix}$$

Now, disintegrating ω:

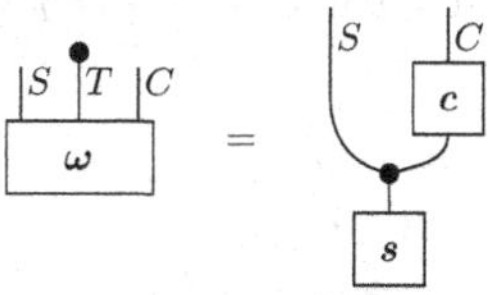

gives $\quad c \approx \begin{pmatrix} 0.81 & 0.32 \\ 0.19 & 0.68 \end{pmatrix}$

The bottom-left element of $\boldsymbol{c}$ is $P(C = 1|S = 0)$, whereas the bottom-right is $P(C = 1|S = 1)$, so this suggests that patients are $\approx$3.5 times as likely to get cancer if they smoke (68% vs. 19%). However, comb-disintegrating ω using Theorem 6.3 gives $\boldsymbol{g}: S \to T$ and a comb $\boldsymbol{f}: T \to S \otimes C$ with the following stochastic matrices:

$$\boldsymbol{f} \approx \begin{pmatrix} 0.53 & 0.21 \\ 0.11 & 0.42 \\ 0.25 & 0.03 \\ 0.12 & 0.34 \end{pmatrix} \qquad \boldsymbol{g} \approx \begin{pmatrix} 0.95 & 0.41 \\ 0.05 & 0.59 \end{pmatrix}$$

Recomposing these with a 'cut' in between, as in the left-hand side of (13), gives the interventional distribution $\omega' \approx (0.38, 0.11, 0.01, 0.02, 0.16, 0.05, 0.07, 0.22)$. Disintegrating:

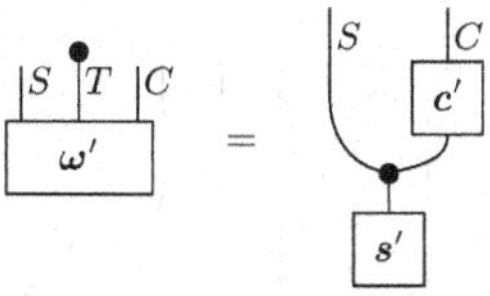

gives $\quad c' \approx \begin{pmatrix} 0.75 & 0.46 \\ 0.25 & 0.54 \end{pmatrix}.$

From the interventional distribution, we conclude that, in a (hypothotetical) clinical trial, patients are about twice as likely to get cancer if they smoke (54% vs. 25%). So, since $54 < 68$, there was *some* confounding influence between S and C in our observational data, but after removing it via comb disintegration, we see there is still a significant causal link between smoking and cancer.

Note this conclusion depends totally on the particular observational data that we picked. For a different interpretation of ω in Stoch, one might conclude that there is *no* causal connection, or even that smoking *decreases* the chance of getting cancer. Interestingly, all three cases can arise even when a naïve analysis of the data shows a strong direct correlation between S and C. To see and/or experiment with these cases, we have provided the Python code[2] used to perform these calculations. See also [19] for a pedagocical overview of this example (using traditional Bayesian network language) with some sample calculations.

8 The General Case for a Single Intervention

While we applied the comb decomposition to a particular example, this technique applies essentially unmodified to many examples where we intervene at a single variable (called X below) within an arbitrary causal structure.

[2] https://gist.github.com/akissinger/aeec1751792a208253bda491ead587b6.

Theorem 8.1. *Let G be a dag with a fixed node X that has corresponding generator $x\colon Y_1 \otimes \ldots \otimes Y_n \to X$ in Syn_G. Then, suppose ω is a morphism in Syn_G of the following form:*

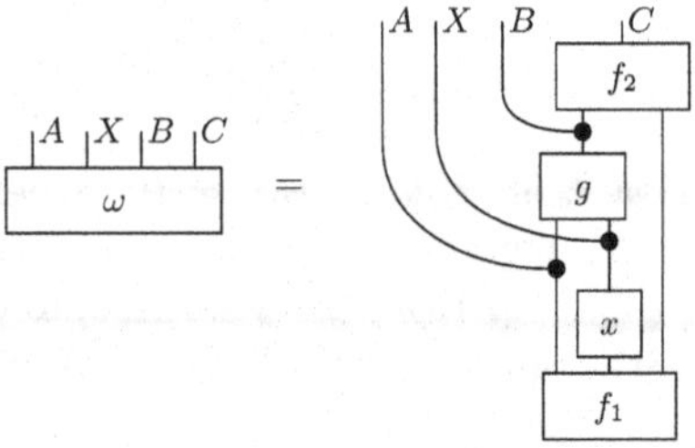

(14)

for some morphisms f_1, f_2 and g in Syn_G not containing x as a subdiagram. Then the interventional distribution $\boldsymbol{\omega}' := \mathcal{F}(\mathsf{cut}_X(\omega))$ is computable from the observational distribution $\boldsymbol{\omega} = \mathcal{F}(\omega)$.

Proof. The proof is very close to the example in the previous section. Interpreting ω into Stoch, we get a diagram of stochastic maps, which we can comb-disintegrate, then recompose with  to produce the interventional distribution:

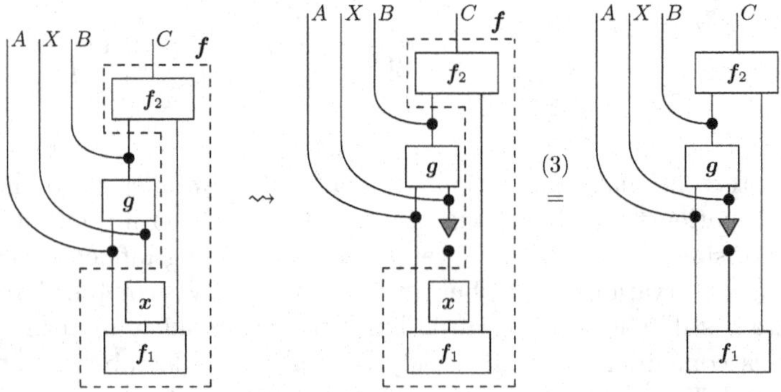

The RHS above is then $\mathcal{F}(\mathsf{cut}_X(\omega))$. □

This is general enough to cover several well-known sufficient conditions from the causality literature, including single-variable versions of the so-called *front-door* and *back-door* criteria, as well as the sufficient condition based on confounding paths given by Pearl and Tian [26]. As the latter subsumes the other two, we will say a few words about the relationship between the Pearl/Tian condition and Theorem 8.1. In [26], the authors focus on *semi-Markovian* models, where the only latent variables have exactly two observed children and no parents. Suppose we write $A \leftrightarrow B$ if two observed variables are connected by a latent common cause, then one can characterise *confounding paths* as the transitive closure of $\leftrightarrow$. They go on to show that the interventional distribution corresponding cutting X is computable whenever there are no confounding paths connecting X to one of its children.

We can compare this to the form of expression ω in Eq. (14). First, note this factorisation implies that all boxes which take X as an input must occur as sub-diagrams of g. Hence, any 'confounding path' connecting X to its children would yield at least one (un-copied) wire from f_1 to g, hence it cannot be factored as (14). Conversely, if there are no confounding paths from X to its children, then we can we can place the boxes involved in any other confounding path either entirely inside of g or entirely outside of g and obtain factorisation (14). Hence, restricting to semi-Markovian models, the no- confounding-path condition from [26] is equivalent to ours. However, Theorem 8.1 is slightly more general: its formulation doesn't rely on the causal structure ω being semi-Markovian.

9 Conclusion and Future Work

This paper takes a fresh, systematic look at the problem of causal identifiability. By clearly distinguishing syntax (string diagram surgery and identification of comb shapes) and semantics (comb-disintegration of joint states) we obtain a clear methodology for computing interventional distributions, and hence causal effects, from observational data.

A natural next step is moving beyond single-variable interventions to the general case, i.e. situations where we allow interventions on multiple variables which may have some arbitrary causal relationships connecting them. This would mean extending the comb factorisation Theorem 6.3 from a 2-comb and a channel to arbitrary n-combs. This seems to be straightforward, via an inductive extension of the proof of Theorem 6.3. A more substantial direction of future work will be the strengthening of Theorem 8.1 from sufficient conditions for causal identifiability to a full characterisation. Indeed, the related condition based on confounding paths from [26] is a necessary and sufficient condition for computing the interventional distribution on a single variable. Hence, it will be interesting to formalise this necessity proof (and more general versions, e.g. [10]) within our framework and investigate, for example, the extent to which it holds beyond the semi-Markovian case.

While we focus exclusively on the case of taking models in Stoch in this paper, the techniques we gave are posed at an abstract level in terms of composition and factorisation. Hence, we are optimistic about their prospects to generalise to other probabilistic (e.g. infinite discrete and continuous variables) and quantum settings. In the latter case, this could provide insights into the emerging field of *quantum causal structures* [6,9,18,22,23], which attempts in part to replay some of the results coming from statistical causal reasoning, but where quantum processes play a role analogous to stochastic ones. A key difficulty in applying our framework to a category of quantum processes, rather than Stoch, is the unavailability of 'copy' morphisms due to the quantum no-cloning theorem [27]. However, a recent proposal for the formulation of 'quantum common causes' [1] suggests a (partially-defined) analogue to the role played by 'copy' in our formulation constructed via multiplication of certain commuting Choi matrices. Hence, it may yet be possible to import results from classical causal reasoning into the quantum case just by changing the category of models.

Acknowledgements. FZ acknowledges support from EPSRC grant EP/R020604/1. AK would like to thank Tom Claassen for useful discussions on causal identification criteria.

References

1. Allen, J.-M.A., Barrett, J., Horsman, D.C., Lee, C.M., Spekkens, R.W.: Quantum common causes and quantum causal models. Phys. Rev. X **7**, 031021 (2017)
2. Bonchi, F., Sobociński, P., Zanasi, F.: Deconstructing Lawvere with distributive laws. J. Log. Algebr. Meth. Program. **95**, 128–146 (2018)
3. Chiribella, G., D'Ariano, G.M., Perinotti, P.: Quantum circuit architecture. Phys. Rev. Lett. **101**, 060401 (2008)
4. Cho, K., Jacobs, B.: Disintegration and Bayesian inversion, both abstractly and concretely (2017). arxiv.org/abs/1709.00322
5. Clerc, F., Danos, V., Dahlqvist, F., Garnier, I.: Pointless learning. In: Esparza, J., Murawski, A.S. (eds.) FoSSaCS 2017. LNCS, vol. 10203, pp. 355–369. Springer, Heidelberg (2017). https://doi.org/10.1007/978-3-662-54458-7_21
6. Costa, F., Shrapnel, S.: Quantum causal modelling. New J. Phys. **18**(6), 063032 (2016)
7. Fong, B.: Causal theories: a categorical perspective on Bayesian networks. Master's thesis, University of Oxford (2012). arxiv.org/abs/1301.6201
8. Gutoski, G., Watrous, J.: Toward a general theory of quantum games. In: Proceedings of the Thirty-Ninth Annual ACM Symposium on Theory of Computing, pp. 565–574. ACM (2007)
9. Henson, J., Lal, R., Pusey, M.F.: Theory-independent limits on correlations from generalized Bayesian networks. New J. Phys. **16**(11), 113043 (2014)
10. Huang, Y., Valtorta, M.: On the completeness of an identifiability algorithm for semi-Markovian models. Ann. Math. Artif. Intell. **54**(4), 363–408 (2008)
11. Huang, Y., Valtorta, M.: Pearl's calculus of intervention is complete. CoRR, abs/1206.6831 (2012)
12. Jacobs, B., Kissinger, A., Zanasi, F.: Causal inference by string diagram surgery. CoRR, abs/1811.08338 (2018)
13. Jacobs, B., Zanasi, F.: A predicate/state transformer semantics for Bayesian learning. Electr. Notes Theor. Comput. Sci. **325**, 185–200 (2016)
14. Jacobs, B., Zanasi, F.: The logical essentials of Bayesian reasoning. CoRR, abs/1804.01193 (2018)
15. Kissinger, A., Uijlen, S.: A categorical semantics for causal structure. In: 32nd Annual ACM/IEEE Symposium on Logic in Computer Science, LICS 2017, Reykjavik, Iceland, 20–23 June 2017, pp. 1–12 (2017)
16. Lawvere, F.W.: Ordinal sums and equational doctrines. In: Eckmann, B. (ed.) Seminar on Triples and Categorical Homology Theory. LNM, vol. 80, pp. 141–155. Springer, Heidelberg (1969). https://doi.org/10.1007/BFb0083085
17. Lawvere, F.W.: Functorial semantics of algebraic theories. Proc. Natl. Acad. Sci. U.S.A. **50**(5), 869 (1963)
18. Leifer, M.S., Spekkens, R.W.: Towards a formulation of quantum theory as a causally neutral theory of Bayesian inference. Phys. Rev. A **88**, 052130 (2013)
19. Nielsen, M.: If correlation doesn't imply causation, then what does? http://www.michaelnielsen.org/ddi/if-correlation-doesnt-imply-causation-then-what-does. Accessed 15 Nov 2018

20. Pearl, J.: Causality: Models, Reasoning and Inference. Cambridge University Press, Cambridge (2000)
21. Perinotti, P.: Causal structures and the classification of higher order quantum computations (2016)
22. Pienaar, J., Brukner, Č.: A graph-separation theorem for quantum causal models. New J. Phys. **17**(7), 073020 (2015)
23. Ried, K., Agnew, M., Vermeyden, L., Janzing, D., Spekkens, R.W., Resch, K.J.: A quantum advantage for inferring causal structure. Nat. Phys. **11**, 1745–2473 (2015)
24. Selinger, P.: A survey of graphical languages for monoidal categories. In: Coecke, B. (ed.) New Structures for Physics. LNP, vol. 813. Springer, Heidelberg (2011)
25. Shpitser, I., Pearl, J.: Identification of joint interventional distributions in recursive semi-Markovian causal models. In: Proceedings of the National Conference on Artificial Intelligence, vol. 21, p. 1219. AAAI Press/MIT Press, Menlo Park/Cambridge (1999/2006)
26. Tian, J., Pearl, J.: A general identification condition for causal effects. In: Proceedings of the Eighteenth National Conference on Artificial Intelligence and Fourteenth Conference on Innovative Applications of Artificial Intelligence, 28 July–1 August 2002, Edmonton, Alberta, Canada, pp. 567–573 (2002)
27. Wootters, W.K., Zurek, W.H.: A single quantum cannot be cloned. Nature **299**(5886), 802–803 (1982)

Higher-Order Distributions for Differential Linear Logic

Marie Kerjean[1(✉)] and Jean-Simon Pacaud Lemay[2]

[1] Équipe Gallinette, Inria, LS2N, Nantes, France
marie.kerjean@inria.fr
[2] University of Oxford, Oxford, UK
jean-simon.lemay@kellogg.ox.ac.uk

Abstract. Linear Logic was introduced as the computational counterpart of the algebraic notion of linearity. Differential Linear Logic refines Linear Logic with a proof-theoretical interpretation of the geometrical process of differentiation. In this article, we construct a polarized model of Differential Linear Logic satisfying computational constraints such as an interpretation for higher-order functions, as well as constraints inherited from physics such as a continuous interpretation for spaces. This extends what was done previously by Kerjean for first order Differential Linear Logic without promotion. Concretely, we follow the previous idea of interpreting the exponential of Differential Linear Logic as a space of higher-order distributions with compact-support, which is constructed as an inductive limit of spaces of distributions on Euclidean spaces. We prove that this exponential is endowed with a co-monadic like structure, with the notable exception that it is functorial only on isomorphisms. Interestingly, as previously argued by Ehrhard, this still allows the interpretation of differential linear logic without promotion.

Keywords: Differential Linear Logic · Categorical semantics · Topological vector spaces

1 Introduction

Denotational semantics interprets programs as functions which focuses not on how data from these programs are computed, but rather focusing on the input/output of programs and on data computed from other data [19]. Through the Curry-Howard-Lambek correspondence, this approach refines into the categorical semantics of type systems. In particular, a study of the denotational model of the λ-calculus for coherent spaces led Girard to Linear Logic [9] and the understanding of the use of resources as the computational counterpart of

Marie Kerjean was supported by the ANR Rapido, and would like to thanks Tom Hirschowitz for many comments and discussions on this work. Jean-Simon Pacaud Lemay would like to thank Kellogg College, the Clarendon Fund, and the Oxford-Google DeepMind Graduate Scholarship for financial support.

M. Bojańczyk and A. Simpson (Eds.): FOSSACS 2019, LNCS 11425, pp. 330–347, 2019.
https://doi.org/10.1007/978-3-030-17127-8_19

linearity in algebra. Differential Linear Logic (DiLL) [7] is a refinement of Linear Logic which allows for a notion of linear approximation of non-linear proofs. As a proof-net calculus, DiLL originated from studying vectorial models of Linear Logic which in general are based on spaces of sequences, such as Köthe spaces and finiteness spaces [5].

Recently the first author argued in [14] that as a sequent calculus DiLL has a "smooth" semantical interpretation where the exponential ! (the central object of Linear Logic) is interpreted as a space of *distributions with compact support* [18]. This semantical interpretation of DiLL (along with the Linear Logic typed phenomena of duality and interaction) provides a strong argument that DiLL should be considered as a foundation for a type theory of differential equations, whose semantics would be based on structures developed for mathematical physics. However one of the many divergences between the theoretical study of physical systems and the theoretical study of programming languages lies in the treatment of input data. In the study of differential equations, one generally only accepts a finite number of parameters: typically time and space [1]. While one of the fundamental aspects of the semantics of functional programming languages is the concept of higher-order types [4], which in particular allows programs to take other programs as inputs. Linking these two concepts together requires that when mathematical physics studies functions with finite dimensional domains, the denotational semantical counterpart will be studying functions whose codomains are spaces of functions (which are in general far from being finite dimensional).

This article gives a higher-order notion of distributions with compact support, following the model without higher order constructed by the first author in [14]. Indeed, only functions whose domains are finite dimensional were defined in [14], while no interpretation was given for functions whose domains are spaces of smooth functions. This latter notion relies on the basic intuition that even with a continuous and infinite set of input data, a program will at each computation use only a *finite* amount of data.

Content and Related Work. In this paper, we interpret the exponential as an inductive limit of spaces of distributions with compact support (Definition 7). Non-linear proofs are thus interpreted as elements of a projective limit of spaces of smooth functions. In [3], Blute, Cockett, and Seely construct a general interpretation of an exponential as a *projective* limit of more basic spaces. In [13], Kriegl and Michor construct the free $\mathcal{C}^\infty$-ring over a set X (thus a space of smooth functions) as a projective limit of spaces of smooth functions between Euclidean spaces. Our work thus differs on the fact that we reverse the use of projective and inductive limits for defining the exponential and that we use a finer indexation than the indexation used in [3,13]. The reverse use of limits compared to the literature is motivated by the fact that we are cautious about *polarities* [16], while the finer indexing is for topological considerations. Indeed, we need to carefully consider the functoriality of the exponential and the topology on the objects.

Context. Differential Linear Logic (DiLL) is a sequent calculus enriching Linear Logic (LL) with the possibility of *linearizing* proofs. This linearization is semantically understood as the differentiation at 0. Motivated by the need to explore the similarities between the differential structures inherited from logic and those inherited from physics, one would like to interpret formulas of DiLL by general topological vector spaces and non-linear proofs by smooth functions. The interpretation of the involutive linear negation of DiLL leads to the requirement of *reflexive* topological vector spaces, that is, topological vector spaces E such that $\mathcal{L}(\mathcal{L}(E,\mathbb{R}),\mathbb{R}) \simeq E$, otherwise expressed as $E'' \simeq E$. In [14], the first author argued that in a classical smooth-linear setting, the exponential ! should be interpreted as a space of distributions with compact support [18], that is, $!E := \mathcal{C}^\infty(E,\mathbb{R})'$. The first author also showed that this defines a strong monoidal functor ! from the category of Euclidean vector spaces to the category of reflexive locally convex and Hausdorff vector spaces. As reflexive spaces typically do not form a $*$-autonomous category (or even a monoidal closed category), in [14] the first author constructs a *polarized* model of DiLL structured as chirality [17]. This polarized structure is also necessary here. In Sect. 5, formulas of DiLL_0 are interpreted in two different categories, depending on whether they interpret a positive or a negative formula.

Main Content. In this paper we construct an interpretation for the exponential ! (Definition 10) which is strong monoidal (Theorem 3). The exponential constructed in this paper is a generalization of the compact-support exponential from [14]. Explicitly, for a reflexive space E, the exponential $!E$ is defined as the inductive limit of spaces $\mathcal{C}^\infty(\mathbb{R}^n,\mathbb{R})'$, indexed by *linear continuous* functions $f : \mathbb{R}^n \multimap E$ (Definition 7),

$$!E := \varinjlim_{f:\mathbb{R}^n \multimap E} \mathcal{C}^\infty(\mathbb{R}^n,\mathbb{R})'.$$

We also consider the "why not" connective ? (Definition 9) where for a reflexive space E, $?E$ is interpreted as the space of smooth scalar functions on E, $\mathcal{C}^\infty(E,\mathbb{R})$. Explicitly, being the dual of $!E$, $?E$ is the projective limit of spaces $\mathcal{C}^\infty(\mathbb{R}^n,\mathbb{R})$, indexed by the injective linear continuous functions $f : \mathbb{R}^n \multimap E'$ (Proposition 4),

$$?E := \varprojlim_{f:\mathbb{R}^n \multimap E'} \mathcal{C}^\infty(\mathbb{R}^n,\mathbb{R}).$$

An important drawback of this work is that the functoriality of ! is ensured only on isomorphisms, that is, ! is an endofunctor on the category REFL_{iso} of reflexive spaces and isomorphisms between them. We use a technique developed by Ehrhard in [6] to show that this still provides a model of *finitary* Differential linear logic (DiLL_0), that is, DiLL without the promotion rule. We also discuss how this construction also leads to a polarized model of DiLL_0 (Sect. 5).

Organization of the Paper. Section 2 gives an overview of the development in DiLL which led to this paper and gives some background in functional analysis. In Sect. 3 we discuss higher-order functions and distributions, and prove strong monoidality. Section 4 provides the interpretation of the dereliction and codereliction and the bialgebraic structure of the exponential. Finally in Sect. 5 we discuss the polarized interpretation of formulas.

Notation. In this article, we borrow notation from Linear Logic. In particular, we use $\multimap$ to distinguish between linear functions and non-linear ones, for example, $f : E \multimap F$ would be *linear continuous* while $g : E \longrightarrow F$ would only be smooth. We also denote elements of $!E$ and $?E$, which are index by linear continuous *injective* indexes $f : \mathbb{R}^n \hookrightarrow E$, in bold with their indexing in subscript: $\mathbf{g}_f \in !E$ or $\mathbf{f}_f \in ?E$.

2 Preliminaries

2.1 Differential Linear Logic and Its Semantics

Linear Logic [9] refines Intuitionistic Logic with a linear negation, $(-)^\perp$, and a notion of linearity of proofs, $\multimap$. More precisely, Linear Logic introduces the fundamental isomorphism between $A \Rightarrow B$, proofs of B from A, and $!A \multimap B$, linear proofs of B from $!A$ the exponential of A. In particular, Linear Logic features a *dereliction* rule d, which allows one to consider linear proofs as particular cases of non-linear proofs:

$$\frac{A^\perp \vdash \Gamma}{!(A^\perp) \vdash \Gamma}\, d$$

Differential Linear Logic (DiLL) brings a notion differentiation to the picture by introducing a *codereliction* rule $\bar{d}$. By cut-elimination, the codereliction rule allows one to *linearize* a non-linear proof:

$$\frac{\vdash \Gamma, A}{\vdash \Gamma, !A}\, \bar{d}$$

In Linear Logic, the exponential group also features weakening and contraction rules. While DiLL adds co-weakening and co-contraction rules, which in the context of this paper correspond respectively as integration and convolution (see [15] for more details). DiLL without promotion, or finitary Differential Linear Logic, is denoted DiLL_0 and is the original version of Differential Linear Logic by Ehrhard and Regnier [7]. Its exponential rules for $\{?, !\}$ can be found in Fig. 1. The other rules of DiLL_0 correspond to the usual ones for the MALL group $\{\otimes, ⅋, \oplus, \times\}$. Non-finitary DiLL can be constructed by adding the promotion rule to DiLL_0, which in particular requires functoriality of the exponential. Cut-elimination in DiLL and DiLL_0 generates *sums of proofs* [7], and therefore the categorical interpretation of proofs must be done in a category enriched over commutative monoids.

$$\frac{\vdash \Gamma}{\vdash \Gamma, ?E}\,w \qquad \frac{\vdash \Gamma, ?E, ?E}{\vdash \Gamma, ?E}\,c \qquad \frac{\vdash \Gamma, E}{\vdash \Gamma, ?E}\,d$$

$$\frac{\vdash \Gamma}{\vdash \Gamma, !E}\,\bar{w} \qquad \frac{\vdash \Gamma, !E \qquad \vdash \Delta, !E}{\vdash \Gamma, \Delta, !E}\,\bar{c} \qquad \frac{\vdash \Gamma, E}{\vdash \Gamma, !E}\,\bar{d}$$

Fig. 1. Exponential rules of DiLL_0

Following Fiore's definition in [8], a categorical model of DiLL is an extension of Seely's axiomatization of categorical models of Linear Logic [20]. Explicitly a model of DiLL consists of a $*$-autonomous category $(\mathcal{L}, \otimes, 1, (\_)^*)$ with a finite biproduct structure $\times$ with zero object 0, a strong monoidal comonad $! : (\mathcal{L}, \times, 0) \longrightarrow (\mathcal{L}, \otimes, 1)$, and a natural transformation $\bar{d} : id_{\mathcal{L}} \Rightarrow !$, called the *codereliction* operator, which interprets differentiation at zero. A particular important coherence for the codereliction is that composing the co-unit of the co-monad $d : ! \Rightarrow id_{\mathcal{L}}$ with $\bar{d}$ results in the identity (the top left triangle of Definition 1). Intuitively, this means that differentiating a linear map results in the same linear map.

Working Without Promotion. The special particularity of our work is that we *do not* interpret promotion and thus only obtain a denotational model of DiLL_0 but not of DiLL. The main reason for this is that in the formula

$$\mathscr{E}'(E) := \varinjlim_{f:\mathbb{R}^n \multimap E} \mathscr{E}'_f(\mathbb{R}^n),$$

injectivity of the indexes $f : \mathbb{R}^n \multimap E$ is needed to have a well-defined order to properly define an inductive limit (Definition 6). Therefore the exponential constructed in this paper cannot be functorial with respect to every linear continuous morphism in TopVec. In the construction of the exponential, one needs to compose injective indexes f with maps ℓ of the category (resp. their dual ℓ'), and these composition $\ell \circ f$ (resp. $\ell' \circ g$) are required to again be injective. As shown by Treves [21, Chapter 23.2], ℓ' is injective if and only if ℓ has a dense image. Therefore we have no choice but to ask for isomorphisms and thus we obtain an endofunctor on Refl_{iso}, the category of reflexive spaces and linear continuous isomorphisms between them.

Models of DiLL_0 in which promotion is not necessarily interpreted were studied by Ehrhard in his survey on Differential Linear Logic [6]. He introduces *exponential structures* which provides a categorical setting which differs from the traditional axiomatization of Seely's models.

Definition 1 *[6, Section 2.5]. Let $\mathcal{L}$ be pre-additive $*$-autonomous category (i.e. a commutative monoid enriched $*$-autonomous category [6, Sect. 2.4]) and let $\mathcal{L}_{iso}$ be the wide subcategory of $\mathcal{L}$ with only isomorphisms as morphisms. An* *exponential structure on a $\mathcal{L}$ is as tuple $(!, w, c, \bar{w}, \bar{c}, d, \bar{d})$ consisting of an endofunctor $! : \mathcal{L}_{iso} \longrightarrow \mathcal{L}_{iso}$, and families of morphisms of $\mathcal{L}$ (not necessarily of $\mathcal{L}_{iso}$) indexed by the objects of $\mathcal{L}$:*

$$w_A : !A \longrightarrow 1 \qquad c_A : !A \longrightarrow !A \otimes !A \qquad \bar{w}_A : 1 \longrightarrow !A \qquad \bar{c}_A : !A \otimes !A \longrightarrow !A$$

$$d_A : !A \longrightarrow A \qquad \bar{d}_A : A \longrightarrow !A$$

which are natural for morphisms of $\mathcal{L}_{iso}$, and such that for each object A, $(!A, w_A, c_A, \bar{w}_A, \bar{c}_A)$ is a commutative bialgebra in $\mathcal{L}$, and that the following diagrams to commute:

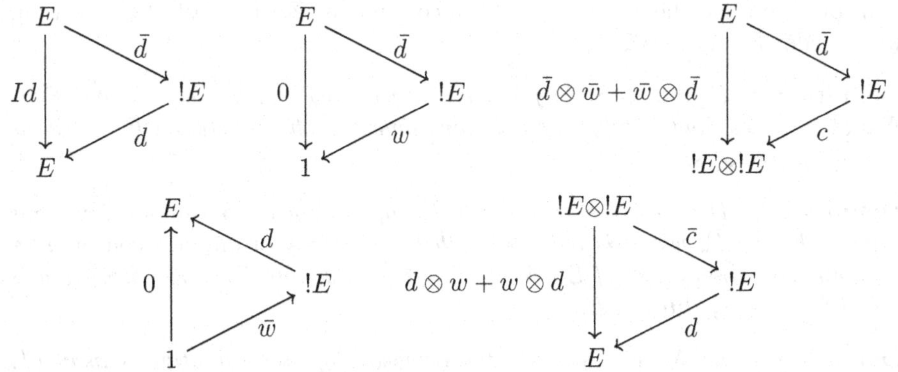

The above commutative diagrams allow for a direct interpretation of the cut-elimination process of DiLL_0. Ehrhard shows in particular that the interpretation of the structural and co-structural rules of DiLL_0 only needs the functoriality of the exponential on the isomorphisms [6, Sect. 2.5]. Indeed, in a *classical* model of DiLL (that is a model in which the interpretation of the linear negation is involutive) functoriality on isomorphisms is needed to guaranty the duality between ? and !. Otherwise, the structural exponential rules are interpreted by natural transformations c, $\bar{c}$, w, $\bar{w}$, d, and $\bar{d}$. These natural transformations can be constructed as in [8], following a co-monadic structure $(!A, w_A, \mu_A)$ on each object $!A$ [7, Sect. 2.6]. To sum up:

Functorality of the exponential on isomorphisms is needed for duality but is not needed to interpret finitary proofs as morphisms of a category.

That we have a model of DiLL_0 and not of DiLL fits well with our motivation, as we are looking for the computational counterpart of type theories modeled by analysis. DiLL_0 is indeed the sequent calculus which is refined into an understanding of Linear Partial Differential Equations in [14] and the meaning of promotion with respect to differential equations remains unclear. However, we are still able to construct a natural promotion-like morphism for our exponential (Definition 13).

2.2 Reflexive Spaces and Distributions

In this paper, we study and use the theory of locally convex topological vector spaces [12] to give concrete models of DiLL. Topological vector spaces are a generalization of normed spaces or metric spaces, in which continuity is only characterized by a collection of open sets (which may not necessarily come from a metric or a norm). In this section, we highlight some key concepts which hopefully will give the reader a better understanding of the difficulties of constructing models of DiLL using smooth spaces. We refer respectively to [12] or [18] for details on topological vector spaces or distribution theory.

By a locally convex topological vector space (lcs), we mean a *locally convex and Hausdorff topological vector space* on $\mathbb{R}$. Briefly, these are vector space endowed with a topology generated by convex open subsets such that the scalar multiplication and the addition are both continuous. For the rest of the section, we consider E and F two lcs.

Definition 2. *Denote $E \sim F$ for a linear isomorphism between E and F as $\mathbb{R}$-vector spaces, and $E \simeq F$ for a linear homeomorphism between E and F as topological vector spaces.*

Definition 3. *Denote $\mathcal{L}_b(E, F)$ as the lcs of all* linear continuous *functions between E and F, which is endowed with the* topology of uniform convergence on bounded subsets *[12] of E. When $F = \mathbb{R}$, we denote $E' = \mathcal{L}_b(E, \mathbb{R})$ and is called the* ***strong dual*** *of E.*

Definition 4. *Let $\delta : E \longrightarrow E''$ be the transpose of the evaluation map in E', which is explicitly defined as follows:*

$$\delta : \begin{cases} E \longrightarrow E'' \\ x \mapsto \delta_x : (f \longrightarrow f(x)) \end{cases}$$

A lcs E is said to be **semi-reflexive** *if δ is a linear isomorphism, that is, $E \sim E''$. A semi-reflexive lcs E is* ***reflexive*** *when δ is a linear homeomorphism, that is, $E \simeq E''$.*

The following proposition is crucial to the constructions of this paper. In terms of polarization, it shows how semi-reflexivity is a negative construction, while reflexivity mixes positives and negative requirements.

Proposition 1 *[12, Chapter 11.4].*

- *Semi-reflexivity is preserved by projective limits, that is, the projective limit of semi-reflexive lcs is a semi-reflexive lcs.*
- *A lcs E is reflexive if and only if it is semi-reflexive and* barrelled, *meaning that every convex, balanced, absorbing and closed subspace of E is a 0-neighbourhood.*
- *Barrelled spaces are preserved by inductive limits, that is, the inductive limit of barrelled spaces is a barrelled space.*

Next we briefly recall a few facts about distributions.

Definition 5. *For each $n \in \mathbb{N}$, a function $f : \mathbb{R}^n \longrightarrow \mathbb{R}$ is said to be smooth if it is infinitely differentiable. Let $\mathcal{E}(\mathbb{R}^n) = \mathcal{C}^\infty(\mathbb{R}^n, \mathbb{R})$ denote the space of all smooth functions $f : \mathbb{R}^n \longrightarrow \mathbb{R}$, and which is endowed with the topology of uniform convergence of all differentials on all compact subsets of $\mathbb{R}^n$ [12]. The strong dual of $\mathcal{E}(\mathbb{R}^n)$, $\mathcal{E}'(\mathbb{R}^n)$, is called the* ***space of distributions with compact support****.*

We now recall the famous Schwartz kernel theorem, which states that the construction of a kernel of $f \otimes g \in \mathscr{E}(\mathbb{R}^n) \otimes \mathscr{E}(\mathbb{R}^m) \mapsto f \cdot g \in \mathscr{E}(\mathbb{R}^{n+m})$ is in fact an isomorphism on the completed tensor product $\mathscr{E}(\mathbb{R}^n)\hat{\otimes}\mathscr{E}(\mathbb{R}^m)$:

Theorem 1 ([18])**.** *For any $n, m \in \mathbb{N}$, we have the following:*

$$\mathcal{E}'(\mathbb{R}^m)\hat{\otimes}_\pi\mathcal{E}'(\mathbb{R}^m) \simeq \mathcal{E}'(\mathbb{R}^{n+m}) \simeq \mathcal{L}_b(\mathcal{E}'(\mathbb{R}^m), \mathcal{E}(\mathbb{R}^n))$$

Theorem 2 ([14])**.** *There is a first-order polarized denotational model of* DiLL$_0$ *in which the exponential is interpreted as a space of distributions: $!(\mathbb{R}^n) := \mathscr{E}'(\mathbb{R}^n)$.*

This interpretation did not generalize to higher-order as we were unable to define $!E$ for an infinite dimensional space E, even for those sharing the topological properties of spaces of smooth functions[1]. For example, the definition of $!!\mathbb{R}$ is in no way obvious. This is the problem we tackle in the following sections.

3 Higher-Order Distributions and Kernel

In this section we define spaces of higher-order functions and distributions, we prove that they are reflexive (Proposition 2) and verify a kernel theorem (Theorem 3).

Definition 6. *Let E be a lcs and $f : \mathbb{R}^n \hookrightarrow E$ and $g : \mathbb{R}^m \hookrightarrow E$ be two linear continuous injective functions. We say that $f \leqslant g$ when $n \leqslant m$ and $f = g_{|\mathbb{R}^n}$, that is, $f = g \circ \iota_{n,m}$ where $\iota_{n,m} : \mathbb{R}^n \longrightarrow \mathbb{R}^m$ is the canonical injection.*

The ordering $\leqslant$ in the above definition provides an order on the set of dependent pairs (n, f) where $n \in \mathbb{N}$ and $f : \mathbb{R}^\mathbb{N} \hookrightarrow E$ is linear injective. This will allow us to construct an inductive limit (a categorical colimit) of lcs.

Definition 7. *Let E any lcs.*

1. *For every linear continuous injective function $f : \mathbb{R}^n \multimap E$, define the lcs $\mathscr{E}'_f(\mathbb{R}^n)$ as follows:*

$$\mathscr{E}'_f(\mathbb{R}^n) := \mathcal{C}^\infty(\mathbb{R}^n)'$$

[1] These spaces are in particular nuclear (F)-spaces, see [14].

2. *Define $\mathscr{E}'(E)$, the* ***space of distributions*** *on E, as follows:*

$$\mathscr{E}'(E) := \varinjlim_{f:\mathbb{R}^n \multimap E} \mathscr{E}'_f(\mathbb{R}^n)$$

that is, the inductive limit [12, Chapter4.5] (or colimit) in the category TopVec *of the family of lcs $\{\mathscr{E}'_f(\mathbb{R}^n) | f : \mathbb{R}^n \multimap E$ linear continuous injective$\}$ directed under the inclusion maps defined as*

$$S_{f,g} : \mathscr{E}'_g(\mathbb{R}^n) \longrightarrow \mathscr{E}'_f(\mathbb{R}^m), \phi \mapsto (h \mapsto \phi(h \circ \iota_{n,m}))$$

when $f \leqslant g$.

Intuitively this definition of $\mathscr{E}'(E)$ says that distributions with compact support on E are the distributions with a finite dimensional compact support $K \subset \mathbb{R}^n$.

Proposition 2. *For any lcs E, $\mathscr{E}'(E)$ is a reflexive lcs.*

The following proposition justifies the notation of $\mathscr{E}'(\mathbb{R}^n)$ from Definition 5.

Proposition 3. *If $E \simeq \mathbb{R}^n$ for some $n \in \mathbb{N}$, then $\mathscr{E}'(E) \simeq \mathcal{C}^\infty(\mathbb{R}^n)'$.*

As $\mathscr{E}'(E)$ is reflexive, we give a special (yet obvious) notation for the strong dual of $\mathscr{E}'(E)$.

Definition 8. *For a reflexive lcs E, let $\mathscr{E}(E)$ denote the strong dual of $\mathscr{E}'(E)$.*

Since the strong dual of a reflexive lcs is again reflexive [12], it follows by Proposition 3 that for any reflexive lcs E, $\mathscr{E}(E)$ is also reflexive.

The strong dual of a projective limit is *linearly isomorphic* to the inductive limit of the duals, however as noted in [12, Chapter 8.8.12], the topologies may not coincide. When E is endowed with its Mackey topology (which is the case in particular when E is reflexive), then the topologies do coincide.

Proposition 4. *Let E be a reflexive lcs. For every linear continuous injective function $f : \mathbb{R}^n \multimap E$, define the lcs $\mathscr{E}_f(\mathbb{R}^n) := \mathcal{C}^\infty(\mathbb{R}^n)$. Then we have the following linear homeomorphism:*

$$\mathscr{E}(E) \simeq \varprojlim_{f:\mathbb{R}^n \multimap E} \mathscr{E}_f(\mathbb{R}^n)$$

where the lcs on the right is the projective limit [12, Chapter 2.6] in TopVec *of the family of lcs $\{\mathscr{E}_f(\mathbb{R}^n) | \ f : \mathbb{R}^n \multimap E$ linear continuous injective$\}$ with projections defined as:*

$$T_{g,f} = S'_{f,g} : \mathscr{E}_g(\mathbb{R}^m) \longrightarrow \mathscr{E}_f(\mathbb{R}^n), \boldsymbol{g} \mapsto \boldsymbol{g} \circ \iota_{n,m}$$

when $f \leqslant g$.

The elements of $\mathbf{f} \in \mathscr{E}(E)$ are families $\mathbf{f} := (\mathbf{f}_f)_{f:\mathbb{R}^n \multimap E}$ such that if $f \leqslant g$, we have that $\mathbf{f}_f = \mathbf{f}_g \circ \iota_{n,m}$. The intuition here is that distributions of a reflexive lcs E are in fact distributions with compact support on a finite dimensional space, or equivalently that smooth functions $E \longrightarrow \mathbb{R}$ are functions which are smooth when restricted to $\mathbb{R}^n$ (viewed as a finite dimensional subspace of E). This makes it possible to define multinomials on E in the following way:

$$P(x \in \mathbb{R}^k) = \sum_{I \subset [|1,n|]} a_\alpha x_1^{\alpha^1} \dots x_n^{\alpha_n^I}$$

where we either embedded or projected $\mathbb{R}^k$ into $\mathbb{R}^n$ in the canonical way.

It also seems possible to provide a setting restricted specifically to higher order spaces of distributions and not to every reflexive space. Indeed, we would like to describe smooth scalar functions on $\mathscr{E}(\mathbb{R}^n)$ as

$$h \in \mathscr{E}(\mathbb{R}^n) \mapsto h(0)^2$$

taking into account that we have as inputs non-linear functions. This seem to indicate another direction of research, where we would construct smooth functions indexed by Dirac functions $\delta : \mathbb{R}^n \multimap E' = \mathscr{E}'(\mathbb{R}^n)$ as defined in Definition 4.

The Kernel Theorem. We now provide the Kernel theorem for spaces $\mathscr{E}(E)$. Indeed, the spaces of functions are the one which can be described as projective limits, and projective limits are the ones which commute with the completed projective tensor product $\hat{\otimes}_\pi$. While we do not provide a proof here, we would like to highlight that the proof of this theorem depends heavily on the fact that the considered spaces of functions are nuclear spaces [12].

Theorem 3. *For every lcs E and F, we have a linear homeomorphism:*

$$\mathscr{E}(E)\hat{\otimes}_\pi\mathscr{E}(F) \simeq \mathscr{E}(E \oplus F).$$

We now give the definitions of functors ? and !, both of which agree with the previous characterization described by the first author in [14] on Euclidean spaces $\mathbb{R}^n$. However, as discussed in the introduction, while these functors can be defined properly on all objects, they will only be defined on isomorphisms. So let REFL_{iso} denote the category of reflexive lcs and linear homeomorphism between them.

Definition 9. *Define the endofunctor* $? : \text{REFL}_{iso} \longrightarrow \text{REFL}_{iso}$ *as follows:*

$$? : \begin{cases} \text{REFL}_{iso} \longrightarrow \text{REFL}_{iso} \\ E \mapsto \mathscr{E}(E') \\ \ell : E \longrightarrow F \mapsto ?\ell : \mathscr{E}(E') \longrightarrow \mathscr{E}(F') \end{cases} \tag{1}$$

where for $\boldsymbol{f} \in \mathscr{E}(E')$*, the* $g : \mathbb{R}^m \multimap F'$ *component of* $?\ell(\boldsymbol{f}) \in \mathscr{E}(F')$ *is defined as:*

$$?\ell(\boldsymbol{f})_g = \boldsymbol{f}_{\ell' \circ g}$$

where $\ell' : F' \multimap E'$ *denotes the transpose of* ℓ*.*

Note that $?\ell : \mathscr{E}(E') \longrightarrow \mathscr{E}(F')$ is defined by the universal property of the projective limit, that is, $?\ell$ is uniquely defined by post-composing by the projections $\pi_g : \mathscr{E}(F') \longrightarrow \mathscr{E}(\mathbb{R}^n)$ for each linear continuous injective function $g :\multimap F'$. We also note that $\mathbf{f}_{\ell' \circ g}$ is well-defined since ℓ' is injective and therefore so is $\ell' \circ g$. The universality of the projective limit also insures that $?\ell$ is an isomorphism and that ? is functorial.

Definition 10. *Define the functor* $! : \text{REFL}_{iso} \longrightarrow \text{REFL}_{iso}$ *on objects as* $!E := (?E')'$ *and on isomorphisms as* $!\ell = (?\ell')'$. *Explicitly,* ! *is defined as follows:*

$$! : \begin{cases} \text{REFL}_{iso} \longrightarrow \text{REFL}_{iso} \\ E \mapsto \mathscr{E}'(E) \\ \ell : E \longrightarrow F \mapsto !\ell \in \mathscr{E}(F') \end{cases} \tag{2}$$

where for the $f : \mathbb{R}^n \multimap E$ *component of* $\boldsymbol{f} \in \mathscr{E}'(E)$, $!\ell(\boldsymbol{f}_f) \in \mathscr{E}'(F)$ *is defined as:*

$$!\ell(\boldsymbol{f}_f) = \boldsymbol{f}_{\ell \circ f : \mathbb{R}^n \multimap F}$$

As before, $!\ell$ is defined by the co-universal property of the inductive limit, that is, $!\ell$ is defined by pre-composition with the injections $\iota_f : \mathscr{E}'_f(\mathbb{R}^n) \hookrightarrow \mathscr{E}'(E)$ for every linear continuous injective function $f : \mathbb{R} \multimap E$. Functoriality of ! is ensured by functoriality of ? and reflexivity of the objects.

4 Structural Morphisms on the Exponential

We consider the exponential from the DiLL model of convenient vector spaces in [2] as a guideline for defining the structural morphisms on $!E$. In that setting, structural operations can be defined on Dirac operations. For example, the codereliction d_{conv} maps δ_x to x. Here the mapping δ_x must be understood as the linear continuous function which maps $x \in E$ to $\left((\mathbf{f}_f)_f \in \mathscr{E}(E') \mapsto \mathbf{f}(f^{-1}(x))\right) \in \mathscr{E}'(E)$, which we show is well defined below.

4.1 Dereliction and Co-dereliction

Definition 11. *For a reflexive lcs* E, *define the following linear continuous morphism:*

$$d_E : \begin{cases} !(E) \longrightarrow E'' \simeq E \\ \phi \mapsto (\ell \in E' \mapsto \phi((\ell \circ f)_{f:\mathbb{R}^n \multimap E} \in \mathscr{E}(E)) \end{cases} \tag{3}$$

We stress that d_E is a map in REFL and not a map in REFL_{iso} (though sufficient for Definition 1). The map d_E is well defined as $\ell \circ f$ is a linear continuous injective function $\mathbb{R}^n \multimap \mathbb{R}$, and thus is smooth and belongs in particular to $\mathscr{E}(\mathbb{R}^n)$. Also, as we are working with reflexive spaces, d_E could have been described equivalently as a map of the following type:

$$\begin{aligned} &E \longrightarrow ?(E) \\ &x \mapsto (ev_x \circ f \in \mathcal{L}(\mathbb{R}^n, \mathbb{R}))_{f:\mathbb{R}^n \multimap E'} \end{aligned} \tag{4}$$

Lemma 1. *The morphisms d_E are natural with respect to linear homeomorphisms, that is, maps of* REFL_{iso}. *Explicitly, if $\ell : E \longrightarrow F \in \text{REFL}_{iso}$ then $d_F \circ !\ell = \ell \circ d_E$.*

We now study the interpretation of the codereliction $\bar{d}$. Let $D_0 : \mathcal{C}^\infty(\mathbb{R}^n) \longrightarrow (\mathbb{R}^n)'$ denote the operator which maps a function to its differential at 0.

$$D_0 : \begin{cases} \mathcal{C}^\infty(\mathbb{R}^n) \longrightarrow (\mathbb{R}^n)' \\ \qquad \mathbf{f} \mapsto \left(v \in \mathbb{R}^n \mapsto \lim_{t \longrightarrow 0} \dfrac{f(tx) - f(0)}{t} = \sum_{i=1}^{n} \dfrac{\partial f}{\partial x_i}(0) v_i \right) \end{cases}$$

The operator D_0 is linear in $\mathbf{f} \in \mathcal{C}^\infty(\mathbb{R}^n)$. It is continuous: the reciprocal image by D_0 of the polar $B_{0,1}$ is the set of all functions $\mathbf{f} \in \mathcal{C}^\infty(\mathbb{R}^n)$ whose partial derivatives of order one have maximal value 1 on the compact $\{0\}$. This contains the set $\{\mathbf{f} | \forall i, |\frac{\partial f}{\partial x_i}(0)| < 1\}$, which is open in the topology described in Definition 5.

Definition 12. *For a reflexive lcs E, define the following linear continuous morphism:*

$$\bar{d}_E : \begin{cases} E \longrightarrow !E \simeq (\mathscr{E}(E))' \\ x \mapsto (\mathbf{f}_f \in \mathcal{C}^\infty_f(\mathbb{R}^n, \mathbb{R}))_{f:\mathbb{R}^n \multimap E'} \mapsto D_0 \mathbf{f}_f(f^{-1}(x)) \\ \textit{where } f \textit{ is injective such that } x \in Im(f). \end{cases} \tag{5}$$

We should explain why the choice of $f^{-1}(x)$ does not matter. Here $f^{-1}(x)$ is the *linear* argument of the differentiation. Indeed suppose that $f \leqslant g$, that is, $f = g \circ \iota_{n,m}$. Thus by definition of the projective limit we have $\mathbf{f}_f = \mathbf{f}_g \circ \iota_{n,m}$ and:

$$\begin{aligned} D_0 \mathbf{f}_f(f^{-1}(x)) &= D_0(\mathbf{f}_g \circ \iota_{n,m})((g \circ \iota_{n,m})^{-1}(x)) \\ &= D_0 \mathbf{f}_g(D_0 \iota_{n,m}(\iota_{n,m}^{-1}(g^{-1}(x)))) \\ &= D_0 \mathbf{f}_g(\iota_{n,m}(\iota_{n,m}^{-1}(g^{-1}(x))) && (\text{as } \iota_{n,m} \text{ is linear}) \\ &= D_0 \mathbf{f}_g(g^{-1}(x))) \end{aligned}$$

As any pair of of linear functions $f : \mathbb{R}^n \multimap E$ and $g : \mathbb{R}^m \longrightarrow E$ is bounded by $f \times g : \mathbb{R}^{n+m} \longrightarrow E$, we obtain the required uniqueness.

Similar to the dereliction, the codereliction could alternatively have been described as a map of the following type:

$$\begin{aligned} \mathscr{E}(E') &\longrightarrow E'' \simeq E \\ (\mathbf{f}_f)_{f:\mathbb{R}^n \multimap E'} &\mapsto (\ell \in E' \mapsto D_0 \mathbf{f}_f(f^{-1}(\ell)) \end{aligned} \tag{6}$$

We again stress that $\bar{d}_E$ is not a map in REFL_{iso}.

Lemma 2. *The morphisms $\bar{d}_E$ are natural with respect to linear homeomorphisms, that is, maps of* REFL_{iso}. *Explicitly, if $\ell : E \longrightarrow F \in \text{REFL}_{iso}$ then $\bar{d}_F \circ \ell = !\ell \circ \bar{d}_E$.*

Finally, we observe that d_E and $\bar{d}_E$ satisfy the all-important coherence condition between derelictions and coderelictions.

Proposition 5. *For a reflexive lcs E, $d_E \circ \bar{d}_E = Id_E$.*

4.2 (Co-)contraction and (Co-)weakening

In this section, we define the interpretation of the other exponential rules: weakening w, co-weakening $\bar{w}$, contraction c, and co-contraction $\bar{c}$, which will be generalized from [14]. We start with weakening and co-weakening, which are fairly straightforward.

$$w : \begin{cases} !E \longrightarrow \mathbb{R} \\ \phi \mapsto \sum_f \phi_f(1) \end{cases} \qquad \bar{w} : \begin{cases} \mathbb{R} \longrightarrow !E \\ 1 \mapsto \delta_0 : ((\mathbf{f}_f)_f \in \mathcal{E}(E) \mapsto \mathbf{f}(0)) \text{ for any } f \end{cases}$$

According to [8], the rules c and $\bar{c}$ are interpreted respectively via the kernel theorem and pre-composition with the diagonal $E \longrightarrow E \times E$ and co-diagonal $E \times E \longrightarrow E$ maps of the biproduct. This is however not defined in a context where ! is functorial only on isomorphisms. Thus we give a direct, componentwise interpretation of contraction and co-contraction.

$$c : \begin{cases} !E \longrightarrow !(E \times E) \simeq !E \otimes !E \\ \phi \mapsto (\mathbf{g}_g)_{g:\mathbb{R}^n \hookrightarrow E \times E} \mapsto \phi((\mathbf{g}_{(x \in \mathbb{R}^n \mapsto (f(x), f(x)))})_{f:\mathbb{R}^n \hookrightarrow E}) \end{cases}$$

$$\bar{c} : \begin{cases} !E \otimes !E \longrightarrow !E \\ \phi \otimes \psi \mapsto (\mathbf{f}_f)_{f:\mathbb{R}^n \hookrightarrow E} \mapsto \phi((x \in \mathbb{R}^n \mapsto \psi((y \in \mathbb{R}^m \mapsto \mathbf{f}_f(x) + \mathbf{f}_{f'}(y))_{f'}))_f) \end{cases}$$

where $f : \mathbb{R}^n \hookrightarrow E$ and $f' : \mathbb{R}^m \hookrightarrow E$.

Theorem 4. *The morphisms $(w, \bar{w}, c, \bar{c}, d, \bar{d})$ satisfy the coherences of exponential structure on $!E$, as detailed in Definition* 1.

We note that this does not give an exponential structure per say since REFL is not a monoidal category, as we will explain in Sect. 5. That said, in Sect. 5 we are still able to construct a *polarized* model of DiLL_0.

4.3 Co-multiplication

The categorical interpretation of the exponential rules of linear logic requires a co-monad $! : \mathcal{L} \longrightarrow \mathcal{L}$. However in the case of this paper, the exponential ! is functorial only on isomorphisms. As such, one cannot interpret the promotion rule of Linear Logic, as this requires functoriality of ! on the interpretation of any proof (and typically on linear continuous maps which are not isomorphisms).

That said, functoriality is the only missing ingredient, and one can still define natural transformations of the same type as the co-multiplication and co-unit of the co-monad. This section details this point, leaving the exploration of a functorial ! for future work.

Definition 13. *For a reflexive lcs E, define the following linear continuous morphism:*

$$\mu_E : \begin{cases} !E \longrightarrow !!E \\ \phi \mapsto \left((\mathbf{g}_g)_g \in \mathscr{E}(!E) \simeq \varinjlim_g \mathcal{C}_g^\infty(\mathbb{R}^m) \right) \mapsto \mathbf{g}_g(g^{-1}(\phi)) \\ \textit{when } \phi \in Im(g) \textit{ and } g \textit{ is injective} \end{cases} \tag{7}$$

This is well defined, as we can show as for the codereliction (5) that the term $\mathbf{g}_g(g^{-1}(\phi))$ is unique when $g : \mathbb{R}^m \longrightarrow !E$ linear and $\mathbf{g}_g \in \mathcal{C}_g^\infty(\mathbb{R}^m)$ varies. Moreover there is at least one linear function $g : \mathbb{R}^m \longrightarrow !E$ which has ϕ in its image.

Lemma 3. *The morphisms μ_E are natural with respect to linear homeomorphisms, that is, maps of* REFL_{iso}. *Explicitly, if $\ell : E \longrightarrow F \in \text{REFL}_{iso}$ then $\mu_F \circ !!\ell = !\ell \circ \mu_E$.*

Proposition 6. *For any reflexive lcs E, $d_{!E} \circ \mu_E = Id_{!E}$*

The identity of Proposition 6 is one of the identities of a comonad. The other comonad identities require applying ! to μ and d, which we cannot do in our context as ! is only defined on isomorphisms.

5 A Model of DiLL_0

In Sect. 4 we defined the structural morphisms on the exponential and proved the equations allowing to interpret proofs of DiLL_0 by morphisms in REFL, independent of cut-elimination. We now detail which categories allow to interpret formulas of MALL. This will be done in a polarized setting generalizing the one of [14].

Polarization. So far we have constructed an exponential $! : \text{REFL}_{iso} \longrightarrow \text{REFL}_{iso}$ which is strong monoidal. However, the category of reflexive spaces is too big to give us a model of DiLL_0. Interpreting the multiplicative connective requires a monoidal setting, and reflexive spaces are not stable by topological tensor products. If we study more closely the definition of spaces of higher-order smooth functions, we see that their reflexivity follows from a more restrictive class of spaces. These spaces are however not stable by duality, thus resulting in a *polarized* model of DiLL_0.

In this section we briefly show how the techniques develop above constructs a *polarized model* of DiLL_0. The syntax of polarized (Differential) Linear Logic

[16] is recalled below. A distinction is made between positive formulas (preserved by $\otimes$ and $\oplus$) and negative formulas (preserved by $\parr$ and $\&$). The same deduction rule apply.

$$\text{Negative Formulas: } N, M := \bot|1||?P|N \parr M|N \times M|P^{\perp}$$
$$\text{Positive Formulas: } P, Q := \top|0|!N|P \otimes Q|P \oplus Q|N^{\perp}$$

Models of polarized linear logic are axiomatized categorically as an *adjunction between a category of positives and a category of negative*, where two interpretations for negation play the role of adjoint functors. These categories obey the axiomatic of chiralities [17].

Additives. Interpreting the additive connectives of linear logic is straightforward. The product $\times$ and coproduct $\oplus$ of lcs are linearly homeomorphic on finite indexes and therefore give biproducts, which leads to the usual commutative monoid enrichment as described in [8].

Multiplicatives. When sticking to finite dimensional spaces or normed spaces, duality is pretty straightforward in the sense that the dual of a normed space is still normed. This, however, is no longer the case when one generalizes to metric spaces. Indeed, the dual of a metric space may not be endowed with a metric. A Fréchet space, or (F)-space, is a complete and metrizable lcs. The duals of these spaces are not metrizable in general, but they are (DF)-spaces (see [10] for the definition):

Proposition 7 ([11] IV.3.1).

- *If E is metrizable, then its strong dual E' is a (DF)-space.*
- *If E is a (DF)-space, then E' is an (F)-space.*

Typical examples of nuclear (F)-spaces are the spaces of smooth functions $\mathscr{E}(\mathbb{R}^n)$, while typical examples of nuclear (DF)-spaces are the spaces of distributions with compact support $\mathscr{E}'(\mathbb{R}^n)$. In particular, all these spaces are reflexive. In [14], the first author interpreted positive formulas as Nuclear (DF)-spaces, while negative formulas were interpreted as (F)-spaces. Following the construction of Sect. 3, we will consider respectively inductive limits and projective limits.

Definition 14. *A lcs is said to be a* LNF*-space if it is a regular projective limit of nuclear Fréchet spaces. The category of* LNF*-spaces and linear continuous injective maps is denoted* LNF. *A lcs E is said to be a* LNDF*-space if it is an inductive limit of nuclear complete (DF)-spaces.*

Proposition 8. *1. A* LNF*-space E is reflexive.*
2. The dual of a LNF*-space is a* LNDF*-space.*

The above proposition can be proven using the same techniques as computing the dual of $\mathscr{E}(E)$.

The difficulty of constructing a model of MLL in topological vector spaces is choosing the topology which will make the tensor product associative and commutative on the already chosen category of lcs. Contrary to what happens in a purely algebraic setting, the definition of a topological tensor product is not straightforward and several topologies can be defined, with each corresponding to a different notion of continuity for bilinear maps [10]. On nuclear spaces, such as $\mathscr{E}(\mathbb{R}^n)$ and $\mathscr{E}'(\mathbb{R}^n)$, most of these tensor product coincide with one another. In [14], both multiplicative connectors ($\otimes$ and $⅋$) were interpreted as the completed projective (equivalently injective) tensor product $\hat{\otimes}_\pi$ (see [12, 15.1 and 21.2]) This property is lost when working with limits. However, there is still a good interpretation of $⅋$ for LNF spaces (which are thus the interpretation of negatives formulas). Indeed, the completed injective tensor product $\hat{\otimes}_\varepsilon$ of a projective limit of lcs is the projective limit of the completed injective tensor products [12, 16.3.2]. Taking the duals of Theorem 3 applied to E' and F' gives the following:

Proposition 9. *For any reflexive spaces E and F we have a linear homeomorphism:*

$$?E\hat{\otimes}_\varepsilon ?F \simeq ?(E \oplus F).$$

and shows that $⅋$ is interpreted by $\hat{\otimes}_\varepsilon$. The multiplicative conjunction $\otimes$ is interpreted as the dual of $\hat{\otimes}_\varepsilon$, which may not be necessarily linearly homeomorphic to $\hat{\otimes}_\pi$.

6 Conclusion

In this paper, we extended the polarized model of DiLL without higher order constructed in [14] to a higher-order polarized model of DiLL_0. The motivating idea was that computation on spaces of functions used only a finite number of arguments. This lead to constructing an exponential on a reflexive lcs as an inductive limit of exponentials of finite dimensional vector spaces. While this exponential is only functorial for linear homeomorphisms we were still able to provide structural morphisms interpreting (co)weakening, (co)contraction, and (co)dereliction, and hints of a co-monad.

The next step would be to extend the definition of the exponential in this paper to an interpretation of the promotion rule and thus of LL – this could be done through epi-mono decomposition of arrows in REFL. Another task is to properly work out which tensor product of reflexive space will provide a model of DiLL. Such a model should use chiralities [17], and underline the similarities between shifts and (co-)dereliction.

More generally, this works highlights again that the interpretation of the exponential in lcs relies on a computing principle. Indeed, it always requires finding a higher-order extension of distributions. While what we have constructed here relies on a finitary principle, the construction of a free exponential in [3] relies on the principle that higher-order operations are computed on Dirac distributions δ_x. That is, the exponential is constructed following a discretization scheme. The appearance of such numerical methods in a semantic study of DiLL provides another link between theoretical computer science and mathematical physics. This opens the door to studying relating numerical schemes of numerical analysis and the theoretical study of programming language.

References

1. Blass, A.: A game semantics for linear logic. Ann. Pure Appl. Logic **56**(1), 183–220 (1992). https://doi.org/10.1016/0168-0072(92)90073-9
2. Blute, R., Ehrhard, T., Tasson, C.: A convenient differential category. Cah. Topol. Géom. Différ. Catég. (2012)
3. Blute, R.F., Cockett, J.R.B., Seely, R.A.G.: Differential categories. Math. Struct. Comput. Sci. **16**, 6 (2006). https://doi.org/10.1017/S0960129506005676
4. Church, A.: A formulation of the simple theory of types. J. Symb. Logic **5**(2), 56–68 (1940)
5. Ehrhard, T.: On Köthe sequence spaces and linear logic. Math. Struct. Comput. Sci. **12**(5), 579–623 (2002)
6. Ehrhard, T.: An introduction to differential linear logic: proof-nets, models and antiderivatives. Math. Struct. Comput. Sci. **28**(7), 995–1060 (2018). https://doi.org/10.1017/S0960129516000372
7. Ehrhard, T., Regnier, L.: Differential interaction nets. Theor. Comput. Sci. **364**(2), 166–195 (2006)
8. Fiore, M.: Differential structure in models of multiplicative biadditive intuitionistic linear logic. In: Della Rocca, S.R. (ed.) TLCA 2007. LNCS, vol. 4583, pp. 163–177. Springer, Heidelberg (2007). https://doi.org/10.1007/978-3-540-73228-0_13
9. Girard, J.Y.: Linear logic. Theor. Comput. Sci. **50**(1), 1–101 (1987). https://doi.org/10.1016/0304-3975(87)90045-4
10. Grothendieck, A.: Produits tensoriels topologiques et espaces nucléaires. Memoirs of the AMS, 16 (1966)
11. Grothendieck, A.: Topological vector spaces. Gordon and Breach Science Publishers (1973). Traducteur: O. Chaljub
12. Jarchow, H.: Locally Convex Spaces. B. G. Teubner, Berlin (1981)
13. Kainz, G., Kriegl, A., Michor, P.: $\mathcal{C}^\infty$-algebras from the functional analytic view point. J. Pure Appl. Algebra **46**(1), 89–107 (1987)
14. Kerjean, M.: A logical account for linear partial differential equations. In: Proceedings of the 33rd Annual ACM/IEEE Symposium on Logic in Computer Science, LICS. ACM (2018). https://doi.org/10.1145/3209108
15. Kerjean, M.: Reflexive spaces of smooth functions: a logical account for linear partial differential equations. Ph.D. thesis, Université Paris Diderot, October 2018
16. Laurent, O.: Etude de la polarisation en logique. Thèse de doctorat, Université Aix-Marseille II (2002)
17. Melliès, P.A.: Dialogue categories and chiralities. Publ. Res. Inst. Math. Sci. **52**(4), 359–412 (2016)
18. Schwartz, L.: Théorie des distributions. Publications de l'Institut de Mathématique de l'Université de Strasbourg, No. IX-X, Hermann, Paris (1966)
19. Scott, D., Strachey, C.: Towards a mathematical semantics for programming languages (1971)
20. Seely, R.: Linear logic, *-autonomous categories and cofree coalgebras. In: Categories in Computer Science and Logic. American Mathematical Society (1989)
21. Trèves, F.: Topological Vector Spaces, Distributions and Kernels. Academic Press, New York, London (1967)

Languages Ordered by the Subword Order

Dietrich Kuske[1(✉)] and Georg Zetzsche[2]

[1] Technische Universität Ilmenau, Ilmenau, Germany
dietrich.kuske@tu-ilmenau.de
[2] Max Planck Institute for Software Systems (MPI-SWS), Kaiserslautern, Germany
georg@mpi-sws.org

Abstract. We consider a language together with the subword relation, the cover relation, and regular predicates. For such structures, we consider the extension of first-order logic by threshold- and modulo-counting quantifiers. Depending on the language, the used predicates, and the fragment of the logic, we determine four new combinations that yield decidable theories. These results extend earlier ones where only the language of all words without the cover relation and fragments of first-order logic were considered.

Keywords: Subword order · First-order logic · Counting quantifiers · Decidable theories

1 Introduction

The subword relation (sometimes called scattered subword relation) is a simple example of a well-quasi ordering [7]. This property allows its prominent use in the verification of infinite-state systems [4]. The subword relation can be understood as embeddability of one word into another. This embeddability relation has been considered for other classes of structures like trees, posets, semilattices, lattices, graphs etc. [8–11,14–16,22,23].

We are interested in logics over the subword order. Prior work on this has concentrated on first-order logic where the universe consists of all words over some alphabet. In this setting, we already have a rather precise picture about the border between decidability and undecidability: For the subword order alone, the $\exists^*$-theory is decidable [17] and the $\exists^*\forall^*$-theory is undecidable [6,12]. If we add constants to the signature, already the $\exists^*$-theory becomes undecidable [6]. With regular predicates, the two-variable theory is decidable, but the three-variable theory is undecidable [12].

Thus, the decidable theories identified so far leave little room to express natural properties. First, the universe is confined to the set of all words and

Part of the results were obtained when the second author was affiliated with the Laboratoire Spécification et Vérification (ENS Paris-Saclay) and supported by a fellowship within the Postdoc-Program of the German Academic Exchange Service (DAAD) and by Labex DigiCosme, Université Paris-Saclay, project VERICONISS.

M. Bojańczyk and A. Simpson (Eds.): FOSSACS 2019, LNCS 11425, pp. 348–364, 2019.
https://doi.org/10.1007/978-3-030-17127-8_20

predicates for subsets quickly incur undecidability. Moreover, neither in the $\exists^*$-, nor in the two-variable fragment of first-order logic, one can express the cover relation $\sqsubset\!\!\cdot$ (i.e., "u is a proper subword of v and there is no word properly between these two"). As another example, one cannot express threshold properties like "there are at most k subwords with a given property" in any of these two logics.

In this paper, we aim to identify decidable logics that are more expressive. To that end, we consider four additions to the expressivity of the logic:

- Instead of all words over some alphabet, the universe is a language L.
- We add regular predicates or constants to the structure.
- Besides the subword order, we also consider the cover relation $\sqsubset\!\!\cdot$.
- We add threshold and modulo counting quantifiers to the logic.

Formally, this means we consider structures of the form

$$(L, \sqsubseteq, \sqsubset\!\!\cdot, (K \cap L)_{K \text{ regular}}, (w)_{w \in L}),$$

where the universe is a language $L \subseteq \Sigma^*$, $\sqsubseteq$ is the subword ordering, $\sqsubset\!\!\cdot$ is the cover relation, there is a predicate $K \cap L$ for each regular $K \subseteq \Sigma^*$, and a constant symbol for each $w \in L$. Moreover, we consider fragments of the logic C+MOD, which extends first-order logic by threshold- and modulo-counting quantifiers.

The key idea of this paper is to find decidable theories by varying the universe L and thereby either (i) simplify the structure $(L, \sqsubseteq)$ enough to obtain decidability even with the extensions above or (ii) generalize existing results that currently only apply to $L = \Sigma^*$. This leads to the following results.

1. First, we require L to be bounded. This means, we have $L \subseteq w_1^* \cdots w_m^*$ for some words $w_1, \ldots, w_m \in \Sigma^*$. Then, as soon as L is context-free, the C+MOD-theory of the whole structure is decidable (Theorem 3.4).
2. To lift the boundedness restriction, we show that if L is regular, we still obtain decidability for the whole structure if we stay within the two-variable fragment C+MOD^2 (Corollary 4.8). This generalizes the decidability of the FO^2-theory without the cover relation as shown in [12, Theorem 5.5].
3. Moreover, we consider a regular universe, but lift the two-variable requirement. To get decidability, we restrict quantifiers and available predicates: We show that for regular L, the Σ_1-theory of the structure $(L, \sqsubseteq)$ is decidable (Theorem 5.1). In the case $L = \Sigma^*$, this had been shown in [17, Prop. 2.2].
4. Finally, we place a further restriction on L, but in return obtain decidability with constants. We show that if L is regular and every letter is "frequent" in L (see Sect. 6), then the Σ_1-theory of the structure $(L, \sqsubseteq, (w)_{w \in L})$ is decidable (Theorem 6.2). Note that, by [6, Theorem 3.3], this theory is undecidable if $L = \Sigma^*$.

Our first result is shown by a first-order interpretation of the structure in $(\mathbb{N}, +)$. Since $L \subseteq w_1^* \cdots w_n^*$, instead of words, one can argue about vectors $(x_1, \ldots, x_n) \in \mathbb{N}^n$ for which $w_1^{x_1} \cdots w_n^{x_n} \in L$. For the interpretation, we use the fact that semilinearity of context-free languages yields a Presburger formula expressing $w_1^{x_1} \cdots w_n^{x_n} \in L$ for $(x_1, \dots, x_n) \in \mathbb{N}^n$. Moreover, Presburger definability of $w_1^{x_1} \cdots w_n^{x_n} \sqsubseteq w_1^{y_1} \cdots w_n^{y_n}$ for $(x_1, \dots, x_n) \in \mathbb{N}^n$ and $(y_1, \dots, y_n) \in \mathbb{N}^n$ is a simple consequence of the subword relation being rational, which was observed in [12]. The first-order interpretation of our structure in $(\mathbb{N}, +)$ then enables us to employ decidability of the C+MOD-theory of the latter structure [1,5,21]. (Note that this decidability does not follow directly from Presburger's result since in first-order logic, one cannot make statements like "the number of witnesses $x \in \mathbb{N}$ satisfying ... is even"). A similar interpretation in $(\mathbb{N}, +)$ was used in [6] for various algorithms concerning $(\Sigma^*, \sqsubseteq, (w)_{w \in \Sigma^*})$ for fragments of FO related to bounded languages.

Our second result extends an approach from [12] for decidability of the FO^2-theory of the structure $(\Sigma^*, \sqsubseteq, (L)_{L \text{ regular}})$. The authors of [12] provide a quantifier elimination procedure showing that every unary relation FO^2-definable in this structure is regular. Our extended quantifier-elimination procedure uses the same invariant, now relying on the following two properties:

- The class of regular languages is closed under *counting* images under *unambiguous* rational relations.
  This can be shown either directly or (as we do here) using weighted automata [20].
- The proper subword, the cover, and the incomparability relation are *unambiguous* rational.

Our third result extends the decidability of the Σ_1-theory of $(\Sigma^*, \sqsubseteq)$ from [17]. In [17], decidability is a consequence of the fact that every finite partial order can be embedded into $(\Sigma^*, \sqsubseteq)$ if $|\Sigma| \geq 2$. This certainly fails for general regular languages: $(a^*, \sqsubseteq)$ can only accomodate linear orders. However, we can distinguish two cases: If L is a bounded language, then decidability of the Σ_1-theory of $(L, \sqsubseteq)$ follows from our first result. If L is not bounded, then we show that again every finite partial order embeds into $(L, \sqsubseteq)$. To this end, we first extend a well-known property of unbounded regular languages, namely that there are $x, u, v, y \in \Sigma^*$ with $x\{u, v\}^* y \subseteq L$ such that $|u| = |v|$ and $u \neq v$. We show that here, u, v can be chosen so that uv is a primitive word. We then observe that for large enough n, any embedding of the word $(uv)^{n-1}$ into $(uv)^n$ must hit either the left-most position or the right-most position in $(uv)^n$. This enables us to argue that for large enough n, sending a tuple $(t_1, \dots, t_m) \in \{0,1\}^m$ to $xv^{t_1}(uv)^n \cdots v^{t_m}(uv)^n y$ is in fact an embedding of $(\{0,1\}^m, \leq)$ into $(L, \sqsubseteq)$, where $\leq$ denotes coordinate-wise comparison. Since any partial order with $\leq m$ elements embeds into $(\{0,1\}^m, \leq)$, this completes the proof.

Regarding our fourth result, we know from [6] that decidability of the Σ_1-theory of $(L, \sqsubseteq, (w)_{w \in L})$ does not hold for every regular L: Undecidability holds already for $L = \{a, b\}^*$. Therefore, we require that every letter is frequent in L, meaning that in some automaton for L, every letter occurs in every cycle. In case L is bounded, we can again invoke our first result. If L is not bounded, we deduce from the frequency condition that for every $w \in \Sigma^*$, there are only finitely many words in L that do not have w as a subword. Removing those finitely many words preserves unboundedness, so that every finite partial order embeds in L above w. We then proceed to show that for such languages, any Σ_1-sentence is effectively equivalent to a sentence where constants are only used to express that all variables take values above a certain word w. Since every finite partial order embeds above w, this implies decidability.

The full version of this work is available as [18].

2 Preliminaries

Throughout this paper, let Σ be some finite alphabet. A word $u = a_1a_2\ldots a_m$ with $a_1, a_2, \ldots, a_m \in \Sigma$ is a *subword* of a word $v \in \Sigma^*$ if there are words $v_0, v_1, \ldots, v_m \in \Sigma^*$ with $v = v_0a_1v_1a_2v_2\cdots a_mv_m$. In that case, we write $u \sqsubseteq v$; if, in addition, $u \neq v$, then we write $u \sqsubset v$ and call u a *proper* subword of v. If $u, w \in \Sigma^*$ such that $u \sqsubset w$ and there is no word v with $u \sqsubset v \sqsubset w$, then we say that w is a *cover* of u and write $u \sqsubset\!\!\cdot\, w$. This is equivalent to saying $u \sqsubseteq w$ and $|u| + 1 = |w|$ where $|u|$ is the length of the word u. If neither u is a subword of v nor *vice versa*, then the words u and v are *incomparable* and we write $u \parallel v$. For instance, $aa \sqsubset babbba$, $aa \sqsubset\!\!\cdot\, aba$, and $aba \parallel aabb$.

Let $\mathcal{S} = (L, (R_i)_{i\in I}, (w_j)_{j\in J})$ be a *structure*, i.e., L is a set, $R_i \subseteq L^{n_i}$ is a relation of arity n_i (for all $i \in I$), and $w_j \in L$ for all $j \in J$. Then, formulas φ of the logic C+MOD are defined by the following grammar:

$$\varphi ::= (s = t) \mid R_i(s_1, \ldots, s_{n_i}) \mid \neg\varphi \mid \varphi \vee \varphi \mid \exists x\, \varphi \mid \exists^{\geq k} x\, \varphi \mid \exists^{p \bmod q} x\, \varphi$$

where $s, t, s_1, \ldots, s_{n_i}$ are variables or constants w_j with $j \in J$, $i \in I$, $k \in \mathbb{N}$, and $p, q \in \mathbb{N}$ with $p < q$. We call $\exists^{\geq k}$ a *threshold counting quantifier* and $\exists^{p \bmod q}$ a *modulo counting quantifier*. The semantics of these quantifiers is defined as follows:

- $\mathcal{S} \models \exists^{\geq k} x\, \alpha$ iff $|\{w \in L \mid \mathcal{S} \models \alpha(w)\}| \geq k$
- $\mathcal{S} \models \exists^{p \bmod q} x\, \alpha$ iff $|\{w \in L \mid \mathcal{S} \models \alpha(w)\}| \in p + q\mathbb{N}$

For instance, $\exists^{0 \bmod 2} x\, \alpha$ expresses that the number of elements of the structure satisfying α is even. Then $(\exists^{0 \bmod 2} x\, \alpha) \vee (\exists^{1 \bmod 2} x\, \alpha)$ holds iff only finitely many elements of the structure satisfy α. The fragment FO+MOD of C+MOD comprises all formulas not containing any threshold counting quantifier. First-order logic FO is the set of formulas from C+MOD not mentioning any counting quantifier. Let Σ_1 denote the set of first-order formulas of the form $\exists x_1 \exists x_2 \ldots \exists x_n \colon \psi$ where ψ is quantifier-free; these formulas are also called *existential*.

The threshold quantifier $\exists^{\geq k}$ can be expressed using the existential quantifier, only. Consequently, the logics FO+MOD and C+MOD are equally expressive. The situation changes when we restrict the number of variables that can be used in a formula: Let FO+MOD^2 and C+MOD^2 denote the set of formulas from FO+MOD and C+MOD, respectively, that use the variables x and y, only. Then, the existence of ≥ 3 elements in the structure is expressible in C+MOD^2, but not in FO+MOD^2.

In this paper, we will consider the following structures:

- The largest one is $(L, \sqsubseteq, \sqsubset, (K \cap L)_{K \text{ regular}}, (w)_{w \in L})$ for some $L \subseteq \Sigma^*$. The universe of this structure is the language L, we have two binary predicates ($\sqsubseteq$ and $\sqsubset$), a unary predicate $K \cap L$ for every regular language K, and we can use every word from L as a constant.
- The other extreme is the structure $(L, \sqsubseteq)$ for some $L \subseteq \Sigma^*$ where we consider only the binary predicate $\sqsubseteq$.
- Finally, we will also prove results on the intermediate structure $(L, \sqsubseteq, (w)_{w \in L})$ that has a binary relation and any word from the language as a constant.

For any structure $\mathcal{S}$ and any of the logics $\mathcal{L}$, the *$\mathcal{L}$-theory* of $\mathcal{S}$ is the set of sentences from $\mathcal{L}$ that hold in $\mathcal{S}$.

A non-deterministic finite automaton is called *non-degenerate* if every state lies on a path from an initial to a final state. A language $L \subseteq \Sigma^*$ is *bounded* if there are a number $n \in \mathbb{N}$ and words $w_1, w_2, \dots, w_n \in \Sigma^*$ such that $L \subseteq w_1^* w_2^* \cdots w_n^*$. Otherwise, it is *unbounded.*

For a monoid M, a subset $S \subseteq M$ is called *rational* if it is a homomorphic image of a regular language. In other words, there exists an alphabet Δ, a regular $R \subseteq \Delta^*$, and a homomorphism $h\colon \Delta^* \to M$ with $S = h(R)$. In particular, if Σ_1, Σ_2 are alphabets and $M = \Sigma_1^* \times \Sigma_2^*$, then a subset $S \subseteq \Sigma_1^* \times \Sigma_2^*$ is rational iff there is an alphabet Δ, a regular $R \subseteq \Delta^*$, and homomorphisms $h_i\colon \Delta^* \to \Sigma_i^*$ with $S = \{(h_1(w), h_2(w)) \mid w \in R\}$. This fact is known as *Nivat's theorem* [2].

For an alphabet Γ, a word $w \in \Gamma^*$, and a letter $a \in \Gamma$, let $|w|_a$ denote the number of occurrences of the letter a in the word w. The *Parikh vector* of w is the tuple $\Psi_\Gamma(w) = (|w|_a)_{a \in \Gamma} \in \mathbb{N}^\Gamma$. Note that Ψ_Γ is a homomorphism from the free monoid Γ^* onto the additive monoid $(\mathbb{N}^\Gamma, +)$.

3 The FO+MOD-Theory with Regular Predicates

The aim of this section is to prove that the full FO+MOD-theory of the structure

$$(L, \sqsubseteq, \sqsubset, (K \cap L)_{K \text{ regular}}, (w)_{w \in L})$$

is decidable for L bounded and context-free. This is achieved by interpreting this structure in $(\mathbb{N}, +)$, i.e., in Presburger arithmetic whose FO+MOD-theory is known to be decidable [1,5,21]. We start with three preparatory lemmas.

Lemma 3.1. *Let $K \subseteq \Sigma^*$ be context-free, $w_1, \dots, w_n \in \Sigma^*$, and $g\colon \mathbb{N}^n \to \Sigma^*$ be defined by $g(\overline{m}) = w_1^{m_1} w_2^{m_2} \cdots w_n^{m_n}$ for all $\overline{m} = (m_1, m_2, \dots, m_n) \in \mathbb{N}^n$. The set $g^{-1}(K) = \{\overline{m} \in \mathbb{N}^n \mid g(\overline{m}) \in K\}$ is effectively semilinear.*

Proof. Let $\Gamma = \{a_1, a_2, \dots, a_n\}$ be an alphabet and define the monoid homomorphism $f\colon \Gamma^* \to \Sigma^*$ by $f(a_i) = w_i$ for all $i \in [1, n]$.

Since the class of context-free languages is effectively closed under inverse homomorphisms and under intersections with regular languages, the language

$$L = f^{-1}(K) \cap a_1^* a_2^* \cdots a_n^* = \{u \in a_1^* a_2^* \cdots a_n^* \mid f(u) \in K\}$$

is effectively context-free. Its Parikh image $\Psi_\Gamma(L) \subseteq \mathbb{N}^n$ is effectively semilinear [19]. Moreover, $\Psi_\Gamma(L)$ equals the set $g^{-1}(K)$ from the lemma. □

Lemma 3.2. *Let $w_1, \dots, w_n \in \Sigma^*$ and $g\colon \mathbb{N}^n \to \Sigma^*$ be defined by $g(\overline{m}) = w_1^{m_1} w_2^{m_2} \cdots w_n^{m_n}$ for all $\overline{m} = (m_1, m_2, \dots, m_n) \in \mathbb{N}^n$. The set $\{(\overline{m}, \overline{n}) \in \mathbb{N}^n \times \mathbb{N}^n \mid g(\overline{m}) \sqsubseteq g(\overline{n})\}$ is semilinear.*

Proof. Let $\Gamma = \{a_1, a_2, \dots, a_n\}$ be an alphabet and define the monoid homomorphism $f\colon \Gamma^* \to \Sigma^*$ by $f(a_i) = w_i$ for all $i \in [1, n]$. One first shows that

$$S_2 = \{(u, v) \mid u, v \in a_1^* a_2^* \dots a_n^*,\ f(v) \sqsubseteq f(v)\}$$

is rational. We now employ Nivat's theorem. It tells us that there are a regular language R over some alphabet Δ and two homomorphisms $h_1, h_2\colon \Delta^* \to \Gamma^*$ so that we can write $S_2 = \{(h_1(w), h_2(w)) \mid w \in R\}$. Since R is regular, its Parikh-image $\Psi_\Delta(R) = \{\Psi_\Delta(w) \mid w \in R\}$ is semilinear [19]. There are monoid homomorphisms $p_1, p_2\colon \mathbb{N}^\Delta \to \mathbb{N}^n$ with $\Psi_\Gamma(h_i(w)) = p_i(\Psi_\Delta(w))$ for all $i \in \{1, 2\}$ and $w \in \Delta^*$. With these, the image $H = \{(p_1(\Psi_\Delta(w)), p_2(\Psi_\Delta(w))) \mid w \in R\}$ of the set $\Psi_\Delta(R)$ under the monoid homomorphism $(p_1, p_2)\colon \mathbb{N}^\Delta \to \mathbb{N}^n \times \mathbb{N}^n$ is semilinear. It turns out that this set equals the set from the lemma. □

Lemma 3.3. *Let $w_1, w_2, \dots, w_n \in \Sigma^*$, $L \subseteq w_1^* w_2^* \cdots w_n^*$ be context-free, and $g\colon \mathbb{N}^n \to \Sigma^*$ be defined by $g(\overline{m}) = w_1^{m_1} w_2^{m_2} \cdots w_n^{m_n}$ for every tuple $\overline{m} = (m_1, m_2, \dots, m_n) \in \mathbb{N}^n$. Then there exists a semilinear set $U \subseteq \mathbb{N}^n$ such that g maps U bijectively onto L.*

Proof. The set U contains, for each $u \in L$, the lexicographically minimal tuple $\overline{m} \in \mathbb{N}^n$ with $g(\overline{m}) = u$. Then, Lemmas 3.1 and 3.2 and the closure of the class of semilinear sets under first-order definitions imply the required properties. □

Now we can prove the main result of this section.

Theorem 3.4. *Let $L \subseteq \Sigma^*$ be context-free and bounded. Then the* FO+MOD*-theory of $(L, \sqsubseteq, \sqsubseteq\!\!\!\cdot, (K \cap L)_{K \text{ regular}}, (w)_{w \in L})$ is decidable.*

Proof. It suffices to prove the decidability for the structure $\mathcal{S} = (L, \sqsubseteq, (K \cap L)_{K \text{ regular}})$ since the theory of the structure from the theorem can be reduced to that of $\mathcal{S}$ ($x \sqsubseteq\!\!\!\cdot\, y$ gets replaced by its definition and $x\theta w$ by $\exists y\colon y \in \{w\} \wedge x\theta y$ where θ is any binary relation symbol).

Since L is bounded, there are words $w_1, w_2, \dots, w_n \in \Sigma^*$ such that L is included in $w_1^* w_2^* \cdots w_n^*$. For an n-tuple $\overline{m} = (m_1, m_2, \dots, m_n) \in \mathbb{N}^n$ we define $g(\overline{m}) = w_1^{m_1} w_2^{m_2} \cdots w_n^{m_n} \in \Sigma^*$.

1. By Lemma 3.3, there is a semilinear set $U \subseteq \mathbb{N}^n$ that is mapped by g bijectively onto L.
2. The set $\{(\overline{m}, \overline{n}) \mid g(\overline{m}) \sqsubseteq g(\overline{n})\}$ is semilinear by Lemma 3.2.
3. For any regular language $K \subseteq \Sigma^*$ the set $\{\overline{m} \in \mathbb{N}^n \mid g(\overline{m}) \in K\} \subseteq \mathbb{N}^n$ is effectively semilinear by Lemma 3.1.

From these semilinear sets, we obtain first-order formulas $\lambda(\overline{x})$, $\sigma(\overline{x},\overline{y})$, and $\kappa_K(\overline{x})$ in the language of $(\mathbb{N},+)$ such that, for any $\overline{m},\overline{n} \in \mathbb{N}^n$, we have

1. $(\mathbb{N},+) \models \lambda(\overline{m}) \iff \overline{m} \in U$,
2. $(\mathbb{N},+) \models \sigma(\overline{m},\overline{n}) \iff g(\overline{m}) \sqsubseteq g(\overline{n})$, and
3. $(\mathbb{N},+) \models \kappa_K(\overline{m}) \iff g(\overline{m}) \in K$.

One then defines, from an FO+MOD-formula $\varphi(x_1,\ldots,x_k)$ in the language of $\mathcal{S}$, an FO+MOD-formula $\varphi'(\overline{x_1},\ldots,\overline{x_k})$ in the language of $(\mathbb{N},+)$ such that

$$(\mathbb{N},+) \models \varphi'(\overline{m_1},\ldots,\overline{m_k}) \iff \mathcal{S} \models \varphi(g(\overline{m_1}),\ldots,g(\overline{m_k})).$$

(This construction can be found in the full version [18] and increases the formula size at least exponentially.)

Consequently, any sentence φ from FO+MOD in the language of $\mathcal{S}$ is translated into an equivalent sentence φ' in the language of $(\mathbb{N},+)$. By [1,5,21], validity of the sentence φ' in $(\mathbb{N},+)$ is decidable. □

4 The C+MOD²-Theory with Regular Predicates

It is the aim of this section to show that the C+MOD^2-theory of the structure $(L, \sqsubseteq, \sqsubset\!\!\!\cdot, (K \cap L)_{K \text{ regular}}, (w)_{w \in L})$ is decidable for any regular language L. To this aim, we first show that the C+MOD^2-theory of

$$\mathcal{S} = (\Sigma^*, \sqsubseteq, \sqsubset\!\!\!\cdot, (L)_{L \text{ regular}})$$

is decidable. This decidability proof extends the proof from [12] for the decidability of the FO^2-theory of $(\Sigma^*, \sqsubseteq, (L)_{L \text{ regular}})$. It provides a quantifier-elimination procedure (see Sect. 4.3) that relies on the following two properties:

1. The class of regular languages is closed under *counting* images under *unambiguous* rational relations (Sect. 4.2) and
2. the proper subword, the cover, and the incomparability relation are *unambiguous* rational (Sect. 4.1).

4.1 Unambiguous Rational Relations

Recall that, by Nivat's theorem, a relation $R \subseteq \Sigma^* \times \Sigma^*$ is rational if there exist an alphabet Γ, a homomorphism $h\colon \Gamma^* \to \Sigma^* \times \Sigma^*$, and a regular language $S \subseteq \Gamma^*$ such that h maps S surjectively onto R. We call R an *unambiguous rational relation* if, in addition, h maps S *injectively* (and therefore bijectively) onto R. Note that these are precisely the relations accepted by unambiguous 2-tape-automata.

While the class of rational relations is closed under unions, this is not the case for unambiguous rational relations (e.g., $R = \{(a^m b a^n, a^m) \mid m,n \in \mathbb{N}\} \cup \{(a^m b a^n, a^n) \mid m,n \in \mathbb{N}\}$ is the union of unambiguous rational relations but not unambiguous). But it is closed under *disjoint* unions.

Lemma 4.1. *For any alphabet Σ, the cover relation $\sqsubset\!\!\cdot$ and the relation $\sqsubset \setminus \sqsubset\!\!\cdot$ are unambiguous rational.*

Proof. For $i \in \{1,2\}$, let $\Sigma_i = \Sigma \times \{i\}$ and $\Gamma = \Sigma_1 \cup \Sigma_2$. Furthermore, let the homomorphism $\mathrm{proj}_i \colon \Gamma^* \to \Sigma^*$ be defined by $\mathrm{proj}_i(a,i) = a$ and $\mathrm{proj}_i(a,3-i) = \varepsilon$ for all $a \in \Sigma$. Finally, let the homomorphism $\mathrm{proj} \colon \Gamma^* \to \Sigma^* \times \Sigma^*$ be defined by $\mathrm{proj}(w) = (\mathrm{proj}_1(w), \mathrm{proj}_2(w))$.

- The regular language

$$\mathrm{Sub} = \left(\bigcup_{a\in\Sigma} \left((\Sigma_2 \setminus \{(a,2)\})^* \, (a,2) \, (a,1) \right) \right)^* \Sigma_2{}^*.$$

  is mapped bijectively onto the subword relation.
- Let S be the regular language of words from Sub with precisely one more occurrence of letters from Σ_2 than from Σ_1. Then S is mapped bijectively onto the relation $\sqsubset\!\!\cdot$, hence this relation is unambiguous rational.
- Similarly, let S' denote the regular language of all words from Sub with at least two more occurrences of letters from Σ_2 than from Σ_1. It is mapped bijectively onto the relation $\sqsubset \setminus \sqsubset\!\!\cdot$, i.e., $\sqsubset \setminus \sqsubset\!\!\cdot$ is unambiguous rational. $\square$

Lemma 4.2. *For any alphabet Σ, the incomparability relation*

$$\parallel = \{(u,v) \in \Sigma^* \times \Sigma^* \mid \textit{neither } u \sqsubseteq v \textit{ nor } v \sqsubseteq u\}$$

is unambiguous rational.

Proof. We will show that the following three relations are unambiguous rational:

1. $R_1 = \{(u,v) \mid |u| < |v| \text{ and not } u \sqsubseteq v\}$,
2. $R_2 = \{(u,v) \mid |u| = |v| \text{ and } u \neq v\}$, and
3. $R_3 = \{(u,v) \mid |u| > |v| \text{ and not } v \sqsubseteq u\}$.

The result follows since $\parallel$ is the disjoint union of these relations. Let Σ_i, Γ, proj_i, and proj be defined as in the previous proof. First, the regular language

$$\mathrm{Inc}_2 = (\Sigma_2\Sigma_1)^* \cdot \{(a,2)(b,1) \mid a,b \in \Sigma, a \neq b\} \cdot (\Sigma_2\Sigma_1)^*.$$

is mapped by proj bijectively onto R_2.

From [12, Lemma 5.2], we learn that $(u,v) \in R_1 \cup R_2$ if, and only if,

- $u = a_1 a_2 \ldots a_\ell u'$ for some $\ell \geq 1$, $a_1, \ldots, a_\ell \in \Sigma$, $u' \in \Sigma^*$, and
- $v \in (\Sigma \setminus \{a_1\})^* a_1 \, (\Sigma \setminus \{a_2\})^* a_2 \cdots (\Sigma \setminus \{a_{\ell-1}\})^* a_{\ell-1} \, (\Sigma \setminus \{a_\ell\})^+ v'$ for some word $v' \in \Sigma^*$ with $|u'| = |v'|$.

Consequently, proj maps the following language bijectively onto $R_1 \cup R_2$:

$$\mathrm{Inc}_{1,2} = \left(\bigcup_{a\in\Sigma} \left((\Sigma_2 \setminus \{(a,2)\})^* (a,2)(a,1) \right) \right)^* \cdot \bigcup_{a\in\Sigma} \left((\Sigma_2 \setminus \{(a,2)\})^+ (a,1) \right) \cdot (\Sigma_2\Sigma_1)^*$$

and since $\mathrm{Inc}_2 \subseteq \mathrm{Inc}_{1,2}$, proj maps $\mathrm{Inc}_1 = \mathrm{Inc}_{1,2} \setminus \mathrm{Inc}_2$ bijectively onto R_1. The claim regarding R_3 follows analogously. $\square$

4.2 Closure Properties of the Class of Regular Languages

Let $R \subseteq \Sigma^* \times \Sigma^*$ be an unambiguous rational relation and $L \subseteq \Sigma^*$ a regular language. We want to show that the languages of all words $u \in \Sigma^*$

$$\text{with } |\{v \in L \mid (u,v) \in R\}| \geq k \quad (1)$$
$$(\text{with } |\{v \in L \mid (u,v) \in R\}| \in p + q\mathbb{N}, \text{ respectively}) \quad (2)$$

are effectively regular for all $k \in \mathbb{N}$ and all $0 \leq p < q$, respectively (this does not hold for arbitrary rational relations). It is straightforward to work out direct automata constructions for this. However, the full details of this are somewhat cumbersome. Instead, we provide a proof via weighted automata, which enables us to split the two constructions into several simple steps.

Let S be a semiring. A function $r\colon \Sigma^* \to S$ is *realizable over* S if there are $n \in \mathbb{N}$, $\lambda \in S^{1\times n}$, a homomorphism $\mu\colon \Sigma^* \to S^{n\times n}$, and $\nu \in S^{n\times 1}$ with $r(w) = \lambda \cdot \mu(w) \cdot \nu$ for all $w \in \Sigma^*$. The triple (λ, μ, ν) is a *presentation of dimension* n or a *weighted automaton for* r.

In the following, we consider the semiring $\mathbb{N}^\infty$, i.e., the set $\mathbb{N} \cup \{\infty\}$ together with the commutative operations $+$ and $\cdot$ (with $x + \infty = \infty$ for all $x \in \mathbb{N} \cup \{\infty\}$, $x \cdot \infty = \infty$ for all $x \in (\mathbb{N} \cup \{\infty\}) \setminus \{0\}$, and $0 \cdot \infty = 0$). Sometimes, we will argue about sums of infinitely many elements from $\mathbb{N}^\infty$, which are defined as expected.

Proposition 4.3. *Let Γ and Σ be alphabets, $f\colon \Gamma^* \to \Sigma^*$ a homomorphism, and $\chi\colon \Gamma^* \to \mathbb{N}^\infty$ a realizable function over $\mathbb{N}^\infty$. Then the following function r is effectively realizable over $\mathbb{N}^\infty$:*

$$r = \chi \circ f^{-1}\colon \Sigma^* \to \mathbb{N}^\infty\colon u \mapsto \sum_{\substack{w \in \Gamma^* \\ f(w) = u}} \chi(w)$$

Proof. The homomorphism f can be written as $f = f_2 \circ f_1$ where $f_1\colon \Gamma^* \to \Gamma^*$ is non-expanding (i.e., $f_1(a) \in \Gamma \cup \{\varepsilon\}$ for all $a \in \Gamma$) and $f_2\colon \Gamma^* \to \Sigma^*$ is non-erasing (i.e., $f_2(a) \in \Sigma^+$ for all $a \in \Gamma$). Then $r = (\chi \circ f_1^{-1}) \circ f_2^{-1}$. Then $\chi' = \chi \circ f_1^{-1}$ is effectively realizable by [3, Lemma 2.2(b)].

Let (λ, μ, ν) be a presentation of dimension n for χ'. For $\sigma \in \Sigma \cup \{\varepsilon\}$, set $\Gamma_\sigma = \{b \in \Gamma \mid f_2(b) = \sigma\}$. Furthermore, define the matrix $M \in (\mathbb{N}^\infty)^{n\times n}$ by

$$M_{ij} = \begin{cases} \infty & \text{if there is } w \in \Gamma_\varepsilon^* \text{ with } n < |w| \leq 2n \text{ and } \mu(w)_{ij} > 0 \\ \sum_{w \in \Gamma_\varepsilon^{\leq n}} \mu(w)_{ij} & \text{otherwise.} \end{cases}$$

Then $M_{ij} = \sum_{w \in \Gamma_\varepsilon^*} \mu(w)_{ij}$ for all $i, j \in [1, n]$. Setting $\lambda' = \lambda \cdot M$ and

$$\mu'(a) = \sum_{b \in \Gamma_a} \big(\mu(b) \cdot M\big) \text{ for all } a \in \Sigma$$

defines the presentation (λ', μ', ν) for the function $r = \chi' \circ f_2^{-1}$. □

Lemma 4.4. *Let $R \subseteq \Sigma^* \times \Sigma^*$ be an unambiguous rational relation and $L \subseteq \Sigma^*$ be regular. Then the following function r is effectively realizable over $\mathbb{N}^\infty$:*

$$r\colon \Sigma^* \to \mathbb{N}^\infty : u \mapsto |\{v \in L \mid (u,v) \in R\}|$$

Proof. Since R is unambiguous rational, so is $R \cap (\Sigma^* \times L)$, i.e., there are an alphabet Γ, homomorphisms $f, g\colon \Gamma^* \to \Sigma^*$, and a regular language $S_L \subseteq \Gamma^*$ such that

$$(f,g)\colon \Gamma^* \to \Sigma^* \times \Sigma^* : w \mapsto \big(f(w), g(w)\big)$$

maps S_L bijectively onto $R \cap (\Sigma^* \times L)$. Since S_L is regular, its characteristic function χ is effectively realizable by [20, Prop. 3.12]. One then shows that r is the function $\chi \circ f^{-1}$ as in Proposition 4.3. □

We now come to the main result of this section.

Proposition 4.5. *Let $R \subseteq \Sigma^* \times \Sigma^*$ be an unambiguous rational relation and $L \subseteq \Sigma^*$ be regular. Then, for $k \in \mathbb{N}$ and for $p, q \in \mathbb{N}$ with $p < q$, the set H of words w satisfying* (1) *and* (2)*, respectively, is effectively regular.*

Let R denote the rational relation mentioned before Lemma 4.1. Then a word $a^m b a^n$ has ≥ 2 "R-partners" iff it has an even number of "R-partners" iff $m \neq n$. Hence, the above proposition does not hold for arbitrary rational relations.

Proof. Let r be the function from Lemma 4.4. Setting $x \equiv y$ iff $x = y$ or $k \leq x, y < \infty$ defines a congruence $\equiv$ on $\mathbb{N}^\infty$. Then $S_k^\infty = \mathbb{N}^\infty/\!\equiv$ is a finite semiring and the function $s\colon \Sigma^* \to S_k^\infty : u \mapsto [r(u)]$ is effectively realizable. Since the semiring S_k^∞ is finite, the "level sets" $s^{-1}([i]) = \{u \in \Sigma^* \mid s(u) \equiv i\}$ are effectively regular by [20, Prop. 4.5]. Since $s^{-1}([k]) \cup s^{-1}([\infty])$ is the language of words u satisfying (1), the first result follows.

For the second language, we consider the congruence $\equiv \subseteq \mathbb{N}^\infty \times \mathbb{N}^\infty$ with $x \equiv y$ iff $x = y$ or $q \leq x, y < \infty$ and $x - y \in q\mathbb{N}$. □

4.3 Quantifier Elimination for C+MOD^2

Our decision algorithm employs a quantifier alternation procedure, i.e., we will transform an arbitrary formula into an equivalent one that is quantifier-free. As usual, the heart of this procedure handles formulas $\psi = \mathsf{Q}y\,\varphi$ where Q is a quantifier and φ is quantifier-free. Since the logic C+MOD^2 has only two variables, any such formula ψ has at most one free variable. In other words, it defines a language K. The following lemma shows that this language is effectively regular, such that ψ is equivalent to the quantifier-free formula $x \in K$.

Lemma 4.6. *Let $\varphi(x,y)$ be a quantifier-free formula from* C+MOD^2 *in the language of the structure $\mathcal{S} = (\Sigma^*, \sqsubseteq, \sqsubset\!\!\cdot, (L)_{L \text{ regular}})$. Then the sets*

$$\{x \in \Sigma^* \mid \mathcal{S} \models \exists^{\geq k} y\,\varphi\} \text{ and } \{x \in \Sigma^* \mid \mathcal{S} \models \exists^{p \bmod q} y\,\varphi\}$$

are effectively regular for all $k \in \mathbb{N}$ and all $p, q \in \mathbb{N}$ with $p < q$.

Proof. Since φ is quantifier-free, we can rewrite it into a Boolean combination of formulas of the form $x \in K$ and $y \in L$ for some regular languages K and L, $x \sqsubseteq y$ and $y \sqsubseteq x$, and $x \mathrel{\sqsubset\!\!\cdot}\, y$ and $y \mathrel{\sqsubset\!\!\cdot}\, x$.

There are six possible relations between the two variables x and y in the partial order: we can have $x = y$, $x \mathrel{\sqsubset\!\!\cdot}\, y$ or *vice versa*, $x \sqsubset y \wedge \neg x \mathrel{\sqsubset\!\!\cdot}\, y$ or *vice versa*, or $x \parallel y$. Let $\theta_i(x, y)$ for $1 \le i \le 6$ be formulas describing these relations.

Hence φ is equivalent to $\bigvee_{1 \le i \le 6}(\theta_i \wedge \varphi)$. In this formula, any occurrence of φ appears in conjunction with precisely one of the formulas θ_i. Depending on this formula θ_i (i.e., the relation between x and y), we can simplify φ to φ_i by replacing the atomic subformulas that compare x and y by true or false. As a result, the formula φ is equivalent to $\bigvee_{1 \le i \le 6}(\theta_i \wedge \varphi_i)$ where the formulas φ_i are Boolean combinations of formulas of the form $x \in K$ and $y \in L$ for some regular languages K and L.

Now let $k \in \mathbb{N}$. Since the formulas θ_i are mutually exclusive, we get

$$\exists^{\ge k} y\, \varphi \equiv \exists^{\ge k} y \bigvee_{1 \le i \le 6} (\theta_i \wedge \varphi_i) \equiv \bigvee_{(*)} \bigwedge_{1 \le i \le 6} \exists^{\ge k_i} y\, (\theta_i \wedge \varphi_i)$$

where the disjunction $(*)$ extends over all $(k_1, \ldots, k_6) \in \mathbb{N}^6$ with $\sum_{1 \le i \le 6} k_i = k$.

Hence it suffices to show that

$$\{x \in \Sigma^* \mid \exists^{\ge k} y\, (\theta_i \wedge \varphi)\} \tag{3}$$

is effectively regular for all $1 \le i \le 6$, all $k \in \mathbb{N}$, and all Boolean combinations φ of formulas of the form $x \in K$ and $y \in L$ where K and L are regular languages. We can find regular languages K_M and L_M and a finite set I such that φ is equivalent to $\bigvee_{M \in I}(x \in K_M \wedge y \in L_M)$ and such that this disjunction is exclusive. Hence the set from (3) equals the union of the sets

$$\{x \in \Sigma^* \mid \exists^{\ge k} y\, (\theta_i \wedge x \in K_M \wedge y \in L_M)\} = K_M \cap \underbrace{\{x \in \Sigma^* \mid \exists^{\ge k} y \in L_M \colon \theta_i\}}_{H_M}$$

for $M \in I$. The set H_M is effectively regular by Proposition 4.5 and Lemmas 4.1 and 4.2. Since the language in the claim of the lemma is a Boolean combination of such sets, the first claim is demonstrated; the second follows similarly. □

The only atomic formulas with a single variable x are $x \in L$ with L regular, $x = x$, $x \sqsubseteq x$ (which are equivalent to $x \in \Sigma^*$), and $x \mathrel{\sqsubset\!\!\cdot}\, x$ (which is equivalent to $x \in \emptyset$). Hence, any quantifier-free formula with a single free variable x is a Boolean combination of statements of the form $x \in L$. Lemma 4.6 thus implies:

Theorem 4.7. *Let $\mathcal{S} = (\Sigma^*, \sqsubseteq, \mathrel{\sqsubset\!\!\cdot}, (L)_{L \text{ regular}})$. Let $\varphi(x)$ be a formula from C+MOD^2. Then the set $\{x \in \Sigma^* \mid \mathcal{S} \models \varphi\}$ is effectively regular.*

Corollary 4.8. *Let $L \subseteq \Sigma^*$ be a regular language. Then the C+MOD^2-theory of the structure $\mathcal{S}_L = (L, \sqsubseteq, \mathrel{\sqsubset\!\!\cdot}, (K \cap L)_{K \text{ regular}}, (w)_{w \in L})$ is decidable.*

Proof. Let $\varphi \in \mathrm{C{+}MOD}^2$ be a sentence. We build φ_L by (1) restricting all quantifications to L, (2) replace $x\theta w$ by $\exists y\colon y \in \{w\} \wedge x\theta y$, and dually for $y\theta w$ for all $w \in L$ and all binary relations θ.

With $\mathcal{S}$ the structure from Theorem 4.7, we obtain $\mathcal{S} \models \varphi_L \iff \mathcal{S}_L \models \varphi$. By Theorem 4.7, the language $\{x \mid \mathcal{S} \models \varphi_L\}$ is regular (since φ_L is a sentence, it is $\emptyset$ or Σ^*). Hence φ_L holds iff this set is nonempty, which is decidable. □

5 The Σ_1-Theory

In this section, we study for which regular languages L the Σ_1-theory of the structure $(L, \sqsubseteq)$ is decidable. If L is bounded, then decidability follows from Theorem 3.4. In the case of $(\Sigma^*, \sqsubseteq)$, decidability is known as well [17]. Here, we prove decidability for every regular language L. Note that in terms of quantifier block alternation, this is optimal: The Σ_2-theory is undecidable already in the simple case of $(\{a, b\}^*, \sqsubseteq)$ [6].

Theorem 5.1. *For every regular $L \subseteq \Sigma^*$, the Σ_1-theory of $(L, \sqsubseteq)$ is decidable.*

Observe that very generally, the Σ_1-theory of a partially ordered set $(P, \leq)$ is decidable if every finite partial order embeds into $(P, \leq)$: In that case, a formula with n variables is satisfied in $(P, \leq)$ if and only if it is satisfied for some finite partial order with at most n elements. This is used to obtain decidability for the case $L = \Sigma^*$ with $|\Sigma| \geq 2$ in [17].

As mentioned above, if L is bounded, decidability follows from Theorem 3.4. If L is unbounded, it is well-known that there is a subset $x\{p, q\}^*y \subseteq L$ such that $|p| = |q|$ and $p \neq q$ (see Lemma 5.2). Since in that case, the monoids $(\{a, b\}^*, \cdot)$ and $(\{p, q\}^*, \cdot)$ are isomorphic, it is tempting to assume that $(\{a, b\}^*, \sqsubseteq)$ embeds into $(\{p, q\}^*, \sqsubseteq)$ and thus into $(x\{p, q\}^*y, \sqsubseteq)$. However, that is not the case. If $L = \{ab, ba\}^*$, then the downward closure of any infinite subset of L includes all of L. Since, on the other hand, $(\{a, b\}^*, \sqsubseteq)$ has infinite downward closed strict subsets such as a^*, it cannot embed into $(L, \sqsubseteq)$. Nevertheless, the rest of this section demonstrates that every finite partial order embeds into $(L, \sqsubseteq)$ whenever L is an unbounded regular language. By the previous paragraph, this implies Theorem 5.1.

We recall a well-known property of unbounded regular languages.

Lemma 5.2. *If $L \subseteq \Sigma^*$ is not bounded, then there are $x, y, p, q \in \Sigma^*$ such that $|p| = |q|$, $p \neq q$, and $x\{p, q\}^*y \subseteq L$.*

Proof. Let A be any non-degenerate deterministic finite automaton accepting L. Then at least one strongly connected component of A is not a cycle since otherwise, L would be bounded. Hence, there is a state s and prefix-incomparable words u, v, each of which is read on a cycle starting in s. Since u and v are prefix-incomparable, the words $p = uv$ and $q = vu$ are distinct, but equally long. Since A is non-degenerate, there are words $x, y \in \Sigma^*$ with $x\{p, q\}^*y \subseteq L$. □

To have some control over how words can embed, we prove a stronger version of Lemma 5.2. Two words $p, q \in \Sigma^*$ are *conjugate* if there are $x, y \in \Sigma^*$ with $p = xy$ and $q = yx$. A word $p \in \Sigma^*$ is *primitive* if there is no $q \in \Sigma^*$ with $p \in qq^+$.

Proposition 5.3. *For every unbounded regular language $L \subseteq \Sigma^*$, there are $x, u, v, y \in \Sigma^*$ such that $|u| = |v|$, the word uv is primitive, and $x\{u, v\}^*y \subseteq L$.*

Proof. Since L is unbounded and regular, Lemma 5.2 yields words $x, y, p, q \in \Sigma^*$ with $|p| = |q|$, $p \neq q$, and $x\{p, q\}^*y \subseteq L$. Then the words $r = pq$ and $s = pp$ are not conjugate, because every conjugate of a square is a square. Moreover, $|r| = |s|$, and $x\{r, s\}^*y \subseteq x\{p, q\}^*y \subseteq L$. Let $n = |r|$, $u = rs^{n-1}$, and $v = s^n$. Towards a contradiction, suppose $uv = rs^{2n-1}$ is not primitive. Then there is a word $w \in \Sigma^*$ with $rs^{2n-1} \in ww^+$. Depending on whether $|w| \geq n$ or $|w| < n$, we have $n \leq |w^t| \leq n^2$ either for $t = 1$ or for $t = n$. It follows that r is a prefix of w^t and that w^t is a suffix of s^n, implying that r is a factor of s^n. Since r and s are not conjugate, this is impossible. □

We are now ready to describe how to embed a finite partial order into $(L, \sqsubseteq)$. Observe that every finite partial order with m elements embeds into $(\{0, 1\}^m, \leq)$ where $\leq$ is the componentwise order. Hence, it suffices to embed this partial order into $(\{u, v\}^*, \sqsubseteq)$. We do this as follows. Let $n = |uv| + m + 3$ and define, for a tuple $t = (t_1, \ldots, t_m) \in \{0, 1\}^m$,

$$\varphi_m(t_1, \ldots, t_m) = v^{t_1}(uv)^n \cdots v^{t_m}(uv)^n.$$

Then, clearly, $s \leq t$ implies $\varphi_m(s) \sqsubseteq \varphi_m(t)$. The converse requires a careful analysis of how prefixes of $\varphi_m(s)$ can embed into prefixes of $\varphi_m(t)$. For $x, y \in \Sigma^*$, we write $x \hookrightarrow y$ if x, but no word xa with $a \in \Sigma$ is a subword of y. In other words, $x \hookrightarrow y$ if x is a *prefix-maximal subword of* y. This gives us a criterion for non-embeddability: If x has a strict prefix x_0 with $x_0 \hookrightarrow y$, then certainly $x \not\sqsubseteq y$. In this case, the word x_1 with $x = x_0x_1$ is called *residue*. We show the following:

Lemma 5.4. *Let $u, v \in \Sigma^*$ be words such that $|u| = |v|$ and uv is primitive. Then, for all $\ell, n \in \mathbb{N}$ with $n > |uv| + \ell + 2$, we have*

(i) $(uv)^n \hookrightarrow v(uv)^n$,
(ii) $(uv)^\ell v(uv)^{n-\ell-1} \hookrightarrow (uv)^n$, *and*
(iii) $(uv)^{1+\ell} v(uv)^{n-\ell-2} \hookrightarrow v(uv)^n$.

For this lemma, it is crucial to observe that for a primitive word w and $n > |w|+1$, any embedding of w^{n-1} into w^n must either hit the left-most or the right-most position in w^n. To conclude that $s \not\leq t$ implies $\varphi_m(s) \not\sqsubseteq \varphi_m(t)$, we argue about prefixes of the form $p_i = v^{s_1}(uv)^n \cdots v^{s_i}(uv)^n$ and $q_i = v^{t_1}(uv)^n \cdots v^{t_i}(uv)^n$ for $i \in [1, m]$. If $s \not\leq t$, let $i \in [1, m]$ be the index with $s_i = 1$, $t_i = 0$ and $s_j \leq t_j$ for all $j \in [1, i-1]$. Then clearly $p_{i-1} \sqsubseteq q_{i-1}$. In fact, Lemma 5.4 (i) implies that even $p_{i-1} \hookrightarrow q_{i-1}$, since $x \hookrightarrow y$ and $x' \hookrightarrow y'$ imply $xy \hookrightarrow x'y'$. Then, by

Lemma 5.4 (ii), $p_i = p_{i-1}v(uv)^{n-1}(uv)$ has a residue of uv in $q_i = q_{i-1}(uv)^n$. To conclude $\varphi_m(s) \not\sqsubseteq \varphi_m(t)$, it remains to be shown that this can never be rectified when considering prefixes p_j and q_j for $j = i+1, \ldots, m$. To this end, Lemma 5.4 (ii) and (iii) tell us that if p_j has a residue of $(uv)^\ell$ in q_j, then the word p_{j+1} has a residue of $(uv)^\ell$ or even $(uv)^{\ell+1}$ in q_{j+1}.

6 The Σ_1-Theory with Constants

In this section, we study for which languages L the structure $(L, \sqsubseteq, (w)_{w \in L})$ has a decidable Σ_1-theory. From Theorem 3.4, we know that this is the case whenever L is bounded. However, there are very simple languages for which decidability is lost: If $|\Sigma| \geq 2$, then the Σ_1-theory of $(\Sigma^*, \sqsubseteq, (w)_{w \in \Sigma^*})$ is undecidable [6]. Here, we present a sufficient condition for the Σ_1-theory of $(L, \sqsubseteq, (w)_{w \in \Sigma^*})$ to be decidable.

Let $L \subseteq \Sigma^*$. We say that a letter $a \in \Sigma$ is *frequent* in L if there is a real constant $\delta > 0$ so that $|w|_a \geq \delta \cdot |w|$ for all but finitely many $w \in L$. Our sufficient condition requires that all letters be frequent in L. If L is regular, this is equivalent to saying that in every non-degenerate automaton for L, every cycle contains every letter. An example of such a language is $\{ab, ba\}^*$.

We shall prove that this condition implies decidability of the Σ_1-theory of $(L, \sqsubseteq, (w)_{w \in \Sigma^*})$. If L is bounded, decidability already follows from Theorem 3.4. In case L is unbounded, we employ our results from Sect. 5 to show another embeddability result. For $w \in \Sigma^*$, let $w{\uparrow} = \{u \in \Sigma^* \mid w \sqsubseteq u\}$ denote the upward closure of $\{w\}$ in $(\Sigma^*, \sqsubseteq)$. We will show that if L is unbounded, then for each $w \in \Sigma^*$, the decomposition of $L = (L \setminus w{\uparrow}) \cup (L \cap w{\uparrow})$ yields two simple parts: The set $L \setminus w{\uparrow}$ is finite and the set $L \cap w{\uparrow}$ embeds every finite partial order. This simplifies the conditions under which a Σ_1-sentence is satisfied.

Lemma 6.1. *Let $L \subseteq \Sigma^*$ be an unbounded regular language where every letter is frequent. For every $w \in \Sigma^*$, the set $L \setminus w{\uparrow}$ is finite and $L \cap w{\uparrow}$ is unbounded.*

Proof. In a non-degenerate automaton A for L, every cycle must contain every letter. Therefore, if A has n states and $v \in L$ has $|v| > n \cdot |w|$, then a computation for v must contain some state more than $|w|$ times, which implies $w \sqsubseteq v$ and hence $v \notin L \setminus w{\uparrow}$. Therefore, $L \setminus w{\uparrow}$ is finite. This implies that $L \cap w{\uparrow}$ is unbounded: Otherwise $L = (L \cap w{\uparrow}) \cup (L \setminus w{\uparrow})$ would be bounded as well. □

Theorem 6.2. *Let $L \subseteq \Sigma^*$ be an unbounded regular language where every letter is frequent. Then the Σ_1-theory of $(L, \sqsubseteq, (w)_{w \in L})$ is decidable.*

Proof. For decidability, we may assume that we are given a formula φ that is a disjunction of conjunctions of literals of the following forms (where x and y are arbitrary variables and w an arbitrary word from L):

(i) $x \sqsubseteq w$	(iii) $w \sqsubseteq x$	(v) $x \sqsubseteq y$
(ii) $x \not\sqsubseteq w$	(iv) $w \not\sqsubseteq x$	(vi) $x \not\sqsubseteq y$

Step 1. We first show that literals of types (i) and (iv) can be eliminated. To this end, we observe that for each $w \in L$, both of the sets $\{u \in L \mid u \sqsubseteq w\}$, and $\{u \in L \mid w \not\sqsubseteq u\}$ are finite (in the latter case, this follows from Lemma 6.1). Thus, every conjunction that contains a literal $x \sqsubseteq w$ or $w \not\sqsubseteq x$, constrains x to finitely many values. Therefore, we can replace this conjunction with a disjunction of conjunctions that result from replacing x by one of these values. (Here, we might obtain literals $u \sqsubseteq v$ or $u \not\sqsubseteq v$, but those can be replaced by other equivalent formulas). We repeat this until there are no more literals of the form (i) and (iv).
Step 2. We now eliminate literals of the form (ii). Note that the language $\{u \in L \mid u \not\sqsubseteq w\}$ is upward closed in $(L, \sqsubseteq)$. Since L is regular, we can compute the finite set of minimal elements of this set. Thus, $x \not\sqsubseteq w$ is equivalent to a finite disjunction of literals of the form $w' \sqsubseteq x$. The resulting formula ψ is a disjunction of conjunction of literals of the form (iii), (v), (vi).
Step 3. To check satisfiability, we may assume that ψ is a conjunction of literals of the form (iii), (v), (vi). We can write ψ as $\gamma_1 \wedge \gamma_2$, where γ_1 is a conjunction of literals of the form (iii) and γ_2 is a conjunction of literals of the form (v) and (vi). We claim that ψ is satisfiable if and only if γ_2 is satisfiable in some partial order. The "only if" direction is trivial, so suppose γ_2 is satisfied by some finite partial order $(P, \leq)$ and let $w \in \Sigma^*$ be a concatenation of all words occurring in γ_1. By Lemma 6.1, $L \cap w\!\uparrow$ is unbounded, which implies that $(P, \leq)$ can be embedded into $(L \cap w\!\uparrow, \sqsubseteq)$ (see Sect. 5). This means, there exists a satisfying assignment where even $w \sqsubseteq x$ for every variable x. In particular, it satisfies $\psi = \gamma_1 \wedge \gamma_2$. □

Open Questions

We did not consider complexity issues. In particular, from [13], we know that the FO^2-theory of the structure $(\Sigma^*, \sqsubseteq, (w)_{w \in \Sigma^*})$ can be decided in elementary time. We are currently working out the details for the extension of this result to the $\mathrm{C{+}MOD}^2$-theory of the structure $(L, \sqsubseteq, (w)_{w \in L})$ for regular languages L. We reduced the FO+MOD-theory of the full structure (for L context-free and bounded) to the FO+MOD-theory of $(\mathbb{N}, +)$, which is known to be decidable in elementary time [5]. Our reduction increases the formula exponentially due to the need of handling statements of the form "there is an even number of pairs $(x, y) \in \mathbb{N}^2$ such that ..." It should be checked whether the proof from [5] can be extended to handle such statements in FO+MOD for $(\mathbb{N}, +)$ directly.

Finally, our results raise an interesting question: For which regular languages L does the structure $(L, \sqsubseteq, (w)_{w \in L})$ have a decidable Σ_1-theory? If every letter is frequent in L, we have decidability. For example, this applies to $L = \{ab, ba\}^*$ or $L = \{ab, baa\}^* \cup bb\{abb\}^*$. If $L = \Sigma^*$ for $|\Sigma| \geq 2$, we have undecidability [6].

References

1. Apelt, H.: Axiomatische Untersuchungen über einige mit der Presburgerschen Arithmetik verwandten Systeme. Z. Math. Logik Grundlagen Math. **12**, 131–168 (1966)
2. Berstel, J.: Transductions and Context-Free Languages. Teubner Studienbücher, Stuttgart (1979)
3. Droste, M., Gastin, P.: Weighted automata and weighted logics. In: Droste, M., Kuich, W., Vogler, H. (eds.) Handbook of Weighted Automata, pp. 176–211. Springer, Heidelberg (2009). https://doi.org/10.1007/978-3-642-01492-5_5
4. Finkel, A., Schnoebelen, Ph.: Well-structured transition systems everywhere! Theor. Comput. Sci. **256**, 63–92 (2001)
5. Habermehl, P., Kuske, D.: On Presburger arithmetic extended with modulo counting quantifiers. In: Pitts, A. (ed.) FoSSaCS 2015. LNCS, vol. 9034, pp. 375–389. Springer, Heidelberg (2015). https://doi.org/10.1007/978-3-662-46678-0_24
6. Halfon, S., Schnoebelen, Ph., Zetzsche, G.: Decidability, complexity, and expressiveness of first-order logic over the subword ordering. In: Proceedings of the Thirty-Second Annual ACM/IEEE Symposium on Logic in Computer Science (LICS 2017), pp. 1–12. IEEE Computer Society (2017)
7. Higman, G.: Ordering by divisibility in abstract algebras. Proc. London Math. Soc. **2**, 326–336 (1952)
8. Ježek, J., McKenzie, R.: Definability in substructure orderings. I: finite semilattices. Algebra Univers. **61**(1), 59–75 (2009)
9. Ježek, J., McKenzie, R.: Definability in substructure orderings. III: finite distributive lattices. Algebra Univers. **61**(3–4), 283–300 (2009)
10. Ježek, J., McKenzie, R.: Definability in substructure orderings. IV: finite lattices. Algebra Univers. **61**(3–4), 301–312 (2009)
11. Ježek, J., McKenzie, R.: Definability in substructure orderings. II: finite ordered sets. Order **27**(2), 115–145 (2010)
12. Karandikar, P., Schnoebelen, Ph.: Decidability in the logic of subsequences and supersequences. In: Harsha, P., Ramalingam, G. (eds.) Proceedings of the 35th Conference on Foundations of Software Technology and Theoretical Computer Science (FSTTCS 2015). Leibniz International Proceedings in Informatics, vol. 45, pp. 84–97. Leibniz-Zentrum für Informatik (2015)
13. Karandikar, P., Schnoebelen, Ph.: The height of piecewise-testable languages with applications in logical complexity. In: Talbot, J.-M., Regnier, L. (eds.) Proceedings of the 25th EACSL Annual Conference on Computer Science Logic (CSL 2016). Leibniz International Proceedings in Informatics, vol. 62, pp. 37:1–37:22 (2016)
14. Kudinov, O.V., Selivanov, V.L.: Undecidability in the homomorphic quasiorder of finite labelled forests. J. Log. Comput. **17**(6), 1135–1151 (2007)
15. Kudinov, O.V., Selivanov, V.L., Yartseva, L.V.: Definability in the subword order. In: Ferreira, F., Löwe, B., Mayordomo, E., Mendes Gomes, L. (eds.) CiE 2010. LNCS, vol. 6158, pp. 246–255. Springer, Heidelberg (2010). https://doi.org/10.1007/978-3-642-13962-8_28
16. Kudinov, O.V., Selivanov, V.L., Zhukov, A.V.: Definability in the h-quasiorder of labeled forests. Ann. Pure Appl. Logic **159**(3), 318–332 (2009)
17. Kuske, D.: Theories of orders on the set of words. Theor. Inf. Appl. **40**, 53–74 (2006)
18. Kuske, D., Zetzsche, G.: Languages ordered by the subword order. CoRR, abs/1901.02194 (2019)

19. Parikh, R.: On context-free languages. J. ACM **13**(4), 570–581 (1966)
20. Sakarovitch, J.: Rational and recognisable power series. In: Droste, M., Kuich, W., Vogler, H. (eds.) Handbook of Weighted Automata, pp. 105–174. Springer, Heidelberg (2009). https://doi.org/10.1007/978-3-642-01492-5_4
21. Schweikardt, N.: Arithmetic, first-order logic, and counting quantifiers. ACM Trans. Comput. Log. **6**(3), 634–671 (2005)
22. Thinniyam, R.S.: Definability of recursive predicates in the induced subgraph order. In: Ghosh, S., Prasad, S. (eds.) ICLA 2017. LNCS, vol. 10119, pp. 211–223. Springer, Heidelberg (2017). https://doi.org/10.1007/978-3-662-54069-5_16
23. Thinniyam, R.S.: Defining recursive predicates in graph orders. Logical Methods Comput. Sci. **14**(3:21), 1–38 (2018)

Strong Adequacy and Untyped Full-Abstraction for Probabilistic Coherence Spaces

Thomas Leventis[1,2(✉)] and Michele Pagani[1]

[1] IRIF UMR 8243, Université Paris Diderot, Sorbonne Paris Cité, CNRS, Paris, France
{leventis,pagani}@irif.fr
[2] University of Bologna, Bologna, Italy

Abstract. We consider the probabilistic untyped lambda-calculus and prove a stronger form of the adequacy property for probabilistic coherence spaces (PCoh), showing how the denotation of a term statistically distributes over the denotations of its head-normal forms.

We use this result to state a precise correspondence between PCoh and a notion of probabilistic Nakajima trees, recently introduced by Leventis in order to prove a separation theorem. As a consequence, we get full abstraction for PCoh. This latter result has already been mentioned as a corollary of Clairambault and Paquet's full abstraction theorem for probabilistic concurrent games. Our approach allows to prove the property directly, without the need of a third model.

Keywords: Lambda-Calculus · Denotational semantics · Probabilistic functional programming

1 Introduction

Full abstraction for the maximal consistent sensible λ-theory $\mathcal{H}^\star$ [1] is a crucial property for a model of the untyped λ-calculus, stating that two terms M, N have the same denotation in the model iff for every context $C[\,]$ the head-reduction sequences of $C[M]$ and $C[N]$ either both terminate or both diverge. The first such result was obtained for Scott's model $\mathcal{D}^\infty$ by Hyland [10] and Wadsworth [15]. More recently, Manzonetto developed a general technique for achieving full abstraction for a large class of models, decomposing it into the *adequacy property* and a notion of *well-stratification* [13]. An adequacy property states that the semantics of a λ-term is different from the bottom element iff its head-reduction terminates. Well-stratification is more technical, basically it means that the semantics of a λ-term can be stratified into different levels, expressing in the model the nesting of the head-normal forms defining the interaction between a λ-term and a context.

Our paper reconsiders these results in the setting of the probabilistic untyped λ-calculus Λ^+. The language extends the untyped λ-calculus with a barycentric

M. Bojańczyk and A. Simpson (Eds.): FOSSACS 2019, LNCS 11425, pp. 365–381, 2019.
https://doi.org/10.1007/978-3-030-17127-8_21

sum constructor allowing for terms like $M +_p N$, with $p \in [0,1]$, reducing to M with probability p and to N with probability $1-p$. In recent years there has been a renewed interest in Λ^+ as a core language for (untyped) discrete probabilistic functional programming. In particular, Leventis proves in [12] a separation property for Λ^+ based on a probabilistic version of *Nakajima trees*, the latter describing a nesting of sub-probability distributions of infinitary η-long head-normal forms (see Sect. 5 and the examples in Fig. 2).

We consider the semantics of Λ^+ given by the probabilistic coherence space $\mathcal{D}$ defined by Danos and Ehrhard in [5] and proved to be adequate in [6]. We show that the denotation $[\![M]\!]$ in $\mathcal{D}$ of a Λ^+ term M enjoys a kind of stratification property (Theorem 1, called here *strong adequacy*) and we use this property to prove that $[\![M]\!]$ is a faithful description of the probabilistic Nakajima tree of M (Corollary 1). As a consequence of this result and the previously mentioned separation theorem, we achieve full abstraction for $\mathcal{D}$ (Theorem 2), thus reconstructing in this setting Manzonetto's reasoning for classical λ-calculus.

Very recently, and independently from this work, Clairambault and Paquet also prove full abstraction for $\mathcal{D}$ [2]. Their proof uses a game semantics model representing in an abstract way the probabilistic Nakajima trees and a faithful functor from this game semantics to the weighted relational semantics of [11]. The latter provides a model having the same equational theory over Λ^+ as the probabilistic coherence space $\mathcal{D}$, so full abstraction for $\mathcal{D}$ follows immediately. By the way, let us emphasise that all results in our paper can be transferred as they are to the weighted relational semantics of [11]. We decided however to consider the probabilistic coherence space model in order to highlight the correspondence between the definition of $\mathcal{D}$ (Eq. (11)) and the definition of the logical relation (Eq. (13)) which is the key ingredient in the proof of our notion of stratification.

Let us give some more intuitions on this latter notion, which has an interest in its own. The model $\mathcal{D}$ is defined as the limit of a chain of probabilistic coherence spaces $(\mathcal{D}_\ell)_{\ell\in\mathbb{N}}$ approximating more and more the denotation of Λ^+ terms. The adequacy property proven in [6] states that the probability of a term M to converge to a head-normal form is given by the mass of the semantics $[\![M]\!]$ restricted to the subspace $\mathcal{D}_2$ [6, Theorem 22]. The natural question is then to understand which kind of operational meaning carries the rest of the mass of $[\![M]\!]$, i.e. the points of order greater than 2. Our Theorem 1 answers this question, showing that the semantics $[\![M]\!]$ distributes over the semantics of its head-normal forms according to the operational semantics of Λ^+. By iterating this reasoning one gets a stratification of $[\![M]\!]$ into a nesting of (η-expanded) head-normal forms which is the key ingredient linking $[\![M]\!]$ and the probabilistic Nakajima trees (Corollary 1).

The fact that our proof of full abstraction is based on the notion of strong adequacy makes very plausible that the proof can be adapted to a more general class of models than only probabilistic coherence spaces and weighted semantics. In particular, we would like to stress that we did not use the property of analyticity of term denotations, which is instead at the core of the proof of full abstraction for probabilistic PCF-like languages [7,8].

Notational convention. We write $\mathbb{N}$ for the set of natural numbers and $\mathbb{R}_{\geq 0}$ for the set of non-negative real numbers. Given any set X we write $\mathcal{M}_f(X)$ for the set of **finite multisets of** X: an element $m \in \mathcal{M}_f(X)$ is a function $X \to \mathbb{N}$ such that the **support** of m $\mathrm{Supp}(m) = \{x \in X \mid m(x) > 0\}$ is finite. We write $[x_1, \ldots, x_n]$ for the multiset m such that $m(x) =$ *number of indices* i *s.t.* $x = x_i$, so $[]$ is the empty multiset and $\uplus$ the disjoint union. The **Kronecker delta** over a set X is defined for $x, y \in X$ by: $\delta_{x,y} = 1$ if $x = y$, and $\delta_{x,y} = 0$ otherwise.

2 The Probabilistic Language Λ^+

We recall the call-by-name untyped probabilistic λ-calculus, following [6]. The set Λ^+ of terms over a set $\mathcal{V}$ of variables is defined inductively by:

$$M, N \in \Lambda^+ ::= x \mid \lambda x.M \mid MN \mid M +_p N, \tag{1}$$

where x ranges over $\mathcal{V}$ and p ranges over $[0, 1]$. Note that we consider probabilities over the whole interval $[0, 1]$ but our proofs still hold if we restrict them to rational numbers. We use the λ-calculus terminology and notations as in [1]: terms are considered modulo *α-equivalence*, i.e. variable renaming; we write $\mathrm{FV}(M)$ for the set of free variables of a term M. For any finite list of variables $\Gamma = x_1, \ldots, x_n$ we write Λ^+_Γ for the set of terms $M \in \Lambda^+$ such that $\mathrm{FV}(M) \subseteq \{x_1, \ldots, x_n\}$. Given two terms $M, N \in \Lambda^+$ and $x \in \mathcal{V}$ we write $M\{N/x\}$ for the term obtained by substituting N for the free occurrences of x in M, subject to the usual proviso of renaming bound variables of M to avoid capture of free variables in N.

Example 1. Some terms useful in giving examples: the duplicator $\delta = \lambda x.xx$, the Turing fixed point combinator $\boldsymbol{\Theta} = (\lambda xy.y(xxy))(\lambda xy.y(xxy))$ and $\boldsymbol{\Omega} = \delta\delta$.

A **context** $C[\,]$ is a term containing a single occurrence of a distinguished variable denoted $[\,]$ and called hole. A **head-context** is of the form $E[\,] = \lambda x_1 \ldots x_n.[\,]M_1 \ldots M_k$, for $n, k \geq 0$ and $M_i \in \Lambda^+$. Given $M \in \Lambda^+$, we write $C[M]$ for the term obtained by replacing M for the hole in $C[\,]$ possibly with capture of free variables. The **operational semantics** is given by a Markov chain over Λ^+, mixing together the standard head-reduction of untyped λ-calculus with the probabilistic choice $+_p$. Precisely, this system is given by the transition matrix Red in Eq. (2). It is well known that any Λ^+-term M can be uniquely decomposed into $E[R]$ for $E[\,]$ a head-context and R either a β-redex, or a $+_p$-redex (for some $p \in [0, 1]$) or a variable in $\mathcal{V}$. This gives the following cases:

$$\mathrm{Red}_{E[R],N} ::= \begin{cases} 1 & \text{if } R = (\lambda x.M')M'' \text{ and } N = E[M'\{M''/x\}] \\ p & \text{if } R = M' +_p M'', M' \neq M'' \text{ and } N = E[M'] \\ 1-p & \text{if } R = M' +_p M'', M' \neq M'' \text{ and } N = E[M''] \\ 1 & \text{if } R = M' +_p M' \text{ and } N = E[M'] \\ 1 & \text{if } R \in \mathcal{V} \text{ and } N = E[R] \\ 0 & \text{otherwise} \end{cases} \tag{2}$$

This matrix is stochastic, i.e. for any term M, $\sum_N \mathrm{Red}_{M,N} = 1$. A **head-normal form** is a term of the form $E[y]$, with $y \in \mathcal{V}$ called its **head-variable**. We write HNF for the set of all head-normal forms. Following [5,6], we consider the head-normal forms as absorbing states of the process. Hence the n-th power Red^n of the matrix Red describes the process of performing *exactly* n steps: $\mathrm{Red}^n_{M,N}$ is the probability that after n process steps M will reach state N.

Example 2. Let $L = (x +_p y)$, we have $\mathrm{Red}_{\delta L, LL} = 1$, and $\mathrm{Red}^n_{\delta L, xL} = p$, $\mathrm{Red}^n_{\delta L, yL} = 1 - p$ for all $n \geq 2$. In fact both xL and yL are head-normal forms, so absorbing states. The term $\boldsymbol{\Omega}$ β-reduces to itself, so $\mathrm{Red}^n_{\Omega,\Omega} = 1$ for any n, giving an example of absorbing state which is not a head-normal form.

The Turing fixed point combinator needs two β-steps to unfold its argument, so, for any term M, $\mathrm{Red}^2_{\boldsymbol{\Theta} M, M(\boldsymbol{\Theta} M)} = 1$. In the case M is a probabilistic function like $M = \lambda f.(f +_p y)$, we get $\mathrm{Red}^{4n}_{\boldsymbol{\Theta} M, \boldsymbol{\Theta} M} = p^n$ and $\mathrm{Red}^{4n}_{\boldsymbol{\Theta} M, y} = 1 - p^n$, for any n. In the case $M = \lambda f.(yf +_p y)$, we get: $\mathrm{Red}^{4(n+1)}_{\boldsymbol{\Theta} M, y^n(\boldsymbol{\Theta} M)} = p^{n+1}$ and $\mathrm{Red}^{4(n+1)}_{\boldsymbol{\Theta} M, y^n(y)} = (1-p)p^n$, where $y^n(...)$ denotes the n-fold application $y(\dots y(...))$.

Notice that for $h \in \mathrm{HNF}$ and $M \in \Lambda^+$, the sequence $\left(\mathrm{Red}^n_{M,h}\right)_{n\in\mathbb{N}}$ is monotone increasing and bounded by 1, so it converges. We define its limit by:

$$\forall M \in \Lambda^+, \forall h \in \mathrm{HNF},\ \mathrm{Red}^{\infty}_{M,h} ::= \sup_{n\in\mathbb{N}}\left(\mathrm{Red}^n_{M,h}\right) \in [0,1]. \tag{3}$$

This quantity gives the total probability of M to reduce to the head-normal form h in any number (possibly infinitely many) of finite reduction sequences.

Example 3. Recall the terms in Example 2. We have $\mathrm{Red}^{\infty}_{\delta L, xL} = p$ and $\mathrm{Red}^{\infty}_{\delta L, yL} = 1 - p$. For any $h \in \mathrm{HNF}$ and $n \in \mathbb{N}$ we have $\mathrm{Red}^n_{\boldsymbol{\Omega}, h} = 0$ so $\mathrm{Red}^{\infty}_{\boldsymbol{\Omega}, h} = 0$. The quantity $\mathrm{Red}^{\infty}_{\boldsymbol{\Theta}(\lambda f.(f +_p y)), y}$ is the first example of limit, being equal to 1 whereas $\mathrm{Red}^n_{\boldsymbol{\Theta}(\lambda f.(f +_p y)), y} < 1$ for all $n \in \mathbb{N}$. Operationally this means that the term $\boldsymbol{\Theta}(\lambda f.(f +_p y))$ reduces to y with probability 1 but the length of these reductions is not bounded. Finally, $\mathrm{Red}^{\infty}_{\boldsymbol{\Theta}(\lambda f.(yf +_p y)), y^n(y)} = (1-p)p^n$, this means that $\boldsymbol{\Theta}(\lambda f.(yf +_p y))$ converges with probability 1 but it can reach infinitely many different head-normal forms.

Given $M, N \in \Lambda^+$, we say that M is **contextually equivalent** to N if, and only if, $\forall C[\,]$, $\sum_{h\in\mathrm{HNF}} \mathrm{Red}^{\infty}_{C[M],h} = \sum_{h\in\mathrm{HNF}} \mathrm{Red}^{\infty}_{C[N],h}$.

An important property in the following is **extensionality**, meaning invariance under η-equivalence. The **η-equivalence** is the smallest congruence such that, for any $M \in \Lambda^+$ and $x \notin \mathrm{FV}(M)$ we have $M =_\eta \lambda x.Mx$. Notice that the contextual equivalence is extensional (see [1] for the classical λ-calculus).

3 Probabilistic Coherence Spaces

Girard introduced probabilistic coherence spaces (PCS) as a "quantitative refinement" of coherence spaces [9]. Danos and Ehrhard considered then the category

Pcoh of linear and Scott-continuous functions between PCS as a model of linear logic and the cartesian closed category $\mathbf{Pcoh}_!$ of entire functions between PCS as the Kleisli category associated with the comonad of **Pcoh** modelling the exponential modality [5]. They proved also that $\mathbf{Pcoh}_!$ provides an adequate model of probabilistic PCF and the reflexive object $\mathcal{D}$ which is our object of study.

The two categories **Pcoh** and $\mathbf{Pcoh}_!$ have been then studied in various papers. In particular, $\mathbf{Pcoh}_!$ is proved to be fully abstract for the call-by-name probabilistic PCF [7]. This result has been also extended to richer languages, e.g. call-by-push-value probabilistic PCF [8]. The untyped model $\mathcal{D}$ is proven adequate for Λ^+ [6]. This paper is the continuation of the latter result, showing full abstraction for $\mathcal{D}$ as a consequence of a stronger form of adequacy.

We briefly recall here the cartesian closed category $\mathbf{Pcoh}_!$ and the reflexive object $\mathcal{D}$. Because of space we omit to consider the linear logic model **Pcoh**, from which $\mathbf{Pcoh}_!$ is derived. We refer the reader to [5,6] for more details.

Probabilistic coherence spaces and entire functions. A **probabilistic coherence space**, or PCS for short, is a pair $\mathcal{X} = (|\mathcal{X}|, \mathrm{P}(\mathcal{X}))$ where $|\mathcal{X}|$ is a countable set called the **web** of $\mathcal{X}$ and $\mathrm{P}(\mathcal{X})$ is a subset of the semi-module $(\mathbb{R}_{\geq 0})^{|\mathcal{X}|}$ such that the following three conditions hold: (i) *closedness*: $\mathrm{P}(\mathcal{X})^{\perp\perp} = \mathrm{P}(\mathcal{X})$, where, given a set $P \subseteq (\mathbb{R}_{\geq 0})^{|\mathcal{X}|}$, the **dual of** P is defined as $P^{\perp} ::= \{y \in (\mathbb{R}_{\geq 0})^{|\mathcal{X}|} \mid \forall x \in P \ \sum_{a \in |\mathcal{X}|} x_a y_a \leq 1\}$; (ii) *boundedness*: $\forall a \in |\mathcal{X}|$, $\exists \mu > 0$, $\forall x \in \mathrm{P}(\mathcal{X})$, $x_a \leq \mu$; (iii) *completeness*: $\forall a \in |\mathcal{X}|$, $\exists x \in \mathrm{P}(\mathcal{X})$, $x_a > 0$.

Given $x, y \in \mathrm{P}(\mathcal{X})$, we write $x \leq y$ for the order defined pointwise, i.e. for every $a \in |\mathcal{X}|$, $x_a \leq y_a$. The closedness condition is equivalent to require that $\mathrm{P}(\mathcal{X})$ is convex and Scott-closed, as stated below.

Proposition 1 (e.g. [4]). *Given an index set I and a subset $P \subset (\mathbb{R}_{\geq 0})^I$ which is bounded and complete, we have $P = P^{\perp\perp}$ iff the following two conditions hold: (i) P is convex, i.e. for every $x, y \in P$ and $\lambda \in [0,1]$, $\lambda x + (1-\lambda) y \in P$; (ii) P is Scott-closed, i.e. for every $x \leq y \in P$, $x \in P$ and for every increasing chain $\{x_i\}_{i \in \mathrm{N}} \subseteq P$, $\sup_i x_i \in P$.*

A data-type is denoted by a PCS $\mathcal{X}$ and its data by vectors in $\mathrm{P}(\mathcal{X})$: convexity allows for probabilistic superposition and Scott-closedness for recursion.

Example 4. A simple example of PCS is $\mathcal{U} = (|\mathcal{U}|, \mathrm{P}(\mathcal{U}))$ with $|\mathcal{U}|$ a singleton set and $\mathrm{P}(\mathcal{U}) = [0,1]$. Notice $\mathrm{P}(\mathcal{U})^{\perp} = \mathrm{P}(\mathcal{U})$. This PCS gives the flat interpretation of the unit type in a typed language. The boolean type is denoted by the two dimensional PCS $\mathcal{B} ::= (\{\mathtt{t}, \mathtt{f}\}, \{(\rho_{\mathtt{t}}, \rho_{\mathtt{f}}) \mid \rho_{\mathtt{t}} + \rho_{\mathtt{f}} \leq 1\})$. Notice that $\mathrm{P}(\mathcal{B})$ can be seen as the set of the probabilistic sub-distributions of the boolean values.

As soon as one consider functional types, the intuitive notion of (discrete) sub-probabilistic distribution is lost. In particular, the reflexive object $\mathcal{D}$ defined below is an example of an infinite dimensional PCS where scalars arbitrarily big may appear in $\mathrm{P}(\mathcal{D})$. One can think of PCS's as a generalisation of the notion of discrete sub-probabilistic distributions allowing a cartesian closed category.

An **entire function from $\mathcal{X}$ to $\mathcal{Y}$** is a matrix $f \in \mathbb{R}_{\geq 0}{}^{\mathcal{M}_f(|\mathcal{X}|)\times|\mathcal{Y}|}$ such that for any $x \in \mathrm{P}(\mathcal{X})$, the image $f(x)$ under f belongs to $\mathrm{P}(\mathcal{Y})$, where $f(x)$ is

$$f(x) ::= \left(\sum_{m\in\mathcal{M}_f(|\mathcal{X}|)} f_{m,b}x^m \right)_{b\in|\mathcal{Y}|} \qquad \text{where } x^m ::= \prod_{a\in\mathrm{Supp}(m)} x_a^{m(a)} \tag{4}$$

Notice that the condition $f(x) \in \mathrm{P}(\mathcal{Y})$ requires that the possibly infinite sum in the previous equation must converge. Recently, Crubillé proves that the entire maps can be characterised independently from their matrix representation as the absolutely monotonic and Scott-continuous maps between PCS's, see [3].

The cartesian closed category. The Kleisli category $\mathbf{Pcoh}_!$ has PCS's as objects and entire maps as morphisms. Given $f \in \mathbf{Pcoh}_!(\mathcal{X}, \mathcal{Y})$ and $g \in \mathbf{Pcoh}_!(\mathcal{Y}, \mathcal{Z})$, the **composition** $g \circ f$ is the usual functional composition, whose matrix can be explicitly given by, for $m \in \mathcal{M}_f(|\mathcal{X}|), c \in |\mathcal{Z}|$:

$$(g\circ f)_{m,c} ::= \sum_{p\in\mathcal{M}_f(|\mathcal{Y}|)} g_{p,c} f^{(m,p)} \qquad \text{where } f^{(m,[b_1,\dots,b_n])} ::= \sum_{\substack{(m_1,\dots,m_n)\\ \text{s.t. } m=\biguplus m_i}} \prod_{i=1}^{n} f_{m_i,b_i} \tag{5}$$

The boundedness condition over $\mathcal{Z}$ and the completeness condition over $\mathcal{X}$ ensure that the possibly infinite sum over $p \in \mathcal{M}_f(|\mathcal{Y}|)$ in Eq. (5) converges. The **identity** is the matrix $\mathrm{id}^{\mathcal{X}}_{m,a} = \delta_{[a],a}$, where δ is the Kronecker delta.

The **cartesian product** of any countable family $(\mathcal{X}_i)_{i\in I}$ of PCS's is:

$$\begin{aligned} \left|\textstyle\prod_{i\in I}\mathcal{X}_i\right| &::= \textstyle\bigcup_{i\in I}\{i\}\times|\mathcal{X}_i|, \\ \mathrm{P}\left(\textstyle\prod_{i\in I}\mathcal{X}_i\right) &::= \{x \in (\mathbb{R}_{\geq 0})^{|\prod_{i\in I}\mathcal{X}_i|} \mid \forall i\in I, \pi_i(x)\in\mathrm{P}(\mathcal{X}_i)\}, \end{aligned} \tag{6}$$

where $\pi_i(x)$ is the vector in $(\mathbb{R}_{\geq 0})^{|\mathcal{X}_i|}$ denoting the i-th component of x, i.e. $\pi_i(x)_a ::= x_{(i,a)}$. This means that $\mathrm{P}\left(\prod_{i\in I}\mathcal{X}_i\right)$ can be seen as the set-theoretical product $\prod_{i\in I}\mathrm{P}(\mathcal{X}_i)$, by mapping $x \in \mathrm{P}\left(\prod_{i\in I}\mathcal{X}_i\right)$ to the sequence $(\pi_i(x))_{i\in I}$. The j-th projection $\mathrm{pr}^j \in \mathbf{Pcoh}_!(\prod_{i\in I}\mathcal{X}_i, \mathcal{X}_j)$ is defined by $\mathrm{pr}^j_{m,b} ::= \delta_{m,[(j,b)]}$. If all components of a product are equal to a PCS $\mathcal{X}$ we can use the exponential notation $\mathcal{X}^I$. Binary products can be written as $\mathcal{X}\times\mathcal{Y}$. In the following, we will often denote the finite multisets in $\mathcal{M}_f\left(\left|\prod_{i\in I}\mathcal{X}_i\right|\right)$ as I-families of finite multisets almost everywhere empty, using the set-theoretical isomorphism:[1]

$$\mathcal{M}_f\left(\left|\prod_{i\in I}\mathcal{X}_i\right|\right) \quad\simeq\quad \{\boldsymbol{m} \in \prod_{i\in I}\mathcal{M}_f(|\mathcal{X}_i|) \mid \mathrm{Supp}(\boldsymbol{m}) \text{ finite}\}. \tag{7}$$

For example, the multi-set $[(0,a),(0,a'),(1,b)] \in \mathcal{M}_f(|\mathcal{X}\times\mathcal{Y}|)$ will be denoted as the pair $([a,a'],[b])$, or the multiset $[(2,a),(4,a'),(4,a'')] \in \mathcal{M}_f\left(\left|\prod_{n\in\mathbb{N}}\mathcal{X}_n\right|\right)$ as the almost everywhere empty sequence $([],[],[a],[],[a',a''],[],\dots)$.

[1] In fact, this isomorphism corresponds, for I finite, to the fundamental exponential isomorphism $!(A \,\&\, B) \simeq\, !A \otimes !B$ of linear logic.

The **object of morphisms** from $\mathcal{X}$ to $\mathcal{Y}$ is $\mathbf{Pcoh}_!(\mathcal{X},\mathcal{Y})$ itself, i.e.:

$$|\mathcal{X} \Rightarrow \mathcal{Y}| ::= \mathcal{M}_f(|\mathcal{X}|) \times |\mathcal{Y}|, \quad \mathrm{P}(\mathcal{X} \Rightarrow \mathcal{Y}) ::= \mathbf{Pcoh}_!(\mathcal{X},\mathcal{Y}). \tag{8}$$

The proof that $\mathrm{P}(\mathcal{X} \Rightarrow \mathcal{Y})$ so defined enjoys the closedness, completeness and boundedness conditions of the definition of a PCS is not trivial and it is argued by the fact that $\mathbf{Pcoh}_!$ is the Kleisli category associated with the exponential comonad of the linear logic model **Pcoh** mentioned in the introduction.

The **evaluation** $\mathrm{Ev}^{\mathcal{X},\mathcal{Y}} \in \mathbf{Pcoh}_!((\mathcal{X} \Rightarrow \mathcal{Y}) \times \mathcal{X}, \mathcal{Y})$ and the **curryfication** $\mathrm{Cur}^{\mathcal{X},\mathcal{Z},\mathcal{Y}}(v) \in \mathbf{Pcoh}_!(\mathcal{Z}, \mathcal{X} \Rightarrow \mathcal{Y})$ of a morphism $v \in \mathbf{Pcoh}_!(\mathcal{X} \times \mathcal{Z}, \mathcal{Y})$ are:

$$\mathrm{Ev}^{\mathcal{X},\mathcal{Y}}_{(m,p),a} ::= \delta_{m,[(p,a)]}, \qquad \mathrm{Cur}^{\mathcal{X},\mathcal{Z},\mathcal{Y}}(v)_{m,(p,a)} ::= v_{(p,m),a}. \tag{9}$$

The reflexive object $\mathcal{D}$. We set $\mathcal{X} \subseteq \mathcal{Y}$ whenever $|\mathcal{X}| \subseteq |\mathcal{Y}|$ and $\mathrm{P}(\mathcal{X}) = \{v|_{|\mathcal{X}|} \text{ s.t. } v \in \mathrm{P}(\mathcal{Y})\}$, where $v|_{|\mathcal{X}|}$ is the vector in $\mathbb{R}^{|\mathcal{X}|}_{\geq 0}$ obtained by restricting $v \in \mathbb{R}^{|\mathcal{Y}|}_{\geq 0}$ to the indexes in $|\mathcal{X}| \subseteq |\mathcal{Y}|$. This defines a complete order over PCS's. The model $\mathcal{D}$ of Λ^+ is then given by the least fix-point of the Scott-continuous functor $\mathcal{X} \mapsto \mathcal{X}^{\mathbb{N}} \Rightarrow \mathcal{U}$ (where $\mathcal{U}$ is the one-dimensional PCS defined in Example 4). We do not detail here its definition, but we give explicitly the chain $\mathcal{D}_0 = (\emptyset, \mathbf{0})$, $\mathcal{D}_{\ell+1} = \mathcal{D}_\ell^{\mathbb{N}} \Rightarrow \mathcal{U}$ whose (co)limit is the least fix-point $\mathcal{D}$ of $\mathcal{X} \mapsto \mathcal{X}^{\mathbb{N}} \Rightarrow \mathcal{U}$ by the Knaster-Tarski theorem. We refer to [5, Sect. 2] for details.

The webs of these spaces are given by:

$$|\mathcal{D}_0| ::= \emptyset, \quad |\mathcal{D}_{\ell+1}| ::= \mathcal{M}_f(|\mathcal{D}_\ell|)^{(\omega)}, \quad |\mathcal{D}| ::= \bigcup_{\ell \in \mathbb{N}} |\mathcal{D}_\ell| \tag{10}$$

where $\mathcal{M}_f(|\mathcal{D}_\ell|)^{(\omega)}$ denotes the set of infinite sequences of multisets of $|\mathcal{D}_\ell|$ that are almost everywhere empty (notice we are using the isomorphism mentioned in Eq. (7)). The set $|\mathcal{D}_1|$ is the singleton containing the infinite sequence $([],[],[]\dots)$ of empty multisets, which we denote by $\star$. Given a multiset $m \in \mathcal{M}_f(|\mathcal{D}_\ell|)$ and a sequence $d \in \mathcal{M}_f(|\mathcal{D}_{\ell+1}|)$, we denote by $m :: d$ the element of $|\mathcal{D}_{\ell+1}|$ having at first position m and then all the multisets of d shifted by one position. Notice that any element of $|\mathcal{D}_{\ell+1}|$ can be written as $m_1 :: \dots m_n :: \star$ for an n sufficiently large and $m_1, \dots, m_n \in \mathcal{M}_f(|\mathcal{D}_\ell|)$. In particular, $[] :: \star = \star$.[2]

The sets of vectors $\mathrm{P}(\mathcal{D}_\ell)$ and $\mathrm{P}(\mathcal{D})$ completing the definition of a PCS are:

$$\begin{aligned}
\mathrm{P}(\mathcal{D}_0) &::= \mathbf{0} \\
\mathrm{P}(\mathcal{D}_{\ell+1}) &::= \left\{ v \in (\mathbb{R}_{\geq 0})^{|\mathcal{D}_{\ell+1}|} \text{ s.t. } \begin{array}{c} \forall n \in \mathbb{N}, \forall u_1, \dots, u_n \in \mathrm{P}(\mathcal{D}_\ell) \\ \displaystyle\sum_{\substack{m_1,\dots,m_n \in \\ \mathcal{M}_f(|\mathcal{D}_\ell|)}} v_{m_1 :: \dots m_n :: \star} u_1^{m_1} \dots u_n^{m_n} \leq 1 \end{array} \right\} \\
\mathrm{P}(\mathcal{D}) &::= \left\{ v \in (\mathbb{R}_{\geq 0})^{|\mathcal{D}|} \text{ s.t. } \forall \ell \in \mathbb{N}, v|_{|\mathcal{D}_\ell|} \in \mathrm{P}(\mathcal{D}_\ell) \right\}
\end{aligned} \tag{11}$$

The above definition of $\mathrm{P}(\mathcal{D}_{\ell+1})$ is actually equivalent to the standard one inferred from the definition of the countable product $\mathcal{D}^{\mathbb{N}}$, which would require

[2] The elements of $|\mathcal{D}|$ can be seen as intersection types generated from the constant $\star$, the :: operation being the arrow and multisets non-idempotent intersections.

$$[\![x]\!]^{\Gamma}_{\boldsymbol{m},d} = \begin{cases} 1 & \boldsymbol{m}_x = [d] \text{ and } \forall y \in \Gamma \setminus x, \boldsymbol{m}_y = [], \\ 0 & \text{otherwise} \end{cases},$$

$$[\![\lambda x.M]\!]^{\Gamma}_{\boldsymbol{m},m::d} = [\![M]\!]^{x,\Gamma}_{(m,\boldsymbol{m}),d},$$

$$[\![MN]\!]^{\Gamma}_{\boldsymbol{m},d} = \sum_{m \in \mathcal{M}_f(|\mathcal{D}|)} \sum_{\substack{(\boldsymbol{m}_1,\boldsymbol{m}_2) \text{ s.t.} \\ \forall x \in \Gamma, \boldsymbol{m}_x = \boldsymbol{m}_{1x} \uplus \boldsymbol{m}_{2x}}} [\![M]\!]^{\Gamma}_{\boldsymbol{m}_1,m::d}([\![N]\!]^{\Gamma})^{\boldsymbol{m}_2,m},$$

$$[\![M +_p N]\!]^{\Gamma}_{\boldsymbol{m},d} = p[\![M]\!]^{\Gamma}_{\boldsymbol{m},d} + (1-p)[\![N]\!]^{\Gamma}_{\boldsymbol{m},d}.$$

Fig. 1. Explicit definition of the denotation of a term in Λ^+_Γ as a matrix in $\mathrm{P}(\mathcal{D}^\Gamma \Rightarrow \mathcal{D})$. Recall Eq. (5) for the notation $([\![N]\!]^\Gamma)^{\boldsymbol{m}_2,m}$.

to apply v to a countable family $(u_i)_{i\in\mathbb{N}}$ of vectors in $\mathrm{P}(\mathcal{D}_\ell)$. The two definitions are equivalent because of the continuity of the scalar multiplication and the sum.

It happens that any solution of $\mathcal{X} = \mathcal{X}^{\mathbb{N}} \Rightarrow \mathcal{U}$ gives also a solution (although not minimal) to $\mathcal{X} = \mathcal{X} \Rightarrow \mathcal{X}$ and hence a reflexive object of $\mathbf{Pcoh}_!$. The isomorphism pair $\lambda \in \mathbf{Pcoh}_!(\mathcal{D} \Rightarrow \mathcal{D}, \mathcal{D})$ and $\mathsf{app} \in \mathbf{Pcoh}_!(\mathcal{D}, \mathcal{D} \Rightarrow \mathcal{D})$ is given by, for any $p \in \mathcal{M}_f(|\mathcal{D} \Rightarrow \mathcal{D}|)$, $m, q \in \mathcal{M}_f(|\mathcal{D}|)$, and $d \in |\mathcal{D}|$,

$$\lambda_{p,m::d} ::= \delta_{p,[(m,d)]}, \qquad \mathsf{app}_{q,(m,d)} ::= \delta_{q,[m::d]}. \tag{12}$$

It is easy to check that $\mathsf{app} \circ \lambda = \mathrm{id}^{\mathcal{D}\Rightarrow\mathcal{D}}$ and $\lambda \circ \mathsf{app} = \mathrm{id}^{\mathcal{D}}$, so $(\mathcal{D}, \lambda, \mathsf{app})$ yields an extensional model of untyped λ-calculus, i.e. $[\![M]\!] = [\![N]\!]$ whenever $M =_\eta N$.

Interpretation of the Terms of Λ^+. Given a term M and a list Γ of pairwise different variables containing $\mathrm{FV}(M)$, the interpretation of M is a morphism $[\![M]\!]^\Gamma \in \mathbf{Pcoh}_!(\mathcal{D}^\Gamma, \mathcal{D})$, i.e. a matrix in $\mathbb{R}_{\geq 0}^{\mathcal{M}_f(|\mathcal{D}^\Gamma|)\times|\mathcal{D}|} = \mathbb{R}_{\geq 0}^{\mathcal{M}_f(|\mathcal{D}|)^\Gamma\times|\mathcal{D}|}$. The definition of $[\![M]\!]^\Gamma$ is the standard one determined by the cartesian closed structure of $\mathbf{Pcoh}_!$ and the reflexive object $(\mathcal{D}, \lambda, \mathsf{app})$: $[\![x]\!]^\Gamma$ is the x-th projection of the product $\mathcal{D}^\Gamma$, $[\![\lambda x.M]\!]^\Gamma = \lambda \circ \mathrm{Cur}\left([\![M]\!]^{x,\Gamma}\right)$ and $[\![MN]\!]^\Gamma = \mathrm{Ev} \circ \langle \mathsf{app} \circ [\![M]\!]^\Gamma, [\![N]\!]^\Gamma \rangle$, where $\langle\ ,\ \rangle$ is the cartesian product of two morphisms. Figure 1 makes explicit the coefficients of the matrix $[\![M]\!]^\Gamma$ by structural induction on M. The only non-standard operation is the barycentric sum $[\![M +_p N]\!]$ which is still a morphism of $\mathbf{Pcoh}_!$ by the convexity of $\mathrm{P}(\mathcal{D}^\Gamma \Rightarrow \mathcal{D})$ (Proposition 1).

Proposition 2 (Soundness, [5,6]**).** *For every term* $M \in \Lambda^+$ *and sequence* $\Gamma \supseteq \mathrm{FV}(M)$*:* $[\![M]\!]^\Gamma = \sum_{N\in\Lambda^+} \mathrm{Red}_{M,N}[\![N]\!]^\Gamma$.

4 Strong Adequacy

In this section we state and prove Theorem 1, enhancing the $\mathbf{Pcoh}_!$ adequacy property given in [6]. This latter explains the computational meaning of the mass of $[\![M]\!]$ restricted to $\mathcal{D}_2 \subseteq \mathcal{D}$, while our generalisation considers the whole $[\![M]\!]$, showing that it encodes the way the operational semantics dispatches the mass into the denotation of the head-normal forms. As in [6], the proof of Theorem 1 adapts a method introduced by Pitts [14], consisting in building a recursively specified relation of formal approximation $\lhd$ (Proposition 3) which satisfies the same recursive equation as $\mathcal{D}$. However, our generalisation requires a subtler definition of $\lhd$ with respect to [6]. In particular, we must consider open terms in order to prove Lemma 7.

The approximation relation. Let us introduce some convenient notation, extending the definition of λ-abstraction and application to general morphisms.

Definition 1. *Given* $v \in \mathrm{P}(\mathcal{D}^{x,\Gamma} \Rightarrow \mathcal{D})$, *let* $\Lambda(v)$ *be the vector* $\lambda \circ \mathrm{Cur}(v) \in \mathrm{P}(\mathcal{D}^{\Gamma} \Rightarrow \mathcal{D})$. *Given* $v, u \in \mathrm{P}(\mathcal{D}^{\Gamma} \Rightarrow \mathcal{D})$ *let* $v @ u$ *be the vector* $\mathrm{Ev} \circ \langle \mathsf{app} \circ v, u\rangle \in \mathrm{P}(\mathcal{D}^{\Gamma} \Rightarrow \mathcal{D})$. *Finally, given a finite sequence* $u_1, \ldots, u_n \in \mathrm{P}(\mathcal{D}^{\Gamma} \Rightarrow \mathcal{D})$, *for* $n \in \mathbb{N}$, *we denote by* $v @ u_1 \ldots u_n$ *the vector* $(v @ u_1) @ \ldots u_n$.

Lemma 1. *The map* $v \mapsto \Lambda(v)$ *is linear, i.e. for any vectors* v, v' *and scalars* $p, p' \in [0,1]$ *such that* $p + p' \leq 1$, *we have* $\Lambda(pv + p'v') = p\Lambda(v) + p'\Lambda(v')$, *and Scott-continuous, i.e. for any countable increasing chain* $(v_n)_{n\in\mathbb{N}}$, $\Lambda(\sup_n(v_n)) = \sup_n(\Lambda(v_n))$. *The map* $(v, u_1, \ldots, u_n) \mapsto v @ u_1 \ldots u_n$ *is Scott-continuous on all of its arguments but linear only on its first argument* v.

Proof. Scott-continuity is because the scalar multiplication and the sum are Scott-continuous. The linearity is because the matrices app, λ are associated with linear maps (namely, they have non-zero coefficients only on singleton multisets, see (12)) as well as the left-most component of Ev, see (9). $\square$

For any $\Gamma \subseteq \Delta$ there exists the projection $\mathrm{pr} : \mathrm{P}(\mathcal{D})^{\Delta} \to \mathrm{P}(\mathcal{D})^{\Gamma}$. Then, given a matrix $v \in \mathrm{P}(\mathcal{D}^{\Gamma} \Rightarrow \mathcal{D})$ we denote by $v{\uparrow^{\Delta}} \in \mathrm{P}(\mathcal{D}^{\Delta} \Rightarrow \mathcal{D})$ the matrix corresponding to the pre-composition of the morphism associated with v with pr. This can be explicitly defined by, for $\boldsymbol{m} \in \mathcal{M}_{\mathrm{f}}(|\mathcal{D}|)^{\Delta}$, $d \in |\mathcal{D}|$, $(v{\uparrow^{\Delta}})_{\boldsymbol{m},d} = v_{(\boldsymbol{m}_x)_{x\in\Gamma},d}$ if $\forall y \in \Delta \setminus \Gamma, \boldsymbol{m}_y = []$, and $(v{\uparrow^{\Delta}})_{\boldsymbol{m},d} = 0$ otherwise.

We define an operation ϕ acting on the relations $R \subseteq \bigcup_{\Gamma} (\mathrm{P}(\mathcal{D}^{\Gamma} \Rightarrow \mathcal{D}) \times \Lambda^{+}_{\Gamma})$. Each component $\phi^{\Gamma}(R) \subseteq (\mathrm{P}(\mathcal{D}^{\Gamma} \Rightarrow \mathcal{D})) \times \Lambda^{+}_{\Gamma}$ is given by:

$$\begin{aligned}(v, M) \in \phi^{\Gamma}(R) \text{ iff } & \forall \Delta \supseteq \Gamma, \forall n \in \mathbb{N}, \forall u_1, \ldots, u_n \in \mathrm{P}(\mathcal{D}^{\Delta} \Rightarrow \mathcal{D}) \\ & \forall N_1, \ldots, N_n \in \Lambda^{+}_{\Delta}, \text{ s.t. } (u_i, N_i) \in R \text{ for all } i \leq n, \\ & v{\uparrow^{\Delta}} @ u_1 \ldots, u_n \leq \textstyle\sum_{h \in \mathrm{HNF}_{\Delta}} \mathrm{Red}^{\infty}_{M\, N_1 \ldots N_n, h} [\![h]\!]^{\Delta}.\end{aligned} \tag{13}$$

The above definition is similar to Eq. (11), giving $\mathcal{D}_{\ell+1}$ from $\mathcal{D}_{\ell}$. In the following we look for a fixed-point of ϕ (Proposition 3). Its quest is not simple because ϕ is not monotone. We derive then from ϕ a monotone operator ψ on a larger space, and we compute its fixed-point by using Tarski's Theorem (Lemma 3).

Given $(R^+, R^-) \in \mathcal{P}\left(\bigcup_{\Gamma} (\mathrm{P}(\mathcal{D}^{\Gamma} \Rightarrow \mathcal{D}) \times \Lambda^{+}_{\Gamma})\right)^2$, we define $\psi(R^+, R^-) = (\phi(R^-), \phi(R^+))$. Given two such pairs $(R_1^+, R_1^-), (R_2^+, R_2^-)$, we define $(R_1^+, R_1^-) \sqsubseteq (R_2^+, R_2^-)$ iff $R_1^+ \subseteq R_2^+$ and $R_1^- \supseteq R_2^-$.

Lemma 2. *The relation $\sqsubseteq$ is an order relation giving a complete lattice on* $\mathcal{P}\left(\bigcup_\Gamma \left(\mathrm{P}(\mathcal{D}^\Gamma \Rightarrow \mathcal{D}) \times \Lambda_\Gamma^+\right)\right)^2$.

Thanks to the previous lemma, we set $(\lhd^+, \lhd-)$ as the glb of the set $\{(R^+, R^-) \mid \psi(R^+, R^-) \sqsubseteq (R^+, R^-)\}$ of the pre-fixed points of ψ.

Lemma 3. $\psi(\lhd^+, \lhd^-) = (\lhd^+, \lhd^-)$, *so* $\lhd^+ = \phi(\lhd^-)$ *and* $\lhd^- = \phi(\lhd^+)$.

Proof. One can check that ψ is monotone increasing wrt $\sqsubseteq$, so the result follows from Tarski's Theorem on fixed points. □

Lemma 4. *For any* $R \subseteq \bigcup_\Gamma \left(\mathrm{P}(\mathcal{D}^\Gamma \Rightarrow \mathcal{D}) \times \Lambda_\Gamma^+\right)$ *and* $M \in \Lambda_\Gamma^+$, *the set* $\{v \in \mathrm{P}(\mathcal{D}^\Gamma \Rightarrow \mathcal{D}) \mid (v, M) \in \phi^\Gamma(R)\}$ *contains* 0, *is downward closed and chain closed.*

Proof. Consequence of the fact that the application $v \mathbin{@} u_1 \ldots u_n$ and the lifting $v{\uparrow}^\Delta$ are Scott-continuous (Lemma 1). Also, $v{\uparrow}^\Delta$ is linear as well as $v \mathbin{@} u_1 \ldots u_n$ on its left argument v (always Lemma 1), so $0{\uparrow}^\Delta \mathbin{@} u_1 \ldots u_n = 0$. □

Proposition 3. *We have* $\lhd^+ = \lhd^-$. *From now on we denote it simply by* $\lhd$. *We note* $\lhd^\Gamma$ *its component on* $\left(\mathrm{P}(\mathcal{D}^\Gamma \Rightarrow \mathcal{D})\right) \times \Lambda_\Gamma^+$.

Proof. First $(\lhd^-, \lhd^+)$ is a (pre-)fixed point of ψ so $(\lhd^+, \lhd^-) \sqsubseteq (\lhd^-, \lhd^+)$, i.e. $\lhd^+ \subseteq \lhd^-$. To prove the converse, we reason by induction on $|\mathcal{D}|$. For $v \in \mathrm{P}(\mathcal{D}^\Gamma \Rightarrow \mathcal{D})$ and $\ell \in \mathbb{N}$, we note $v_{|\ell}$ its restriction to $\left|\mathcal{D}^\Gamma \Rightarrow \mathcal{D}_\ell\right|$, i.e.: $(v_{|\ell})_{m,d} = v_{m,d}$ if $d \in |\mathcal{D}_\ell|$, and $(v_{|\ell})_{m,d} = 0$ otherwise. Notice that $v_{|\ell}$ is a morphism $\mathrm{P}(\mathcal{D}^\Gamma \Rightarrow \mathcal{D})$, since $v_{|\ell} \leq v \in \mathrm{P}(\mathcal{D}^\Gamma \Rightarrow \mathcal{D})$. We prove by induction on ℓ that:

$$\forall v \in \mathrm{P}(\mathcal{D}^\Gamma \Rightarrow \mathcal{D}), \forall M \in \Lambda_\Gamma^+, (v, M) \in \lhd^- \text{ implies } (v_{|\ell}, M) \in \lhd^+.$$

For $\ell = 0$ we have $v_{|0} = 0$ so by Lemma 4 $(v_{|0}, M) \in \lhd^+ = \phi(\lhd^-)$. At level $\ell + 1$ we want to prove $(v_{|\ell+1}, M) \in \lhd^+ = \phi(\lhd^-)$. Let $\Delta \supseteq \Gamma$, $u_1, \ldots, u_n \in \mathrm{P}(\mathcal{D}^\Delta \Rightarrow \mathcal{D})$, $N_1, \ldots, N_n \in \Lambda_\Delta^+$ such that for all $i \leq n$, $(u_i, N_i) \in \lhd^-$. By induction hypothesis we have $((u_i)_{|\ell}, N_i) \in \lhd^+$ for all $i \leq n$. Besides by hypothesis $(v, M) \in \lhd^- = \phi(\lhd^+)$ and we have $v_{|\ell+1} \leq v$ so Lemma 4 gives $(v_{|\ell+1}, M) \in \phi(\lhd^+)$. Hence $v_{|\ell+1}{\uparrow}^\Delta \mathbin{@} (u_1)_{|\ell} \cdots (u_n)_{|\ell} \leq \sum_{h \in \mathrm{HNF}_\Delta} \mathrm{Red}^\infty_{MN_1 \ldots N_n, h} [\![h]\!]^\Delta$. We conclude by observing that $v_{|\ell+1}{\uparrow}^\Delta \mathbin{@} (u_1)_{|\ell} \cdots (u_n)_{|\ell} = v_{|\ell+1}{\uparrow}^\Delta \mathbin{@} u_1 \ldots u_n$.

Now if $(v, M) \in \lhd^-$ then for all $\ell \in \mathbb{N}$, $(v_{|\ell}, M) \in \lhd^+$, but we have $v = \sup_{\ell \in \mathbb{N}} v_{|\ell}$ so Lemma 4 gives $(v, M) \in \lhd^+$. □

The key lemma. Lemma 9 is the so-called key-lemma for the relation $\lhd$. The reasoning is standard, except for the proof of Lemma 8 allowing strong adequacy.

Lemma 5. *For* $M \in \Lambda_{x,\Gamma}^+$, $N \in \Lambda_\Gamma^+$, $(v, (\lambda x.M)N) \in \lhd^\Gamma$ *iff* $(v, M\{N/x\}) \in \lhd^\Gamma$.

Proof. Observe that for all $n \in \mathbb{N}$, $N_1, \ldots, N_n \in \Lambda^+$ and $h \in \mathrm{HNF}$ we have $\mathrm{Red}^\infty_{(\lambda x.M)NN_1 \ldots N_n, h} = \mathrm{Red}^\infty_{M\{N/x\}N_1 \ldots N_n, h}$. □

Lemma 6. *Let* (v, M) *and* (r, L) *in* $\lhd^\Gamma$, *then* $(pv + (1-p)r, M +_p L) \in \lhd^\Gamma$.

Proof. Simply observe that for all $h \in \mathrm{HNF}$ and $N_1, \ldots, N_n \in \Lambda^+$ we have $\mathrm{Red}^{\infty}_{(M+_pL)N_1\ldots N_n,h} = p\mathrm{Red}^{\infty}_{MN_1\ldots N_n,h} + (1-p)\mathrm{Red}^{\infty}_{LN_1\ldots N_n,h}$. □

Lemma 7. *For all $x \in \Gamma$, $(\mathrm{pr}^{\Gamma}_x, x) \in \lhd^{\Gamma}$.*

Proof. Given any $\Delta \supseteq \Gamma$, $n \in \mathbb{N}$ and $(u_1, N_1), \ldots, (u_n, N_n) \in \lhd^{\Delta}$, we have:

$$\sum_{h \in \mathrm{HNF}_{\Delta}} \mathrm{Red}^{\infty}_{xN_1\ldots N_n,h} [\![h]\!]^{\Delta} = [\![xN_1 \ldots N_n]\!]^{\Delta} = \mathrm{pr}^{\Delta}_x @ [\![N_1]\!]^{\Delta} \ldots [\![N_n]\!]^{\Delta}$$

Besides for all $i \leq n$, as $(u_i, N_i) \in \lhd^{\Delta}$ we have $u_i \leq \sum_{h\in\mathrm{HNF}_{\Delta}} \mathrm{Red}^{\infty}_{N_i,h}[\![h]\!]^{\Delta} \leq [\![N_i]\!]^{\Delta}$. The latter inequality is because Proposition 2 implies that for all $k \in \mathbb{N}$, $\sum_{h\in\mathrm{HNF}_{\Delta}} \mathrm{Red}^{k}_{N_i,h}[\![h]\!] \leq [\![N_i]\!]$. The application @ being increasing in both its arguments we have $\mathrm{pr}^{\Gamma}_x {\uparrow}^{\Delta} @ u_1 \ldots u_n \leq \mathrm{pr}^{\Delta}_x @ [\![N_1]\!]^{\Delta} \ldots [\![N_n]\!]^{\Delta}$. □

Lemma 8. *Let $(v, M) \in \big(\mathrm{P}(\mathcal{D}^{\Gamma} \Rightarrow \mathcal{D})\big) \times \Lambda^+_{\Gamma}$, we have $(v, M) \in \lhd^{\Gamma}$ iff for all $(r, L) \in \lhd^{\Delta}$ with $\Delta \supseteq \Gamma$, $(v{\uparrow}^{\Delta} @ r, ML) \in \lhd^{\Delta}$.*

Proof. If $(v, M) \in \lhd^{\Gamma} = \phi^{\Gamma}(\lhd)$ and $(r, L) \in \lhd^{\Delta}$ then using the definition of ϕ it is easy to check that $(v{\uparrow}^{\Delta} @ r, ML) \in \lhd^{\Delta}$. Conversely if for all $(r, L) \in \lhd^{\Delta}$ we have $(v{\uparrow}^{\Delta} @ r, ML) \in \lhd^{\Delta}$ and we want to prove that $(v, M) \in \phi^{\Gamma}(\lhd)$ then the conditions of Eq. (13) trivially holds whenever $n \geq 1$, so we need to consider only the case for $n = 0$.

Suppose that for all $(r, L) \in \lhd^{\Delta}$, $(v{\uparrow}^{\Delta} @ r, ML) \in \lhd^{\Delta}$, let us prove that $v \leq \sum_{h\in\mathrm{HNF}_{\Gamma}} \mathrm{Red}^{\infty}_{M,h}[\![h]\!]^{\Gamma}$. Let x be a fresh variable, according to Lemma 7 we have $(\mathrm{pr}^{x,\Gamma}_x, x) \in \lhd^{x,\Gamma}$ so $v{\uparrow}^{x,\Gamma} @ \mathrm{pr}^{x,\Gamma}_x \leq \sum_{h\in\mathrm{HNF}_{x,\Gamma}} \mathrm{Red}^{\infty}_{Mx,h}[\![h]\!]^{x,\Gamma}$. Then:

$$\begin{aligned}
v &= \Lambda(v{\uparrow}^{x,\Gamma} @ \mathrm{pr}^{x,\Gamma}_x) && \text{extensionality of } \mathcal{D}\\
&\leq \Lambda\Big(\sum_{h\in\mathrm{HNF}_{x,\Gamma}} \mathrm{Red}^{\infty}_{Mx,h}[\![h]\!]^{x,\Gamma}\Big) && \text{monotonicity } \Lambda(\,), \text{Lemma 1}\\
&= \sum_{h\in\mathrm{HNF}_{x,\Gamma}} \mathrm{Red}^{\infty}_{Mx,h}\Lambda([\![h]\!]^{x,\Gamma}) && \text{linearity and contin.} \Lambda(\,), \text{Lemma 1}\\
&= \sum_{h\in\mathrm{HNF}_{x,\Gamma}} \mathrm{Red}^{\infty}_{Mx,h}[\![\lambda x.h]\!]^{\Gamma} && \text{def. of } \Lambda(\,).
\end{aligned}$$

One can check that for $h \in \mathrm{HNF}_{x,\Gamma}$, $\mathrm{Red}^{\infty}_{Mx,h} = \sum_{h_0\in\mathrm{HNF}_{\Gamma}} \mathrm{Red}^{\infty}_{M,h_0}\mathrm{Red}^{\infty}_{h_0x,h}$ (recall that x is not free in M). If h_0 is a head-normal form $yP_1 \ldots P_m$ then $\mathrm{Red}^{\infty}_{h_0x,h} \neq 0$ only if $h = yP_1 \ldots P_m x$ with $x \notin \mathrm{FV}(yP_1 \ldots P_m)$ (and $\mathrm{Red}^{\infty}_{h_0x,h} = 1$). If $h_0 = \lambda x_0.h'$ then $\mathrm{Red}^{\infty}_{h_0x,h} \neq 0$ only if $h = h'\{x/x_0\}$ (and $\mathrm{Red}^{\infty}_{h_0x,h} = 1$). In the first case we have $[\![\lambda x.h]\!]^{\Gamma} = [\![\lambda x.(h_0x)]\!]^{\Gamma} = [\![h_0]\!]^{\Gamma}$. In the second case we have $\lambda x.h = h_0$ modulo α-equivalence and $[\![\lambda x.h]\!]^{\Gamma} = [\![h_0]\!]^{\Gamma}$. Therefore: $v \leq \sum_{h_0\in\mathrm{HNF}_{\Gamma}} \mathrm{Red}^{\infty}_{M,h_0}[\![h_0]\!]^{\Gamma}$. □

Lemma 9 (Key Lemma). *For all $M \in \Lambda^+_{\Gamma}$ with $\Gamma = \{y_1, \ldots, y_n\}$, for all $\Delta \supseteq \Gamma$, for all $u_1, \ldots, u_n$ in $\mathrm{P}(\mathcal{D}^{\Delta} \Rightarrow \mathcal{D})$ and $N_1, \ldots, N_n$ in Λ^+_{Δ} with $(u_i, N_i) \in \lhd^{\Delta}$,*

$$[\![M]\!]^{\Gamma} \circ (u_1, \ldots, u_n) \lhd^{\Delta} M\{N_1/y_1, \ldots, N_n/y_n\}$$

Proof. The proof is by induction on M. The abstraction uses Lemmas 5 and 8, the application uses Lemma 8 and the barycentric sum Lemma 6. □

Theorem 1 (Strong adequacy). *For all $M \in \Lambda_\Gamma^+$ we have:*

$$[\![M]\!]^\Gamma = \sum_{h \in \mathrm{HNF}_\Gamma} \mathrm{Red}^\infty_{M,h} [\![h]\!]^\Gamma.$$

Proof. The invariance of the interpretation by reduction (Proposition 2) gives that for all $n \in \mathbb{N}$, $[\![M]\!]^\Gamma = \sum_{N \in \Lambda_\Gamma^+} \mathrm{Red}^n_{M,N} [\![N]\!]^\Gamma \geq \sum_{h \in \mathrm{HNF}_\Gamma} \mathrm{Red}^n [\![h]\!]^\Gamma$. When $n \to \infty$ we get $[\![M]\!]^\Gamma \geq \sum_{h \in \mathrm{HNF}_\Gamma} \mathrm{Red}^\infty_{M,h} [\![h]\!]^\Gamma$.

Conversely using Lemma 9 with $\Delta = \Gamma$ and $(u_i, N_i) = (\pi^\Gamma_{y_i}, y_i)$, which is in $\lhd^\Gamma$ thanks to Lemma 7, we get $([\![M]\!]^\Gamma, M) \in \lhd^\Gamma$. The definition of $\lhd = \phi(\lhd)$ with $\Delta = \Gamma$ and $n = 0$ gives $[\![M]\!]^\Gamma \leq \sum_{h \in \mathrm{HNF}_\Gamma} \mathrm{Red}^\infty_{M,h} [\![h]\!]^\Gamma$. □

5 Nakajima Trees and Full Abstraction

We apply our strong adequacy to infer full abstraction (Theorem 2). As mentioned in the Introduction, the bridge linking syntax and semantics is given by the notion of probabilistic Nakajima tree defined by Leventis [12] (here Definitions 2 and 3) in order to prove a separation theorem for Λ^+. Lemma 11 shows that the equality of Nakajima trees implies the denotational equality. The proof of this lemma uses the strong adequacy property.

Definition 2. *The set $\mathcal{PT}^\eta_\ell$ of* **Nakajima trees** *with depth at most $\ell \in \mathbb{N}$ is the set of subprobability distributions over* **value Nakajima trees** *$\mathcal{VT}^\eta_\ell$. These sets are defined by mutual recursion as follows:*

$$\mathcal{VT}^\eta_0 = \emptyset, \qquad \mathcal{VT}^\eta_{\ell+1} = \left\{ \lambda \boldsymbol{x}.y\, \boldsymbol{T} \mid \boldsymbol{x} \in \mathcal{V}^{\mathbb{N}}, y \in \mathcal{V}, \boldsymbol{T} \in (\mathcal{PT}^\eta_\ell)^{\mathbb{N}} \right\},$$

$$\mathcal{PT}^\eta_0 = \{\bot\}, \qquad \mathcal{PT}^\eta_{\ell+1} = \left\{ T \in [0,1]^{\mathcal{VT}^\eta_{\ell+1}} \mid \sum_{t \in \mathcal{VT}^\eta_{\ell+1}} T(t) \leq 1 \right\}.$$

The notation $\bot$ represents the empty function (i.e. the distribution with empty support), encoding undefinedness and allowing directed sets of approximants.

Value Nakajima trees represent infinitary η-long head-normal forms: up to η-equivalence every head-normal form $h = \lambda x_1 \dots x_n.y\, M_1 \, \dots \, M_m$ is equal to $\lambda x_1 \dots x_{n+k}.y\, M_1 \, \dots \, M_m\, x_{n+1} \, \dots \, x_{n+k}$ for any $k \in \mathbb{N}$ and $x_{n+1}, \dots, x_{n+k}$ fresh, and value Nakajima trees are infinitary variants of such η-expansions.

Definition 3. *By mutual recursion we associate value trees VT^η with head-normal forms and general trees PT^η with general Λ^+ terms:*

$$\begin{aligned} &VT^\eta_{\ell+1}(\lambda x_1 \dots x_n.y\, M_1 \, \dots \, M_m) \\ &\qquad = \lambda x_1 \dots x_n x_{n+1} \dots .y\, PT^\eta_\ell(M_1) \, \dots \, PT^\eta_\ell(M_m)\, PT^\eta_\ell(x_{n+1}) \, \dots \end{aligned}$$

where the x_is are pairwise distinct variables and, for $i > m$, the x_i's are fresh;

$$PT_0^\eta(M) = \bot, \qquad PT_{\ell+1}^\eta(M) = t \mapsto \sum_{h\in(VT_{\ell+1}^\eta)^{-1}(t)} \mathrm{Red}_{M,h}^\infty$$

Remark 1. In [12], following the definition of deterministic Nakajima trees in [1], the value tree $VT_{\ell+1}^\eta(\lambda x_1 \dots x_n.y\, M_1 \dots M_m)$ includes explicitly the difference $n-m$. This yields a heavier but somewhat more convenient definition, as then Lemma 10 also holds for $\ell = 1$. In this paper we chose to use the lighter definition. This choice does not influence the Nakajima tree equality by Lemma 10.

Example 5. Figure 2(a) depicts some examples of value Nakajima trees associated with the head-normal form $\lambda x_1.y(\mathbf{\Omega}x_1)x_1$. Notice that these trees are equivalent to the Nakajima trees associated with $y(\mathbf{\Omega}x_1)$ as well as $y\mathbf{\Omega}$. In fact, all these terms are contextually equivalent.

Figure 2(b) shows the Nakajima tree of depth 2 associated with the term $y(u +_q v) +_p (y' +_{p'} \mathbf{\Omega})$. Notice that the two sums $+_p$ and $+_{p'}$ contribute to the same subprobability distribution, whereas they are kept distinct from the sum $+_q$ on the argument side of an application.

Figure 2(c) gives some examples of the Nakajima trees associated with the term $\mathbf{\Theta}(\lambda f.(y +_p y(f))$, discussed also in Examples 2 and 3. Notice that the more the depth ℓ increases, the more the top-level distribution's support grows.

It is clear that the family $(\mathcal{PT}_\ell^\eta(M))_{\ell\in\mathbb{N}}$ converges to a limit, but we do not need to make it explicit for our purposes, so we avoid defining the topology over $\bigcup_\ell \mathcal{PT}_\ell^\eta$ yielding the convergence of $(\mathcal{PT}_\ell^\eta(M))_{\ell\in\mathbb{N}}$.

The next lemma shows that the first levels of a VT^η of a head-normal form h give a lot of information about the shape of h.

Lemma 10. *Given two head-normal forms $h = \lambda x_1 \dots x_n.yM_1 \dots M_m$ and $h' = \lambda x_1 \dots x_{n'}.y'M_1' \dots M_{m'}'$ and any $\ell \geq 2$, if $VT_\ell^\eta(h) = VT_\ell^\eta(h')$, then $y = y'$ and $n - m = n' - m'$.*

Proof. The fact $y = y'$ follows immediately from the definition of VT^η. Concerning the second equality, one can assume $n = n'$ by η-expanding one of the two terms, in fact VT^η is invariant under η-expansion. Modulo α-equivalence, we can then restrict ourselves to consider the case of $h = \lambda x_1 \dots x_n.yM_1 \dots M_m$ and $h' = \lambda x_1 \dots x_n.yM_1' \dots M_{m'}'$.

Suppose, by the sake of contradiction, that $m > m'$. Then we should have $PT_{\ell-1}^\eta(M_{m'+1}) = PT_{\ell-1}^\eta(x_{n+1})$, where x_{n+1} is a fresh variable, in particular $x_{n+1} \notin \mathrm{FV}(M_{m'+1})$. Since $\ell-1 > 0$, we have that $PT_{\ell-1}^\eta(x_{n+1})(t) = 1$ only if t is equal to $\lambda z_1 z_2 \dots .x_{n+1} PT_{\ell-2}^\eta(z_1) PT_{\ell-2}^\eta(z_2) \dots$, otherwise $PT_{\ell-1}^\eta(x_{n+1})(t) = 0$. So, $PT_{\ell-1}^\eta(M_{m'+1}) = PT_{\ell-1}^\eta(x_{n+1})$ implies that $\mathrm{Red}_{M_{m'+1},h}^\infty > 0$ for some h having x_{n+1} as free variable, which is impossible since $x_{n+1} \notin \mathrm{FV}(M_{m'+1})$. □

Thanks to the strong adequacy property we can prove that for $M \in \Lambda_\Gamma^+$ each coefficient of $[\![M]\!]^\Gamma$ is entirely defined by $PT_\ell^\eta(M)$ for ℓ large enough. To do so we define the following size on $|\mathcal{D}|$, $\mathcal{M}_\mathrm{f}(|\mathcal{D}|)$ and $\mathcal{M}_\mathrm{f}(|\mathcal{D}|)^\Gamma \times |\mathcal{D}|$:

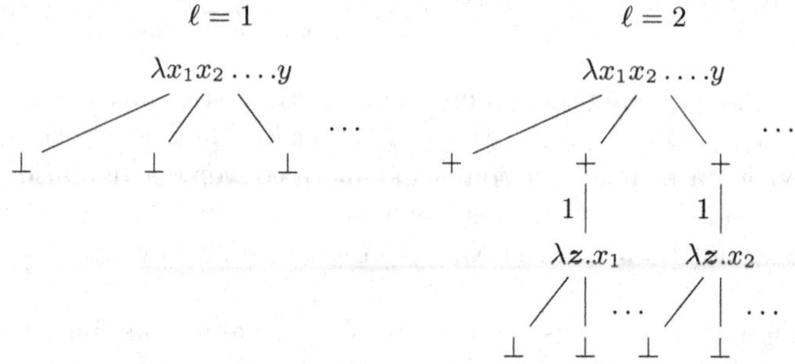

(a) $VT^{\eta}_{\ell}(\lambda x_1.y(\boldsymbol{\Omega} x_1)x_1)$ for some ℓ, also equal to $VT^{\eta}_{\ell}(y\boldsymbol{\Omega})$.

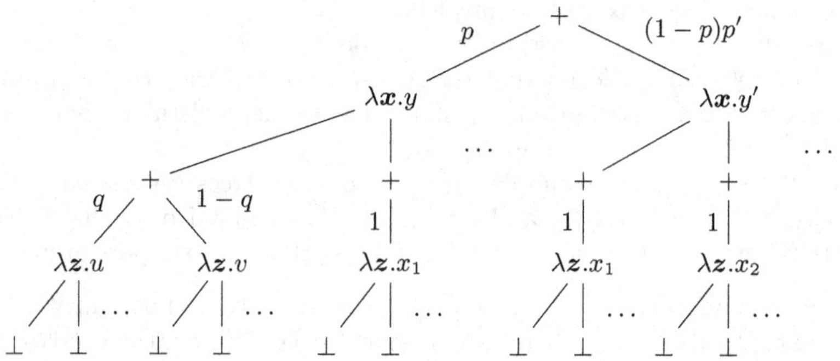

(b) $PT^{\eta}_{2}(y(u +_q v) +_p (y' +_{p'} \boldsymbol{\Omega}))$. Notice the layers of distributions.

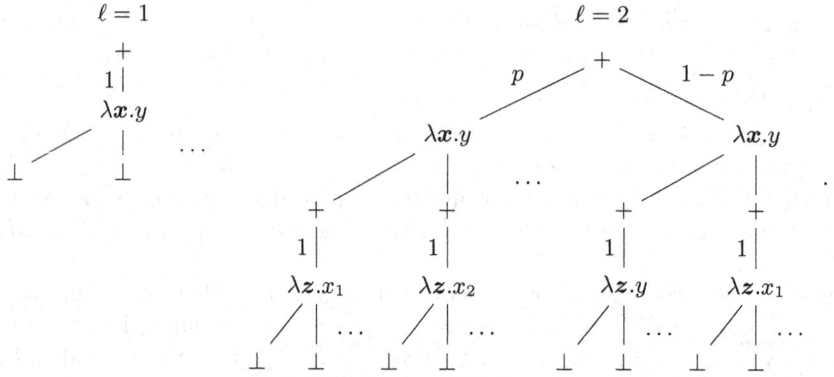

(c) $PT^{\eta}_{\ell}(\boldsymbol{\Theta}(\lambda f.(y +_p y(f))))$ for some ℓ

Fig. 2. Examples of Nakajima trees. Distributions are represented by barycentric sums, depicted as + nodes whose outgoing edges are weighted by probabilities.

- $\#(\star) = 0$ for the base element,
- $\#(m :: d) = \#(m) + \#(d)$ for $m \in \mathcal{M}_f(|\mathcal{D}|)$ and $d \in |\mathcal{D}|$,
- $\#([d_1, \dots, d_n]) = n + \sum_{i=1}^{n} \#(d_i)$ for $d_1, \dots, d_n \in |\mathcal{D}|$,
- $\#(\boldsymbol{m}, d) = \#(d) + \sum_{x \in \Gamma}(\#(\boldsymbol{m}_x))$ for $\boldsymbol{m} \in \mathcal{M}_f(|\mathcal{D}|)^{\Gamma}$ and $d \in |\mathcal{D}|$.

Lemma 11. *Given $\ell \in \mathbb{N}$ and $M, N \in \Lambda_\Gamma^+$, if $PT_\ell^\eta(M) = PT_\ell^\eta(N)$ then for any $(\boldsymbol{m}, d) \in \mathcal{M}_f(|\mathcal{D}|)^\Gamma \times |\mathcal{D}|$ with $\#(\boldsymbol{m}, d) < \ell$, we have $[\![M]\!]^\Gamma_{\boldsymbol{m},d} = [\![N]\!]^\Gamma_{\boldsymbol{m},d}$.*

Proof. We do induction on ℓ. If $\ell \leq 1$, then $\#(\boldsymbol{m}, d) = 0$ implies $d = \star$ and for every $x \in \Gamma$, $\boldsymbol{m}_x = [\,]$. In this case we remark that both $[\![M]\!]^\Gamma_{\boldsymbol{m},d}, [\![N]\!]^\Gamma_{\boldsymbol{m},d}$ are null. This in fact can be easily checked by inspecting the rules of Fig. 1, computing the matrix denoting a term by structural induction over the term.

Otherwise, by Theorem 1, we have: $[\![M]\!]^\Gamma_{\boldsymbol{m},d} = \sum_{h \in \mathrm{HNF}_\Gamma} \mathrm{Red}^\infty_{M,h} [\![h]\!]^\Gamma_{\boldsymbol{m},d}$. This last sum can be refactored as $\sum_{t \in VT_\ell^\eta} \sum_{h \in (VT_\ell^\eta)^{-1}(t)} \mathrm{Red}^\infty_{M,h} [\![h]\!]^\Gamma_{\boldsymbol{m},d}$. A similar reasoning for N gives $[\![N]\!]^\Gamma_{\boldsymbol{m},d} = \sum_{t \in VT_\ell^\eta} \sum_{h \in (VT_\ell^\eta)^{-1}(t)} \mathrm{Red}^\infty_{N,h} [\![h]\!]^\Gamma_{\boldsymbol{m},d}$.

Let us fix a $t \in VT_\ell^\eta$ and $(\boldsymbol{m}, d) \in \mathcal{M}_f(|\mathcal{D}|)^\Gamma \times |\mathcal{D}|$ with $\#(\boldsymbol{m}, d) < \ell$. Let us prove that:

$\square$ for any $h, h' \in (VT_\ell^\eta)^{-1}(t)$, we have $[\![h]\!]^\Gamma_{\boldsymbol{m},d} = [\![h']\!]^\Gamma_{\boldsymbol{m},d}$.

Notice that $\square$ implies $[\![M]\!]^\Gamma_{\boldsymbol{m},d} = [\![N]\!]^\Gamma_{\boldsymbol{m},d}$, since the hypothesis $PT_\ell^\eta(M) = PT_\ell^\eta(N)$ gives $\sum_{h \in (VT_\ell^\eta)^{-1}(t)} \mathrm{Red}^\infty_{M,h} = \sum_{h \in (VT_\ell^\eta)^{-1}(t)} \mathrm{Red}^\infty_{N,h}$, for any $t \in VT_\ell^\eta$.

Let then $h = \lambda x_1 \dots x_n.y M_1 \dots M_k$ and $h' = \lambda x_1 \dots x_{n'}.y' M'_1 \dots M'_{k'}$. Since $\ell \geq 2$, $VT_\ell^\eta(h) = VT_\ell^\eta(h')$ implies by Lemma 10 that $y = y'$ and $n - k = n' - k'$. Since $\mathcal{D}$ is extensional (see Sect. 3), by η-expanding one of the two terms, we can suppose $n = n'$ and, then, $k = k'$. Besides if $n > 0$ let us write $d = m :: d'$, we have $[\![h]\!]^\Gamma_{\boldsymbol{m},d} = [\![\lambda x_2 \dots x_n.y M_1 \dots M_k]\!]^{x_1,\Gamma}_{(m,\boldsymbol{m}),d'}$ with $\#((m, \boldsymbol{m}), d') = \#(\boldsymbol{m}, d)$, and similarly for $[\![h']\!]^\Gamma_{\boldsymbol{m},d}$. So, we can reduce to consider the case: $h = y M_1 \dots M_k$ and $h' = y M'_1 \dots M'_k$. If $k = 0$ the claim $\square$ is trivial, otherwise by unfolding the applications of h using the applicative case in Fig. 1, we have that:

$$[\![h]\!]^\Gamma_{\boldsymbol{m},d} = \sum_{\substack{(\boldsymbol{m}_0,\dots,\boldsymbol{m}_k)\\ \text{s.t. } \boldsymbol{m} = \biguplus_i \boldsymbol{m}_i}} \sum_{\substack{m_1,\dots,m_k\\ \in \mathcal{M}_f(|\mathcal{D}|)}} [\![y]\!]^\Gamma_{\boldsymbol{m}_0, m_1 :: \dots :: m_k :: d} ([\![M_1]\!]^\Gamma)^{\boldsymbol{m}_1, m_1} \dots ([\![M_k]\!]^\Gamma)^{\boldsymbol{m}_k, m_k}$$

and the same for h', replacing each M_i with M'_i. Notice that $[\![y]\!]^\Gamma_{\boldsymbol{m}_0, m_1 :: \dots :: m_k :: d} \neq 0$ implies $(\boldsymbol{m}_0)_y = [m_1 :: \dots :: m_k :: d]$, hence $\#(m_i) < \#(\boldsymbol{m}_0)$ for any $i \leq k$, thus $\#(\boldsymbol{m}_i, m_i) < \#(\boldsymbol{m}_i) + \#(\boldsymbol{m}_0) \leq \#(\boldsymbol{m}) \leq \#(\boldsymbol{m}, d) < \ell$ and $\#(\boldsymbol{m}_i, m_i) < \ell - 1$. Moreover, the hypothesis $VT_\ell^\eta(h) = VT_\ell^\eta(h')$, implies $PT_{\ell-1}^\eta(M_i) = PT_{\ell-1}^\eta(M'_i)$ for any $i \leq k$, so we conclude by induction hypothesis on each term in the sums appearing in $([\![M_i]\!]^\Gamma)^{\boldsymbol{m}_i, m_i}$ and $([\![M'_i]\!]^\Gamma)^{\boldsymbol{m}_i, m_i}$. $\square$

Corollary 1. *Let $M, N \in \Lambda_\Gamma^+$, $\forall \ell \in \mathbb{N}, PT_\ell^\eta(M) = PT_\ell^\eta(N)$ implies $[\![M]\!]^\Gamma = [\![N]\!]^\Gamma$.*

Theorem 2. *For any two terms $M, N \in \lambda_\Gamma^+$, the following are equivalent:*

1. *M and N are contextually equivalent;*
2. *M and N have the same Nakajima trees;*
3. *M and N have the same interpretation in $\mathcal{D}$.*

Proof. (1) to (2) is given by [12, Theorem 10.1]. From (2) and Corollary 1, we get (3). Finally, (3) implies (1) by the adequacy of probabilistic coherence spaces, proven in [6, Corollary 25]. □

References

1. Barendregt, H.: The Lambda-Calculus: Its Syntax and Semantics. Studies in Logic and the Foundations of Mathematics, vol. 103. North-Holland, Amsterdam (1984)
2. Clairambault, P., Paquet, H.: Fully abstract models of the probabilistic lambda-calculus. In: Ghica, D.R., Jung, A. (eds.) 27th EACSL Annual Conference on Computer Science Logic, CSL 2018, 4–7 September 2018, LIPIcs, Birmingham, UK, vol. 119, pp. 16:1–16:17. Schloss Dagstuhl - Leibniz-Zentrum fuer Informatik (2018). https://doi.org/10.4230/LIPIcs.CSL.2018.16
3. Crubillé, R.: Probabilistic stable functions on discrete cones are power series. In: Dawar, A., Grädel, E. (eds.) Proceedings of the 33rd Annual ACM/IEEE Symposium on Logic in Computer Science, LICS 2018, Oxford, UK, 09–12 July 2018, pp. 275–284. ACM (2018). https://doi.org/10.1145/3209108.3209198, http://doi.acm.org/10.1145/3209108.3209198
4. Crubillé, R., Ehrhard, T., Pagani, M., Tasson, C.: The free exponential modality of probabilistic coherence spaces. In: Esparza, J., Murawski, A.S. (eds.) FoSSaCS 2017. LNCS, vol. 10203, pp. 20–35. Springer, Heidelberg (2017). https://doi.org/10.1007/978-3-662-54458-7_2
5. Danos, V., Ehrhard, T.: Probabilistic coherence spaces as a model of higher-order probabilistic computation. Inf. Comput. **209**(6), 966–991 (2011)
6. Ehrhard, T., Pagani, M., Tasson, C.: The computational meaning of probabilistic coherence spaces. In: Grohe, M. (ed.) Proceedings of the 26th Annual IEEE Symposium on Logic in Computer Science (LICS 2011), pp. 87–96. IEEE Computer Society Press (2011)
7. Ehrhard, T., Pagani, M., Tasson, C.: Probabilistic coherence spaces are fully abstract for probabilistic PCF. In: Sewell, P. (ed.) The 41th Annual ACM SIGPLAN-SIGACT Symposium on Principles of Programming Languages, POPL 2014, San Diego, USA. ACM (2014)
8. Ehrhard, T., Tasson, C.: Probabilistic call by push value (2016). http://arxiv.org/abs/1607.04690
9. Girard, J.Y.: Between logic and quantic: a tract. In: Ehrhard, T., Girard, J.Y., Ruet, P., Scott, P. (eds.) Linear Logic in Computer Science. London Mathematical Society Lecture Note Series, vol. 316. CUP, Cambridge (2004)
10. Hyland, M.: A syntactic characterization of the equality in some models for the lambda calculus. J. London Math. Soc. **12**, 361–370 (1976)
11. Laird, J., Manzonetto, G., McCusker, G., Pagani, M.: Weighted relational models of typed lambda-calculi. In: 28th Annual ACM/IEEE Symposium on Logic in Computer Science, LICS 2013, New Orleans, LA, USA, 25–28 June 2013. IEEE Computer Society, June 2013

12. Leventis, T.: Probabilistic Böhm trees and probabilistic separation. In: Proceedings of the 33rd Annual ACM/IEEE Symposium on Logic in Computer Science, LICS 2018, Oxford, UK, 09–12 July 2018, pp. 649–658 (2018). https://doi.org/10.1145/3209108.3209126
13. Manzonetto, G.: A general class of models of $\mathcal{H}^*$. In: Královič, R., Niwiński, D. (eds.) MFCS 2009. LNCS, vol. 5734, pp. 574–586. Springer, Heidelberg (2009). https://doi.org/10.1007/978-3-642-03816-7_49
14. Pitts, A.M.: Computational adequacy via 'mixed' inductive definitions. In: Brookes, S., Main, M., Melton, A., Mislove, M., Schmidt, D. (eds.) MFPS 1993. LNCS, vol. 802, pp. 72–82. Springer, Heidelberg (1994). https://doi.org/10.1007/3-540-58027-1_3
15. Wadsworth, C.P.: The relation between computational and denotational properties for scott's D_∞-models of the lambda-calculus. SIAM J. Comput. **5**, 488–521 (1976)

A Sound and Complete Logic for Algebraic Effects

Cristina Matache(✉) and Sam Staton

University of Oxford, Oxford, UK
cristina.matache@balliol.ox.ac.uk

Abstract. This work investigates three notions of program equivalence for a higher-order functional language with recursion and general algebraic effects, in which programs are written in continuation-passing style. Our main contribution is the following: we define a logic whose formulas express program properties and show that, under certain conditions which we identify, the induced program equivalence coincides with a contextual equivalence. Moreover, we show that this logical equivalence also coincides with an applicative bisimilarity. We exemplify our general results with the nondeterminism, probabilistic choice, global store and I/O effects.

1 Introduction

Logic is a fundamental tool for specifying the behaviour of programs. A general approach is to consider that a logical formula ϕ encodes a program property, and the satisfaction relation of the logic, $t \models \phi$, asserts that program t enjoys property ϕ. An example is Hennessy-Milner logic [12] used to model concurrency and nondeterminism. Other program logics include Hoare logic [13], which describes imperative programs with state, and more recently separation logic [28]. Both state and nondeterminism are examples of *computational effects* [25], which represent impure behaviour in a functional programming language. The logics mentioned so far concern languages with first-order functions, so as a natural extension, we are interested in finding a logic which describes higher-order programs with general effects.

The particular flavour of effects we consider is that of *algebraic effects* developed by Plotkin and Power [32–34]. This is a unified framework in which effectful computation is triggered by a set of operations whose behaviour is axiomatized by a set of equations. For example, nondeterminism is given by a binary choice operation $or(-,-)$ that satisfies the equations of a semilattice. Thus, general effectful programs have multiple possible execution paths, which can be visualized as an (effect) tree, with effect operations labelling the nodes. Consider the following function `or_suc` which has three possible return values, and the effect tree of (`or_suc` 2):

M. Bojańczyk and A. Simpson (Eds.): FOSSACS 2019, LNCS 11425, pp. 382–399, 2019.
https://doi.org/10.1007/978-3-030-17127-8_22

$$\texttt{or\_suc} = \lambda x{:}\texttt{nat}.\quad or(x,\ or(x+1, x+2))$$

$(\texttt{or\_suc}\ 2) \longmapsto$ 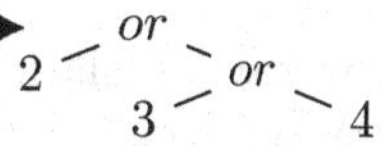

Apart from state and nondeterminism, examples of algebraic effects include probabilistic choice and input and output operations.

Apart from providing a specification language for programs, a logic can also be used to compare two different programs. This leads to a notion of program equivalence: two programs are equivalent when they satisfy exactly the same formulas in the logic.

Many other definitions of program equivalence for higher-order languages exist. An early notion is contextual equivalence [26], which asserts that two programs are equivalent if they have the same observable behaviour in all program contexts. However, this is hard to establish in practice due to the quantification over all contexts. Another approach, which relies on the existence of a suitable denotational model of the language, is checking equality of denotations. Yet another notion, meant to address the shortcomings of the previous two, is that of applicative bisimilarity [1].

Given the wide range of definitions of program equivalence, comparing them is an interesting question. For example, the equivalence induced by Hennessy-Milner logic is known to coincide with bisimilarity for CCS. Thus, we not only aim to find a logic describing general algebraic effects, but also to compare it to existing notions of program equivalence.

Program equivalence for general algebraic effects has been studied by Johann, Simpson and Voigtländer [17] who define a notion of contextual equivalence and a corresponding logical relation. Dal Lago, Gavazzo and Levy [7] provide an abstract treatment of applicative bisimilarity in the presence of algebraic effects. Working in a typed, call-by-value setting, Simpson and Voorneveld [38] propose a modal logic for effectful programs whose induced program equivalence coincides with applicative bisimilarity, but not with contextual equivalence (see counterexample in Sect. 5). Dal Lago, Gavazzo and Tanaka [8] propose a notion of applicative similarity that coincides with contextual equivalence for an untyped, call-by-name effectful calculus.

These papers provide the main starting point for our work. Our goal is to find a logic of program properties which characterizes contextual equivalence for a higher-order language with algebraic effects. We study a typed call-by-value language in which programs are written in continuation-passing style (CPS). CPS is known to simplify contextual equivalence, through the addition of control operators (e.g. [5]), but it also implies that all notions of program equivalence we define can only use continuations to test return values. Contextual equivalence and bisimilarity for lambda-calculi with control, but without general effects, have been studied extensively (e.g. [4,15,23,41]).

In CPS, functions receive as argument the continuation (which is itself a function) to which they pass their return value. Consider the function that adds two natural numbers. This usually has type $\texttt{nat} \to \texttt{nat} \to \texttt{nat}$, but its CPS version is defined as: $\texttt{addk} = \lambda(n{:}\texttt{nat}, m{:}\texttt{nat}, k{:}\texttt{nat}{\to}\texttt{R}).\ k\ (n+m)$ for some fixed return type $\texttt{R}$. The function $\texttt{or\_suc}$ becomes in CPS:

$$\texttt{or\_succ} = \lambda(x{:}\texttt{nat}, k{:}\texttt{nat}{\to}\texttt{R}).\ \mathit{or}(k\ x,\ \mathit{or}(\texttt{addk}\ (x,\ 1,\ k),\ \texttt{addk}\ (x,\ 2,\ k))).$$

A general translation of direct-style functions into CPS can be found in Sect. 5.

We fix a calculus named ECPS (Sect. 2), in which programs are not expected to return, except through a call to the continuation. Contextual equivalence is defined using a custom set of observations $\mathfrak{P}$, where the elements of $\mathfrak{P}$ are sets of effect trees. We consider a logic $\mathcal{F}$ whose formulas express properties of ECPS programs (Sect. 3). For example, `or_succ` satisfies the following formula: $\phi = (\{2\},\ (\{3\} \vee \{4\}) \mapsto \Box) \mapsto \Diamond$.

Here, $\Diamond$ is the set of all effect trees for which at least one execution path succeeds and $\Box$ is the set of trees that always succeed. So `or_succ` $\models_{\mathcal{F}} \phi$ says that, when given arguments 2 and a continuation that always succeeds for input 3 or 4, then `or_succ` *may* succeed. In other words, `or_succ` may 'return' 3 or 4 to the continuation. In contrast, `or_succ` $\models_{\mathcal{F}} \phi' = (\{2\},\ (\{3\} \vee \{4\}) \mapsto \Box) \mapsto \Box$ says that the program `or_succ` *must* return 3 or 4 to the continuation. Thus `or_succ` $\not\models_{\mathcal{F}} \phi'$ because the continuation k might diverge on 2.

Another example can be obtained by generalizing the `or_succ` function to take a function as a parameter, rather than using `addk`:

$$\begin{aligned}\texttt{or\_succ'} = \lambda(x : \texttt{nat},\ k : \texttt{nat}{\to}\texttt{R},\ f : (\texttt{nat},\ \texttt{nat},\ \texttt{nat}{\to}\texttt{R}){\to}\texttt{R}).\\ \mathit{or}(k\ x,\ \mathit{or}(f\ (x, 1, k),\ f\ (x, 2, k)))\\ \models_{\mathcal{F}} \Big(\{2\},\ \{4\} \mapsto \Diamond,\ ((\{2\},\ \{2\},\ \{4\} \mapsto \Diamond) \mapsto \Diamond)\Big) \mapsto \Diamond.\end{aligned}$$

The formula above says that `or_succ'` may call f with arguments 2, 2 and k.

The main theorem concerning the logic $\mathcal{F}$ (Theorem 1) is that, under certain restrictions on the observations in $\mathfrak{P}$, logical equivalence coincides with contextual equivalence. In other words, $\mathcal{F}$ is sound and complete with respect to contextual equivalence. The proof of this theorem, outlined in Sect. 4, involves applicative bisimilarity as an intermediate step. Thus, we show in fact that three notions of program equivalence for ECPS are the same: logical equivalence, contextual equivalence and applicative bisimilarity. Due to space constraints, proofs are omitted but they can be found in [21].

2 Programming Language – ECPS

We consider a simply-typed functional programming language with general recursion, a datatype of natural numbers and general algebraic effects as introduced by Plotkin and Power [32]. We will refer to this language as ECPS because programs are written in continuation-passing style.

ECPS distinguishes between terms which can reduce further, named computations, and values, which cannot reduce. ECPS is a variant of both Plotkin's PCF [31] and Levy's Jump-With-Argument language [20], extended with algebraic effects. A fragment of ECPS is discussed in [18] in connection with logic.

Types	$A, A_1, B := (A_1, \ldots, A_n){\to}\texttt{R} \mid \texttt{nat}$	$(n \geq 0)$
Typing contexts	$\Gamma := \emptyset \mid \Gamma, x : A.$	

The only base type in ECPS is $\mathtt{nat}$. The return type of functions, $\mathtt{R}$, is fixed and is *not* a first-class type. Intuitively, we consider that functions are not expected to return. A type in direct style $A \to B$ becomes in ECPS: $(A, B {\to} \mathtt{R}) {\to} \mathtt{R}$. In the typing context $(\Gamma, x : A)$, the free variable x does not appear in Γ.

First, consider the pure fragment of ECPS, without effects, named CPS:

$$\begin{array}{lll} \text{Values} & v, w := \mathtt{zero} \mid \mathtt{succ}(v) \mid \lambda(x_1{:}A_1, \ldots, x_n{:}A_n).t \mid x & (n \geq 0) \\ \text{Computations} & s, t := v(w_1, \ldots, w_n) \mid \mathtt{case}\ v\ \mathtt{of}\ \{\mathtt{zero} \Rightarrow s,\ \mathtt{succ}(x) \Rightarrow t\} \mid & \\ & \quad (\mathtt{rec}\ x.v)(w_1, \ldots, w_n). & \end{array}$$

Variables, natural numbers and lambdas are values. Computations include function application and an eliminator for natural numbers. The expression $\mathtt{rec}\ x.v$ is a recursive definition of the function v, which must be applied. If exactly one argument appears in a lambda abstraction or an application term, we will sometimes omit the parentheses around that argument.

There are two typing relations in CPS, one for values $\Gamma \vdash v : A$, which says that value v has type A in the context Γ, and one for computations $\Gamma \vdash t : \mathtt{R}$. This says that t is well-formed given the context Γ. All computations have the same return type $\mathtt{R}$. We also define the *order of a type* recursively, which roughly speaking counts the number of function arrows $\to$ in a type.

$$\frac{}{\Gamma, x : A \vdash x : A} \qquad \frac{\Gamma, \overrightarrow{x : A} \vdash t : \mathtt{R}}{\Gamma \vdash \lambda(\overrightarrow{x{:}A}).t : (\overrightarrow{A}){\to}\mathtt{R}} \qquad \frac{}{\Gamma \vdash \mathtt{zero} : \mathtt{nat}} \qquad \frac{\Gamma \vdash v : \mathtt{nat}}{\Gamma \vdash \mathtt{succ}(v) : \mathtt{nat}}$$

$$\frac{\Gamma \vdash v : (\overrightarrow{A}){\to}\mathtt{R} \quad (\Gamma \vdash w_i : A_i)_i}{\Gamma \vdash v\ (\overrightarrow{w}) : \mathtt{R}} \qquad \frac{\Gamma, x : (\overrightarrow{A}){\to}\mathtt{R} \vdash v : (\overrightarrow{A}){\to}\mathtt{R} \quad (\Gamma \vdash w_i : A_i)_i}{\Gamma \vdash (\mathtt{rec}\ x.v)(\overrightarrow{w}) : \mathtt{R}}$$

$$\frac{\Gamma \vdash v : \mathtt{nat} \qquad \Gamma \vdash t : \mathtt{R} \qquad \Gamma, x : \mathtt{nat} \vdash s : \mathtt{R}}{\Gamma \vdash \mathtt{case}\ v\ \mathtt{of}\ \{\mathtt{zero} \Rightarrow t,\ \mathtt{succ}(x) \Rightarrow s\} : \mathtt{R}}$$

$$ord(\mathtt{nat}) = 0 \qquad ord(() {\to} \mathtt{R}) = 1$$

$$ord((A_1, \ldots, A_n) {\to} \mathtt{R}) = max_{1 \leq i \leq n}(ord(A_i)) + 1 \qquad (\text{if } n > 0)$$

To introduce algebraic effects into our language, we consider a new kind of context Σ, disjoint from Γ, which we call an *effect context*. The symbols σ appearing in Σ stand for effect operations and their type must have either order 1 or 2. For example, the binary choice operation $or : (() {\to} \mathtt{R},\ () {\to} \mathtt{R}) {\to} \mathtt{R}$ expects two thunked computations. The output operation $output : (\mathtt{nat},\ () {\to} \mathtt{R}) {\to} \mathtt{R}$ expects a parameter and a continuation. An operation signifying success, which takes no arguments, is $\downarrow : () {\to} \mathtt{R}$. Roughly, Σ could be regarded as a countable algebraic signature.

We extend the syntax of CPS with effectful computations. The typing relations now carry a Σ context: $\Gamma \vdash_\Sigma v : A$ and $\Gamma \vdash_\Sigma t : \mathtt{R}$. Otherwise, the typing judgements remain unchanged; we have a new rule for typing effect operations:

$$s,t := \ldots \mid \sigma(\overrightarrow{v}, \overrightarrow{k}) \qquad \frac{\sigma : (\overrightarrow{A}, \overrightarrow{B}) \to \mathtt{R} \in \Sigma \quad (\Gamma \vdash_\Sigma v_i : A_i)_i \quad (\Gamma \vdash_\Sigma k_j : B_j)_j}{\Gamma \vdash_\Sigma \sigma(\overrightarrow{v}, \overrightarrow{k}) : \mathtt{R}}$$

In ECPS, the only type with order 0 is $\mathtt{nat}$, so in fact $A_i = \mathtt{nat}$ for all i. Notice that the grammar does not allow function abstraction over a symbol from Σ and that σ is not a first-class term. So we can assume that Σ is fixed, as in the examples from Sect. 2.1.

As usual, we identify terms up to alpha-equivalence. Substitution of values for free variables that are not operations, $v[w/x]$ and $t[w/x]$, is defined in the standard way by induction on the structure of v and t. We use $\overline{n}$ to denote the term $\mathtt{succ}^n(\mathtt{zero})$. Let $(\vdash_\Sigma)$ be the set of well-formed closed computations and $(\vdash_\Sigma A)$ the set of closed values of type A.

2.1 Operational Semantics

We define a family of relations on closed computation terms $(\longrightarrow) \subseteq (\vdash_\Sigma) \times (\vdash_\Sigma)$ for any effect context Σ:

$$(\lambda(\overrightarrow{x{:}A}).t)\ (\overrightarrow{w}) \longrightarrow t[\overrightarrow{w}/\overrightarrow{x}]$$
$$(\mathtt{rec}\ x.v)\ (\overrightarrow{w}) \longrightarrow (v[(\lambda(\overrightarrow{y{:}A}).(\mathtt{rec}\ x.v)(\overrightarrow{y}))/x])\ (\overrightarrow{w})$$
$$\mathtt{case\ zero\ of}\ \{\mathtt{zero} \Rightarrow s,\ \mathtt{succ}(x) \Rightarrow t\} \longrightarrow s$$
$$\mathtt{case\ succ}(v)\ \mathtt{of}\ \{\mathtt{zero} \Rightarrow s,\ \mathtt{succ}(x) \Rightarrow t\} \longrightarrow t[v/x].$$

Observe that the reduction given by $\longrightarrow$ can either run forever or terminate with an effect operation. If the effect operation does not take any arguments of order 1 (i.e. continuations), the computation stops. If the reduction reaches $\sigma(\overrightarrow{v}, \overrightarrow{k})$, the intuition is that any continuation k_i may be chosen, and executed with the results of operation σ. Thus, repeatedly evaluating effect operations leads to the construction of an infinitely branching tree (similar to that in [32]), as we now explain, which we call an *effect tree*. A path in the tree represents a possible execution path of the program.

An effect tree, of possibly infinite depth and width, can contain:

- leaves labelled $\bot$, which signifies nontermination of $\longrightarrow$;
- leaves labelled $\sigma_{\overrightarrow{v}}$, where $\sigma : (\overrightarrow{A}) \to \mathtt{R} \in \Sigma$ and $(\vdash_\Sigma v_i : A_i)_i$;
- nodes labelled $\sigma_{\overrightarrow{v}}$, where $\sigma : (\overrightarrow{A}, \overrightarrow{B}) \to \mathtt{R} \in \Sigma$ and each $\vdash_\Sigma v_i : A_i$; such a node has an infinite number of children $t_0, t_1, \ldots$.

Denote the set of all effect trees by Trees_Σ. This set has a partial order: $tr_1 \leq tr_2$ if and only if tr_1 can be obtained by replacing subtrees of tr_2 by $\bot$. Every ascending chain $t_1 \leq t_2 \leq \ldots$ has a least upper bound $\bigsqcup_n t_n$. In fact Trees_Σ is the free pointed Σ-algebra [2] and therefore also has a coinductive property [9].

Next, we define a sequence of effect trees associated with each well-formed closed computation. Each element in the sequence can be seen as evaluating the computation one step further. Let $[\![-]\!]_{(-)} : (\vdash_\Sigma) \times \mathbb{N} \longrightarrow \mathit{Trees}_\Sigma$:

$$[\![t]\!]_0 = \bot$$

$$[\![t]\!]_{m+1} = \begin{cases} [\![s]\!]_m & \text{if } t \longrightarrow s \\ \sigma_{\overrightarrow{v}}\left(\left(\left([\![k_i\ (\overline{n_1}, \ldots, \overline{n_{p_i}})]\!]_m\right)_{n_1,\ldots,n_{p_i} \in \mathbb{N}}\right)_i\right) & \text{if } t = \sigma(\overrightarrow{v}, \overrightarrow{k}) \end{cases}$$

These are all the cases since well-formed computations do not get stuck. We define the function $[\![-]\!] : (\vdash_\Sigma) \longrightarrow \mathit{Trees}_\Sigma$ as the least upper bound of the chain $\{[\![t_n]\!]\}_{n\in\mathbb{N}}$: $[\![t]\!] = \bigsqcup_{n\in\mathbb{N}}[\![t]\!]_n$.

We now give examples of effect contexts Σ for different algebraic effects, and of some computations and their associated effect trees.

Example 1 (Pure functional computation). $\Sigma = \{\downarrow : ()\to\mathtt{R}\}$. Intuitively, $\downarrow$ is a top-level success flag, analogous to a 'barb' in process algebra. This is to ensure a reasonable contextual equivalence for CPS programs, which never actually return results. For example, $\mathit{loop} = (\mathtt{rec}\ f.\lambda().(f\ x))\ ()$ runs forever, and

$$\mathtt{test\_zero} = \lambda(y{:}\mathtt{nat}).\ \mathtt{case}\ y\ \mathtt{of}\ \{\mathtt{zero} \Rightarrow\ \downarrow (),\ \mathtt{succ}(x) \Rightarrow \mathit{loop}\}$$

is a continuation that succeeds just when it is passed zero. Generally, an effect tree for a pure computation is either $\downarrow$ if it succeeds or $\bot$ otherwise.

Example 2 (Nondeterminism). $\Sigma = \{\mathit{or} : (()\to\mathtt{R},\ ()\to\mathtt{R})\to\mathtt{R},\ \downarrow : ()\to\mathtt{R}\}$. Intuitively, $\mathit{or}(k_1, k_2)$ performs a nondeterministic choice between computations k_1 () and k_2 (). Consider a continuation $\mathtt{test\_3} : \mathtt{nat}\to\mathtt{R}$ that diverges on 3 and succeeds otherwise. The program `or_succ` from the introduction is in ECPS:

$$\begin{aligned}\mathtt{or\_succ} = \lambda(x{:}\mathtt{nat}, k{:}\mathtt{nat}\to\mathtt{R}).\ &\mathit{or}(\lambda().\ k\ x,\\ &\lambda().\ \mathit{or}(\lambda().k\ (\mathtt{succ}(x)),\\ &\qquad \lambda().k\ (\mathtt{succ}(\mathtt{succ}(x)))))\end{aligned}$$

$$[\![\mathtt{or\_succ}\ (\overline{2},\ \mathtt{test\_3})]\!] = \begin{array}{ccccc} & \mathit{or} & & & \\ \swarrow & & \searrow & & \\ \downarrow & & & \mathit{or} & \\ & & \swarrow & & \searrow \\ & & \bot & & \downarrow \end{array}$$

Example 3 (Probabilistic choice). $\Sigma = \{\mathit{p\text{-}or} : (()\to\mathtt{R},\ ()\to\mathtt{R})\to\mathtt{R},\ \downarrow : ()\to\mathtt{R}\}$. Intuitively, the operation $\mathit{p\text{-}or}(k_1, k_2)$ chooses between k_1 () and k_2 () with probability 0.5. Consider the following term which encodes the geometric distribution:

$$\begin{aligned}&\mathtt{geom} = \lambda k{:}\mathtt{nat}\to\mathtt{R}.\\ &\quad(\mathtt{rec}\ f.\ \lambda(n{:}\mathtt{nat}, k'{:}\mathtt{nat}\to\mathtt{R}).\mathit{p\text{-}or}(\lambda().k'\ n,\ \lambda().f\ (\mathtt{succ}(n), k')))\ (\overline{1}, k).\end{aligned}$$

The probability that `geom` passes a number $n > 0$ to its continuation is 2^{-n}. To test it, consider $k = (\lambda x{:}\mathtt{nat}.\downarrow ())$; then $[\![\mathtt{geom}\ k]\!]$ is an infinite tree:

$$[\![\mathtt{geom}\ k]\!] = \begin{array}{ccc} & \mathit{p\text{-}or} & \\ \swarrow & & \searrow \\ \downarrow & & [\![\mathtt{geom}\ k]\!] \end{array}$$

Example 4 (Global store). $\mathbb{L}$ is a finite set of locations storing natural numbers and $\Sigma = \{\mathit{lookup}_l : (\mathtt{nat}\to\mathtt{R})\to\mathtt{R},\ \mathit{update}_l : (\mathtt{nat}, ()\to\mathtt{R})\to\mathtt{R} \mid l \in \mathbb{L}\} \cup \{\downarrow : ()\to\mathtt{R}\}$. Intuitively, $\mathit{lookup}_l(k)$ looks up the value at storage location l, if this is $\overline{n}$ it continues with $k\,(\overline{n})$. For $update_l(v, k)$ the intuition is: write the number v in location l then continue with the computation $k\,()$. For example:

$$[\![update_{l_0}(\overline{1},\ \lambda().lookup_{l_0}(\lambda x{:}\mathtt{nat}.\mathtt{case}\ x\ \mathtt{of}\ \{\mathtt{zero} \Rightarrow \downarrow\ (),\ \mathtt{succ}(y) \Rightarrow loop\}))]\!] =$$

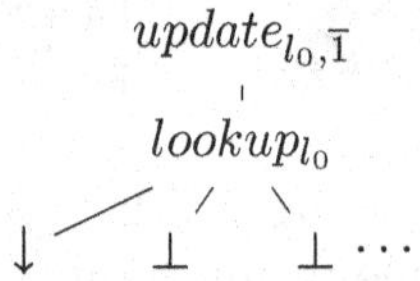

Only the second branch of $lookup_{l_0}$ can occur. The other branches are still present in the tree because $[\![-]\!]$ treats effect operations as uninterpreted syntax.

Example 5 (Interactive input/output). $\Sigma = \{\downarrow : ()\to\mathtt{R},\ output : (\mathtt{nat}, ()\to\mathtt{R})\to\mathtt{R},\ input : (\mathtt{nat}\to\mathtt{R})\to\mathtt{R}\}$. Intuitively, the computation $input(k)$ accepts number $\overline{n}$ from the input channel and continues with $k\,(\overline{n})$. The computation $output(v, k)$ writes v to the output channel then continues with computation $k\,()$. Below is a computation that inputs a number $\overline{n}$ then outputs it immediately, and repeats.

$$[\![\mathtt{echo}]\!] = [\![(\mathtt{rec}\ f.\ \lambda().\ input(\lambda x{:}\mathtt{nat}.\ output(x,\ \lambda().f\ ())))\ ()]\!] =$$

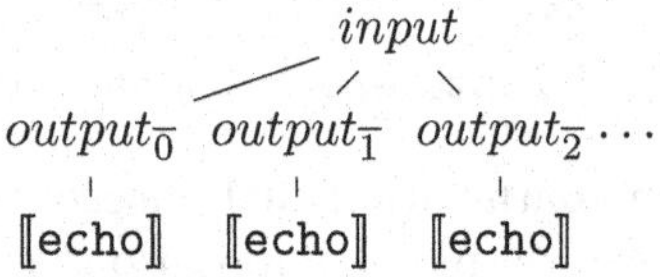

2.2 Contextual Equivalence

Informally, two terms are contextually equivalent if they have the same *observable behaviour* in all program contexts. The definition of observable behaviour depends on the programming language under consideration. In ECPS, we can observe effectful behaviour such as interactive output values or the probability with which a computation succeeds. This behaviour is encoded by the effect tree of a computation. Therefore, we represent an ECPS observation as a set of effect trees P. A computation t exhibits observation P if $[\![t]\!] \in P$.

For a fixed set of effect operations Σ, we define the set $\mathfrak{P}$ of possible *observations.* The elements of $\mathfrak{P}$ are subsets of $Trees_\Sigma$. Observations play a similar role to the modalities from [38]. For our running examples of effects, $\mathfrak{P}$ is defined as follows:

Example 6 (Pure functional computation). Define $\mathfrak{P} = \{\Downarrow\}$ where $\Downarrow = \{\downarrow\}$. There are no effect operations so the $\Downarrow$ observation only checks for success.

Example 7 (Nondeterminism). Define $\mathfrak{P} = \{\Diamond, \Box\}$ where:

$$\Diamond = \{tr \in Trees_\Sigma \mid \text{at least one of the paths in } tr \text{ has a } \downarrow \text{ leaf}\}$$
$$\Box = \{tr \in Trees_\Sigma \mid \text{the paths in } tr \text{ are all finite and finish with a } \downarrow\}.$$

The intuition is that, if $[\![t]\!] \in \Diamond$, then computation t *may* succeed, whereas if $[\![t]\!] \in \Box$, then t *must* succeed.

Example 8 (Probabilistic choice). Define $\mathbb{P} : \mathit{Trees}_\Sigma \longrightarrow [0,1]$ to be the least function, by the pointwise order, such that:

$$\mathbb{P}(\downarrow) = 1 \qquad \mathbb{P}(\mathit{p\text{-}or}(tr_0, tr_1)) = \frac{1}{2}\mathbb{P}(tr_0) + \frac{1}{2}\mathbb{P}(tr_1).$$

Notice that $\mathbb{P}(\bot) = 0$. Then observations are defined as:

$$\mathbf{P}_{>q} = \{tr \in \mathit{Trees}_\Sigma \mid \mathbb{P}(tr) > q\} \qquad \mathfrak{P} = \{\mathbf{P}_{>q} \mid q \in \mathbb{Q},\ 0 \leq q < 1\}.$$

This means that $[\![t]\!] \in \mathbf{P}_{>q}$ if the probability that t succeeds is greater than q.

Example 9 (Global store). Define the set of states as the set of functions from storage locations to natural numbers: $\mathit{State} = \mathbb{L} \longrightarrow \mathbb{N}$. Given a state S, we write $[S\!\downarrow] \subseteq \mathit{Trees}_\Sigma$ for the set of effect trees that terminate when starting in state S. More precisely, $[-]$ is the least State-indexed family of sets satisfying the following:

$$\frac{-}{\downarrow \in [S\!\downarrow]} \qquad \frac{l \in \mathbb{L} \qquad tr_{S(l)} \in [S\!\downarrow]}{\mathit{lookup}_l(tr_0, tr_1, tr_2, \ldots) \in [S\!\downarrow]} \qquad \frac{l \in \mathbb{L} \qquad tr \in [S[l := n]\!\downarrow]}{\mathit{update}_{l,\overline{n}}(tr) \in [S\!\downarrow]}$$

The set of observations is: $\mathfrak{P} = \{[S\!\downarrow] \mid S \in \mathit{State}\}$.

Example 10 (Interactive input/output). An I/O-trace is a finite word w over the alphabet $\{?n \mid n \in \mathbb{N}\} \cup \{!n \mid n \in \mathbb{N}\}$. For example, $?1\,!1\,?2\,!2\,?3\,!3$. The set of observations is: $\mathfrak{P} = \{\langle W\rangle_{\ldots},\ \langle W\rangle\!\downarrow \mid W$ an I/O-trace$\}$. Observations are defined as the least sets satisfying the following rules:

$$\frac{-}{tr \in \langle\epsilon\rangle_{\ldots}} \quad \frac{tr = \downarrow}{tr \in \langle\epsilon\rangle\!\downarrow} \quad \frac{tr_n \in \langle W\rangle_{\ldots}}{\mathit{input}(tr_0, tr_1, \ldots) \in \langle(?n)W\rangle_{\ldots}} \quad \frac{tr' \in \langle W\rangle_{\ldots}}{\mathit{output}_{\overline{n}}(tr') \in \langle(!n)W\rangle_{\ldots}}$$

and the analogous rules for $\langle(?n)W\rangle\!\downarrow$ and $\langle(!n)W\rangle\!\downarrow$. Thus, $[\![t]\!] \in \langle W\rangle_{\ldots}$ if computation t produces I/O trace W, and $[\![t]\!] \in \langle W\rangle\!\downarrow$ if additionally t succeeds immediately after producing W.

Using the set of observations $\mathfrak{P}$, we can now define contextual equivalence as the greatest compatible and adequate equivalence relation between possibly open terms of the same type. Adequacy specifies a necessary condition for two *closed* computations to be related, namely producing the same observations.

Definition 1. *A well-typed relation $\mathcal{R} = (\mathcal{R}^{\mathfrak{v}}_A, \mathcal{R}^{\mathfrak{c}})$ (i.e. a family of relations indexed by ECPS types where $\mathcal{R}^{\mathfrak{c}}$ relates computations) on possibly open terms is adequate if:*

$$\forall s, t.\ \vdash_\Sigma s\ \mathcal{R}^{\mathfrak{c}}\ t \implies \forall P \in \mathfrak{P}.\ [\![s]\!] \in P \iff [\![t]\!] \in P.$$

Relation $\mathcal{R}$ is compatible if it is closed under the rules in [21, Page 57]. As an example, the rules for application and lambda abstraction are:

$$\frac{\Gamma \vdash_\Sigma v\ \mathcal{R}^{\mathfrak{v}}_{(\overrightarrow{A})\to\mathtt{R}}\ v' \qquad (\Gamma \vdash_\Sigma w_i\ \mathcal{R}^{\mathfrak{v}}_{A_i}\ w'_i)_i}{\Gamma \vdash_\Sigma v(\overrightarrow{w})\ \mathcal{R}^{\mathfrak{c}}\ v'(\overrightarrow{w'})} \qquad \frac{\Gamma, \overrightarrow{x : A} \vdash_\Sigma s\ \mathcal{R}^{\mathfrak{c}}\ t}{\Gamma \vdash_\Sigma \lambda(\overrightarrow{x{:}A}).s\ \mathcal{R}^{\mathfrak{v}}_{(\overrightarrow{A})\to\mathtt{R}}\ \lambda(\overrightarrow{x{:}A}).t}$$

Definition 2 (Contextual equivalence). *Let $\mathbb{CA}$ be the set of well-typed relations on possibly open terms that are both compatible and adequate. Define contextual equivalence $\equiv_{\text{ctx}}$ to be $\bigcup \mathbb{CA}$.*

Proposition 1. *Contextual equivalence $\equiv_{\text{ctx}}$ is an equivalence relation, and is moreover compatible and adequate.*

This definition of contextual equivalence, originally proposed in [11,19], can be easily proved equivalent to the traditional definition involving program contexts (see [21, §7]). As Pitts observes [30], reasoning about program contexts directly is inconvenient because they cannot be defined up to alpha-equivalence, hence we prefer using Definition 2.

For example, in the pure setting (Example 1), we have $\overline{0} \not\equiv_{\text{ctx}} \overline{1}$, because $\texttt{test\_zero}(\overline{0}) \not\equiv_{\text{ctx}} \texttt{test\_zero}(\overline{1})$; they are distinguished by the observation $\Downarrow$. In the state example, $lookup_{l_1}(k) \not\equiv_{\text{ctx}} lookup_{l_2}(k)$, because they are distinguished by the context $(\lambda k{:}\texttt{nat}{\to}\texttt{R}.\,[-])\ (\texttt{test\_zero})$ and the observation $[S{\downarrow}]$ where $S(l_1) = \overline{0}$ and $S(l_2) = \overline{1}$. In the case of probabilistic choice (Example 3), $\texttt{geom}\ (\lambda x{:}\texttt{nat}.\downarrow ()) \equiv_{\text{ctx}} \downarrow ()$ because $(\texttt{geom}\ (\lambda x{:}\texttt{nat}.\downarrow ()))$ succeeds with probability 1 ('almost surely').

3 A Program Logic for ECPS – $\mathcal{F}$

This section contains the main contribution of the paper: a logic $\mathcal{F}$ of program properties for ECPS which characterizes contextual equivalence. Crucially, the logic makes use of the observations in $\mathfrak{P}$ to express properties of computations.

In $\mathcal{F}$, there is a distinction between formulas that describe values and those that describe computations. Each value formula is associated an ECPS type A. Value formulas are constructed from the basic formulas $(\phi_1, \ldots, \phi_n) \mapsto P$ and $\phi = \{n\}$, where $n \in \mathbb{N}$ and $P \in \mathfrak{P}$, as below. The indexing set I can be infinite, even uncountable. Computation formulas are simply the elements of $\mathfrak{P}$.

$$\frac{n \in \mathbb{N}}{\{n\} : \texttt{nat}} \qquad \frac{\phi_1 : A_1 \ldots \phi_n : A_n \qquad P \in \mathfrak{P}}{(\phi_1, \ldots, \phi_n) \mapsto P : (A_1, \ldots, A_n) \to \texttt{R}}\ (\text{VAL}) \qquad \frac{(\phi_i : A)_{i \in I}}{\bigvee_{i \in I} \phi_i : A} \qquad \frac{(\phi_i : A)_{i \in I}}{\bigwedge_{i \in I} \phi_i : A} \qquad \frac{\phi : A}{\neg \phi : A}$$

The satisfaction relation $\models_{\mathcal{F}}$ relates a closed value $\vdash_{\Sigma} v : A$ to a value formula $\phi : A$ of the same type, or a closed computation t to an observation P. Relation $t \models_{\mathcal{F}} P$ tests the shape of the effect tree of t.

$$\begin{aligned}
v \models_{\mathcal{F}} \{n\} \quad &\iff \quad v = \overline{n} \\
v \models_{\mathcal{F}} (\phi_1, \ldots, \phi_n) \mapsto P \quad &\iff \quad \text{for all closed values } w_1, \ldots, w_n \text{ such that} \\
&\qquad\quad \forall i.\ w_i \models_{\mathcal{F}} \phi_i \text{ then } v(w_1, \ldots, w_n) \models_{\mathcal{F}} P \\
v \models_{\mathcal{F}} \bigvee_{i \in I} \phi_i \quad &\iff \quad \text{there exists } j \in I \text{ such that } v \models_{\mathcal{F}} \phi_j \\
v \models_{\mathcal{F}} \bigwedge_{i \in I} \phi_i \quad &\iff \quad \text{for all } j \in I,\ v \models_{\mathcal{F}} \phi_j \\
v \models_{\mathcal{F}} \neg \phi \quad &\iff \quad \text{it is false that } v \models_{\mathcal{F}} \phi \\
t \models_{\mathcal{F}} P \quad &\iff \quad [\![t]\!] \in P.
\end{aligned}$$

Example 11. Consider the following formulas, where only ϕ_3 and ϕ_4 refer to the same effect context:

$$\begin{aligned}
&\phi_1 = ((\{3\} \mapsto \Diamond) \mapsto \Diamond) \wedge ((\{4\} \mapsto \Diamond) \mapsto \Diamond) \wedge ((\{3\} \mapsto \Box \wedge \{4\} \mapsto \Box) \mapsto \Box)\\
&\phi_2 = ((\vee_{n>1}\{n\}) \mapsto \mathbf{P}_{>q}) \mapsto \mathbf{P}_{>q/2}\\
&\phi_3 = \wedge_{S\in State}((\{S(l)\} \mapsto [S\downarrow]) \mapsto [S\downarrow])\\
&\phi_4 = \wedge_{S\in State} \wedge_{n\in\mathbb{N}} ((\{n\},\ () \mapsto [S[l_0 := n, l_1 := n+1]\downarrow]) \mapsto [S[l_0 := n]\downarrow])\\
&\phi_5 = \wedge_{k\in\mathbb{N}} \vee_{n_1,\ldots,n_k\in\mathbb{N}} (() \mapsto \langle ?n_1!n_1?n_2!n_2\ \ldots\ ?n_k!n_k\rangle_{\ldots}).
\end{aligned}$$

Given a function $v : (\texttt{nat}\rightarrow\texttt{R})\rightarrow\texttt{R}$, $v \models_{\mathcal{F}} \phi_1$ means that v is guaranteed to call its argument only with $\overline{3}$ or $\overline{4}$. The function `geom` from Example 3 satisfies ϕ_2 because with probability $1/2$ it passes to the continuation a number $n > 1$.

For example, the following satisfactions hold: $\lambda k{:}\texttt{nat}\rightarrow\texttt{R}.\, lookup_l(k) \models_{\mathcal{F}} \phi_3$ and $f = \lambda(x{:}\texttt{nat}, k{:}()\rightarrow\texttt{R}).\, update_{l_1}(\texttt{succ}(x), k) \models_{\mathcal{F}} \phi_4$. The latter formula says that, either f always succeeds, or f evaluated with $\overline{n}$ changes the state from $S[l_0 := n]$ to $S[l_0 := n, l_1 := n+1]$ before calling its continuation. This is similar to a total correctness assertion $[S[l_0 := n]](-)[S[l_0 := n, l_1 := n+1]]$ from Hoare logic, for a direct style program. Formula ϕ_5 is satisfied by $\lambda().\texttt{echo}$, where `echo` is the computation defined in Example 5.

Even though the indexing set I in $\wedge_{i\in I}$ and $\vee_{i\in I}$ may be uncountable, the sets of values and computations are countable. Since logical formulas are interpreted over values and computations, all conjunctions and disjunctions are logically equivalent to countable ones.

Definition 3 (Logical equivalence). *For any closed values $\vdash_\Sigma v_1 : A$ and $\vdash_\Sigma v_2 : A$, and for any closed computations $\vdash_\Sigma s_1$ and $\vdash_\Sigma s_2$:*

$$\begin{aligned}
v_1 \equiv_{\mathcal{F}} v_2 \quad &\Longleftrightarrow \quad \forall \phi : A \text{ in } \mathcal{F}.\ (v_1 \models_{\mathcal{F}} \phi \Longleftrightarrow v_2 \models_{\mathcal{F}} \phi)\\
s_1 \equiv_{\mathcal{F}} s_2 \quad &\Longleftrightarrow \quad \forall P \text{ in } \mathcal{F}.\ (s_1 \models_{\mathcal{F}} P \Longleftrightarrow s_2 \models_{\mathcal{F}} P).
\end{aligned}$$

To facilitate equational reasoning, logical equivalence should be compatible, a property proved in the next section (Proposition 3). Compatibility allows substitution of related programs for a free variable that appears on both sides of a program equation. Notice that logical equivalence would not be changed if we added conjunction, disjunction and negation at the level of computation formulas. We have omitted such connectives for simplicity.

To state our main theorem, first define the open extension of a well-typed relation $\mathcal{R}$ on closed terms as: $\overrightarrow{x : A} \vdash_\Sigma t\, \mathcal{R}^\circ\, s$ if and only if for any closed values $(\vdash_\Sigma v_i : A_i)_i$, $t[\overrightarrow{v/x}]\ \mathcal{R}\ s[\overrightarrow{v/x}]$. Three sufficient conditions that we impose on the set of observations $\mathfrak{P}$ are defined below. The first one, consistency, ensures that contextual equivalence can distinguish at least two programs.

Definition 4 (Consistency). *A set of observations $\mathfrak{P}$ is consistent if there exists at least one observation $P_0 \in \mathfrak{P}$ such that:*

1. $P_0 \neq \mathit{Trees}_\Sigma$ and
2. there exists at least one computation t_0 such that $[\![t_0]\!] \in P_0$.

Definition 5 (Scott-openness). *A set of trees X is Scott-open if:*

1. *It is upwards closed, that is: $tr \in X$ and $tr \leq tr'$ imply $tr' \in X$.*
2. *Whenever $tr_1 \leq tr_2 \leq \ldots$ is an ascending chain with least upper bound $\bigsqcup tr_i \in X$, then $tr_j \in X$ for some j.*

Definition 6 (Decomposability). *The set of observations $\mathfrak{P}$ is decomposable if for any $P \in \mathfrak{P}$, and for any $tr \in P$:*

$$\forall \sigma \in \Sigma.\ (tr = \sigma_{\overrightarrow{v}}(\overrightarrow{tr'}) \implies$$
$$\exists \overrightarrow{P'} \in \mathfrak{P} \cup \{\mathit{Trees}_\Sigma\}.\ \overrightarrow{tr'} \in \overrightarrow{P'} \text{ and } \forall \overrightarrow{p'} \in \overrightarrow{P'}.\ \sigma_{\overrightarrow{v}}(\overrightarrow{p'}) \in P).$$

Theorem 1 (Soundness and Completeness of $\mathcal{F}$). *For a decomposable set of Scott-open observations $\mathfrak{P}$ that is consistent, the open extension of $\mathcal{F}$-logical equivalence coincides with contextual equivalence: $(\equiv^\circ_\mathcal{F}) = (\equiv_{\text{ctx}})$.*

The proof of this theorem is outlined in Sect. 4. It is easy to see that for all running examples of effects the set $\mathfrak{P}$ is consistent. The proof that each $P \in \mathfrak{P}$ is Scott-open is similar to that for modalities from [38]. It remains to show that for all our examples $\mathfrak{P}$ is decomposable. Intuitively, decomposability can be understood as saying that logical equivalence is a congruence for the effect context Σ.

Example 12 (Pure functional computation). The only observation is $\Downarrow = \{\downarrow\}$. There are no trees in $\Downarrow$ whose root has children, so decomposability is satisfied.

Example 13 (Nondeterminism). Consider $tr \in \Diamond$. Either $tr = \downarrow$, in which case we are done, or $tr = or(tr'_0, tr'_1)$. It must be the case that either tr'_0 or tr'_1 have a $\downarrow$-leaf. Without loss of generality, assume this is the case for tr'_0. Then we know $tr'_0 \in \Diamond$ so we can choose $P'_0 = \Diamond, P'_1 = \mathit{Trees}_\Sigma$. For any $\overrightarrow{p'} \in \overrightarrow{P'}$ we know $or(\overrightarrow{p'}) \in \Diamond$ because p'_0 has a $\downarrow$-leaf, so decomposability holds. The argument for $tr \in \Box$ is analogous: $P'_0 = P'_1 = \Box$.

Example 14 (Probabilistic choice). Consider $tr = p\text{-}or(tr'_0, tr'_1) \in \mathbf{P}_{>q}$. Choose: $q_0 = \frac{\mathbb{P}(tr'_0)}{\mathbb{P}(tr'_0)+\mathbb{P}(tr'_1)} \cdot 2q$ and $q_1 = \frac{\mathbb{P}(tr'_1)}{\mathbb{P}(tr'_0)+\mathbb{P}(tr'_1)} \cdot 2q$. From $\mathbb{P}(tr) = \frac{1}{2}(\mathbb{P}(tr'_0)+\mathbb{P}(tr'_1)) > q$ we can deduce that: $1 \geq \mathbb{P}(tr'_0) > q_0$ and $1 \geq \mathbb{P}(tr'_1) > q_1$. So we can choose $P'_0 = \mathbf{P}_{>q_0}, P'_1 = \mathbf{P}_{>q_1}$ to satisfy decomposability.

Example 15 (Global store). Consider a tree $tr = \sigma_{\overrightarrow{v}}(tr'_0, tr'_1, tr'_2, \ldots) \in [S\downarrow]$. If $\sigma = \mathit{lookup}_l$, then we know $tr'_{S(l)} \in [S\downarrow]$. In the definition of decomposability, choose $P'_{S(l)} = [S\downarrow]$ and $P'_{k \neq S(l)} = \mathit{Trees}_\Sigma$ and we are done. If $\sigma_{\overrightarrow{v}} = \mathit{update}_{l,\overline{n}}$, then $tr'_0 \in [S[l := n]\downarrow]$. Choose $P'_0 = [S[l := n]\downarrow]$.

Example 16 (Interactive input/output). Consider an I/O trace $W \neq \epsilon$ and a tree $tr = \sigma_{\overrightarrow{v}}(tr'_0, tr'_1, tr'_2, \ldots) \in \langle W \rangle_{\ldots}$. If $\sigma = input$, it must be the case that $W = (?k)W'$ and $tr'_k \in \langle W' \rangle_{\ldots}$. We can choose $P'_k = \langle W' \rangle_{\ldots}$ and $P'_{m \neq k} = \langle \epsilon \rangle_{\ldots}$ and we are done. If $\sigma_{\overrightarrow{v}} = output_{\overline{n}}$, then $W = (!n)W'$ and $tr'_0 \in \langle W' \rangle_{\ldots}$. Choose $P'_0 = \langle W' \rangle_{\ldots}$ and we are done. The proof for $\langle W \rangle\downarrow$ is analogous.

4 Soundness and Completeness of the Logic $\mathcal{F}$

This section outlines the proof of Theorem 1, which says that $\mathcal{F}$-logical equivalence coincides with contextual equivalence. The full proof can be found in [21]. First, we define applicative bisimilarity for ECPS, similarly to the way Simpson and Voorneveld [38] define it for PCF with algebraic effects. Then, we prove in turn that $\mathcal{F}$-logical equivalence coincides with applicative bisimilarity, and that applicative bisimilarity coincides with contextual equivalence. Thus, three notions of program equivalence for ECPS are in fact the same.

Definition 7 (Applicative $\mathfrak{P}$-bisimilarity). *A collection of relations $\mathcal{R}^{\mathfrak{v}}_A \subseteq (\vdash_\Sigma A)^2$ for each type A and $\mathcal{R}^{\mathfrak{c}} \subseteq (\vdash_\Sigma)^2$ is an applicative* $\mathfrak{P}$-simulation *if:*

1. $v \, \mathcal{R}^{\mathfrak{v}}_{\mathtt{nat}} \, w \implies v = w$.
2. $s \, \mathcal{R}^{\mathfrak{c}} \, t \implies \forall P \in \mathfrak{P}. \, (\llbracket s \rrbracket \in P \implies \llbracket t \rrbracket \in P)$.
3. $v \, \mathcal{R}^{\mathfrak{v}}_{(\overrightarrow{A}) \to \mathtt{R}} \, u \implies \forall (\vdash_\Sigma w_i : A_i)_i. \, v(\overrightarrow{w}) \, \mathcal{R}^{\mathfrak{c}} \, u(\overrightarrow{w})$.

An applicative $\mathfrak{P}$-bisimulation *is a symmetric simulation.* Bisimilarity, *denoted by $\sim$, is the union of all bisimulations. Therefore, it is the greatest applicative $\mathfrak{P}$-bisimulation.*

Notice that applicative bisimilarity uses the set of observations $\mathfrak{P}$ to relate computations, just as contextual and logical equivalence do. It is easy to show that bisimilarity is an equivalence relation.

Proposition 2. *Given a decomposable set of Scott-open observations $\mathfrak{P}$, the open extension of applicative $\mathfrak{P}$-bisimilarity, $\sim^\circ$, is compatible.*

Proof (notes). This is proved using Howe's method [14], following the structure of the corresponding proof from [38]. Scott-openness is used to show that the observations P interact well with the sequence of trees $\llbracket - \rrbracket_{(-)}$ associated with each computation. For details see [21, §5.4]. □

Proposition 3. *Given a decomposable set of Scott-open observations $\mathfrak{P}$, applicative $\mathfrak{P}$-bisimilarity $\sim$ coincides with $\mathcal{F}$-logical equivalence $\equiv_\mathcal{F}$. Hence, the open extension of $\mathcal{F}$-logical equivalence $\equiv^\circ_\mathcal{F}$ is compatible.*

Proof (sketch). We define a new logic $\mathcal{V}$ which is almost the same as $\mathcal{F}$ except that the (VAL) rule is replaced by:

$$\frac{\vdash_\Sigma w_1 : A_1 \ldots \vdash_\Sigma w_n : A_n \quad P \in \mathfrak{P}}{(w_1, \ldots, w_n) \mapsto P : (A_1, \ldots, A_n) \to \mathtt{R}} \qquad v \models_\mathcal{V} (\overrightarrow{w}) \mapsto P \iff v(\overrightarrow{w}) \models_\mathcal{V} P.$$

That is, formulas of function type are now constructed using ECPS values. It is relatively straightforward to show that $\mathcal{V}$-logical equivalence coincides with applicative $\mathfrak{P}$-bisimilarity [21, Prop. 6.3.1]. However, we do not know of a similar direct proof for the logic $\mathcal{F}$. From Proposition 2, we deduce that $\mathcal{V}$-logical equivalence is compatible.

Next, we prove that the logics $\mathcal{F}$ and $\mathcal{V}$ are in fact equi-expressive, so they induce the same relation of logical equivalence on ECPS programs [21, Prop. 6.3.4]. Define a translation of formulas from $\mathcal{F}$ to $\mathcal{V}$, $(-)^\flat$, and one from $\mathcal{V}$ to $\mathcal{F}$, $(-)^\sharp$. The most interesting cases are those for formulas of function type:

$$((\phi_1,\ldots,\phi_n) \mapsto P)^\flat = \bigwedge\{(w_1,\ldots,w_n) \mapsto P \mid w_1 \models_{\mathcal{V}} \phi_1^\flat,\ldots,w_n \models_{\mathcal{V}} \phi_n^\flat\}$$
$$((w_1,\ldots,w_n) \mapsto P)^\sharp = (\chi_{w_1},\ldots,\chi_{w_n}) \mapsto P$$

where χ_{w_i} is the characteristic formula of w_i, that is $\chi_{w_i} = \bigwedge\{\phi \mid w_i \models_{\mathcal{F}} \phi\}$. Equi-expressivity means that the satisfaction relation remains unchanged under both translations, for example $v \models_{\mathcal{V}} \phi \iff v \models_{\mathcal{F}} \phi^\sharp$. Most importantly, the proof of equi-expressivity makes use of compatibility of $\equiv_{\mathcal{V}}$, which we established previously. For a full proof see [21, Prop. 6.2.3]). □

Finally, to prove Theorem 1 we show that applicative $\mathfrak{P}$-bisimilarity coincides with contextual equivalence [21, Prop. 7.2.2]:

Proposition 4. *Consider a decomposable set $\mathfrak{P}$ of Scott-open observations that is consistent. The open extension of applicative $\mathfrak{P}$-bisimilarity $\sim^\circ$ coincides with contextual equivalence $\equiv_{\text{ctx}}$.*

Proof (sketch). Prove $(\equiv_{\text{ctx}}) \subseteq (\sim^\circ)$ in two stages: first we show it holds for closed terms by showing $\equiv_{\text{ctx}}$ for them is a bisimulation; we make use of consistency of $\mathfrak{P}$ in the case of natural numbers. Then we extend to open terms using compatibility of $\equiv_{\text{ctx}}$. The opposite inclusion follows immediately by compatibility and adequacy of $\sim^\circ$. □

5 Related Work

The work closest to ours is that by Simpson and Voorneveld [38]. In the context of a direct-style language with algebraic effects, EPCF, they propose a modal logic which characterizes applicative bisimilarity but not contextual equivalence. Consider the following example from [19] (we use simplified EPCF syntax):

$$M = \lambda().\texttt{?nat} \qquad N = \textbf{let } y \Rightarrow \texttt{?nat} \textbf{ in } \lambda().\texttt{min(?nat, y)} \tag{1}$$

where ?`nat` is a computation, defined using *or*, which returns a natural number nondeterministically. Term M satisfies the formula $\Phi = \Diamond(\mathit{true} \mapsto \wedge_{n\in\mathbb{N}}\Diamond\{n\})$ in the logic of [38], which says that M may return a function which in turn may return any natural number. However, N does not satisfy Φ because it always returns a *bounded* number generator G. The bound on G is arbitrarily high so M and N are contextually equivalent, since a context can only test a finite number of outcomes of G.

EPCF can be translated into ECPS via a continuation-passing translation that preserves the shape of computation trees. The translation maps a value $\Gamma \vdash V : \tau$ to a value $\Gamma^* \vdash V^* : \tau^*$. An EPCF computation $\Gamma \vdash M : \tau$ becomes an ECPS value $\Gamma^* \vdash M^* : (\tau^* \to \mathtt{R}) \to \mathtt{R}$, which intuitively is waiting for a continuation k to pass its return result to (see [21, §4]). As an example, consider the cases for functions and application, where k stands for a continuation:

$$(\Gamma \vdash \lambda x{:}\tau.M : \tau \to \rho)^* = \Gamma^* \vdash \lambda(x{:}\tau^*, k{:}\rho^* \to \mathtt{R}).\,(M^*\ k) : (\tau^*, (\rho^* \to \mathtt{R})) \to \mathtt{R}$$
$$(\Gamma \vdash V\ W : \rho)^* = \Gamma^* \vdash \lambda k{:}\rho^* \to \mathtt{R}.\, V^*\ (W^*, k) : (\rho^* \to \mathtt{R}) \to \mathtt{R}.$$

This translation suggests that ECPS functions of type $(A_1, \ldots, A_n) \to \mathtt{R}$ can be regarded as continuations that never return. In EPCF the CPS-style algebraic operations can be replaced by direct-style generic effects [34], e.g. $\mathit{input}() : \mathtt{nat}$.

One way to understand this CPS translation is that it arises from the fact that $((-) \to \mathtt{R}) \to \mathtt{R}$ is a monad on the multicategory of values (in a suitable sense, e.g. [40]), which means that we can use the standard monadic interpretation of a call-by-value language. As usual, the algebraic structure on the return type $\mathtt{R}$ induces an algebraic structure on the entire monad (see e.g. [16], [24, §8]). We have not taken a denotational perspective in this paper, but for the reader with this perspective, a first step is to note that the quotient set $Q \stackrel{\text{def}}{=} (\mathit{Trees}_\Sigma)/_{\equiv_\mathfrak{P}}$ is a Σ-algebra, where $(tr \equiv_\mathfrak{P} tr')$ if $\forall P \in \mathfrak{P}, (tr \in P \iff tr' \in P)$; decomposability implies that $(\equiv_\mathfrak{P})$ is a Σ-congruence. This thus induces a CPS monad $Q^{(Q^-)}$ on the category of cpos.

Note that the terms in (1) above are an example of programs that are not bisimilar in EPCF but become bisimilar when translated to ECPS. This is because in ECPS bisimilarity, like contextual and logical equivalence, uses continuations to test return results. Therefore, in ECPS we cannot test for all natural numbers, like formula Φ does. This example provides an intuition of why we were able to show that all three notions of equivalence coincide, while [38] was not.

The modalities in Simpson's and Voorneveld's logic are similar to the observations from $\mathfrak{P}$, because they also specify shapes of effect trees. Since EPCF computations have a return value, a modality is used to *lift* a formula about the return values to a computation formula. In contrast, in the logic $\mathcal{F}$ observations alone suffice to specify properties of computations. From this point of view, our use of observations is closer to that found in the work of Johann et al. [17]. This use of observations also leads to a much simpler notion of decomposability (Definition 6) than that found in [38].

It can easily be shown that for the running examples of effects, $\mathcal{F}$-logical equivalence induces the program equations which are usually used to axiomatize algebraic effects, for example the equations for global store from [33]. Thus our choice of observations is justified further.

A different logic for algebraic effects was proposed by Plotkin and Pretnar [35]. It has a modality for each effect operation, whereas observations in $\mathfrak{P}$ are determined by the behaviour of effects, rather than by the syntax of their operations. Plotkin and Pretnar prove that their logic is sound for establishing several notions of program equivalence, but not complete in general. Refinement types are yet another approach to specifying the behaviour of algebraic effects, (e.g. [3]). Several monadic-based logics for computational effects have been proposed, such as [10], [29], although without the focus on contextual equivalence.

A logic describing a higher-order language with local store is the Hoare logic of Yoshida, Honda and Berger [42]. Hoare logic has also been integrated into a type system for a higher-order functional language with dependent types, in the form of Hoare type theory [27]. Although we do not yet know how to deal with local state or dependent types in the logic $\mathcal{F}$, an advantage of our logic over the previous two is that we describe different algebraic effects in a uniform manner.

Another aspect worth noticing is that some (non-trivial) $\mathcal{F}$-formulas are not inhabited by any program. For example, there is no function $v : (()\to\mathtt{R})\to\mathtt{R}$ satisfying: $\psi = (() \mapsto \langle !0\rangle_{\ldots}) \mapsto \langle !1\rangle_{\ldots} \;\wedge\; (() \mapsto \langle !1\rangle_{\ldots}) \mapsto \langle !0\rangle_{\ldots}$.

Formula ψ says that, if the first operation of a continuation is $output(\overline{0})$, this is replaced by $output(\overline{1})$ and vice-versa. But in ECPS, one cannot check whether an argument outputs something without also causing the output observation, and so the formula is never satisfied.

However, ψ could be inhabited if we extended ECPS to allow λ-abstraction over the symbols in the effect context Σ, and allowed such symbols to be *captured* during substitution (dynamic scoping). Consider the following example in an imaginary extended ECPS where we abstract over *output*:

$$\begin{aligned} h = \lambda(x{:}\mathtt{nat}, k{:}()\to\mathtt{R}).\ &\mathtt{case}\ x\ \mathtt{of}\ \{\mathtt{zero} \Rightarrow output(\overline{1}, k),\ \mathtt{succ}(y) \Rightarrow \\ &\qquad \mathtt{case}\ y\ \mathtt{of}\ \{\mathtt{zero} \Rightarrow output(\overline{0}, k),\ \mathtt{succ}(z) \Rightarrow k\ ()\}\} \\ v = \lambda f{:}()\to\mathtt{R}.&\big((\lambda output{:}(\mathtt{nat}, ()\to\mathtt{R})\to\mathtt{R}.\ (f\ ()))\ h\big). \end{aligned}$$

The idea is that during reduction of $(v\ f)$, the *output* operations in f are captured by $\lambda output$. Thus, $output(\overline{0})$ operations from $(f\ ())$ are replaced by $output(\overline{1})$ and vice-versa, and all other writes are skipped; so in particular $v \models_{\mathcal{F}} \psi$. This behaviour is similar to that of *effect handlers* [36]: computation $(f\ ())$ is being handled by handler h. We leave for future work the study of handlers in ECPS and of their corresponding logic.

6 Concluding Remarks

To summarize, we have studied program equivalence for a higher-order CPS language with general algebraic effects and general recursion (Sect. 2). Our main contribution is a logic $\mathcal{F}$ of program properties (Sect. 3) whose induced program equivalence coincides with contextual equivalence (Theorem 1; Sect. 4). Previous work on algebraic effects concentrated on logics that are sound for contextual equivalence, but not complete [35,38]. Moreover, $\mathcal{F}$-logical equivalence also coincides with applicative bisimilarity for our language. We exemplified our results for nondeterminism, probabilistic choice, global store and I/O. A next step would be to consider local effects (e.g. [22,33,37,39]) or normal form bisimulation (e.g. [6]).

Acknowledgements. This research was supported by an EPSRC studentship, a Balliol College Jowett scholarship, and the Royal Society. We would like to thank Niels Voorneveld for pointing out example (1), Alex Simpson and Ohad Kammar for useful discussions, and the anonymous reviewers for comments and suggestions.

References

1. Abramsky, S.: The lazy λ-calculus. In: Turner, D. (ed.) Research Topics in Functional Programming. Chapter 4, pp. 65–117. Addison Wesley, Boston (1990)
2. Abramsky, S., Jung, A.: Domain theory. In: Abramsky, S., Gabbay, D.M., Maibaum, T.S.E. (eds.) Handbook of Logic in Computer Science, Chap. 1, vol. 3, pp. 1–168. Oxford University Press, Oxford (1994)
3. Ahman, D., Plotkin, G.: Refinement types for algebraic effects. In: TYPES (2015)
4. Biernacki, D., Lenglet, S.: Applicative bisimulations for delimited-control operators. In: Birkedal, L. (ed.) FoSSaCS 2012. LNCS, vol. 7213, pp. 119–134. Springer, Heidelberg (2012). https://doi.org/10.1007/978-3-642-28729-9_8
5. Cartwright, R., Curien, P., Felleisen, M.: Fully abstract semantics for observably sequential languages. Inf. Comput. **111**(2), 297–401 (1994)
6. Dal Lago, U., Gavazzo, F.: Effectful normal form bisimulation. In: Proceedings of ESOP 2019 (2019)
7. Dal Lago, U., Gavazzo, F., Levy, P.: Effectful applicative bisimilarity: monads, relators, and Howe's method. In: LICS (2017)
8. Dal Lago, U., Gavazzo, F., Tanaka, R.: Effectful applicative similarity for call-by-name lambda calculi. In: ICTCS/CILC (2017)
9. Freyd, P.: Algebraically complete categories. In: Carboni, A., Pedicchio, M.C., Rosolini, G. (eds.) Category Theory. LNM, vol. 1488, pp. 95–104. Springer, Heidelberg (1991). https://doi.org/10.1007/BFb0084215
10. Goncharov, S., Schröder, L.: A relatively complete generic Hoare logic for order-enriched effects. In: LICS (2013)
11. Gordon, A.: Operational equivalences for untyped and polymorphic object calculi. In: Gordon, A., Pitts, A. (eds.) Higher Order Operational Techniques in Semantics, pp. 9–54. Cambridge University Press, Cambridge (1998)
12. Hennessy, M., Milner, R.: Algebraic laws for nondeterminism and concurrency. J. ACM **32**(1), 137–161 (1985)
13. Hoare, C.: An axiomatic basis for computer programming. Commun. ACM **12**(10), 576–580 (1969)
14. Howe, D.: Proving congruence of bisimulation in functional programming languages. Inf. Comput. **124**(2), 103–112 (1996)
15. Hur, C.K., Neis, G., Dreyer, D., Vafeiadis, V.: A logical step forward in parametric bisimulations. Technical report MPI-SWS-2014-003, January 2014
16. Hyland, M., Levy, P.B., Plotkin, G., Power, J.: Combining algebraic effects with continuations. Theoret. Comput. Sci. **375**, 20–40 (2007)
17. Johann, P., Simpson, A., Voigtländer, J.: A generic operational metatheory for algebraic effects. In: LICS (2010)
18. Lafont, Y., Reus, B., Streicher, T.: Continuations semantics or expressing implication by negation. Technical report 9321, Ludwig-Maximilians-Universität, München (1993)
19. Lassen, S.: Relational reasoning about functions and nondeterminism. Ph.D. thesis, University of Aarhus, BRICS, December 1998

20. Lassen, S.B., Levy, P.B.: Typed normal form bisimulation. In: Duparc, J., Henzinger, T.A. (eds.) CSL 2007. LNCS, vol. 4646, pp. 283–297. Springer, Heidelberg (2007). https://doi.org/10.1007/978-3-540-74915-8_23
21. Matache, C.: Program equivalence for algebraic effects via modalities. Master's thesis, University of Oxford, September 2018. https://arxiv.org/abs/1902.04645
22. Melliès, P.-A.: Local states in string diagrams. In: Dowek, G. (ed.) RTA 2014. LNCS, vol. 8560, pp. 334–348. Springer, Cham (2014). https://doi.org/10.1007/978-3-319-08918-8_23
23. Merro, M.: On the observational theory of the CPS-calculus. Acta Inf. **47**(2), 111–132 (2010)
24. Møgelberg, R.E., Staton, S.: Linear usage of state. Log. Meth. Comput. Sci. **10** (2014)
25. Moggi, E.: Notions of computation and monads. Inf. Comput. **93**(1), 55–92 (1991)
26. Morris, J.: Lambda calculus models of programming languages. Ph.D. thesis, MIT (1969)
27. Nanevski, A., Morrisett, J., Birkedal, L.: Hoare type theory, polymorphism and separation. J. Funct. Program. **18**(5–6), 865–911 (2008)
28. O'Hearn, P., Reynolds, J., Yang, H.: Local reasoning about programs that alter data structures. In: Fribourg, L. (ed.) CSL 2001. LNCS, vol. 2142, pp. 1–19. Springer, Heidelberg (2001). https://doi.org/10.1007/3-540-44802-0_1
29. Pitts, A.: Evaluation logic. In: Birtwistle, G. (ed.) IVth Higher Order Workshop, Banff 1990. Springer, Heidelberg (1991). https://doi.org/10.1007/978-1-4471-3182-3_11
30. Pitts, A.: Howe's method for higher-order languages. In: Sangiorgi, D., Rutten, J. (eds.) Advanced Topics in Bisimulation and Coinduction. Chapter 5, pp. 197–232. Cambridge University Press, Cambridge (2011)
31. Plotkin, G.: LCF considered as a programming language. Theor. Comput. Sci. **5**(3), 223–255 (1977)
32. Plotkin, G., Power, J.: Adequacy for algebraic effects. In: Honsell, F., Miculan, M. (eds.) FoSSaCS 2001. LNCS, vol. 2030, pp. 1–24. Springer, Heidelberg (2001). https://doi.org/10.1007/3-540-45315-6_1
33. Plotkin, G., Power, J.: Notions of computation determine monads. In: Nielsen, M., Engberg, U. (eds.) FoSSaCS 2002. LNCS, vol. 2303, pp. 342–356. Springer, Heidelberg (2002). https://doi.org/10.1007/3-540-45931-6_24
34. Plotkin, G., Power, J.: Algebraic operations and generic effects. Appl. Categ. Struct. **11**(1), 69–94 (2003)
35. Plotkin, G., Pretnar, M.: A logic for algebraic effects. In: LICS (2008)
36. Plotkin, G., Pretnar, M.: Handling Algebraic Effects. Log. Methods Comput. Sci. 9(4) (2013)
37. Power, J.: Indexed Lawvere theories for local state. In: Models, Logics and Higher-Dimensional Categories, pp. 268–282. AMS (2011)
38. Simpson, A., Voorneveld, N.: Behavioural equivalence via modalities for algebraic effects. In: Ahmed, A. (ed.) ESOP 2018. LNCS, vol. 10801, pp. 300–326. Springer, Cham (2018). https://doi.org/10.1007/978-3-319-89884-1_11
39. Staton, S.: Instances of computational effects. In: Proceedings of LICS 2013 (2013)
40. Staton, S., Levy, P.B.: Universal properties for impure programming languages. In: Proceedings of POPL 2013 (2013)

41. Yachi, T., Sumii, E.: A sound and complete bisimulation for contextual equivalence in λ-calculus with call/cc. In: Igarashi, A. (ed.) APLAS 2016. LNCS, vol. 10017, pp. 171–186. Springer, Heidelberg (2016). https://doi.org/10.1007/978-3-319-47958-3_10
42. Yoshida, N., Honda, K., Berger, M.: Logical reasoning for higher-order functions with local state. Log. Methods Comput. Sci. **4**(4) (2008)

Equational Axiomatization of Algebras with Structure

Stefan Milius and Henning Urbat(✉)

Friedrich-Alexander-Universität Erlangen-Nürnberg, Erlangen, Germany
henning.urbat@fau.de

Abstract. This paper proposes a new category theoretic account of equationally axiomatizable classes of algebras. Our approach is well-suited for the treatment of algebras equipped with additional computationally relevant structure, such as ordered algebras, continuous algebras, quantitative algebras, nominal algebras, or profinite algebras. Our main contributions are a generic HSP theorem and a sound and complete equational logic, which are shown to encompass numerous flavors of equational axiomizations studied in the literature.

1 Introduction

A key tool in the algebraic theory of data structures is their specification by operations (constructors) and equations that they ought to satisfy. Hence, the study of models of equational specifications has been of long standing interest both in mathematics and computer science. The seminal result in this field is Birkhoff's celebrated HSP theorem [7]. It states that a class of algebras over a signature Σ is a *variety* (i.e. closed under homomorphic images, subalgebras, and products) iff it is axiomatizable by equations $s = t$ between Σ-terms. Birkhoff also introduced a complete deduction system for reasoning about equations.

In algebraic approaches to the semantics of programming languages and computational effects, it is often natural to study algebras whose underlying sets are equipped with additional computationally relevant structure and whose operations preserve that structure. An important line of research thus concerns extensions of Birkhoff's theory of equational axiomatization beyond ordinary Σ-algebras. On the syntactic level, this requires to enrich Birkhoff's notion of an equation in ways that reflect the extra structure. Let us mention a few examples:

(1) *Ordered algebras* (given by a poset and monotone operations) and *continuous algebras* (given by a complete partial order and continuous operations) were identified by the ADJ group [14] as an important tool in denotational semantics. Subsequently, Bloom [8] and Adámek, Nelson, and Reiterman [2,3]

S. Milius—Supported by Deutsche Forschungsgemeinschaft (DFG) under project MI 717/5-1.
H. Urbat—Supported by Deutsche Forschungsgemeinschaft (DFG) under project SCHR 1118/8-2.

M. Bojańczyk and A. Simpson (Eds.): FOSSACS 2019, LNCS 11425, pp. 400–417, 2019.
https://doi.org/10.1007/978-3-030-17127-8_23

established ordered versions of the HSP theorem along with complete deduction systems. Here, the role of equations $s = t$ is taken over by inequations $s \leq t$.

(2) *Quantitative algebras* (given by an extended metric space and nonexpansive operations) naturally arise as semantic domains in the theory of probabilistic computation. In recent work, Mardare, Panangaden, and Plotkin [18,19] presented an HSP theorem for quantitative algebras and a complete deduction system. In the quantitative setting, equations $s =_\varepsilon t$ are equipped with a non-negative real number ε, interpreted as "s and t have distance at most ε".

(3) *Nominal algebras* (given by a nominal set and equivariant operations) are used in the theory of name binding [24] and have proven useful for characterizing logics for data languages [9,11]. Varieties of nominal algebras were studied by Gabbay [13] and Kurz and Petrişan [16]. Here, the appropriate syntactic concept involves equations $s = t$ with constraints on the support of their variables.

(4) *Profinite algebras* (given by a profinite topological space and continuous operations) play a central role in the algebraic theory of formal languages [22]. They serve as a technical tool in the investigation of *pseudovarieties* (i.e. classes of finite algebras closed under homomorphic images, subalgebras, and finite products). As shown by Reiterman [25] and Eilenberg and Schützenberger [12], pseudovarieties can be axiomatized by *profinite equations* (formed over free profinite algebras) or, equivalently, by sequences of ordinary equations $(s_i = t_i)_{i<\omega}$, interpreted as "all but finitely many of the equations $s_i = t_i$ hold".

The present paper proposes a general category theoretic framework that allows to study classes of algebras with extra structure in a systematic way. Our overall goal is to isolate the domain-specific part of any theory of equational axiomatization from its generic core. Our framework is parametric in the following data:

- a category $\mathscr{A}$ with a factorization system $(\mathcal{E}, \mathcal{M})$;
- a full subcategory $\mathscr{A}_0 \subseteq \mathscr{A}$;
- a class Λ of cardinal numbers;
- a class $\mathscr{X} \subseteq \mathscr{A}$ of objects.

Here, $\mathscr{A}$ is the category of algebras under consideration (e.g. ordered algebras, quantitative algebras, nominal algebras). Varieties are formed within $\mathscr{A}_0$, and the cardinal numbers in Λ determine the arities of products under which the varieties are closed. Thus, the choice $\mathscr{A}_0 =$ finite algebras and $\Lambda =$ finite cardinals corresponds to pseudovarieties, and $\mathscr{A}_0 = \mathscr{A}$ and $\Lambda =$ all cardinals to varieties. The crucial ingredient of our setting is the parameter $\mathscr{X}$, which is the class of objects over which equations are formed; thus, typically, $\mathscr{X}$ is chosen to be some class of freely generated algebras in $\mathscr{A}$. Equations are modeled as $\mathcal{E}$-quotients $e\colon X \twoheadrightarrow E$ (more generally, filters of such quotients) with domain $X \in \mathscr{X}$.

The choice of $\mathscr{X}$ reflects the desired expressivity of equations in a given setting. Furthermore, it determines the type of quotients under which equationally axiomatizable classes are closed. More precisely, in our general framework a *variety* is defined to be a subclass of $\mathscr{A}_0$ closed under $\mathcal{E}_{\mathscr{X}}$-quotients, $\mathcal{M}$-subobjects, and Λ-products, where $\mathcal{E}_{\mathscr{X}}$ is a subclass of $\mathcal{E}$ derived from $\mathscr{X}$. Due to its parametric nature, this concept of a variety is widely applicable and turns out to specialize to many interesting cases. The main result of our paper is the

General HSP Theorem. *A subclass of $\mathscr{A}_0$ forms a variety if and only if it is axiomatizable by equations.*

In addition, we introduce a generic deduction system for equations, based on two simple proof rules (see Sect. 4), and establish a

General Completeness Theorem. *The generic deduction system for equations is sound and complete.*

The above two theorems can be seen as the generic building blocks of the model theory of algebras with structure. They form the common core of numerous Birkhoff-type results and give rise to a systematic recipe for deriving concrete HSP and completeness theorems in settings such as (1)–(4). In fact, all that needs to be done is to translate our abstract notion of equation and equational deduction, which involves (filters of) quotients, into an appropriate syntactic concept. This is the domain-specific task to fulfill, and usually amounts to identifying an "exactness" property for the category $\mathscr{A}$. Subsequently, one can apply our general results to obtain HSP and completeness theorems for the type of algebras under consideration. Several instances of this approach are shown in Sect. 5. Omitted proofs and details for the examples can be found in [20].

Related work. Generic approaches to universal algebra have a long tradition in category theory. They aim to replace syntactic notions like terms and equations by suitable categorical abstractions, most prominently Lawvere theories and monads [4,17]. Our present work draws much of its inspiration from the classical paper of Banaschewski and Herrlich [6] on HSP classes in $(\mathcal{E}, \mathcal{M})$-structured categories. These authors were the first to model equations as quotients $e\colon X \twoheadrightarrow E$. However, their approach does not feature the parameter $\mathscr{X}$ and assumes that equations are formed over $\mathcal{E}$-projective objects X. This limits the scope of their results to categories with enough projectives, a property that typically fails in categories of algebras with structure (including continuous, quantitative or nominal algebras). The identification of the parameter $\mathscr{X}$ and of the derived parameter $\mathcal{E}_{\mathscr{X}}$ as a key concept is thus a crucial step towards a categorical view of such structures.

Equational logics on the level of abstraction of Banaschewski and Herrlich's work were studied by Roşu [26,27] and Adámek, Hébert, and Sousa [1]. These authors work under assumptions on the category $\mathscr{A}$ different from our framework, e.g. they require existence of pushouts. Hence, the proof rules and completeness results in *loc. cit.* are not directly comparable to our approach in Sect. 4.

In the present paper, we model equations as filters of quotients rather than single quotients, which allows us to encompass several HSP theorems for finite algebras [12,23,25]. The first categorical generalization of such results was given by Adámek, Chen, Milius, and Urbat [10,29] who considered algebras for a monad $\mathbb{T}$ on an algebraic category and modeled equations as filters of finite quotients of free $\mathbb{T}$-algebras (equivalently, as profinite quotients of free profinite $\mathbb{T}$-algebras). This idea was generalized by Salamánca [28] to monads on concrete categories. However, again, this work only applies to categories with enough projectives.

2 Preliminaries

We start by recalling some notions from category theory. A *factorization system* $(\mathcal{E}, \mathcal{M})$ in a category $\mathscr{A}$ consists of two classes $\mathcal{E}, \mathcal{M}$ of morphisms in $\mathscr{A}$ such that (1) both $\mathcal{E}$ and $\mathcal{M}$ contain all isomorphisms and are closed under composition, (2) every morphism f has a factorization $f = m \cdot e$ with $e \in \mathcal{E}$ and $m \in \mathcal{M}$, and (3) the *diagonal fill-in* property holds: for every commutative square $g \cdot e = m \cdot f$ with $e \in \mathcal{E}$ and $m \in \mathcal{M}$, there exists a unique d with $m \cdot d = g$ and $d \cdot e = f$. The morphisms m and e in (2) are unique up to isomorphism and are called the *image* and *coimage* of f, resp. The factorization system is *proper* if all morphisms in $\mathcal{E}$ are epic and all morphisms in $\mathcal{M}$ are monic. From now on, we will assume that $\mathscr{A}$ is a category equipped with a proper factorization system $(\mathcal{E}, \mathcal{M})$. Quotients and subobjects in $\mathscr{A}$ are taken with respect to $\mathcal{E}$ and $\mathcal{M}$. That is, a *quotient* of an object X is represented by a morphism $e\colon X \twoheadrightarrow E$ in $\mathcal{E}$ and a *subobject* by a morphism $m\colon M \rightarrowtail X$ in $\mathcal{M}$. The quotients of X are ordered by $e \le e'$ iff e' factorizes through e, i.e. there exists a morphism h with $e' = h \cdot e$. Identifying quotients e and e' which are isomorphic (i.e. $e \le e'$ and $e' \le e$), this makes the quotients of X a partially ordered class. Given a full subcategory $\mathscr{A}_0 \subseteq \mathscr{A}$ we denote by $X {\downarrow} \mathscr{A}_0$ the class of all quotients of X represented by $\mathcal{E}$-morphisms with codomain in $\mathscr{A}_0$. The category $\mathscr{A}$ is *$\mathcal{E}$-co-wellpowered* if for every object $X \in \mathscr{A}$ there is only a set of quotients with domain X. In particular, $X {\downarrow} \mathscr{A}_0$ is then a poset. Finally, an object $X \in \mathscr{A}$ is called *projective* w.r.t. a morphism $e\colon A \to B$ if for every $h\colon X \to B$, there exists a morphism $g\colon X \to A$ with $h = e \cdot g$.

3 The Generalized Variety Theorem

In this section, we introduce our categorical notions of equation and variety, and derive the HSP theorem. Fix a category $\mathscr{A}$ with a proper factorization system $(\mathcal{E}, \mathcal{M})$, a full subcategory $\mathscr{A}_0 \subseteq \mathscr{A}$, a class Λ of cardinal numbers, and a class $\mathscr{X} \subseteq \mathscr{A}$ of objects. An object of $\mathscr{A}$ is called *$\mathscr{X}$-generated* if it is a quotient of some object in $\mathscr{X}$. A key role will be played by the subclass $\mathcal{E}_{\mathscr{X}} \subseteq \mathcal{E}$ defined by

$$\mathcal{E}_{\mathscr{X}} = \{e \in \mathcal{E} \ : \ \text{every } X \in \mathscr{X} \text{ is projective w.r.t. } e\}.$$

Note that $\mathscr{X} \subseteq \mathscr{X}'$ implies $\mathcal{E}_{\mathscr{X}'} \subseteq \mathcal{E}_{\mathscr{X}}$. The choice of $\mathscr{X}$ is a trade-off between "having enough equations" (that is, $\mathscr{X}$ needs to be rich enough to make equations sufficiently expressive) and "having enough projectives" (cf. (3) below).

Assumptions 3.1. Our data is required to satisfy the following properties:

(1) $\mathscr{A}$ has Λ-products, i.e. for every $\lambda \in \Lambda$ and every family $(A_i)_{i<\lambda}$ of objects in $\mathscr{A}$, the product $\prod_{i<\lambda} A_i$ exists.
(2) $\mathscr{A}_0$ is closed under isomorphisms, Λ-products and $\mathscr{X}$-generated subobjects. The last statement means that for every subobject $m\colon A \rightarrowtail B$ in $\mathcal{M}$ where $B \in \mathscr{A}_0$ and A is $\mathscr{X}$-generated, one has $A \in \mathscr{A}_0$.
(3) Every object of $\mathscr{A}_0$ is an $\mathcal{E}_{\mathscr{X}}$-quotient of some object of $\mathscr{X}$, that is, for every object $A \in \mathscr{A}_0$ there exists some $e\colon X \twoheadrightarrow A$ in $\mathcal{E}_{\mathscr{X}}$ with domain $X \in \mathscr{X}$.

Example 3.2. Throughout this section, we will use the following three running examples to illustrate our concepts. For further applications, see Sect. 5.

(1) *Classical Σ-algebras.* The setting of Birkhoff's seminal work [7] in general algebra is that of algebras for a signature. Recall that a *(finitary) signature* is a set Σ of operation symbols each with a prescribed finite arity, and a *Σ-algebra* is a set A equipped with operations $\sigma\colon A^n \to A$ for each n-ary $\sigma \in \Sigma$. A *morphism* of Σ-algebras (or a *Σ-homomorphism*) is a map preserving all Σ-operations. The forgetful functor from the category $\mathbf{Alg}(\Sigma)$ of Σ-algebras and Σ-homomorphisms to $\mathbf{Set}$ has a left adjoint assigning to each set X the *free Σ-algebra* $T_\Sigma X$, carried by the set of all Σ-terms in variables from X. To treat Birkhoff's results in our categorical setting, we choose the following parameters:
   - $\mathscr{A} = \mathscr{A}_0 = \mathbf{Alg}(\Sigma)$;
   - $(\mathcal{E}, \mathcal{M}) =$ (surjective morphisms, injective morphisms);
   - $\Lambda =$ all cardinal numbers;
   - $\mathscr{X} =$ all free Σ-algebras $T_\Sigma X$ with $X \in \mathbf{Set}$.

   One easily verifies that $\mathcal{E}_{\mathscr{X}}$ consists of all surjective morphisms, that is, $\mathcal{E}_{\mathscr{X}} = \mathcal{E}$.
(2) *Finite Σ-algebras.* Eilenberg and Schützenberger [12] considered classes of finite Σ-algebras, where Σ is assumed to be a signature with only finitely many operation symbols. In our framework, this amounts to choosing
   - $\mathscr{A} = \mathbf{Alg}(\Sigma)$ and $\mathscr{A}_0 = \mathbf{Alg}_\mathsf{f}(\Sigma)$, the full subcategory of finite Σ-algebras;
   - $(\mathcal{E}, \mathcal{M}) =$ (surjective morphisms, injective morphisms);
   - $\Lambda =$ all finite cardinal numbers;
   - $\mathscr{X} =$ all free Σ-algebras $T_\Sigma X$ with $X \in \mathbf{Set}_\mathsf{f}$.

   As in (1), the class $\mathcal{E}_{\mathscr{X}}$ consists of all surjective morphisms.
(3) *Quantitative Σ-algebras.* In recent work, Mardare, Panangaden, and Plotkin [18,19] extended Birkhoff's theory to algebras endowed with a metric. Recall that an *extended metric space* is a set A with a map $d_A\colon A \times A \to [0, \infty]$ (assigning to any two points a possibly infinite distance), subject to the axioms (i) $d_A(a, b) = 0$ iff $a = b$, (ii) $d_A(a, b) = d_A(b, a)$, and (iii) $d_A(a, c) \leq d_A(a, b) + d_A(b, c)$ for all $a, b, c \in A$. A map $h\colon A \to B$ between extended metric spaces is *nonexpansive* if $d_B(h(a), h(a')) \leq d_A(a, a')$ for $a, a' \in A$. Let $\mathbf{Met}_\infty$ denote the category of extended metric spaces and nonexpansive maps. Fix a, not necessarily finitary, signature Σ, that is, the arity of an operation symbol $\sigma \in \Sigma$ is any cardinal number. A *quantitative Σ-algebra*

is a Σ-algebra A endowed with an extended metric d_A such that all Σ-operations $\sigma\colon A^n \to A$ are nonexpansive. Here, the product A^n is equipped with the sup-metric $d_{A^n}((a_i)_{i<n}, (b_i)_{i<n}) = \sup_{i<n} d_A(a_i, b_i)$. The forgetful functor from the category $\mathbf{QAlg}(\Sigma)$ of quantitative Σ-algebras and nonexpansive Σ-homomorphisms to $\mathbf{Met}_\infty$ has a left adjoint assigning to each space X the free quantitative Σ-algebra $T_\Sigma X$. The latter is carried by the set of all Σ-terms (equivalently, well-founded Σ-trees) over X, with metric inherited from X as follows: if s and t are Σ-terms of the same shape, i.e. they differ only in the variables, their distance is the supremum of the distances of the variables in corresponding positions of s and t; otherwise, it is ∞.

We aim to derive the HSP theorem for quantitative algebras proved by Mardare et al. as an instance of our general results. The theorem is parametric in a regular cardinal number $c > 1$. In the following, an extended metric space is called *c-clustered* if it is a coproduct of spaces of size $< c$. Note that coproducts in $\mathbf{Met}_\infty$ are formed on the level of underlying sets. Choose the parameters

- $\mathscr{A} = \mathscr{A}_0 = \mathbf{QAlg}(\Sigma)$;
- $(\mathcal{E}, \mathcal{M})$ given by morphisms carried by surjections and subspaces, resp.;
- $\Lambda =$ all cardinal numbers;
- $\mathscr{X} =$ all free algebras $T_\Sigma X$ with $X \in \mathbf{Met}_\infty$ a c-clustered space.

One can verify that a quotient $e\colon A \twoheadrightarrow B$ belongs to $\mathcal{E}_\mathscr{X}$ if and only if for each subset $B_0 \subseteq B$ of cardinality $< c$ there exists a subset $A_0 \subseteq A$ such that $e[A_0] = B_0$ and the restriction $e\colon A_0 \to B_0$ is isometric (that is, $d_B(e(a), e(a')) = d_A(a, a')$ for $a, a' \in A_0$). Following the terminology of Mardare et al., such a quotient is called *c-reflexive.* Note that for $c = 2$ every quotient is c-reflexive, so $\mathcal{E}_\mathscr{X} = \mathcal{E}$. If c is infinite, $\mathcal{E}_\mathscr{X}$ is a proper subclass of $\mathcal{E}$.

Definition 3.3. An *equation over* $X \in \mathscr{X}$ is a class $\mathscr{T}_X \subseteq X{\downarrow}\mathscr{A}_0$ that is

(1) *Λ-codirected:* every subset $F \subseteq \mathscr{T}_X$ with $|F| \in \Lambda$ has a lower bound in F;
(2) *closed under $\mathcal{E}_\mathscr{X}$-quotients:* for every $e\colon X \twoheadrightarrow E$ in $\mathscr{T}_X$ and $q\colon E \twoheadrightarrow E'$ in $\mathcal{E}_\mathscr{X}$ with $E' \in \mathscr{A}_0$, one has $q \cdot e \in \mathscr{T}_X$.

An object $A \in \mathscr{A}$ *satisfies* the equation $\mathscr{T}_X$ if every morphism $h\colon X \to A$ factorizes through some $e \in \mathscr{T}_X$. In this case, we write

$$A \models \mathscr{T}_X.$$

Remark 3.4. In many of our applications, one can simplify the above definition and replace classes of quotients by single quotients. Specifically, if $\mathscr{A}$ is $\mathcal{E}$-cowellpowered (so that every equation is a set, not a class) and $\Lambda =$ all cardinal numbers, then every equation $\mathscr{T}_X \subseteq X{\downarrow}\mathscr{A}_0$ contains a least element $e_X\colon X \twoheadrightarrow E_X$, viz. the lower bound of all elements in $\mathscr{T}_X$. Then an object A satisfies $\mathscr{T}_X$ iff it satisfies e_X, in the sense that every morphism $h\colon X \to A$ factorizes through e_X. Therefore, in this case, one may equivalently define an equation to be a morphism $e_X\colon X \twoheadrightarrow E_X$ with $X \in \mathscr{X}$. This is the concept of equation investigated by Banaschewski and Herrlich [6].

Example 3.5. In our running examples, we obtain the following concepts:

(1) *Classical Σ-algebras.* By Remark 3.4, an equation corresponds to a quotient $e_X: T_\Sigma X \twoheadrightarrow E_X$ in $\mathbf{Alg}(\Sigma)$, where X is a set of variables.
(2) *Finite Σ-algebras.* An equation $\mathscr{T}_X$ over a finite set X is precisely a filter (i.e. a codirected and upwards closed subset) in the poset $T_\Sigma X{\downarrow}\mathbf{Alg}_\mathsf{f}(\Sigma)$.
(3) *Quantitative Σ-algebras.* By Remark 3.4, an equation can be presented as a quotient $e_X: T_\Sigma X \twoheadrightarrow E_X$ in $\mathbf{QAlg}(\Sigma)$, where X is a c-clustered space.

We shall demonstrate in Sect. 5 how to interpret the above abstract notions of equations, i.e. (filters of) quotients of free algebras, in terms of concrete syntax.

Definition 3.6. A *variety* is a full subcategory $\mathcal{V} \subseteq \mathscr{A}_0$ closed under $\mathcal{E}_\mathscr{X}$-quotients, subobjects, and Λ-products. More precisely,

(1) for every $\mathcal{E}_\mathscr{X}$-quotient $e: A \twoheadrightarrow B$ in $\mathscr{A}_0$ with $A \in \mathcal{V}$ one has $B \in \mathcal{V}$,
(2) for every $\mathcal{M}$-morphism $m: A \rightarrowtail B$ in $\mathscr{A}_0$ with $B \in \mathcal{V}$ one has $A \in \mathcal{V}$, and
(3) for every family of objects A_i $(i < \lambda)$ in $\mathcal{V}$ with $\lambda \in \Lambda$ one has $\prod_{i<\lambda} A_i \in \mathcal{V}$.

Example 3.7. In our examples, we obtain the following notions of varieties:

(1) *Classical Σ-algebras.* A *variety of Σ-algebras* is a class of Σ-algebras closed under quotient algebras, subalgebras, and products. This is Birkhoff's original concept [7].
(2) *Finite Σ-algebras.* A *pseudovariety of Σ-algebras* is a class of finite Σ-algebras closed under quotient algebras, subalgebras, and finite products. This concept was studied by Eilenberg and Schützenberger [12].
(3) *Quantitative Σ-algebras.* For any regular cardinal number $c > 1$, a *c-variety of quantitative Σ-algebras* is a class of quantitative Σ-algebras closed under c-reflexive quotients, subalgebras, and products. This notion of a variety was introduced by Mardare et al. [19].

Construction 3.8. Given a class $\mathbb{E}$ of equations, put

$$\mathcal{V}(\mathbb{E}) = \{A \in \mathscr{A}_0 : A \models \mathscr{T}_X \text{ for each } \mathscr{T}_X \in \mathbb{E}\}.$$

A subclass $\mathcal{V} \subseteq \mathscr{A}_0$ is called *equationally presentable* if $\mathcal{V} = \mathcal{V}(\mathbb{E})$ for some $\mathbb{E}$.

We aim to show that varieties coincide with the equationally presentable classes (see Theorem 3.16 below). The "easy" part of the correspondence is established by the following lemma, which is proved by a straightforward verification.

Lemma 3.9. *For every class $\mathbb{E}$ of equations, $\mathcal{V}(\mathbb{E})$ is a variety.*

As a technical tool for establishing the general HSP theorem and the corresponding sound and complete equational logic, we introduce the following concept:

Definition 3.10. An *equational theory* is a family of equations

$$\mathscr{T} = (\mathscr{T}_X \subseteq X{\downarrow}\mathscr{A}_0)_{X \in \mathscr{X}}$$

with the following two properties (illustrated by the diagrams below):

(1) *Substitution invariance.* For every morphism $h\colon X \to Y$ with $X,Y \in \mathcal{X}$ and every $e_Y\colon Y \twoheadrightarrow E_Y$ in $\mathcal{T}_Y$, the coimage $e_X\colon X \twoheadrightarrow E_X$ of $e_Y \cdot h$ lies in $\mathcal{T}_X$.

(2) $\mathcal{E}_{\mathcal{X}}$-*completeness.* For every $Y \in \mathcal{X}$ and every quotient $e\colon Y \twoheadrightarrow E_Y$ in $\mathcal{T}_Y$, there exists an $X \in \mathcal{X}$ and a quotient $e_X\colon X \twoheadrightarrow E_X$ in $\mathcal{T}_X \cap \mathcal{E}_{\mathcal{X}}$ with $E_X = E_Y$.

$$
\begin{array}{ccc}
X \xrightarrow{\;\forall h\;} Y & \qquad & X \qquad Y \\
{\scriptstyle e_X}\downarrow \qquad \downarrow{\scriptstyle \forall e_Y} & & {\scriptstyle \exists e_X}\downarrow \qquad \downarrow{\scriptstyle \forall e_Y} \\
E_X \rightarrowtail E_Y & & E_X == E_Y
\end{array}
$$

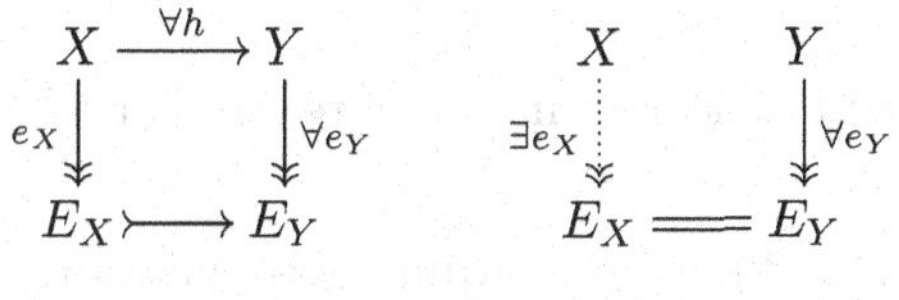

Remark 3.11. In many settings, the slightly technical concept of an equational theory can be simplified. First, note that $\mathcal{E}_{\mathcal{X}}$-completeness is trivially satisfied whenever $\mathcal{E}_{\mathcal{X}} = \mathcal{E}$. If, additionally, every equation contains a least element (e.g. in the setting of Remark 3.4), an equational theory corresponds exactly to a family of quotients $(e_X\colon X \twoheadrightarrow E_X)_{X \in \mathcal{X}}$ such that $E_X \in \mathscr{A}_0$ for all $X \in \mathcal{X}$, and for every $h\colon X \to Y$ with $X,Y \in \mathcal{X}$ the morphism $e_Y \cdot h$ factorizes through e_X.

Example 3.12 (Classical Σ-algebras). Recall that a *congruence* on a Σ-algebra A is an equivalence relation $\equiv\, \subseteq A \times A$ that forms a subalgebra of $A \times A$. It is well-known that there is an isomorphism of complete lattices

$$
\text{quotient algebras of } A \quad \cong \quad \text{congruences on } A \tag{3.1}
$$

assigning to a quotient $e\colon A \twoheadrightarrow B$ its *kernel*, given by $a \equiv_e a'$ iff $e(a) = e(a')$. Consequently, in the setting of Example 3.2(1), an equational theory – presented as a family of single quotients as in Remark 3.11 – corresponds precisely to a family of congruences $(\equiv_X\, \subseteq T_\Sigma X \times T_\Sigma X)_{X \in \mathbf{Set}}$ closed under substitution, that is, for every $s,t \in T_\Sigma X$ and every morphism $h\colon T_\Sigma X \to T_\Sigma Y$ in $\mathbf{Alg}(\Sigma)$,

$$
s \equiv_X t \quad \text{implies} \quad h(s) \equiv_Y h(t).
$$

We saw in Lemma 3.9 that every class of equations, so in particular every equational theory $\mathcal{T}$, yields a variety $\mathcal{V}(\mathcal{T})$ consisting of all objects of $\mathscr{A}_0$ that satisfy every equation in $\mathcal{T}$. Conversely, to every variety one can associate an equational theory as follows:

Construction 3.13. Given a variety $\mathcal{V}$, form the family of equations

$$
\mathcal{T}(\mathcal{V}) = (\mathcal{T}_X \subseteq X{\downarrow}\mathscr{A}_0)_{X \in \mathcal{X}},
$$

where $\mathcal{T}_X$ consists of all quotients $e_X\colon X \twoheadrightarrow E_X$ with codomain $E_X \in \mathcal{V}$.

Lemma 3.14. *For every variety $\mathcal{V}$, the family $\mathcal{T}(\mathcal{V})$ is an equational theory.*

We are ready to state the first main result of our paper, the HSP Theorem. Given two equations $\mathscr{T}_X$ and $\mathscr{T}'_X$ over $X \in \mathscr{X}$, we put $\mathscr{T}_X \leq \mathscr{T}'_X$ if every quotient in $\mathscr{T}'_X$ factorizes through some quotient in $\mathscr{T}_X$. Theories form a poset with respect to the order $\mathscr{T} \leq \mathscr{T}'$ iff $\mathscr{T}_X \leq \mathscr{T}'_X$ for all $X \in \mathscr{X}$. Similarly, varieties form a poset (in fact, a complete lattice) ordered by inclusion.

Theorem 3.15 (HSP Theorem). *The complete lattices of equational theories and varieties are dually isomorphic. The isomorphism is given by*

$$\mathcal{V} \mapsto \mathscr{T}(\mathcal{V}) \quad \textit{and} \quad \mathscr{T} \mapsto \mathcal{V}(\mathscr{T}).$$

One can recast the HSP Theorem into a more familiar form, using equations in lieu of equational theories:

Theorem 3.16 (HSP Theorem, equational version). *A class $\mathcal{V} \subseteq \mathscr{A}_0$ is equationally presentable if and only if it forms a variety.*

Proof. By Lemma 3.9, every equationally presentable class $\mathcal{V}(\mathbb{E})$ is a variety. Conversely, for every variety $\mathcal{V}$ one has $\mathcal{V} = \mathcal{V}(\mathscr{T}(\mathcal{V}))$ by Theorem 3.15, so $\mathcal{V}$ is presented by the equations $\mathbb{E} = \{\mathscr{T}_X : X \in \mathscr{X}\}$ where $\mathscr{T} = \mathscr{T}(\mathcal{V})$. □

4 Equational Logic

The correspondence between theories and varieties gives rise to the second main result of our paper, a generic sound and complete deduction system for reasoning about equations. The corresponding semantic concept is the following:

Definition 4.1. An equation $\mathscr{T}_X \subseteq X{\downarrow}\mathscr{A}_0$ *semantically entails* the equation $\mathscr{T}'_Y \subseteq Y{\downarrow}\mathscr{A}_0$ if every $\mathscr{A}_0$-object satisfying $\mathscr{T}_X$ also satisfies $\mathscr{T}'_Y$ (that is, if $\mathcal{V}(\mathscr{T}_X) \subseteq \mathcal{V}(\mathscr{T}_Y)$). In this case, we write $\mathscr{T}_X \models \mathscr{T}'_Y$.

The key to our proof system is a categorical formulation of term substitution:

Definition 4.2. Let $\mathscr{T}_X \subseteq X{\downarrow}\mathscr{A}_0$ be an equation over $X \in \mathscr{X}$. The *substitution closure* of $\mathscr{T}_X$ is the smallest theory $\overline{\mathscr{T}} = (\overline{\mathscr{T}}_Y)_{Y\in\mathscr{X}}$ such that $\mathscr{T}_X \leq \overline{\mathscr{T}}_X$.

The substitution closure of an equation can be computed as follows:

Lemma 4.3. *For every equation $\mathscr{T}_X \subseteq X{\downarrow}\mathscr{A}_0$ one has $\overline{\mathscr{T}} = \mathscr{T}(\mathcal{V}(\mathscr{T}_X))$.*

The deduction system for semantic entailment consists of two proof rules:

(Weakening) $\mathscr{T}_X \vdash \mathscr{T}'_X$ for all equations $\mathscr{T}'_X \leq \mathscr{T}_X$ over $X \in \mathscr{X}$.
(Substitution) $\mathscr{T}_X \vdash \overline{\mathscr{T}}_Y$ for all equations $\mathscr{T}_X$ over $X \in \mathscr{X}$ and all $Y \in \mathscr{X}$.

Given equations $\mathscr{T}_X$ and $\mathscr{T}'_Y$ over X and Y, respectively, we write $\mathscr{T}_X \vdash \mathscr{T}'_Y$ if $\mathscr{T}'_Y$ arises from $\mathscr{T}_X$ by a finite chain of applications of the above rules.

Theorem 4.4 (Completeness Theorem). *The deduction system for semantic entailment is sound and complete: for every pair of equations $\mathscr{T}_X$ and $\mathscr{T}'_Y$,*

$$\mathscr{T}_X \models \mathscr{T}'_Y \quad \textit{iff} \quad \mathscr{T}_X \vdash \mathscr{T}'_Y.$$

5 Applications

In this section, we present some of the applications of our categorical results (see [20] for full details). Transferring the general HSP theorem of Sect. 3 into a concrete setting requires to perform the following four-step procedure:

Step 1. Instantiate the parameters $\mathscr{A}$, $(\mathcal{E}, \mathcal{M})$, $\mathscr{A}_0$, Λ and $\mathscr{X}$ of our categorical framework, and characterize the quotients in $\mathcal{E}_{\mathscr{X}}$.
Step 2. Establish an *exactness property* for the category $\mathscr{A}$, i.e. a correspondence between quotients $e\colon A \twoheadrightarrow B$ in $\mathscr{A}$ and suitable relations between elements of A.
Step 3. Infer a suitable syntactic notion of equation, and prove it to be expressively equivalent to the categorical notion of equation given by Definition 3.3.
Step 4. Invoke Theorem 3.15 to deduce an HSP theorem.

The details of Steps 2 and 3 are application-specific, but typically straightforward. In each case, the bulk of the usual work required for establishing the HSP theorem is moved to our general categorical results and thus comes for free.

Similarly, to obtain a complete deduction system in a concrete application, it suffices to phrase the two proof rules of our generic equational logic in syntactic terms, using the correspondence of quotients and relations from Step 2; then Theorem 4.4 gives the completeness result.

5.1 Classical Σ-Algebras

The classical Birkhoff theorem emerges from our general results as follows.

Step 1. Choose the parameters of Example 3.2(1), and recall that $\mathcal{E}_{\mathscr{X}} = \mathcal{E}$.
Step 2. The exactness property of $\mathbf{Alg}(\Sigma)$ is given by the correspondence (3.1).
Step 3. Recall from Example 3.5(1) that equations can be presented as single quotients $e\colon T_\Sigma X \twoheadrightarrow E_X$. The exactness property (3.1) leads to the following classical syntactic concept: a *term equation* over a set X of variables is a pair $(s,t) \in T_\Sigma X \times T_\Sigma X$, denoted as $s = t$. It is *satisfied* by a Σ-algebra A if for every map $h\colon X \to A$ we have $h^\sharp(s) = h^\sharp(t)$. Here, $h^\sharp\colon T_\Sigma X \to A$ denotes the unique extension of h to a Σ-homomorphism. Equations and term equations are expressively equivalent in the following sense:

(1) For every equation $e\colon T_\Sigma X \twoheadrightarrow E_X$, the kernel $\equiv_e \subseteq T_\Sigma X \times T_\Sigma X$ is a set of term equations equivalent to e, that is, a Σ-algebra satisfies the equation e iff it satisfies all term equations in $\equiv_e$. This follows immediately from (3.1).
(2) Conversely, given a term equation $(s,t) \in T_\Sigma X \times T_\Sigma X$, form the smallest congruence $\equiv$ on $T_\Sigma X$ with $s \equiv t$ (viz. the intersection of all such congruences) and let $e\colon T_\Sigma X \twoheadrightarrow E_X$ be the corresponding quotient. Then a Σ-algebra satisfies $s = t$ iff it satisfies e. Again, this is a consequence of (3.1).

Step 4. From Theorem 3.16 and Example 3.7(1), we deduce the classical

Theorem 5.1 (Birkhoff [7]**).** *A class of Σ-algebras is a variety (i.e. closed under quotients, subalgebras, products) iff it is axiomatizable by term equations.*

Similarly, one can obtain Birkhoff's complete deduction system for term equations as an instance of Theorem 4.4; see [20, Section B.1] for details.

5.2 Finite Σ-Algebras

Next, we derive Eilenberg and Schützenberger's equational characterization of pseudovarieties of algebras over a finite signature Σ using our four-step plan:

Step 1. Choose the parameters of Example 3.2(2), and recall that $\mathcal{E}_{\mathscr{X}} = \mathcal{E}$.
Step 2. The exactness property of $\mathbf{Alg}(\Sigma)$ is given by (3.1).
Step 3. By Example 3.2(2), an equational theory is given by a family of filters $\mathscr{T}_n \subseteq T_\Sigma n{\downarrow}\mathbf{Alg}_\mathsf{f}(\Sigma)$ $(n < \omega)$. The corresponding syntactic concept involves sequences $(s_i = t_i)_{i<\omega}$ of term equations. We say that a finite Σ-algebra A *eventually satisfies* such a sequence if there exists $i_0 < \omega$ such that A satisfies all equations $s_i = t_i$ with $i \geq i_0$. Equational theories and sequences of term equations are expressively equivalent:

(1) Let $\mathscr{T} = (\mathscr{T}_n)_{n<\omega}$ be a theory. Since Σ is a finite signature, for each finite quotient $e\colon T_\Sigma n \twoheadrightarrow E$ the kernel $\equiv_e$ is a finitely generated congruence [12, Prop. 2]. Consequently, for each $n < \omega$ the algebra $T_\Sigma n$ has only countably many finite quotients. In particular, the codirected poset $\mathscr{T}_n$ is countable, so it contains an ω^{op}-chain $e_0^n \geq e_1^n \geq e_2^n \geq \cdots$ that is *cofinal*, i.e., each $e \in \mathscr{T}_n$ is above some e_i^n. The e_i^n can be chosen in such a way that, for each $m > n$ and $q\colon m \to n$, the morphism $e_i^n \cdot T_\Sigma q$ factorizes through e_i^m. For each $n < \omega$, choose a finite subset $W_n \subseteq T_\Sigma n \times T_\Sigma n$ generating the kernel of e_n^n. Let $(s_i = t_i)_{i<\omega}$ be a sequence of term equations where (s_i, t_i) ranges over $\bigcup_{n<\omega} W_n$. One can verify that a finite Σ-algebra lies in $\mathcal{V}(\mathscr{T})$ iff it eventually satisfies $(s_i = t_i)_{i<\omega}$.

(2) Conversely, given a sequence of term equations $(s_i = t_i)_{i<\omega}$ with $(s_i, t_i) \in T_\Sigma m_i \times T_\Sigma m_i$, form the theory $\mathscr{T} = (\mathscr{T}_n)_{n<\omega}$ where $\mathscr{T}_n$ consists of all finite quotients $e\colon T_\Sigma n \twoheadrightarrow E$ with the following property:

$$\exists i_0 < \omega : \forall i \geq i_0 : \forall (g\colon T_\Sigma m_i \to T_\Sigma n) : e \cdot g(s_i) = e \cdot g(t_i).$$

Then a finite Σ-algebra eventually satisfies $(s_i = t_i)_{i<\omega}$ iff it lies in $\mathcal{V}(\mathscr{T})$.

Step 4. The theory version of our HSP theorem (Theorem 3.16) now implies:

Theorem 5.2 (Eilenberg-Schützenberger [12]**).** *A class of finite Σ-algebras is a pseudovariety (i.e. closed under quotients, subalgebras, and finite products) iff it is axiomatizable by a sequence of term equations.*

In an alternative characterization of pseudovarieties due to Reiterman [25], where the restriction to finite signatures Σ can be dropped, sequences of term equations are replaced by the topological concept of a *profinite equation*. This result can also be derived from our general HSP theorem, see [20, Section B.4].

5.3 Quantitative Algebras

In this section, we derive an HSP theorem for quantitative algebras.

Step 1. Choose the parameters of Example 3.2(3). Recall that we work with fixed regular cardinal $c > 1$ and that $\mathcal{E}_{\mathscr{X}}$ consists of all c-reflexive quotients.
Step 2. To state the exactness property of $\mathbf{QAlg}(\Sigma)$, recall that an *(extended) pseudometric* on a set A is a map $p\colon A\times A \to [0,\infty]$ satisfying all axioms of an extended metric except possibly the implication $p(a,b) = 0 \Rightarrow a = b$. Given a quantitative Σ-algebra A, a pseudometric p on A is called a *congruence* if (i) $p(a,a') \le d_A(a,a')$ for all $a, a' \in A$, and (ii) every Σ-operation $\sigma\colon A^n \to A$ ($\sigma \in \Sigma$) is nonexpansive w.r.t. p. Congruences are ordered by $p \le q$ iff $p(a,a') \le q(a,a')$ for all $a, a' \in A$. There is a dual isomorphism of complete lattices

$$\text{quotient algebras of } A \quad \cong \quad \text{congruences on } A \tag{5.1}$$

mapping $e\colon A \twoheadrightarrow B$ to the congruence p_e on A given by $p_e(a,b) = d_B(e(a),e(b))$.
Step 3. By Example 3.5(3), equations can be presented as single quotients $e\colon T_\Sigma X \twoheadrightarrow E$, where X is a c-clustered space. The exactness property (5.1) suggests to replace equations by the following syntactic concept. A *c-clustered equation* over the set X of variables is an expression

$$x_i =_{\varepsilon_i} y_i \ (i \in I) \ \vdash \ s =_\varepsilon t \tag{5.2}$$

where (i) I is a set, (ii) $x_i, y_i \in X$ for all $i \in I$, (iii) s and t are Σ-terms over X, (iv) $\varepsilon_i, \varepsilon \in [0,\infty]$, and (v) the equivalence relation on X generated by the pairs (x_i, y_i) $(i \in I)$ has all equivalence classes of cardinality $< c$. In other words, the set of variables can be partitioned into subsets of size $< c$ such that only relations between variables in the same subset appear on the left-hand side of (5.2). A quantitative Σ-algebra A *satisfies* (5.2) if for every map $h\colon X \to A$ with $d_A(h(x_i), h(y_i)) \le \varepsilon_i$ for all $i \in I$, one has $d_A(h^\sharp(s), h^\sharp(t)) \le \varepsilon$. Here $h^\sharp\colon T_\Sigma X \to A$ denotes the unique Σ-homomorphism extending h.

Equations and c-clustered equations are expressively equivalent:
(1) Let X be a c-clustered space, i.e. $X = \coprod_{j\in J} X_j$ with $|X_j| < c$. Every equation $e\colon T_\Sigma X \twoheadrightarrow E$ induces a set of c-clustered equations over X given by

$$x =_{\varepsilon_{x,y}} y \ (j \in J,\ x, y \in X_j) \ \vdash \ s =_{\varepsilon_{s,t}} t \quad (s,t \in T_\Sigma X), \tag{5.3}$$

with $\varepsilon_{x,y} = d_X(x,y)$ and $\varepsilon_{s,t} = d_E(e(s), e(t))$. It is not difficult to show that e and (5.3) are equivalent: an algebra satisfies e iff it satisfies all equations (5.3).
(2) Conversely, to every c-clustered equation (5.2) over a set X of variables, we associate an equation in two steps:
   - Let p the largest pseudometric on X with $p(x_i, y_i) \le \varepsilon_i$ for all i (that is, the pointwise supremum of all such pseudometrics). Form the corresponding quotient $e_p\colon X \twoheadrightarrow X_p$, see (5.1). It is easy to see that X_p is c-clustered.

- Let q be the largest congruence on $T_\Sigma(X_p)$ with $q(T_\Sigma e_p(s), T_\Sigma e_p(t)) \leq \varepsilon$ (that is, the pointwise supremum of all such congruences). Form the corresponding quotient $e_q\colon T_\Sigma(X_p) \twoheadrightarrow E_q$.

A routine verification shows that (5.2) and e_q are expressively equivalent, i.e. satisfied by the same quantitative Σ-algebras.

Step 4. From Theorem 3.16 and Example 3.7(3), we deduce the following

Theorem 5.3 (Quantitative HSP Theorem). *A class of quantitative Σ-algebras is a c-variety (i.e. closed under c-reflexive quotients, subalgebras, and products) iff it is axiomatizable by c-clustered equations.*

The above theorem generalizes a recent result of Mardare, Panangaden, and Plotkin [19] who considered only signatures Σ with operations of finite or countably infinite arity and cardinal numbers $c \leq \aleph_1$. Theorem 5.3 holds without any restrictions on Σ and c. In addition to the quantitative HSP theorem, one can also derive the completeness of quantitative equational logic [18] from our general completeness theorem, see [20, Section B.5] for details.

5.4 Nominal Algebras

In this section, we derive an HSP theorem for algebras in the category **Nom** of nominal sets and equivariant maps; see Pitts [24] for the required terminology. We denote by $\mathbb{A}$ the countably infinite set of atoms, by $\mathrm{Perm}(\mathbb{A})$ the group of finite permutations of $\mathbb{A}$, and by $\mathsf{supp}_X(x)$ the least support of an element x of a nominal set X. Recall that X is *strong* if, for all $x \in X$ and $\pi \in \mathrm{Perm}(\mathbb{A})$,

$$[\forall a \in \mathsf{supp}_X(x) : \pi(a) = a] \iff \pi \cdot x = x.$$

A *supported set* is a set X equipped with a map $\mathsf{supp}_X\colon X \to \mathcal{P}_f(\mathbb{A})$. A *morphism* $f\colon X \to Y$ of supported sets is a function with $\mathsf{supp}_Y(f(x)) \subseteq \mathsf{supp}_X(x)$ for all $x \in X$. Every nominal set X is a supported set w.r.t. its least-support map supp_X. The following lemma, whose first part is a reformulation of [21, Prop. 5.10], gives a useful description of strong nominal sets in terms of supported sets.

Lemma 5.4. *The forgetful functor from* **Nom** *to* **SuppSet** *has a left adjoint* $F\colon$ **SuppSet** $\to$ **Nom**. *The nominal sets of the form FY ($Y \in$* **SuppSet***) are up to isomorphism exactly the strong nominal sets.*

Fix a finitary signature Σ. A *nominal Σ-algebra* is a Σ-algebra A carrying the structure of a nominal set such that all Σ-operations $\sigma\colon A^n \to A$ are equivariant. The forgetful functor from the category **NomAlg**(Σ) of nominal Σ-algebras and equivariant Σ-homomorphisms to **Nom** has a left adjoint assigning to each nominal set X the *free nominal Σ-algebra* $T_\Sigma X$, carried by the set of Σ-terms and with group action inherited from X. To derive a nominal HSP theorem from our general categorical results, we proceed as follows.

Step 1. Choose the parameters of our setting as follows:
- $\mathscr{A} = \mathscr{A}_0 = \mathbf{NomAlg}(\Sigma)$;
- $(\mathcal{E}, \mathcal{M}) =$ (surjective morphisms, injective morphisms);
- $\Lambda =$ all cardinal numbers;
- $\mathscr{X} = \{T_\Sigma X \,:\, X \text{ is a strong nominal set}\}$.

One can show that a quotient $e\colon A \twoheadrightarrow B$ belongs to $\mathcal{E}_{\mathscr{X}}$ iff it is *support-reflecting*: for every $b \in B$ there exists $a \in A$ with $e(a) = b$ and $\mathsf{supp}_A(a) = \mathsf{supp}_B(b)$.

Step 2. A *nominal congruence* on a nominal Σ-algebra A is a Σ-algebra congruence $\equiv\, \subseteq A \times A$ that forms an equivariant subset of $A \times A$. In analogy to (3.1), there is an isomorphism of complete lattices

$$\text{quotient algebras of } A \quad \cong \quad \text{nominal congruences on } A. \tag{5.4}$$

Step 3. By Remark 3.4, an equation can be presented as a single quotient $e\colon T_\Sigma X \twoheadrightarrow E$, where X is a strong nominal set. Equations can be described by syntactic means as follows. A *nominal Σ-term* over a set Y of variables is an element of $T_\Sigma(\mathrm{Perm}(\mathbb{A}) \times Y)$. Every map $h\colon Y \to A$ into a nominal Σ-algebra A extends to the Σ-homomorphism

$$\hat{h} = (T_\Sigma(\mathrm{Perm}(\mathbb{A}) \times Y) \xrightarrow{T_\Sigma(\mathrm{Perm}(\mathbb{A})\times h)} T_\Sigma(\mathrm{Perm}(\mathbb{A}) \times A) \xrightarrow{T_\Sigma(-\cdot-)} T_\Sigma A \xrightarrow{id^\sharp} A)$$

where $id^\sharp$ is the unique Σ-homomorphism extending the identity map $id\colon A \to A$. A *nominal equation* over Y is an expression of the form

$$\mathsf{supp}_Y \vdash s = t, \tag{5.5}$$

where $\mathsf{supp}_Y\colon Y \to \mathcal{P}_f(\mathbb{A})$ is a function and s and t are nominal Σ-terms over Y. A nominal Σ-algebra A *satisfies* the equation $\mathsf{supp}_Y \vdash s = t$ if for every map $h\colon Y \to A$ with $\mathsf{supp}_A(h(y)) \subseteq \mathsf{supp}_Y(y)$ for all $y \in Y$ one has $\hat{h}(s) = \hat{h}(t)$. Equations and nominal equations are expressively equivalent:

(1) Given an equation $e\colon T_\Sigma X \twoheadrightarrow E$ with X a strong nominal set, choose a supported set Y with $X = FY$, and denote by $\eta_Y\colon Y \to FY$ the universal map (see Lemma 5.4). Form the nominal equations over Y given by

$$\mathsf{supp}_Y \vdash s = t \quad (s, t \in T_\Sigma(\mathrm{Perm}(\mathbb{A}) \times Y) \text{ and } e \cdot T_\Sigma m(s) = e \cdot T_\Sigma m(t)) \tag{5.6}$$

where m is the composite $\mathrm{Perm}(\mathbb{A}) \times Y \xrightarrow{\mathrm{Perm}(\mathbb{A})\times\eta_Y} \mathrm{Perm}(\mathbb{A}) \times X \xrightarrow{-\cdot-} X$. It is not difficult to see that a nominal Σ-algebra satisfies e iff it satisfies (5.6).

(2) Conversely, given a nominal equation (5.5) over the set Y, let $X = FY$ and form the nominal congruence on $T_\Sigma X$ generated by the pair $(T_\Sigma m(s), T_\Sigma m(t))$, with m defined as above. Let $e\colon T_\Sigma X \twoheadrightarrow E$ be the corresponding quotient, see (5.4). One can show that a nominal Σ-algebra satisfies e iff it satisfies (5.5).

Step 4. We thus deduce the following result as an instance of Theorem 3.16:

Theorem 5.5 (Kurz and Petrişan [16]). *A class of nominal Σ-algebras is a variety (i.e. closed under support-reflecting quotients, subalgebras, and products) iff it is axiomatizable by nominal equations.*

For brevity and simplicity, in this section we restricted ourselves to algebras for a signature. Kurz and Petrişan proved a more general HSP theorem for algebras over an endofunctor on **Nom** with a suitable finitary presentation. This extra generality allows to incorporate, for instance, algebras for binding signatures.

5.5 Further Applications

Let us briefly mention some additional instances of our framework, all of which are given a detailed treatment in the full arXiv paper [20].

Ordered Algebras. Bloom [8] proved an HSP theorem for Σ-algebras in the category of posets: a class of such algebras is closed under homomorphic images, subalgebras, and products, iff it is axiomatizable by inequations $s \leq t$ between Σ-terms. This result can be derived much like the unordered case in Sect. 5.1.

Continuous Algebras. A more intricate ordered version of Birkhoff's theorem concerns *continuous algebras*, i.e. Σ-algebras with an ω-cpo structure on their underlying set and continuous Σ-operations. Adámek, Nelson, and Reiterman [3] proved that a class of continuous algebras is closed under homomorphic images, subalgebras, and products, iff it axiomatizable by inequations between terms with formal suprema (e.g. $\sigma(x) \leq \bigvee_{i<\omega} c_i$). This result again emerges as an instance of our general HSP theorem. A somewhat curious feature of this application is that the appropriate factorization system $(\mathcal{E}, \mathcal{M})$ takes as $\mathcal{E}$ the class of dense morphisms, i.e. morphisms of $\mathcal{E}$ are not necessarily surjective. However, one has $\mathcal{E}_{\mathscr{X}} =$ surjections, so homomorphic images are formed in the usual sense.

Abstract HSP Theorems. Our results subsume several existing categorical generalizations of Birkhoff's theorem. For instance, Theorem 3.15 yields Manes' [17] correspondence between quotient monads $\mathbb{T} \twoheadrightarrow \mathbb{T}'$ and varieties of $\mathbb{T}$-algebras for any monad $\mathbb{T}$ on **Set**. Similarly, Banaschewski and Herrlich's [6] HSP theorem for objects in categories with enough projectives is a special case of Theorem 3.16.

6 Conclusions and Future Work

We have presented a categorical approach to the model theory of algebras with additional structure. Our framework applies to a broad range of different settings and greatly simplifies the derivation of HSP-type theorems and completeness results for equational deduction systems, as the generic part of such derivations now comes for free using our Theorems 3.15, 3.16 and 4.4. There remain a number of interesting directions and open questions for future work.

As shown in Sect. 5, the key to arrive at a syntactic notion of equation lies in identifying a correspondence between quotients and suitable relations, which we informally coined "exactness". The similarity of these correspondences in our applications suggests that there should be a (possibly enriched) notion of *exact category* that covers our examples; cf. Kurz and Velebil's [15] 2-categorical view of ordered algebras. This would allow to move more work to the generic theory.

Theorem 4.4 can be used to recover several known sound and complete equational logics, but it also applies to settings where no such logic is known, for instance, a logic of profinite equations (however, cf. recent work of Almeida and Klíma [5]). In each case, the challenge is to translate our two abstract proof rules into concrete syntax, which requires the identification of a syntactic equivalent of the two properties of an equational theory. While substitution invariance always translates into a syntactic substitution rule in a straightforward manner, $\mathcal{E}_{\mathscr{X}}$-completeness does not appear to have an obvious syntactic counterpart. In most of the cases where a concrete equational logic is known, this issue is obfuscated by the fact that one has $\mathcal{E}_{\mathscr{X}} = \mathcal{E}$, so $\mathcal{E}_{\mathscr{X}}$-completeness becomes a trivial property. Finding a syntactic account of $\mathcal{E}_{\mathscr{X}}$-completeness remains an open problem. One notable case where $\mathcal{E}_{\mathscr{X}} \neq \mathcal{E}$ is the one of nominal algebras. Gabbay's work [13] does provide an HSP theorem and a sound and complete equational logic in a setting slightly different from Sect. 5.4, and it should be interesting to see whether this can be obtained as an instance of our framework.

Finally, in previous work [29] we have introduced the notion of a *profinite theory* (a special case of the equational theories in the present paper) and shown how the dual concept can be used to derive Eilenberg-type correspondences between varieties of languages and pseudovarieties of finite algebras. Our present results pave the way to an extension of this method to new settings, such as nominal sets. Indeed, a simple modification of the parameters in Sect. 5.4 yields a new HSP theorem for *orbit-finite* nominal Σ-algebras. We expect that a dualization of this result in the spirit of *loc. cit.* leads to a correspondence between varieties of data languages and varieties of orbit-finite nominal monoids, an important step towards an algebraic theory of data languages.

Acknowledgement. The authors would like to thank Thorsten Wißmann for insightful discussions on nominal sets.

References

1. Adámek, J., Hébert, M., Sousa, L.: A logic of injectivity. J. Homotopy Relat. Struct. **2**(2), 13–47 (2007)
2. Adámek, J., Mekler, A.H., Nelson, E., Reiterman, J.: On the logic of continuous algebras. Notre Dame J. Formal Logic **29**(3), 365–380 (1988)
3. Adámek, J., Nelson, E., Reiterman, J.: The Birkhoff variety theorem for continuous algebras. Algebra Univers. **20**(3), 328–350 (1985)
4. Adámek, J., Rosický, J., Vitale, E.M.: Algebraic Theories: A Categorical Introduction to General Algebra. Cambridge Tracts in Mathematics. Cambridge University Press, Cambridge (2010)

5. Almeida, J., Klíma, O.: Towards a pseudoequational proof theory. arXiv preprint arXiv:1708.09681 (2017)
6. Banaschewski, B., Herrlich, H.: Subcategories defined by implications. Houston J. Math. **2**(2), 149–171 (1976)
7. Birkhoff, G.: On the structure of abstract algebras. Proc. Camb. Philos. Soc. **10**, 433–454 (1935)
8. Bloom, S.L.: Varieties of ordered algebras. J. Comput. Syst. Sci. **2**(13), 200–212 (1976)
9. Bojańczyk, M.: Nominal monoids. Theory Comput. Syst. **53**(2), 194–222 (2013)
10. Chen, L.-T., Adámek, J., Milius, S., Urbat, H.: Profinite monads, profinite equations, and Reiterman's theorem. In: Jacobs, B., Löding, C. (eds.) FoSSaCS 2016. LNCS, vol. 9634, pp. 531–547. Springer, Heidelberg (2016). https://doi.org/10.1007/978-3-662-49630-5_31
11. Colcombet, T., Ley, C., Puppis, G.: Logics with rigidly guarded data tests. Log. Methods Comput. Sci. **11**(3) (2015)
12. Eilenberg, S., Schützenberger, M.P.: On pseudovarieties. Adv. Math. **10**, 413–418 (1976)
13. Gabbay, M.J.: Nominal algebra and the HSP theorem. J. Logic Comput. **19**, 341–367 (2009)
14. Goguen, J.A., Thatcher, J.W., Wagner, E.G., Wright, J.B.: Initial algebra semantics and continuous algebras. J. ACM **24**(1), 68–95 (1977)
15. Kurz, A., Velebil, J.: Quasivarieties and varieties of ordered algebras: regularity and exactness. Math. Struct. Comput. Sci. **27**, 1153–1194 (2017)
16. Kurz, A., Petrisan, D.: On universal algebra over nominal sets. Math. Struct. Comput. Sci. **20**(2), 285–318 (2010)
17. Manes, E.G.: Algebraic Theories. Graduate Texts in Mathematics, vol. 26. Springer, New York (1976). https://doi.org/10.1007/978-1-4612-9860-1
18. Mardare, R., Panangaden, P., Plotkin, G.: Quantitative algebraic reasoning. In: Proceedings of the 31st Annual ACM/IEEE Symposium on Logic in Computer Science, LICS 2016, pp. 700–709. ACM (2016)
19. Mardare, R., Panangaden, P., Plotkin, G.: On the axiomatizability of quantitative algebras. In: 32nd Annual ACM/IEEE Symposium on Logic in Computer Science, LICS 2017, Reykjavik, Iceland, 20–23 June 2017, pp. 1–12. IEEE Computer Society (2017). https://doi.org/10.1109/LICS.2017.8005102
20. Milius, S., Urbat, H.: Equational axiomatization of algebras with structure. CoRR abs/1812.02016 (2018). http://arxiv.org/abs/1812.02016
21. Milius, S., Schröder, L., Wißmann, T.: Regular behaviours with names. Appl. Categorical Struct. **24**(5), 663–701 (2016)
22. Pin, J.É.: Profinite methods in automata theory. In: Albers, S., Marion, J.Y. (eds.) 26th International Symposium on Theoretical Aspects of Computer Science STACS 2009, pp. 31–50. IBFI Schloss Dagstuhl (2009)
23. Pin, J.É., Weil, P.: A Reiterman theorem for pseudovarieties of finite first-order structures. Algebra Univers. **35**, 577–595 (1996)
24. Pitts, A.M.: Nominal Sets: Names and Symmetry in Computer Science. Cambridge University Press, Cambridge (2013)
25. Reiterman, J.: The Birkhoff theorem for finite algebras. Algebra Univers. **14**(1), 1–10 (1982)
26. Roşu, G.: Complete categorical equational deduction. In: Fribourg, L. (ed.) CSL 2001. LNCS, vol. 2142, pp. 528–538. Springer, Heidelberg (2001). https://doi.org/10.1007/3-540-44802-0_37

27. Roşu, G.: Complete categorical deduction for satisfaction as injectivity. In: Futatsugi, K., Jouannaud, J.-P., Meseguer, J. (eds.) Algebra, Meaning, and Computation. LNCS, vol. 4060, pp. 157–172. Springer, Heidelberg (2006). https://doi.org/10.1007/11780274_9
28. Salamánca, J.: Unveiling Eilenberg-type Correspondences: Birkhoff's Theorem for (finite) Algebras + Duality (2017). https://arxiv.org/abs/1702.02822
29. Urbat, H., Adámek, J., Chen, L., Milius, S.: Eilenberg theorems for free. CoRR abs/1602.05831 (2017). http://arxiv.org/abs/1602.05831

Towards a Structural Proof Theory of Probabilistic μ-Calculi

Christophe Lucas[1(✉)] and Matteo Mio[2(✉)]

[1] ENS–Lyon, Lyon, France
christophe.lucas@ens-lyon.fr
[2] CNRS and ENS–Lyon, Lyon, France
matteo.mio@ens-lyon.fr

Abstract. We present a structural proof system, based on the machinery of hypersequent calculi, for a simple probabilistic modal logic underlying very expressive probabilistic μ-calculi. We prove the soundness and completeness of the proof system with respect to an equational axiomatisation and the fundamental cut-elimination theorem.

1 Introduction

Modal and temporal logics are formalisms designed to express properties of mathematical structures representing the behaviour of computing systems, such as, e.g., Kripke frames, trees and labeled transition systems. A fundamental problem regarding such logics is the *equivalence problem*: given two formulas ϕ and ψ, establish whether ϕ and ψ are semantically equivalent. For many temporal logics, including the basic modal logic K (see, e.g., [BdRV02]) and its many extensions such as the *modal μ-calculus* [Koz83], the equivalence problem is decidable and can be answered automatically. This is, of course, a very desirable fact. However, a fully automatic approach is not always viable due to the high complexity of the algorithms involved. An alternative and complementary approach is to use *human-aided* proof systems for constructing *formal proofs* of the desired equalities. As a concrete example, the well-known equational axioms of Boolean algebras together with two axioms for the $\Diamond$ modality:

$$\Diamond\bot = \bot \qquad \Diamond(x \vee y) = \Diamond(x) \vee \Diamond(y)$$

can be used to construct formal proofs of all valid equalities between formulas of modal logic using the familiar deductive rules of *equational logic* (see Definition 3). The simplicity of equational logic is a great feature of this kind of system but sometimes comes at a cost because even seemingly trivial equalities often require significant human ingenuity to be proved.[1] The problem lies in

[1] Example: the law of idempotence $x \vee x = x$ can be derived from the standard axioms of Boolean algebras (i.e., complemented distributive lattices) as: $x \vee x = (x \vee x) \wedge \top = (x \vee x) \wedge (x \vee \neg x) = x \vee (x \wedge \neg x) = x \vee \bot = x$.

The authors were supported by the French project ANR-16-CE25-0011 REPAS.

M. Bojańczyk and A. Simpson (Eds.): FOSSACS 2019, LNCS 11425, pp. 418–435, 2019.
https://doi.org/10.1007/978-3-030-17127-8_24

the *transitivity rule* ($a = b$ & $b = c \Rightarrow a = c$) which requires to guess, among infinitely many possibilities, an interpolant formula b to prove the equality $a = c$.

The field of *structural proof theory* (see [Bus98]), originated with the seminal work of Gentzen on his *sequent calculus* proof system LK for classical propositional (first-order) logic [Gen34], investigates proof systems which, roughly speaking, require less human ingenuity. The key technical result regarding the sequent calculus, the *cut-elimination theorem*, implies that when searching for a proof of a statement, only certain formulas need to be considered: the so-called *sub-formula property*. This simplifies significantly, in practice, the *proof search* endeavour. The original system LK of Gentzen has been extensively investigated and generalised and, for example, it can be extended with rules for the $\Diamond$ modality and becomes a convenient proof system for modal logic [Wan96]. Furthermore, it is possible to extend it with rules for dealing with (co)inductive definitions and it becomes a proof system for the modal μ-calculus (see, e.g., [Stu07]). Research on the structural proof theory of the modal μ-calculus is an active area of research (see, e.g., recent [Dou17]).

Probabilistic Logics and the Riesz Modal Logic. Probabilistic logics are temporal logics specifically designed to express properties of mathematical structures (e.g., Markov chains and Markov decision processes) representing the behaviour of computing systems using probabilistic features such as random bit generation. Unlike the non-probabilistic case, the equivalence problem for most expressive probabilistic logics (e.g., *pCTL* [LS82,HJ94], see also [BK08,BBLM17]) is not known to be decidable. Hence, human-aided proof systems are currently the only viable approach to establish equalities of formulas of expressive probabilistic logics. To the best of our knowledge, however, all the proof systems proposed in the literature (see, e.g., [DFHM16] for the logic pCTL, [BGZB09,Hsu17] for pRHL and [Koz85] for pPDL) are not entirely satisfactory because they include rules, such as the transitivity rule discussed above, violating the sub-formula property.

Another line of work on probabilistic logics has focused on *probabilistic μ-calculi* ([MM07,HK97,DGJP00,dA03,MS17,Mio11,Mio12a,Mio14]). These logical formalisms are, similarly to Kozen's modal μ-calculus, obtained by extending a base *real-valued* modal logic with (co)inductively defined operators. Recently, in [MFM17], a base real-valued modal logic called *Riesz modal logic* ($\mathcal{R}$) has been defined and a sound and complete equational axiomatisation has been obtained (see Definition 2). Importantly, the logic $\mathcal{R}$ extended with (co)inductively defined operators is sufficiently expressive to interpret most other probabilistic logics, including pCTL [Mio12b,Mio18,MS13a]. Hence, the Riesz modal logic appears to be a convenient base for developing the theory of probabilistic μ-calculi and, more generally, probabilistic logics.

Contributions of This Work. This work is a first step towards the development of the structural proof theory of probabilistic μ-calculi. We introduce a *hypersequent calculus* called MGA (read *modal* GA) for a version of the Riesz modal logic (the *scalar-free fragment*, see Sect. 2 for details) and by proving the cut-elimination theorem. Formally we prove:

Theorem 1. *The hypersequent calculus MGA is sound and complete with respect to the equational axioms of Fig. 1 and the CUT rule is eliminable.*

The machinery of hypersequent calculi has been introduced by Avron in [Avr87] and, independently, by Pottinger in [Pot83]. Our calculus extends the hypersequent calculus GA of Metcalfe, Olivetti and Gabbay [MOG05] (see also the book [MOG09] and the related [CM03] and [DMS18]) which is a sound and complete structural proof system for the equational theory of lattice-ordered abelian groups (axioms (1) in Fig. 1, see [Vul67] for an overview). The main contributions of this work are:

1. The careful extension of the system GA of [MOG05] with appropriate proof rules for the modality ($\Diamond$) and the proof of soundness and completeness.
2. The non-trivial adaptation of the proof-technique used in [MOG09, §5.2] to prove the cut-elimination theorem for GA.
3. The formalisation using the theorem prover Agda of our key technical results: Theorems 4 and 9. The code is freely available at [Agd].

In particular, the last point above guarantees the correctness of the proofs of all our novel technical results which, as it is often the case in proof theory, involve complex and long induction arguments. Given the availability of formalised proofs, in this work we focus on illustrating the main ideas behind our arguments rather than spelling out all technical details.

Organisation of the Paper. In Sect. 2 we provide the necessary definitions about the Riesz modal logic from [MFM17, Mio18] and about the hypersequent calculus GA of [MOG05, MOG09]. In Sect. 3 we present our hypersequent calculus MGA and state the main theorems. In Sect. 4 we sketch the main ideas behind our proof of cut-elimination. Lastly, in Sect. 5 we discuss some directions for future work.

2 Technical Background

2.1 The Riesz Modal Logic and Its Scalar-free Fragment

The Riesz modal logic $\mathcal{R}$ introduced in [MFM17] is a probabilistic logic for expressing properties of discrete or continuous Markov chains. We refer to [MFM17] for a detailed introduction. Here we just restrict ourselves to the purely *syntactical* aspects of this logic: its syntax and its axiomatisation.

Definition 1 (Syntax). *The set of formulas of the Riesz modal logic is generated by the following grammar:* $\phi, \psi ::= x \mid 0 \mid 1 \mid \phi + \psi \mid r\phi \mid \phi \sqcup \psi \mid \phi \sqcap \psi \mid \Diamond\phi$ *where* r*, called a* scalar*, ranges over the set* $\mathbb{R}$ *of real numbers. We just write* $-\phi$ *in place of* $(-1)\phi$.

A main result of [MFM17] is that two formulas ϕ and ψ are semantically equivalent if and only if the identity $\phi = \psi$ holds in all *modal Riesz spaces*.

Definition 2. *A modal Riesz space is an algebraic structure R over the signature $\Sigma = \{0, 1, +, r, \sqcup, \sqcap, \Diamond\}_{r\in\mathbb{R}}$ such that the following set $\mathcal{R}$ of axioms hold:*

1. *$\{R, 0, +, r, \sqcup, \sqcap\}_{r\in\mathbb{R}}$ is a Riesz space (see, e.g., [LZ71]), i.e.,*
   - *$(R, 0, +, r)_{r\in\mathbb{R}}$ is an $\mathbb{R}$-vector space,*
   - *$(R, \sqcup, \sqcap)$ is a lattice,*
   - *the lattice order ($x \leq y \Leftrightarrow x \sqcap y = x$) is compatible with addition, i.e.:*
     (a) $x \leq y$ implies $x + z \leq y + z$ (i.e., $(x \sqcap y) + z = ((x \sqcap y) + z) \sqcap (y + z)$),
     (b) $x \geq 0$ implies $rx \geq 0$ (i.e., $0 = 0 \sqcap r(x \sqcup 0)$) for every $r \in \mathbb{R}_{\geq 0}$,
2. *$0 \leq 1$ (i.e., $0 = 0 \sqcap 1$),*
3. *the $\Diamond$ operation is linear, positive and 1-decreasing, i.e.:*
   - *$\Diamond(x + y) = \Diamond(x) + \Diamond(y)$ and $\Diamond(rx) = r\Diamond(x)$,*
   - *if $x \geq 0$ then $\Diamond(x) \geq 0$ (i.e., $0 = 0 \sqcap \Diamond(x \sqcup 0)$),*
   - *$\Diamond(1) \leq 1$ (i.e., $\Diamond 1 = \Diamond 1 \sqcap 1$).*

Note that the definition of modal Riesz spaces is purely equational: all axioms of Riesz spaces (1) can be expressed equationally and so can the axioms (2) and (3). This means, by Birkoff completeness theorem, that two formulas are semantically equivalent if and only if the identity $\phi = \psi$ can be derived using the familiar deductive rules of equational logic, written as $\mathcal{R} \vdash \phi = \psi$.

Definition 3 (Deductive Rules of Equational Logic). *Rules for deriving identities from a set $\mathcal{A}$ of equational axioms:*

$$\frac{(t_1 = t_2) \in \mathcal{A}}{\mathcal{A} \vdash t_1 = t_2}\,Ax \qquad \frac{}{\mathcal{A} \vdash t = t}\,\mathit{refl} \qquad \frac{\mathcal{A} \vdash t_2 = t_1}{\mathcal{A} \vdash t_1 = t_2}\,\mathit{sym} \qquad \frac{\mathcal{A} \vdash t_1 = t_2}{\mathcal{A} \vdash C[t_1] = C[t_2]}\,\mathit{ctxt}$$

$$\frac{\mathcal{A} \vdash t_1 = t_2 \quad \mathcal{A} \vdash t_2 = t_3}{\mathcal{A} \vdash t_1 = t_3}\,\mathit{trans} \qquad \frac{\mathcal{A} \vdash f(\boldsymbol{s}, x, \boldsymbol{u}) = g(\boldsymbol{w}, x, \boldsymbol{z})}{\mathcal{A} \vdash f(\boldsymbol{s}, t, \boldsymbol{u}) = g(\boldsymbol{w}, t, \boldsymbol{z})}\,\mathit{subst}$$

where $C[\cdot]$ is a context and f, g are function symbols of the fixed signature.

In what follows we denote with $\mathcal{R} \vdash \phi \leq \psi$ the judgment $\mathcal{R} \vdash \phi = \phi \sqcap \psi$. The following elementary facts from the theory of Riesz spaces (see, e.g., [LZ71, §2.12]) will be useful.

Proposition 1. *The following assertions hold:*

- *$\mathcal{R} \vdash \phi = \psi$ iff $\mathcal{R} \vdash \phi - \psi = 0$,*
- *$\mathcal{R} \vdash \phi = \psi$ iff $(\mathcal{R} \vdash \phi \leq \psi$ and $\mathcal{R} \vdash \psi \leq \phi)$.*
- *$\mathcal{R} \vdash r(x \sqcup y) = rx \sqcup ry$, $\mathcal{R} \vdash r(x \sqcap y) = rx \sqcap ry$.*

The first point says that an equality $\phi = \psi$ can always be expressed as an identity with 0. The second point says that we can express equalities with inequalities and *vice versa*. The third point, together will the other axioms, implies that scalar multiplication distributes over all other operations $\{+, \sqcup, \sqcap, \Diamond\}$.

For most practical purposes (when expressing properties of probabilistic models) the scalars in the Riesz modal logic can be restricted to be rational numbers.

Definition 4 (Rational and Scalar-free formulas). *A formula ϕ is* rational *if all its scalars are rational numbers. Similarly, ϕ is* scalar-free *if its scalars are all equal to (-1). Equivalently, the set of scalar-free formulas is generated by the following grammar:* $A, B ::= x \mid 0 \mid 1 \mid A+B \mid -A \mid A \sqcup B \mid A \sqcap B \mid \Diamond(A)$.

Note how we have switched to the letters A and B to range over scalar-free formulas to highlight this distinction.

Proposition 2. *Let ϕ be a rational formula. Then there exists a scalar-free formula A such that $\mathcal{R} \vdash \phi = 0$ iff $\mathcal{R} \vdash A = 0$.*

Proof. Let $\{r_i\}_{i \in I}$ be the list of rational scalars in ϕ, with $r_i = \frac{n_i}{m_i}$ and let $d = \prod_i m_i$ be the product of all denominators. Since scalar multiplication distributes with all operations it is easy to show that $\mathcal{R} \vdash d\phi = \psi$, for a formula ψ whose scalars are all integers. We can then obtain A from ψ by inductively replacing any sub-formula of ψ the form nB with $(B + B + \cdots + B)$ (n times) if n is positive, with $-(B + B + \cdots + B)$ if n is negative and with 0 if $n = 0$. □

For this reason in this work we restrict attention to scalar-free formulas and we consider the restricted set of axioms $\mathbb{T}$ of Fig. 1. The axioms of Riesz spaces, when scalar multiplication is omitted, reduce to the axioms of *lattice ordered abelian groups* (see, e.g., [Vul67]). The axiom $0 \leq 1$ is unaltered and the axioms for the $\Diamond$ modality are naturally adapted. For these reasons we refer to these axioms as of those of *lattice-ordered modal abelian groups.*

1. **Axioms of Lattice–ordered abelian groups:**
   - Abelian Group: $x + (y + z) = (x + y) + z$, $x + y = y + x$, $x + 0 = x$, $x - x = 0$.
   - Lattice axioms: (associativity) $x \sqcup (y \sqcup z) = (x \sqcup y) \sqcup z$, $x \sqcap (y \sqcap z) = (x \sqcap y) \sqcap z$, (commutativity) $z \sqcup y = y \sqcup z$, $z \sqcap y = y \sqcap z$, (absorption) $z \sqcup (z \sqcap y) = z$, $z \sqcap (z \sqcup y) = z$, (idempotency) $x \sqcup x = x$, $x \sqcap x = x$.
   - Compatibility: $(x \sqcap y) + z = ((x \sqcap y) + z) \sqcap (y + z)$
2. **Axioms for the unit:** $0 = 0 \sqcap 1$,
3. **Modal axioms:**
   - $\Diamond(x + y) = \Diamond(x) + \Diamond(y)$, $\Diamond(-x) = -\Diamond(x)$ and $\Diamond(0) = 0$,
   - $0 = 0 \sqcap \Diamond(x \sqcup 0)$,
   - $\Diamond 1 = \Diamond 1 \sqcap 1$.

Fig. 1. Set of axioms $\mathbb{T}$ of lattice-ordered modal Abelian groups.

Remark 1. Note that from the previous discussion it does not follow directly that $\mathcal{R} \vdash A = B$ implies $\mathbb{T} \vdash A = B$. We indeed conjecture that $\mathcal{R}$ is a conservative extension of $\mathbb{T}$ but we have not proved this fact so far. In any case, this is not required for results of this work.

The main contribution of this work is the design of a sound and complete hypersequent calculus for the theory $\mathbb{T}$ and the proof of cut-elimination.

2.2 The Hypersequent Calculus GA

Our starting point is the hypersequent calculus GA of [MOG05, MOG09] for the theory of lattice-ordered abelian groups (set of axioms (1) in Fig. 1).

Definition 5 (Formulas, Sequents and hypersequents). *A* formula *A is a term built from a set of variables (ranged over by x, y, z) over the signature $\{0, +, -, \sqcap, \sqcup\}$. A* sequent *S is a pair of two (possibly empty) multisets of formulas $\Gamma = A_0, \ldots, A_n$ and $\Delta = B_0, \ldots, B_m$, denoted as $\Gamma \vdash \Delta$. A* hypersequent *G is a nonempty multiset $S_1, \ldots, S_n$ of sequents, denoted as $S_1 | \ldots | S_n$.*

Following [MOG05, MOG09], with some abuse of notation, we denote with S both the sequent and the hypersequent consisting of only the sequent S. The system GA is a deductive system for deriving hypersequents consisting of the rules of Fig. 2. The system GA without the CUT rule is denoted by GA*.

Another convention we adopt from [MOG05, MOG09] is to write $d \vDash_{\mathrm{GA}} G$ to express the fact that d is a valid GA-derivation of the hypersequent G. We write $\vDash_{\mathrm{GA}} G$ to express the existence of a GA-derivation d such that $d \vDash_{\mathrm{GA}} G$. Similarly, we write $d \vDash_{\mathrm{GA}^*} G$ and $\vDash_{\mathrm{GA}^*} G$ when referring to the subsystem GA*.

Axioms:

$$\frac{}{\vdash}\ \Delta\text{-ax} \qquad \frac{}{A \vdash A}\ \text{ID-ax}$$

Structural rules:

$$\frac{G}{G | \Gamma \vdash \Delta}\ \text{Weakening (W)} \qquad \frac{G | \Gamma \vdash \Delta | \Gamma \vdash \Delta}{G | \Gamma \vdash \Delta}\ \text{Contraction (C)}$$

$$\frac{G | \Gamma_1, \Gamma_2 \vdash \Delta_1, \Delta_2}{G | \Gamma_1 \vdash \Delta_1 | \Gamma_2 \vdash \Delta_2}\ \text{Split (S)} \qquad \frac{G | \Gamma_1 \vdash \Delta_1 \quad G | \Gamma_2 \vdash \Delta_2}{G | \Gamma_1, \Gamma_2 \vdash \Delta_1, \Delta_2}\ \text{Mix (M)}$$

Logical rules:

$$\frac{G | \Gamma \vdash \Delta}{G | \Gamma, 0 \vdash \Delta}\ 0_L \qquad \frac{G | \Gamma \vdash \Delta}{G | \Gamma \vdash \Delta, 0}\ 0_R$$

$$\frac{G | \Gamma, A, B \vdash \Delta}{G | \Gamma, A + B \vdash \Delta}\ +_L \qquad \frac{G | \Gamma \vdash \Delta, A, B}{G | \Gamma \vdash \Delta, A + B}\ +_R$$

$$\frac{G | \Gamma \vdash \Delta, A}{G | \Gamma, -A \vdash \Delta}\ -_L \qquad \frac{G | \Gamma, A \vdash \Delta}{G | \Gamma \vdash \Delta, -A}\ -_R$$

$$\frac{G | \Gamma, A \vdash \Delta \quad G | \Gamma, B \vdash \Delta}{G | \Gamma, A \sqcup B \vdash \Delta}\ \sqcup_L \qquad \frac{G | \Gamma \vdash \Delta, A | \Gamma \vdash \Delta, B}{G | \Gamma \vdash \Delta, A \sqcup B}\ \sqcup_R$$

$$\frac{G | \Gamma, A \vdash \Delta | \Gamma, B \vdash \Delta}{G | \Gamma, A \sqcap B \vdash \Delta}\ \sqcap_L \qquad \frac{G | \Gamma \vdash \Delta, A \quad G | \Gamma \vdash \Delta, B}{G | \Gamma \vdash \Delta, A \sqcap B}\ \sqcap_R$$

CUT rule:

$$\frac{G \mid \Gamma_1 \vdash \Delta_1, A \quad G \mid \Gamma_2, A \vdash \Delta_2}{G \mid \Gamma_1, \Gamma_2 \vdash \Delta_1, \Delta_2}\ \mathit{Cut}$$

Fig. 2. Inference rules of the hypersequent system GA of [MOG05].

Multisets of formulas, sequents and hypersequents are interpreted as a single formula as follows:

Definition 6 (Interpretation). *A multiset of formulas $\Gamma = \phi_1, \ldots, \phi_n$ is interpreted as the formula $[\![\Gamma]\!] = \phi_1 + \phi_1 + \cdots + \phi_n$ if $n \geq 1$ and as $[\![\Gamma]\!] = 0$ if $\Gamma = \emptyset$. A sequent $S = \Gamma \vdash \Delta$ is interpreted as the formula $[\![S]\!] = [\![\Delta]\!] - [\![\Gamma]\!]$. Finally, a hypersequent $G = S_0 \mid \cdots \mid S_n$ is interpreted as the formula $[\![G]\!] = [\![S_0]\!] \sqcup \cdots \sqcup [\![S_n]\!]$.*

Example 1. Consider the hypersequent $G = (0 \sqcup x, y \vdash y) \mid (-y \vdash)$ consisting of two sequents. Then $[\![G]\!] = (y - ((0 \sqcup x) + y)) \sqcup (0 - (-y))$.

The soundness and completeness of the hypersequent system GA with respect to the theory of lattice-ordered abelian groups (axioms (1) of Fig. 1, written as $\mathbb{T}_{(1)}$) is expressed by the following theorem.

Theorem 2 ([MOG05]). *For all formulas A and hypersequents G:*

Soundness: if $\vDash_{GA} G$ then $\mathbb{T}_{(1)} \vdash [\![G]\!] \geq 0$.
Completeness: if $\mathbb{T}_{(1)} \vdash A \geq 0$ then $\vDash_{GA} (\vdash A)$

Proof. The proofs presented in [MOG05] exploit the following well-known fact (see, e.g., [Vul67]): the equality $A = B$ holds in all lattice-ordered abelian groups if and only if it holds in $(\mathbb{R}, 0, +, -, \max, \min)$ under any interpretation of the variables as real numbers. In other words, $\mathbb{R}$ generates the variety of lattice-ordered abelian groups. □

The main result of [MOG05] regarding GA is that the CUT rule is eliminable.

Theorem 3 (Cut-elimination [MOG05]**).** *Any GA-derivation of a hypersequent G can be effectively transformed into a GA*-derivation of G.*

3 The Hypersequent System MGA

In this section we introduce our hypersequent calculus system MGA, a modal extension of the GA system of [MOG05]. The system MGA deals with formulas over the signature of modal lattice-ordered abelian groups (see Fig. 1) thus including the constant 1 and the unary modality $\Diamond$.

Definition 7 (Formulas of MGA). *A* formula *A is a term built from a set of variables (ranged over by x, y, z) over the signature $\{0, 1, +, -, \sqcap, \sqcup, \Diamond\}$.*

The definitions of sequents and hypersequents are given exactly as for the system GA in Definition 5 of Sect. 2.2. Similarly, multisets of formulas, sequents and hypersequents are interpreted as formulas exactly as already specified in Definition 6 of Sect. 2.2 for the system GA. Before presenting the deduction rules of MGA, it is useful to introduce the following abbreviations.

- For $n \in \mathbb{N}_{\geq 0}$, we denote with nF the multiset of formulas $F, F, \ldots, F$. So for example we write $2A, 1B \vdash 0C, D$ to denote the sequent $A, A, B \vdash D$.
- Given a multiset of formulas $\Gamma = F_0, \ldots, F_k$ and $n \in \mathbb{N}_{\geq 0}$ we denote with $n\Gamma$ the multiset of formulas $nF_0, \ldots, nF_k$. If $\Gamma = \emptyset$ then also $n\Gamma = \emptyset$.
- Given a multiset of formulas $\Gamma = F_0, \ldots, F_n$ we denote with $\Diamond\Gamma$ the multiset of formulas $\Diamond F_0, \ldots, \Diamond F_n$. Consistently, if $\Gamma = \emptyset$ then also $\Diamond\Gamma = \emptyset$.

The rules of the system MGA consist of all rules of GA (see Fig. 2) together with the additional rules of Fig. 3.

Axiom for 1:	**Rule for $\Diamond$:**
$\dfrac{}{\vdash 1}\ \text{1–ax}$	$\dfrac{\Gamma \vdash \Delta, n\mathbf{1}}{\Diamond\Gamma \vdash \Diamond\Delta, n\mathbf{1}}\ \Diamond\text{-rule}$

Fig. 3. Additional inference rules of the hypersequent system MGA

The axiom (1-ax) for the constant 1 is straightforward and it simply expresses the axiom $0 \leq 1$ from Fig. 1 (i.e., $\mathbb{T} \vdash [\![\vdash 1]\!] \geq 0$).

The rule ($\Diamond$-rule) for the modality is more subtle as it imposes strong constraints on the shape of its premise and conclusion. First, both the conclusion and the premise are required to be hypersequents consisting of exactly one sequent. Furthermore, in the conclusion, all formulas, except those of the form 1 on the right side, need to be of the form $\Diamond C$ for some C.

The following is an illustrative example of derivation in the system MGA:

$$\dfrac{\dfrac{\dfrac{\dfrac{\dfrac{\dfrac{\dfrac{\dfrac{\dfrac{}{1 \vdash 1}\ \text{ID-ax} \qquad \dfrac{}{A \vdash A}\ \text{ID-ax}}{A, 1 \vdash 1, A}\ \text{M}}{A, 1, -(A) \vdash 1}\ -_L}{A, 1 - A \vdash 1}\ +_L}{\boldsymbol{A, A \vdash 1} \mid A, 1 - A \vdash 1}\ \text{W}}{A, \boldsymbol{A \sqcap (1 - A)} \vdash 1}\ \sqcap_L}{A, A \sqcap (1 - A) \vdash 1 \mid \boldsymbol{1 - A, A \sqcap (1 - A) \vdash 1}}\ \text{W}}{\boldsymbol{A \sqcap (1 - A)}, A \sqcap (1 - A) \vdash 1}\ \sqcap_L}{\Diamond((A \sqcap (1 - A))), \Diamond((A \sqcap (1 - A))) \vdash 1}\ \Diamond\text{-rule}}{\boldsymbol{\Diamond((A \sqcap (1 - A))) + \Diamond((A \sqcap (1 - A)))} \vdash 1}\ +_L$$

Our first theorem regarding MGA states its soundness and completeness with respect to the theory of modal lattice-ordered abelian groups (see Fig. 1). The proof of [MOG05] of Theorem 2 cannot be directly adapted here because, unlike the case for lattice-ordered abelian groups and $\mathbb{R}$, we are not aware of any simple modal lattice-order abelian group which generates the whole variety.

Theorem 4. *For all formulas A and hypersequents G:*

Soundness: if $\vDash_{MGA} G$ *then* $\mathbb{T} \vdash [\![G]\!] \geq 0$.

Completeness: if $\mathbb{T} \vdash A \geq 0$ *then* $\vDash_{MGA} (\vdash A)$.

Proof. Soundness is proven by translating every MGA derivation d of G to a derivation in equational logic π of $[\![G]\!] \geq 0$. This is done by induction on the complexity of d. The difficult cases correspond to when d ends by applications of either the S-rule, the M-rule or the $\sqcup_L$ rule. The formalised proof is implemented in the agda file `Syntax/Agda/MGA-Cut/Soundness.agda` in [Agd] and the type of the function is: soundness : $(G$: HSeq$) \rightarrow$ (MGA G) $\rightarrow$ botAG $\leq$S $[\![$ G $]\!]$.

Conversely, completeness is proven by translating every equational logic derivation π of $A = B$ to the MGA derivations d_1 and d_2 of the (hyper)sequents $A \vdash B$ and $B \vdash A$ respectively. The proof goes by induction on π. First, MGA derivations are obtained for all axioms of Fig. 1. For example, for the axiom $\Diamond(x+y) = \Diamond(x) + \Diamond(y)$ we can derive the (hyper)sequent $\Diamond(x+y) \vdash \Diamond(x) + \Diamond(y)$ as showed below (left-side). Translating applications of the rules *refl* and *sym* is simple. Translating applications of the *trans* rules is immediate using the *CUT* rule of MGA. To translate applications of the *ctxt* rule, it is sufficient to prove (by induction) a simple context-lemma that states that if $A \vdash B$ is MGA derivable then also $C[A] \vdash C[B]$ is MGA derivable. Similarly, to translate applications of the *subst* rule, it is sufficient to prove (by induction) a simple substitution-lemma stating that if G is MGA derivable then $G[A/x]$ is also derivable, where $G[A/x]$ is the hypersequent where every occurrence of x is replaced by A.

Note that $\mathbb{T} \vdash A \geq 0$ means that $\mathbb{T} \vdash 0 = 0 \sqcap A$. By the translation method outlined above, the (hyper)sequent $0 \vdash 0 \sqcap A$ is MGA derivable. We can then get a MGA derivation of $\vdash A$ as follows (right-side):

$$
\dfrac{\dfrac{\dfrac{\dfrac{\dfrac{}{x \vdash x}\ \text{ID-ax} \qquad \dfrac{}{y \vdash y}\ \text{ID-ax}}{x, y \vdash x, y}\ \text{M}}{x + y \vdash x, y}\ +_L}{\Diamond(x+y) \vdash \Diamond(x), \Diamond(y)}\ \Diamond}{\Diamond(x+y) \vdash \Diamond(x) + \Diamond(y)}\ +_R
\qquad\Bigg|\qquad
\dfrac{\dfrac{\dfrac{\dfrac{}{A \vdash A}\ \text{ID-ax}}{0 \vdash A \mid A \vdash A}\ \text{W}}{0 \sqcap A \vdash A}\ \sqcap_L \qquad \dfrac{0 \vdash 0 \sqcap A \qquad \dfrac{\dfrac{}{\vdash}\ \Delta\text{-ax}}{\vdash 0}\ 0_R}{\vdash 0 \sqcap A}\ \text{cut}}{\vdash A}\ \text{cut}
$$

The file `Syntax/Agda/MGA-Cut/Completeness.agda` in [Agd] contains the formalised proof and the type of the function is: completeness : $(A$: Term$) \rightarrow$ botAG $\leq$S $A \rightarrow$ MGA (head (\[\], \[\] :: A)). □

Remark 2. The following natural looking variant of the ($\Diamond$-rule), allowing hypersequents with more than one component, is unsound:

$$\frac{G \mid \Gamma \vdash \Delta, n1}{G \mid \Diamond\Gamma \vdash \Diamond\Delta, n1}$$

Our main theorem regarding the system MGA is the cut-elimination theorem. We denote with MGA* the system without the CUT rule.

Theorem 5 (Cut-elimination). *Any MGA-derivation of a hypersequent G can be effectively transformed into a* MGA**-derivation of G.*

Theorems 4 and 5 imply the statement of Theorem 1 in the Introduction.

4 Overview of the Proof of the Cut-Elimination Theorem

In this section we illustrate the structure of our proof of the cut-elimination theorem. We first explain the main ideas behind the proof of cut-elimination for GA of [MOG09, §5.2]. We then explain why these idea are not directly applicable to the system MGA. Lastly, we discuss our key technical contribution which makes it possible to adapt the proof method of [MOG09, §5.2] to prove the cut-elimination theorem for the MGA system.

4.1 The CAN-Elimination Theorem for the System GA

A key idea of [MOG09, §5.2] is to replace the CUT rule with an easier to handle rule called *cancellation* (CAN) rule. The CAN rule can derive the CUT rule in the basic cut-free system GA* as follows (right-side):

$$\frac{G|\Gamma, A \vdash A, \Delta}{G|\Gamma \vdash \Delta}\ \text{CAN} \qquad\Bigg|\qquad \frac{\dfrac{\dfrac{d_1}{G|\Gamma_1, A \vdash \Delta_1} \quad \dfrac{d_2}{G|\Gamma_2 \vdash A, \Delta_2}}{G|\Gamma_1, \Gamma_2, A \vdash A, \Delta_1, \Delta_2}\ \text{M}}{G|\Gamma_1, \Gamma_2 \vdash \Delta_1, \Delta_2}\ \text{CAN}$$

The cut-elimination theorem is obtained in [MOG09, §5.2] by proving a CAN-elimination theorem expressed as: if $\vDash_{\text{GA}^*} G|\Gamma, A \vdash A, \Delta$ then $\vDash_{\text{GA}^*} G|\Gamma \vdash \Delta$.

The CAN-elimination theorem for the system GA is proved in three steps:

Step A: proving the invertibility of all the logical rules ([MOG09, Lemma 5.18]). The invertibility states that if the conclusion of a logical rule (for instance, $G|\Gamma, A+B \vdash \Delta$ for the $+_L$ rule) is derivable without the CAN-rule, then all the premises (in this case $G|\Gamma, A, B \vdash \Delta$) are derivable too without the CAN-rule.

Step B: proving the atomic CAN-elimination theorem ([MOG09, Lemma 5.17]). This theorem deals with the special case of A being a variable and states that if $d \vDash_{\text{GA}^*} G|\Gamma, x \vdash x, \Delta$ then $\vDash_{\text{GA}^*} G|\Gamma \vdash \Delta$. This theorem is proven by induction on d and is mostly straightforward: the only difficult case is when d finishes with an application of the M-rule. A separate technical result ([MOG09, Lemma 5.16]) is used to take care of this difficult case.

Step C: proving the CAN-elimination theorem ([MOG09, Theorem 5.19]). The CAN-elimination theorem states that if $\vDash_{\text{GA}^*} G|\Gamma, A \vdash A, \Delta$ then $\vDash_{\text{GA}^*} G|\Gamma \vdash \Delta$. This proof is by induction on A:

- If A is a variable, we can conclude with the atomic CAN-elimination theorem.
- Otherwise we use the invertibility of the logical rules and we can conclude with the induction hypothesis. For instance, if $A = B + C$, then by invertibility of the $+_L$ and $+_R$ rules we have a GA*-derivation of $\vDash_{\text{GA}^*} G|\Gamma, B, C \vdash \Delta, B, C$ and, from it, we can obtain a GA*-derivation of $G|\Gamma \vdash \Delta$ by using twice the induction hypothesis, first on B then on C.

4.2 Issues in Adapting the Proof for the System MGA

The proofs of [MOG09] can be adapted to the context of MGA without much difficulty to perform the first two steps:

Theorem 6 (Invertibility of the logical rules). *All logical rules (including the $\Diamond$-rule) are invertible in the system MGA*.*

Proof. The same proof technique used in [MOG09] works. The main idea is, in order to deal easily with the (S) and the (C) rules, to prove a slightly stronger statement about the invertibility of more general rules. For instance, the generalisation of the rule $+_L$ is:

$$\frac{[\Gamma_i, n_i A, n_i B \vdash \Delta_i]_{i=1}^{k}}{[\Gamma_i, n_i(A+B) \vdash \Delta_i]_{i=1}^{n}}$$

□

Theorem 7 (Atomic CAN-elimination theorem). *If $\vDash_{MGA^*} \Gamma, x \vdash x, \Delta$ then $\vDash_{MGA^*} \Gamma \vdash \Delta$.*

The complication comes from the third and last Step C. We want to prove that if $\vDash_{\text{MGA}^*} G|\Gamma, A \vdash A, \Delta$ then $\vDash_{\text{MGA}^*} G|\Gamma \vdash \Delta$. An ordinary proof by induction on A could get stuck when $A = \Diamond B$. For instance, if the hypersequent is $x, \Diamond B \vdash \Diamond B, x$, the invertibility of the $\Diamond$-rule can not be used because of the syntactic constraints the $\Diamond$-rule imposes on its conclusion. Indeed the invertibility of the $\Diamond$-rule states that if $\vDash_{\text{MGA}^*} \Diamond\Gamma \vdash \Diamond\Delta$ then $\vDash_{\text{MGA}^*} \Gamma \vdash \Delta$, but $x, \Diamond A \vdash \Diamond A, x$ is not of this form because it contains the variable x.

For this reason, we deal with the case $A = \Diamond B$ in a different way, using an induction argument on the derivation of $G|\Gamma, A \vdash A, \Delta$. In this argument, however, the M-rule is hard to deal with (as already remarked it is a main source of complications also on the proof of atomic CAN-elimination of [MOG09, §5.2]).

Our main technical result is that the M-rule can be eliminated from a simple variant of the system MGA called MGA-SR (which stands MGA with *scalar rules*). The system MGA-SR is obtained by modifying MGA as follows:

- The logical left-rules and right-rules for the connectives $\{0, -, +, \sqcup, \sqcap\}$ are generalised to deal with scalar coefficients (syntactic sugaring introduced in Sect. 3). For instance, the rules $+_L$ and $\sqcup_L$ become:

$$\frac{G \mid \Gamma, nA, nB \vdash \Delta}{G \mid \Gamma, n(A+B) \vdash \Delta} +_L \qquad \frac{G|\Gamma, nA \vdash \Delta \quad G|\Gamma, nB \vdash \Delta}{G|\Gamma, n(A \sqcup B) \vdash \Delta} \sqcup_L$$

- The axioms ID-ax and 1-ax are replaced by the rules

$$\frac{G|\Gamma \vdash \Delta}{G|\Gamma, nA \vdash nA, \Delta} \text{ID-rule} \qquad \frac{G \mid \Gamma \vdash \Delta}{G \mid \Gamma \vdash \Delta, n1} \text{1-rule}$$

- All structural rules (C, W, S, M), the $\Diamond$-rule and the CAN rule remain exactly as in MGA (see Fig. 2).

It is possible to verify that MGR and MGR-SR are equivalent, i.e., they can derive exactly the same hypersequents (Theorem 8 below). The first modification (scalar rules) is technically motivated because it simplifies several proofs: in fact scalar rules are also implicitly considered in several of the proofs of [MOG09] for the system GA. The second modification (ID-rule and 1-rule) is essential. Indeed in the system MGA (and also in GA) the (hyper)sequent $x, y \vdash x, y$ is not derivable without applying the M-rule. Hence M-elimination in MGA is impossible. On the other hand the (hyper)sequent $x, y \vdash x, y$ is easily derivable in MGA-SR without requiring applications of the M rule

$$\dfrac{\dfrac{\overline{\vdash}\ \Delta\text{-ax}}{y \vdash y}\ \text{ID-rule}}{x, y \vdash x, y}\ \text{ID-rule}$$

and, as we will prove (Theorem 12), it is indeed possible to eliminate all applications of the M-rule from MGA-SR.

As outlined above, the presence of the M-rule was the main source of complications in adapting Step C. Once the equivalence between MGA-SR and MGA-SR without the M-rule is established, most complications disappear and the CAN-elimination proof can be obtained by performing Steps A–B–C for the system MGA-SR.

4.3 The System MGA-SR and the M-Elimination Theorem

In this subsection we introduce the system MGA-SR (MGA with *scalar rules*) for which we will prove the M-elimination theorem.

Definition 8 (MGA-SR). *The inference rules of MGA-SR are the rules of MGA modified as discussed previously. We denote by MGA-SR***, MGA-SR*† *and MGA-SR*$^{\dagger *}$ *the systems without the CUT rule, the M-rule and both the CUT and M-rules, respectively.*

Theorem 8. *The two systems MGA and MGA-SR are equivalent:* $\vDash_{MGA} G$ *if and only if* $\vDash_{MGA-SR} G$.

The two systems MGA* *and MGA-SR** *are equivalent:* $\vDash_{MGA^*} G$ *if and only if* $\vDash_{MGA-SR^*} G$.

Proof. Translating MGA proofs to MGA-SR proofs is straightforward. All rules of MGA are specific instances of the scalar rules of MGA-SR (taking the scalar $n = 1$) and the the axioms 1-Axiom and ID-axioms are easily derivable in MGA-SR (without the need of the CAN rule) by using the id-rule and 1-rule (again, using the scalar $n = 1$). Translating MGA-SR to MGA is also mostly straightforward. Some care is needed to translate instances of the scalar-rules $\sqcup_L$ and $\sqcap_R$ from MGA-SR to MGA. This can be done by induction on the scalar n using the fact that the two premises $G|\Gamma, nA, B \vdash \Delta$ and $G|\Gamma, nB, A \vdash \Delta$ are derivable from $G|\Gamma, (n+1)A \vdash \Delta$ and $G|\Gamma, (n+1)B \vdash \Delta$. We remark that this derivation may require the usage of the M rule. □

We now state our main technical contribution: the M-elimination theorem for the system MGA-SR.

Theorem 9 (M-elimination). *If* $d_1 \vDash_{MGA\text{-}SR^\dagger} G_1 \mid \Gamma \vdash \Delta$ *and* $d_2 \vDash_{MGA\text{-}SR^\dagger} G_2 \mid \Sigma \vdash \Pi$ *then* $\vDash_{MGA\text{-}SR^\dagger} G_1 \mid G_2 \mid \Gamma, \Sigma \vdash \Delta, \Pi$.

If $d_1 \vDash_{MGA\text{-}SR^{\dagger *}} G_1 \mid \Gamma \vdash \Delta$ *and* $d_2 \vDash_{MGA\text{-}SR^{\dagger *}} G_2 \mid \Sigma \vdash \Pi$ *then* $\vDash_{MGA\text{-}SR^{\dagger *}} G_1 \mid G_2 \mid \Gamma, \Sigma \vdash \Delta, \Pi$.

We now give a sketch of our proof argument. A formalised proof in Agda is available in [Agd] and is contained in the files `Syntax/MGA-SR/M-Elim.agda` and `Syntax/MGA-SR-CAN/M-Elim-CAN.agda`.

The general idea is to combine the derivations d_1 and d_2 in a *sequential way*. We first consider the case when no applications of the $\Diamond$-rule appear in d_1 nor d_2. First the proof d_1 is transformed into a pre-proof (i.e., where the derivation is left incomplete at some leaves) d_1' of $G_1 \mid G_2 \mid \Gamma, \Sigma \vdash \Delta, \Pi$. The pre-proof d_1' is structurally identically to d_1 and it essentially just ignores the G_2, Σ and Π components of the hypersequent. While the leaves of d_1 are all of the form $(\vdash)$ because Δ-ax is the only axiom of MGA-SR, the leaves of the pre-proof d_1' are of the form $G_2 \mid n\Sigma \vdash n\Pi$ (the ignored part carried out until the end, which can get multiplied by applications of the C and S rules). We can now proceed with the second step and provide derivations for these leaves using (easily modified versions of) the proof d_2.

When occurrences of the $\Diamond$-rule appear in d_1 or d_2 the argument requires more care. Indeed an application of the $\Diamond$-rule on d_1 acting on some hypersequent (necessarily) of the form:

$$\Diamond\Gamma_1 \vdash \Diamond\Delta_1, k1$$

cannot turned into an application of $\Diamond$-rule on:

$$G_2 \mid \Sigma, \Diamond\Gamma_1 \vdash \Diamond\Delta_1, k_1, \Pi$$

because this hypersequent violates the structural constraints of the $\Diamond$-rule. For this reason, we stop the construction of d_1' at these points and, as a results, the leaves of the pre-proof d_1' are generally of the form: $G_2 \mid n\Sigma, \Diamond\Gamma_1 \vdash \Diamond\Delta_1, k1, n\Pi$, for some Γ_1, Δ_1 and scalars n, k.

The idea now is, following the same kind of procedure, to modify the proof d_2 and turn it to a pre-proof d_2' of $G_2 \mid n\Sigma, \Diamond\Gamma_1 \vdash \Diamond\Delta_1, k1, n\Pi$. Crucially, the previous issue disappears. Indeed proof steps in d_2 acting on hypersequents of the form:

$$\Diamond\Sigma_1 \vdash \Diamond\Pi_1, m1$$

using the $\Diamond$-rule, can be turned into valid $\Diamond$-rule steps for the extended hypersequent:

$$\Diamond\Sigma_1, \Diamond\Gamma_1 \vdash \Diamond\Delta_1, k1, \Diamond\Pi_1, m_1 1$$

because the shape of the sequent is compatible with the constraint of the $\Diamond$ rule. Note that the hypersequent resulting from the application of the $\Diamond$-rule is $\Sigma_1, \Gamma_1 \vdash \Gamma_1, k_1 1, \Pi_1, m_1 1$ and has a lower modal-depth than the starting one. Hence an inductive argument on modal-complexity can be arranged to recursively reduce the general M-elimination procedure to the simpler case where d_1 and d_2 do not have occurrences of the $\Diamond$-rule (Fig. 4).

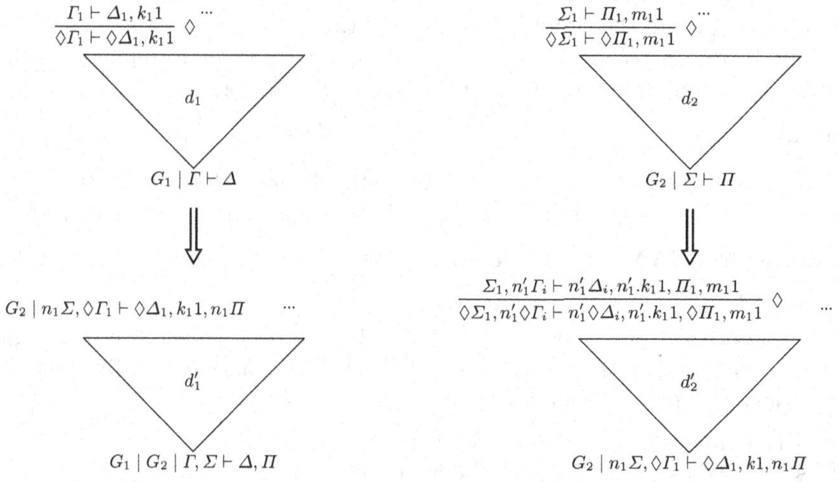

Fig. 4. Sequentially composing d_1 and d_2 in the M-elimination proof.

The following is a direct consequence Theorems 8 and 9.

Corollary 1. *The two systems MGA and MGA-SR† are equivalent:* $\vDash_{MGA} G$ *if and only if* $\vDash_{MGA-SR^\dagger} G$.

The two systems MGA* and MGA-SR$^{\dagger *}$ are equivalent: $\vDash_{MGA^*} G$ *if and only if* $\vDash_{MGA-SR^{\dagger *}} G$.

4.4 Cut-Elimination Theorem for the System MGA

We have already remarked that the cut-elimination theorem for the system MGA follows from the CAN-elimination theorem. By Corollary 1, the CAN-elimination theorem for the system MGA-SR† implies the CAN-elimination for MGA. Since there is no M-rule in MGA-SR†, the proof of CAN-elimination can follow the three Steps A–B–C outlined in Subsect. 4.1. As for Step A, we need to prove the invertibility of the logical rules in the system MGA-SR$^{\dagger *}$.

Theorem 10 (Invertibility of the logical rules). *The logical rules of the system MGA-SR$^{\dagger *}$,* $\{0_L, 0_R, +_L, +_R, \sqcup_L, \sqcup_R, \sqcap_L, \sqcup_R\}$, *are invertible.*

Remark 3. We note that, just as in [MOG09, §5.2], it is in fact possible and indeed technically useful to prove the invertibility of generalised scalar rules dealing with scalar rules, as in the proof of Theorem 6.

As for Step B we prove the atomic CAN-elimination theorem. Following the previous remark, we prove the following stronger version of the statement.

Theorem 11 (Atomic CAN-elimination). *If* $\vDash_{MGA\text{-}SR^{\dagger *}} [\Gamma_i, k_i x \vdash k_i x, \Delta_i]_{i=1}^n$ *then* $\vDash_{MGA\text{-}SR^{\dagger *}} [\Gamma_i \vdash \Delta_i]_{i=1}^n$.

Since we removed the M-rule, there are no significant difficulties in the induction arguments, and the proof is quite straightforward.

We also need a technical lemma regarding the constant formula 1 which is provable by a simple induction on the length of derivations.

Lemma 1. *If* $\vDash_{MGA\text{-}SR^{\dagger *}} [\Gamma_i, n_i 1 \vdash n_i 1, \Delta_i]_{i=1}^n$ *then* $\vdash_{MGA\text{-}SR^{\dagger *}} [\Gamma_i \vdash \Delta_i]_{i=1}^n$.

We can now prove the CAN-elimination theorem for MGA-SR†. This, together with Corollary 1 implies the cut-elimination (Theorem 5) for MGA.

Theorem 12 (CAN-elimination). *If* $d \vDash_{MGA\text{-}SR^{\dagger *}} G \mid \Gamma, A \vdash A, \Delta$ *then* $\vDash_{MGA\text{-}SR^{\dagger *}} G \mid \Gamma \vdash \Delta$.

Proof. Again, it is convenient to prove the stronger statement: If $d \vDash_{\text{MGA-SR}^{\dagger *}} [\Gamma_i, k_i A \vdash, k_i A, \Delta_i]_{i=1}^n$ then $\vDash_{\text{MGA-SR}^{\dagger *}} [\Gamma_i \vdash \Delta_i]_{i=1}^n$. This is done by induction on the (lexicographical) complexity of the pair (A, d):

- If A is a variable, we can conclude with Theorem 11.
- If $A = 1$, we can conclude with Lemma 1.
- If $A = \Diamond B$, we look at d.
  - If d finished with the $\Diamond$-rule, then the end hypersequent is necessarily of the form: $[\Gamma_i, k_i A \vdash, k_i A, \Delta_i]_{i=1}^n = \Diamond\Gamma_1, n_1 \Diamond B \vdash n_1 \Diamond B, \Diamond\Delta_1, k1$, and is derived from the hypersequent $\vDash_{\text{MGA-SR}^{\dagger *}} \Gamma_1, n_1 B \vdash n_1 B, \Delta_1, k1$. By induction hypotheses (B has smaller complexity than A), we have that $\vDash_{\text{MGA-SR}^{\dagger *}} \Gamma_1 \vdash \Delta_1, k1$. Hence we can derive $\vDash_{\text{MGA-SR}^{\dagger *}} \Diamond\Gamma_1 \vdash \Diamond\Delta_1, k1$ by application of the $\Diamond$-rule.
  - Otherwise, the hypersequent is derived by application of some other rule (not active on $A = \Diamond B$) from some premises. In this case, we simply apply the inductive hypothesis on the premises (the formula A is unchanged but the complexity of the premise derivation has decreased) and use the same rule to construct a derivation of the desired hypersequent.
- Otherwise, using the same argument of [MOG09, §5.2] discussed in Sect. 4.1, we make progress in the inductive proof (reducing the complexity of A) by using the invertibility of the logical rules (Theorem 10). □

5 Conclusions and Future Work

We have presented a structural proof system called MGA for the scalar-free fragment of the Riesz modal logic. A natural direction of research is to extend the system MGA to deal with the full Riesz modal logic, thus handling arbitrary scalars $r \in \mathbb{R}$. The (integer-)scalar rules of the system MGA-SR could be naturally generalised to handle real-scalars but it is not clear, at the present moment, if the resulting system would satisfy a reasonable formulation of the sub-formula property. Another interesting topic of research is to consider extensions of MGA for fixed-point extensions of the Riesz modal logic (e.g., [MS17, Mio18]). In this direction, the machinery of *cyclic proofs* (see, e.g., [Stu07, MS13b, BS11, Dou17]) appears to be particularly promising.

References

[Agd] Repository containing the proofs formalised in Agda. https://github.com/clucas26e4/M-elimination

[Avr87] Avron, A.: A constructive analysis of RM. J. Symbolic Logic **52**(4), 939–951 (1987)

[BBLM17] Bacci, G., Bacci, G., Larsen, K.G., Mardare, R.: On the metric-based approximate minimization of Markov chains. In: Proceedings of 44th ICALP. LIPIcs, vol. 80, pp. 104:1–104:14. Schloss Dagstuhl - Leibniz-Zentrum fuer Informatik (2017)

[BdRV02] Blackburn, P., de Rijke, M., Venema, Y.: Modal Logic. Cambridge Tracts in Theoretical Computer Science. Cambridge University Press, Cambridge (2002)

[BGZB09] Barthe, G., Grégoire, B., Zanella-Beguelin, S.: Formal certification of code-based cryptographic proofs. In: Proceedings of POPL, pp. 90–101 (2009)

[BK08] Baier, C., Katoen, J.P.: Principles of Model Checking. The MIT Press, Cambridge (2008)

[BS11] Brotherston, J., Simpson, A.: Sequent calculi for induction and infinite descent. J. Log. Comput. **21**(6), 1177–1216 (2011)

[Bus98] Buss, S.R.: An introduction to proof theory. In: Handbook of Proof Theory, pp. 1–78. Elsevier (1998)

[CM03] Ciabattoni, A., Metcalfe, G.: Bounded Łukasiewicz logics. In: Cialdea Mayer, M., Pirri, F. (eds.) TABLEAUX 2003. LNCS (LNAI), vol. 2796, pp. 32–47. Springer, Heidelberg (2003). https://doi.org/10.1007/978-3-540-45206-5_6

[dA03] Alfaro, L.: Quantitative verification and control via the μ-calculus. In: Amadio, R., Lugiez, D. (eds.) CONCUR 2003. LNCS, vol. 2761, pp. 103–127. Springer, Heidelberg (2003). https://doi.org/10.1007/978-3-540-45187-7_7

[DFHM16] Dimitrova, R., Ferrer Fioriti, L.M., Hermanns, H., Majumdar, R.: Probabilistic CTL*: the deductive way. In: Chechik, M., Raskin, J.-F. (eds.) TACAS 2016. LNCS, vol. 9636, pp. 280–296. Springer, Heidelberg (2016). https://doi.org/10.1007/978-3-662-49674-9_16

[DGJP00] Desharnais, J., Gupta, V., Jagadeesan, R., Panangaden, P.: Approximating labelled Markov processes. In: Proceedings of LICS (2000)

[DMS18] Diaconescu, D., Metcalfe, G., Schnüriger, L.: A real-valued modal logic. Log. Methods Comput. Sci. **14**(1), 1–27 (2018)

[Dou17] Doumane, A.: On the infinitary proof theory of logics with fixed points. Ph.D. thesis, University Paris Diderot (2017)

[Gen34] Gentzen, G.: Untersuchungen über das logische schließen. Math. Z. **39**, 405–431 (1934)

[HJ94] Hansson, H., Jonsson, B.: A logic for reasoning about time and reliability. Form. Asp. Comput. **6**, 512–535 (1994)

[HK97] Huth, M., Kwiatkowska, M.: Quantitative analysis and model checking. In: Proceedings of LICS (1997)

[Hsu17] Hsu, J.: Probabilistic couplings for probabilistic reasoning. Ph.D. thesis, University of Pennsylvania (2017)

[Koz83] Kozen, D.: Results on the propositional μ-calculus. Theor. Comput. Sci. **27**, 333–354 (1983)

[Koz85] Kozen, D.: A probabilistic PDL. J. Comput. Syst. Sci. **30**(2), 162–178 (1985)

[LS82] Lehmann, D., Shelah, S.: Reasoning with time and chance. Inf. Control **53**(3), 165–1983 (1982)

[LZ71] Luxemburg, W.A.J., Zaanen, A.C.: Riesz Spaces. North-Holland Mathematical Library, vol. 1. Elsevier, Amsterdam (1971)

[MFM17] Mio, M., Furber, R., Mardare, R.: Riesz modal logic for Markov processes. In: 32nd ACM/IEEE Symposium on Logic in Computer Science (LICS), pp. 1–12. IEEE (2017). https://doi.org/10.1109/LICS.2017.8005091

[Mio11] Mio, M.: Probabilistic modal μ-calculus with independent product. In: Hofmann, M. (ed.) FoSSaCS 2011. LNCS, vol. 6604, pp. 290–304. Springer, Heidelberg (2011). https://doi.org/10.1007/978-3-642-19805-2_20

[Mio12a] Mio, M.: On the equivalence of denotational and game semantics for the probabilistic μ-calculus. Log. Methods Comput. Sci. **8**(2) (2012). https://lmcs.episciences.org/787, https://doi.org/10.2168/LMCS-8(2:7)2012

[Mio12b] Mio, M.: Probabilistic modal μ-calculus with independent product. Log. Methods Comput. Sci. **8**(4) (2012). https://lmcs.episciences.org/789, https://doi.org/10.2168/LMCS-8(4:18)2012

[Mio14] Mio, M.: Upper-expectation bisimilarity and Łukasiewicz μ-calculus. In: Muscholl, A. (ed.) FoSSaCS 2014. LNCS, vol. 8412, pp. 335–350. Springer, Heidelberg (2014). https://doi.org/10.1007/978-3-642-54830-7_22

[Mio18] Mio, M.: Riesz modal logic with threshold operators. In: 33rd ACM/IEEE Symposium on Logic in Computer Science (LICS), pp. 710–719. ACM (2018). https://doi.org/10.1145/3209108.3209118

[MM07] McIver, A., Morgan, C.: Results on the quantitative μ-calculus q$M\mu$. ACM Trans. Comput. Log. **8**(1) (2007). https://dl.acm.org/citation.cfm?doid=1182613.1182616

[MOG05] Metcalfe, G., Olivetti, N., Gabbay, D.M.: Sequent and hypersequent calculi for Abelian and Łukasiewicz logics. ACM Trans. Comput. Log. **6**(3), 578–613 (2005)

[MOG09] Metcalfe, G., Olivetti, N., Gabbay, D.M.: Proof Theory for Fuzzy Logics. Applied Logic Series, vol. 36. Springer, Dordrecht (2009). https://doi.org/10.1007/978-1-4020-9409-5

[MS13a] Mio, M., Simpson, A.: Łukasiewicz μ-calculus. In: Proceedings Workshop on Fixed Points in Computer Science, FICS. EPTCS, vol. 126, pp. 87–104 (2013). https://doi.org/10.4204/EPTCS.126.7

[MS13b] Mio, M., Simpson, A.: A proof system for compositional verification of probabilistic concurrent processes. In: Pfenning, F. (ed.) FoSSaCS 2013. LNCS, vol. 7794, pp. 161–176. Springer, Heidelberg (2013). https://doi.org/10.1007/978-3-642-37075-5_11

[MS17] Mio, M., Simpson, A.: Łukasiewicz μ-calculus. Fundam. Informaticae **150**(3–4), 317–346 (2017). https://doi.org/10.3233/FI-2017-1472

[Pot83] Pottinger, G.: Uniform, cut-free formulations of T, S4 and S5 (abstract). J. Symbolic Logic **48**(3), 898–910 (1983)

[Stu07] Studer, T.: On the proof theory of the modal μ-calculus. Stud. Logica. **89**(3), 343–363 (2007)

[Vul67] Vulikh, B.Z.: Introduction to the Theory of Partially Ordered Spaces. Wolters-Noordhoff Scientific Publications LTD., Groningen (1967)

[Wan96] Wansing, H. (ed.): Proof Theory of Modal Logic. Applied Logic Series, vol. 2. Springer, Dordrecht (1996). https://doi.org/10.1007/978-94-017-2798-3

Partial and Conditional Expectations in Markov Decision Processes with Integer Weights

Jakob Piribauer(✉) and Christel Baier

Technische Universität Dresden, Dresden, Germany
{jakob.piribauer,christel.baier}@tu-dresden.de

Abstract. The paper addresses two variants of the stochastic shortest path problem ("optimize the accumulated weight until reaching a goal state") in Markov decision processes (MDPs) with integer weights. The first variant optimizes partial expected accumulated weights, where paths not leading to a goal state are assigned weight 0, while the second variant considers conditional expected accumulated weights, where the probability mass is redistributed to paths reaching the goal. Both variants constitute useful approaches to the analysis of systems without guarantees on the occurrence of an event of interest (reaching a goal state), but have only been studied in structures with non-negative weights. Our main results are as follows. There are polynomial-time algorithms to check the finiteness of the supremum of the partial or conditional expectations in MDPs with arbitrary integer weights. If finite, then optimal weight-based deterministic schedulers exist. In contrast to the setting of non-negative weights, optimal schedulers can need infinite memory and their value can be irrational. However, the optimal value can be approximated up to an absolute error of ϵ in time exponential in the size of the MDP and polynomial in $\log(1/\epsilon)$.

1 Introduction

Stochastic shortest path (SSP) problems generalize the shortest path problem on graphs with weighted edges. The SSP problem is formalized using finite state Markov decision processes (MDPs), which are a prominent model combining probabilistic and nondeterministic choices. In each state of an MDP, one is allowed to choose nondeterministically from a set of actions, each of them is augmented with probability distributions over the successor states and a weight (cost or reward). The SSP problem asks for a policy to choose actions (here called a scheduler) maximizing or minimizing the expected accumulated weight until reaching a goal state. In the classical setting, one seeks an optimal *proper* scheduler where proper means that a goal state is reached almost surely. Polynomial-time solutions exist exploiting the fact that optimal memoryless deterministic

The authors are supported by the DFG through the Research Training Group QuantLA (GRK 1763), the DFG-project BA-1679/11-1, the Collaborative Research Center HAEC (SFB 912), and the cluster of excellence CeTI.

M. Bojańczyk and A. Simpson (Eds.): FOSSACS 2019, LNCS 11425, pp. 436–452, 2019.
https://doi.org/10.1007/978-3-030-17127-8_25

schedulers exist (provided the optimal value is finite) and can be computed using linear programming techniques, possibly in combination with model transformations (see [1,5,10]). The restriction to proper schedulers, however, is often too restrictive. First, there are models that have no proper scheduler. Second, even if proper schedulers exist, the expectation of the accumulated weight of schedulers missing the goal with a positive probability should be taken into account as well. Important such applications include the semantics of probabilistic programs (see e.g. [4,7,12,14,16]) where no guarantee for almost sure termination can be given and the analysis of program properties at termination time gives rise to stochastic shortest (longest) path problems in which the goal (halting configuration) is not reached almost surely. Other examples are the fault-tolerance analysis (e.g., expected costs of repair mechanisms) in selected error scenarios that can appear with some positive, but small probability or the trade-off analysis with conjunctions of utility and cost constraints that are achievable with positive probability, but not almost surely (see e.g. [2]).

This motivates the switch to variants of classical SSP problems where the restriction to proper schedulers is relaxed. One option (e.g., considered in [8]) is to seek a scheduler optimizing the expectation of the random variable that assigns weight 0 to all paths not reaching the goal and the accumulated weight of the shortest prefix reaching the goal to all other paths. We refer to this expectation as *partial expectation*. Second, we consider the *conditional expectation* of the accumulated weight until reaching the goal under the condition that the goal is reached. In general, partial expectations describe situations in which some reward (positive and negative) is accumulated but only retrieved if a certain goal is met. In particular, partial expectations can be an appropriate replacement for the classical expected weight before reaching the goal if we want to include schedulers which miss the goal with some (possibly very small) probability. In contrast to conditional expectations, the resulting scheduler still has an incentive to reach the goal with a high probability, while schedulers maximizing the conditional expectation might reach the goal with a very small positive probability.

Previous work on partial or conditional expected accumulated weights was restricted to the case of non-negative weights. More precisely, partial expectations have been studied in the setting of stochastic multiplayer games with non-negative weights [8]. Conditional expectations in MDPs with non-negative weights have been addressed in [3]. In both cases, optimal values are achieved by weight-based deterministic schedulers that depend on the current state and the weight that has been accumulated so far, while memoryless schedulers are not sufficient. Both [8] and [3] prove the existence of a *saturation point* for the accumulated weight from which on optimal schedulers behave memoryless and maximize the probability to reach a goal state. This yields exponential-time algorithms for computing optimal schedulers using an iterative linear programming approach. Moreover, [3] proves that the threshold problem for conditional expectations ("does there exist a scheduler $\mathfrak{S}$ such that the conditional expectation under $\mathfrak{S}$ exceeds a given threshold?") is PSPACE-hard even for acyclic MDPs.

The purpose of the paper is to study partial and conditional expected accumulated weights for MDPs with integer weights. The switch from non-negative to integer weights indeed causes several additional difficulties. We start with the following observation. While optimal partial or conditional expectations in non-negative MDPs are rational, they can be irrational in the general setting:

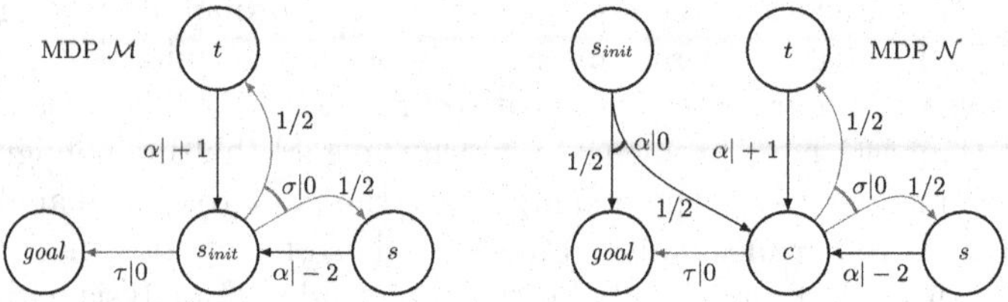

Fig. 1. Enabled actions are denoted by Greek letters and the weight associated to the action is stated after the bar. Probabilistic choices are marked by a bold arc and transition probabilities are denoted next to the arrows.

Example 1. Consider the MDP $\mathcal{M}$ depicted on the left in Fig. 1. In the initial state s_{init}, two actions are enabled. Action τ leads to *goal* with probability 1 and weight 0. Action σ leads to the states s and t with probability 1/2 from where we will return to s_{init} with weight -2 or $+1$, respectively. The scheduler choosing τ immediately leads to an expected weight of 0 and is optimal among schedulers reaching the goal almost surely. As long as we choose σ in s_{init}, the accumulated weight follows an asymmetric random walk increasing by 1 or decreasing by 2 with probability 1/2 before we return to s_{init}. It is well known that the probability to ever reach accumulated weight $+1$ in this random walk is $1/\Phi$ where $\Phi = \frac{1+\sqrt{5}}{2}$ is the golden ratio. Likewise, ever reaching accumulated weight n has probability $1/\Phi^n$ for all $n \in \mathbb{N}$. Consider the scheduler $\mathfrak{S}_k$ choosing τ as soon as the accumulated weight reaches k in s_{init}. Its partial expectation is k/Φ^k as the paths which never reach weight k are assigned weight 0. The maximum is reached at $k = 2$. In Sect. 4, we prove that there are optimal schedulers whose decisions only depend on the current state and the weight accumulated so far. With this result we can conclude that the maximal partial expectation is indeed $2/\Phi^2$, an irrational number.

The conditional expectation of $\mathfrak{S}_k$ in $\mathcal{M}$ is k as $\mathfrak{S}_k$ reaches the goal with accumulated weight k if it reaches the goal. So, the conditional expectation is not bounded. If we add a new initial state making sure that the goal is reached with positive probability as in the MDP $\mathcal{N}$, we can obtain an irrational maximal conditional expectation as well: The scheduler $\mathfrak{T}_k$ choosing τ in c as soon as the weight reaches k has conditional expectation $\frac{k/2\Phi^k}{1/2+1/2\Phi^k}$. The maximum is obtained for $k = 3$; the maximal conditional expectation is $\frac{3/\Phi^3}{1+1/\Phi^3} = \frac{3}{3+\sqrt{5}}$.

Moreover, while the proposed algorithms of [3,8] crucially rely on the monotonicity of the accumulated weights along the prefixes of paths, the accumulated weights of prefixes of path can oscillate when there are positive and negative weights. As we will see later, this implies that the existence of saturation points is no longer ensured and optimal schedulers might require infinite memory (more precisely, a counter for the accumulated weight). These observations provide evidence why linear-programming techniques as used in the case of non-negative MDPs [3,8] cannot be expected to be applicable for the general setting.

Contributions. We study the problem of maximizing the partial and conditional expected accumulated weight in MDPs with integer weights. Our first result is that the finiteness of the supremum of partial and conditional expectations in MDPs with integer weights can be checked in polynomial time (Sect. 3). For both variants we show that there are optimal weight-based deterministic schedulers if the supremum is finite (Sect. 4). Although the suprema might be irrational and optimal schedulers might need infinite memory, the suprema can be ϵ-approximated in time exponential in the size of the MDP and polynomial in $\log(1/\epsilon)$ (Sect. 5). By duality of maximal and minimal expectations, analogous results hold for the problem of minimizing the partial or conditional expected accumulated weight. (Note that we can multiply all weights by -1 and then apply the results for maximal partial resp. conditional expectations.)

Related Work. Closest to our contribution is the above mentioned work on partial expected accumulated weights in stochastic multiplayer games with non-negative weights in [8] and on computation schemes for maximal conditional expected accumulated weights in non-negative MDPs [3]. Conditional expected termination time in probabilistic push-down automata has been studied in [11], which can be seen as analogous considerations for a class of infinite-state Markov chains with non-negative weights. The recent work on notions of conditional value at risk in MDPs [15] also studies conditional expectations, but the considered random variables are limit averages and a notion of (non-accumulated) weight-bounded reachability.

2 Preliminaries

We give basic definitions and present our notation. More details can be found in textbooks, e.g. [18].

Notations for Markov Decision Processes. A *Markov decision process* (MDP) is a tuple $\mathcal{M} = (S, Act, P, s_{init}, wgt)$ where S is a finite set of states, Act a finite set of actions, $s_{init} \in S$ the initial state, $P : S \times Act \times S \to [0,1] \cap \mathbb{Q}$ is the transition probability function and $wgt : S \times Act \to \mathbb{Z}$ the weight function. We require that $\sum_{t \in S} P(s, \alpha, t) \in \{0, 1\}$ for all $(s, \alpha) \in S \times Act$. We write $Act(s)$ for the set of actions that are enabled in s, i.e., $\alpha \in Act(s)$ iff $\sum_{t \in S} P(s, \alpha, t) = 1$. We assume that $Act(s)$ is non-empty for all s and that all states are reachable from s_{init}. We call a state absorbing if the only enabled action leads to the state itself with probability 1 and weight 0. The paths of $\mathcal{M}$ are finite or infinite sequences $s_0\, \alpha_0\, s_1\, \alpha_1\, s_2\, \alpha_2 \ldots$ where states and actions alternate such that $P(s_i, \alpha_i, s_{i+1}) > 0$ for all $i \geq 0$. If $\pi = s_0\, \alpha_0\, s_1\, \alpha_1 \ldots \alpha_{k-1}\, s_k$ is finite, then

$wgt(\pi) = wgt(s_0, \alpha_0) + \ldots + wgt(s_{k-1}, \alpha_{k-1})$ denotes the accumulated weight of π, $P(\pi) = P(s_0, \alpha_0, s_1) \cdot \ldots \cdot P(s_{k-1}, \alpha_{k-1}, s_k)$ its probability, and $last(\pi) = s_k$ its last state. The *size* of $\mathcal{M}$, denoted $size(\mathcal{M})$, is the sum of the number of states plus the total sum of the logarithmic lengths of the non-zero probability values $P(s, \alpha, s')$ as fractions of co-prime integers and the weight values $wgt(s, \alpha)$.

Scheduler. A *(history-dependent, randomized) scheduler* for $\mathcal{M}$ is a function $\mathfrak{S}$ that assigns to each finite path π a probability distribution over $Act(last(\pi))$. $\mathfrak{S}$ is called memoryless if $\mathfrak{S}(\pi) = \mathfrak{S}(\pi')$ for all finite paths π, π' with $last(\pi) = last(\pi')$, in which case $\mathfrak{S}$ can be viewed as a function that assigns to each state s a distribution over $Act(s)$. $\mathfrak{S}$ is called deterministic if $\mathfrak{S}(\pi)$ is a Dirac distribution for each path π, in which case $\mathfrak{S}$ can be viewed as a function that assigns an action to each finite path π. Scheduler $\mathfrak{S}$ is said to be *weight-based* if $\mathfrak{S}(\pi) = \mathfrak{S}(\pi')$ for all finite paths π, π' with $wgt(\pi) = wgt(\pi')$ and $last(\pi) = last(\pi')$. Thus, deterministic weight-based schedulers can be viewed as functions that assign actions to state-weight-pairs. By $HR^{\mathcal{M}}$ we denote the class of all schedulers, by $WR^{\mathcal{M}}$ the class of weight-based schedulers, by $WD^{\mathcal{M}}$ the class of weight-based, deterministic schedulers, and by $MD^{\mathcal{M}}$ the class of memoryless deterministic schedulers. Given a scheduler $\mathfrak{S}$, $\varsigma = s_0\, \alpha_0\, s_1\, \alpha_1 \ldots$ is a $\mathfrak{S}$-path iff ς is a path and $\mathfrak{S}(s_0\, \alpha_0\, s_1\, \alpha_1 \ldots \alpha_{k-1}\, s_k)(\alpha_k) > 0$ for all $k \geq 0$.

Probability Measure. We write $\Pr^{\mathfrak{S}}_{\mathcal{M},s}$ or briefly $\Pr^{\mathfrak{S}}_{s}$ to denote the probability measure induced by $\mathfrak{S}$ and s. For details, see [18]. We will use LTL-like formulas to denote measurable sets of paths and also write $\Diamond(wgt \bowtie x)$ to describe the set of infinite paths having a prefix π with $wgt(\pi) \bowtie x$ for $x \in \mathbb{Z}$ and $\bowtie \in \{<, \leq, =, \geq, >\}$. Given a measurable set ψ of infinite paths, we define $\Pr^{\min}_{\mathcal{M},s}(\psi) = \inf_{\mathfrak{S}} \Pr^{\mathfrak{S}}_{\mathcal{M},s}(\psi)$ and $\Pr^{\max}_{\mathcal{M},s}(\psi) = \sup_{\mathfrak{S}} \Pr^{\mathfrak{S}}_{\mathcal{M},s}(\psi)$ where $\mathfrak{S}$ ranges over all schedulers for $\mathcal{M}$. Throughout the paper, we suppose that the given MDP has a designated state *goal*. Then, $p^{\max}_s$ and $p^{\min}_s$ denote the maximal resp. minimal probability of reaching *goal* from s. That is, $p^{\max}_s = \sup_{\mathfrak{S}} \Pr^{\mathfrak{S}}_s(\Diamond goal)$ and $p^{\min}_s = \inf_{\mathfrak{S}} \Pr^{\mathfrak{S}}_s(\Diamond goal)$. Let $Act^{\max}(s) = \{\alpha \in Act(s) | \sum_{t \in S} P(s, \alpha, t) \cdot p^{\max}_t = p^{\max}_s\}$, and $Act^{\min}(s) = \{\alpha \in Act(s) | \sum_{t \in S} P(s, \alpha, t) \cdot p^{\min}_t = p^{\min}_s\}$.

Mean Payoff. A well-known measure for the long-run behavior of a scheduler $\mathfrak{S}$ in an MDP $\mathcal{M}$ is the *mean payoff.* Intuitively, the mean payoff is the amount of weight accumulated per step on average in the long run. Formally, we define the mean payoff as the following random variable on infinite paths $\zeta = s_0\alpha_0 s_1\alpha_1 \ldots$: $MP(\zeta) := \liminf\limits_{k \to \infty} \frac{\sum_{i=0}^{k} wgt(s_i, \alpha_i)}{k+1}$. The mean payoff of the scheduler $\mathfrak{S}$ starting in s_{init} is then defined as the expected value $\mathbb{E}^{\mathfrak{S}}_{s_{init}}(MP)$. The maximal mean payoff is the supremum over all schedulers which is equal to the maximum over all MD-schedulers: $\mathbb{E}^{\max}_{s_{init}}(MP) = \max_{\mathfrak{S} \in MD} \mathbb{E}^{\mathfrak{S}}_{s_{init}}(MP)$. In strongly connected MDPs, the maximal mean payoff does not depend on the initial state.

End Components, MEC-Quotient. An *end component* of $\mathcal{M}$ is a strongly connected sub-MDP. End components can be formalized as pairs $\mathcal{E} = (E, \mathfrak{A})$ where E is a nonempty subset of S and $\mathfrak{A}$ a function that assigns to each state $s \in E$ a nonempty subset of $Act(s)$ such that the graph induced by $\mathcal{E}$ is strongly connected. $\mathcal{E}$ is called *maximal* if there is no end component $\mathcal{E}' = (E', \mathfrak{A}')$ with $\mathcal{E} \neq \mathcal{E}'$, $E \subseteq E'$ and $\mathfrak{A}(s) \subseteq \mathfrak{A}'(s)$ for all $s \in E$. The *MEC-quotient* of an MDP $\mathcal{M}$ is the MDP $MEC(\mathcal{M})$ arising from $\mathcal{M}$ by collapsing all states that belong to the same maximal end component $\mathcal{E}$ to a state $s_{\mathcal{E}}$. All actions enabled in some state in $\mathcal{E}$ not belonging to $\mathcal{E}$ are enabled in $s_{\mathcal{E}}$. Details and the formal construction can be found in [9]. We call an end component $\mathcal{E}$ *positively weight-divergent* if there is a scheduler $\mathfrak{S}$ for $\mathcal{E}$ such that $\Pr^{\mathfrak{S}}_{\mathcal{E},s}(\Diamond(wgt \geq n)) = 1$ for all $s \in \mathcal{E}$ and $n \in \mathbb{N}$. In [1], it is shown that the existence of positively weight-divergent end components can be decided in polynomial time.

3 Partial and Conditional Expectations in MDPs

We define *partial* and *conditional expectations* in MDPs. We extend the definition of [8] by introducing partial expectations with *bias* which are closely related to conditional expectations. Afterwards, we sketch the computation of maximal partial expectations in MDPs with non-negative weights and in Markov chains.

Partial and Conditional Expectation. In the sequel, let $\mathcal{M}$ be an MDP with a designated absorbing goal state *goal*. Furthermore, we collapse all states from which *goal* is not reachable to one absorbing state *fail*. Let $b \in \mathbb{R}$. We define the random variable $\oplus^b goal$ on infinite paths ζ by

$$\oplus^b goal(\zeta) = \begin{cases} wgt(\zeta) + b & \text{if } \zeta \vDash \Diamond goal, \\ 0 & \text{if } \zeta \nvDash \Diamond goal. \end{cases}$$

We call the expectation of this random variable under a scheduler $\mathfrak{S}$ the *partial expectation with bias b* of $\mathfrak{S}$ and write $PE^{\mathfrak{S}}_{\mathcal{M},s_{init}}[b] := \mathbb{E}^{\mathfrak{S}}_{\mathcal{M},s_{init}}(\oplus^b goal)$ as well as $PE^{\sup}_{\mathcal{M},s_{init}}[b] := \sup_{\mathfrak{S}\in HR^{\mathcal{M}}} PE^{\mathfrak{S}}_{\mathcal{M},s_{init}}[b]$. If $b = 0$, we sometimes drop the argument b; if $\mathcal{M}$ is clear from the context, we drop the subscript. In order to maximize the partial expectation, intuitively one has to find the right balance between reaching *goal* with high probability and accumulating a high positive amount of weight before reaching *goal*. The bias can be used to shift this balance by additionally rewarding or penalizing a high probability to reach *goal*.

The *conditional expectation* of $\mathfrak{S}$ is defined as the expectation of $\oplus^0 goal$ under the condition that *goal* is reached. It is defined if $\Pr^{\mathfrak{S}}_{\mathcal{M},s_{init}}(\Diamond goal) > 0$. We write $CE^{\mathfrak{S}}_{\mathcal{M},s_{init}} := \mathbb{E}^{\mathfrak{S}}_{\mathcal{M},s_{init}}(\oplus^0 goal|\Diamond goal)$ and $CE^{\sup}_{\mathcal{M},s_{init}} = \sup_{\mathfrak{S}} CE^{\mathfrak{S}}_{\mathcal{M},s_{init}}$ where the supremum is taken over all schedulers $\mathfrak{S}$ with $\Pr^{\mathfrak{S}}_{\mathcal{M},s_{init}}(\Diamond goal) > 0$. We can express the conditional expectation as $CE^{\mathfrak{S}}_{\mathcal{M},s_{init}} = PE^{\mathfrak{S}}_{\mathcal{M},s_{init}}/\Pr^{\mathfrak{S}}_{\mathcal{M},s_{init}}(\Diamond goal)$. The following proposition establishes a close connection between conditional expectations and partial expectations with bias.

Proposition 2. *Let $\mathcal{M}$ be an MDP, $\mathfrak{S}$ a scheduler with $\Pr^{\mathfrak{S}}_{s_{init}}(\Diamond goal) > 0$, $\theta \in \mathbb{Q}$, and $\bowtie \in \{<, \leq, \geq, >\}$. Then we have $PE^{\mathfrak{S}}_{s_{init}}[-\theta] \bowtie 0$ iff $CE^{\mathfrak{S}}_{s_{init}} \bowtie \theta$. Further, if $\Pr^{\min}_{s_{init}}(\Diamond goal) > 0$, then $PE^{\sup}_{s_{init}}[-\theta] \bowtie 0$ iff $CE^{\sup}_{s_{init}} \bowtie \theta$.*

Proof. The first claim follows from $PE^{\mathfrak{S}}_{s_{init}}[-\theta] = PE^{\mathfrak{S}}_{s_{init}}[0] - \Pr^{\mathfrak{S}}_{s_{init}}(\Diamond goal) \cdot \theta$. The second claim follows by quantification over all schedulers. □

In [3], it is shown that deciding whether $CE^{\sup}_{s_{init}} \bowtie \theta$ for $\bowtie \in \{<, \leq, \geq, >\}$ and $\theta \in \mathbb{Q}$ is PSPACE-hard even for acyclic MDPs. We conclude:

Corollary 3. *Given an MDP $\mathcal{M}$, $\bowtie \in \{<, \leq, \geq, >\}$, and $\theta \in \mathbb{Q}$, deciding whether $PE^{\sup}_{\mathcal{M},s_{init}} \bowtie \theta$ is PSPACE-hard.*

Finiteness. We present criteria for the finiteness of $PE^{\sup}_{s_{init}}[b]$ and $CE^{\sup}_{s_{init}}$. Detailed proofs can be found in Appendix A.1 of [17]. By slightly modifying the construction from [1] which removes end components only containing 0-weight cycles, we obtain the following result.

Proposition 4. *Let $\mathcal{M}$ be an MDP which does not contain positively weight-divergent end components and let $b \in \mathbb{Q}$. Then there is a polynomial time transformation to an MDP $\mathcal{N}$ containing all states from $\mathcal{M}$ and possibly an additional absorbing state fail such that*

- *all end components of $\mathcal{N}$ have negative maximal expected mean payoff,*
- *for any scheduler $\mathfrak{S}$ for $\mathcal{M}$ there is a scheduler $\mathfrak{S}'$ for $\mathcal{N}$ with $\Pr^{\mathfrak{S}}_{\mathcal{M},s}(\Diamond goal) = \Pr^{\mathfrak{S}'}_{\mathcal{N},s}(\Diamond goal)$ and $PE^{\mathfrak{S}}_{\mathcal{M},s}[b] = PE^{\mathfrak{S}'}_{\mathcal{N},s}[b]$ for any state s in $\mathcal{M}$, and vice versa.*

Hence, we can restrict ourselves to MDPs in which all end components have negative maximal expected mean payoff if there are no positively weight divergent end components. The following result is now analogous to the result in [1] for the classical SSP problem.

Proposition 5. *Let $\mathcal{M}$ be an MDP and $b \in \mathbb{R}$ arbitrary. The optimal partial expectation $PE^{\sup}_{s_{init}}[b]$ is finite if and only if there are no positively weight-divergent end components in $\mathcal{M}$.*

To obtain an analogous result for conditional expectations, we observe that the finiteness of the maximal partial expectation is necessary for the finiteness of the maximal conditional expectation. However, this is not sufficient. In [3], a *critical scheduler* is defined as a scheduler $\mathfrak{S}$ for which there is a path containing a positive cycle and for which $\Pr^{\mathfrak{S}}_{s_{init}}(\Diamond goal) = 0$. Given a critical scheduler, it is easy to construct a sequence of schedulers with unbounded conditional expectation (see Appendix A.1 of [17] and [3]). On the other hand, if $\Pr^{\min}_{\mathcal{M},s_{init}}(\Diamond goal) > 0$, then $CE^{\sup}_{s_{init}}$ is finite if and only if $PE^{\sup}_{s_{init}}$ is finite. We will show how we can restrict ourselves to this case if there are no critical schedulers:

So, let $\mathcal{M}$ be an MDP with $\Pr^{\min}_{\mathcal{M},s_{init}}(\Diamond goal) = 0$ and suppose there are no critical schedulers for $\mathcal{M}$. Let S_0 be the set of all states reachable from s_{init} while only choosing actions in $Act^{\min}$. As there are no critical schedulers, $(S_0, Act^{\min})$ does not contain positive cycles. So, there is a finite maximal weight w_s among paths leading from s_{init} to s in S_0. Consider the following MDP $\mathcal{N}$: It contains the MDP $\mathcal{M}$ and a new initial state t_{init}. For each $s \in S_0$ and each $\alpha \in Act(s) \setminus Act^{\min}(s)$, $\mathcal{N}$ also contains a new state $t_{s,\alpha}$ which is reachable from

t_{init} via an action $\beta_{s,\alpha}$ with weight w_s and probability 1. In $t_{s,\alpha}$, only action α with the same probability distribution over successors and the same weight as in s is enabled. So in $\mathcal{N}$, one has to decide immediately in which state to leave S_0 and one accumulates the maximal weight which can be accumulated in $\mathcal{M}$ to reach this state in S_0. In this way, we ensure that $\Pr^{\min}_{\mathcal{N},t_{init}}(\Diamond goal) > 0$.

Proposition 6. *The constructed MDP $\mathcal{N}$ satisfies $CE^{\sup}_{\mathcal{N},t_{init}} = CE^{\sup}_{\mathcal{M},s_{init}}$.*

We can rely on this reduction to an MDP in which *goal* is reached with positive probability for ϵ-approximations and the exact computation of the optimal conditional expectation. In particular, the values w_s for $s \in S_0$ are easy to compute by classical shortest path algorithms on weighted graphs. Furthermore, we can now decide the finiteness of the maximal conditional expectation.

Proposition 7. *For an arbitrary MDP $\mathcal{M}$, $CE^{\sup}_{\mathcal{M},s_{init}}$ is finite if and only if there are no positively weight-divergent end components and no critical schedulers.*

Partial and Conditional Expectations in Markov Chains. Markov chains with integer weights can be seen as MDPs with only one action α enabled in every state. Consequently, there is only one scheduler for a Markov chain. Hence, we drop the superscripts in $p^{\max}$ and $PE^{\sup}$.

Proposition 8. *The partial and conditional expectation in a Markov chain $\mathcal{C}$ are computable in polynomial time.*

Proof. Let α be the only action available in $\mathcal{C}$. Assume that all states from which *goal* is not reachable have been collapsed to an absorbing state *fail*. Then $PE_{\mathcal{C},s_{init}}$ is the value of $x_{s_{init}}$ in the unique solution to the following system of linear equations with one variable x_s for each state s:

$$x_{goal} = x_{fail} = 0,$$
$$x_s = wgt(s,\alpha) \cdot p_s + \sum_t P(s,\alpha,t) \cdot x_t \text{ for } s \in S \setminus \{goal, fail\}.$$

The existence of a unique solution follows from the fact that $\{goal\}$ and $\{fail\}$ are the only end components (see [18]). It is straight-forward to check that $(PE_{\mathcal{C},s})_{s\in S}$ is this unique solution. The conditional expectation is obtained from the partial expectation by dividing by the probability $p_{s_{init}}$ to reach the goal. □

This result can be seen as a special case of the following result. Restricting ourselves to schedulers which reach the goal with maximal or minimal probability in an MDP without positively weight-divergent end components, linear programming allows us to compute the following two memoryless deterministic schedulers (see [3,8]).

Proposition 9. *Let $\mathcal{M}$ be an MDP without positively weight-divergent end components. There is a scheduler $\mathfrak{Max} \in MD^{\mathcal{M}}$ such that for each $s \in S$ we have $\Pr_s^{\mathfrak{Max}}(\Diamond goal) = p_s^{\max}$ and $PE_s^{\mathfrak{Max}} = \sup_{\mathfrak{S}} PE_s^{\mathfrak{S}}$ where the supremum is taken over all schedulers $\mathfrak{S}$ with $\Pr_s^{\mathfrak{S}}(\Diamond goal) = p_s^{\max}$. Similarly, there is a scheduler $\mathfrak{Min} \in MD^{\mathcal{M}}$ maximizing the partial expectation among all schedulers reaching the goal with minimal probability. Both these schedulers and their partial expectations are computable in polynomial time.*

These schedulers will play a crucial role for the approximation of the maximal partial expectation and the exact computation of maximal partial expectations in MDPs with non-negative weights.

Partial Expectations in MDPs with Non-negative Weights. In [8], the computation of maximal partial expectations in stochastic multiplayer games with non-negative weights is presented. We adapt this approach to MDPs with non-negative weights. A key result is the existence of a *saturation point*, a bound on the accumulated weight above which optimal schedulers do not need memory.

In the sequel, let $R \in \mathbb{Q}$ be arbitrary, let $\mathcal{M}$ be an MDP with non-negative weights, $PE_{s_{init}}^{\sup} < \infty$, and assume that end components have negative maximal mean payoff (see Proposition 4). A saturation point for bias R is a natural number $\mathfrak{p}$ such that there is a scheduler $\mathfrak{S}$ with $PE_{s_{init}}^{\mathfrak{S}}[R] = PE_{s_{init}}^{\sup}[R]$ which is memoryless and deterministic as soon as the accumulated weight reaches $\mathfrak{p}$. I.e. for any two paths π and π', with $last(\pi) = last(\pi')$ and $wgt(\pi), wgt(\pi') > \mathfrak{p}$, $\mathfrak{S}(\pi) = \mathfrak{S}(\pi')$.

Transferring the idea behind the saturation point for conditional expectations given in [3], we provide the following saturation point which can be considerably smaller than the saturation point given in [8] in stochastic multiplayer games. Detailed proofs to this section are given in Appendix A.2 of [17].

Proposition 10. *We define $p_{s,\alpha}^{\max} := \sum_{t\in S} P(s,\alpha,t)\cdot p_t^{\max}$ and $PE_{s,\alpha}^{\mathfrak{Max}} := p_{s,\alpha}^{\max} \cdot wgt(s,\alpha) + \sum_{t\in S} P(s,\alpha,t)\cdot PE_t^{\mathfrak{Max}}$. Then,*

$$\mathfrak{p}_R := \sup \left\{ \frac{PE_{s,\alpha}^{\mathfrak{Max}} - PE_s^{\mathfrak{Max}}}{p_s^{\max} - p_{s,\alpha}^{\max}} \;\middle|\; s \in S, \alpha \in Act(s) \setminus Act^{\max}(s) \right\} - R$$

is an upper saturation point for bias R in $\mathcal{M}$.

The saturation point $\mathfrak{p}_R$ is chosen such that, as soon as the accumulated weight exceeds $\mathfrak{p}_R$, the scheduler $\mathfrak{Max}$ is better than any scheduler deviating from $\mathfrak{Max}$ for only one step. So, the proposition states that $\mathfrak{Max}$ is then also better than any other scheduler.

As all values involved in the computation can be determined by linear programming, the saturation point $\mathfrak{p}_R$ is computable in polynomial time. This also means that the logarithmic length of $\mathfrak{p}_R$ is polynomial in the size of $\mathcal{M}$ and hence $\mathfrak{p}_R$ itself is at most exponential in the size of $\mathcal{M}$.

Proposition 11. *Let $R \in \mathbb{Q}$ and let B_R be the least integer greater or equal to $\mathfrak{p}_R + \max_{s\in S,\alpha\in Act(s)} wgt(s,\alpha)$ and let $S' := S \setminus \{goal, fail\}$. The values*

$(PE^{\sup}_{s_{init}}[r{+}R])_{s\in S',0\leq r\leq B_R}$ *form the unique solution to the following linear program in the variables* $(x_{s,r})_{s\in S',0\leq r\leq B_R}$ *(r ranges over integers):*

Minimize $\sum_{s\in S',0\leq r\leq B_R} x_{s,r}$ *under the following constraints:*

$$\text{For } r \geq \mathfrak{p}_R : x_{s,r} = p^{\max}_s \cdot (r{+}R) + E^{\mathfrak{Max}}_s,$$
$$\text{for } r < \mathfrak{p}_R \text{ and } \alpha \in Act(s):$$
$$x_{s,r} \geq P(s,\alpha,goal)\cdot(r{+}R{+}wgt(s,\alpha)) + \sum_{t\in S'} P(s,\alpha,t)\cdot x_{t,r+wgt(s,\alpha)}.$$

From a solution x to the linear program, we can easily extract an optimal weight-based deterministic scheduler. This scheduler only needs finite memory because the accumulated weight increases monotonically along paths and as soon as the saturation point is reached $\mathfrak{Max}$ provides the optimal decisions. As B_R is exponential in the size of $\mathcal{M}$, the computation of the optimal partial expectation via this linear program runs in time exponential in the size of $\mathcal{M}$.

4 Existence of Optimal Schedulers

We prove that there are optimal weight-based deterministic schedulers for partial and conditional expectations. After showing that, if finite, $PE^{\sup}_{s_{init}}$ is equal to $\sup_{\mathfrak{S}\in WD^{\mathcal{M}}} PE^{\mathfrak{S}}_{s_{init}}$, we take an analytic approach to show that there is a weight-based deterministic scheduler maximizing the partial expectation. We define a metric on $WD^{\mathcal{M}}$ turning it into a compact space. Then, we prove that the function assigning the partial expectation to schedulers is upper semi-continuous. We conclude that there is a weight-based deterministic scheduler obtaining the maximum. Proofs to this section can be found in Appendix B of [17].

Proposition 12. *Let* $\mathcal{M}$ *be an MDP with* $PE^{\sup}_{s_{init}} < \infty$*. Then we have* $PE^{\sup}_{s_{init}} = \sup_{\mathfrak{S}\in WD^{\mathcal{M}}} PE^{\mathfrak{S}}_{s_{init}}$.

Proof sketch. We can assume that all end components have negative maximal expected mean payoff (see Proposition 4). Given a scheduler $\mathfrak{S} \in HR^{\mathcal{M}}$, we take the expected number of times $\theta_{s,w}$ that s is visited with accumulated weight w under $\mathfrak{S}$ for each state-weight pair (s,w), and the expected number of times $\theta_{s,w,\alpha}$ that $\mathfrak{S}$ then chooses α. These values are finite due to the negative maximal mean payoff in end components. We define the scheduler $\mathfrak{T} \in WR^{\mathcal{M}}$ choosing α in s with probability $\theta_{s,w,\alpha}/\theta_{s,w}$ when weight w has been accumulated. Then, we show by standard arguments that we can replace all probability distributions that $\mathfrak{T}$ chooses by Dirac distributions to obtain a scheduler $\mathfrak{T}' \in WD^{\mathcal{M}}$ such that $PE^{\mathfrak{T}'}_{s_{init}} \geq PE^{\mathfrak{S}}_{s_{init}}$. □

It remains to show that the supremum is obtained by a weight-based deterministic scheduler. Given an MDP $\mathcal{M}$ with arbitrary integer weights, we define the following metric $d^{\mathcal{M}}$ on the set of weight-based deterministic schedulers, i.e. on the set of functions from $S \times \mathbb{Z} \to Act$: For two such schedulers $\mathfrak{S}$ and

$\mathfrak{T}$, we let $d^{\mathcal{M}}(\mathfrak{S},\mathfrak{T}) := 2^{-R}$ where R is the greatest natural number such that $\mathfrak{S} \upharpoonright S \times \{-(R-1),\ldots,R-1\} = \mathfrak{T} \upharpoonright S \times \{-(R-1),\ldots,R-1\}$ or ∞ if there is no greatest such natural number.

Lemma 13. *The metric space* $(Act^{S\times\mathbb{Z}}, d^{\mathcal{M}})$ *is compact.*

Having defined this compact space of schedulers, we can rely on the analytic notion of upper semi-continuity.

Lemma 14 (Upper Semi-Continuity of Partial Expectations). *If* $PE^{\sup}_{s_{init}}$ *is finite in* $\mathcal{M}$*, then the function* $PE : (WD, d^{WD}) \to (\mathbb{R}_{\infty}, d^{euclid})$ *assigning* $PE^{\mathfrak{S}}_{s_{init}}$ *to a weight-based deterministic scheduler* $\mathfrak{S}$ *is upper semi-continuous.*

The technical proof of this lemma can be found in Appendix B of [17]. We arrive at the main result of this section.

Theorem 15 (Existence of Optimal Schedulers for Partial Expectations). *If* $PE^{\sup}_{s_{init}}$ *is finite in an MDP* $\mathcal{M}$*, then there is a weight-based deterministic scheduler* $\mathfrak{S}$ *with* $PE^{\sup}_{s_{init}} = PE^{\mathfrak{S}}_{s_{init}}$.

Proof. If $PE^{\sup}_{s_{init}}$ is finite, then the map $PE : (WD, d^{WD}) \to (\mathbb{R}_{\infty}, d^{euclid})$ is upper semi-continuous. So, this map has a maximum because (WD, d^{WD}) is a compact metric space. □

Corollary 16 (Existence of Optimal Schedulers for Conditional Expectations). *If* $CE^{\sup}_{s_{init}}$ *is finite in an MDP* $\mathcal{M}$*, then there is a weight-based deterministic scheduler* $\mathfrak{S}$ *with* $CE^{\sup}_{s_{init}} = CE^{\mathfrak{S}}_{s_{init}}$.

Proof. By Proposition 6, we can assume that $\Pr^{\min}_{s_{init}}(\Diamond goal) > 0$. We know that $PE^{\sup}_{s_{init}}[-CE^{\sup}_{s_{init}}] = 0$ and that there is a weight-based deterministic scheduler $\mathfrak{S}$ with $PE^{\mathfrak{S}}_{s_{init}}[-CE^{\sup}_{s_{init}}] = 0$. By Proposition 2, $\mathfrak{S}$ maximizes the conditional expectation as it reaches *goal* with positive probability. □

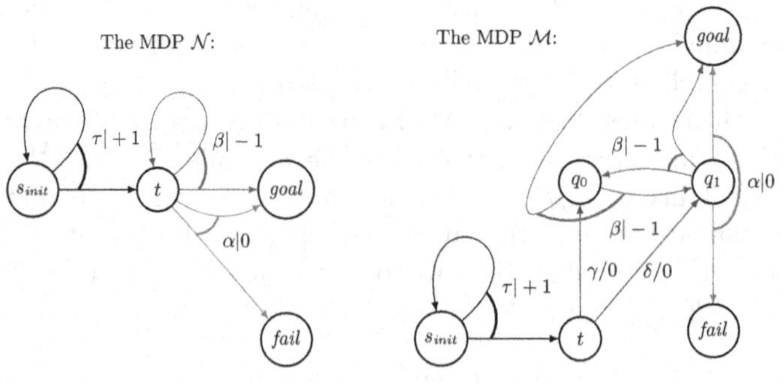

Fig. 2. All non-trivial transition probabilities are 1/2. In the MDP $\mathcal{M}$, the optimal choice to maximize the partial expectation in t depends on the parity of the accumulated weight.

In MDPs with non-negative weights, the optimal decision in a state s only depends on s as soon as the accumulated weight exceeds a saturation point. In MDPs with arbitrary integer weights, it is possible that the optimal choice of action does not become stable for increasing values of accumulated weight as we see in the following example.

Example 17. Let us first consider the MDP $\mathcal{N}$ depicted in Fig. 2. Let π be a path reaching t for the first time with accumulated weight r. Consider a scheduler which chooses β for the first k times and then α. In this situation, the partial expectation from this point on is:

$$\frac{1}{2^{k+1}}(r-k)+\sum_{i=1}^{k}\frac{1}{2^i}(r-i)=\frac{1}{2^{k+1}}+\sum_{i=1}^{k+1}\frac{1}{2^i}(r-i)=\frac{k-r+4}{2^{k+1}}+r-2.$$

For $r \geq 2$, this partial expectation has its unique maximum for the choice $k = r-2$. This already shows that an optimal scheduler needs infinite memory. No matter how much weight r has been accumulated when reaching t, the optimal scheduler has to count the $r-2$ times it chooses β.

Furthermore, we can transfer the optimal scheduler for the MDP $\mathcal{N}$ to the MDP $\mathcal{M}$. In state t, we have to make a nondeterministic choice between two action leading to the states q_0 and q_1, respectively. In both of these states, action β is enabled which behaves like the same action in the MDP $\mathcal{N}$ except that it moves between the two states if *goal* is not reached. So, the action α is only enabled every other step. As in $\mathcal{N}$, we want to choose α after choosing β $r-2$ times if we arrived in t with accumulated weight $r \geq 2$. So, the choice in t depends on the parity of r: For $r = 1$ or r even, we choose δ. For odd $r \geq 3$, we choose γ. This shows that the optimal scheduler in the MDP $\mathcal{M}$ needs specific information about the accumulated weight, in this case the parity, no matter how much weight has been accumulated.

In the example, the optimal scheduler has a periodic behavior when fixing a state and looking at optimal decisions for increasing values of accumulated weight. The question whether an optimal scheduler always has such a periodic behavior remains open.

5 Approximation

As the optimal values for partial and conditional expectation can be irrational, there is no hope to compute these values by linear programming as in the case of non-negative weights. In this section, we show how we can nevertheless approximate the values. The main result is the following.

Theorem 18. *Let $\mathcal{M}$ be an MDP with $PE^{\sup}_{\mathcal{M},s_{init}} < \infty$ and $\epsilon > 0$. The maximal partial expectation $PE^{\sup}_{\mathcal{M},s_{init}}$ can be approximated up to an absolute error of ϵ in time exponential in the size of $\mathcal{M}$ and polynomial in $\log(1/\epsilon)$. If further, $CE^{\sup}_{\mathcal{M},s_{init}} < \infty$, also $CE^{\sup}_{\mathcal{M},s_{init}}$ can be approximated up to an absolute error of ϵ in time exponential in the size of $\mathcal{M}$ and polynomial in $\log(1/\epsilon)$.*

We first prove that upper bounds for $PE^{\sup}_{\mathcal{M},s_{init}}$ and $CE^{\sup}_{\mathcal{M},s_{init}}$ can be computed in polynomial time. Then, we show that there are ϵ-optimal schedulers for the partial expectation which become memoryless as soon as the accumulated weight leaves a sufficiently large weight window around 0. We compute the optimal partial expectation of such a scheduler by linear programming. The result can then be extended to conditional expectations.

Upper Bounds. Let $\mathcal{M}$ be an MDP in which all end components have negative maximal mean payoff. Let δ be the minimal non-zero transition probability in $\mathcal{M}$ and $W := \max_{s\in S,\alpha\in Act(s)} |wgt(s,\alpha)|$. Moving through the MEC-quotient, the probability to reach an accumulated weight of $|S|\cdot W$ is bounded by $1-\delta^{|S|}$ as *goal* or *fail* is reached within S steps with probability at least $1-\delta^{|S|}$. It remains to show similar bounds inside an end component.

We will use the characterization of the maximal mean payoff in terms of super-harmonic vectors due to Hordijk and Kallenberg [13] to define a super-martingale controlling the growth of the accumulated weight in an end component under any scheduler. As the value vector for the maximal mean payoff in an end component is constant and negative in our case, the results of [13] yield:

Proposition 19 (Hordijk, Kallenberg). *Let $\mathcal{E} = (S, Act)$ be an end component with maximal mean payoff $-t$ for some $t > 0$. Then there is a vector $(u_s)_{s\in S}$ such that $-t + u_s \geq wgt(s,\alpha) + \sum_{s'\in S} P(s,\alpha,s')\cdot u_{s'}$.*

Furthermore, let v be the vector $(-t,\ldots,-t)$ in $\mathbb{R}^S$. Then, (v,u) is the solution to a linear program with $2|S|$ variables, $2|S||Act|$ inequalities, and coefficients formed from the transition probabilities and weights in $\mathcal{E}$.

We will call the vector u a *super-potential* because the expected accumulated weight after i steps is at most $u_s - \min_{t\in S} u_t - i\cdot t$ when starting in state s. Let $\mathfrak{S}$ be a scheduler for $\mathcal{E}$ starting in some state s. We define the following random variables on $\mathfrak{S}$-runs in $\mathcal{E}$: let $s(i)\in S$ be the state after i steps, let $\alpha(i)$ be the action chosen after i steps, let $w(i)$ be the accumulated weight after i steps, and let $\pi(i)$ be the history, i.e. the finite path after i steps.

Lemma 20. *The sequence $m(i) := w(i) + u_{s(i)}$ satisfies $\mathbb{E}(m(i+1)|\pi(0),\ldots,\pi(i)) \leq m(i) - t$ for all i.*[1]

Proof. By Proposition 19, $\mathbb{E}(m(i+1)|\pi(0),\ldots,\pi(i)) - m(i) = wgt(s(i),\mathfrak{S}(\pi(i))) + \sum_{s'\in S} P(s(i),\mathfrak{S}(\pi(i)),s')\cdot u_{s'} - u_{s(i)} \leq -t$. □

We are going to apply the following theorem by Blackwell [6].

Theorem 21 (Blackwell [6]). *Let $X_1, X_2,\ldots$ be random variables, and let $S_n := \sum_{k=1}^{n} X_k$. Assume that $|X_i| \leq 1$ for all i and that there is a $u > 0$ such that $\mathbb{E}(X_{n+1}|X_1,\ldots,X_n) \leq -u$. Then, $\Pr(\sup_{n\in\mathbb{N}} S_n \geq t) \leq \left(\frac{1-u}{1+u}\right)^t$.*

[1] This means that $m(i) + i\cdot t$ is a super-martingale with respect to the history $\pi(i)$.

We denote $\max_{s'\in S} u_{s'} - \min_{s'\in S} u_{s'}$ by $\|u\|$. Observe that $|m(i+1)-m(i)| \leq \|u\| + W =: c_{\mathcal{E}}$. We can rescale the sequence $m(i)$ by defining $m'(i) := (m(i) - m(0))/c_{\mathcal{E}}$. This ensures that $m'(0) = 0$, $|m'(i+1) - m'(i)| \leq 1$ and $\mathbb{E}(m'(i+1)|m'(0),\ldots,m'(i)) \leq -t/c_{\mathcal{E}}$ for all i. In this way, we arrive at the following conclusion, putting $\lambda_{\mathcal{E}} := \frac{1-t/c_{\mathcal{E}}}{1+t/c_{\mathcal{E}}}$.

Corollary 22. *For any scheduler $\mathfrak{S}$ and any starting state s in $\mathcal{E}$, we have* $\Pr_s^{\mathfrak{S}}(\Diamond wgt \geq (k+1)\cdot c_{\mathcal{E}}) \leq \lambda_{\mathcal{E}}^k$.

Proof. By Theorem 21, $\Pr_s^{\mathfrak{S}}(\Diamond wgt \geq (k+1)\cdot c_{\mathcal{E}}) \leq \Pr_s^{\mathfrak{S}}(\Diamond wgt \geq \|u\| + k\cdot c_{\mathcal{E}}) \leq \Pr_s^{\mathfrak{S}}(\exists i : m(i) - m(0) \geq k\cdot c_{\mathcal{E}}) = \Pr_s^{\mathfrak{S}}(\sup_{i\in\mathbb{N}} m'(i) \geq k) \leq \left(\frac{1-t/c_{\mathcal{E}}}{1+t/c_{\mathcal{E}}}\right)^k$. □

Let *MEC* be the set of maximal end components in $\mathcal{M}$. For each $\mathcal{E} \in MEC$, let $\lambda_{\mathcal{E}}$ and $c_{\mathcal{E}}$ be as in Corollary 22. Define $\lambda_{\mathcal{M}} := 1 - (\delta^{|S|}\cdot \prod_{\mathcal{E}\in MEC}(1-\lambda_{\mathcal{E}}))$, and $c_{\mathcal{M}} := |S|\cdot W + \sum_{\mathcal{E}\in MEC} c_{\mathcal{E}}$. Then an accumulated weight of $c_{\mathcal{M}}$ cannot be reached with a probability greater than $\lambda_{\mathcal{M}}$ because reaching accumulated weight $c_{\mathcal{M}}$ would require reaching weight $c_{\mathcal{E}}$ in some end component $\mathcal{E}$ or reaching weight $|S|\cdot W$ in the MEC-quotient and $1-\lambda_{\mathcal{M}}$ is a lower bound on the probability that none of this happens (under any scheduler).

Proposition 23. *Let $\mathcal{M}$ be an MDP with $PE_{s_{init}}^{\sup} < \infty$. There is an upper bound PE^{ub} for the partial expectation in $\mathcal{M}$ computable in polynomial time.*

Proof. In any end component $\mathcal{E}$, the maximal mean payoff $-t$ and the super-potential u are computable in polynomial time. Hence, $c_{\mathcal{E}}$ and $\lambda_{\mathcal{E}}$, and in turn also $c_{\mathcal{M}}$ and $\lambda_{\mathcal{M}}$ are also computable in polynomial time. When we reach accumulated weight $c_{\mathcal{M}}$ for the first time, the actual accumulated weight is at most $c_{\mathcal{M}} + W$. So, we conclude that $\Pr_s^{\max}(\Diamond wgt \geq k\cdot(c_{\mathcal{M}} + W)) \leq \lambda_{\mathcal{M}}^k$. The partial expectation can now be bounded by $\sum_{k=0}^{\infty}(k+1)\cdot(c_{\mathcal{M}}+W)\cdot\lambda_{\mathcal{M}}^k = \frac{c_{\mathcal{M}}+W}{(1-\lambda_{\mathcal{M}})^2}$. □

Corollary 24. *Let $\mathcal{M}$ be an MDP with $CE_{\mathcal{M},s_{init}}^{\sup} < \infty$. There is an upper bound CE^{ub} for the conditional expectation in $\mathcal{M}$ computable in polynomial time.*

Proof. By Proposition 6, we can construct an MDP $\mathcal{N}$ in which *goal* is reached with probability $q > 0$ in polynomial time with $CE_{\mathcal{M},s_{init}}^{\sup} = CE_{\mathcal{N},s_{init}}^{\sup}$. Now, $CE^{ub} := PE^{ub}/q$ is an upper bound for the conditional expectation in $\mathcal{M}$. □

Approximating Optimal Partial Expectations. The idea for the approximation is to assume that the partial expectation is $PE_{s_{init}}^{\mathfrak{Max}} + w\cdot p_s^{\max}$ if a high weight w has been accumulated in state s. Similarly, for small weights w', we use the value $PE_{s_{init}}^{\mathfrak{Min}} + w\cdot p_s^{\min}$. We will first provide a lower "saturation point" making sure that only actions minimizing the probability to reach the goal are used by an optimal scheduler as soon as the accumulated weight drops below this saturation point. For the proofs to this section, see Appendix C.1 of [17].

Proposition 25. *Let $\mathcal{M}$ be an MDP with $PE^{\sup}_{s_{init}} < \infty$. Let $s \in S$ and let $\mathfrak{q}_s := \frac{PE^{ub} - PE^{\mathfrak{Min}}_s}{p^{\min}_s - \min_{\alpha \notin Act^{\min}(s)} p^{\min}_{s,\alpha}}$. Then any weight-based deterministic scheduler $\mathfrak{S}$ maximizing the partial expectation in $\mathcal{M}$ satisfies $\mathfrak{S}(s,w) \in Act^{\min}(s)$ if $w \leq \mathfrak{q}_s$.*

Let $\mathfrak{q} := \min_{s \in S} \mathfrak{q}_s$ and let $D := PE^{ub} - \min\{PE^{\mathfrak{Max}}_s, PE^{\mathfrak{Min}}_s | s \in S\}$. Given $\epsilon > 0$, we define $R^+_\epsilon := (c_\mathcal{M} + W) \cdot \left\lceil \frac{\log(2D)+\log(1/\epsilon)}{\log(1/\lambda_\mathcal{M})} \right\rceil$ and $R^-_\epsilon := \mathfrak{q} - R^+_\epsilon$.

Theorem 26. *There is a weight-based deterministic scheduler $\mathfrak{S}$ such that the scheduler $\mathfrak{T}$ defined by*

$$\mathfrak{T}(\pi) = \begin{cases} \mathfrak{S}(\pi) & \text{if any prefix } \pi' \text{ of } \pi \text{ satisfies } R^-_\epsilon \leq wgt(\pi') \leq R^+_\epsilon, \\ \mathfrak{Max}(\pi) & \text{if the shortest prefix } \pi' \text{ of } \pi \text{ with } wgt(\pi') \notin [R^-_\epsilon, R^+_\epsilon] \\ & \text{satisfies } wgt(\pi') > R^+_\epsilon, \\ \mathfrak{Min}(\pi) & \text{otherwise,} \end{cases}$$

satisfies $PE^{\mathfrak{T}}_{s_{init}} \geq PE^{\sup}_{s_{init}} - \epsilon$.

This result now allows us to compute an ϵ-approximation and an ϵ-optimal scheduler with finite memory by linear programming, similar to the case of non-negative weights, in a linear program with $R^+_\epsilon + R^-_\epsilon$ many variables and $|Act|$-times as many inequalities.

Corollary 27. *$PE^{\sup}_{s_{init}}$ can be approximated up to an absolute error of ϵ in time exponential in the size of $\mathcal{M}$ and polynomial in $\log(1/\epsilon)$.*

If the logarithmic length of $\theta \in \mathbb{Q}$ is polynomial in the size of $\mathcal{M}$, we can also approximate $PE^{\sup}_{s_{init}}[\theta]$ up to an absolute error of ϵ in time exponential in the size of $\mathcal{M}$ and polynomial in $\log(1/\epsilon)$: We can add a new initial state s with a transition to s_{init} with weight θ and approximate $PE^{\sup}_s$ in the new MDP.

Transfer to Conditional Expectations. Let $\mathcal{M}$ be an MDP with $CE^{sup}_{s_{init}} < \infty$ and $\epsilon > 0$. By Proposition 6, we can assume that $\Pr^{\min}_{\mathcal{M},s_{init}}(\Diamond goal) =: p$ is positive. Clearly, $CE^{sup}_{s_{init}} \in [CE^{\mathfrak{Max}}_{s_{init}}, CE^{ub}]$. We perform a binary search to approximate $CE^{sup}_{s_{init}}$: We put $A_0 := CE^{\mathfrak{Max}}_{s_{init}}$ and $B_0 := CE^{ub}$. Given A_i and B_i, let $\theta_i := (A_i + B_i)/2$. Then, we approximate $PE^{\sup}_{s_{init}}[-\theta_i]$ up to an absolute error of $p \cdot \epsilon$. Let E_i be the value of this approximation. If $E_i \in [-2p \cdot \epsilon, 2p \cdot \epsilon]$, terminate and return θ_i as the approximation for $CE^{\sup}_{s_{init}}$. If $E_i < -2p \cdot \epsilon$, put $A_{i+1} := A_i$ and $B_{i+1} := \theta_i$, and repeat. If $E_i > 2p \cdot \epsilon$, put $A_{i+1} := \theta_i$ and $B_{i+1} := B_i$, and repeat.

Proposition 28. *The procedure terminates after at most $\lceil \log((A_0 - B_0)/(p \cdot \epsilon)) \rceil$ iterations and returns an 3ϵ-approximation of $CE^{\sup}_{s_{init}}$ in time exponential in the size of $\mathcal{M}$ and polynomial in $\log(1/\epsilon)$.*

The proof can be found in Appendix C.2 of [17]. This finishes the proof of Theorem 18.

6 Conclusion

Compared to the setting of non-negative weights, the optimization of partial and conditional expectations faces substantial new difficulties in the setting of integer weights. The optimal values can be irrational showing that the linear programming approaches from the setting of non-negative weights cannot be applied for the computation of optimal values. We showed that this approach can nevertheless be adapted for approximation algorithms. Further, we were able to show that there are optimal weight-based deterministic schedulers. These schedulers, however, can require infinite memory and it remains open whether we can further restrict the class of schedulers necessary for the optimization. In examples, we have seen that optimal schedulers can switch periodically between actions they choose for increasing values of accumulated weight. Further insights on the behavior of optimal schedulers would be helpful to address threshold problems ("Is $PE^{\sup}_{s_{init}} \geq \theta$?").

References

1. Baier, C., Bertrand, N., Dubslaff, C., Gburek, D., Sankur, O.: Stochastic shortest paths and weight-bounded properties in Markov decision processes. In: Proceedings of the 33rd Annual ACM/IEEE Symposium on Logic in Computer Science (LICS), pp. 86–94. ACM (2018)
2. Baier, C., Dubslaff, C., Klein, J., Klüppelholz, S., Wunderlich, S.: Probabilistic model checking for energy-utility analysis. In: van Breugel, F., Kashefi, E., Palamidessi, C., Rutten, J. (eds.) Horizons of the Mind. A Tribute to Prakash Panangaden. LNCS, vol. 8464, pp. 96–123. Springer, Cham (2014). https://doi.org/10.1007/978-3-319-06880-0_5
3. Baier, C., Klein, J., Klüppelholz, S., Wunderlich, S.: Maximizing the conditional expected reward for reaching the goal. In: Legay, A., Margaria, T. (eds.) TACAS 2017. LNCS, vol. 10206, pp. 269–285. Springer, Heidelberg (2017). https://doi.org/10.1007/978-3-662-54580-5_16
4. Barthe, G., Espitau, T., Ferrer Fioriti, L.M., Hsu, J.: Synthesizing probabilistic invariants via Doob's decomposition. In: Chaudhuri, S., Farzan, A. (eds.) CAV 2016. LNCS, vol. 9779, pp. 43–61. Springer, Cham (2016). https://doi.org/10.1007/978-3-319-41528-4_3
5. Bertsekas, D.P., Tsitsiklis, J.N.: An analysis of stochastic shortest path problems. Math. Oper. Res. **16**(3), 580–595 (1991)
6. Blackwell, D.: On optimal systems. Ann. Math. Stat. **25**, 394–397 (1954)
7. Chatterjee, K., Fu, H., Goharshady, A.K.: Termination analysis of probabilistic programs through Positivstellensatz's. In: Chaudhuri, S., Farzan, A. (eds.) CAV 2016. LNCS, vol. 9779, pp. 3–22. Springer, Cham (2016). https://doi.org/10.1007/978-3-319-41528-4_1
8. Chen, T., Forejt, V., Kwiatkowska, M., Parker, D., Simaitis, A.: Automatic verification of competitive stochastic systems. Formal Methods Syst. Des. **43**(1), 61–92 (2013)
9. Ciesinski, F., Baier, C., Größer, M., Klein, J.: Reduction techniques for model checking Markov decision processes. In: Proceedings of the Fifth International Conference on Quantitative Evaluation of Systems (QEST), pp. 45–54. IEEE (2008)

10. de Alfaro, L.: Computing minimum and maximum reachability times in probabilistic systems. In: Baeten, J.C.M., Mauw, S. (eds.) CONCUR 1999. LNCS, vol. 1664, pp. 66–81. Springer, Heidelberg (1999). https://doi.org/10.1007/3-540-48320-9_7
11. Esparza, J., Kucera, A., Mayr, R.: Quantitative analysis of probabilistic pushdown automata: expectations and variances. In: Proceedings of the 20th Annual IEEE Symposium on Logic in Computer Science (LICS), pp. 117–126. IEEE (2005)
12. Gretz, F., Katoen, J.-P., McIver, A.: Operational versus weakest pre-expectation semantics for the probabilistic guarded command language. Perform. Eval. **73**, 110–132 (2014)
13. Hordijk, A., Kallenberg, L.: Linear programming and Markov decision chains. Manage. Sci. **25**(4), 352–362 (1979)
14. Katoen, J.-P., Gretz, F., Jansen, N., Kaminski, B.L., Olmedo, F.: Understanding probabilistic programs. In: Meyer, R., Platzer, A., Wehrheim, H. (eds.) Correct System Design. LNCS, vol. 9360, pp. 15–32. Springer, Cham (2015). https://doi.org/10.1007/978-3-319-23506-6_4
15. Kretínský, J., Meggendorfer, T.: Conditional value-at-risk for reachability and mean payoff in Markov decision processes. In: Proceedings of the 33rd Annual ACM/IEEE Symposium on Logic in Computer Science (LICS), pp. 609–618. ACM (2018)
16. Olmedo, F., Gretz, F., Jansen, N., Kaminski, B.L., Katoen, J.-P., Mciver, A.: Conditioning in probabilistic programming. ACM Trans. Program. Lang. Syst. (TOPLAS) **40**(1), 4:1–4:50 (2018)
17. Piribauer, J., Baier, C.: Partial and conditional expectations in Markov decision processes with integer weights (extended version). arXiv:1902.04538 (2019)
18. Puterman, M.L.: Markov Decision Processes: Discrete Stochastic Dynamic Programming. Wiley, New York (1994)

Equational Theories and Monads from Polynomial Cayley Representations

Maciej Piróg(✉), Piotr Polesiuk, and Filip Sieczkowski

University of Wrocław, Wrocław, Poland
mpirog@cs.uni.wroc.pl

Abstract. We generalise Cayley's theorem for monoids by providing an explicit formula for a (multi-sorted) equational theory represented by the type $PX \to X$, where P is an arbitrary polynomial endofunctor with natural coefficients. From the computational perspective, examples of effects given by such theories include backtracking nondeterminism (obtained with the original Cayley representation $X \to X$), finite mutable state (obtained with $n \to X$, for a constant n), and their different combinations (via $n \times X \to X$ or $X^n \to X$). Moreover, we show that monads induced by such theories are implementable using the type formers available in programming languages based on a polymorphic λ-calculus, both as compositions of algebraic datatypes and as continuation-like monads. We give a set-theoretic model of the latter in terms of Barr-dinatural transformations. We also introduce CayMon, a tool that takes a polynomial as an input and generates the corresponding equational theory together with the two implementations of the induced monad in Haskell.

1 Introduction

The relationship between universal algebra and monads has been studied at least since Linton [13] and Eilenberg and Moore [4], while the relationship between monads and the general theory of computational effects (exceptions, mutable state, nondeterminism, and such) has been observed by Moggi [14]. By transitivity, one can study computational effects using concepts from universal algebra, which is the main theme of Plotkin and Power's prolific research programme (see [10,20–24] among many others).

The simplest possible case of this approach is to describe an effect via a finitary equational theory: a finite set of operations (of finite arities), together with a finite set of equations. One such example is the theory of monoids:

Operations: $\gamma, \quad \varepsilon$

Equations: $\gamma(x,\varepsilon) = x, \quad \gamma(\varepsilon, x) = x, \quad \gamma(\gamma(x,y),z) = \gamma(x,\gamma(y,z))$

M. Bojańczyk and A. Simpson (Eds.): FOSSACS 2019, LNCS 11425, pp. 453–469, 2019.
https://doi.org/10.1007/978-3-030-17127-8_26

The above reads that the signature of the theory consists of two operations: binary γ and nullary ε. The equations state that γ is associative, with ε being its left and right unit.[1] One can also read this theory as a specification of backtracking nondeterminism, in which the order of results matters, where γ is an operation that creates a new computation as a choice between two subcomputations, while ε denotes failure. The connection between the equational theory and the computational effect becomes apparent when we consider the monad of free monoids (that is, the list monad), which is in fact used to form backtracking computations in programming; see, for example, Bird's pearl [1].

This suggests a simple recipe for computational effects: it is enough to come up with an equational theory, and out of the hat comes the induced monad of free algebras that implements the corresponding effect. Such an approach is indeed possible in the category **Set**, where every finitary equational theory admits a free monad, constructed by quotienting terms over the signature by the congruence induced by the equations. However, if we want to implement this monad in a programming language, the situation is not so simple, since in most programming languages (in particular, those without higher inductive types) we cannot generally express this kind of quotients. For instance, to describe a variant of nondeterminism that does not admit duplicate results, we may extend the theory of monoids with an equation stating that γ is idempotent, that is, $\gamma(x, x) = x$. But, unlike in the case of general monoids, the monad induced by the theory of idempotent monoids seems to be no longer directly expressible in, say, Haskell. This means that there is no implementation that satisfies all the equations of the theory "on the nose"—one informal argument is that the representations of $\gamma(x, x)$ and x should be the same whatever the type of x, and this would require a decidable equality test on every type, which is not possible.

Thus, both from the practical viewpoint of programming and as a question on the general nature of equational theories, it makes sense to ask which theories are "simple" enough to induce monads expressible using only the basic type formers, such as (co)products, function spaces, algebraic datatypes, universal and existential quantification. This question seems difficult in general, and to our knowledge there is little work that addresses it. In this paper, we focus on a small piece of this problem: we study a certain subset of such implementable equational theories, and conjure some novel extensions.

The monads that we consider arise from Cayley representations. The overall idea is that if a theory has an expressible, well-behaved (in a sense that we make precise in the paper) Cayley representation, the induced monad also has an expressible implementation. The well-known Cayley theorem for monoids states that every monoid with a carrier X embeds in the monoid of endofunctions $X \to X$. In this paper, we generalise this result: given a polynomial **Set**-endofunctor P with natural coefficients, we provide an explicit formula for an equational theory such that its every algebra with a carrier X embeds in a certain algebra with the carrier given by $PX \to X$. Then, we show that the monad of

[1] Although one usually writes γ as an infix operation, we use a "functional" syntax, since, in the following, the arity of corresponding operations may vary.

free algebras of such a theory can be implemented as a continuation-like monad with the endofunctor given at a set A as:

$$\forall X.(A \to PX \to X) \to PX \to X \tag{1}$$

This type is certainly expressible in programming languages based on polymorphic λ-calculi, such as Haskell.

However, before we can give the details of this construction, we need to address some technical issues. It is easy to notice that there may be more than one "Cayley representation" of a given theory: for example, a monoid X embeds not only in $X \to X$, but also in a "smaller" monoid $X \overset{\gamma}{\rightsquigarrow} X$, by which we mean the monoid of functions of the type $X \to X$ of the shape $a \mapsto \gamma(b, a)$, where $b \in X$. The same monoid X embeds also in a "bigger" monoid $X^2 \to X$, in which we interpret the operations as $\gamma(f, g) = (x, y) \mapsto f(g(x, y), y)$ and $\varepsilon = (x, y) \mapsto x$. What makes $X \to X$ special is that instantiating (1) with $PX = X$ gives a monad that is *isomorphic* to the list monad (note that in this case, the type (1) is simply the Church representation of lists). At the same time, we cannot use $X \overset{\gamma}{\rightsquigarrow} X$ instead of $X \to X$, since (1) quantifies over sets, and thus there is no natural candidate for γ. Moreover, even though we may use the instantiation $PX = X^2$, this choice yields a *different* monad (which we describe in more detail in Sect. 5.4). To sort this out, in Sect. 2, we introduce the notion of *tight Cayley representation*. This notion gives rise to the monad of the following shape, which is a strict generalisation of (1), where R is a **Set**-bifunctor of mixed variance:

$$\forall X.(A \to R(X, X)) \to R(X, X) \tag{2}$$

Formally, all our constructions are set-theoretic—to focus the presentation, the connection with programming languages and type theory is left implicit. Thus, the second issue that we discuss in Sect. 2 is the meaning of the universal quantifier $\forall$ in (1). It is known [27] that polymorphic functions of this shape enjoy a form of dinaturality proposed by Michael Barr (see Paré and Román [16]), called by Mulry *strong* dinaturality [15]. We model the universally quantified types above as collections of Barr-dinatural transformations, and prove that if R is a tight representation, the collection (2) is always a set.

In Sect. 4, we give the formula that defines an equational theory given a polynomial functor P. In general, the theories we construct can be multi-sorted, which is useful for avoiding a combinatory explosion of the induced theories, hence a brief discussion of such theories in Sect. 3. We show that $PX \to X$ is indeed a tight representation of the generated theory. Then, in Sect. 5, we study a number of examples in order to discover what effects are denoted by the generated theories. It turns out that each theory can be seen as a (rather complex, for nontrivial polynomial functors) composition of backtracking nondeterminism and finite mutable state. Moreover, in Sect. 6, we show that the corresponding monads can be implemented not only as continuation-like monads (1), but also in "direct style", using algebraic datatypes.

Since they are parametrised by a polynomial, both the equational theory and its representation consist of many indexed components, so it is not necessarily trivial to get much intuition simply by looking at the formulas. To facilitate this, we have implemented a tool, called CayMon, that generates the theory from a given polynomial, and produces two implementations in Haskell: as a composition of algebraic datatypes and as a continuation-like ("Cayley") monad (1). The tool can be run in a web browser, and is available at http://pl-uwr.bitbucket.io/caymon.

2 Tight Cayley Representations

In this section, we take a more abstract view on the concept of "Cayley representation". In the literature (for example, [2,5,17,25]), authors usually define Cayley representations of different forms of algebraic structures in terms of embeddings. This means that given an object X, there is a homomorphism $\sigma : X \to Y$ to a different object Y, and moreover σ has a retraction (not necessarily a homomorphism) $\rho : Y \to X$ (meaning $\rho \cdot \sigma = \mathsf{id}$). One important fact, which is usually left implicit, is that the construction of Y from X is in some sense functorial. Since we are interested in coming up with representations for many different equational theories, we first identify sufficient properties of such a representation needed to carry out the construction of the monad (2) sketched in the introduction. In particular, we introduce the notion of *tight Cayley representation*, which characterises the functoriality and naturality conditions for the components of the representation.

As for notation, we use $A \to B$ to denote both the type of a morphism in a category, and the set of all functions from A to B (the exponential object in **Set**). Also, for brevity, we write the application of a bifunctor to two arguments, e.g., $G(X, Y)$, without parentheses, as GXY. We begin with the following definition:

Definition 1 (see [16]). *Let $\mathscr{C}, \mathscr{D}$ be categories, and $G, H : \mathscr{C}^{\mathsf{op}} \times \mathscr{C} \to \mathscr{D}$ be functors. Then, a collection of $\mathscr{D}$-morphisms $\theta_X : GXX \to HXX$ indexed by $\mathscr{C}$-objects is called a* Barr-dinatural transformation *if it is the case that for all objects A in $\mathscr{D}$, objects X, Y in $\mathscr{C}$, morphisms $f_1 : A \to GXX$, $f_2 : A \to GYY$ in $\mathscr{D}$, and a morphism $g : X \to Y$ in $\mathscr{C}$,*

if $\begin{array}{ccc} & GXX & \\ {}^{f_1}\nearrow & & \searrow^{GXg} \\ A & & GXY \\ {}_{f_2}\searrow & & \nearrow_{GgY} \\ & GYY & \end{array}$ *commutes, then* $\begin{array}{cccc} & GXX \xrightarrow{\theta_X} HXX & & \\ {}^{f_1}\nearrow & & \searrow^{HXg} & \\ A & & HXY & \\ {}_{f_2}\searrow & & \nearrow_{HgY} & \\ & GYY \xrightarrow{\theta_Y} HYY & & \end{array}$ *commutes.*

An important property of Barr-dinaturality is that the component-wise composition gives a well-behaved notion of vertical composition of two such transformations. The connection between Barr-dinatural transformations and Cayley representations is suggested by the fact, shown by Paré and Román [16], that the collection of such transformations of type $H \to H$ for the **Set**-bifunctor $H(X, Y) = X \to Y$ is isomorphic to the set of natural numbers. The latter, equipped with addition and zero (or the former with composition and the identity transformation, respectively), is simply the free monoid with a single generator, that is, an instance of (1) with $PX = X$ and $A = 1$.

For the remainder of this section, assume that $\mathscr{T}$ is a category, while F : **Set** $\to \mathscr{T}$ is a functor with a right adjoint $U : \mathscr{T} \to$ **Set**. Intuitively, $\mathscr{T}$ is a category of algebras of some theory, and U is the forgetful functor. Then, the monad generated by the theory is given by the composition UF. For a function $f : A \to UX$, we write $\widehat{f} = Uf' : UFA \to UX$, where $f' : FA \to X$ is the contraposition of f via the adjunction (intuitively, the unique homomorphism induced by the freeness of the algebra FA).

Definition 2. *A* tight Cayley representation *of $\mathscr{T}$ with respect to $F \dashv U$ consists of the following components:*

(a) A bifunctor $R : \mathbf{Set}^{\mathsf{op}} \times \mathbf{Set} \to \mathbf{Set}$,
(b) For each set X, an object $\mathbb{R}X$ in $\mathscr{T}$, such that $U\mathbb{R}X = RXX$,
(c) For all sets A, X, Y and functions $f_1 : A \to RXX$, $f_2 : A \to RYY$, $g : X \to Y$, it is the case that

if 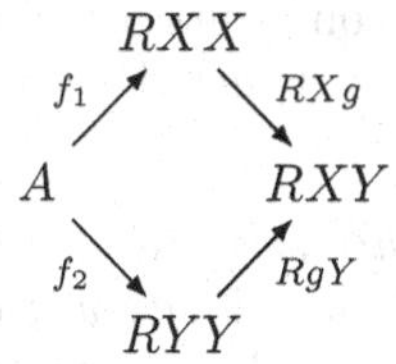

commutes, then 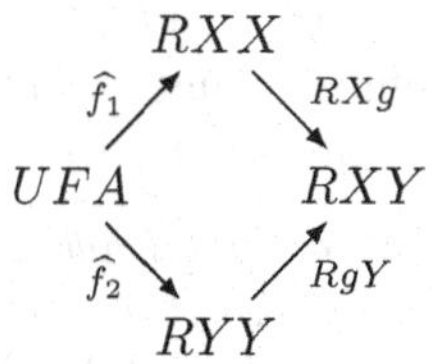

commutes.

(d) For each object M in $\mathscr{T}$, a morphism $\sigma_M : M \to \mathbb{R}(UM)$ in $\mathscr{T}$, such that $U\sigma_M : UM \to R(UM)(UM)$ is Barr-dinatural in M,
(e) A Barr-dinatural transformation $\rho_M : R(UM)(UM) \to UM$, such that $\rho_M \cdot U\sigma_M = \mathsf{id}$,
(f) For each set X, a set of indices I_X and a family of functions $\mathsf{run}_{X,i}$: $RXX \to X$, where $i \in I_X$, such that $R(RXX)\mathsf{run}_X$ is a jointly monic family, and the following diagram commutes for all X and $i \in I_X$:

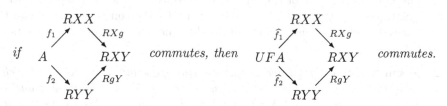

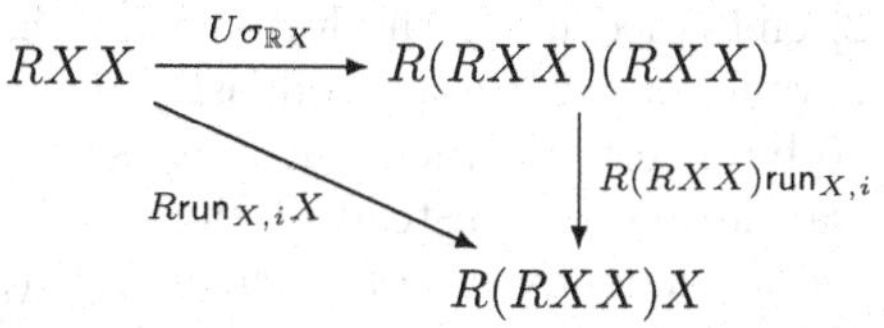

Note that the condition (c) states that the objects $\mathbb{R}$ are, in a sense, natural. Intuitively, understanding an object $\mathbb{R}X$ as an algebra, the condition states that the algebraic structure of $\mathbb{R}X$ does not really depend on the set X. The condition (f) may seem rather complicated: the intuition behind the technical formulation is that RXY behaves like a form of a function space (after all, we are interested in abstract *Cayley* representations), and $\mathsf{run}_{X,i}$ is an application to an argument specified by i, as in the example below. In such a case, the joint monicity becomes the extensional equality of functions.

Example 3. Let us check how Cayley representation for monoids fits the definition above: (a) The bifunctor is $RXY = X \to Y$. (b) The $\mathscr{T}$-object for a monoid M is the monoid $M \to M$ with $\gamma(f, g) = f \circ g$ and $\varepsilon = \mathsf{id}$. (c) Given some elements $a, b, \ldots, c \in A$, we need to see that $g \circ f_1(a) \circ f_1(b) \circ \cdots \circ f_1(c) = f_2(a) \circ f_2(b) \circ \cdots \circ f_2(c) \circ g$. Fortunately, the assumption, which in this case becomes $g \circ f_1(a) = f_2(a) \circ g$ for all $a \in A$, allows us to "commute" g from one side of the chain of function compositions to the other. (d) $\sigma_M(a) = b \mapsto \gamma(a, b)$. It is easy to verify that it is a homomorphism. The Barr-dinaturality condition: assuming $f(m) = n$ for some $m \in M$ and $n \in N$, and a homomorphism $f : M \to N$, it is the case that, omitting the U functor, $RfN(\sigma_N(n)) = RfN(\sigma_N(f(m))) = b \mapsto \gamma(f(m), f(b)) = b \mapsto f(\gamma(m, b)) = RMf(\sigma_M(m))$, where the equalities can be explained respectively as: assumption in the definition of Barr-dinaturality, unfolding definitions, homomorphism, unfolding definitions. (e) $\rho_M(f) = f(\varepsilon)$. It is easy to show that it is Barr-dinatural; note that we need to use the fact that $\mathscr{T}$-morphisms (that is, monoid homomorphisms) preserve ε. (f) We define $I_X = X$, while $\mathsf{run}_{X,i}(f) = f(i)$.

The first main result of this paper states that given a tight representation of $\mathscr{T}$ with respect to $F \dashv U$, the monad given by the composition UF can be alternatively defined using a continuation-like monad constructed with sets of Barr-dinatural transformations:

Theorem 4. *For a tight Cayley representation R with respect to $F \dashv U$, elements of the set UFA are in 1-1 correspondence with Barr-dinatural transformations of the type $(A \to RXX) \to RXX$. In particular, this means that the latter form a set. Moreover, this correspondence gives a monad isomorphism between UF and the evident continuation-like structure on (2), given by the unit $(\eta_A(a))_X(f) = f(a)$ and the Kleisli extension $(f^*(k))_X(g) = k_X(a \mapsto (f(a))_X(g))$.*

We denote the set of all Barr-dinatural transformations from the bifunctor $(X, Y) \mapsto A \to RXY$ to R as $\forall X.(A \to RXX) \to RXX$. This gives us a monad similar in shape to the continuation monad, or, more generally, Kock's codensity monad [12] embodied using the formula for right Kan extensions as ends. One important difference with the codensity monad (except, of course, the fact that we have bifunctors on the right-hand sides of arrows) is that we use Barr-dinatural transformations instead of the usual dinatural transformations [3]. Indeed, if we use ends instead of $\forall$, the end $\int_X (A \to RXX) \to RXX$ is given as the collection of all dinatural transformations of the given shape. It is known, however, that even in the simple case when $A = 1$ and $RXY = X \to Y$, this collection is too big to be a set (see the discussion in [16]), hence such end does not exist.

3 Multi-sorted Equational Theories with a Main Sort

The equational theories that we generate in Sect. 4 are multi-sorted, which is useful for trimming down the combinatorial complexity of the result. This turns out to be, in our view, essential in understanding what computational effects they actually represent. In this section, we give a quick overview of what kind of equational theories we work with, and discuss the construction of their free algebras.

We need to discuss the free algebras here, since we want the freeness to be with respect to a forgetful functor to **Set**, rather than to the usual category of sorted sets; compare [26]. This is because we want the equational theories to generate monads on **Set**, as described in the previous section. In particular, we are interested in theories in which one of the sorts is chosen as the *main* one, and work with the functor that forgets not only the structure, but also the carriers of all the other sorts, only preserving the main one. Luckily, this functor can be factored as a composition of two forgetful functors, each with an obvious left adjoint.

In detail, assume a finite set of sorts $S = \{\Omega, K_1, \ldots, K_d\}$ for some $d \in \mathbb{N}$, where Ω is the main sort. The category of sorted sets is simply the category $\mathbf{Set}^{|S|}$, where $|S|$ is the discrete category generated by the set S. More explicitly, the objects of $\mathbf{Set}^{|S|}$ are tuples of sets (one for each sort), while morphisms are tuples of functions. Given an S-sorted finitary theory $\mathfrak{T}$, we denote the category of its algebras as $\mathfrak{T}$-Alg. To see that the forgetful functor from $\mathfrak{T}$-Alg to **Set** has a left adjoint, consider the following composition of adjunctions:

$$X \mapsto (X, \emptyset, \ldots, \emptyset) \qquad\qquad \text{free}$$

$$\mathbf{Set} \;\rightleftharpoons\; \mathbf{Set}^{|S|} \;\rightleftharpoons\; \mathfrak{T}\text{-Alg}$$

$$(X, A_1, \ldots, A_d) \mapsto X \qquad\qquad \text{carriers}$$

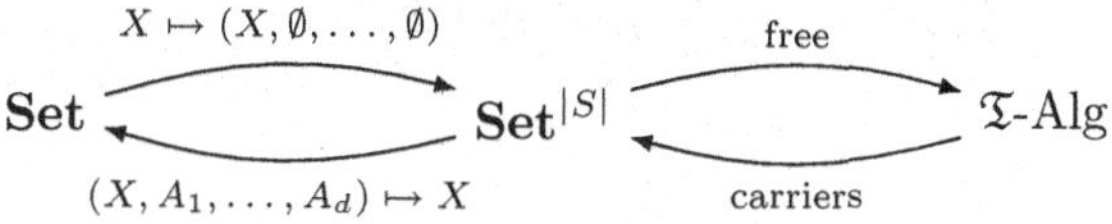

This means that the free algebra for each sort has the carrier given by the set of terms of the given sort (with variables appearing only at positions intended for the main sort Ω) quotiented by the congruence induced by the equations. This kind of composition of adjunctions is similar to [18], but in this case the compound right adjoints of the theories given in the next section are monadic.

4   Theories from Polynomial Cayley Representations

In this section, we introduce algebraic theories that are tightly Cayley-represented by $PX \to X$ for a polynomial functor P. Notation-wise, whenever we write $i \leq k$ for a fixed $k \in \mathbb{N}$, we mean that i is a natural number in the range $1, \ldots, k$, and use $[x_i]_{i \leq k}$ to denote a sequence $x_1, \ldots, x_k$. The latter notation is used also in arguments of functions and operations, so $f([x_i]_{i \leq k})$ means $f(x_1, \ldots, x_k)$, while $f(x, [y_i]_{i \leq k})$ means $f(x, y_1, \ldots, y_k)$. We sometimes use double indexing; for example, by $\prod_{i=1}^{k} \prod_{j=1}^{t_i} X_{i,j} \to Y$ for some $[t_i]_{i \leq k}$, we mean the type $X_{1,1} \times \cdots \times X_{1,t_1} \times \cdots \times X_{k,1} \times \cdots \times X_{k,t_k} \to Y$. This is matched by a double-nested notation in arguments, that is, $f([[x_i^j]_{j \leq t_i}]_{i \leq k})$ means $f(x_1^1, \ldots, x_1^{t_1}, \ldots, x_k^1, \ldots, x_k^{t_k})$. Also, whenever we want to repeat an argument k-times, we write $[x]_k$; for example, $f([x]_3)$ means $f(x, x, x)$. Because we use a lot of sub- and superscripts as indices, we do not use the usual notation for exponentiation. This means that x^i always denotes some x at index i, while a k-fold product of some type X, ordinarily denoted X^k, is written as $\prod^k X$. We use the $[\![\text{-}]\!]$ brackets to denote the interpretation of sorts and operations in an algebra (that is, a model of the theory). If the algebra is clear from the context, we skip the brackets in the interpretation of operations.

For the rest of the paper, let $d \in \mathbb{N}$ (the number of monomials in the polynomial) and sequences of natural numbers $[c_i]_{i \leq d}$ and $[e_i]_{i \leq d}$ (the coeffcients and exponents respectively) define the following polynomial endofunctor on **Set**:

$$PX = \sum_{i=1}^{d} c_i \times \prod^{e_i} X, \tag{3}$$

where c_i is an overloaded notation for the set $\{1, \ldots, c_i\}$. With this data, we define the following equational theory:

Definition 5. *Assuming d, $[c_i]_{i \leq d}$, and $[e_i]_{i \leq d}$ as above, we define the following equational theory $\mathfrak{T}$:*

- *Sorts:*

$$\begin{aligned} &\Omega && \text{(main sort)} \\ &K_i, \text{ for all } i \leq d \end{aligned}$$

- *Operations:*

$$\begin{aligned} &\mathsf{cons} : \textstyle\prod_{i=1}^{d} \prod^{c_i} K_i \to \Omega \\ &\pi_i^j : \Omega \to K_i, \text{ for } i \leq d \text{ and } j \leq c_i \\ &\varepsilon_i^j : K_i, \text{ for } i \leq d \text{ and } j \leq e_i \\ &\gamma_i^j : K_j \times \textstyle\prod^{e_j} K_i \to K_i, \text{ for } i, j \leq d \end{aligned}$$

- *Equations:*

$$\pi_i^j(\mathsf{cons}([[x_i^j]_{j \leq c_i}]_{i \leq d})) = x_i^j \tag{beta-π}$$

$$\mathsf{cons}([[\pi_i^j(x)]_{j \leq c_i}]_{i \leq d}) = x \tag{eta-π}$$

$$\gamma_i^j(\varepsilon_j^k, [x_t]_{t \leq e_j}) = x_k \tag{beta-ε}$$

$$\gamma_i^i(x, [\varepsilon_i^j]_{j \leq e_i}) = x \tag{eta-ε}$$

$$\gamma_i^j(\gamma_j^k(x, [y_t]_{t \leq e_k}), [z_s]_{s \leq e_j}) = \gamma_i^k(x, [\gamma_i^j(y_t, [z_s]_{s \leq e_j})]_{t \leq e_k}) \tag{assoc-γ}$$

Thus, in the theory $\mathfrak{T}$, there is a main sort Ω, which we think of as corresponding to the entire functor, and one sort K_i for each "monomial" $\prod^{e_i} X$. Then, we can think of Ω as a tuple containing elements of each sort, where each sort K_i has exactly c_i occurrences. The fact that Ω is a tuple, which is witnessed by the cons and π operations equipped with the standard equations for tupling and projections, is not too surprising—one should keep in mind that $\mathfrak{T}$ is a theory represented by the type $PX \to X$, which can be equivalently given as the *product* of function spaces $c_i \times \prod^{e_i} X \to X$ for all $i \le d$.

Each operation γ_i^j can be used to compose an element of K_j and e_j elements of K_i to obtain an element of K_i. The ε constants can be seen as selectors: in (beta-ε), ε_j^k in the first argument of γ_i^j selects the k-th argument of the sort K_i, while the (eta-ε) equation states that composing a value of K_i with the successive selectors of K_i gives back the original value. The equation (assoc-γ) states that the composition of values is associative in an appropriate sense. In Sect. 5, we provide a reading of the theory $\mathfrak{T}$ as a specification of a computational effect for different choices of d, c_i, and e_i.

Remark 6. If it is the case that $e_i = e_j$ for some $i, j \le d$, then the sorts K_i and K_j are isomorphic. This means that in every algebra of such a theory, there is an isomorphism of sorts $\varphi : [\![K_i]\!] \to [\![K_j]\!]$, given by $\varphi(x) = \gamma_j^i(x, [\varepsilon_j^k]_{k \le e_i})$. This suggests an alternative setting, in which instead of having a single $c_i \times \prod^{e_i} X$ comoponent, we can have c_i components of the shape $\prod^{e_i} X$. In such a setting, the equational theory $\mathfrak{T}$ in Definition 5 would be slightly simpler—specifically, there would be no need for double-indexing in the types of cons and π. On the downside, this would obfuscate the connection with computational effects described in Sect. 5 and some conjured extensions in Sect. 7.

The theory $\mathfrak{T}$ has a tight Cayley representation using functions from P, as detailed in the following theorem. This gives us the second main result of this paper: by Theorem 4, the theory $\mathfrak{T}$ is the equational theory of the monad (1). The notation in_i means the i-th inclusion of the coproduct in the functor P.

Theorem 7. *The equational theory $\mathfrak{T}$ from Definition 5 is tightly Cayley-represented by the following data:*

- *The bifunctor $RXY = PX \to Y$,*
- *For a set X, the following algebra:*
  - *Carriers of sorts:*

$$[\![\Omega]\!] = RXX$$
$$[\![K_i]\!] = \prod\nolimits^{e_i} X \to X$$

  - *Interpretation of operations:*

$$[\![\mathsf{cons}]\!]([[f_k^j]_{j \le c_k}]_{k \le d})(\mathsf{in}_i(c, [x_t]_{t \le e_i})) = f_i^c([x_t]_{t \le e_i})$$
$$[\![\pi_i^j]\!](f)([x_t]_{t \le e_i}) = f(\mathsf{in}_i(j, [x_t]_{t \le e_i}))$$
$$[\![\varepsilon_i^j]\!]([x_t]_{t \le e_i}) = x_j$$
$$[\![\gamma_i^j]\!](f, [g_k]_{k \le e_j})([x_t]_{t \le e_i}) = f([g_k([x_t]_{t \le e_i})]_{k \le e_j})$$

- *The homomorphism σ_M for the main sort and sorts K_i:*

$$\sigma_M^\Omega(m)(\mathsf{in}_i(c, [x_t]_{t \le e_i})) = \mathsf{cons}([[\gamma_k^i(\pi_i^c(m), [\pi_k^j(x_t)]_{t \le e_i})]_{j \le e_k}]_{k \le d})$$
$$\sigma_M^i(s)([x_t]_{t \le e_i}) = \mathsf{cons}([[\gamma_k^i(s, [\pi_k^j(x_t)]_{t \le e_i})]_{j \le e_k}]_{k \le d})$$

- *The transformation* ρ_M*:*

$$\rho_M(f) = \mathsf{cons}([[\pi^j_k(f(\mathsf{in}_k(j, [\mathsf{cons}([w^f_r]_{r<k}, [\varepsilon^t_k]_{c_k}, [w^f_r]_{k<r\leq d})]_{t\leq e_k})))]_{j\leq c_k}]_{k\leq d})$$
$$\text{where } w^f_r = [\pi^c_r(f(\mathsf{in}_r(c, [\varepsilon^j_r]_{j\leq e_r})))]_{c\leq c_r}$$

- *The set of indices* $I_X = PX$ *and the functions* $\mathsf{run}_{X,i}(f) = f(i)$.

In the representing algebra, it is the case that each $[\![K_i]\!]$ represents one monomial, as mentioned in the description of $\mathfrak{T}$, while $[\![\Omega]\!]$ is the appropriate tuple of representations of monomials, which is encoded as a single function from a coproduct (in our opinion, this encoding turns out to be much more readable on paper), while cons and π are indeed given by tupling and projections. For each $i \leq d$, the function ε^j_i simply returns its j-th argument, while γ is interpreted as the usual composition of multi-argument functions.

Homomorphisms between multi-sorted algebras are defined as operation-preserving functions for each sort, so σ is defined for the sort Ω and for each sort K_i. In general, the point of Cayley representations is to encode an element m of an algebra M using its possible behaviours with other elements of the algebra. It is no different here: for each sort K_i at the c-th occurrence in the tuple, the function σ^Ω packs (using cons) all possible compositions (by means of γ) of values of K_i with the "components" of m (extracted using π). The same happens for each $s \in [\![K_i]\!]$ in $\sigma^i_M(s)$, but there is no need to unpack s, as it is already a value of a single sort.

The transformation ρ_M is a bit more complicated. The argument f is, in general, a function from a coproduct to M, but we cannot simply apply f to one value $\mathsf{in}_i(\ldots)$ for some sort K_i, as we would obviously lose the information about the components in different sorts. This is why we need to apply f to all possible sorts with ε in the right place to ensure that we recover the original value. We extract the information about particular sorts from such values, and combine them using cons. Interestingly, the elements of w^f_r could actually be replaced by any expression of the appropriate sort that is preserved by homomorphisms, assuming that f is also preserved. This is needed to ensure that ρ is Barr-dinatural (the fact that f is preserved by homomorphisms is exactly the assumption in the definition of Barr-dinaturality). For example, if $e_r > 0$ for some $r \leq d$, one can define w^f_r simply as $[\varepsilon^j_r]_{c_r}$ for some $j \leq e_r$. The complicated expression in the definition of w^f_r is a way to produce values also for sorts K_r with $e_r = 0$, which do not have any ε constants.

5 Effects Modeled by Polynomial Representations

Now we describe what kind of computational effects are captured by the theories introduced in the previous section. It turns out that they all are different compositions of finite mutable state and backtracking nondeterminism. These compositions include the two most basic ones: when the state is *local* for each nondeterministic branch, and when it is *global* to the entire computation.

In the following, if there is only one object of a given kind, we skip the indices. For example, if for some i, it is the case that $e_i = 1$, we write ε_i instead of ε_i^1. If $d = 1$, we skip the subscripts altogether.

5.1 Backtracking Nondeterminism via Monoids

We recover the original Cayley theorem for monoids instantiating Theorem 7 with $PX = X$, that is, $d = 1$ and $c_1 = e_1 = 1$. In this case, we obtain two sorts, Ω and K, while the equations (beta-π) and (eta-π) instantiate respectively as follows:

$$\pi(\mathsf{cons}(x)) = x, \quad \mathsf{cons}(\pi(x)) = x$$

This means that both sorts are isomorphic, so one can think of this theory as being single-sorted. Of course, this is always the case if $d = 1$ and $c_1 = 1$. Since $e_1 = 1$, the operation γ is binary and there is a single ε constant. The equations (beta-ε) and (eta-ε) say, respectively, that ε is the left and right unit of γ, that is:

$$\gamma(\varepsilon, x) = x, \quad \gamma(x, \varepsilon) = x$$

Interestingly, the two unit laws for monoids are symmetrical, but in general the (beta-ε) and (eta-ε) equations are not. One should note that the symmetry is already broken when one implements free monoids (that is, lists) in a programming language: in the usual right-nested implementation, the "beta" rule is part of the definition of the **append** function, while the "eta" rule is a theorem. The (assoc-γ) equation instantiates as the associativity of γ:

$$\gamma(\gamma(x, y), z) = \gamma(x, \gamma(y, z))$$

5.2 Finite Mutable State

For $n \in \mathbb{N}$, if we take $PX = n$, that is, $d = 1$, $c_1 = n$ and $e_1 = 0$, we obtain the equational theory of a single mutable cell in which the set of possible states is $\{1, \ldots, n\}$. There are two sorts in the theory: Ω and K. The sort K does not have any interesting structure on its own, as there are no constants ε, and the equation (eta-ε) instantiates to

$$\gamma(x) = x,$$

which means that γ is necessarily an identity. The fact that this theory is indeed the theory of state becomes apparent when we identify Ω as a sort of computations that require some initial state to proceed, and K as computations that produce a final state. Then, the operations $\pi^j : \Omega \to K$ ($j \leq n$) are the "update" operations, where π^j sets the current state to j, while $\mathsf{cons} : \prod^n K \to \Omega$ is the "lookup" operation, in which the j-th argument is the computation to be executed if the current state is j. The equations (beta-π), for all $j \leq n$, and (eta-π) state respectively:

$$\pi^j(\mathsf{cons}([x_i]_{i \leq n})) = x_j, \quad \mathsf{cons}([\pi^i(x)]_{i \leq n}) = x$$

These equations embody the natural behaviour rules for this limited form of state. The former reads that setting the current state to j and then proceeding with the computation x_i if the current state is i is the same thing as simply proceeding with x_j (note that x_j is of the sort K, hence it does not use the information that the current state has just been updated to j, so there is no need to keep the π^j operation on the right-hand side of the equation). The latter states that if the current state is i and we set the current state to i, it is the same thing as not changing the state at all (note that x does not depend on the current state, as it is the same in every argument of cons).

Interestingly, the presentations of equational theories for state in the literature (for example, [7,23]) are all single-sorted. Such a setting can be recovered by defining the following macro-operations on the sort Ω:

$$\mathsf{put}^j : \Omega \to \Omega \qquad\qquad \mathsf{get} : \textstyle\prod^n \Omega \to \Omega$$
$$\mathsf{put}^j(x) = \mathsf{cons}([\pi^j(x)]_n) \qquad\qquad \mathsf{get}([x_i]_{i \le n}) = \mathsf{cons}([\pi^i(x_i)]_{i \le n})$$

The trick here is that the get operation does not change the state (by setting the new state to the current one), while put does not depend on the current state (by having the same computation in every argument of cons). The usual four equations for the interaction of put and get can be obtained by unfolding the definitions and using the (beta-π) and (eta-π) equations:

$$\mathsf{put}^j(\mathsf{put}^k(x)) = \mathsf{put}^k(x) \qquad\qquad \mathsf{put}^j(\mathsf{get}([x_i]_{i \le n})) = \mathsf{put}^j(x_j)$$
$$\mathsf{get}([\mathsf{get}([x_i]_{i \le n})]_n) = \mathsf{get}([x_i]_{i \le n}) \qquad\qquad \mathsf{get}([\mathsf{put}^i(x_i)]_{i \le n}) = \mathsf{get}([x_i]_{i \le n})$$

The connection with the implementation of state in programming becomes evident when we take a closer look at the endofunctor of the induced monad from Theorem 4. Consider the following informal calculation:

$$\begin{aligned}
&\forall X.(A \to n \to X) \to n \to X \\
&\cong \forall X.n \to (A \to n \to X) \to X && \text{(flipping the arguments)} \\
&\cong n \to \forall X.(A \to n \to X) \to X && \text{($\forall$ commutes with arrows)} \\
&\cong n \to \forall X.(A \times n \to X) \to X && \text{(Curry)} \\
&\cong n \to A \times n && \text{(Church)}
\end{aligned}$$

This means that not only do we prove that the equational theory corresponds to the usual state monad, but we can actually *derive* the implementation of state as the endofunctor $A \mapsto (n \to A \times n)$.

5.3 Backtracking with Local State

We obtain one way to combine nondeterminism with state using the functor $PX = n \times X$, for $n \in \mathbb{N}$, that is, $d = 1$, $c_1 = n$ and $e_1 = 1$. It has two sorts, Ω and K, which play roles similar to those detailed in the previous section. However, this time K additionally has the structure of a monoid. This gives us the theory of backtracking with *local* state, which means that whenever we make a choice using the γ operation, the computations in each argument carry separate, non-interfering states. In particular, in a computation $\gamma(x, y)$, both subcomputations x and y start with the same state, which is the initial state of the entire computation. This non-interference is guaranteed simply by the system of sorts: the arguments of γ are of the sort K, which means that the stateful computations inside the arguments begin with π, which sets a new state.

We can also obtain a single-sorted theory, similar to the case of the pure state. To the put and get macro-operations, we add choice and failure as follows:

$$\mathsf{choose} : \Omega \times \Omega \to \Omega \qquad\qquad \mathsf{fail} : \Omega$$

$$\mathsf{choose}(x, y) = \mathsf{cons}([\gamma(\pi^j(x), \pi^j(y))]_{j \le n}) \qquad\qquad \mathsf{fail} = \mathsf{cons}([\varepsilon]_n)$$

Then, the locality of state can be summarised by the following equality, which is easy to show using the (beta-π) and (eta-π) equations:

$$\mathsf{put}^k(\mathsf{choose}(x, y)) = \mathsf{choose}(\mathsf{put}^k(x), \mathsf{put}^k(y))$$

5.4 Backtracking with Global State

Another way to compose nondeterminism and state is by using *global* state, which is obtained for $n \in \mathbb{N}$ and $PX = X^n$, that is, $d = 1$, $c_1 = 1$, and $e_1 = n$. As in the case of pure backtracking nondeterminism, it means that the sorts Ω and K are isomorphic. The intuitive understanding of the expression $\gamma(x, [y_i]_{i \le n})$ is: first perform the computation x, and then the computation y_i, where i is the final state of the computation x. The operation ε^j is: fail, but set the current state to j. In this case, the equations (beta-ε) instantiate to the following for all $j \le n$:

$$\gamma(\varepsilon^j, [y_i]_{i \le n}) = y_j$$

It states that if the first computation fails but sets the state to j, the next step is to try the computation y_j. Note that there is no other way to give a new state than via failure, but this can be circumvented using $\gamma(x, [\varepsilon^k]_n)$ to set the state to k after performing x. The (eta-ε) instantiates to:

$$\gamma(x, [\varepsilon^j]_{j \le n}) = x$$

This reads that if we execute x and then set the current state to the resulting state of x, it is the same as just executing x.

6 Direct-Style Implementation

Free algebras of the theory $\mathfrak{T}$ from Definition 5 can also be presented as terms of a certain shape. They are best described as terms built using the operations from $\mathfrak{T}$ that are well-typed according to the following typing rules, where the types are called Ω, K_i, and P_i for $i \leq d$. The type of the entire term is Ω, and $\textsc{Var}(x)$ means that x is a variable.

$$\frac{[[t_i^j : K_i]_{j\leq c_i}]_{i\leq d}}{\mathsf{cons}([[t_i^j]_{j\leq c_i}]_{i\leq d}) : \Omega} \qquad \varepsilon_i^j : K_i \qquad \frac{t : P_j \qquad [w_k : K_i]_{k\leq e_j}}{\gamma_i^j(t, [w_k]_{k\leq e_j}) : K_i} \qquad \frac{\textsc{Var}(x)}{\pi_i^j(x) : P_i}$$

Note that even though variables appear as arguments to the operations π, they are not of the type Ω. This means that the entire term cannot be a variable, as it is always constructed with cons as the outermost operation. Each argument of cons is a term of the type K_i for an appropriate i, which is built out of the operations ε and γ. Note that the first argument of γ is always a variable wrapped in π, while all the other arguments are again terms of the type K_i. Overall, such terms can be captured as the following endofunctors on **Set**, where W^i represents terms of the type K_i, while W^Ω represents terms of the type Ω. By $\mu Y.GY$ we mean the carrier of the initial algebra of an endofunctor G.

$$W^i X = \mu Y.e_i + \textstyle\sum_{j=1}^{d} (\sum^{c_i} X) \times \prod^{e_j} Y$$
$$W^\Omega X = \textstyle\prod_{i=1}^{d} \prod^{c_i} W^i X$$

Clearly, e_i in the definition of W^i represents the ε_i constants, while the second component of the coproduct is a choice between the γ_i operations with appropriate arguments.

It is the case that every term of the sort Ω can be normalised to a term of the type Ω by a term-rewriting system obtained by orienting the "beta" and "assoc" equations left to right, and eta-expanding variables at the top-level:

$$\pi_i^j(\mathsf{cons}([[x_i^j]_{j\leq c_i}]_{i\leq d})) \rightsquigarrow x_i^j$$
$$\gamma_i^j(\varepsilon_j^k, [x_t]_{t\leq e_j}) \rightsquigarrow x_k$$
$$\gamma_i^j(\gamma_j^k(x, [y_t]_{t\leq e_k}), [z_s]_{s\leq e_j}) \rightsquigarrow \gamma_i^k(x, [\gamma_i^j(y_t, [z_s]_{s\leq e_j})]_{t\leq e_k})$$
$$x \rightsquigarrow \mathsf{cons}([[\gamma_i^i(\pi_i^j(x), [\varepsilon_i^k]_{k\leq e_i})]_{j\leq c_i}]_{i\leq d})$$

This term rewriting system gives rise to a natural implementation of the monadic structure, where the "beta" and "assoc" rules normalise the two-level term structure, thus implementing the monadic multiplication, while the eta-expansion rule implements the monadic unit.

7 Discussion

The idea for employing Cayley representations to explore implementations of monads induced by equational theories is inspired by Hinze [8], who suggested a connection between codensity monads, Church representation of lists, and the Cayley theorem for monoids. We note that Hinze's discussion is informal, but he suggests using ends, which, as we discuss in Sect. 2, is not sound.

Most of related work follows one of two main paths: it either concentrates on algebraic explanation of monads already used in programming and semantics (for example, [11,19,23]), or on the general connection between different kinds of algebraic theories and computational effects, but without much interest in whether it leads to structures implementable in a programming language. Some exceptions are the construction of the sum of a theory and a free theory [9] or the sum of ideal monads [6]. What we propose in Sect. 4 is a form of a "functional combinatorics": given a type, what kind of algebra describes the possible values?

As our approach veers off the main paths of the recent work on effects, there are many possible directions of future work. One interesting direction would be to generalise **Set**, the base category used throughout this paper, to more abstract categories. After all, we want to talk about structures definable only in terms of (co)products, exponentials, and quantifiers—which are all constructions whose universal properties are singled out and explored using (co)cartesian (or even monoidal) closed categories. However, the current development relies heavily on the particular properties of **Set**, such as extensional equality of functions, which appears in disguise in the condition (f) in Definition 2.

One can also try to extend the type used as a Cayley representation. For example, we could consider the polynomial P in (3) to range over the space of all sets, that is, allow the coefficients c_i to vary over sets rather than natural numbers. In the Cayley representation, it would be enough to consider functions from c_i in place of c_i-fold products. We would immediately gain expressiveness, as the obtained state monad would no longer need to be defined only for a finite set of possible states. On the flip side, this would make the resulting theory infinitary – which, of course, is not uncommon in the field of algebraic treatment of computational effects. However, we decide to stick to the simplest possible setting in this paper, which greatly simplifies the presentation, but still gives us some novel observations, like the fact that the theory of finite state is simply the theory of 2-sorted tuples in Sect. 5.2, or the novel theory of backtracking nondeterminism with global state in Sect. 5.4. Other future extensions that we believe are worth exploring include iterating the construction to obtain a from of a distributive tensor (compare Rivas *et al.*'s [25] "double" representation of near-semirings) or quantifying over more variables, leading to less interaction between sorts.

Acknowledgements. We thank the reviewers for their insightful comments and suggestions.

Maciej Piróg was supported by the National Science Centre, Poland under POLONEZ 3 grant "Algebraic Effects and Continuations" no. 2016/23/P/ST6/02217. This project has received funding from the European Union's Horizon 2020 research and innovation programme under the Marie Skłodowska-Curie grant agreement No 665778.

Piotr Polesiuk was supported by the National Science Centre, Poland, under grant no. 2014/15/B/ST6/00619.

Filip Sieczkowski was supported by the National Science Centre, Poland, under grant no. 2016/23/D/ST6/01387.

References

1. Bird, R.: Functional pearl: a program to solve Sudoku. J. Funct. Program. **16**(6), 671–679 (2006). http://dx.doi.org/10.1017/S0956796806006058
2. Bloom, S.L., Ésik, Z., Manes, E.G.: A Cayley theorem for Boolean algebras. Am. Math. Monthly **97**(9), 831–833 (1990). http://dx.doi.org/10.2307/2324751
3. Dubuc, E., Street, R.: Dinatural transformations. In: MacLane, S., et al. (eds.) Reports of the Midwest Category Seminar IV, pp. 126–137. Springer, Heidelberg (1970). https://doi.org/10.1007/BFb0060443
4. Eilenberg, S., Moore, J.C.: Adjoint functors and triples. Illinois J. Math. **9**(3), 381–398 (1965). https://projecteuclid.org:443/euclid.ijm/1256068141
5. Ésik, Z.: A Cayley theorem for ternary algebras. Int. J. Algebra Comput. **8**, 311–316 (1998)
6. Ghani, N., Uustalu, T.: Coproducts of ideal monads. ITA **38**(4), 321–342 (2004). https://doi.org/10.1051/ita:2004016
7. Gibbons, J., Hinze, R.: Just do it: simple monadic equational reasoning. In: Chakravarty, M.M.T., Hu, Z., Danvy, O. (eds.) Proceeding of the 16th ACM SIGPLAN international conference on Functional Programming, ICFP 2011, Tokyo, Japan, 19–21 September 2011, pp. 2–14. ACM (2011). http://doi.acm.org/10.1145/2034773.2034777
8. Hinze, R.: Kan extensions for program optimisation *Or*: Art and Dan explain an old trick. In: Gibbons, J., Nogueira, P. (eds.) MPC 2012. LNCS, vol. 7342, pp. 324–362. Springer, Heidelberg (2012). https://doi.org/10.1007/978-3-642-31113-0_16
9. Hyland, M., Plotkin, G.D., Power, J.: Combining effects: sum and tensor. Theor. Comput. Sci. **357**(1–3), 70–99 (2006). https://doi.org/10.1016/j.tcs.2006.03.013
10. Hyland, M., Power, J.: The category theoretic understanding of universal algebra: Lawvere theories and monads. Electron. Notes Theor. Comput. Sci. **172**, 437–458 (2007). https://doi.org/10.1016/j.entcs.2007.02.019
11. Jaskelioff, M., Moggi, E.: Monad transformers as monoid transformers. Theor. Comput. Sci. **411**(51–52), 4441–4466 (2010). https://doi.org/10.1016/j.tcs.2010.09.011
12. Kock, A.: Continuous Yoneda representation of a small category (1966). Aarhus University preprint. http://home.math.au.dk/kock/CYRSC.pdf
13. Linton, F.: Some aspects of equational categories. In: Eilenberg, S., Harrison, D.K., MacLane, S., Röhrl, H. (eds.) Proceedings of the Conference on Categorical Algebra, pp. 84–94. Springer, Heidelberg (1966). https://doi.org/10.1007/978-3-642-99902-4_3
14. Moggi, E.: Notions of computation and monads. Inf. Comput. **93**(1), 55–92 (1991). https://doi.org/10.1016/0890-5401(91)90052-4
15. Mulry, P.S.: Strong monads, algebras and fixed points. London Mathematical Society Lecture Note Series, pp. 202–216. Cambridge University Press, New York (1992)
16. Paré, R., Román, L.: Dinatural numbers. J. Pure Appl. Algebra **128**(1), 33–92 (1998). http://www.sciencedirect.com/science/article/pii/S0022404997000364
17. Piróg, M.: Eilenberg-Moore monoids and backtracking monad transformers. In: Atkey, R., Krishnaswami, N.R. (eds.) Proceedings of 6th Workshop on Mathematically Structured Functional Programming, MSFP@ETAPS 2016, Eindhoven, Netherlands, 8th April 2016. EPTCS, vol. 207, pp. 23–56 (2016). https://doi.org/10.4204/EPTCS.207.2

18. Piróg, M., Schrijvers, T., Wu, N., Jaskelioff, M.: Syntax and semantics for operations with scopes. In: Proceedings of the 33rd Annual ACM/IEEE Symposium on Logic in Computer Science. LICS 2018, pp. 809–818. ACM, New York (2018). http://doi.acm.org/10.1145/3209108.3209166
19. Piróg, M., Staton, S.: Backtracking with cut via a distributive law and left-zero monoids. J. Funct. Program. **27**, e17 (2017). https://doi.org/10.1017/S0956796817000077
20. Plotkin, G.: Adequacy for algebraic effects with state. In: Fiadeiro, J.L., Harman, N., Roggenbach, M., Rutten, J. (eds.) CALCO 2005. LNCS, vol. 3629, pp. 51–51. Springer, Heidelberg (2005). https://doi.org/10.1007/11548133_3
21. Plotkin, G., Power, J.: Adequacy for algebraic effects. In: Honsell, F., Miculan, M. (eds.) FoSSaCS 2001. LNCS, vol. 2030, pp. 1–24. Springer, Heidelberg (2001). https://doi.org/10.1007/3-540-45315-6_1
22. Plotkin, G.D., Power, J.: Semantics for algebraic operations. Electron. Notes Theor. Comput. Sci. **45**, 332–345 (2001). https://doi.org/10.1016/S1571-0661(04)80970-8
23. Plotkin, G., Power, J.: Notions of computation determine monads. In: Nielsen, M., Engberg, U. (eds.) FoSSaCS 2002. LNCS, vol. 2303, pp. 342–356. Springer, Heidelberg (2002). https://doi.org/10.1007/3-540-45931-6_24
24. Plotkin, G.D., Power, J.: Computational effects and operations: an overview. Electron. Notes Theor. Comput. Sci. **73**, 149–163 (2004). http://dx.doi.org/10.1016/j.entcs.2004.08.008
25. Rivas, E., Jaskelioff, M., Schrijvers, T.: From monoids to near-semirings: the essence of MonadPlus and alternative. In: Falaschi, M., Albert, E. (eds.) Proceedings of the 17th International Symposium on Principles and Practice of Declarative Programming, Siena, Italy, 14–16 July 2015. pp. 196–207. ACM (2015). http://doi.acm.org/10.1145/2790449.2790514
26. Tarlecki, A.: Some nuances of many-sorted universal algebra: a review. Bull. EATCS **104**, 89–111 (2011)
27. Vene, V., Ghani, N., Johann, P., Uustalu, T.: Parametricity and strong dinaturality (2006). https://www.ioc.ee/~tarmo/tday-voore/vene-slides.pdf

A Dialectica-Like Interpretation of a Linear MSO on Infinite Words

Pierre Pradic[1,2] and Colin Riba[1(✉)]

[1] ENS de Lyon, Université de Lyon,
LIP, UMR 5668 CNRS ENS Lyon UCBL Inria, Lyon, France
colin.riba@ens-lyon.fr

[2] Faculty of Mathematics, Informatics and Mechanics,
University of Warsaw, Warsaw, Poland

Abstract. We devise a variant of Dialectica interpretation of intuitionistic linear logic for LMSO, a linear logic-based version MSO over infinite words. LMSO was known to be correct and complete w.r.t. Church's synthesis, thanks to an automata-based realizability model. Invoking Büchi-Landweber Theorem and building on a complete axiomatization of MSO on infinite words, our interpretation provides us with a syntactic approach, without any further construction of automata on infinite words. Via Dialectica, as linear negation directly corresponds to switching players in games, we furthermore obtain a complete logic: either a closed formula or its linear negation is provable. This completely axiomatizes the theory of the realizability model of LMSO. Besides, this shows that in principle, one can solve Church's synthesis for a given ∀∃-formula by only looking for proofs of either that formula or its linear negation.

Keywords: Linear logic · Dialectica interpretation · MSO on Infinite Words

1 Introduction

Monadic Second-Order Logic (MSO) over ω-words is a simple yet expressive language for reasoning on non-terminating systems which subsumes non-trivial logics used in verification such as LTL (see e.g. [2,30]). MSO on ω-words is decidable by Büchi's Theorem [6] (see e.g. [24,29]), and can be completely axiomatized as a subsystem of second-order Peano's arithmetic [28]. While MSO admits an effective translation to finite-state (Büchi) automata, it is a non-constructive logic, in the sense that it has true (*i.e.*provable) ∀∃-statements which can be witnessed by no continuous stream function.

On the other hand, Church's synthesis [8] can be seen as a decision problem for a strong form of constructivity in MSO. More precisely (see e.g. [12,32]),

This work was partially supported by the ANR-14-CE25-0007 - RAPIDO and Polish National Science Centre grant no. 2014/13/B/ST6/03595.

M. Bojańczyk and A. Simpson (Eds.): FOSSACS 2019, LNCS 11425, pp. 470–487, 2019.
https://doi.org/10.1007/978-3-030-17127-8_27

Church's synthesis takes as input a $\forall\exists$-formula of MSO and asks whether it can be realized by a finite-state causal stream transducer. Church's synthesis is known to be decidable since Büchi-Landweber Theorem [7], which gives an effective solution to ω-regular games on finite graphs generated by $\forall\exists$-formulae. In traditional (theoretical) solutions to Church's synthesis, the game graphs are induced from deterministic (say parity) automata obtained by McNaughton's Theorem [19]. Despite its long history, Church's synthesis has not yet been amenable to tractable solutions for the full language of MSO (see e.g. [12]).

In recent works [25,26], the authors suggested a Curry-Howard approach to Church's synthesis based on intuitionistic and linear variants of MSO. In particular, [26] proposed a system LMSO based on (intuitionistic) linear logic [13], in which via a translation $(-)^L$: MSO $\to$ LMSO, the provable $\forall\exists(-)^L$-statements exactly correspond to the realizable instances of Church's synthesis. Realizer extraction for LMSO is done via an external realizability model based on alternating automata, which amounts to see every formula $\varphi(a)$ as a formula of the form $(\exists u)(\forall x)\varphi_{\mathcal{D}}(u, x, a)$, where $\varphi_{\mathcal{D}}$ represents a deterministic automaton.

In this paper, we use a variant of Gödel's "Dialectica" functional interpretation as a syntactic formulation of the automata-based realizability model of [26]. Dialectica associates to $\varphi(a)$ a formula $\varphi^D(a)$ of the form $(\exists u)(\forall x)\varphi_D(u, x, a)$. In usual versions formulated in higher-types arithmetic (see e.g. [1,16]), the formula φ_D is quantifier-free, so that φ^D is a prenex form of φ. This prenex form is constructive, and a constructive proof of φ can be turned to a proof of φ^D with an explicit witness for $\exists u$. Even if Dialectica originally interprets intuitionistic arithmetic, it is structurally linear, and linear versions of Dialectica were formulated at the very beginning of linear logic [21–23] (see also [14,27]).

We show that the automata-based realizability model of [26] can be obtained by a suitable modification of the usual linear Dialectica interpretation, in which the formula φ_D essentially represents a deterministic automaton on ω-words and is in general not quantifier-free, and whose realizers are exactly the finite-state accepting strategies in the model of [26]. In addition to provide a syntactic extraction procedure with internalized and automata-free correctness proof, this reformulation has a striking consequence, namely that there exists an extension LMSO($\mathfrak{C}$) of LMSO which is complete in the sense that for each closed formula φ, it either proves φ or its linear negation $\varphi \multimap \bot$. Since LMSO($\mathfrak{C}$) has realizers for all provable $\forall\exists(-)^L$-statements, its completeness contrasts with the classical setting, in which due to provable non-constructive statements, one can not decide Church's synthesis by only looking for proofs of $\forall\exists$-statements or their negations. Besides, LMSO($\mathfrak{C}$) has a linear choice axiom which is realizable in the sense of both $(-)^D$ and [26], but whose naive MSO counterpart is false.

The paper is organized as follows. We present our basic setting in Sect. 2, with a particular emphasis on particularities of (finite-state) causal functions to model strategies and realizers. Our variant of Dialectica and the corresponding linear system are discussed in Sect. 3, while Sect. 4 defines the systems LMSO and LMSO($\mathfrak{C}$) and shows the completeness of LMSO($\mathfrak{C}$).

2 Preliminaries

Alphabets (denoted Σ, Γ, etc) are finite non-empty sets of the form $\mathbf{2}^p$ for some $p \in \mathbb{N}$. We let $\mathbf{1} := \mathbf{2}^0$. Note that alphabets are closed under Cartesian products and set-theoretic function spaces. It follows that taking $[\![o]\!] := \mathbf{2}$, we have an alphabet $[\![\tau]\!]$ for each simple type $\tau \in \mathrm{ST}$, where

$$\sigma, \tau \in \mathrm{ST} \quad ::= \quad \mathbf{1} \mid o \mid \sigma \times \tau \mid \sigma \to \tau$$

We often write $(\tau)\sigma$ for the type $\sigma \to \tau$. Given an ω-word (or stream) $B \in \Sigma^\omega$ and $n \in \mathbb{N}$, we write $B{\restriction}n$ for the finite word $B(0).\cdots.B(n-1) \in \Sigma^*$.

Church's Synthesis and Causal Functions. Church's synthesis consists in the automatic extraction of stream functions from input-output specifications (see e.g. [12,31]). These specifications are in general asked to be ω-regular, or equivalently definable in MSO over ω-words. In practice, proper subsets of MSO (and even of LTL) are assumed (see e.g. [5,11,12]). As an example, the relation

$$(\exists^\infty k)B(k) \;\Rightarrow\; (\exists^\infty k)C(k) \qquad \text{resp.} \qquad (\forall^\infty k)B(k) \;\Rightarrow\; (\exists^\infty k)C(k) \quad (1)$$

with input $B \in \mathbf{2}^\omega$ and output $C \in \mathbf{2}^\omega$ specifies functions $F : \mathbf{2}^\omega \to \mathbf{2}^\omega$ such that $F(B) \in \mathbf{2}^\omega \simeq \mathcal{P}(\mathbb{N})$ is infinite whenever $B \in \mathbf{2}^\omega \simeq \mathcal{P}(\mathbb{N})$ is infinite (resp. the complement of B is finite). One may also additionally require to respect the transitions of some automaton. For instance, following [31], in addition to either case of (1) one can ask $C \subseteq B$ and C not to contain two consecutive positions:

$$(\forall n)(C(n) \;\Rightarrow\; B(n)) \qquad \text{and} \qquad (\forall n)(C(n) \;\Rightarrow\; \neg C(n+1)) \qquad (2)$$

In any case, the realizers must be (finite-state) causal functions. A stream function $F : \Sigma^\omega \to \Gamma^\omega$ is causal (notation $F : \Sigma \to_{\mathbb{S}} \Gamma$) if it can produce a prefix of length n of its output from a prefix of length n of its input. Hence F is causal if it is induced by a map $f : \Sigma^+ \to \Gamma$ as follows:

$$F(B)(n) \quad = \quad f(B(0) \cdot \ldots \cdot B(n)) \qquad \text{(for all } B \in \Sigma^\omega \text{ and all } n \in \mathbb{N})$$

The finite-state (f.s.) causal functions are those induced by Mealy machines. A Mealy machine $\mathcal{M} : \Sigma \to \Gamma$ is a DFA over input alphabet Σ equipped with an output function $\lambda : Q_{\mathcal{M}} \times \Sigma \to \Gamma$ (where $Q_{\mathcal{M}}$ is the state set of $\mathcal{M}$). Writing $\partial^* : \Sigma^* \to Q_{\mathcal{M}}$ for the iteration of the transition function ∂ of $\mathcal{M}$ from its initial state, $\mathcal{M}$ induces a causal function via $(\bar{\mathsf{a}}.\mathsf{a} \in \Sigma^+) \;\mapsto\; (\lambda(\partial^*(\bar{\mathsf{a}}), \mathsf{a}) \in \Gamma)$.

Causal and f.s. causal functions form categories with finite products. Let $\mathbb{S}$ be the category whose objects are alphabets and whose maps from Σ to Γ are causal functions $F : \Sigma^\omega \to \Gamma^\omega$. Let $\mathbb{M}$ be the wide subcategory of $\mathbb{S}$ whose maps are finite-state causal functions.[1]

[1] A subcategory $\mathbb{D}$ of $\mathbb{C}$ is *wide* if $\mathbb{D}$ has the same objects as $\mathbb{C}$.

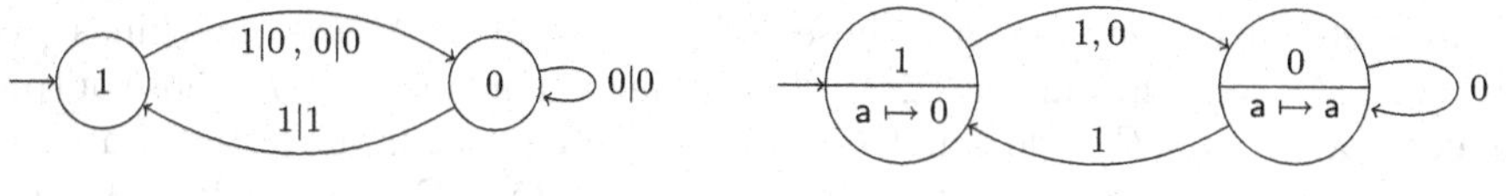

Fig. 1. A Mealy machine (left) and an equivalent eager (Moore) machine (right).

Example 1. (a) Usual functions $\Sigma \to \Gamma$ lift to (pointwise, one-state) maps $\Sigma \to_{\mathbb{M}} \Gamma$. For instance, the identity $\Sigma \to_{\mathbb{M}} \Sigma$ is induced by the Mealy machine with $\langle \partial, \lambda \rangle : (-, \mathsf{a}) \mapsto (-, \mathsf{a})$.
(b) Causal functions $\mathbf{1} \to_{\mathbb{S}} \Sigma$ correspond exactly to ω-words $B \in \Sigma^\omega$.
(c) The conjunction of (2) with either side of (1) is realized by the causal function $F : \mathbf{2} \to_{\mathbb{M}} \mathbf{2}$ induced by the machine $\mathcal{M} : \mathbf{2} \to \mathbf{2}$ displayed on Fig. 1 (left, where a transition $\mathsf{a}|\mathsf{b}$ outputs b from input a), taken from [31].

Proposition 1. *The Cartesian product of $\Sigma_1, \ldots, \Sigma_n$ (for $n \geq 0$) in $\mathbb{S}, \mathbb{M}$ is given by the product of sets $\Sigma_1 \times \cdots \times \Sigma_n$ (so that $\mathbf{1}$ is terminal).*

The Logic MSO(M). Our specification language MSO(**M**) is an extension of MSO on ω-words with one function symbol for each f.s. causal function. More precisely, MSO(**M**) is a many-sorted first-order logic, with one sort for each simple type $\tau \in \mathrm{ST}$, and with one function symbol of arity $(\sigma_1, \ldots, \sigma_n; \tau)$ for each map $[\![\sigma_1]\!] \times \cdots \times [\![\sigma_n]\!] \to_{\mathbb{M}} [\![\tau]\!]$. A term $\mathtt{t}$ of sort τ (notation $\mathtt{t}^\tau$) with free variables among $x_1^{\sigma_1}, \ldots, x_n^{\sigma_n}$ (we say that $\mathtt{t}$ is of arity $(\sigma_1, \ldots, \sigma_n; \tau)$) thus induces a map $[\![\mathtt{t}]\!] : [\![\sigma_1]\!] \times \cdots \times [\![\sigma_n]\!] \to_{\mathbb{M}} [\![\tau]\!]$. Given a valuation $x_i \mapsto B_i \in [\![\sigma_i]\!]^\omega \simeq \mathbb{S}[\mathbf{1}, [\![\sigma_i]\!]]$ for $i \in \{1, \ldots, n\}$, we then obtain an ω-word

$$[\![\mathtt{t}]\!] \circ \langle B_1, \ldots, B_n \rangle \quad \in \quad \mathbb{S}[\mathbf{1}, [\![\tau]\!]] \quad \simeq \quad [\![\tau]\!]^\omega$$

MSO(**M**) extends MSO with $\exists x^\tau$ and $\forall x^\tau$ ranging over $\mathbb{S}[\mathbf{1}, [\![\tau]\!]] \simeq [\![\tau]\!]^\omega$ and with sorted equalities $\mathtt{t}^\tau \doteq \mathtt{u}^\tau$ interpreted as equality over $\mathbb{S}[\mathbf{1}, [\![\tau]\!]] \simeq [\![\tau]\!]^\omega$. Write $\models \varphi$ when φ holds in this model, called the *standard* model. The full definition of MSO(**M**) is deferred to Sect. 4.1.

An instance of Church's synthesis problem is given by a closed formula $(\forall x^\sigma)(\exists u^\tau)\varphi(u, x)$. A positive solution (or realizer) of this instance is a term $\mathtt{t}(x)$ of arity $(\sigma; \tau)$ such that $(\forall x^\sigma)\varphi(\mathtt{t}(x), x)$ holds.

Proposition 1 implies that MSO(**M**) proves the following equations:

$$\pi_i(\langle \mathtt{t}_1, \ldots, \mathtt{t}_n \rangle) \doteq_{\sigma_i} \mathtt{t}_i \qquad \text{and} \qquad \mathtt{t} \doteq_{\sigma_1 \times \cdots \times \sigma_n} \langle \pi_1(\mathtt{t}), \ldots, \pi_n(\mathtt{t}) \rangle \tag{3}$$

Hence each formula $\varphi(a_1^{\sigma_1}, \ldots, a_n^{\sigma_n})$ can be seen as a formula $\varphi(a^{\sigma_1 \times \cdots \times \sigma_n})$.

Eager Functions. A causal function $\Sigma \to_{\mathbb{S}} \Gamma$ is eager if it can produce a prefix of length $n+1$ of its output from a prefix of length n of its input. More precisely, an eager $F : \Sigma \to_{\mathbb{S}} \Gamma$ is induced by a map $f : \Sigma^* \to \Gamma$ as

$$F(B)(n) \quad = \quad f(B(0) \cdot \ldots \cdot B(n-1)) \qquad (\text{for all } B \in \Sigma^\omega \text{ and all } n \in \mathbb{N})$$

Finite-state eager functions are those induced by eager (Moore) machines (see also [11]). An eager machine $\mathcal{E} : \Sigma \to \Gamma$ is a Mealy machine $\Sigma \to \Gamma$ whose output function $\lambda : Q_{\mathcal{E}} \to \Gamma$ is does not depend on the current input letter. An eager $\mathcal{E} : \Sigma \to \Gamma$ induces an eager function via the map $(\bar{\mathsf{a}} \in \Sigma^*) \mapsto (\lambda_{\mathcal{E}}(\partial^*_{\mathcal{E}}(\bar{\mathsf{a}})) \in \Gamma)$.

We write $F : \Sigma \to_{\mathbb{E}} \Gamma$ when $F : \Sigma \to_{\mathbb{S}} \Gamma$ is eager and $F : \Sigma \to_{\mathbb{EM}} \Gamma$ when F is f.s. eager. All functions $F : \Sigma \to_{\mathbb{M}} \mathbf{1}$, and more generally, constants functions $F : \Sigma \to_{\mathbb{S}} \Gamma$ are eager. Note also that if $F : \Sigma \to_{\mathbb{S}} \Gamma$ is eager, then $F : \Sigma \to_{\mathbb{EM}} \Gamma$. On the other hand, if $F : \Sigma \to_{\mathbb{EM}} \Gamma$ is induced by an eager machine $\mathcal{E}$ then F is finite-state causal as being induced by the Mealy machine with same states and transitions as $\mathcal{E}$, but with output function $(q, \mathsf{a}) \mapsto \lambda_{\mathcal{E}}(q)$.

Eager functions do not form a category since the identity of $\mathbb{S}$ is not eager. On the other hand, eager functions are closed under composition with causal functions.

Proposition 2. *If F is eager and G, H are causal then $H \circ F \circ G$ is eager.*

Isolating eager functions allows a proper treatment of strategies in games and realizers w.r.t. the Dialectica interpretation. Since $\Sigma^+ \to \Gamma \simeq \Sigma^* \to \Gamma^\Sigma$, maps $\Sigma \to_{\mathbb{E}} \Gamma^\Sigma$ are in bijection with maps $\Sigma \to_{\mathbb{S}} \Gamma$. This easily extends to machines. Given a Mealy machine $\mathcal{M} : \Sigma \to \Gamma$, let $\Lambda(\mathcal{M}) : \Sigma \to \Gamma^\Sigma$ be the eager machine defined as $\mathcal{M}$ but with output map taking $q \in Q_{\mathcal{M}}$ to $(\mathsf{a} \mapsto \lambda_{\mathcal{M}}(q, \mathsf{a})) \in \Gamma^\Sigma$.

Example 2. Recall the Mealy machine $\mathcal{M} : \mathbf{2} \to \mathbf{2}$ of Ex. 1.(c). Then $\Lambda(\mathcal{M}) : \mathbf{2} \to \mathbf{2}^\mathbf{2}$ is the eager machine displayed in Fig. 1 (right, where the output is indicated within states).

Eager f.s. functions will often be used with the following notations. First, let @ be the pointwise lift to $\mathbb{M}$ of the usual application function $\Gamma^\Sigma \times \Sigma \to \Gamma$. We often write $(F)G$ for $@(F, G)$. Consider a Mealy machine $\mathcal{M} : \Sigma \to \Gamma$ and the induced eager machine $\Lambda(\mathcal{M}) : \Sigma \to \Gamma^\Sigma$. We have

$$F_{\mathcal{M}}(B) \quad = \quad @(F_{\Lambda(\mathcal{M})}(B), B) \qquad \qquad (\text{for all B} \in \Sigma^\omega)$$

Given $F : \Gamma \to_{\mathbb{E}} \Sigma^\Gamma$, we write $\mathbf{e}(F)$ for the causal $@(F(-), -) : \Gamma \to_{\mathbb{S}} \Sigma$. Given $F : \Gamma \to_{\mathbb{S}} \Sigma$, we write $\Lambda(F)$ for the eager $\Gamma \to_{\mathbb{E}} \Sigma^\Gamma$ such that $F = \mathbf{e}(\Lambda(F))$. We extend these notations to terms.

Eager functions admit fixpoints similar to those of contractive maps in the topos of tree (see e.g. [4, Thm. 2.4]).

Proposition 3. *For each $F : \Sigma \times \Gamma \to_{\mathbb{E}} \Sigma^\Gamma$ there is a $\mathrm{fix}(F) : \Gamma \to_{\mathbb{E}} \Sigma^\Gamma$ s.t.*

$$\mathrm{fix}(F)(C) \quad = \quad F\big(\mathbf{e}(\mathrm{fix}(F))(C)\,,\, C\big) \qquad \qquad (\text{for all } C \in \Gamma^\omega)$$

If F is induced by the eager machine $\mathcal{E} : \Sigma \times \Gamma \to \Sigma^\Gamma$, then $\mathrm{fix}(F)$ is induced by the eager $\mathcal{H} : \Gamma \to \Sigma^\Gamma$ defined as $\mathcal{E}$ but with $\partial_{\mathcal{H}} : (q, \mathsf{b}) \mapsto \partial_{\mathcal{E}}\big(q,\, ((\lambda_{\mathcal{E}}(q))\mathsf{b}, \mathsf{b})\big)$.

Games. Traditional solutions to Church's synthesis turn specifications to infinite two-player games with ω-regular winning conditions. Consider an $\mathsf{MSO}(\mathbf{M})$ formula $\varphi(u^\tau, x^\sigma)$ with no free variable other than u, x. We see this formula as defining a two-player infinite game $\mathcal{G}(\varphi)(u^\tau, x^\sigma)$ between the *Proponent* P ($\exists$loïse), playing moves in $[\![\tau]\!]$ and the *Opponent* O ($\forall$bélard), playing moves in $[\![\sigma]\!]$. The Proponent begins, and then the two players alternate, producing an infinite play of the form

$$\chi \quad := \quad \mathsf{u}_0\mathsf{x}_0 \cdots \mathsf{u}_n\mathsf{x}_n \cdots \quad \simeq \quad ((\mathsf{u}_k)_k, (\mathsf{x}_k)_k) \in [\![\tau]\!]^\omega \times [\![\sigma]\!]^\omega$$

The play χ is winning for P if $\varphi((\mathsf{u}_k)_k, (\mathsf{x})_k)$ holds. Otherwise χ is winning for O. Strategies for P resp. O in this game are functions

$$[\![\sigma]\!]^* \longrightarrow [\![\tau]\!] \qquad \text{resp.} \qquad [\![\tau]\!]^+ \longrightarrow [\![\sigma]\!] \;\simeq\; [\![\tau]\!]^* \longrightarrow [\![\sigma]\!]^{[\![\tau]\!]}$$

Hence finite-state strategies are represented by f.s. eager functions. In particular, a realizer of $(\forall x^\sigma)(\exists u^\tau)\varphi(u, x)$ in the sense of Church is a f.s. P-strategy in

$$\mathcal{G}\big(\varphi((u)x, x)\big)\big(u^{(\tau)\sigma}, x^\sigma\big)$$

Most approaches to Church's synthesis reduce to Büchi-Landweber Theorem [7], stating that games with ω-regular winning conditions are effectively determined, and that the winner always has a finite-state winning strategy. We will use Büchi-Landweber Theorem in following form. Note that an O-strategy in the game $\mathcal{G}(\varphi)(u^\tau, x^\sigma)$ is a P-strategy in the game $\mathcal{G}\big(\neg\varphi(u, (x)u)\big)\big(x^{(\sigma)\tau}, u^\tau\big)$.

Theorem 1 ([7]). *Let $\varphi(u^\tau, x^\sigma)$ be an $\mathsf{MSO}(\mathbf{M})$-formula with only u, x free. Then either there is an eager term $\mathtt{u}(x)$ of arity $(\sigma;\tau)$ such that $\models (\forall x)\varphi(\mathtt{u}(x), x)$ or there is an eager term $\mathtt{x}(u)$ of arity $(\tau;(\sigma)\tau)$ such that $\models (\forall u)\neg\varphi(u, \mathbf{e}(\mathtt{x})(u))$. It is decidable which case holds and the terms are computable from φ.*

Curry-Howard Approaches. Following the complete axiomatization of MSO on ω-words of [28] (see also [26]), one can axiomatize $\mathsf{MSO}(\mathbf{M})$ with a deduction system based on arithmetic (see Sect. 4.1). Consider an instance of Church's synthesis $(\forall x^\sigma)(\exists u^\tau)\varphi(u, x)$. Then we get from Theorem 1 the alternative

$$\vdash_{\mathsf{MSO}(\mathbf{M})} (\forall x)\varphi\big(\mathbf{e}(\mathtt{u})(x), x\big) \quad \text{or} \quad \vdash_{\mathsf{MSO}(\mathbf{M})} (\forall u)\neg\varphi\big((u)(\mathtt{x}(u)), \mathtt{x}(u)\big) \tag{4}$$

for an eager term $\mathtt{u}(x)$ or a causal term $\mathtt{x}(u)$. By enumerating proofs and machines, one thus gets a (naive) syntactic algorithm for Church's synthesis. But it seems however unlikely to obtain a complete classical system in which the provable $\forall\exists$-statements do correspond to the realizable instances of Church's synthesis, because $\mathsf{MSO}(\mathbf{M})$ has true but unrealizable $\forall\exists$-statements. Besides, note that

$$\begin{array}{rcl}
(\forall x^\sigma)\varphi\big(\mathbf{e}(\mathtt{u})(x), x\big) & \vdash_{\mathsf{MSO}(\mathbf{M})} & (\forall x^\sigma)(\exists u^\tau)\varphi(u, x) \\
(\forall u^{(\tau)\sigma})\neg\varphi\big((u)(\mathtt{x}(u)), \mathtt{x}(u)\big) & \vdash_{\mathsf{MSO}(\mathbf{M})} & (\forall u^{(\tau)\sigma})(\exists x^\sigma)\neg\varphi\big((u)x, x\big) \\
\neg(\forall x^\sigma)(\exists u^\tau)\varphi(u, x) & \vdash_{\mathsf{MSO}(\mathbf{M})} & (\forall u^{(\tau)\sigma})(\exists x^\sigma)\neg\varphi\big((u)x, x\big)
\end{array}$$

while it is possible both for realizable and unrealizable instances to have

$$\vdash_{\mathsf{MSO(M)}} \quad (\forall x^\sigma)(\exists u^\tau)\varphi(u,x) \ \wedge \ (\forall u^{(\tau)\sigma})(\exists x^\sigma)\neg\varphi((u)x, x) \tag{5}$$

In previous works [25,26], the authors devised intuitionistic and linear variants of MSO on ω-words in which, thanks to automata-based polarity systems, proofs of suitably polarized existential statements correspond exactly to realizers for Church's synthesis. In particular, [26] proposed a system LMSO based on (intuitionistic) linear logic [13], such that via a translation $(-)^L$: MSO $\to$ LMSO, provable $\forall\exists(-)^L$-statements exactly correspond to realizable instances of Church's synthesis, while (4) exactly corresponds to alternatives of the form

$$\vdash_{\mathsf{LMSO}} (\forall x^\sigma)(\exists u^\tau)\left[\varphi((u)x, x)\right]^L \ \text{or} \ \vdash_{\mathsf{LMSO}} (\forall u^{(\tau)\sigma})(\exists x^\sigma)\left[\neg\varphi((u)x, x)\right]^L \tag{6}$$

This paper goes further. We show that the automata-based realizability model of [26] can be obtained in a syntactic way, thanks to a (linear) Dialectica-like interpretation of a variant of LMSO, which turns a formula φ to a formula φ^D of the form $(\exists u)(\forall x)\varphi_D(u,x)$, where $\varphi_D(u,x)$ essentially represents a deterministic automaton. While the correctness of the extraction procedure of [25,26] relied on automata-theoretic techniques, we show here that it can be performed syntactically. Second, by extending LMSO with realizable axioms, we obtain a system LMSO($\mathfrak{C}$) in which, using an adaptation of the usual *Characterization Theorem* for Dialectica stating that $\varphi \multimap\!\!\circ\, \varphi^D$ (see e.g. [16]), alternatives of the form (6) imply that for a closed φ,

$$\vdash_{\mathsf{LMSO}(\mathfrak{C})} \varphi \qquad \text{or} \qquad \vdash_{\mathsf{LMSO}(\mathfrak{C})} \varphi \multimap \bot$$

where $(-) \multimap \bot$ is a *linear* negation. We thus get a complete *linear* system with extraction of suitably polarized $\forall\exists$-statements. Such a system can of course not have a standard semantics, and indeed, LMSO($\mathfrak{C}$) has a functional choice axiom

$$(\forall x^\sigma)(\exists y^\tau)\varphi(x,y) \quad \multimap \quad (\exists f^{(\tau)\sigma})(\forall x^\sigma)\varphi(x,(f)x) \tag{LAC}$$

which is realizable in the sense of both $(-)^D$ and [26], but whose translation to MSO(M) (which precludes (5)) is false in the standard model.

3 A Monadic Linear Dialectica-Like Interpretation

Gödel's "Dialectica" functional interpretation associates to $\varphi(a)$ a formula $\varphi^D(a)$ of the form $(\exists u^\tau)(\forall x^\sigma)\varphi_D(u,x,a)$. In usual versions formulated in higher-types arithmetic (see e.g. [1,16]), the formula φ_D is quantifier-free, so that φ^D is a prenex form of φ. This prenex form is constructive, and a constructive proof of φ can be turned to a proof of φ^D with an explicit (closed) witness for $\exists u$. We call such witnesses *realizers* of φ. Even if Dialectica originally interprets intuitionistic arithmetic, it is structurally linear: in general, realizers of contraction

$$\varphi(a) \quad \longrightarrow \quad \varphi(a) \wedge \varphi(a)$$

$$\frac{}{\varphi \vdash \varphi} \qquad \frac{\overline{\varphi} \vdash \gamma, \overline{\varphi}' \qquad \overline{\psi}, \gamma \vdash \overline{\psi}'}{\overline{\varphi}, \overline{\psi} \vdash \overline{\varphi}', \overline{\psi}'} \qquad \frac{\overline{\varphi}, \varphi, \psi, \overline{\psi} \vdash \overline{\varphi}'}{\overline{\varphi}, \psi, \varphi, \overline{\psi} \vdash \overline{\varphi}'} \qquad \frac{\overline{\varphi} \vdash \overline{\varphi}', \varphi, \psi, \overline{\psi}'}{\overline{\varphi} \vdash \overline{\varphi}', \psi, \varphi, \overline{\psi}'}$$

$$\frac{\overline{\varphi} \vdash \overline{\psi}}{\overline{\varphi}, \mathbf{I} \vdash \overline{\psi}} \qquad \frac{}{\vdash \mathbf{I}} \qquad \frac{\overline{\varphi}, \varphi_0, \varphi_1 \vdash \overline{\varphi}'}{\overline{\varphi}, \varphi_0 \otimes \varphi_1 \vdash \overline{\varphi}'} \qquad \frac{\overline{\varphi} \vdash \varphi, \overline{\varphi}' \qquad \overline{\psi} \vdash \psi, \overline{\psi}'}{\overline{\varphi}, \overline{\psi} \vdash \varphi \otimes \psi, \overline{\varphi}', \overline{\psi}'} \qquad \frac{\overline{\varphi}, \varphi \vdash \psi}{\overline{\varphi} \vdash \varphi \multimap \psi}$$

$$\frac{}{\bot \vdash} \qquad \frac{\overline{\varphi} \vdash \overline{\psi}}{\overline{\varphi} \vdash \bot, \overline{\psi}} \qquad \frac{\overline{\varphi}, \varphi \vdash \overline{\varphi}' \qquad \overline{\psi}, \psi \vdash \overline{\psi}'}{\overline{\varphi}, \overline{\psi}, \varphi \parr \psi \vdash \overline{\varphi}', \overline{\psi}'} \qquad \frac{\overline{\varphi} \vdash \varphi_0, \varphi_1, \overline{\varphi}'}{\overline{\varphi} \vdash \varphi_0 \parr \varphi_1, \overline{\varphi}'} \qquad \frac{\overline{\varphi} \vdash \varphi, \overline{\varphi}' \qquad \overline{\psi}, \psi \vdash \overline{\psi}'}{\overline{\varphi}, \overline{\psi}, \varphi \multimap \psi \vdash \overline{\varphi}', \overline{\psi}'}$$

$$\frac{\overline{\varphi}, \varphi \vdash \overline{\varphi}'}{\overline{\varphi}, (\exists z^\tau)\varphi \vdash \overline{\varphi}'} \qquad \frac{\overline{\varphi} \vdash \varphi[\mathtt{t}^\tau/x^\tau], \overline{\varphi}'}{\overline{\varphi} \vdash (\exists x^\tau)\varphi, \overline{\varphi}'} \qquad \frac{\overline{\varphi}, \varphi[\mathtt{t}^\tau/x^\tau] \vdash \overline{\varphi}'}{\overline{\varphi}, (\forall x^\tau)\varphi \vdash \overline{\varphi}'} \qquad \frac{\overline{\varphi} \vdash \varphi}{\overline{\varphi} \vdash (\forall z^\tau)\varphi}$$

Fig. 2. Deduction for MF (where z^τ is fresh).

only exist when the term language can decide $\varphi_D(u, x, a)$, which is possible in arithmetic but not in all settings. Besides, linear versions of Dialectica were formulated at the very beginning of linear logic [21–23] (see also [14,27]).

In this paper, we use a variant of Dialectica as a syntactic formulation of the automata-based realizability model of [26]. The formula φ_D essentially represents a deterministic automaton on ω-words and is in general not quantifier-free. Moreover, we extract f.s. causal functions, while the category $\mathbb{M}$ is not closed. As a result, a realizer of φ is an *open* (eager) term $\mathtt{u}(x)$ of arity $(\sigma; \tau)$ satisfying $\varphi_D(\mathtt{u}(x), x)$. While it is possible to exhibit realizers for contraction on closed φ thanks to the Büchi-Landweber Theorem, this is generally not the case for open $\varphi(a)$. We therefore resort to working in a linear system, in which we obtain witnesses for $\forall\exists(-)^L$-statements (and thus for realizable instances of Church's synthesis), but not for all $\forall\exists$-statements.

Fix a set of atomic formulae At containing all $(\mathtt{t}^\tau \doteq \mathtt{u}^\tau)$, and a standard interpretation extending Sect. 2 for each $\alpha \in$ At.

3.1 The Multiplicative Fragment

Our linear system is based on *full intuitionistic linear logic* (see [15]). The formulae of the multiplicative fragment MF are given by the grammar:

$$\varphi, \psi \ ::= \ \mathbf{I} \mid \bot \mid \alpha \mid \varphi \multimap \psi \mid \varphi \otimes \psi \mid \varphi \parr \psi \mid (\exists x^\tau)\varphi \mid (\forall x^\tau)\varphi$$

(where $\alpha \in$ At). Deduction is given by the rules of Fig. 2 and the axioms

$$\frac{}{\vdash \mathtt{t}^\tau \doteq \mathtt{t}^\tau} \qquad \frac{}{\mathtt{t}^\tau \doteq \mathtt{u}^\tau,\ \varphi[\mathtt{t}^\tau/x^\tau] \vdash \varphi[\mathtt{u}^\tau/x^\tau]} \qquad \frac{[\![\mathtt{t}^\tau]\!] = [\![\mathtt{u}^\tau]\!]}{\vdash \mathtt{t}^\tau \doteq \mathtt{u}^\tau} \tag{7}$$

Each formula φ of MF can be mapped to a classical formula $\lfloor \varphi \rfloor$ (where $\mathbf{I}$, $\multimap$, $\otimes$, $\parr$ are replaced resp. by $\top, \to, \wedge, \vee$). Hence $\lfloor \varphi \rfloor$ holds whenever $\vdash \varphi$

The Dialectica interpretation of MF is the usual one rewritten with the connectives of MF, but for the disjunction $\parr$ that we treat similarly as $\otimes$. To each

$$
\begin{array}{lll}
(\varphi \otimes \psi)^D(a) & := & \exists\langle u,v\rangle\forall\langle x,y\rangle.\ (\varphi\otimes\psi)_D(\langle u,v\rangle,\langle x,y\rangle,a) \quad := \\
& & \qquad \exists\langle u,v\rangle\forall\langle x,y\rangle.\ \varphi_D(u,x,a)\otimes\psi_D(v,y,a) \\
(\varphi \parr \psi)^D(a) & := & \exists\langle u,v\rangle\forall\langle x,y\rangle.\ (\varphi\parr\psi)_D(\langle u,v\rangle,\langle x,y\rangle,a) \quad := \\
& & \qquad \exists\langle u,v\rangle\forall\langle x,y\rangle.\ \varphi_D(u,x,a)\parr\psi_D(v,y,a) \\
(\varphi \multimap \psi)^D(a) & := & \exists\langle f,F\rangle\forall\langle u,y\rangle.\ (\varphi\multimap\psi)_D(\langle f,F\rangle,\langle u,y\rangle,a) \quad := \\
& & \qquad \exists\langle f,F\rangle\forall\langle u,y\rangle.\ \varphi_D(u,(F)uy,a)\multimap\psi_D((f)u,y,a) \\
(\exists w.\varphi)^D(a) & := & \exists\langle u,w\rangle\forall x.\ (\exists w.\varphi)_D(\langle u,w\rangle,x,a) \quad := \ \exists\langle u,w\rangle\forall x.\ \varphi_D(u,x,\langle a,w\rangle) \\
(\forall w.\varphi)^D(a) & := & \exists f\,\forall\langle x,w\rangle.\ (\forall w.\varphi)_D(f,\langle x,w\rangle,a) \quad := \ \exists f\,\forall\langle x,w\rangle.\ \varphi_D((f)w,x,\langle a,w\rangle)
\end{array}
$$

Fig. 3. The Dialectica Interpretation of MF (where types are leaved implicit).

formula $\varphi(a)$ with only a free, we associate a formula $\varphi^D(a)$ with only a free, as well as a formula φ_D with possibly other free variables. For atomic formulae we let $\varphi^D(a) := \varphi_D(a) := \varphi(a)$. The inductive cases are given on Fig. 3, where $\varphi^D(a) = (\exists u)(\forall x)\varphi_D(u,x,a)$ and $\psi^D(a) = (\exists v)(\forall y)\psi_D(v,y,a)$.

Dialectica is such that φ^D is equivalent to φ via possibly non-intuitionistic but constructive principles. The tricky connectives are implication and universal quantification. Similarly as in the intuitionistic case (see e.g. [1,16,33]), $(\varphi \multimap \psi)^D$ is prenex a form of $\varphi^D \multimap \psi^D$ obtained using (LAC) together with linear variants of the *Markov* and *Independence of premises* principles. In our case, the equivalence $\varphi \multimapboth \varphi^D$ also requires additional axioms for $\otimes$ and $\parr$. We give details for the full system in Sect. 3.3.

The soundness of $(-)^D$ goes as usual, excepted that we extract *open eager* terms: from a proof of $\varphi(a^\kappa)$ we extract a realizer of $(\forall a)\varphi(a)$, that is an open eager term $\mathtt{u}(x,a)$ s.t. $\vdash \varphi_D(@(\mathtt{u}(x,a),a),x,a)$. Composition of realizers (in part. required for the cut rule) is given by the fixpoints of Proposition 3. Note that a realizer of a closed φ is a finite-state winning P-strategy in $\mathcal{G}(\lfloor\varphi_D\rfloor)(u,x)$.

3.2 Polarized Exponentials

It is well-known that the structure of Dialectica is linear, as it makes problematic the interpretation of contraction:

$$\varphi(a) \quad\multimap\quad \varphi(a)\otimes\varphi(a) \qquad \text{and} \qquad \varphi(a)\parr\varphi(a) \quad\multimap\quad \varphi(a)$$

In our case, the Büchi-Landweber Theorem implies that all closed instances of contraction have realizers which are correct in the standard model. But this is in general not true for open instances.

Example 3. Realizers of $\varphi \multimap \varphi\otimes\varphi$ for a closed φ are given by eager terms $\mathtt{U}_1(u,x_1,x_2)$, $\mathtt{U}_2(u,x_1,x_2)$, $\mathtt{X}(u,x_1,x_2)$ which must represent P-strategies in the game $\mathcal{G}(\Phi)(\langle U_1,U_2,X\rangle,\langle u,x_1,x_2\rangle)$, where Φ is

$$\lfloor\varphi_D(u,(X)ux_1x_2)\rfloor \quad\longrightarrow\quad \lfloor\varphi_D((U_1)u,x_1)\rfloor \ \wedge\ \lfloor\varphi_D((U_2)u,x_2)\rfloor$$

By the Büchi-Landweber Theorem 1, either there is an eager term $\mathtt{U}(x)$ such that $\lfloor\varphi_D(\mathtt{U}(x), x)\rfloor$ holds, so that

$$\lfloor\varphi_D(u, x_1)\rfloor \quad\longrightarrow\quad \lfloor\varphi_D(\mathbf{e}(\mathtt{U})(x_1), x_1)\rfloor \;\wedge\; \lfloor\varphi_D(\mathbf{e}(\mathtt{U})(x_2), x_2)\rfloor$$

or there is an eager term $\mathtt{X}(u)$ such that $\neg\lfloor\varphi_D(u, \mathbf{e}(\mathtt{X})(u))\rfloor$ holds, so that

$$\lfloor\varphi_D(u, \mathbf{e}(\mathtt{X})(u))\rfloor \quad\longrightarrow\quad \lfloor\varphi_D(u, x_1)\rfloor \;\wedge\; \lfloor\varphi_D(u, x_2)\rfloor$$

Example 4. Consider the open formula $\varphi(a^o) := (\forall x^o)(\mathtt{t}(x, a) \doteq 0^\omega)$ where $[\![\mathtt{t}]\!](B, C) = 0^{n+1}1^\omega$ for the first $n \in \mathbb{N}$ with $C(n+1) = B(0)$ if such n exists, and such that $[\![\mathtt{t}]\!](B, C) = 0^\omega$ otherwise. The game induced by $((\forall a)(\varphi \multimap \varphi \otimes \varphi))_D$ is $\mathcal{G}(\Phi)(X, \langle x_1, x_1, a\rangle)$, where Φ is

$$\mathtt{t}((X)x_1x_2a, a) \doteq 0^\omega \quad\longrightarrow\quad \mathtt{t}(x_1, a) \doteq 0^\omega \;\wedge\; \mathtt{t}(x_2, a) \doteq 0^\omega$$

In this game, P begins by playing a function $\mathbf{2}^3 \to \mathbf{2}$, O replies in $\mathbf{2}^3$, and then P and O keep on alternatively playing moves of the expected type. A finite-state winning strategy for O is easy to find. Let P begin with the function X. Fix some $\mathsf{a} \in \mathbf{2}$ and let $i := \mathsf{X}(0, 1, \mathsf{a})$. O replies $(0, 1, \mathsf{a})$ to X. The further moves of P are irrelevant, and O keeps on playing $(-, -, 1 - i)$ (the values of x_1 and x_2 are irrelevant after the first round). This strategy ensures

$$\mathtt{t}((X)x_1x_2a, a) \doteq 0^\omega \quad\wedge\quad \neg(\mathtt{t}(x_1, a) \doteq 0^\omega \;\wedge\; \mathtt{t}(x_2, a) \doteq 0^\omega)$$

Hence we can not realize contraction while remaining correct w.r.t. the standard model. On the other hand, Dialectica induces polarities generalizing the usual polarities of linear logic (see e.g. [17]). Say that $\varphi(a)$ is *positive* (resp. *negative*) if $\varphi^D(a)$ is of the form $\varphi^D(a) = (\exists u^\tau)\varphi_D(u, -, a)$ (resp. $\varphi^D(a) = (\forall x^\sigma)\varphi_D(-, x, a)$). Quantifier-free formulae are thus both positive and negative.

Example 5. Polarized contraction

$$\varphi^+ \multimap \varphi^+ \otimes \varphi^+ \qquad\text{and}\qquad \psi^- \mathbin{⅋} \psi^- \multimap \psi^- \qquad (\varphi^+ \text{ positive, } \psi^- \text{ negative})$$

gives realizers of all instances of itself. Indeed, with say $\varphi^D(a) = (\exists u)\varphi_D(u, -, a)$ and $\psi^D(a) = (\forall y)\psi_D(-, y, a)$, $\boldsymbol{\Lambda}(\pi_1)$ (for π_1 a $\mathbb{M}$-projection on suitable types) gives eager terms $\mathtt{U}(u, a)$ and $\mathtt{Y}(y, a)$ such that

$$\begin{array}{rcl} \varphi_D(u, -, a) & \multimap & \Big(\varphi_D\big(\mathbf{e}(\mathtt{U})(u, a), -, a\big) \;\otimes\; \varphi_D\big(\mathbf{e}(\mathtt{U})(u, a), -, a\big)\Big) \\ \text{and} \quad \Big(\psi_D\big(-, \mathbf{e}(\mathtt{Y})(y, a), a\big) \mathbin{⅋} \psi_D\big(-, \mathbf{e}(\mathtt{Y})(y, a), a\big)\Big) & \multimap & \psi_D(-, y, a) \end{array}$$

We only have exponentials for polarized formulae. First, following the usual polarities of linear logic, we can let

$$\begin{array}{rcccl} (!(\varphi^+))^D(a) & := & (\exists u)(!(\varphi^+))_D(u, -, a) & := & (\exists u)!\varphi_D(u, -, a) \\ (?(\psi^-))^D(a) & := & (\forall y)(?(\psi^-))_D(-, y, a) & := & (\forall x)?\psi_D(-, y, a) \end{array} \tag{8}$$

$$\frac{\overline{\psi} \vdash \overline{\psi}'}{\overline{\psi}, !\varphi \vdash \overline{\psi}'} \qquad \frac{\overline{\psi}, !\varphi, !\varphi \vdash \overline{\psi}'}{\overline{\psi}, !\varphi \vdash \overline{\psi}'} \qquad \frac{\overline{\varphi}, \varphi \vdash \overline{\varphi}'}{\overline{\varphi}, !\varphi \vdash \overline{\varphi}'} \qquad \frac{!\overline{\varphi} \vdash \varphi, ?\overline{\psi}}{!\overline{\varphi} \vdash !\varphi, ?\overline{\psi}} \qquad \frac{\overline{\varphi}, !\varphi \vdash \psi, ?\overline{\psi}}{\overline{\varphi} \vdash !\varphi \multimap \psi, ?\overline{\psi}}$$

$$\frac{\overline{\psi} \vdash \overline{\psi}'}{\overline{\psi} \vdash ?\varphi, \overline{\psi}'} \qquad \frac{\overline{\psi} \vdash ?\varphi, ?\varphi, \overline{\psi}'}{\overline{\psi} \vdash ?\varphi, \overline{\psi}'} \qquad \frac{\overline{\varphi} \vdash \varphi, \overline{\psi}}{\overline{\varphi} \vdash ?\varphi, \overline{\psi}} \qquad \frac{!\overline{\varphi}, \varphi \vdash ?\overline{\psi}}{!\overline{\varphi}, ?\varphi \vdash ?\overline{\psi}} \qquad \frac{\overline{\varphi} \vdash \varphi, ?\overline{\psi}}{\overline{\varphi} \vdash (\forall z)\varphi, ?\overline{\psi}}$$

Fig. 4. Exponential rules of PF.

Hence $!\varphi$ is positive for a positive φ and $?\psi$ is negative for a negative ψ. The following exponential contraction axioms are then interpreted by themselves:

$$!(\varphi^+) \quad \multimap \quad !(\varphi^+) \otimes !(\varphi^+) \qquad \text{and} \qquad ?(\psi^-) \parr ?(\psi^-) \quad \multimap \quad ?(\psi^-)$$

Second, we can have exponentials $!(\psi^-)$ and $?(\varphi^+)$ with the automata-based reading of [26]. Positive formulae are seen as non-deterministic automata, and $?(-)$ on positive formulae is determinization on ω-words (McNaughton's Theorem [19]). Negative formulae are seen as universal automata, and $!(-)$ on negative formulae is co-determinization (an instance of the *Simulation Theorem* [10,20]). Formulae which are both positive and negative (notation $(-)^{\pm}$) correspond to deterministic automata, and are called *deterministic*. We let

$$\begin{array}{lllll} (!(\psi^-))^D(a) & := & (!(\psi^-))_D(-,-,a) & := & !(\forall x)\psi_D(-,x,a) \\ (?(\varphi^+))^D(a) & := & (?(\varphi^+))_D(-,-,a) & := & ?(\exists u)\varphi_D(u,-,a) \end{array} \qquad (9)$$

So $!(\psi^-)$ and $?(\varphi^+)$ are always deterministic. The corresponding exponential contraction axioms are interpreted by themselves. This leads to the following polarized fragment PF (the deduction rules for exponentials are given on Fig. 4):

$$\begin{array}{lll} \varphi^{\pm}, \psi^{\pm} & ::= & \mathbf{I} \mid \bot \mid \alpha \mid !(\varphi^-) \mid ?(\varphi^+) \mid \varphi^{\pm} \otimes \psi^{\pm} \mid \varphi^{\pm} \parr \psi^{\pm} \mid \varphi^{\pm} \multimap \psi^{\pm} \\ \varphi^+, \psi^+ & ::= & \varphi^{\pm} \mid !(\varphi^+) \mid (\exists x^{\sigma})\varphi^+ \mid \varphi^+ \otimes \psi^+ \mid \varphi^+ \parr \psi^+ \mid \varphi^- \multimap \psi^+ \\ \varphi^-, \psi^- & ::= & \varphi^{\pm} \mid ?(\varphi^-) \mid (\forall x^{\sigma})\varphi^- \mid \varphi^- \otimes \psi^- \mid \varphi^- \parr \psi^- \mid \varphi^+ \multimap \psi^- \end{array}$$

3.3 The Full System

The formulae of the full system FS are given by the following grammar:

$$\varphi, \psi \ ::= \ \varphi^+ \mid \varphi^- \mid \varphi \multimap \psi \mid \varphi \otimes \psi \mid \varphi \parr \psi \mid (\exists x^{\tau})\varphi \mid (\forall x^{\tau})\varphi$$

Deduction in FS is given by Figs. 2, 4 and (7). We extend $\lfloor - \rfloor$ to FS with $\lfloor !\varphi \rfloor := \lfloor ?\varphi \rfloor := \lfloor \varphi \rfloor$. Hence $\lfloor \varphi \rfloor$ holds when $\vdash \varphi$ is derivable. The Dialectica interpretation of FS is given by Fig. 3 and (8), (9) (still taking $\varphi^D(a) := \varphi_D(a) := \varphi(a)$ for atoms). Note that $(-)^D$ preserves and reflects polarities.

Theorem 2 (Soundness). *Let φ be closed with $\varphi^D = (\exists u^{\tau})(\forall x^{\sigma})\varphi_D(u,x)$. From a proof of φ in* FS *one can extract an eager term* $\mathrm{u}(x)$ *such that* FS *proves* $(\forall x^{\sigma})\varphi_D(\mathrm{u}(x), x)$.

As usual, proving $\varphi \multimap\!\circ\, \varphi^D$ requires extra axioms. Besides (LAC), we use the following *(linear) semi-intuitionistic principles* (LSIP), with polarities as shown:

$$\begin{array}{rcl} (\forall a)(\varphi^-(a) \otimes \psi^-) & \multimap & (\forall a)\varphi^-(a) \otimes \psi^- \\ (\forall a)(\varphi^-(a) \parr \psi^-) & \multimap & (\forall a)\varphi^-(a) \parr \psi^- \\ (\exists a)\varphi^-(a) \parr \psi & \multimap & (\exists a)(\varphi^-(a) \parr \psi) \\ (\psi^- \multimap (\exists a)\varphi^-(a)) & \multimap & (\exists a)(\psi^- \multimap \varphi^-(a)) \\ ((\forall a)\varphi^\pm(a) \multimap \psi^\pm) & \multimap & (\exists a)(\varphi^\pm(a) \multimap \psi^\pm) \end{array} \qquad (\mathsf{LSIP})$$

as well as the following *deterministic exponential* axioms (DEXP):

$$\delta \multimap\, !\delta \qquad \text{and} \qquad ?\delta \multimap \delta \qquad (\delta \text{ deterministic})$$

All these axioms but (LAC) are true in the standard model (via $\lfloor - \rfloor$). Moreover:

Proposition 4. *The axioms* (LAC) *and* (LSIP) *are realized in* FS. *The axioms* (DEXP) *are realized in* FS + (DEXP).

Theorem 3 (Characterization). *We have*

$$\begin{array}{rccll} \vdash_{\mathsf{FS+(LAC)+(LSIP)+(DEXP)}} & \varphi(a) & \multimap\!\circ & \varphi^D(a) & (\varphi \text{ FS-formula}) \\ \vdash_{\mathsf{FS+(LSIP)+(DEXP)}} & \varphi(a) & \multimap\!\circ & \varphi^D(a) & (\varphi \text{ PF-formula}) \end{array}$$

Corollary 1 (Extraction). *Consider a closed formula* $\varphi := (\forall x^\sigma)(\exists u^\tau)\delta(u, x)$ *with* δ *deterministic. From a proof of* φ *in* FS + (LAC) + (LSIP) + (DEXP) *one can extract a term* $\mathtt{t}(x)$ *such that* $\models (\forall x^\sigma)\lfloor \delta(\mathtt{t}(x), x) \rfloor$.

Note that FS + (DEXP) proves $\delta \parr (\delta \multimap \bot)$ for all deterministic δ.

3.4 Translations of Classical Logic

There are many translations from classical to linear logic. Two canonical possibilities are the $(-)^T$ and $(-)^Q$-translation of [9] (see also [17,18]) targeting resp. negative and positive formulae. Both take classical sequents to linear sequents of the form $!(-) \vdash ?(-)$, which are provable in FS thanks to the PF rules

$$\frac{\overline{\varphi}, !\varphi \vdash \psi, ?\overline{\psi}}{\overline{\varphi} \vdash !\varphi \multimap \psi, ?\overline{\psi}} \qquad\qquad \frac{\overline{\varphi} \vdash \varphi, ?\overline{\psi}}{\overline{\varphi} \vdash (\forall z)\varphi, ?\overline{\psi}}$$

For the completeness of LMSO($\mathfrak{C}$) (Theorem 6, Sect. 4), we shall actually require a translation $(-)^L$ such that the linear equivalences (with polarities as displayed)

$$?\varphi^+ \multimap\!\circ \lfloor \varphi^+ \rfloor^L \qquad\qquad \delta^\pm \multimap\!\circ \lfloor \delta^\pm \rfloor^L \qquad\qquad !\psi^- \multimap\!\circ \lfloor \psi^- \rfloor^L \qquad (10)$$

are provable possibly with extra axioms that we require to realize themselves. In part., (10) implies (DEXP), and $(-)^L$ should give deterministic formulae. While $(-)^T$ and $(-)^Q$ can be adapted accordingly, (10) induces axioms which make the resulting translations equivalent to the deterministic $(-)^L$-translation of [26]:

$$\bot^L := \bot \quad \top^L := \mathbf{I} \quad \alpha^L := \alpha \qquad (\varphi \vee \psi)^L := \varphi^L \parr \psi^L \qquad (\exists x^\sigma.\varphi)^L := ?(\exists x^\sigma)\varphi^L$$
$$(\varphi \to \psi)^L := \varphi^L \multimap \psi^L \qquad (\varphi \wedge \psi)^L := \varphi^L \otimes \psi^L \qquad (\forall x^\sigma.\varphi)^L := !(\forall x^\sigma)\varphi^L$$

Proposition 5. *The scheme (10) is equivalent in* FS *to* (DEXP)+(PEXP)*, where* (PEXP) *are the following* polarized exponential *axioms, with polarities as shown:*

$$\begin{array}{rclrcl}
?(\varphi^+) & \multimap & ?!(\varphi^+) & !?(\psi^-) & \multimap & !(\psi^-) \\
!(\varphi^-) \multimap ?(\psi^+) & \multimap & ?(\varphi^- \multimap \psi^+) & ?(\varphi^+) \multimap !(\psi^-) & \multimap & !(\varphi^+ \multimap \psi^-) \\
?(\varphi^+) \otimes ?(\psi^+) & \multimap & ?(\varphi^+ \otimes \psi^+) & !(\varphi^- \otimes \psi^-) & \multimap & !(\varphi^-) \otimes !(\psi^-) \\
?(\varphi^+) \parr ?(\psi^+) & \multimap & ?(\varphi^+ \parr \psi^+) & !(\varphi^- \parr \psi^-) & \multimap & !(\varphi^-) \parr !(\psi^-)
\end{array}$$

Proposition 6. *If φ is provable in many-sorted classical logic with equality then* FS + (DEXP) *proves φ^L.*

Proposition 7. *The axioms* (PEXP) *are realized in* FS + (LSIP) + (DEXP) + (PEXP)*. Corollary 1 thus extends to* FS + (LAC) + (LSIP) + (DEXP) + (PEXP).

Note that φ^L is deterministic and that $\lfloor \varphi^L \rfloor = \varphi$.

4 Completeness

In Sect. 3 we devised a Dialectica-like $(-)^D$ providing a syntactic extraction procedure for $\forall\exists(-)^L$-statements. In this Section, building on an axiomatic treatment of MSO($\mathbf{M}$), we show that LMSO, an arithmetic extension of FS + (LSIP) + (DEXP) + (PEXP) adapted from [26], is correct and complete w.r.t. Church's synthesis, in the sense that the provable $\forall\exists(-)^L$-statements are exactly the realizable ones. We then turn to the main result of this paper, namely the completeness of LMSO($\mathfrak{C}$) := LMSO + (LAC). We fix the set of atomic formulae

$$\alpha \in \mathrm{At} \quad ::= \quad \mathtt{t}^\tau \doteq \mathtt{u}^\tau \mid \mathtt{t}^o \,\dot\subseteq\, \mathtt{u}^o \mid \mathsf{E}(\mathtt{t}^o) \mid \mathsf{N}(\mathtt{t}^o) \mid \mathsf{S}(\mathtt{t}^o, \mathtt{u}^o) \mid \mathsf{0}(\mathtt{t}^o) \mid \mathtt{t}^o \,\dot\leq\, \mathtt{u}^o$$

4.1 The Logic MSO(M)

MSO($\mathbf{M}$) is many-sorted first-order logic with atomic formulae $\alpha \in \mathrm{At}$. Its sorts and terms are those given in Sect. 2, and standard interpretation extends that of Sect. 2 as follows: $\dot\subseteq$ is set inclusion, E holds on B iff B is empty, N (resp. 0) holds on B iff B is a singleton $\{n\}$ (resp. the singleton $\{0\}$), and $\mathsf{S}(B, C)$ (resp. $B \,\dot\leq\, C$) holds iff $B = \{n\}$ and $C = \{n+1\}$ for some $n \in \mathbb{N}$ (resp. $B = \{n\}$ and $C = \{m\}$ for some $n \leq m$). We write x^ι for variables x^o relativized to N, so that $(\exists x^\iota)\varphi$ and $(\forall x^\iota)\varphi$ stand resp. for $(\exists x^o)(\mathsf{N}(x) \wedge \varphi)$ and $(\forall x^o)(\mathsf{N}(x) \to \varphi)$. Moreover, $x^\iota \,\dot\in\, \mathtt{t}$ stands for $x^\iota \,\dot\subseteq\, \mathtt{t}$, so that $\mathtt{t}^o \,\dot\subseteq\, \mathtt{u}^o$ is equivalent to $(\forall x^\iota)(x \,\dot\in\, \mathtt{t} \to x \,\dot\in\, \mathtt{u})$.

The logic MSO$^+$ [26] is MSO($\mathbf{M}$) restricted to the type o, hence with only terms for Mealy machines of sort $(\mathbf{2}, \ldots, \mathbf{2}; \mathbf{2})$. The MSO of [26] is the purely relational (term-free) restriction of MSO$^+$. Recall from [26, Prop. 2.6], that for

$$\frac{}{\mathsf{E}(\mathtt{t}) \vdash \mathtt{t} \mathrel{\dot{\subseteq}} \mathtt{u}} \quad \frac{\overline{\varphi} \vdash \mathtt{t} \mathrel{\dot{\subseteq}} z, \overline{\varphi}'}{\overline{\varphi} \vdash \mathsf{E}(\mathtt{t}), \overline{\varphi}'} \quad \frac{\overline{\varphi}, z \mathrel{\dot{\subseteq}} \mathtt{t} \vdash \mathsf{E}(z), z \doteq \mathtt{t}, \overline{\varphi}'}{\overline{\varphi} \vdash \mathsf{N}(\mathtt{t}), \mathsf{E}(\mathtt{t}), \overline{\varphi}'} \quad \frac{\overline{\varphi}, \mathsf{N}(z), z \mathrel{\dot{\subseteq}} \mathtt{t} \vdash z \mathrel{\dot{\subseteq}} \mathtt{u}, \overline{\varphi}'}{\overline{\varphi} \vdash \mathtt{t} \mathrel{\dot{\subseteq}} \mathtt{u}, \overline{\varphi}'} \quad \frac{}{\mathsf{N}(\mathtt{t}), \mathsf{E}(\mathtt{t}) \vdash}$$

$$\frac{}{\vdash \mathtt{t} \mathrel{\dot{\subseteq}} \mathtt{t}} \quad \frac{}{\mathtt{t} \mathrel{\dot{\subseteq}} \mathtt{u}, \mathtt{u} \mathrel{\dot{\subseteq}} \mathtt{v} \vdash \mathtt{t} \mathrel{\dot{\subseteq}} \mathtt{v}} \quad \frac{}{\mathtt{t} \mathrel{\dot{\subseteq}} \mathtt{u}, \mathtt{u} \mathrel{\dot{\subseteq}} \mathtt{t} \vdash \mathtt{t} \doteq \mathtt{u}} \quad \frac{}{\mathsf{N}(\mathtt{t}), \mathtt{u} \mathrel{\dot{\subseteq}} \mathtt{t} \vdash \mathsf{E}(\mathtt{u}), \mathtt{u} \doteq \mathtt{t}} \quad \frac{}{\mathsf{S}(\mathtt{t}, \mathtt{u}), \mathsf{0}(\mathtt{u}) \vdash}$$

$$\frac{}{\mathsf{N}(\mathtt{t}) \vdash \mathtt{t} \mathrel{\dot{\leq}} \mathtt{t}} \quad \frac{}{\mathtt{t} \mathrel{\dot{\leq}} \mathtt{u}, \mathtt{u} \mathrel{\dot{\leq}} \mathtt{v} \vdash \mathtt{t} \mathrel{\dot{\leq}} \mathtt{v}} \quad \frac{}{\mathtt{t} \mathrel{\dot{\leq}} \mathtt{u}, \mathtt{u} \mathrel{\dot{\leq}} \mathtt{t} \vdash \mathtt{t} \doteq \mathtt{u}} \quad \frac{}{\mathsf{S}(\mathtt{t}, \mathtt{u}) \vdash \mathtt{t} \mathrel{\dot{\leq}} \mathtt{u}} \quad \frac{}{\mathsf{0}(\mathtt{t}) \vdash \mathsf{N}(\mathtt{t})}$$

$$\frac{\overline{\varphi}, \mathsf{0}(z) \vdash \overline{\varphi}'}{\overline{\varphi} \vdash \overline{\varphi}'} \quad \frac{}{\mathsf{S}(\mathtt{u}, \mathtt{v}), \mathtt{t} \mathrel{\dot{\leq}} \mathtt{v} \vdash \mathtt{t} \doteq \mathtt{v}, \mathtt{t} \mathrel{\dot{\leq}} \mathtt{u}} \quad \frac{}{\mathtt{t} \mathrel{\dot{\leq}} \mathtt{u} \vdash \mathsf{N}(t)} \quad \frac{}{\mathtt{t} \mathrel{\dot{\leq}} \mathtt{u} \vdash \mathsf{N}(u)} \quad \frac{}{\mathsf{S}(\mathtt{t}, \mathtt{u}) \vdash \mathsf{N}(\mathtt{t})}$$

$$\frac{\overline{\varphi}, \mathsf{S}(\mathtt{t}, z) \vdash \overline{\varphi}'}{\overline{\varphi} \vdash \overline{\varphi}'} \quad \frac{}{\mathsf{0}(\mathtt{t}), \mathsf{0}(\mathtt{u}) \vdash \mathtt{t} \doteq \mathtt{u}} \quad \frac{}{\mathsf{S}(\mathtt{t}, \mathtt{u}), \mathsf{S}(\mathtt{t}, \mathtt{v}) \vdash \mathtt{u} \doteq \mathtt{v}} \quad \frac{}{\mathsf{S}(\mathtt{u}, \mathtt{t}), \mathsf{S}(\mathtt{v}, \mathtt{t}) \vdash \mathtt{u} \doteq \mathtt{v}} \quad \frac{}{\mathsf{S}(\mathtt{t}, \mathtt{u}) \vdash \mathsf{N}(\mathtt{u})}$$

Fig. 5. The Arithmetic Rules of MSO(**M**) and LMSO (with terms of sort o and z fresh).

each Mealy machine $\mathcal{M} : \mathbf{2}^p \to \mathbf{2}$, there is an MSO-formula $\delta_{\mathcal{M}}(\overline{X}, x)$ such that for all $n \in \mathbb{N}$ and all $\overline{B} \in (\mathbf{2}^\omega)^p$, we have $F_{\mathcal{M}}(\overline{B})(n) = 1$ iff $\delta_{\mathcal{M}}(\{n\}, \overline{B})$ holds.

The axioms of MSO(**M**) are the arithmetic rules of Fig. 5, the axioms (7) and the following, where $\mathcal{M} : \mathbf{2}^p \to \mathbf{2}$ and y, z, X are fresh.

$$\frac{}{\vdash (\forall \overline{X}^o)(\forall x^\iota)\left(x \mathrel{\dot{\in}} \mathtt{f}_{\mathcal{M}}(\overline{X}) \;\leftrightarrow\; \delta_{\mathcal{M}}(x, \overline{X})\right)} \qquad \frac{}{\vdash (\exists X^o)(\forall x^\iota)\,(x \mathrel{\dot{\in}} X \;\leftrightarrow\; \varphi)}$$

$$\frac{\overline{\varphi}, \mathsf{0}(z) \vdash \varphi[z/x], \overline{\varphi}' \qquad \overline{\varphi}, \mathsf{S}(y, z), \varphi[y/x] \vdash \varphi[z/x], \overline{\varphi}'}{\overline{\varphi} \vdash (\forall x^\iota)\varphi, \overline{\varphi}'}$$

The theory MSO(**M**) is complete. Thus provability in MSO(**M**) and validity in the standard model coincide. This extends [26, Thm. 2.11 (via [28])].

Theorem 4 (Completeness of MSO(M)). *For closed MSO(**M**)-formulae φ, we have $\models \varphi$ if and only if $\vdash_{\mathsf{MSO}(\mathbf{M})} \varphi$.*

4.2 The Logic LMSO

The system LMSO is FS + (LSIP) + (DEXP) + (PEXP) extended with Fig. 5 and

$$\frac{}{\vdash (\forall \overline{X}^o)(\forall x^\iota)\left(x \mathrel{\dot{\in}} \mathtt{f}_{\mathcal{M}}(\overline{X}) \;\multimap\!\circ\; \delta^L_{\mathcal{M}}(x, \overline{X})\right)} \qquad \frac{}{\vdash ?(\exists X^o)!(\forall x^\iota)\,(x \mathrel{\dot{\in}} X \;\multimap\!\circ\; \delta^\pm)}$$

$$\frac{!\overline{\varphi}, \mathsf{0}(z) \vdash \varphi^-[z/x], ?\overline{\varphi}' \qquad !\overline{\varphi}, \mathsf{S}(y, z), !\varphi^-[y/x] \vdash \varphi^-[z/x], ?\overline{\varphi}'}{!\overline{\varphi} \vdash (\forall x^\iota)\varphi^-, ?\overline{\varphi}'}$$

Let $\mathsf{LMSO}(\mathfrak{C}) := \mathsf{LMSO} + (\mathsf{LAC})$. Note that $\vdash_{\mathsf{MSO}(\mathbf{M})} \lfloor \varphi \rfloor$ whenever $\vdash_{\mathsf{LMSO}} \varphi$. Proposition 6 extends so that similarly as in [26] we have

Proposition 8. *If $\vdash_{\mathsf{MSO}(\mathbf{M})} \varphi$ then $\vdash_{\mathsf{LMSO}} \varphi^L$. In part., for a realizable instance of Church's synthesis $(\forall x^\sigma)(\exists u^\tau)\varphi(u, x)$, we have $\vdash_{\mathsf{LMSO}} (\forall x^\sigma)(\exists u^\tau)\varphi^L(u, x)$.*

Moreover, the soundness of $(-)^D$ extends to LMSO. It follows that $\mathsf{LMSO}(\mathfrak{C})$ is coherent and proves exactly the realizable $\forall\exists(-)^L$-statements.

Theorem 5 (Soundness). *Let φ be closed with $\varphi^D = (\exists u^\tau)(\forall x^\sigma)\varphi_D(u,x)$. From a proof of φ in* $\mathsf{LMSO}(\mathfrak{C})$ *one can extract an eager term $\mathtt{u}(x)$ such that* LMSO *proves $(\forall x^\sigma)\varphi_D(\mathtt{u}(x),x)$.*

Corollary 2 (Extraction). *Consider a closed formula $\varphi := (\forall x^\sigma)(\exists u^\tau)\delta(u,x)$ with δ deterministic. From a proof of φ in* $\mathsf{LMSO}(\mathfrak{C})$ *one can extract a term $\mathtt{t}(x)$ such that $\models (\forall x^\sigma)\lfloor\delta(\mathtt{t}(x),x)\rfloor$.*

4.3 Completeness of $\mathsf{LMSO}(\mathfrak{C})$

The completeness of $\mathsf{LMSO}(\mathfrak{C})$ follows from a couple of important facts. First, $\mathsf{LMSO}(\mathfrak{C})$ proves the elimination of linear double negation, using (via Theorem 3) the same trick as in [26].

Lemma 1. *For all* LMSO*-formula φ, we have $(\varphi \multimap \bot) \multimap \bot \vdash_{\mathsf{LMSO}(\mathfrak{C})} \varphi$.*

Combining Lemma 1 with (LAC) gives classical linear choice.

Corollary 3. $(\forall f)(\exists x)\varphi(x,(f)x) \vdash_{\mathsf{LMSO}(\mathfrak{C})} (\exists x)(\forall y)\varphi(x,y)$.

The key to the completeness of $\mathsf{LMSO}(\mathfrak{C})$ is the following quantifier inversion.

Lemma 2. $(\forall x^\sigma)\varphi(\mathtt{t}^\tau(x),x) \vdash_{\mathsf{LMSO}(\mathfrak{C})} (\exists u^\tau)(\forall x^\sigma)\varphi(u,x)$, *where $\mathtt{t}(x)$ is eager.*

Lemma 2 follows (via Corollary 3) from the fixpoints on eager machines (Proposition 3). Fix an eager $\mathtt{t}^\tau(x^\sigma)$. Taking the fixpoint of $[\![(f)\mathtt{t}(x)]\!] : [\![\sigma]\!] \times [\![(\sigma)\tau]\!] \to_{\mathrm{EM}} [\![\sigma]\!]^{[\![(\sigma)\tau]\!]}$ gives a term $\mathtt{v}^\sigma(f^{(\sigma)\tau})$ such that $\mathtt{v}(f) \doteq @(f,\mathtt{t}(\mathtt{v}(f)))$. Then conclude with

$$\begin{array}{rll} (\forall x^\sigma)\varphi(\mathtt{t}(x),x) & \vdash_{\mathsf{LMSO}} & \varphi(\mathtt{t}(\mathtt{v}(f)),\mathtt{v}(f)) \\ & \vdash_{\mathsf{LMSO}} & \varphi(\mathtt{t}(\mathtt{v}(f)),@(f,\mathtt{t}(\mathtt{v}(f)))) \\ & \vdash_{\mathsf{LMSO}} & (\exists u^\tau)\varphi(u,(f)u) \\ & \vdash_{\mathsf{LMSO}} & (\forall f^{(\sigma)\tau})(\exists u^\tau)\varphi(u,(f)u) \\ & \vdash_{\mathsf{LMSO}(\mathfrak{C})} & (\exists u^\tau)(\forall x^\sigma)\varphi(u,x) \end{array}$$

Completeness of $\mathsf{LMSO}(\mathfrak{C})$ then follows via $(-)^D$, Proposition 5, completeness of $\mathsf{MSO}(\mathbf{M})$ and Büchi-Landweber Theorem 1. The idea is to lift a f.s. winning P-strat. in $\mathcal{G}(\lfloor\varphi_D(u,x)\rfloor)(u,x)$ to a realizer of $\varphi^D = (\exists u)(\forall x)\varphi_D(u,x)$ in $\mathsf{LMSO}(\mathfrak{C})$.

Theorem 6 (Completeness of $\mathsf{LMSO}(\mathfrak{C})$). *For each closed formula φ, either $\vdash_{\mathsf{LMSO}(\mathfrak{C})} \varphi$ or $\vdash_{\mathsf{LMSO}(\mathfrak{C})} \varphi \multimap \bot$.*

5 Conclusion

We provided a linear Dialectica-like interpretation of $\mathsf{LMSO}(\mathfrak{C})$, a linear variant of MSO on ω-words based on [26]. Our interpretation is correct and complete w.r.t. Church's synthesis, in the sense that it proves exactly the realizable $\forall\exists(-)^L$-statements. We thus obtain a syntactic extraction procedure with correctness proof internalized in $\mathsf{LMSO}(\mathfrak{C})$. The system $\mathsf{LMSO}(\mathfrak{C})$ is moreover complete in the sense that for every closed formula φ, it proves either φ or its linear negation. While completeness for a linear logic necessarily collapse some linear structure, the corresponding axioms (DEXP) and (PEXP) do respect the structural constraints allowing for realizer extraction from proofs. The completeness of $\mathsf{LMSO}(\mathfrak{C})$ contrasts with that of the classical system $\mathsf{MSO}(\mathbf{M})$, since the latter has provable unrealizable $\forall\exists$-statements. In particular, proof search in $\mathsf{LMSO}(\mathfrak{C})$ for $\forall\exists(-)^L$-formulae and their negation is correct and complete w.r.t. Church's synthesis. The design of the Dialectica interpretation also clarified the linear structure of LMSO, as it allowed us to decompose it starting from a system based on usual full intuitionistic linear logic (see e.g. [3] for recent references on the subject).

An outcome of witness extraction for $\mathsf{LMSO}(\mathfrak{C})$ is the realization of a simple version of the fan rule (in the usual sense of e.g. [16]). We plan to investigate monotone variants of Dialectica for our setting. Thanks to the compactness of Σ^ω, we expect this to allow extraction of uniform bounds, possibly with translations to stronger constructive logics than LMSO.

References

1. Avigad, J., Feferman, S.: Gödel's functional ("Dialectica") interpretation. In: Buss, S. (ed.) Handbook Proof Theory. Studies in Logic and the Foundations of Mathematics, vol. 137, pp. 337–405. Elsevier, Amsterdam (1998)
2. Baier, C., Katoen, J.P.: Principles of Model Checking. MIT Press, New York (2008)
3. Bellin, G., Heijltjes, W.: Proof nets for bi-intuitionistic linear logic. In: Kirchner, H. (ed.) FSCD. LIPIcs, vol. 108, pp. 10:1–10:18. Schloss Dagstuhl-Leibniz-Zentrum fuer Informatik, Dagstuhl, Germany (2018)
4. Birkedal, L., Møgelberg, R.E., Schwinghammer, J., Støvring, K.: First steps in synthetic guarded domain theory: step-indexing in the topos of trees. Logical Methods Comput. Sci. **8**(4), 1–45 (2012)
5. Bloem, R., Jobstmann, B., Piterman, N., Pnueli, A., Sa'ar, Y.: Synthesis of reactive (1) designs. J. Comput. Syst. Sci. **78**(3), 911–938 (2012)
6. Büchi, J.R.: On a decision method in restricted second-order arithmetic. In: Nagel, E., Suppes, P., Tarski, A. (eds.) Logic, Methodology and Philosophy of Science (Proc. 1960 Intern. Congr.), pp. 1–11. Stanford University Press, Stanford (1962)
7. Büchi, J.R., Landweber, L.H.: Solving sequential conditions by finite-state strategies. Transation Am. Math. Soc. **138**, 367–378 (1969)
8. Church, A.: Applications of recursive arithmetic to the problem of circuit synthesis. In: Summaries of the SISL. vol. 1, pp. 3–50. Cornell University, Ithaca (1957)
9. Danos, V., Joinet, J.B., Schellinx, H.: A new deconstructive logic: linear logic. J. Symb. Log. **62**(3), 755–807 (1997)

10. Emerson, E.A., Jutla, C.S.: Tree automata, mu-calculus and determinacy (extended abstract). In: FOCS. pp. 368–377. IEEE Computer Society (1991)
11. Filiot, E., Jin, N., Raskin, J.F.: Antichains and compositional algorithms for LTL synthesis. Form. Methods Syst. Des. **39**(3), 261–296 (2011)
12. Finkbeiner, B.: Synthesis of reactive systems. In: Esparza, J., Grumberg, O., Sickert, S. (eds.) Dependable Software Systems Engineering, NATO Science for Peace and Security Series, D: Information and Communication Security, vol. 45, pp. 72–98. IOS Press, Amsterdam (2016)
13. Girard, J.Y.: Linear logic. Theor. Comput. Sci. **50**, 1–102 (1987)
14. Hyland, J.M.E.: Proof theory in the abstract. Ann. Pure Appl. Logic **114**(1–3), 43–78 (2002)
15. Hyland, M., de Paiva, V.C.V.: Full intuitionistic linear logic (extended abstract). Ann. Pure Appl. Logic **64**(3), 273–291 (1993)
16. Kohlenbach, U.: Applied Proof Theory: Proof Interpretations and their Use in Mathematics. Springer Monographs in Mathematics. Springer, Heidelberg (2008). https://doi.org/10.1007/978-3-540-77533-1
17. Laurent, O., Regnier, L.: About translations of classical logic into polarized linear logic. In: Proceedings of LICS 2003, pp. 11–20. IEEE Computer Society Press (2003)
18. LLWiki: LLWiki (2008). http://llwiki.ens-lyon.fr/mediawiki/
19. McNaughton, R.: Testing and generating infinite sequences by a finite automaton. Inf. Control **9**(5), 521–530 (1966)
20. Muller, D.E., Schupp, P.E.: Simulating alternating tree automata by nondeterministic automata: new results and new proofs of theorems of Rabin, McNaughton and Safra. Theor. Comput. Sci. **141**(1&2), 69–107 (1995)
21. de Paiva, V.C.V.: The DIalectica categories. In: Proceedings of Categories in Computer Science and Logic, Boulder, CO, Contemporary Mathematics, vol. 92. American Mathematical Society (1987)
22. de Paiva, V.C.V.: A DIalectica-like model of linear logic. In: Pitt, D.H., Rydeheard, D.E., Dybjer, P., Pitts, A.M., Poigné, A. (eds.) Category Theory and Computer Science. LNCS, vol. 389, pp. 341–356. Springer, Heidelberg (1989). https://doi.org/10.1007/BFb0018360
23. de Paiva, V.C.V.: The DIalectica categories. Technical report 213, University of Cambridge Computer Laboratory, January 1991
24. Perrin, D., Pin, J.É.: Infinite Words: Automata, Semigroups, Logic and Games. Pure and Applied Mathematics. Elsevier (2004)
25. Pradic, P., Riba, C.: A Curry-Howard approach to Church's synthesis. In: Proceedings of FSCD 2017. LIPIcs, vol. 84, pp. 30:1–30:16. Schloss Dagstuhl - Leibniz-Zentrum fuer Informatik (2017)
26. Pradic, P., Riba, C.: LMSO: a Curry-Howard approach to Church's synthesis via linear logic. In: Proceedings of LICS 2018. ACM (2018)
27. Shirahata, M.: The DIalectica interpretation of first-order classical affine logic. Theory Appl. Categ. **17**(4), 49–79 (2006)
28. Siefkes, D.: Decidable Theories I: Büchi's Monadic Second Order Successor Arithmetic. LNM, vol. 120. Springer, Heidelberg (1970). https://doi.org/10.1007/BFb0061047
29. Thomas, W.: Automata on infinite objects. In: van Leeuwen, J. (ed.) Handbook of Theoretical Computer Science, vol. B: Formal Models and Semantics, pp. 133–192. Elsevier Science Publishers (1990)

30. Thomas, W.: Languages, automata, and logic. In: Rozenberg, G., Salomaa, A. (eds.) Handbook of Formal Languages, vol. III, pp. 389–455. Springer, Heidelberg (1997). https://doi.org/10.1007/978-3-642-59126-6_7
31. Thomas, W.: Solution of Church's problem: a tutorial. New Perspect. Games Interact. **5**, 23 (2008)
32. Thomas, W.: Facets of synthesis: revisiting Church's problem. In: de Alfaro, L. (ed.) FoSSaCS 2009. LNCS, vol. 5504, pp. 1–14. Springer, Heidelberg (2009). https://doi.org/10.1007/978-3-642-00596-1_1
33. Troelstra, A.: Metamathematical Investigation of Intuitionistic Arithmetic and Analysis. LNM, vol. 344. Springer, Heidelberg (1973). https://doi.org/10.1007/BFb0066739

Deciding Equivalence of Separated Non-nested Attribute Systems in Polynomial Time

Helmut Seidl[1], Raphaela Palenta[1(✉)], and Sebastian Maneth[2]

[1] Fakultät für Informatik, TU München, Munich, Germany
{seidl,palenta}@in.tum.de
[2] FB3 - Informatik, Universität Bremen, Bremen, Germany
maneth@uni-bremen.de

Abstract. In 1982, Courcelle and Franchi-Zannettacci showed that the equivalence problem of separated non-nested attribute systems can be reduced to the equivalence problem of total deterministic separated basic macro tree transducers. They also gave a procedure for deciding equivalence of transducer in the latter class. Here, we reconsider this equivalence problem. We present a new alternative decision procedure and prove that it runs in polynomial time. We also consider extensions of this result to partial transducers and to the case where parameters of transducers accumulate strings instead of trees.

1 Introduction

Attribute grammars are a well-established formalism for realizing computations on syntax trees [20,21], and implementations are available for various programming languages, see, e.g. [12,28,29]. A fundamental question for any such specification formalism is whether two specifications are semantically equivalent. As a particular case, attribute grammars have been considered which compute uninterpreted trees. Such devices that translate input trees (viz. the parse trees of a context-free grammar) into output trees, have also been studied under the name "attributed tree transducer" [14] (see also [15]). In 1982, Courcelle and Franchi-Zannettacci showed that the equivalence problem for (strongly noncircular) attribute systems reduces to the equivalence problem for primitive recursive schemes with parameters [3]; the latter model is also known under the name *macro tree transducer* [9]. Whether or not equivalence of attributed tree transducers (ATTs) or of (deterministic) macro tree transducers (MTTs) is decidable remain two intriguing (and very difficult) open problems.

For several subclasses of ATTs it has been proven that equivalence is decidable. The most general and very recent result that covers almost all other known ones about deterministic tree transducers is that "deterministic top-down tree-to-string transducers" have decidable equivalence [27]. Notice that the complexity of this problem remains unknown (the decidability is proved via two semi-decision procedures). The only result concerning deterministic tree transducers

M. Bojańczyk and A. Simpson (Eds.): FOSSACS 2019, LNCS 11425, pp. 488–504, 2019.
https://doi.org/10.1007/978-3-030-17127-8_28

that we are aware of and that is *not* covered by this general result, is the one by Courcelle and Franchi-Zannettacci about decidability of equivalence of "separated non-nested" ATTs (which they reduce to the same problem for "separated non-nested" MTTs). However, in their paper no statement is given concerning the complexity of the problem. In this paper we close this gap and study the complexity of deciding equivalence of separated non-nested MTTs. To do so we propose a new approach that we feel is simpler and easier to understand than the one of [3]. Using our approach we can prove that the problem can be solved in polynomial time.

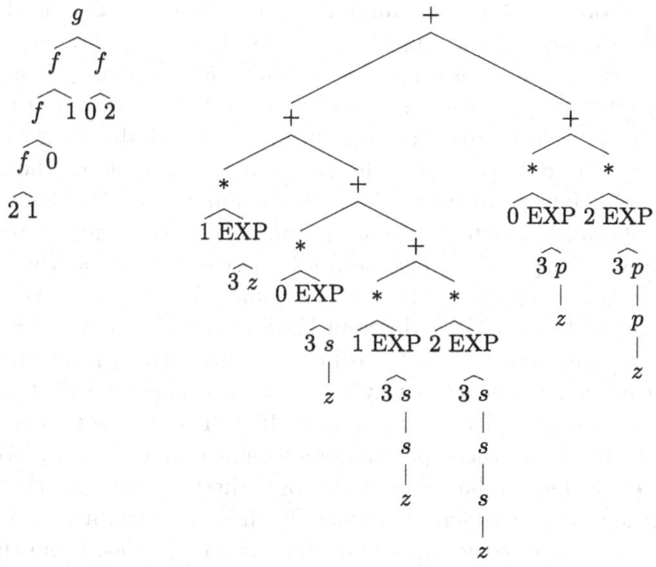

Fig. 1. Input tree for 2101.01 (in ternary) and corresponding output tree of M_{tern}.

In a separated non-nested attribute system, distinct sets of operators are used for the construction of inherited and synthesized attributes, respectively, and inherited attributes may depend on inherited attributes only. Courcelle and Franchi-Zannettacci's algorithm first translates separated non-nested attribute grammars into separated total deterministic non-nested macro tree transducers. In the sequel we will use the more established term *basic* macro-tree transducers instead of non-nested MTTs. Here, a macro tree transducer is called *separated* if the alphabets used for the construction of parameter values and outside of parameter positions are disjoint. And the MTT is *basic* if there is no nesting of state calls, i.e., there are no state calls inside of parameter positions. Let us consider an example. We want to translate ternary numbers into expressions over $+$, $*$, EXP, plus the constants 0, 1, and 2. Additionally, operators s, p, and z are used to represent integers in unary. The ternary numbers are parsed into particular binary trees; e.g., the left of Fig. 1 shows the binary tree for the

$$\begin{array}{lll} q_0(g(x_1,x_2)) & \rightarrow & +(q(x_1,z),q'(x_2,p(z))) \\ q(f(x_1,x_2),y) & \rightarrow & +(r(x_2,y),q(x_1,s(y))) \\ q'(f(x_1,x_2),y) & \rightarrow & +(r(x_1,y),q'(x_2,p(y))) \\ \phi(i,y) & \rightarrow & *(i,\mathrm{EXP}(3,y)) \quad \text{for } i \in \{0,1,2\}, \phi \in \{q,q',r\} \end{array}$$

Fig. 2. Rules of the transducer M_{tern}.

number 2101.02. This tree is translated by our MTT into the tree in the right of Fig. 1 (which indeed evaluates to $64.\overline{2}$ in decimal). The rules of our transducer M_{tern} are shown in Fig. 2. The example is similar to the one used by Knuth [20] in order to introduce attribute grammars. The transducer is indeed basic and separated: the operators p, s, and z are only used in parameter positions.

Our polynomial time decision procedure works in two phases: first, the transducer is converted into an "earliest" normal form. In this form, output symbols that are not produced within parameter positions are produced as early as possible. In particular it means that the root output symbols of the right-hand sides of rules for one state must differ. For instance, our transducer M_{tern} is *not* earliest, because all three r-rules produce the same output root symbol ($*$). Intuitively, this symbol should be produced earlier, e.g., at the place when the state r is called. The earliest form is a common technique used for normal forms and equivalence testing of different kinds of tree transducers [8,13,22]. We show that equivalent states of a transducer in this earliest form produce their state-output exactly in the same way. This means especially that the output of parameters is produced in the same places. It is therefore left to check, in the second phase, that also these parameter outputs are equivalent. To this end, we build an equivalence relation on states of earliest transducers that combines the two equivalence tests described before. Technically speaking, the equivalence relation is tested by constructing sets of Herbrand equalities. From these equalities, a fixed point algorithm can, after polynomially many iterations, produce a stable set of equalities.

The proofs of Lemmata 1 and 2 can be found in the appendix of an extended version at http://arxiv.org/abs/1902.03858.

2 Separated Basic Macro Tree Transducers

Let Σ be a ranked alphabet, i.e., every symbol of the finite set Σ has associated with it a fixed rank $k \in \mathbb{N}$. Generally, we assume that the input alphabet Σ is *non-trivial*, i.e., Σ has cardinality at least 2, and contains at least one symbol of rank 0 and at least one symbol of rank > 0. The set $\mathcal{T}_\Sigma$ is the set of all (finite, ordered, rooted) trees over the alphabet Σ. We denote a tree as a string over Σ and parenthesis and commas, i.e., $f(a, f(a, b))$ is a tree over Σ, where f is of rank 2 and a, b are of rank zero. We use Dewey dotted decimal notation to refer to a node of a tree: The root node is denoted ε, and for a node u, its i-th child is denoted by $u.i$. For instance, in the tree $f(a, f(a, b))$ the b-node is at position 2.2. A *pattern* (or k-pattern) (over Δ) is a tree $p \in \mathcal{T}_{\Delta\cup\{\top\}}$ over a ranked alphabet Δ and a disjoint symbol $\top$ (with exactly k occurrences of the symbol $\top$). The occurrences of the dedicated symbol $\top$ serve as place holders for other patterns. Assume that p is a k-pattern and that $p_1, \ldots, p_k$ are patterns; then $p[p_1, \ldots, p_k]$ denotes the pattern obtained from p by replacing, for $i = 1, \ldots, k$, the i-th occurrence (from left-to-right) of $\top$ by the pattern p_i.

A *macro tree transducer* (*MTT*) M is a tuple $(Q, \Sigma, \Delta, \delta)$ where Q is a ranked alphabet of states, Σ and Δ are the ranked input and output alphabets, respectively, and δ is a finite set of rules of the form:

$$q(f(x_1, \ldots, x_k), y_1, \ldots, y_l) \to T \tag{1}$$

where $q \in Q$ is a state of rank $l+1$, $l \geq 0$, $f \in \Sigma$ is an input symbol of rank $k \geq 0$, $x_1, \ldots, x_k$ and $y_1, \ldots, y_l$ are the formal input and output parameters, respectively, and T is a tree built up according to the following grammar:

$$T ::= a(T_1, \ldots, T_m) \mid q'(x_i, T_1, \ldots, T_n) \mid y_j$$

for output symbols $a \in \Delta$ of rank $m \geq 0$ and states $q' \in Q$ of rank $n+1$, input parameter x_i with $1 \leq i \leq k$, and output parameter y_j with $1 \leq j \leq l$. For simplicity, we assume that all states q have the same number l of parameters. Our definition of an MTT does not contain an initial state. We therefore consider an MTT always together with an axiom $A = p[q_1(x_1, \underline{T_1}), \ldots, q_m(x_1, \underline{T_m})]$ where $\underline{T_1}, \ldots, \underline{T_m} \in \mathcal{T}_\Delta^l$ are vectors of output trees (of length l each). Sometimes we only use an MTT M without explicitly mentioning an axiom A, then some A is assumed implicitly. Intuitively, the state q of an MTT corresponds to a function in a functional language which is defined through pattern matching over its first argument, and which constructs tree output using tree top-concatenation only; the second to $(l+1)$-th arguments of state q are its accumulating output parameters. The output produced by a state for a given input tree is determined by the right-hand side T of a rule of the transducer which matches the root symbol f of the current input tree. This right-hand side is built up from accumulating output parameters and calls to states for subtrees of the input and applications of output symbols from Δ. In general MTTs are nondeterministic and only partially defined. Here, however, we concentrate on total deterministic transducers. The MTT M is *deterministic*, if for every $(q, f) \in Q \times \Sigma$ there is at most one rule of the form (1). The MTT M is *total*, if for every $(q, f) \in Q \times \Sigma$ there is at least one rule of the form (1). For total deterministic transducers, the semantics of a state $q \in Q$ with the rule $q(f(x_1, \ldots, x_k), y_1, \ldots, y_l) \to T$ can be considered as a function

$$[\![q]\!] : \mathcal{T}_\Sigma \times \mathcal{T}_\Delta^l \to \mathcal{T}_\Delta$$

which inductively is defined by:

$$[\![q]\!](f(t_1, \ldots, t_k), \underline{S}) = [\![T]\!]\,(t_1, \ldots, t_k)\,\underline{S}$$

where

$$[\![a(T_1, \ldots, T_m)]\!]\,\underline{t}\,\underline{S} = a([\![T_1]\!]\,\underline{t}\,\underline{S}, \ldots, [\![T_m]\!]\,\underline{t}\,\underline{S})$$

$$[\![y_j]\!]\,\underline{t}\,\underline{S} = S_j$$
$$[\![q'(x_i, T_1, \ldots, T_l)]\!]\,\underline{t}\,\underline{S} = [\![q']\!](t_i, [\![T_1]\!]\,\underline{t}\,\underline{S}, \ldots, [\![T_l]\!]\,\underline{t}\,\underline{S})$$

where $\underline{S} = (S_1, \ldots, S_l) \in \mathcal{T}_\Delta^l$ is a vector of output trees. The semantics of a pair (M, A) with MTT M and axiom $A = p[q_1(x_1, \underline{T_1}), \ldots, q_m(x_1, \underline{T_m})]$ is defined by $[\![(M, A)]\!](t) = p[[\![q_1]\!](t, \underline{T_1}), \ldots, [\![q_m]\!](t, \underline{T_m})]$. Two pairs (M_1, A_1), (M_2, A_2) consisting of MTTs M_1, M_2 and corresponding axioms A_1, A_2 are *equivalent*, $(M_1, A_1) \equiv (M_2, A_2)$, iff for all input trees $t \in \mathcal{T}_\Sigma$, and parameter values $\underline{T} \in \mathcal{T}^l_{\Delta_{in}}$, $[\![(M_1, A_1)]\!](t, \underline{T}) = [\![(M_2, A_2)]\!](t, \underline{T})$.

The MTT M is *basic*, if each argument tree T_j of a subtree $q'(x_i, T_1, \ldots, T_n)$ of right-hand sides T of rules (1) may not contain further occurrences of states, i.e., is in $\mathcal{T}_{\Delta \cup Y}$. The MTT M is *separated basic*, if M is basic, and Δ is the disjoint union of ranked alphabets Δ_{out} and Δ_{in} so that the argument trees T_j of subtrees $q'(x_i, T_1, \ldots, T_n)$ are in $\mathcal{T}_{\Delta_{in} \cup Y}$, while the output symbols a outside of such subtrees are from Δ_{out}. The same must hold for the axiom. Thus, letters directly produced by a state call are in Δ_{out} while letters produced in the parameters are in Δ_{in}. The MTT M_{tern} from the Introduction is separated basic with $\Delta_{out} = \{0, 1, 2, 3, *, +, \text{EXP}\}$ and $\Delta_{in} = \{p, s, z\}$.

As separated basic MTTs are in the focus of our interests, we make the grammar for their right-hand side trees T explicit:

$$\begin{aligned} T &::= a(T_1, \ldots, T_m) \mid y_j \mid q'(x_i, T'_1, \ldots, T'_n) \\ T' &::= b(T'_1, \ldots, T'_{m'}) \mid y_j \end{aligned}$$

where $a \in \Delta_{out}$, $q' \in Q$, $b \in \Delta_{in}$ of ranks $m, n+1$ and m', respectively, and p is an n-pattern over Δ. For separated basic MTTs only axioms $A = p[q_1(x_1, \underline{T_1}), \ldots, q_m(x_1, \underline{T_m})]$ with $T_1, \ldots, T_m \in \mathcal{T}^l_{\Delta_{in}}$ are considered.

Note that equivalence of nondeterministic transducers is undecidable (even already for very small subclasses of transductions [18]). Therefore, we assume for the rest of the paper that all MTTs are deterministic and separated basic. We will also assume that all MTTs are total, with the exception of Sect. 5 where we also consider partial MTTs.

Example 1. We reconsider the example from the Introduction and adjust it to our formal definition. The transducer was given without an axiom (but with a tacitly assumed "start state" q_0). Let us now remove the state q_0 and add the axiom $A = q(x_1, z)$. The new q rule for g is:

$$q(g(x_1, x_2), y) \to +(q(x_1, y), q'(x_2, p(y))).$$

To make the transducer total, we add for state q' the rule

$$q'(g(x_1, x_2), y) \to +(*(0, \text{EXP}(3, y)), *(0, \text{EXP}(3, y))).$$

For state r we add rules $q(\alpha(x_1, x_2), y) \to *(0, \text{EXP}(3, y))$ with $\alpha = f, g$. The MTT is separated basic with $\Delta_{out} = \{0, 1, 2, 3, *, +, \text{EXP}\}$ and $\Delta_{in} = \{p, s, z\}$. □

We restricted ourselves to *total* separated basic MTTs. However, we would like to be able to decide equivalence for *partial* transducers as well. For this reason we define now top-down tree automata, and will then decide equivalence of MTTs relative to some given DTA D. A *deterministic top-down tree automaton* (*DTA*) D is a tuple $(B, \Sigma, b_0, \delta_D)$ where B is a finite set of states, Σ is a ranked alphabet of input symbols, $b_0 \in B$ is the initial state, and δ_D is the partial transition function with rules of the form $b(f(x_1, \dots, x_k)) \to (b_1(x_1), \dots, b_k(x_k))$, where $b, b_1, \dots, b_k \in B$ and $f \in \Sigma$ of rank k. W.l.o.g. we always assume that all states b of a DTA are productive, i.e., $\mathsf{dom}(b) \neq \emptyset$. If we consider a MTT M relative to a DTA D we implicitly assume a mapping $\pi : Q \to B$, that maps each state of M to a state of D, then we consider for q only input trees in $\mathsf{dom}(\pi(q))$.

3 Top-Down Normalization of Transducers

In this section we show that each total deterministic basic separated MTT can be put into an "earliest" normal form relative to a fixed DTA D. Intuitively, state output (in Δ_{out}) is produced as early as possible for a transducer in the normal form. It can then be shown that two equivalent transducers in normal form produce their state output in exactly the same way.

Recall the definition of patterns as trees over $\mathcal{T}_{\Delta \cup \{\top\}}$. Substitution of $\top$-symbols by other patterns induces a partial ordering $\sqsubseteq$ over patterns by $p \sqsubseteq p'$ if and only if $p = p'[p_1, \dots, p_m]$ for some patterns $p_1, \dots, p_m$. W.r.t. this ordering, $\top$ is the *largest* element, while all patterns without occurrences of $\top$ are minimal. By adding an artificial *least* element $\bot$, the resulting partial ordering is in fact a *complete lattice*. Let us denote this complete lattice by $\mathcal{P}_\Delta$.

Let $\Delta = \Delta_{in} \cup \Delta_{out}$. For $T \in \mathcal{T}_{\Delta \cup Y}$, we define the *$\Delta_{out}$-prefix* as the pattern $p \in \mathcal{T}_{\Delta_{out} \cup \{\top\}}$ as follows. Assume that $T = a(T_1, \dots, T_m)$.

- If $a \in \Delta_{out}$, then $p = a(p_1, \dots, p_m)$ where for $j = 1, \dots, m$, p_j is the Δ_{out}-prefix of T_j.
- If $a \in \Delta_{in} \cup Y$, then $p = \top$.

By this definition, each tree $t \in \mathcal{T}_{\Delta \cup Y}$ can be uniquely decomposed into a Δ_{out}-prefix p and subtrees $t_1, \dots, t_m$ whose root symbols all are contained in $\Delta_{in} \cup Y$ such that $t = p[t_1, \dots, t_m]$.

Let M be a total separated basic MTT M, D be a given DTA. We define the Δ_{out}-prefix of a state q of M relative to D as the minimal pattern $p \in \mathcal{T}_{\Delta_{out} \cup \{\top\}}$ so that each tree $[\![q]\!](t, \underline{T})$, $t \in \mathsf{dom}(\pi(q)), \underline{T} \in \mathcal{T}_\Delta^l$, is of the form $p[T_1, \dots, T_m]$ for some sequence of subtrees $T_1, \dots, T_m \in \mathcal{T}_\Delta$. Let us denote this unique pattern p by $\mathsf{pref}_o(q)$. If $q(f, y_1, \dots, y_l) \to T$ is a rule of a separated basic MTT and there is an input tree $f(t_1, \dots, t_k) \in \mathsf{dom}(\pi(q))$ then $|\mathsf{pref}_o(q)| \leq |T|$.

Lemma 1. *Let M be a total separated basic MTT and D a given DTA. Let $t \in \mathsf{dom}(\pi(q))$ be a smallest input tree of a state q of M. The Δ_{out}-prefix of every state q of M relative to D can be computed in time $\mathcal{O}(|t| \cdot |M|)$.*

The proof is similar to the one of [8, Theorem 8] for top-down tree transducers. This construction can be carried over as, for the computation of Δ_{out}-prefixes, the precise contents of the output parameters y_j can be ignored.

Example 2. We compute the Δ_{out}-prefix of the MTT M from Example 1. We consider M relative to the trivial DTA D that consists only of one state b with $dom(b) = \mathcal{T}_\Sigma$. We therefore omit D in our example. We get the following system of in-equations: from the rules of state r we obtain $Y_r \sqsubseteq *(i, \text{EXP}(3, \top))$ with $i \in \{0, 1, 2\}$. From the rules of state q we obtain $Y_q \sqsubseteq +(Y_q, Y_{q'})$, $Y_q \sqsubseteq +(Y_r, Y_q)$ and $Y_q \sqsubseteq *(i, \text{EXP}(3, \top))$ with $i \in \{0, 1, 2\}$. From the rules of state q' we obtain $Y_{q'} \sqsubseteq +(*(0, \text{EXP}(3, \top)), *(0, \text{EXP}(3, \top)))$, $Y_{q'} \sqsubseteq +(Y_r, Y_{q'})$ and $Y_{q'} \sqsubseteq *(i, \text{EXP}(3, \top))$ with $i \in \{0, 1, 2\}$. For the fixpoint iteration we initialize $Y_r^{(0)}$, $Y_q^{(0)}$, $Y_{q'}^{(0)}$ with $\bot$ each. Then $Y_r^{(1)} = *(\top, \text{EXP}(3, \top)) = Y_r^{(2)}$ and $Y_q^{(1)} = \top$, $Y_{q'}^{(1)} = \top$. Thus, the fixpoint iteration ends after two rounds with the solution $\mathsf{pref}_o(q) = \top$. □

Let M be a separated basic MTT M and D be a given DTA D. M is called D-earliest if for every state $q \in Q$ the Δ_{out}-prefix with respect to $\pi(q)$ is $\top$.

Lemma 2. *For every pair (M, A) consisting of a total separated basic MTT M and axiom A and a given DTA D, an equivalent pair (M', A') can be constructed so that M' is a total separated basic MTT that is D-earliest. Let t be an output tree of (M, A) for a smallest input tree $t' \in \mathsf{dom}(\pi(q))$ where q is the state occurring in A. Then the construction runs in time $\mathcal{O}(|t| \cdot |(M, A)|)$.*

The construction follows the same line as the one for the earliest form of top-down tree transducer, cf. [8, Theorem 11]. Note that for partial separated basic MTTs the size of the Δ_{out}-prefixes is at most exponential in the size of the transducer. However for total transducer that we consider here the Δ_{out}-prefixes are linear in the size of the transducer and can be computed in quadratic time, cf. [8].

Corollary 1. *For (M, A) consisting of a total deterministic separated basic MTT M and axiom A and the trivial DTA D accepting $\mathcal{T}_\Sigma$ an equivalent pair (M', A') can be constructed in quadratic time such that M' is an D-earliest total deterministic separated basic MTT.*

Example 3. We construct an equivalent earliest MTT M' for the transducer from Example 1. In Example 2 we already computed the Δ_{out}-prefixes of states q, q', r; $\mathsf{pref}_o(q) = \top$, $\mathsf{pref}_o(q') = \top$ and $\mathsf{pref}_o(r) = *(\top, \text{EXP}(3, \top))$. As there is only one occurrence of symbol $\top$ in the Δ_{out}-prefixes of q and q' we call states $\langle q, 1\rangle$ and $\langle q', 1\rangle$ by q and q', respectively. Hence, a corresponding earliest transducer has axiom $A = q(x_1, z)$. The rules of q and q' for input symbol g do not change. For input symbol f we obtain

$$q(f(x_1, x_2), y) \to +(*(r(x_2, y), \text{EXP}(3, y)), q(x1, s(y))) \quad \text{and}$$
$$q'(f(x_1, x_2), y) \to +(*(r(x_1, y), \text{EXP}(3, y), q'(x_2, p(y))).$$

As there is only one occurrence of symbol $\top$ related to a recursive call in $\mathsf{pref}_o(r)$ we call $\langle r, 1\rangle$ by r. For state r we obtain new rules $r(\alpha(x_1, x_2), y) \to 0$ with $\alpha \in \{f, g\}$ and $r(i, y) \to i$ with $i \in \{0, 1, 2\}$. □

We define a family of equivalence relation by induction, $\cong_b \subseteq ((Q, \mathcal{T}^k_{\Delta_{in}}) \cup \mathcal{T}_{\Delta_{in}}) \times ((Q, \mathcal{T}^k_{\Delta_{in}}) \cup \mathcal{T}_{\Delta_{in}})$ with b a state of a given DTA is the intersection of the equivalence relations $\cong_b^{(h)}$, i.e., $X \cong_b Z$ if and only if for all $h \geq 0$, $X \cong_b^{(h)} Z$. We let $(q, \underline{T}) \cong_b^{(h+1)} (q', \underline{T'})$ if for all $f \in \mathsf{dom}(b)$ with $b(f(x_1, \dots, x_k) \to (b_1, \dots, b_k)$, there is a pattern p such that $q(f(x_1, \dots, x_k), \underline{y}) \to p[t_1, \dots, t_m]$ and $q'(f(x_1, \dots, x_k), \underline{y'}) \to p[t'_1, \dots, t'_m]$ with

- if t_i and t'_i are both recursive calls to the same subtree, i.e., $t_i = q_i(x_{j_i}, \underline{T_i})$, $t'_i = q'_i(x_{j'_i}, \underline{T'_i})$ and $j_i = j'_i$, then $(q_i, \underline{T_i})[\underline{T}/\underline{y}] \cong^h_{b_{j_i}} (q'_i, \underline{T'_i})[\underline{T'}/\underline{y'}]$
- if t_i and t'_i are both recursive calls but on different subtrees, i.e., $t_i = q_i(x_{j_i}, \underline{T_i})$, $t'_i = q'_i(x_{j'_i}, \underline{T'_i})$ and $j_i \neq j'_i$, then $\hat{t} := [\![q_i]\!](s, \underline{T_i})[\underline{T}/\underline{y}] = [\![q'_i]\!](s, \underline{T'_i})[\underline{T}/\underline{y}]$ for some $s \in \Sigma^{(0)}$ and $(q_i, \underline{T_i})[\underline{T}/\underline{y}] \cong^{(h)}_{b_{j_i}} \hat{t} \cong^{(h)}_{b_{j'_i}} (q'_i, \underline{T'_i})[\underline{T}/\underline{y}]$
- if t_i and t'_i are both parameter calls, i.e., $t_i = y_{j_i}$ and $t'_i = y'_{j'_i}$, then $T_{j_i} = T'_{j'_i}$
- if t_i is a parameter call and t'_i a recursive call, i.e., $t_i = y_{j_i}$ and $t'_i = q'_i(x_{j'_i}, \underline{T'_i})$, then $T_{j_i} \cong^{(h)}_{b_{j'_i}} (q'_i, \underline{T'_i})[\underline{T'}/\underline{y'}]$
- (symmetric to the latter case) if t_i is a recursive call and t'_i a parameter call, i.e., $t_i = q_i(x_{j_i}, \underline{T_i})$ and $t'_i = y'_{j'_i}$, then $(q_i, \underline{T_i})[\underline{T}/\underline{y}] \cong^{(h)}_{b_{j_i}} T'_{j'_i}$.

We let $T \cong_b^{(h+1)} (q', \underline{T'})$ if for all $f \in \mathsf{dom}(b)$ with $r(f(x_1, \dots, x_k)) \to (b_1, \dots, b_k)$, $q'(f(\underline{x}), \underline{y}) \to t'$,

- if $t' = y_j$ then $T = T'_j$
- if $t' = q'_1(x_i, \underline{T'_1})$ then $T \cong^{(h)}_{b_i} (q'_1, \underline{T'_1})[\underline{T'}/\underline{y'}]$.

Intuitively, $(q, \underline{T}) \cong^h_b (q', \underline{T'})$ if for all input trees $t \in \mathsf{dom}(b)$ of height h, $[\![q]\!](t, \underline{T}) = [\![q']\!](t, \underline{T'})$. Then $(q, \underline{T}) \cong_b (q', \underline{T'})$ if for *all* input trees $t \in \mathsf{dom}(b)$ (independent of the height), $[\![q]\!](t, \underline{T}) = [\![q']\!](t, \underline{T'})$.

Theorem 1. *For a given DTA D with initial state b, let M, M' be D-earliest total deterministic separated basic MTTs with axioms A and A', respectively. Then (M, A) is equivalent to (M', A') relative to D, iff there is a pattern p such that $A = p[q_1(x_1, \underline{T_1}), \dots, q_m(x_1, \underline{T_m})]$, and $A' = p[q'_1(x_1, \underline{T'_1}), \dots, q'_m(x_1, \underline{T'_m})]$ and for $j = 1, \dots, m$, $(q_j, \underline{T_j}) \cong_b (q'_j, \underline{T'_j})$, i.e., q_j and q'_j are equivalent on the values of output parameters $\underline{T}_j$ and $\underline{T'_j}$.*

Proof. Let Δ be the output alphabet of M and M'. Assume that $(M, A) \cong_b (M', A')$. As M and M' are earliest, the Δ_{out}-prefix of $[\![(M, A)]\!](t)$ and $[\![(M', A')]\!](t)$, for $t \in \mathsf{dom}(b)$ is the same pattern p and therefore $A = p[q_1(x_1, \underline{T_1}), \dots, q_m(x_1, \underline{T_m})]$ and $A' = p[q'_1(x_1, \underline{T'_1}), \dots, q'_m(x_1, \underline{T'_m})]$. To show that $(q_i, \underline{T_i}) \cong_b (q'_i, \underline{T'_i})$ let u_i be the position of the i-th $\top$-node in the pattern p. For some $t \in \mathsf{dom}(b)$ and $\underline{T} \in \mathcal{T}_{\Delta_{in}}$ let t_i and t'_i be the subtree of $[\![(M, A)]\!](t, \underline{T})$ and $[\![(M', A')]\!](t, \underline{T})$, respectively. Then $t_i = t'_i$ and therefore $(q_i, \underline{T_i}) \cong_b (q'_i, \underline{T'_i})$.

Now, assume that the axioms $A = p[q_1(x_1, \underline{T_1}), \dots, q_m(x_1, \underline{T_m})]$ and $A' = p[q'_1(x_1, \underline{T'_1}), \dots, q'_m(x_1, \underline{T'_m})]$ consist of the same pattern p and for $i = 1, \dots, m$, $(q_i, \underline{T_i}) \cong_b (q'_i, \underline{T'_i})$. Let $t \in \mathsf{dom}(b)$ be an input tree then

$$\begin{aligned}[\![(M,A)]\!](t) &= p[[\![q_1]\!](t,\underline{T_1}),\ldots,[\![q_m]\!](t,\underline{T_m})]\\ &= p[[\![q'_1]\!](t,\underline{T'_1}),\ldots,[\![q'_m]\!](t,\underline{T'_m})]\\ &= [\![(M',A')]\!](t).\end{aligned}$$

4 Polynomial Time

In this section we prove the main result of this paper, namely, that for each fixed DTA D, equivalence of total deterministic basic separated MTTs (relative to D) can be decided in polynomial time. This is achieved by taking as input two D-earliest such transducers, and then collecting conditions on the parameters of pairs of states of the respective transducers for their produced outputs to be equal.

Example 4. Consider a DTA D with a single state only which accepts all inputs, and states q, q' with

$$q(a, y_1, y_2) \to g(y_1) \qquad q'(a, y'_1, y'_2) \to g(y'_2)$$

Then q and q' can only produce identical outputs for the input a (in $\mathsf{dom}(b)$) if parameter y'_2 of q' contains the same output tree as parameter y_1 of q. This precondition can be formalized by the equality $y'_2 \doteq y_1$. Note that in order to distinguish the output parameters of q' from those of q we have used primed copies y'_i for q'. □

It turns out that *conjunctions* of equalities such as in Example 4 are sufficient for proving equivalence of states. For states q, q' of total separated basic MTTs M, M', respectively, that are both D-earliest for some fixed DTA D, $h \geq 0$ and some fresh variable z, we define

$$\Psi^{(h)}_{b,q}(z) = \bigwedge_{b(f\underline{x})\to(b_1,\ldots,b_k)} \quad \begin{array}{ll} \displaystyle\bigwedge_{q(f\underline{x},\underline{y})\to y_j} (z \doteq y_j) & \wedge \\ \displaystyle\bigwedge_{q(f\underline{x},\underline{y})\to \hat{q}(x_i,\underline{T})} \Psi^{(h-1)}_{b_i,\hat{q}}(z)[\underline{T}/\underline{y}] & \wedge \\ \displaystyle\bigwedge_{\substack{q(f\underline{x},\underline{y})\to p[\ldots]\\ p\neq\top}} \bot & \end{array}$$

where $\bot$ is the boolean value *false*. We denote the output parameters in $\Psi^{(h)}_{b,q}(z)$ by $\underline{y}$, we define $\Psi'^{(h)}_{b,q'}(z)$ in the same lines as $\Psi^{(h)}_{b,q}(z)$ but using $\underline{y'}$ for the output parameters. To substitute the output parameters with trees $\underline{T}$, $\underline{T'}$, we therefore use $\Psi^{(h)}_{b,q}(z)[\underline{T}/\underline{y}]$ and $\Psi'^{(h)}_{b,q'}(z)[\underline{T'}/\underline{y'}]$. Assuming that q is a state of the D-earliest separated basic MTT M then $\Psi^{(h)}_{b,q}(z)$ is true for all ground parameter values $\underline{s}$ and some $T \in \mathcal{T}_{\Delta\cup Y}$ if $[\![q]\!](t,\underline{s}) = T[\underline{s}/\underline{y}]$ for all input trees $t \in \mathsf{dom}(b)$ of height at most h. Note that, since M is D-earliest, T is necessarily in $\mathcal{T}_{\Delta_{in}\cup Y}$. W.l.o.g., we assume that every state b of D is productive, i.e., $\mathsf{dom}(b) \neq \emptyset$. For each state b of D, we therefore may choose some input tree $t_b \in \mathsf{dom}(b)$ of minimal depth. We define $s_{b,q}$ to be the output of q for a minimal input tree $t_r \in \mathsf{dom}(b)$ and parameter values $\underline{y}$—when considering formal output parameters as output symbols in Δ_{in}, i.e., $s_{b,q} = [\![q]\!](t_r, \underline{y})$.

Example 5. We consider again the trivial DTA D with only one state b that accepts all $t \in \mathcal{T}_\Sigma$. Thus, we may choose $t_b = a$. For a state q with the following two rules $q(a, y_1, y_2) \to y_1$ and $q(f(x), y_1, y_2) \to q(x, h(y_2), b)$, we have $s_{b,q} = y_1$. Moreover, we obtain

$$\begin{aligned}
\Psi_{b,q}^{(0)}(z) &= z \doteq y_1 \\
\Psi_{b,q}^{(1)}(z) &= (z \doteq y_1) \wedge (z \doteq h(y_2)) \\
\Psi_{b,q}^{(2)}(z) &= (z \doteq y_1) \wedge (z \doteq h(y_2)) \wedge (z \doteq h(b)) \\
&\equiv (y_2 \doteq b) \wedge (y_1 \doteq h(b)) \wedge (z \doteq h(b)) \\
\Psi_{b,q}^{(3)}(z) &= (z \doteq y_1) \wedge (b \doteq b) \wedge (h(y_2) \doteq h(b)) \wedge (z \doteq h(b)) \\
&\equiv (y_2 \doteq b) \wedge (y_1 \doteq h(b)) \wedge (z \doteq h(b))
\end{aligned}$$

We observe that $\Psi_{b,q}^{(2)}(z) = \Psi_{b,q}^{(3)}(z)$ and therefore for every $h \geq 2$, $\Psi_{b,q}^{(h)}(z) = \Psi_{b,q}^{(3)}(z)$. □

According to our equivalence relation $\cong_b$, b state of the DTA D, we define for states q, q' of D-earliest total deterministic separated basic MTTs M, M', and $h \geq 0$, the conjunction $\Phi_{b,(q,q')}^{(h)}$ by

$$\begin{aligned}
\bigwedge_{\substack{b(f\underline{x}) \to (b_1,\dots,b_k) \\ q(f\underline{x},\underline{y}) \to p[\underline{t}] \\ q'(f\underline{x},\underline{y'}) \to p[\underline{t'}]}} \Bigg(&\bigwedge_{\substack{t_i = y_{j_i}, \\ t'_i = y'_{j'_i}}} (y_{j_i} \doteq y'_{j'_i}) && \wedge \\
&\bigwedge_{\substack{t_i = y_{j_i}, \\ t'_i = q'_i(x_{j'_i}, \underline{T'})}} \Psi'^{(h-1)}_{b_{j'_i}, q'_i}(y_{j_i})[\underline{T'}/\underline{y'}] && \wedge \\
&\bigwedge_{\substack{t'_i = y'_{j'_i}, \\ t_i = q_i(x_{j_i}, \underline{T})}} \Psi^{(h-1)}_{b_{j_i}, q_i}(y'_{j'_i})[\underline{T}/\underline{y}] && \wedge \\
&\bigwedge_{\substack{t_i = q_i(x_{j_i}, \underline{T}), \\ t'_i = q'_i(x_{j'_i}, \underline{T'}) \\ j_i = j'_i}} \Phi^{(h-1)}_{b_{j_i}, (q_i, q'_i)}[\underline{T}/\underline{y}, \underline{T'}/\underline{y'}] && \wedge \\
&\bigwedge_{\substack{t_i = q_i(x_{j_i}, \underline{T}), \\ t'_i = q'_i(x_{j'_i}, \underline{T'}) \\ j_i \neq j'_i}} (\Psi^{(h-1)}_{b_{j_i}, q_i}(s_{b,q_i})[\underline{T}/\underline{y}] \wedge \Psi'^{(h-1)}_{b_{j'_i}, q'_i}(s_{b,q_i}[\underline{T}/\underline{y}])[\underline{T'}/\underline{y'}]) \Bigg) && \wedge \\
\bigwedge_{\substack{b(f) \to (b_1,\dots,b_k) \\ p \neq p', q(f\underline{x},\underline{y}) \to p[\underline{t}] \\ q'(f\underline{x},\underline{q}) \to p'[\underline{t'}]}} &\bot
\end{aligned}$$

$\Phi_{b,(q,q')}^{(h)}$ is defined in the same lines as the equivalence relation $\cong_b^{(h)}$. $\Phi_{b,(q,q')}^{(h)}$ is true for all values of output parameters $\underline{T}$, $\underline{T'}$ such that $[\![q]\!](t, \underline{T}) = [\![q']\!](t, \underline{T'})$ for $t \in \mathsf{dom}(b)$ of height at most h. By induction on $h \geq 0$, we obtain:

Lemma 3. *For a given DTA D, states q, q' of D-earliest total separated basic MTTs, vectors of trees $\underline{T}, \underline{T'}$ over Δ_{in}, b a state of D. $s \in \mathsf{dom}(b)$, and $h \geq 0$ the following two statements hold:*

$$(q, \underline{T}) \cong_b^{(h)} (q', \underline{T'}) \Leftrightarrow \Phi_{b,(q,q')}^{(h)}[\underline{T}/\underline{y}, \underline{T'}/\underline{y'}] \equiv \text{true}$$

$$s \cong_b^{(h)} (q', \underline{T'}) \Leftrightarrow \Psi_{b,q'}^{(h)}(t)[\underline{T'}/\underline{y}] \equiv \text{true}$$

□

$\Phi_{b,(q,q')}^{(h)}$ is a conjunction of equations of the form $y_i \doteq y_j$, $y_i \doteq t$ with $t \in \Delta_{in}$. Every satisfiable conjunction of equalities is equivalent to a (possible empty) finite conjunction of equations of the form $y_i \doteq t_i$, $t_i \in \mathcal{T}_{\Delta_{in} \cup Y}$ where the y_i are distinct and no equation is of the form $y_j \doteq y_j$. We call such conjunctions *reduced.* If we have two inequivalent reduced conjunctions ϕ_1 and ϕ_2 with $\phi_1 \Rightarrow \phi_2$ then ϕ_1 contains strictly more equations. From that follows that for every sequence $\phi_0 \Rightarrow \dots \phi_m$ of pairwise inequivalent reduced conjunctions ϕ_j with k variables, $m \leq k + 1$ holds. This observation is crucial for the termination of the fixpoint iteration we will use to compute $\Phi_{b,(q,q')}^{(h)}$.

For $h \geq 0$ we have:

$$\Psi_{b,q}^{(h)}(z) \Rightarrow \Psi_{b,q}^{(h-1)}(z) \tag{2}$$

$$\Phi_{b,(q,q')}^{(h)} \Rightarrow \Phi_{b,(q,q')}^{(h-1)} \tag{3}$$

As we fixed the number of output parameters to the number l, for each pair (q, q') the conjunction $\Phi_{b,(q,q')}^{(h)}$ contains at most $2l$ variables y_i, y_i'. Assuming that the MTTs to which state q and q' belong have n states each, we conclude that $\Phi_{b,(q,q')}^{(n^2(2l+1))} \equiv \Phi_{b,(q,q')}^{(n^2(2l+1)+i)}$ and $\Psi_{b,q}^{(n(l+1))} \equiv \Psi_{b,q}^{(n(l+1)+i)}$ for all $i \geq 0$. Thus, we can define $\Phi_{b,(q,q')} := \Phi_{b,(q,q')}^{(n^2(2l+1))}$ and $\Psi_{b,q} := \Psi_{b,q}^{(n(l+1))}$. As $(q, \underline{T}) \cong_b (q', \underline{T'})$ iff for all $h \geq 0$, $(q, \underline{T}) \cong_b^{(h)} (q', \underline{T'})$ holds, observation (3) implies that

$$(q, \underline{T}) \cong_b (q', \underline{T'}) \Leftrightarrow \Phi_{b,(q,q')}[\underline{T}/\underline{y}][\underline{T'}/\underline{y'}] \equiv \text{true}$$

Therefore, we have:

Lemma 4. *For a DTA D, states q, q' of D-earliest separated basic MTTs M, M' and states b of D, the formula $\Phi_{b,(q,q')}$ can be computed in time polynomial in the sizes of M and M'.*

Proof. We successively compute the conjunctions $\Psi_{b,q}^{(h)}(z), \Psi_{b,q'}^{(h)}(z), \Phi_{b,(q,q')}^{(h)}$, $h \geq 0$, for all states b, q, q'. As discussed before, some $h \leq n^2(2l+1)$ exists such that the conjunctions for $h+1$ are equivalent to the corresponding conjunctions for h—in which case, we terminate. It remains to prove that the conjunctions for h can be computed from the conjunctions for $h-1$ in polynomial time. For that, it is crucial that we maintain *reduced* conjunctions. Nonetheless, the *sizes* of occurring right-hand sides of equalities may be quite large. Consider for example the conjunction $x_1 \doteq a \wedge x_2 \doteq f(x_1, x_1) \wedge \ldots \wedge x_n \doteq f(x_{n-1}, x_{n-1})$. The corresponding reduced conjunction is then given by $x_1 \doteq a \wedge x_2 \doteq f(a, a) \wedge \ldots \wedge x_n \doteq f(f(f(\ldots(f(a,a))\ldots)$ where the sizes of right-hand sides grow exponentially. In order to arrive at a polynomial-size representation, we therefore rely on compact representations where isomorphic subtrees are represented only once. W.r.t. this representation, reduction of a non-reduced conjunction, implications between reduced conjunctions as well as substitution of variables in conjunctions can all be realized in polynomial time. From that, the assertion of the lemma follows.

Example 6. Let D be a DTA with the following rules $b(f(x)) \to (b)$, $b(g) \to ()$ and $b(h) \to ()$. Let q and q' be states of separated basic MTTs M, M', respectively, that are D-earliest and π, π' be the mappings from the states of D to the states of M, M' with $(b, q) \in \pi$ and $(b, q') \in \pi'$.

$$\begin{aligned} q(f(x), y_1, y_2) &\to a(q(x, b(y_1, y_1), c(y_2), d)) \\ q(g, y_1, y_2) &\to y_1 \\ q(h, y_1, y_2) &\to y_2 \end{aligned}$$

$$\begin{aligned} q'(f(x), y_1', y_2') &\to a(q'(x, c(y_1'), b(y_2', y_2'), d)) \\ q'(g, y_1', y_2') &\to y_2' \\ q'(h, y_1', y_2') &\to y_1' \end{aligned}$$

$$\begin{aligned} \Phi^{(0)}_{r,(q,q')} &= (y_1 \doteq y_2') \wedge (y_2 \doteq y_1') \\ \Phi^{(1)}_{r,(q,q')} &= (y_1 \doteq y_2') \wedge (y_2 \doteq y_1') \wedge (b(y_1, y_1) \doteq b(y_2', y_2')) \wedge (c(y_2) \doteq c(y_1')) \\ &\equiv (y_1 \doteq y_2') \wedge (y_2 \doteq y_1') = \Phi^{(0)}_{r,(q,q')} \end{aligned}$$

□

In summary, we obtain the main theorem of our paper.

Theorem 2. *Let (M, A) and (M', A') be pairs consisting of total deterministic separated basic MTTs M, M' and corresponding axioms A, A' and D a DTA. Then the equivalence of (M, A) and (M', A') relative to D is decidable. If D accepts all input trees, equivalence can be decided in polynomial time.*

Proof. By Lemma 2 we build pairs (M_1, A_1) and (M_1', A_1') that are equivalent to (M, A) and (M', A') where M_1, M_1' are D-earliest separated basic MTTs. If D is trivial the construction is in polynomial time, cf. Corollary 1. Let the axioms be $A_1 = p[q_1(x_{i_1}, \underline{T_1}), \ldots, q_k(x_{i_k}, \underline{T_k})]$ and $A_1' = p'[q_1'(x_{i_1'}, \underline{T_1}), \ldots, q_k'(x_{i_{k'}'}, \underline{T_{k'}})]$. According to Lemma 3 (M_1, A_1) and (M_1', A_1') are equivalent iff

- $p = p'$, $k = k'$ and
- for all $j = 1, \ldots, k$, $\Phi_{b,(q_j, q_j')}[\underline{T_j}/\underline{y}, \underline{T_j}/\underline{y'}]$ is equivalent to true.

By Lemma 4 we can decide the second statements in time polynomial in the sizes of M_1 and M_1'.

5 Applications

In this section we show several applications of our equivalence result. First, we consider partial transductions of separated basic MTTs. To decide the equivalence of partial transductions we need to decide (*a*) whether the domain of two given MTTs is the same and if so, (*b*) whether the transductions on this domain are the same. How the second part of the decision procedure is done was shown in detail in this paper if the domain is given by a DTA. It therefore remains to discuss how this DTA can be obtained. It was shown in [4, Theorem 3.1] that the domain of every top-down tree transducer T can be accepted by some DTA B_T and this automaton can be constructed from T in exponential time. This construction can easily be extended to basic MTTs. The decidability of equivalence of DTAs is well-known and can be done in polynomial time [16,17]. To obtain a total transducer we add for each pair (q, f), $q \in Q$ and $f \in \Sigma$ that has no rule a new rule $q(f(\underline{x}), \underline{y}) \to \bot$, where $\bot$ is an arbitrary symbol in Δ_{out} of rank zero.

Example 7. In Example 1 we discussed how to adjust the transducer from the introduction to our formal definition. We therefore had to introduce additional rules to obtain a total transducer. Now we still add rules for the same pairs (q, f) but only with right-hand sides $\bot$. Therefore the original domain of the transducer is given by a DTA $D = (R, \Sigma, r_0, \delta_D)$ with the rules $r_0(g(x_1, x_2)) \to (r(x_1), r(x_2))$, $r(f(x_1, x_2)) \to (r(x_1), r(x_2))$ and $r(i) \to (\,)$ for $i = 1, 2, 3$. □

Corollary 2. *The equivalence of deterministic separated basic MTTs with a* partial *transition function is decidable.*

Next, we show that our result can be used to decide the equivalence of total separated basic MTTs with look-ahead. A total macro tree transducer with regular look-ahead (MTT^{R}) is a tuple $(Q, \Sigma, \Delta, \delta, R, \delta_R)$ where R is a finite set of look-ahead states and δ_R is a total function from $R^k \to R$ for every $f \in \Sigma^{(k)}$. Additionally we have a deterministic bottom-up tree automaton $(P, \Sigma, \delta, -)$ (without final states). A rule of the MTT is of the form

$$q(f(t_1, \ldots, t_k), y_1, \ldots, y_k) \to t \qquad \langle p_1, \ldots, p_k \rangle$$

and is applicable to an input tree $f(t_1, \ldots, t_k)$ if the look-ahead automaton accepts t_i in state p_i for all $i = 1, \ldots, k$. For every $q, f, p_1, \ldots, p_k$ there is exactly one such rule. Let $N_1 = (Q_1, \Sigma_1, \Delta_1, \delta_1, R_1, \delta_{R1}), N_2 = (Q_2, \Sigma_2, \Delta_2, \delta_2, R_2, \delta_{R2})$ be two total separated basic MTTs with look-ahead. We construct total separated basic MTTs M_1, M_2 *without* look-ahead as follows. The input alphabet contains for every $f \in \Sigma$ and $r_1, \ldots, r_k \in R_1$, $r'_1, \ldots, r'_k \in R_2$ the symbols $\langle f, r_1, \ldots, r_k, r'_1, \ldots, r'_k \rangle$. For $q(f(x_1, \ldots, x_k), \underline{y}) \to p[T_1, \ldots, T_m]\ \langle r_1, \ldots, r_k \rangle$ and $q'(f(x_1, \ldots, x_k), \underline{y'}) \to p'[T'_1, \ldots, T'_m]\ \langle r'_1, \ldots, r'_k \rangle$ we obtain for M_1 the rules

$$\hat{q}(\langle f(x_1, \ldots, x_k), r_1, \ldots, r_k, r'_1, \ldots, r'_k \rangle, \underline{y}) \to p[\hat{T}_1, \ldots, \hat{T}_m]$$

with $\hat{T}_i = \hat{q}_i(\langle x_{j_i}, \hat{r_1}, \ldots, \hat{r_l}, \hat{r'_1}, \ldots, \hat{r'_l}\rangle, Z_i)$ if $T_i = q_i(x_{j_i}, Z_i)$ and $q_i(x_{j_i}, \underline{y}) \to \hat{T}_i \ \langle \hat{r_1}, \ldots, \hat{r_l}\rangle$ and $q'_i(x_{j_i}, \underline{y'}) \to \hat{T'_i} \ \langle \hat{r'_1}, \ldots, \hat{r'_l}\rangle$. If $T_i = y_{j_i}$ then $\hat{T}_i = y_{j_i}$. The total separated basic MTT M_2 is constructed in the same lines. Thus, N_i, $i = 1, 2$ can be simulated by M_i, $i = 1, 2$, respectively, if the input is restricted to the regular tree language of new input trees that represent correct runs of the look-ahead automata.

Corollary 3. *The equivalence of total separated basic MTTs with regular look-ahead is decidable in polynomial time.*

Last, we consider separated basic MTTs that concatenate strings instead of trees in the parameters. We abbreviate this class of transducers by $\mathrm{MTT}^{\mathrm{yp}}$. Thus, the alphabet Δ_{in} is not longer a ranked alphabet but a unranked alphabet which elements/letters can be concatenated to words. The procedure to decide equivalence of $\mathrm{MTT}^{\mathrm{yp}}$ is essentially the same as we discussed in this paper but instead of conjunctions of equations of trees over $\Delta_{in} \cup Y$ we obtain conjunctions equations of words. Equations of words is a well studied problem [23,24,26]. In particular, the confirmed Ehrenfeucht conjecture states that each conjunction of a set of word equations over a finite alphabet and using a finite number of variables, is equivalent to the conjunction of a finite subset of word equations [19]. Accordingly, by a similar argument as in Sect. 4, the sequences of conjunctions $\Psi^{(h)}_{b,q}(z), \Psi'^{(h)}_{b,q'}(z), \Phi^{(h)}_{b,(q,q')}, h \geq 0$, are ultimately stable. Using an encoding of words by integer matrices and applying techniques as in [19], we obtain:

Theorem 3. *The equivalence of total separated basic MTTs that concatenate words instead of trees in the parameters (Δ_{in} is unranked) is decidable.*

6 Related Work

For several subclasses of attribute systems equivalence is known to be decidable. For instance, attributed grammars without inherited attributes are equivalent to deterministic top-down tree transducers (DT) [3,5]. For this class equivalence was shown to be decidable by Esik [10]. Later, a simplified algorithm was provided in [8]. If the tree translation of an attribute grammar is of linear size increase, then equivalence is decidable, because it is decidable for deterministic macro tree transducers (DMTT) of linear size increase. This follows from the fact that the latter class coincides with the class of (deterministic) MSO definable tree translations (DMSOTT) [6] for which equivalence is decidable [7]. Figure 3 shows a Hasse diagram of classes of translations realized by certain deterministic tree transducers. The prefixes "l", "n", "sn", "b" and "sb" mean "linear size increase", "non-nested", "separated non-nested", "basic" and "separated basic", respectively. A minimal class where it is still open whether equivalence is decidable is the class of *non-nested* attribute systems (nATT) which, on the macro tree transducer side, is included in the class of *basic* deterministic macro tree transducers (bDMTT).

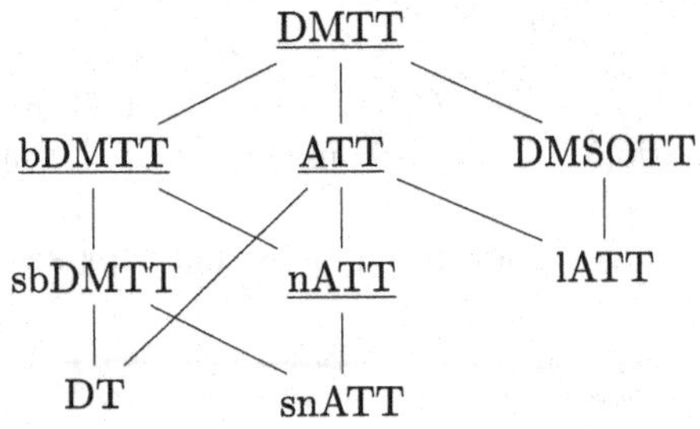

Fig. 3. Classes with and without (underlined) known decidability of equivalence

For deterministic top-down tree transducers, equivalence can be decided in EXPSPACE, and in NLOGSPACE if the transducers are total [25]. For the latter class of transducers, one can decide equivalence in polynomial time by transforming the transducer into a canonical normal form (called "earliest normal form") and then checking isomorphism of the resulting transducers [8]. In terms of hardness, we know that equivalence of deterministic top-down tree transducers is EXPTIME-hard. For linear size increase deterministic macro tree transducers the precise complexity is not known (but is at least NP-hard). More complexity results are known for other models of tree transducers such as streaming tree transducers [1], see [25] for a summary.

7 Conclusion

We have proved that the equivalence problem for separated non-nested attribute systems can be decided in polynomial time. In fact, we have shown a stronger statement, namely that in polynomial time equivalence of *separated basic total deterministic macro tree transducers* can be decided. To see that the latter is a strict superclass of the former, consider the translation that takes a binary tree as input, and outputs the same tree, but under each leaf a new monadic tree is output which represents the inverse Dewey path of that node. For instance, the tree $f(f(a,a),a)$ is translated into the tree $f(f(a(1(1(e))),a(2(1(e)))),a(2(e)))$. A macro tree transducer of the desired class can easily realize this translation using a rule of the form $q(f(x_{1,2}),y) \rightarrow f(q(x_1,1(y)),q(x_2,2(y)))$. In contrast, no attribute system can realize this translation. The reason is that for every attribute system, the number of distinct output subtrees is linearly bounded by the size of the input tree. For the given translation there is no linear such bound (it is bounded by $|s|\log(|s|)$).

The idea of "separated" to use different output alphabets, is related to the idea of transducers "with origin" [2,11]. In future work we would like to define adequate notions of origin for macro tree transducer, and prove that equivalence of such (deterministic) transducers with origin is decidable.

References

1. Alur, R., D'Antoni, L.: Streaming tree transducers. J. ACM **64**(5):31:1–31:55 (2017)
2. Bojańczyk, M.: Transducers with origin information. In: Esparza, J., Fraigniaud, P., Husfeldt, T., Koutsoupias, E. (eds.) ICALP 2014, Part II. LNCS, vol. 8573, pp. 26–37. Springer, Heidelberg (2014). https://doi.org/10.1007/978-3-662-43951-7_3
3. Courcelle, B., Franchi-Zannettacci, P.: Attribute grammars and recursive program schemes I. Theor. Comput. Sci. **17**(2), 163–191 (1982)
4. Engelfriet, J.: Top-down tree transducers with regular look-ahead. Math. Syst. Theory **10**, 289–303 (1977)
5. Engelfriet, J.: Some open questions and recent results on tree transducers and tree languages. In: Formal Language Theory, pp. 241–286. Elsevier (1980)
6. Engelfriet, J., Maneth, S.: Macro tree translations of linear size increase are MSO definable. SIAM J. Comput. **32**(4), 950–1006 (2003)
7. Engelfriet, J., Maneth, S.: The equivalence problem for deterministic MSO tree transducers is decidable. Inf. Process. Lett. **100**(5), 206–212 (2006)
8. Engelfriet, J., Maneth, S., Seidl, H.: Deciding equivalence of top-down XML transformations in polynomial time. J. Comput. Syst. Sci. **75**(5), 271–286 (2009)
9. Engelfriet, J., Vogler, H.: Macro tree transducers. J. Comput. Syst. Sci. **31**(1), 71–146 (1985)
10. Ésik, Z.: Decidability results concerning tree transducers I. Acta Cybern. **5**(1), 1–20 (1980)
11. Filiot, E., Maneth, S., Reynier, P., Talbot, J.: Decision problems of tree transducers with origin. Inf. Comput. 261(Part), 311–335 (2018)
12. Fors, N., Cedersjö, G., Hedin, G.: JavaRAG: a Java library for reference attribute grammars. In: Proceedings of the 14th International Conference on Modularity, MODULARITY 2015, pp. 55–67. ACM, New York (2015)
13. Friese, S., Seidl, H., Maneth, S.: Earliest normal form and minimization for bottom-up tree transducers. Int. J. Found. Comput. Sci. **22**(7), 1607–1623 (2011)
14. Fülöp, Z.: On attributed tree transducers. Acta Cybern. **5**(3), 261–279 (1981)
15. Fülöp, Z., Vogler, H.: Syntax-Directed Semantics - Formal Models Based on Tree Transducers. Monographs in Theoretical Computer Science. An EATCS Series. Springer, Heidelberg (1998). https://doi.org/10.1007/978-3-642-72248-6
16. Gécseg, F., Steinby, M.: Minimal ascending tree automata. Acta Cybern. **4**(1), 37–44 (1978)
17. Gécseg, F., Steinby, M.: Tree Automata. Akadéniai Kiadó, Budapest (1984)
18. Griffiths, T.V.: The unsolvability of the equivalence problem for lambda-free nondeterministic generalized machines. J. ACM **15**(3), 409–413 (1968)
19. Honkala, J.: A short solution for the HDT0L sequence equivalence problem. Theor. Comput. Sci. **244**(1–2), 267–270 (2000)
20. Knuth, D.E.: Semantics of context-free languages. Math. Syst. Theory **2**(2), 127–145 (1968)
21. Knuth, D.E.: Correction: semantics of context-free languages. Math. Syst. Theory **5**(1), 95–96 (1971)
22. Laurence, G., Lemay, A., Niehren, J., Staworko, S., Tommasi, M.: Normalization of sequential top-down tree-to-word transducers. In: Dediu, A.-H., Inenaga, S., Martín-Vide, C. (eds.) LATA 2011. LNCS, vol. 6638, pp. 354–365. Springer, Heidelberg (2011). https://doi.org/10.1007/978-3-642-21254-3_28

23. Lothaire, M.: Algebraic Combinatorics on Words. Cambridge University Press, Cambridge (2002)
24. Makanin, G.S.: The problem of solvability of equations in a free semigroup. Math. USSR-Sb. **32**(2), 129 (1977)
25. Maneth, S.: A survey on decidable equivalence problems for tree transducers. Int. J. Found. Comput. Sci. **26**(8), 1069–1100 (2015)
26. Plandowski, W.: Satisfiability of word equations with constants is in PSPACE. In: 40th Annual Symposium on Foundations of Computer Science, FOCS 1999, New York, NY, USA, 17–18 October 1999, pp. 495–500. IEEE Computer Society (1999)
27. Seidl, H., Maneth, S., Kemper, G.: Equivalence of deterministic top-down tree-to-string transducers is decidable. J. ACM **65**(4), 21:1–21:30 (2018)
28. Sloane, A.M., Kats, L.C., Visser, E.: A pure embedding of attribute grammars. Sci. Comput. Program. **78**(10), 1752–1769 (2013). Special Section on Language Descriptions Tools and Applications (LDTA 2008 and 2009) & Special Section on Software Engineering Aspects of Ubiquitous Computing and Ambient Intelligence (UCAm I 2011)
29. Van Wyk, E., Bodin, D., Gao, J., Krishnan, L.: Silver: an extensible attribute grammar system. Sci. Comput. Program. **75**(1–2), 39–54 (2010)

Justness
A Completeness Criterion for Capturing Liveness Properties (Extended Abstract)

Rob van Glabbeek[1,2(✉)]

[1] Data61, CSIRO, Sydney, Australia
[2] Computer Science and Engineering, University of New South Wales, Sydney, Australia
rvg@cs.stanford.edu

Abstract. This paper poses that transition systems constitute a good model of distributed systems only in combination with a criterion telling which paths model complete runs of the represented systems. Among such criteria, progress is too weak to capture relevant liveness properties, and fairness is often too strong; for typical applications we advocate the intermediate criterion of justness. Previously, we proposed a definition of justness in terms of an asymmetric concurrency relation between transitions. Here we define such a concurrency relation for the transition systems associated to the process algebra CCS as well as its extensions with broadcast communication and signals, thereby making these process algebras suitable for capturing liveness properties requiring justness.

1 Introduction

Transition systems are a common model for distributed systems. They consist of sets of states, also called *processes*, and transitions—each transition going from a source state to a target state. A given distributed system $\mathcal{D}$ corresponds to a state P in a transition system $\mathbb{T}$—the initial state of $\mathcal{D}$. The other states of $\mathcal{D}$ are the processes in $\mathbb{T}$ that are reachable from P by following the transitions. A run of $\mathcal{D}$ corresponds with a *path* in $\mathbb{T}$: a finite or infinite alternating sequence of states and transitions, starting with P, such that each transition goes from the state before to the state after it. Whereas each finite path in $\mathbb{T}$ starting from P models a *partial run* of $\mathcal{D}$, i.e., an initial segment of a (complete) run, typically not each path models a run. Therefore a transition system constitutes a good model of distributed systems only in combination with what we here call a *completeness criterion*: a selection of a subset of all paths as *complete paths*, modelling runs of the represented system.

A *liveness property* says that "something [good] must happen" eventually [18]. Such a property holds for a distributed system if the [good] thing happens in each of its possible runs. One of the ways to formalise this in terms of transition systems is to postulate a set of good states $\mathscr{G}$, and say that the liveness property $\mathscr{G}$ holds for the process P if all complete paths starting in P pass through a state

M. Bojańczyk and A. Simpson (Eds.): FOSSACS 2019, LNCS 11425, pp. 505–522, 2019.
https://doi.org/10.1007/978-3-030-17127-8_29

of $\mathscr{G}$ [16]. Without a completeness criterion the concept of a liveness property appears to be meaningless.

Example 1. The transition system on the right models Cataline eating a croissant in Paris. It abstracts from all activity in the world except the eating of that croissant, and thus has two states only—the states of the world before and after this event—and one transition t. We depict states by circles and transitions by arrows between them. An initial state is indicated by a short arrow without a source state. A possible liveness property says that the croissant will be eaten. It corresponds with the set of states $\mathscr{G}$ consisting of state 2 only. The states of $\mathscr{G}$ are indicated by shading.

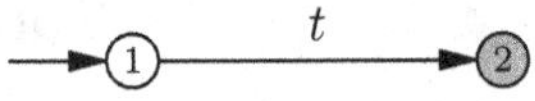

The depicted transition system has three paths starting with state 1: 1, $1\,t$ and $1\,t\,2$. The path $1\,t\,2$ models the run in which Cataline finishes the croissant. The path 1 models a run in which Cataline never starts eating the croissant, and the path $1\,t$ models a run in which Cataline starts eating it, but never finishes. The liveness property $\mathscr{G}$ holds only when using a completeness criterion that rules out the paths 1 and $1\,t$ as modelling actual runs of the system, leaving $1\,t\,2$ as the sole complete path.

The transitions of transition systems can be understood to model atomic actions that can be performed by the represented systems. Although we allow these actions to be instantaneous or durational, in the remainder of this paper we adopt the assumption that "atomic actions always terminate" [23]. This is a partial completeness criterion. It rules out the path $1\,t$ in Example 1. We build in this assumption in the definition of a path by henceforth requiring that finite paths should end with a state.

Progress. The most widely employed completeness criterion is *progress.*[1] In the context of *closed systems*, having no run-time interactions with the environment, it is the assumption that a run will never get stuck in a state with outgoing transitions. This rules out the path 1 in Example 1, as t is outgoing. When adopting progress as completeness criterion, the liveness property $\mathscr{G}$ holds for the system modelled in Example 1.

Progress is assumed in almost all work on process algebra that deals with liveness properties, mostly implicitly. Milner makes an explicit progress assumption for the process algebra CCS in [20]. A progress assumption is built into the temporal logics LTL [24], CTL [7] and CTL* [8], namely by disallowing states without outgoing transitions and evaluating temporal formulas by quantifying over infinite paths only.[2] In [17] the 'multiprogramming axiom' is a progress assumption, whereas in [1] progress is assumed as a 'fundamental liveness property'.

[1] Misra [21,22] calls this the 'minimal progress assumption'. In [22] he uses 'progress' as a synonym for 'liveness'. In session types, 'progress' and 'global progress' are used as names of particular liveness properties [4]; this use has no relation with ours.

[2] Exceptionally, states without outgoing transitions are allowed, and then quantification is over all *maximal* paths, i.e. paths that are infinite or end in a state without outgoing transitions [5].

As we argued in [10,15,16], a progress assumption as above is too strong in the context of reactive systems, meaning that it rules out as incomplete too many paths. There, a transition typically represents an interaction between the distributed system being modelled and its environment. In many cases a transition can occur only if both the modelled system *and* the environment are ready to engage in it. We therefore distinguish *blocking* and *non-blocking* transitions. A transition is non-blocking if the environment cannot or will not block it, so that its execution is entirely under the control of the system under consideration. A blocking transition on the other hand may fail to occur because the environment is not ready for it. The same was done earlier in the setting of Petri nets [26], where blocking and non-blocking transitions are called *cold* and *hot*, respectively.

In [10,15,16] we worked with transition systems that are equipped with a partitioning of the transitions into blocking and non-blocking ones, and reformulated the progress assumption as follows:

> *a (transition) system in a state that admits a non-blocking transition will eventually progress, i.e., perform a transition.*

In other words, a run will never get stuck in a state with outgoing non-blocking transitions. In Example 1, when adopting progress as our completeness criterion, we assume that Cataline actually wants to eat the croissant, and does not willingly remain in State 1 forever. When that assumption is unwarranted, one would model her behaviour by a transition system different from that of Example 1. However, she may still be stuck in State 1 by lack of any croissant to eat. If we want to model the capability of the environment to withhold a croissant, we classify t as a blocking transition, and the liveness property $\mathcal{G}$ does not hold. If we abstract from a possible shortage of croissants, t is deemed a non-blocking transition, and, when assuming progress, $\mathcal{G}$ holds.

As an alternative approach to a dogmatic division of transitions in a transition system, we could shift the status of transitions to the progress property, and speak of B-progress when B is the set of blocking transitions. In that approach, $\mathcal{G}$ holds for State 1 of Example 1 under the assumption of B-progress when $t \notin B$, but not when $t \in B$.

Justness. Justness is a completeness criterion proposed in [10,15,16]. It strengthens progress. It can be argued that once one adopts progress it makes sense to go a step further and adopt even justness.

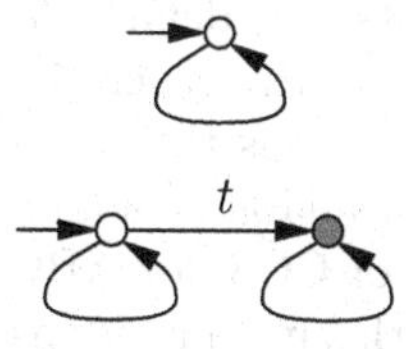

Example 2. The transition system on the top right models Alice making an unending sequence of phone calls in London. There is no interaction of any kind between Alice and Cataline. Yet, we may chose to abstracts from all activity in the world except the eating of the croissant by Cataline, and the making of calls by Alice. This yields the combined transition system on the bottom right. Even when taking the transition t to be non-blocking, progress is not a strong enough completeness criterion to ensure that Cataline will ever eat the croissant. For the infinite path

that loops in the first state is complete. Nevertheless, as nothing stops Cataline from making progress, in reality t will occur [16].

This example is not a contrived corner case, but a rather typical illustration of an issue that is central to the study of distributed systems. Other illustrations of this phenomena occur in [10, Section 9.1], [14, Section 10], [11, Section 1.4], [12] and [6, Section 4]. The criterion of justness aims to ensure the liveness property occurring in these examples. In [16] it is formulated as follows:

> *Once a non-blocking transition is enabled that stems from a set of parallel components, one (or more) of these components will eventually partake in a transition.*

In Example 2, t is a non-blocking transition enabled in the initial state. It stems from the single parallel component Cataline of the distributed system under consideration. Justness therefore requires that Cataline must partake in a transition. This can only be t, as all other transitions involve component Alice only. Hence justness says that t must occur. The infinite path starting in the initial state and not containing t is ruled out as unjust, and thereby incomplete.

In [13,16] we explain how justness is fundamentally different from fairness, and why fairness is too strong a completeness criterion for many applications.

Unlike progress, the concept of justness as formulated above is in need of some formalisation, i.e., to formally define a component, to make precise for concrete transition systems what it means for a transition to stem from a set of components, and to define when a component partakes in a transition.

A formalisation of justness for the transition system generated by the process algebra AWN, the *Algebra for Wireless Networks* [9], was provided in [10]. In the same vain, [15] offered a formalisation for the transition systems generated by CCS [20], and its extension ABC, the *Algebra of Broadcast Communication* [15], a variant of CBS, the *Calculus of Broadcasting Systems* [25]. The same was done for CCS extended with *signals* in [6]. These formalisations coinductively define *B-justness*, where B ranges over sets of transitions deemed to be blocking, as a family of predicates on paths, and proceed by a case distinction on the operators in the language. Although these definitions *do* capture the concept of justness formulated above, it is not easy to see why.

A more syntax-independent formalisation of justness occurs in [16]. There it is defined directly on transition systems equipped with a, possibly asymmetric, concurrency relation between transitions. However, the concurrency relation itself is defined only for the transition system generated by a fragment of CCS, and the generalisation to full CCS, and other process algebras, is non-trivial.

It is the purpose of this paper to make the definition of justness from [16] available to a large range of process algebras by defining the concurrency relation for CCS, for ABC, and for the extension of CCS with signals used in [6]. We do this in a precise as well as in an approximate way, and show that both approaches lead to the same concept of justness. Moreover, in all cases we establish a closure property on the concurrency relation ensuring that justness is a meaningful notion. We show that for all these algebras justness is *feasible*. Here feasibility is a requirement on completeness criteria advocated in [1,16,19]. Finally, we establish agreement between the formalisation of justness from [16] and the present paper, and the original coinductive ones from [15] and [6].

2 Labelled Transition Systems with Concurrency

We start with the formal definitions of a labelled transition system, a path, and the completeness criterion *progress*, which is parametrised by the choice of a collection B of blocking actions. Then we define the completeness criterion *justness* on labelled transition system upgraded with a concurrency relation.

Definition 1. A *labelled transition system* (LTS) is a tuple $(S, Tr, src, target, \ell)$ with S and Tr sets (of *states* and *transitions*), $src, target : Tr \to S$ and $\ell : Tr \to \mathscr{L}$, for some set of transition labels $\mathscr{L}$.

Here we work with LTSs labelled over a structured set of labels $(\mathscr{L}, Act, Rec)$, where $Rec \subseteq Act \subseteq \mathscr{L}$. Labels in Act are *actions*; the ones in $\mathscr{L} \setminus Act$ are *signals*. Transitions labelled with actions model a state chance in the represented system; signal transitions do not—they satisfy $src(t) = target(t)$ and merely convey a property of a state. $Rec \subseteq Act$ is the set of *receptive* actions; sets $B \subseteq Act$ of blocking actions must always contain Rec. In CCS and most other process algebras $Rec = \emptyset$ and $Act = \mathscr{L}$. Let $Tr^\bullet = \{t \in Tr \mid \ell(t) \in Act \setminus Rec\}$ be the set of transitions that are neither signals nor receptive.

Definition 2. A *path* in a transition system $(S, Tr, src, target)$ is an alternating sequence $s_0\, t_1\, s_1\, t_2\, s_2 \cdots$ of states and non-signal transitions, starting with a state and either being infinite or ending with a state, such that $src(t_i) = s_{i-1}$ and $target(t_i) = s_i$ for all relevant i.

A *completeness criterion* is a unary predicate on the paths in a transition system.

Definition 3. Let $B \subseteq Act$ be a set of actions with $Rec \subseteq B$—the *blocking* ones. Then $Tr^\bullet_{\neg B} := \{t \in Tr^\bullet \mid \ell(t) \notin B\}$ is the set of *non-blocking* transitions. A path in $\mathbb{T}$ is *B-progressing* if either it is infinite or its last state is the source of no non-blocking transition $t \in Tr^\bullet_{\neg B}$.

B-progress is a completeness criterion for any choice of $B \subseteq Act$ with $Rec \subseteq B$.

Definition 4. A *labelled transition system with concurrency* (LTSC) is a tuple $(S, Tr, src, target, \ell, \smile^\bullet)$ consisting of a LTS $(S, Tr, src, target, \ell)$ and a *concurrency relation* $\smile^\bullet \subseteq Tr^\bullet \times Tr$, such that:

$$t \not\smile^\bullet t \text{ for all } t \in Tr^\bullet, \tag{1}$$

if $t \in Tr^\bullet$ and π is a path from $src(t)$ to $s \in S$ such that $t \smile^\bullet v$ for all transitions v occurring in π, then there is a $u \in Tr^\bullet$ such that $src(u) = s$, $\ell(u) = \ell(t)$ and $t \not\smile^\bullet u$. (2)

Informally, $t \smile^{\bullet} v$ means that the transition v does not interfere with t, in the sense that it does not affect any resources that are needed by t, so that in a state where t and v are both possible, after doing v one can still do (a future variant u of) t. In many transition systems $\smile^{\bullet}$ is a symmetric relation, denoted $\smile$.

The transition relation in a labelled transition system is often defined as a relation $Tr \subseteq S \times \mathscr{L} \times S$. This approach is not suitable here, as we will encounter multiple transitions with the same source, target and label that ought to be distinguished based on their concurrency relations with other transitions.

Definition 5. A path π in an LTSC is *B-just*, for $Rec \subseteq B \subseteq Act$, if for each transition $t \in Tr^{\bullet}_{\neg B}$ with $s := src(t) \in \pi$, a transition u occurs in the suffix of π starting at s, such that $t \not\smile^{\bullet} u$.

Informally, justness requires that once a non-blocking non-signal transition t is enabled, sooner or later a transition u will occur that interferes with it, possibly t itself. Note that, for any $Rec \subseteq B \subseteq Act$, B-justness is a completeness criterion stronger than B-progress.

Components. Instead of introducing $\smile^{\bullet}$ as a primitive, it is possible to obtain it as a notion derived from two functions $npc, afc : Tr \rightarrow \mathscr{P}(\mathscr{C})$, for a given set of *components* $\mathscr{C}$. These functions could then be added as primitives to the definition of an LTS. They are based on the idea that a process represents a system built from parallel components. Each transition is obtained as a synchronisation of activities from some of these components. Now $npc(t)$ describes the (nonempty) set of components that are *necessary participants* in the execution of t, whereas $afc(t)$ describes the components that are *affected* by the execution of t. The concurrency relation is then defined by

$$t \smile^{\bullet} u \quad \Leftrightarrow \quad npc(t) \cap afc(u) = \emptyset$$

saying that u interferes with t iff a necessary participant in t is affected by u.

Most material above stems from [16]. However, there $Tr^{\bullet} = Tr$, so that $\smile^{\bullet}$ is irreflexive, i.e., $npc(t) \cap afc(t) \neq \emptyset$ for all $t \in Tr$. Moreover, a fixed set B is postulated, so that the notions of progress and justness are not explicitly parametrised with the choice of B. Furthermore, property (2) is new here; it is the weakest closure property that supports Theorem 1 below. In [16] only the model in which $\smile^{\bullet}$ is derived from npc and afc comes with a closure property:

$$\begin{array}{l} \text{If } t, v \in Tr^{\bullet} \text{ with } src(t) = src(v) \text{ and } npc(t) \cap afc(v) = \emptyset, \text{ then} \\ \exists u \in Tr^{\bullet} \text{ with } src(u) = target(v),\ \ell(u) = \ell(t) \text{ and } npc(u) = npc(t). \end{array} \tag{3}$$

Trivially (3) implies (2).

An important requirement on completeness criteria is that any finite path can be extended into a complete path. This requirement was proposed by Apt, Francez and Katz in [1] and called *feasibility*. It also appears in Lamport [19] under the name *machine closure*. The theorem below list conditions under which B-justness is feasible. Its proof is a variant of a similar theorem from [16] showing conditions under which notions of strong and weak fairness are feasible.

Table 1. Structural operational semantics of CCS

$$\alpha.P \xrightarrow{\alpha} P \ (\textsc{Act}) \qquad \frac{P \xrightarrow{\alpha} P'}{P+Q \xrightarrow{\alpha} P'}\ (\textsc{Sum-l}) \qquad \frac{Q \xrightarrow{\alpha} Q'}{P+Q \xrightarrow{\alpha} Q'}\ (\textsc{Sum-r})$$

$$\frac{P \xrightarrow{\eta} P'}{P|Q \xrightarrow{\eta} P'|Q}\ (\textsc{Par-l}) \qquad \frac{P \xrightarrow{c} P',\ Q \xrightarrow{\bar{c}} Q'}{P|Q \xrightarrow{\tau} P'|Q'}\ (\textsc{Comm}) \qquad \frac{Q \xrightarrow{\eta} Q'}{P|Q \xrightarrow{\eta} P|Q'}\ (\textsc{Par-r})$$

$$\frac{P \xrightarrow{\ell} P'}{P\backslash L \xrightarrow{\ell} P'\backslash L}\ (\ell,\bar{\ell} \notin L)\ (\textsc{Res}) \qquad \frac{P \xrightarrow{\ell} P'}{P[f] \xrightarrow{f(\ell)} P'[f]}\ (\textsc{Rel}) \qquad \frac{P \xrightarrow{\alpha} P'}{A \xrightarrow{\alpha} P'}\ (A \stackrel{def}{=} P)\ (\textsc{Rec})$$

Theorem 1. If, in an LTSC with set of blocking actions B, only countably many transitions from $Tr^{\bullet}_{\neg B}$ are enabled in each state, then B-justness is feasible.

All proofs can found in the full version of this paper [13].

3 CCS and Its Extensions with Broadcast and Signals

This section presents four process algebras: Milner's *Calculus of Communicating Systems* (CCS) [20], its extensions with broadcast communication ABC [15] and signals CCSS [6], and an alternative presentation of ABC that avoids negative premises in favour of *discard* transitions.

3.1 CCS

CCS [20] is parametrised with sets $\mathscr{A}$ of *agent identifiers* and $\mathscr{C}_h$ of *(handshake communication) names*; each $A \in \mathscr{A}$ comes with a defining equation $A \stackrel{def}{=} P$ with P being a CCS expression as defined below. $\bar{\mathscr{C}}_h := \{\bar{c} \mid c \in \mathscr{C}_h\}$ is the set of *co-names*. Complementation is extended to $\bar{\mathscr{C}}_h$ by setting $\bar{\bar{c}} = c$. $Act := \mathscr{C}_h \dot{\cup} \bar{\mathscr{C}}_h \dot{\cup} \{\tau\}$ is the set of *actions*, where τ is a special *internal action*. Below, c ranges over $\mathscr{C}_h \cup \bar{\mathscr{C}}_h$, η, α, ℓ over Act, and A, B over $\mathscr{A}$. A *relabelling* is a function $f: \mathscr{C}_h \rightarrow \mathscr{C}_h$; it extends to Act by $f(\bar{c}) = \overline{f(c)}$ and $f(\tau) := \tau$. The set $\mathbb{P}_{\mathrm{CCS}}$ of CCS expressions or *processes* is the smallest set including:

$\mathbf{0}$		*inaction*
$\alpha.P$	for $\alpha \in Act$ and $P \in \mathbb{P}_{\mathrm{CCS}}$	*action prefixing*
$P+Q$	for $P, Q \in \mathbb{P}_{\mathrm{CCS}}$	*choice*
$P\|Q$	for $P, Q \in \mathbb{P}_{\mathrm{CCS}}$	*parallel composition*
$P\backslash L$	for $L \subseteq \mathscr{C}_h$ and $P \in \mathbb{P}_{\mathrm{CCS}}$	*restriction*
$P[f]$	for f a relabelling and $P \in \mathbb{P}_{\mathrm{CCS}}$	*relabelling*
A	for $A \in \mathscr{A}$	*agent identifier*

One often abbreviates $\alpha.\mathbf{0}$ by α, and $P\backslash\{c\}$ by $P\backslash c$. The traditional semantics of CCS is given by the labelled transition relation $\rightarrow \subseteq \mathbb{P}_{\mathrm{CCS}} \times Act \times \mathbb{P}_{\mathrm{CCS}}$, where transitions $P \xrightarrow{\ell} Q$ are derived from the rules of Table 1.

Table 2. Structural operational semantics of ABC broadcast communication

(Bro-l)
$$\frac{P \stackrel{b\sharp_1}{\longrightarrow} P',\ Q \stackrel{b?}{\not\longrightarrow}}{P|Q \stackrel{b\sharp_1}{\longrightarrow} P'|Q}$$

(Bro-c)
$$\frac{P \stackrel{b\sharp_1}{\longrightarrow} P',\ Q \stackrel{b\sharp_2}{\longrightarrow} Q'}{P|Q \stackrel{b\sharp}{\longrightarrow} P'|Q'}\ \sharp_1\circ\sharp_2=\sharp\neq\_ \quad \text{with} \quad \begin{array}{c|cc} \circ & ! & ? \\ \hline ! & \_ & ! \\ ? & ! & ? \end{array}$$

(Bro-r)
$$\frac{P \stackrel{b?}{\not\longrightarrow},\ Q \stackrel{b\sharp_2}{\longrightarrow} Q'}{P|Q \stackrel{b\sharp_2}{\longrightarrow} P|Q'}$$

3.2 ABC—The Algebra of Broadcast Communication

The Algebra of Broadcast Communication (ABC) [15] is parametrised with sets $\mathscr{A}$ of *agent identifiers*, $\mathscr{B}$ of *broadcast names* and $\mathscr{C}_h$ of *handshake communication names*; each $A \in \mathscr{A}$ comes with a defining equation $A \stackrel{def}{=} P$ with P being a guarded ABC expression as defined below.

The collections $\mathscr{B}!$ and $\mathscr{B}?$ of *broadcast* and *receive* actions are given by $\mathscr{B}\sharp := \{b\sharp \mid b \in \mathscr{B}\}$ for $\sharp \in \{!, ?\}$. $Act := \mathscr{B}! \dot\cup \mathscr{B}? \dot\cup \mathscr{C}_h \dot\cup \bar{\mathscr{C}}_h \dot\cup \{\tau\}$ is the set of *actions*. Below, A ranges over $\mathscr{A}$, b over $\mathscr{B}$, c over $\mathscr{C}_h \cup \bar{\mathscr{C}}_h$, η over $\mathscr{C}_h \cup \bar{\mathscr{C}}_h \cup \{\tau\}$ and α, ℓ over Act. A *relabelling* is a function $f : (\mathscr{B} \to \mathscr{B}) \cup (\mathscr{C}_h \to \mathscr{C}_h)$. It extends to Act by $f(\bar{c}) = \overline{f(c)}$, $f(b\sharp) = f(b)\sharp$ and $f(\tau) := \tau$. The set $\mathbb{P}_{\rm ABC}$ of ABC expressions is defined exactly as $\mathbb{P}_{\rm CCS}$. An expression is guarded if each agent identifier occurs within the scope of a prefixing operator. The structural operational semantics of ABC is the same as the one for CCS (see Table 1) but augmented with the rules for broadcast communication in Table 2.

ABC is CCS augmented with a formalism for broadcast communication taken from the Calculus of Broadcasting Systems (CBS) [25]. The syntax without the broadcast and receive actions and all rules except (Bro-l), (Bro-c) and (Bro-r) are taken verbatim from CCS. However, the rules now cover the different name spaces; (Act) for example allows labels of broadcast and receive actions. The rule (Bro-c)—without rules like (Par-l) and (Par-r) with label $b!$—implements a form of broadcast communication where any broadcast $b!$ performed by a component in a parallel composition is guaranteed to be received by any other component that is ready to do so, i.e., in a state that admits a $b?$-transition. In order to ensure associativity of the parallel composition, one also needs this rule for components receiving at the same time ($\sharp_1=\sharp_2=?$). The rules (Bro-l) and (Bro-r) are added to make broadcast communication *non-blocking*: without them a component could be delayed in performing a broadcast simply because one of the other components is not ready to receive it.

3.3 CCS with Signals

CCS with signals (CCSS) [6] is CCS extended with a signalling operator $P\hat{}s$. Informally, $P\hat{}s$ emits the signal s to be read by another process. $P\hat{}s$ could for instance be a traffic light emitting the signal *red*. The reading of the signal emitted by $P\hat{}s$ does not interfere with any transition of P, such as jumping to *green*. Formally, CCS is extended with a set $\mathscr{S}$ of *signals*, ranged over by s and r. In CCSS the set of actions is defined as $Act := \mathscr{S} \dot\cup \mathscr{C}_h \cup \bar{\mathscr{C}}_h \dot\cup \{\tau\}$, and the set

Table 3. Structural operational semantics for signals of CCSS

$$
\begin{array}{ccc}
P\hat{\ }s \xrightarrow{\bar{s}} P\hat{\ }s & \dfrac{P \xrightarrow{\bar{s}} P'}{P+Q \xrightarrow{\bar{s}} P'+Q} & \dfrac{Q \xrightarrow{\bar{s}} Q'}{P+Q \xrightarrow{\bar{s}} P+Q'} \\[2ex]
\dfrac{P \xrightarrow{\alpha} P'}{P\hat{\ }r \xrightarrow{\alpha} P'} & \dfrac{P \xrightarrow{\bar{s}} P'}{P\hat{\ }r \xrightarrow{\bar{s}} P'\hat{\ }r} & \dfrac{P \xrightarrow{\bar{s}} P'}{A \xrightarrow{\bar{s}} A}\ (A \stackrel{def}{=} P)
\end{array}
$$

of labels by $\mathscr{L} := Act \,\dot{\cup}\, \bar{\mathscr{S}}$, where $\bar{\mathscr{S}} := \{\bar{s} \mid s \in \mathscr{S}\}$. A relabelling is a function $f : (\mathscr{S} \to \mathscr{S}) \cup (\mathscr{C}_h \to \mathscr{C}_h)$. It extends to $\mathscr{L}$ by $f(\bar{c}) = \overline{f(c)}$ for $c \in \mathscr{C}_h \cup \mathscr{S}$ and $f(\tau) := \tau$. The set $\mathbb{P}_{\text{CCSS}}$ of CCSS expressions is defined just as $\mathbb{P}_{\text{CCS}}$, but now also $P\hat{\ }s$ is a process for $P \in \mathbb{P}_{\text{CCSS}}$ and $s \in \mathscr{S}$, and restriction also covers signals.

The semantics of CCSS is given by the labelled transition relation $\to\ \subseteq \mathbb{P}_{\text{CCSS}} \times \mathscr{L} \times \mathbb{P}_{\text{CCSS}}$ derived from the rules of CCS (Table 1), where now η, ℓ range over $\mathscr{L}$, α over Act, c over $\mathscr{C}_h \cup \mathscr{S}$ and $L \subseteq \mathscr{C}_h \cup \mathscr{S}$, augmented with the rules of Table 3. The first rule is the base case showing that a process $P\hat{\ }s$ emits the signal s. The rule below models the fact that signalling cannot prevent a process from making progress.

The original semantics of CCSS [6] featured unary predicates $P^{\curvearrowright s}$ on processes to model that P emits the signal s; here, inspired by [3], these predicates are represented as transitions $P \xrightarrow{\bar{s}} P$. Whereas this leads to a simpler operational semantics, the price paid is that these new *signal transitions* need special treatment in the definition of justness—cf. Definitions 2 and 5.

3.4 Using Signals to Avoid Negative Premises in ABC

Finally, we present an alternative operational semantics ABCd of ABC that avoids negative premises. The price to be paid is the introduction of signals that indicate when a state does not admit a receive action.[3] To this end, let $\mathscr{B}{:} := \{b{:} \mid b \in \mathscr{B}\}$ be the set of *broadcast discards*, and $\mathscr{L} := \mathscr{B}{:} \,\dot{\cup}\, Act$ the set of *transition labels*, with Act as in Sect. 3.2. The semantics is given by the labelled transition relation $\to\ \subseteq \mathbb{P}_{\text{ABC}} \times \mathscr{L} \times \mathbb{P}_{\text{ABC}}$ derived from the rules of CCS (Table 1), where now c ranges over $\mathscr{C}_h \cup \bar{\mathscr{C}}_h$, η over $\mathscr{C}_h \cup \bar{\mathscr{C}}_h \cup \{\tau\}$, α over Act and ℓ over $\mathscr{L}$, augmented with the rules of Table 4.

Lemma 1. [25] $P \xrightarrow{b:} Q$ iff $Q = P \wedge P \stackrel{b?}{\not\longrightarrow}$, for $P, Q \in \mathbb{P}_{\text{ABC}}$ and $b \in \mathscr{B}$.

So the structural operational semantics of ABC from Sects. 3.2 and 3.4 yield the same labelled transition relation $\longrightarrow$ when transitions labelled $b{:}$ are ignored. This approach stems from the Calculus of Broadcasting Systems (CBS) [25].

---

[3] A state P admits an action $\alpha \in Act$ if there exists a transition $P \xrightarrow{\alpha} Q$.

Table 4. SOS of ABC broadcast communication with discard transitions

$$\mathbf{0} \xrightarrow{b:} \mathbf{0} \qquad \alpha.P \xrightarrow{b:} \alpha.P \quad (\alpha{\neq}b?) \qquad \frac{P \xrightarrow{b:} P',\ Q \xrightarrow{b:} Q'}{P+Q \xrightarrow{b:} P'+Q'}$$

$$\frac{P \xrightarrow{b\sharp_1} P',\ Q \xrightarrow{b\sharp_2} Q'}{P|Q \xrightarrow{b\sharp} P'|Q'} \ \sharp_1{\circ}\sharp_2{=}\sharp{\neq}\_ \ \text{ with } \begin{array}{c|ccc} \circ & ! & ? & : \\ \hline ! & \_ & ! & ! \\ ? & ! & ? & ? \\ : & ! & ? & : \end{array} \qquad \frac{P \xrightarrow{b:} P'}{A \xrightarrow{b:} A} \ (A \stackrel{def}{=} P)$$

4 An LTS with Concurrency for CCS and Its Extensions

The forthcoming material applies to each of the process algebras from Sect. 3, or combinations thereof. Let $\mathbb{P}$ be the set of processes in the language.

We allocate an LTS as in Definition 1 to these languages by taking S to be the set $\mathbb{P}$ of processes, and Tr the set of *derivations* t of transitions $P \xrightarrow{\ell} Q$ with $P, Q \in \mathbb{P}$. Of course $src(t) = P$, $target(t) = Q$ and $\ell(t) = \ell$. Here a *derivation* of a transition $P \xrightarrow{\ell} Q$ is a well-founded tree with the nodes labelled by transitions, such that the root has label $P \xrightarrow{\ell} Q$, and if μ is the label of a node and K is the set of labels of the children of this node then $\frac{K}{\mu}$ is an instance of a rule of Tables 1, 2, 3 and 4.

We take $Rec := \mathscr{B}?$ in ABC and ABCd: broadcast receipts can always be blocked by the environment, namely by not broadcasting the requested message. For CCS and CCSS we take $Rec := \emptyset$, thus allowing environments that can always participate in certain handshakes, and/or always emit certain signals.

Following [15], we give a name to any derivation of a transition: The unique derivation of the transition $\alpha.P \xrightarrow{\alpha} P$ using the rule (Act) is called $\xrightarrow{\alpha} P$. The unique derivation of the transition $P\hat{}s \xrightarrow{\bar{s}} P\hat{}s$ is called $P^{\rightarrow s}$. The derivation obtained by application of (Comm) or (Bro-c) on the derivations t and u of the premises of that rule is called $t|u$. The derivation obtained by application of (Par-l) or (Bro-l) on the derivation t of the (positive) premise of that rule, and using process Q at the right of $|$, is $t|Q$. In the same way, (Par-r) and (Bro-r) yield $P|u$, whereas (Sum-l), (Sum-r), (Res), (Rel) and (Rec) yield $t{+}Q$, $P{+}t$, $t\backslash L$, $t[f]$ and $A{:}t$. These names reflect syntactic structure: $t|P \neq P|t$ and $(t|u)|v \neq t|(u|v)$.

Table 3, moreover, contributes derivations $t\hat{}r$. The derivations obtained by application of the rules of Table 4 are called $b{:}\mathbf{0}$, $b{:}\alpha.P$, $t+u$, $t|u$ and $A{:}t$, where t and u are the derivations of the premises.

Synchrons. Let $Arg := \{+_{\rm L}, +_{\rm R}, |_{\rm L}, |_{\rm R}, \backslash L, [f], A{:}, \hat{}r \mid L \subseteq \mathscr{C}_h \wedge f \text{ a relabelling} \wedge A \in \mathscr{A} \wedge r \in \mathscr{S}\}$. A *synchron* is an expression $\sigma(\xrightarrow{\alpha} P)$ or $\sigma(P^{\rightarrow s})$ or $\sigma(b{:})$ with $\sigma \in Arg^*$, $\alpha \in Act$, $s \in \mathscr{S}$, $P \in \mathbb{P}$ and $b \in \mathscr{B}$. An *argument* $\iota \in Arg$ is applied componentwise to a set Σ of synchrons: $\iota(\Sigma) := \{\iota\varsigma \mid \varsigma \in \Sigma\}$.

The set of synchrons $\varsigma(t)$ of a derivation t of a transition is defined by

$$\begin{array}{lll} \varsigma(\xrightarrow{\alpha}P) = \{(\xrightarrow{\alpha}P)\} & \varsigma(t+Q) = +_L\varsigma(t) & \varsigma(P+t) = +_R\varsigma(t) \\ \varsigma(t|Q) = |_L\varsigma(t) & \varsigma(t|u) = |_L\varsigma(t) \cup |_R\varsigma(u) & \varsigma(P|u) = |_R\varsigma(u) \\ \varsigma(t\backslash L) = \backslash L\,\varsigma(t) & \varsigma(t[f]) = [f]\varsigma(t) & \varsigma(A{:}t) = A{:}\varsigma(t) \\ \varsigma(P^{\rightarrow s}) = \{(P^{\rightarrow s})\} & \varsigma(t\hat{r}) = \hat{r}\,\varsigma(t) & \\ \varsigma(b{:}\mathbf{0}) = \{(b{:})\} & \varsigma(b{:}\alpha.P) = \{(b{:})\} & \varsigma(t+v) = +_L\varsigma(t) \cup +_R\varsigma(v) \end{array}$$

Thus, a synchron of t represents a path in the proof-tree t from its root to a leaf. Each transition derivation can be seen as the synchronisation of one or more synchrons. Note that we use the symbol ς as a variable ranging over synchrons, and as the name of a function—context disambiguates.

Example 3. The CCS process $P = \big((c.Q + (d.R|e.S))|\bar{c}.T\big)\backslash c$ has 3 outgoing transitions: $P \xrightarrow{\tau} (Q|T)\backslash c$, $P \xrightarrow{d} ((R|e.S)|\bar{c}.T)\backslash c$ and $P \xrightarrow{e} ((d.R|S)|\bar{c}.T)\backslash c$. Let t_τ, t_d and $t_e \in \mathit{Tr}$ be the unique derivations of these transitions. Then t_τ is a synchronisation of two synchrons, whereas t_d and $t_e \in \mathit{Tr}$ have only one each: $\varsigma(t_\tau) = \{\backslash c\,|_L +_L(\xrightarrow{c}Q), \backslash c\,|_R(\xrightarrow{\bar{c}}T)\}$, $\varsigma(t_d) = \{\backslash c\,|_L +_R |_L(\xrightarrow{d}R)\}$ and $\varsigma(t_e) = \{\backslash c\,|_L +_R |_R(\xrightarrow{e}S)\}$. The derivations t_d and $t_e \in \mathit{Tr}$ can be seen as *concurrent*, because their synchrons come from opposite sides of the same parallel composition; one would expect that after one of them occurs, a variant of the other is still possible. Indeed, there is a transition $((d.R|S)|\bar{c}.T)\backslash c \xrightarrow{d} ((R|S)|\bar{c}.T)\backslash c$. Let t'_d be its unique derivation. The derivation t_d and t'_d are surely different, for they have a different source state. Even their synchrons are different: $\varsigma(t'_d) = \{\backslash c\,|_L\,|_L(\xrightarrow{d}R)\}$. Nevertheless, t'_d can be recognised as a future variant of t_d: its only synchron has merely lost an argument $+_R$. This choice got resolved when taking the transition t_e.

We proceed to formalise the concepts "future variant" and "concurrent" that occur above, by defining two binary relations $\leadsto \subseteq \mathit{Tr}^\bullet \times \mathit{Tr}^\bullet$ and $\smile^\bullet \subseteq \mathit{Tr}^\bullet \times \mathit{Tr}$ such that the following properties hold:

$$\text{The relation } \leadsto \text{ is reflexive and transitive.} \tag{4}$$
$$\text{If } t \leadsto t' \text{ and } t \smile^\bullet v, \text{ then } t' \smile^\bullet v. \tag{5}$$
$$\text{If } t \smile^\bullet v \text{ with } \mathit{src}(t) = \mathit{src}(v) \text{ then } \exists t' \text{ with } \mathit{src}(t') = \mathit{target}(v) \text{ and } t \leadsto t'. \tag{6}$$
$$\text{If } t \leadsto t' \text{ then } \ell(t') = \ell(t) \text{ and } t \not\smile^\bullet t'. \tag{7}$$

With $t \smile^\bullet v$ we mean that the possible occurrence of t is unaffected by the occurrence of v. Although for CCS the relation $\smile^\bullet$ is symmetric (and $\mathit{Tr}^\bullet = \mathit{Tr}$), for ABC and CCSS it is not:

Example 4 ([15]). Let P be the process $b!|(b? + c)$, and let t and v be the derivations of the $b!$- and c-transitions of P. The broadcast $b!$ is in our view completely under the control of the left component; it will occur regardless of whether the right component listens to it or not. It so happens that if $b!$ occurs in state P, the right component will listen to it, thereby disabling the possible occurrence of c. For this reason we have $t \smile^\bullet v$ but $v \not\smile^\bullet t$.

Example 5. Let P be the process $a\hat{\ }s|s$, and let t and v be the derivations of the a- and τ-transitions of P. The occurrence of a disrupts the emission of the signal s, thereby disabling the τ-transition. However, reading the signal does not affect the possible occurrence of a. For this reason we have $t \smile^{\bullet} v$ but $v \not\smile^{\bullet} t$.

Proposition 1. Assume (4)–(7). Then the LTS ($\mathbb{P}$, *Tr*, *src*, *target*, ℓ), augmented with the concurrency relation $\smile^{\bullet}$, is an LTSC in the sense of Definition 4.

We now proceed to define the relations $\leadsto$ and $\smile^{\bullet}$ on synchrons, and then lift them to derivations. Subsequently, we establish (4)–(7).

The elements $+_L$, $+_R$, A: and $\hat{\ }r$ of *Arg* are called *dynamic* [20]; the others are *static*. (Static operators stay around when their arguments perform transitions.) For $\sigma \in Arg^*$ let $static(\sigma)$ be the result of removing all dynamic elements from σ. For $\varsigma = \sigma\upsilon$ with $\upsilon \in \{(\stackrel{\alpha}{\rightarrow}P), (P^{\rightarrow s}), (b{:})\}$ let $static(\varsigma) := static(\sigma)\upsilon$.

Definition 6. A synchron ς' is a *possible successor* of a synchron ς, notation $\varsigma \leadsto \varsigma'$, if either $\varsigma' = \varsigma$, or ς has the form $\sigma_1|_D\varsigma_2$ for some $\sigma_1 \in Arg^*$, $D \in \{L, R\}$ and ς_2 a synchron, and $\varsigma' = static(\sigma_1)|_D\varsigma_2$.

Definition 7. Two synchrons ς and υ are *directly concurrent*, notation $\varsigma \smile_d \upsilon$, if ς has the form $\sigma_1|_D\varsigma_2$ and $\upsilon = \sigma_1|_E\upsilon_2$ with $\{D, E\} = \{L, R\}$. Two synchrons ς' and υ' are *concurrent*, notation $\varsigma' \smile \upsilon'$, if $\exists\varsigma,\upsilon.\, \varsigma' \leftsquigarrow \varsigma \smile_d \upsilon \leadsto \upsilon'$.

Necessary and Active Synchrons. All synchrons of the form $\sigma(\stackrel{\alpha}{\rightarrow}P)$ are *active*; their execution causes a transition $\alpha.P \stackrel{\alpha}{\longrightarrow} P$ in the relevant component of the represented system. Synchrons $\sigma(P^{\rightarrow s})$ and $\sigma(b{:})$ are passive; they are not affecting any state change. Let $a\varsigma(t)$ denote the set of active synchrons of a derivation t. So a transition t is labelled by a signal, i.e. $\ell(t) \notin Act$, iff $a\varsigma(t) = \emptyset$.

Whether a synchron $\varsigma \in \varsigma(t)$ is *necessary* for t to occur is defined only for $t \in Tr^{\bullet}$. If t is the derivation of a broadcast transition, i.e., $\ell(t) = b!$ for some $b \in \mathscr{B}$, then exactly one synchron $\upsilon \in \varsigma(t)$ is of the form $\sigma(\stackrel{b!}{\rightarrow}P)$, while all the other $\varsigma \in \varsigma(t)$ are of the form $\sigma'(\stackrel{b?}{\rightarrow}Q)$ (or possibly $\sigma'(b{:})$ in ABCd). Only the synchron υ is necessary for the broadcast to occur, as a broadcast is unaffected by whether or not someone listens to it. Hence we define $n\varsigma(t) := \{\upsilon\}$. For all $t \in Tr^{\bullet}$ with $\ell(t) \notin \mathscr{B}!$ (i.e. $\ell(t) \in \mathscr{S} \cup \mathscr{C}_h \cup \bar{\mathscr{C}}_h \cup \{\tau\}$) we set $n\varsigma(t) := \varsigma(t)$, thereby declaring all synchrons of the derivation necessary.

Definition 8. A derivation $t' \in Tr^{\bullet}$ is a *possible successor* of a derivation $t \in Tr^{\bullet}$, notation $t \leadsto t'$, if t and t' have equally many necessary synchrons and each necessary synchron of t' is a possible successor of one of t; i.e., if $|n\varsigma(t)| = |n\varsigma(t')|$ and $\forall\varsigma' \in n\varsigma(t').\, \exists\varsigma \in n\varsigma(t).\, \varsigma \leadsto \varsigma'$.

This implies that the relation $\leadsto$ between $n\varsigma(t)$ and $n\varsigma(u)$ is a bijection.

Definition 9. Derivation $t \in Tr^{\bullet}$ is *unaffected by* u, notation $t \smile^{\bullet} u$, if $\forall\varsigma \in n\varsigma(t).\, \forall\upsilon \in a\varsigma(u).\, \varsigma \smile \upsilon$.

So t is unaffected by u if no active synchron of u interferes with a necessary synchron of t. Passive synchrons do not interfere at all.

In Example 3 one has $t_d \smile t_e$, $t_d \rightsquigarrow t'_d$ and $t'_d \smile t_e$. Here $t \smile u$ denotes $t \smile^{\bullet} u \wedge u \smile^{\bullet} t$.

Proposition 2. The relations $\rightsquigarrow$ and $\smile^{\bullet}$ satisfy the properties (4)–(7).

5 Components

This section proposes a concept of system components associated to a transition, with a classification of components as necessary and/or affected. We then define a concurrency relation $\smile^{\bullet}_s$ in terms of these components closely mirroring Definition 9 in Sect. 4 of the concurrency relation $\smile^{\bullet}$ in terms of synchrons. We show that $\smile^{\bullet}$ and $\smile^{\bullet}_s$, as well as the concurrency relation defined in terms of components in Sect. 2, give rise to the same concept of justness.

A *static component* is a string $\sigma \in \mathit{Arg}^*$ of static arguments. Let $\mathscr{C}$ be the set of static components. The *static component* $c(\varsigma)$ of a synchron ς is defined to be the largest prefix γ of ς that is a static component.

Let $\mathit{comp}(t) := \{c(\varsigma) \mid \varsigma \in \varsigma(t)\}$ be the set of *static components* of t. Moreover, $\mathit{npc}(t) := \{c(\varsigma) \mid \varsigma \in n\varsigma(t)\}$ and $\mathit{afc}(t) := \{c(\varsigma) \mid \varsigma \in a\varsigma(t)\}$ are the *necessary* and *affected* static components of $t \in \mathit{Tr}$. Since $n\varsigma(t) \subseteq \varsigma(t)$ and $a\varsigma(t) \subseteq \varsigma(t)$, we have $\mathit{npc}(t) \subseteq \mathit{comp}(t)$ and $\mathit{afc}(t) \subseteq \mathit{comp}(t)$.

Two static components γ and δ are *concurrent*, notation $\gamma \smile \delta$, if $\gamma = \sigma_1|_D\gamma_2$ and $\delta = \sigma_1|_E\delta_2$ with $\{D, E\} = \{L, R\}$.

Definition 10. Derivation $t \in \mathit{Tr}^{\bullet}$ is *statically unaffected by* u, $t \smile^{\bullet}_s u$, iff $\forall \gamma \in \mathit{npc}(t).\, \forall \delta \in \mathit{afc}(u).\, \gamma \smile \delta$.

Proposition 3. If $t \smile^{\bullet}_s u$ then $t \smile^{\bullet} u$.

In Example 3 we have $t_d \smile t_e$ but $t_d \not\smile_s t_e$, for $\mathit{npc}(t_e) = \mathit{comp}(t_e) = \mathit{comp}(t_d) = \mathit{afc}(t_d) = \{\backslash c|_L\}$. Here $t \smile_s u$ denotes $t \smile^{\bullet}_s u \wedge u \smile^{\bullet}_s t$. Hence the implication of Proposition 3 is strict.

Proposition 4. The functions npc and $\mathit{afc} : \mathit{Tr} \rightarrow \mathscr{P}(\mathscr{C})$ satisfy closure property (3) of Sect. 2.

The concurrency relation $\smile^{\bullet}_c$ defined in terms of static components according to the template in [16], recalled in Sect. 2, is not identical to $\smile^{\bullet}_s$:

Definition 11. Let t, u be derivations. Write $t \smile^{\bullet}_c u$ iff $\mathit{npc}(t) \cap \mathit{afc}(u) = \emptyset$.

Nevertheless, we show that for the study of justness it makes no difference whether justness is defined using the concurrency relation $\smile^{\bullet}$, $\smile^{\bullet}_s$ or $\smile^{\bullet}_c$.

Theorem 2. A path is $\smile^{\bullet}$-B-just iff it is $\smile^{\bullet}_c$-B-just iff it is $\smile^{\bullet}_s$-B-just.

6 A Coinductive Characterisation of Justness

In this section we show that the $\rightarrowtail$-based concept of justness defined in this paper coincides with a coinductively defined concept of justness, for CCS and ABC originating from [15]. To state the coinductive definition of justness, we need to define the notion of the decomposition of a path starting from a process with a leading static operator.

Any derivation $t \in Tr$ of a transition with $src(t) = P|Q$ has the shape

- $u|Q$, with $target(t) = target(u)|Q$,
- $u|v$, with $target(t) = target(u)|target(v)$,
- or $P|v$, with $target(t) = P|target(v)$.

Let a path *of* a process P be a path as in Definition 2 starting with P. Now the *decomposition* of a path π of $P|Q$ into paths π_1 and π_2 of P and Q, respectively, is obtained by concatenating all left-projections of the states and transitions of π into a path of P and all right-projections into a path of Q—notation $\pi \Rrightarrow \pi_1|\pi_2$. Here it could be that π is infinite, yet either π_1 or π_2 (but not both) are finite.

Likewise, $t \in Tr$ with $src(t) = P[f]$ has the shape $u[f]$ with $target(t) = target(u)[f]$. The *decomposition* π' of a path π of $P[f]$ is the path obtained by leaving out the outermost $[f]$ of all states and transitions in π, notation $\pi \Rrightarrow \pi'[f]$. In the same way one defines the decomposition of a path of $P\backslash c$.

The following co-inductive definition of the family B-justness of predicates on paths, with one family member of each choice of a set B of blocking actions, stems from [15, Appendix E]—here $\bar{D} := \{\bar{c} \mid c \in D\}$.

Definition 12. *B-justness*, for $\mathscr{B}? \subseteq B \subseteq Act$, is the largest family of predicates on the paths in the LTS of ABC such that

- a finite B-just path ends in a state that admits actions from B only;
- a B-just path of a process $P|Q$ can be decomposed into a C-just path of P and a D-just path of Q, for some $C, D \subseteq B$ such that $\tau \in B \vee C \cap \bar{D} = \emptyset$;
- a B-just path of $P\backslash L$ can be decomposed into a $B \cup L \cup \bar{L}$-just path of P;
- a B-just path of $P[f]$ can be decomposed into an $f^{-1}(B)$-just path of P;
- and each suffix of a B-just path is B-just.

Intuitively, justness is a completeness criterion, telling which paths can actually occur as runs of the represented system. A path is B-just if it can occur in an environment that may block the actions in B. In this light, the first, third, fourth and fifth requirements above are intuitively plausible. The second requirement first of all says that if $\pi \Rrightarrow \pi_1|\pi_2$ and π can occur in the environment that may block the actions in B, then π_1 and π_2 must be able to occur in such an environment as well, or in environments blocking less. The last clause in this requirement prevents a C-just path of P and a D-just path of Q to compose into a B-just path of $P|Q$ when C contains an action c and D the complementary action $\bar{c}$ (except when $\tau \in B$). The reason is that no environment (except one that can block τ-actions) can block both actions for their respective components, as nothing can prevent them from synchronising with each other.

The fifth requirement helps characterising processes of the form $b+(A|b)$ and $a.(A|b)$, with $A \stackrel{def}{=} a.A$. Here, the first transition 'gets rid of' the choice and of the leading action a, respectively, and this requirement reduces the justness of paths of such processes to their suffixes.

Example 6. To illustrate Definition 12 consider the unique infinite path of the process Alice|Cataline of Example 2 in which the transition t does not occur. Taking the empty set of blocking actions, we ask whether this path is $\emptyset$-just. If it were, then by the second requirement of Definition 12 the projection of this path on the process Cataline would need to be $\emptyset$-just as well. This is the path 1 (without any transitions) in Example 1. It is not $\emptyset$-just by the first requirement of Definition 12, because its last state 1 admits a transition.

We now establish that the concept of justness from Definition 12 agrees with the concept of justness defined earlier in this paper.

Theorem 3. A path is $\smile\!\!\!\bullet_s$-B-just iff it is B-just in the sense of Definition 12.

If a path π is B-just then it is C-just for any $C \supseteq B$. Moreover, the collection of sets B such that a given path π is B-just is closed under arbitrary intersection, and thus there is a least set B_π such that π is B-just. Actions $\alpha \in \mathscr{B}_\pi$ are called *π-enabled* [14]. A path is called *just* (without a predicate B) iff it is B-just for some $\mathscr{B}? \subseteq B \subseteq \mathscr{B}? \,\dot\cup\, \mathscr{C}_h \,\dot\cup\, \bar{\mathscr{C}}_h \,\dot\cup\, \mathscr{S}$ [3,6,14,15], which is the case iff it is $\mathscr{B}? \,\dot\cup\, \mathscr{C}_h \,\dot\cup\, \bar{\mathscr{C}}_h \,\dot\cup\, \mathscr{S}$-just.

In [3] a definition of justness for CCS with signal transition appears, very similar to Definition 12; it also applies to CCSS as presented here. Generalising Theorem 3, one can show that a path is ($\smile\!\!\!\bullet_s$ or $\smile\!\!\!\bullet_c$ or) $\smile\!\!\!\bullet$-just iff it is just in this sense. The same holds for the coinductive definition of justness from [6].

7 Conclusion

We advocate justness as a reasonable completeness criterion for formalising liveness properties when modelling distributed systems by means of transition systems. In [16] we proposed a definition of justness in terms of a, possibly asymmetric, concurrency relation between transitions. The current paper defined such a concurrency relation for the transition systems associated to CCS, as well as its extensions with broadcast communication and signals, thereby making the definition of justness from [16] available to these languages. In fact, we provided three versions of the concurrency relation, and showed that they all give rise to the same concept of justness. We expect that this style of definition will carry over to many other process algebras. We showed that justness satisfies the criterion of feasibility, and proved that our formalisation agrees with previous coinductive formalisations of justness for these languages.

Concurrency relations between transitions in transition systems have been studied in [28]. Our concurrency relation $\smile\!\!\!\bullet$ follows the same computational intuition. However, in [28] transitions are classified as concurrent or not only when they have the same source, whereas as a basis for the definition of justness here we need to compare transitions with different sources. Apart from that, our concurrency relation is more general in that it satisfies fewer closure properties, and moreover is allowed to be asymmetric.

Concurrency is represented explicitly in models like Petri nets [26], event structures [29], or asynchronous transition systems [2,27,30]. We believe that the semantics of CCS in terms of such models agrees with its semantics in terms of labelled transition systems with a concurrency relation as given here. However, formalising such a claim requires a choice of an adequate justness-preserving semantic equivalence defined on the compared models. Development of such semantic equivalences is a topic for future research.

Acknowledgement. I am grateful to Peter Höfner, Victor Dyseryn and Filippo de Bortoli for valuable feedback.

References

1. Apt, K.R., Francez, N., Katz, S.: Appraising fairness in languages for distributed programming. Distrib. Comput. **2**(4), 226–241 (1988). https://doi.org/10.1007/BF01872848
2. Bednarczyk, M.: Categories of asynchronous systems. Ph.D. thesis, Computer Science, University of Sussex, Brighton (1987)
3. Bouwman, M.S.: Liveness analysis in process algebra: simpler techniques to model mutex algorithms. Technical report, Eindhoven University of Technology (2018). http://www.win.tue.nl/~timw/downloads/bouwman_seminar.pdf
4. Coppo, M., Dezani-Ciancaglini, M., Padovani, L., Yoshida, N.: Inference of global progress properties for dynamically interleaved multiparty sessions. In: De Nicola, R., Julien, C. (eds.) COORDINATION 2013. LNCS, vol. 7890, pp. 45–59. Springer, Heidelberg (2013). https://doi.org/10.1007/978-3-642-38493-6_4
5. De Nicola, R., Vaandrager, F.W.: Three logics for branching bisimulation. J. ACM **42**(2), 458–487 (1995). https://doi.org/10.1145/201019.201032
6. Dyseryn, V., van Glabbeek, R.J., Höfner, P.: Analysing mutual exclusion using process algebra with signals. In: Peters, K., Tini, S. (eds.) Proceedings of the Combined 24th International Workshop on Expressiveness in Concurrency and 14th Workshop on Structural Operational Semantics, Electronic Proceedings in Theoretical Computer Science 255. Open Publishing Association, pp. 18–34 (2017). https://doi.org/10.4204/EPTCS.255.2
7. Emerson, E.A., Clarke, E.M.: Using branching time temporal logic to synthesize synchronization skeletons. Sci. Comput. Program. **2**(3), 241–266 (1982). https://doi.org/10.1016/0167-6423(83)90017-5
8. Emerson, E.A., Halpern, J.Y.: 'Sometimes' and 'Not Never' revisited: on branching time versus linear time temporal logic. J. ACM **33**(1), 151–178 (1986). https://doi.org/10.1145/4904.4999
9. Fehnker, A., van Glabbeek, R.J., Höfner, P., McIver, A., Portmann, M., Tan, W.L.: A process algebra for wireless mesh networks. In: Seidl, H. (ed.) ESOP 2012. LNCS, vol. 7211, pp. 295–315. Springer, Heidelberg (2012). https://doi.org/10.1007/978-3-642-28869-2_15

10. Fehnker, A., van Glabbeek, R.J., Höfner, P., McIver, A.K., Portmann, M., Tan, W.L.: A process algebra for wireless mesh networks used for modelling, verifying and analysing AODV. Technical report 5513, NICTA (2013). http://arxiv.org/abs/1312.7645
11. van Glabbeek, R.J.: Structure preserving bisimilarity, supporting an operational petri net semantics of CCSP. In: Meyer, R., Platzer, A., Wehrheim, H. (eds.) Correct System Design. LNCS, vol. 9360, pp. 99–130. Springer, Cham (2015). https://doi.org/10.1007/978-3-319-23506-6_9. http://arxiv.org/abs/1509.05842
12. van Glabbeek, R.J.: Ensuring Liveness Properties of Distributed Systems (A Research Agenda). Position paper (2016). http://arxiv.org/abs/org/abs/1711.04240
13. van Glabbeek, R.J.: Justness: a completeness criterion for capturing liveness properties. Technical report, Data61, CSIRO (2018). http://www.cse.unsw.edu.au/~rvg/synchrons.pdf. Full version of the present paper
14. van Glabbeek, R.J., Höfner, P.: CCS: It's not fair! Acta Inform. **52**(2–3), 175–205 (2015). https://doi.org/10.1007/s00236-015-0221-6
15. van Glabbeek, R.J., Höfner, P.: Progress, fairness and justness in process algebra. Technical report 8501, NICTA (2015). http://arxiv.org/abs/1501.03268
16. van Glabbeek, R.J., Höfner, P.: Progress, justness and fairness. Survey paper, Data61, CSIRO, Sydney, Australia (2018). https://arxiv.org/abs/1810.07414
17. Kuiper, R., de Roever, W.-P.: Fairness assumptions for CSP in a temporal logic framework. In: Bjørner, D. (ed.) Formal Description of Programming Concepts II, North-Holland, pp. 159–170 (1983)
18. Lamport, L.: Proving the correctness of multiprocess programs. IEEE Trans. Softw. Eng. **3**(2), 125–143 (1977). https://doi.org/10.1109/TSE.1977.229904
19. Lamport, L.: Fairness and hyperfairness. Distrib. Comput. **13**(4), 239–245 (2000). https://doi.org/10.1007/PL00008921
20. Milner, R. (ed.): A Calculus of Communicating Systems. LNCS, vol. 92. Springer, Heidelberg (1980). https://doi.org/10.1007/3-540-10235-3
21. Misra, J.: A Rebuttal of Dijkstra's position on fairness (1988). http://www.cs.utexas.edu/users/misra/Notes.dir/fairness.pdf
22. Misra, J.: A Discipline of Multiprogramming—Programming Theory for Distributed Applications. Springer, New York (2001). https://doi.org/10.1007/978-1-4419-8528-6
23. Owicki, S.S., Lamport, L.: Proving liveness properties of concurrent programs. ACM TOPLAS **4**(3), 455–495 (1982). https://doi.org/10.1145/357172.357178
24. Pnueli, A.: The temporal logic of programs. In: Proceedings of the 18th Annual Symposium on Foundations of Computer Science (FOCS 1977), pp. 46–57. IEEE (1977). https://doi.org/10.1109/SFCS.1977.32
25. Prasad, K.V.S.: A calculus of broadcasting systems. In: Abramsky, S., Maibaum, T.S.E. (eds.) CAAP 1991. LNCS, vol. 493, pp. 338–358. Springer, Heidelberg (1991). https://doi.org/10.1007/3-540-53982-4_19
26. Reisig, W.: Understanding Petri Nets—Modeling Techniques, Analysis Methods, Case Studies. Springer, Heidelberg (2013). https://doi.org/10.1007/978-3-642-33278-4
27. Shields, M.W.: Concurrent machines. Comput. J. **28**(5), 449–465 (1985). https://doi.org/10.1093/comjnl/28.5.449
28. Stark, E.W.: Concurrent transition systems. Theor. Comput. Sci. **64**(3), 221–269 (1989). https://doi.org/10.1016/0304-3975(89)90050-9

29. Winskel, G.: Event structures. In: Brauer, W., Reisig, W., Rozenberg, G. (eds.) ACPN 1986, Part II. LNCS, vol. 255, pp. 325–392. Springer, Heidelberg (1987). https://doi.org/10.1007/3-540-17906-2_31
30. Winskel, G., Nielsen, M.: Models for concurrency. In: Abramsky, S., Gabbay, D., Maibaum, T. (eds.) Handbook of Logic in Computer Science, Chap. 1, 4: Semantic Modelling, pp. 1–148. Oxford University Press, Oxford (1995)

Path Category for Free

Open Morphisms from Coalgebras with Non-deterministic Branching

Thorsten Wißmann[1(✉)], Jérémy Dubut[2,3], Shin-ya Katsumata[2], and Ichiro Hasuo[2,4]

[1] Friedrich-Alexander-Universität Erlangen-Nürnberg, Erlangen, Germany
thorsten.wissmann@fau.de
[2] National Institute of Informatics, Tokyo, Japan
{dubut,s-katsumata,hasuo}@nii.ac.jp
[3] Japanese-French Laboratory for Informatics, Tokyo, Japan
[4] SOKENDAI, Hayama, Kanagawa, Japan

Abstract. There are different categorical approaches to variations of transition systems and their bisimulations. One is coalgebra for a functor G, where a bisimulation is defined as a span of G-coalgebra homomorphism. Another one is in terms of path categories and open morphisms, where a bisimulation is defined as a span of open morphisms. This similarity is no coincidence: given a functor G, fulfilling certain conditions, we derive a path-category for pointed G-coalgebras and lax homomorphisms, such that the open morphisms turn out to be precisely the G-coalgebra homomorphisms. The above construction provides path-categories and trace semantics for free for different flavours of transition systems: (1) non-deterministic tree automata (2) regular nondeterministic nominal automata (RNNA), an expressive automata notion living in nominal sets (3) multisorted transition systems. This last instance relates to Lasota's construction, which is in the converse direction.

Keywords: Coalgebra · Open maps · Categories · Nominal sets

1 Introduction

Coalgebras [25] and *open maps* [16] are two main categorical approaches to transition systems and bisimulations. The former describes the branching type of systems as an endofunctor, a system becoming a coalgebra and bisimulations being spans of coalgebra homomorphisms. Coalgebra theory makes it easy to consider state space types in different settings, e.g. nominal sets [17,18] or algebraic categories [5,11,20]. The latter, open maps, describes systems as objects of

This research was supported by ERATO HASUO Metamathematics for Systems Design Project (No. JPMJER1603), JST. The first author was supported by the DFG project MI 717/5-1. He expresses his gratitude for having been invited to Tokyo, which initiated the present work.

M. Bojańczyk and A. Simpson (Eds.): FOSSACS 2019, LNCS 11425, pp. 523–540, 2019.
https://doi.org/10.1007/978-3-030-17127-8_30

Table 1. Two approaches to categorical (bi)simulations

worlds	data	systems	func. sim.	func. bisim.	(bi)simulation
open maps	$J\colon \mathbb{P} \longrightarrow \mathbb{M}$ Def. 2.4	$\mathbf{obj}\,(\mathbb{M})$	$\mathbf{mor}\,(\mathbb{M})$	open maps Def. 2.5	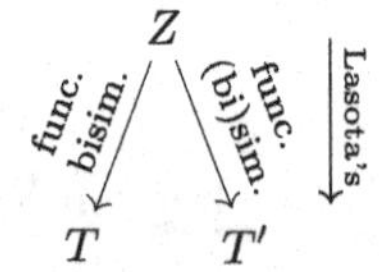
coalgebra	$G\colon \mathbb{C} \longrightarrow \mathbb{C}$ Def. 2.7	pointed G-coalg. Sec. 2.2	lax hom. Def. 2.8	coalg. hom. Def. 2.6	

a category and the execution types as particular objects called paths. In this case, bisimulations are spans of open morphisms. Open maps are particularly adapted to extend bisimilarity to history dependent behaviors, e.g. true concurrency [7,8], timed systems [22] and weak (bi)similarity [9]. Coalgebra homomorphisms and open maps are then key concepts to describe bisimilarity categorically. They intuitively correspond to functional bisimulations, that is, those maps between states whose graph is a bisimulation.

We are naturally interested in the relationship between those two categorical approaches to transition systems and bisimulations. A reduction of open maps situations to coalgebra was given by Lasota using multi-sorted transition systems [19]. In this paper, we give the reduction in the other direction: from the category $\mathsf{Coalg}_l(TF)$ of pointed TF-coalgebras and lax homomorphisms, we construct the path-category Path and a functor $J : \mathsf{Path} \longrightarrow \mathsf{Coalg}_l(TF)$ such that Path-open morphisms coincide with strict homomorphisms, hence functional bisimulations. Here, T is a functor describing the branching behaviour and F describes the input type, i.e. the type of data that is processed (e.g. words or trees). This development is carried out with the case where T is a powerset-like functor, and covers transition systems allowing non-deterministic branching.

The key concept in the construction of Path are *F-precise maps*. Roughly speaking in set, a map $f\colon X \longrightarrow FY$ is F-precise if every $y \in Y$ is used precisely once in f, i.e. there is a unique x such that y appears in $f(x)$ and additionally y appears precisely once in $f(x)$. Such an F-precise map represents one deterministic step (of shape F). Then a path $P \in \mathsf{Path}$ is a finite sequence of deterministic steps, i.e. finitely many precise maps. J converts such a data into a pointed TF-coalgebra. There are many existing notions of paths and traces in coalgebra [4,12,13,21], which lack the notion of *precise* map, which is crucial for the present work.

Once we set up the situation $J\colon \mathsf{Path} \longrightarrow \mathsf{Coalg}_l(TF)$, we are on the framework of open map bisimulations. Our construction of Path using precise maps is justified by the characterisation theorem: Path-open morphisms and strict coalgebra homomorphisms coincide (Theorems 3.20 and 3.24). This coincidence relies on the concept of path-reachable coalgebras, namely, coalgebras such that every state can be reached by a path. Under mild conditions, path-reachability is equivalent to an existing notion in coalgebra, defined as the non-existence of a proper sub-coalgebra (Sect. 3.5). Additionally, this characterization produces a canonical trace semantics for free, given in terms of paths (Sect. 3.6).

We illustrate our reduction with several concrete situations: different classes of non-deterministic top-down tree automata using analytic functors (Sect. 4.1), Regular Nondeterministic Nominal Automata (RNNA), an expressive automata notion living in nominal sets (Sect. 4.2), multisorted transition systems, used in Lasota's work to construct a coalgebra situation from an open map situation (Sect. 4.3).

Notation. We assume basic categorical knowledge and notation (see e.g. [1,3]). The cotupling of morphisms $f\colon A \to C$, $g\colon B \to C$ is denoted by $[f,g]\colon A+B \to C$, and the unique morphsim to the terminal object is $!\colon X \to 1$ for every X.

2 Two Categorical Approaches for Bisimulations

We introduce the two formalisms involved in the present paper: the open maps (Sect. 2.1) and the coalgebras (Sect. 2.2). Those formalisms will be illustrated on the classic example of Labelled Transition Systems (LTSs).

Definition 2.1. *Fix a set A, called the alphabet. A* labelled transition system *is a triple (S, i, Δ) with S a set of* states, *$i \in S$ the* initial state, *and $\Delta \subseteq S \times A \times S$ the* transition relation. *When Δ is obvious from the context, we write $s \xrightarrow{a} s'$ to mean $(s, a, s') \in \Delta$.*

For instance, the tuple $(\{0, \cdots, n\}, 0, \{(k-1, a_k, k) \mid 1 \leq k \leq n\})$ is an LTS, and called the *linear system* over the word $a_1 \cdots a_n \in A^\star$. To relate LTSs, one considers functions that preserves the structure of LTSs:

Definition 2.2. *A* morphism of LTSs *from (S, i, Δ) to (S', i', Δ') is a function $f\colon S \longrightarrow S'$ such that $f(i) = i'$ and for every $(s, a, s') \in \Delta$, $(f(s), a, f(s')) \in \Delta'$. LTSs and morphisms of LTSs form a category, which we denote by LTS_A.*

Some authors choose other notions of morphisms (e.g. [16]), allowing them to operate between LTSs with different alphabets for example. The usual way of comparing LTSs is by using simulations and bisimulations [23]. The former describes what it means for a system to have at least the behaviours of another, the latter describes that two systems have exactly the same behaviours. Concretely:

Definition 2.3. *A* simulation *from (S, i, Δ) to (S', i', Δ') is a relation $R \subseteq S \times S'$ such that (1) $(i, i') \in R$, and (2) for every $s \xrightarrow{a} t$ and $(s, s') \in R$, there is $t' \in S'$ such that $s' \xrightarrow{a} t'$ and $(t, t') \in R$. Such a relation R is a* bisimulation *if $R^{-1} = \{(s', s) \mid (s, s') \in R\}$ is also a simulation.*

Morphisms of LTSs are functional simulations, i.e. functions between states whose graph is a simulation. So how to model (1) systems, (2) functional simulations and (3) functional bisimulations categorically? In the next two sections, we will describe known answers to this question, with open maps and coalgebra. In both cases, it is possible to capture similarity and bisimilarity of two LTSs T

and T'. Generally, a simulation is a (jointly monic) span of a functional bisimulation and a functional simulation, and a bisimulation is a simulation whose converse is also a simulation, as depicted in Table 1. Consequently, to understand similarity and bisimilarity on a general level, it is enough to understand functional simulations and bisimulations.

2.1 Open Maps

The categorical framework of open maps [16] assumes functional simulations to be already modeled as a category $\mathbb{M}$. For example, for $\mathbb{M} := \mathsf{LTS}_A$, objects are LTSs, and morphisms are functional simulations. Furthermore, the open maps framework assumes another category $\mathbb{P}$ of 'paths' or 'linear systems', together with a functor J that tells how a 'path' is to be understood as a system:

Definition 2.4 [16]. *An* open map situation *is given by categories* $\mathbb{M}$ *('systems' with 'functional simulations') and* $\mathbb{P}$ *('paths') together with a functor* $J\colon \mathbb{P} \to \mathbb{M}$.

For example with $\mathbb{M} := \mathsf{LTS}_A$, we pick $\mathbb{P} := (A^\star, \leq)$ to be the poset of words over A with prefix order. Here, the functor J maps a word $w \in A^\star$ to the linear system over w, and $w \leq v$ to the evident functional simulation $J(w \leq v)\colon Jw \longrightarrow Jv$.

In an open map situation $J\colon \mathbb{P} \longrightarrow \mathbb{M}$, we can abstractly represent the concept of a *run* in a system. A run of a path $w \in \mathbb{P}$ in a system $T \in \mathbb{M}$ is simply defined to be an $\mathbb{M}$-morphism of type $Jw \longrightarrow T$. With this definition, each $\mathbb{M}$-morphism $h\colon T \longrightarrow T'$ (i.e. functional simulation) inherently transfers runs: given a run $x\colon Jw \longrightarrow T$, the morphism $h \cdot x\colon Jw \longrightarrow T'$ is a run of w in T'. In the example open map situation $J\colon (A^\star, \leq) \longrightarrow \mathsf{LTS}_A$, a run of a path $w = a_1 \cdots a_n \in A^\star$ in an LTS $T = (S, i, \Delta)$ is nothing but a sequence of states $x_0, \ldots, x_n \in S$ such that $x_0 = i$ and $x_{k-1} \xrightarrow{a_k} x_k$ holds for all $1 \leq k \leq n$.

We introduce the concept of open map [16]. This is an abstraction of the property posessed by *functional bisimulations.* For LTSs $T = (S, i, \Delta)$ and $T' = (S', i', \Delta')$, an LTS_A-morphism $h\colon T \longrightarrow T'$ is a functional bisimulation if the graph of h is a bisimulation. This implies the following relationship between runs in T and runs in T'. Suppose that $w \leq w'$ holds in $A^\star$, and a run x of w in T is given as in (1); here n, m are lengths of w, w' respectively. Then for any run y' of w' in T' extending $h \cdot x$ as in (2), there is a run x' of w' extending x, and moreover its image by h coincides with y' (that is, $h \cdot x' = y'$). Such x' is obtained by repetitively applying the condition of functional bisimulation.

$$\underbrace{\overbrace{\to i \xrightarrow{w_1} x_1 \xrightarrow{w_2} \cdots \xrightarrow{w_n} x_n}^{x} \xrightarrow{w'_{n+1}} x'_{n+1} \xrightarrow{w'_{n+2}} \cdots \xrightarrow{w'_m} x'_m}_{x'} \quad (\text{in } T) \qquad (1)$$

$$\underbrace{\to i' \xrightarrow{w_1} h(x_1) \xrightarrow{w_2} \cdots \xrightarrow{w_n} h(x_n) \xrightarrow{w'_{n+1}} y'_{n+1} \xrightarrow{w_{n+2}} \cdots \xrightarrow{w'_m} y'_m}_{y'} \quad (\text{in } T') \qquad (2)$$

Observe that y' extending $h \cdot x$ can be represented as $y' \cdot J(w \leq w') = h \cdot x$, and x' extending x as $x' \cdot J(w \leq w') = x$. From these, we conclude that if an

LTS_A-morphism $h\colon T \longrightarrow T'$ is a functional bisimulation, then for any $w \leq w'$ in $A^\star$ and run $x\colon Jw \longrightarrow T$ and $y'\colon Jw' \longrightarrow T'$ such that $y' \cdot J(w \leq w') = h \cdot x$, there is a run $x'\colon Jw' \longrightarrow T$ such that $x' \cdot J(w \leq w') = x$ and $h \cdot x' = y'$ (the converse also holds if all states of T are reachable). This necessary condition of functional bisimulation can be rephrased in any open map situation, leading us to the definition of open map.

Definition 2.5 [16]**.** *Let $J\colon \mathbb{P} \longrightarrow \mathbb{M}$ be an open map situation. An $\mathbb{M}$-morphism $h\colon T \longrightarrow T'$ is said to be* open *if for every morphism $\Phi\colon w \longrightarrow w' \in \mathbb{P}$ making the square on the right commute, there is x' making the two triangles commute.*

$$\begin{array}{ccc} Jw & \xrightarrow{x} & T \\ {\scriptstyle J\Phi}\downarrow & \overset{\exists x'}{\nearrow} & \downarrow{\scriptstyle h} \\ Jw' & \xrightarrow{y'} & T' \end{array}$$

Open maps are closed under composition and stable under pullback [16].

2.2 Coalgebras

The theory of G-coalgebras is another categorical framework to study bisimulations. The type of systems is modelled using an endofunctor $G\colon \mathbb{C} \longrightarrow \mathbb{C}$ and a system is then a coalgebra for this functor, that is, a pair of an object S of $\mathbb{C}$ (modeling the state space), and of a morphism of type $S \longrightarrow GS$ (modeling the transitions). For example for LTSs, the transition relation is of type $\Delta \subseteq S \times A \times S$. Equivalently, this can be defined as a function $\Delta\colon S \longrightarrow \mathcal{P}(A \times S)$, where $\mathcal{P}$ is the powerset. In other words, the transition relation is a coalgebra for the Set-functor $\mathcal{P}(A \times \_)$. Intuitively, this coalgebra gives the one-step behaviour of an LTS: S describes the state space of the system, $\mathcal{P}$ describes the 'branching type' as being non-deterministic, $A \times S$ describe the 'computation type' as being linear, and the function itself lists all possible futures after one-step of computation of the system. Now, changing the underlying category or the endofunctor allows to model different types of systems. This is the usual framework of coalgebra, as described for example in [25].

Initial states are modelled coalgebraically by a pointing to the carrier $i\colon I \longrightarrow S$ for a fixed object I in $\mathbb{C}$, describing the 'type of initial states' (see e.g. [2, Sec. 3B]). For example, an initial state of an LTS is the same as a function from the singleton set $I := \{*\}$ to the state space S. This object I will often be the final object of $\mathbb{C}$, but we will see other examples later. In total, an *I-pointed G-coalgebra* is a $\mathbb{C}$-object S together with morphisms $\alpha\colon S \longrightarrow GS$ and $i\colon I \longrightarrow S$. E.g. an LTS is an I-pointed G-coalgebra for $I = \{*\}$ and $GX = \mathcal{P}(A \times X)$.

In coalgebra, functional bisimulations are the first class citizens to be modelled as homomorphisms. The intuition is that those preserve the initial state, and preserve and reflect the one-step relation.

Definition 2.6. *An I-*pointed *G-*coalgebra homomorphism *from $I \xrightarrow{i} S \xrightarrow{\alpha} GS$ to $I \xrightarrow{i'} S' \xrightarrow{\alpha'} GS'$ is a morphism $f\colon S \longrightarrow S'$ making the right-hand diagram commute.*

$$\begin{array}{ccccc} I & \xrightarrow{i} & S & \xrightarrow{\alpha} & GS \\ & {\scriptstyle i'}\searrow & \downarrow{\scriptstyle f} & & \downarrow{\scriptstyle Gf} \\ & & S' & \xrightarrow{\alpha'} & GS' \end{array}$$

For instance, when $G = \mathcal{P}(A \times \_)$, one can easily see that a function f is a G-coalgebra homomorphism iff it is a functional bisimulation. Thus, if we want to capture functional simulations in LTSs, we need to weaken the condition of homomorphism to the inequality $Gf(\alpha(s)) \subseteq \alpha'(f(s))$ (instead of equality). To express this condition for general G-coalgebras, we introduce a partial order $\sqsubseteq_{X,Y}$ on each homset $\mathbb{C}(X, GY)$ in a functorial manner.

Definition 2.7. *A* partial order on G-homsets *is a functor* $\sqsubseteq\colon \mathbb{C}^{\mathsf{op}} \times \mathbb{C} \longrightarrow \mathsf{Pos}$ *such that* $U \cdot \sqsubseteq = \mathbb{C}(\_, G\_)$*; here,* $U\colon \mathsf{Pos} \longrightarrow \mathsf{Set}$ *is the forgetful functor from the category* Pos *of posets and monotone functions.*

The functoriality of $\sqsubseteq$ amounts to that $f_1 \sqsubseteq f_2$ implies $Gh \cdot f_1 \cdot g \sqsubseteq Gh \cdot f_2 \cdot g$.

Definition 2.8. *Given a partial order on G-homsets, an* I-pointed lax G-coalgebra homomorphism $f\colon (S, \alpha, i) \longrightarrow (S', \alpha', i')$ *is a morphism* $f\colon S \longrightarrow S'$ *making the right-hand diagram commute. The* I*-pointed* G*-coalgebras and lax homomorphisms form a category, denoted by* $\mathsf{Coalg}_l(I, G)$.

$$\begin{array}{ccccc} I & \xrightarrow{i} & S & \xrightarrow{\alpha} & GS \\ & \searrow^{i'} & \downarrow f & \sqsubseteq & \downarrow Gf \\ & & S' & \xrightarrow[\alpha']{} & GS' \end{array}$$

Conclusion 2.9. In Set, with $I = \{*\}$, $G = \mathcal{P}(A \times \_)$, define the order $f \sqsubseteq g$ in $\mathsf{Set}(X, \mathcal{P}(A \times Y))$ iff for every $x \in X$, $f(x) \subseteq g(x)$. Then $\mathsf{Coalg}_l(\{*\}, \mathcal{P}(A \times \_)) = \mathsf{LTS}_A$. In particular, we have an open map situation

$$\mathbb{P} = (A^\star, \leq) \quad \xrightarrow{J} \quad \mathbb{M} = \mathsf{LTS}_A = \mathsf{Coalg}_l(\{*\}, \mathcal{P}(A \times \_))$$

and the open maps are precisely the coalgebra homomorphisms (for reachable LTSs). In this paper, we will construct a path category $\mathbb{P}$ for more general I and G, such that the open morphisms are precisely the coalgebra homomorphisms.

3 The Open Map Situation in Coalgebras

Lasota's construction [19] transforms an open map situation $J\colon \mathbb{P} \longrightarrow \mathbb{M}$ into a functor G (with a partial order on G-homsets), together with a functor $\mathsf{Beh}\colon \mathbb{M} \longrightarrow \mathsf{Coalg}_l(I, G)$ that sends open maps to G-coalgebra homomorphisms (see Sect. 4.3 for details). In this paper, we provide a construction in the converse direction for functors G of a certain shape.

As exemplified by LTSs, it is a common pattern that G is the composition $G = TF$ of two functors [12], where T is the branching type (e.g. partial, or non-deterministic) and F is the data type, or the 'linear behaviour' (words, trees, words modulo α-equivalence). If we instantiate our path-construction to $T = \mathcal{P}$ and $F = A \times \_$, we obtain the known open map situation for LTSs (Conclusion 2.9).

Fix a category $\mathbb{C}$ with pullbacks, functors $T, F : \mathbb{C} \longrightarrow \mathbb{C}$, an object $I \in \mathbb{C}$ and a partial order $\sqsubseteq^T$ on T-homsets. They determine a coalgebra situation $(\mathbb{C}, I, TF, \sqsubseteq)$ where $\sqsubseteq$ is the partial order on TF-homsets defined by $\sqsubseteq_{X,Y} = \sqsubseteq^T_{X,FY}$. Under some conditions on T and F, we construct a path-category $\mathsf{Path}(I, F+1)$ and an open map situation $\mathsf{Path}(I, F+1) \hookrightarrow \mathsf{Coalg}_l(I, TF)$ where TF-coalgebra homomorphisms and $\mathsf{Path}(I, F+1)$-open morphisms coincide.

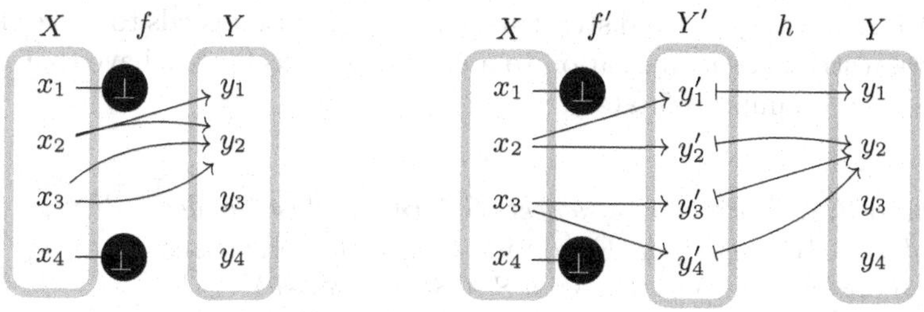

Fig. 1. A non-precise map f that factors through the F-precise $f'\colon X \longrightarrow Y' \times Y' + \{\bot\}$

3.1 Precise Morphisms

While the path category is intuitively clear for $FX = A \times X$, it is not for inner functors F that model tree languages. For example for $FX = A + X \times X$, a $\mathcal{P}F$-coalgebra models transition systems over binary trees with leaves labelled in A, instead of over words. Hence, the paths should be these kind of binary trees. We capture the notion of tree like shape ("every node in a tree has precisely one route to the root") by the following abstract definition:

Definition 3.1. *For a functor $F\colon \mathbb{C} \longrightarrow \mathbb{C}$, a morphism $s\colon S \longrightarrow FR$ is called* F-precise *if for all f, g, h the following implication holds:*

$$\begin{array}{ccc} S & \xrightarrow{f} & FC \\ {\scriptstyle s}\downarrow & & \downarrow{\scriptstyle Fh} \\ FR & \xrightarrow{Fg} & FD \end{array} \quad \overset{\exists d}{\Longrightarrow} \quad \begin{array}{ccc} S & \xrightarrow{f} & FC \\ {\scriptstyle s}\downarrow & \nearrow{\scriptstyle Fd} & \\ FR & & \end{array} \quad \& \quad \begin{array}{ccc} & & C \\ & {\scriptstyle d}\nearrow & \downarrow{\scriptstyle h} \\ R & \xrightarrow{g} & D \end{array}$$

Remark 3.2. If F preserves weak pullbacks, then a morphism s is F-precise iff it fulfils the above definition for $g = \text{id}$.

Example 3.3. Intuitively speaking, for a polynomial Set-functor F, a map $s\colon S \to FR$ is F-precise iff every element of R is mentioned precisely once in the definition of the map f. For example, for $FX = A \times X + \{\bot\}$, the case needed later for LTSs, a map $f\colon X \longrightarrow FY$ is precise iff for every $y \in Y$, there is a unique pair $(x, a) \in X \times A$ such that $f(x) = (a, y)$. For $FX = X \times X + \{\bot\}$ on Set, the map $f\colon X \longrightarrow FY$ in Fig. 1 is not F-precise, because y_2 is used three times (once in $f(x_2)$ and twice in $f(x_3)$), and y_3 and y_4 do not occur in f at all. However, $f'\colon X \longrightarrow FY'$ is F-precise because every element of Y' is used precisely once in f', and we have that $Fh \cdot f' = f$. Also note that f' defines a forest where X is the set of roots, which is closely connected to the intuition that, in the F-precise map f', from every element of Y', there is precisely one edge up to a root in X.

So when transforming a non-precise map into a precise map, one duplicates elements that are used multiple times and drops elements that are not used. We will cover functors F for which this factorization pattern provides F-precise maps. If F involves unordered structure, this factorization needs to make choices, and so we restrict the factorization to a class $\mathcal{S}$ of objects that have that choice-principle (see Example 4.5 later):

Definition 3.4. *Fix a class of objects $\mathcal{S} \subseteq \mathbf{obj}\,\mathbb{C}$ closed under isomorphism. We say that F admits precise factorizations w.r.t. $\mathcal{S}$ if for every $f\colon S \to FY$ with $S \in \mathcal{S}$, there exist $Y' \in \mathcal{S}$, $h\colon Y' \to Y$ and $f'\colon S \to FY'$ F-precise with $Fh \cdot f' = f$.*

$$
\begin{array}{ccc}
S & \xrightarrow{\ \exists f'\ } & FY' \\
& \searrow{\scriptstyle \forall f} & \downarrow{\scriptstyle Fh} \\
& & FY
\end{array}
$$

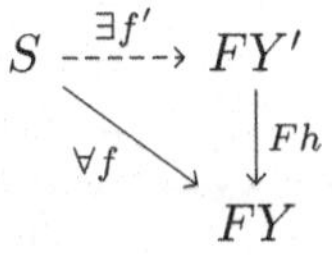

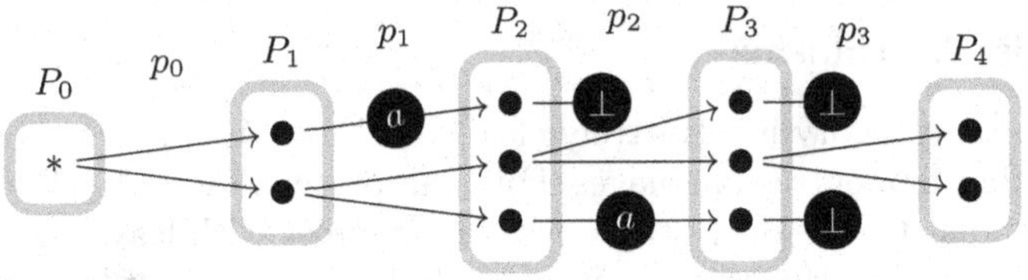

Fig. 2. A path of length 4 for $FX = \{a\} \times X + X \times X + \{\bot\}$ with $I = \{*\}$.

For $\mathbb{C} = \mathsf{Set}$, $\mathcal{S}$ contains all sets. However for the category of nominal sets, $\mathcal{S}$ will only contain the strong nominal sets (see details in Subsect. 4.2).

Remark 3.5. Precise morphisms are essentially unique. If $f_1\colon X \longrightarrow FY_1$ and $f_2\colon X \longrightarrow FY_2$ are F-precise and if there is some $h\colon Y_1 \longrightarrow Y_2$ with $Fh \cdot f_1 = f_2$, then h is an isomorphism. Consequently, if $f\colon S \longrightarrow FY$ with $S \in \mathcal{S}$ is F-precise and F-admits precise factorizations, then $Y \in \mathcal{S}$.

Functors admitting precise factorizations are closed under basic constructions:

Proposition 3.6. *The following functors admit precise factorizations w.r.t. $\mathcal{S}$:*

1. *Constant functors, if $\mathbb{C}$ has an initial object 0 and $0 \in \mathcal{S}$.*
2. *$F \cdot F'$ if $F\colon \mathbb{C} \longrightarrow \mathbb{C}$ and $F'\colon \mathbb{C} \longrightarrow \mathbb{C}$ do so.*
3. *$\prod_{i \in I} F_i$, if all $(F_i)_{i \in I}$ do so and $\mathcal{S}$ is closed under I-coproducts.*
4. *$\coprod_{i \in I} F_i$, if all $(F_i)_{i \in I}$ do so, $\mathbb{C}$ is I-extensive and $\mathcal{S}$ is closed under I-coproducts.*
5. *Right-adjoint functors, if and only if its left-adjoint preserves $\mathcal{S}$-objects.*

Example 3.7. When $\mathbb{C}$ is infinitary extensive and $\mathcal{S}$ is closed under coproducts, every polynomial endofunctor $F\colon \mathbb{C} \longrightarrow \mathbb{C}$ admits precise factorizations w.r.t. $\mathcal{S}$. This is in particular the case for $\mathbb{C} = \mathcal{S} = \mathsf{Set}$. In this case, we shall see later (Sect. 4.1) that many other Set-functors, e.g. the bag functor $\mathcal{B}$, where $\mathcal{B}(X)$ is the set of finite multisets, have precise factorizations. In contrast, $F = \mathcal{P}$ does not admit precise factorizations, and if $f\colon X \longrightarrow \mathcal{P}Y$ is $\mathcal{P}$-precise, then $f(x) = \emptyset$ for all $x \in X$.

3.2   Path Categories in Pointed Coalgebras

We define a path for I-pointed TF-coalgebras as a tree according to F. Following the observation in Example 3.3, one layer of the tree is modelled by a F-precise morphism and hence a path in a TF-coalgebra is defined to be a finite sequence of $(F + 1)$-precise maps, where the $\_ + 1$ comes from the dead states w.r.t. T; the argument is given later in Remark 3.23 when reachability is discussed. Since the $\_ + 1$ is not relevant yet, we define $\mathsf{Path}(I, F)$ in the following and will use $\mathsf{Path}(I, F + 1)$ later. For simplicity, we write $\boldsymbol{X}_n$ for finite families $(X_k)_{0 \le k < n}$.

Definition 3.8. *The category* $\mathsf{Path}(I, F)$ *consists of the following. An object is* $(\boldsymbol{P}_{n+1}, \boldsymbol{p}_n)$ *for an* $n \in \mathbb{N}$ *with* $P_0 = I$ *and* $\boldsymbol{p}_n$ *a family of F-precise maps* $(p_k \colon P_k \longrightarrow FP_{k+1})_{k<n}$. *We say that* $(\boldsymbol{P}_{n+1}, \boldsymbol{p}_n)$ *is a path of length n. A morphism* $\phi_{n+1} \colon (\boldsymbol{P}_{n+1}, \boldsymbol{p}_n) \longrightarrow (\boldsymbol{Q}_{m+1}, \boldsymbol{q}_m)$, $m \ge n$, *is a family* $(\phi_k \colon P_k \longrightarrow Q_k)_{k \le n}$ *with* $\phi_0 = \mathrm{id}_I$ *and* $q_k \cdot \phi_k = F\phi_{k+1} \cdot p_k$ *for all* $0 \le k \le n$.

Example 3.9. Paths for $FX = A \times X + 1$ and $I = \{*\}$ singleton are as follows. First, a map $f \colon I \longrightarrow FX$ is precise iff (up-to isomorphism) either $X = I$ and $f(*) = (a, *)$ for some $a \in A$; or $X = \varnothing$ and $f(*) = \bot$. Then a path is isomorphic to an object of the form: $P_i = I$ for $i \le k$, $P_i = \varnothing$ for $i > k$, $p_i(*) = (a_i, *)$ for $i < k$, and $p_k(*) = \bot$. A path is the same as a word, plus some "junk", concretely, a word in $A^*.\bot^*$. For LTSs, an object in $\mathsf{Path}(I, F)$ with $FX = A \times X$ is simply a word in A^*. For a more complicated functor, Fig. 2 depicts a path of length 4, which is a tree for the signature with one unary, one binary symbol, and a constant. The layers of the tree are the sets $\boldsymbol{P}_4$. Also note that since every p_i is F-precise, there is precisely one route to go from every element of a P_k to $*$.

Remark 3.10. The inductive continuation of Remark 3.5 is as follows. Given a morphism ϕ_{n+1} in $\mathsf{Path}(I, F)$, since ϕ_0 is an isomorphism, then ϕ_k is an isomorphism for all $0 \le k \le n$. If F admits precise factorizations and if $I \in \mathcal{S}$, then for every path $(\boldsymbol{P}_{n+1}, \boldsymbol{p}_n)$, all P_k, $0 \le k \le n$, are in $\mathcal{S}$.

Remark 3.11. If in Definition 3.4, the connecting morphism $h \colon Y' \longrightarrow Y$ uniquely exists, then it follows by induction that the hom-sets of $\mathsf{Path}(I, F)$ are at most singleton. This is the case for all polynomial functors, but not the case for the bag functor on sets (discussed in Subsect. 4.1).

Definition 3.12. *The* path poset $\mathsf{PathOrd}(I, F)$ *is the set* $\coprod_{0 \le n} \mathbb{C}(I, F^n 1)$ *equipped with the order: for* $u \colon I \longrightarrow F^n 1$ *and* $v \colon I \longrightarrow F^m 1$, *we define* $u \le v$ *if* $n \le m$ *and* $F^n(!) \cdot v = u$.

$$
\begin{array}{ccc}
 & & F^n F^{m-n} 1 \\
 & \nearrow^{v} & \downarrow F^n ! \\
I & \xrightarrow{\;u\;} & F^n 1
\end{array}
$$

So $u \le v$ if u is the truncation of v to n levels. This matches the morphisms in $\mathsf{Path}(I, F)$ that witnesses that one path is prefix of another:

Proposition 3.13. *1. The functor* $\mathsf{Comp} \colon \mathsf{Path}(I, F) \longrightarrow \mathsf{PathOrd}(I, F)$ *defined by* $I = P_0 \xrightarrow{p_0} FP_1 \cdots \rightarrow F^n P_n \xrightarrow{F^n !} F^n 1$ *on* $(\boldsymbol{P}_{n+1}, \boldsymbol{p}_n)$ *is full, and reflects isos.*
2. If F admits precise factorizations w.r.t. $\mathcal{S}$ and $I \in \mathcal{S}$, then Comp is sujective.
3. If additionally h in Definition 3.4 is unique, then Comp has a right-inverse.

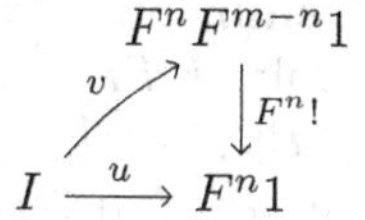

In particular, $\mathsf{PathOrd}(I, F)$ is $\mathsf{Path}(I, F)$ up to isomorphism. In the instances, it is often easier to characterize $\mathsf{PathOrd}(I, F)$. This also shows that $\mathsf{Path}(I, F)$ contains the elements – understood as morphisms from I – of the finite start of the final chain of F: $1 \xleftarrow{!} F1 \xleftarrow{F!} F^2 1 \xleftarrow{F^2!} F^3 1 \leftarrow \cdots$.

Example 3.14. When $FX = A \times X + 1$, $F^n 1$ is isomorphic to the set of words in $A^\star.\bot^\star$ of length n. Consequently, $\mathsf{PathOrd}(I, F)$ is the set of words in $A^\star.\bot^\star$, equipped with the prefix order. In this case, Comp is an equivalence of categories.

3.3 Embedding Paths into Pointed Coalgebras

The paths $(\boldsymbol{P}_{n+1}, \boldsymbol{p}_n)$ embed into $\mathsf{Coalg}_l(I, TF)$ as one expects it for examples like Fig. 2: one takes the disjoint union of the P_k, one has the pointing $I = P_0$ and the linear structure of F is embedded into the branching type T.

During the presentation of the results, we require T, F, and I to have certain properties, which will be introduced one after the other. The full list of assumptions is summarized in Table 2:

(Ax1) – The main theorem will show that coalgebra homomorphisms in $\mathsf{Coalg}_l(I, TF)$ are the open maps for the path category $\mathsf{Path}(I, F + 1)$. So from now on, we assume that $\mathbb{C}$ has finite coproducts and to use the results from the previous sections, we fix a class $\mathcal{S} \subseteq \mathbf{obj}\,\mathbb{C}$ such that $F + 1$ admits precise factorizations w.r.t. $\mathcal{S}$ and that $I \in \mathcal{S}$.

(Ax2) – Recall, that a family of morphisms $(e_i\colon X_i \longrightarrow Y)_{i \in I}$ with common codomain is called jointly epic if for $f, g\colon Y \longrightarrow Z$ we have that $f \cdot e_i = g \cdot e_i$ $\forall i \in I$ implies $f = g$. For Set, this means, that every element $y \in Y$ is in the image of some e_i. Since we work with partial orders on T-homsets, we also need the generalization of this property if $f \sqsubseteq g$ are of the form $Y \longrightarrow TZ'$.

(Ax3) – In this section, we encode paths as a pointed coalgebra by constructing a functor $J\colon \mathsf{Path}(I, F+1) \hookrightarrow \mathsf{Coalg}_l(I, TF)$. For that we need to embed the linear behaviour $FX + 1$ into TFX. This is done by a natural transformation $[\eta, \bot]\colon \mathrm{Id} + 1 \longrightarrow T$, and we require that $\bot\colon 1 \longrightarrow T$ is a bottom element for $\sqsubseteq$.

Example 3.15. For the case where T is the powerset functor $\mathcal{P}$, η is given by the unit $\eta_X(x) = \{x\}$, and $\bot$ is given by empty sets $\bot_X(*) = \varnothing$.

Definition 3.16. *We have an inclusion functor* $J\colon \mathsf{Path}(I, F + 1) \hookrightarrow \mathsf{Coalg}_l(I, TF)$ *that maps a path* $(\boldsymbol{P}_{n+1}, \boldsymbol{p}_n)$ *to an* I*-pointed* TF*-coalgebra on* $\coprod \boldsymbol{P}_{n+1} := \coprod_{0 \le k \le n} P_k$. *The pointing is given by* $\mathsf{in}_0\colon I = P_0 \longrightarrow \coprod \boldsymbol{P}_{n+1}$ *and the structure by:*

$$\coprod_{0 \le k < n} P_k + P_n \xrightarrow{[(F\mathsf{in}_{k+1}+1)\cdot p_k]_{0 \le k < n} + !} F \coprod \boldsymbol{P}_{n+1} + 1 \xrightarrow{[\eta, \bot]} TF \coprod \boldsymbol{P}_{n+1}.$$

Example 3.17. In the case of LTSs, a path, or equivalently a word $a_1 ... a_k.\bot ... \bot \in A^\star.\bot^\star$, is mapped to the finite linear system over $a_1 ... a_k$ (see Sect. 2.1), seen as a coalgebra (see Sect. 2.2).

Proposition 3.18. *Given a morphism* $[x_k]_{k\leq n}\colon \coprod \boldsymbol{P}_{n+1} \longrightarrow X$ *for some system* (X,ξ,x_0) *and a path* $(\boldsymbol{P}_{n+1},\boldsymbol{p}_n)$*, we have*

$$\begin{array}{c} J(\boldsymbol{P}_{n+1},\boldsymbol{p}_n) \xrightarrow{[x_k]_{k\leq n}} (X,\xi,x_0) \\ \text{a run in } \mathsf{Coalg}_l(I,TF) \end{array} \iff \forall k<n\colon \begin{array}{ccccc} P_k & & \xrightarrow{\quad x_k \quad} & & X \\ {\scriptstyle p_k}\downarrow & & Fx_{k+1}+1 \sqsubseteq [\eta,\bot]_X & & \downarrow{\scriptstyle \xi} \\ FP_{k+1}+1 & \longrightarrow & FX+1 & \longrightarrow & TFX. \end{array}$$

Also note that the pointing x_0 of the coalgebra is necessarily the first component of any run in it. In a run $[x_k]_{k\leq n}$, p_k corresponds to an edge from x_k to x_{k+1}.

Example 3.19. For LTSs, since the P_k are singletons, x_k just picks the kth state of the run. The right-hand side of this lemma describes that this is a run iff there is a transition from the kth state and the $(k+1)$–th state.

3.4 Open Morphisms Are Exactly Coalgebra Homomorphisms

In this section, we prove our main contribution, namely that $\mathsf{Path}(I,F+1)$-open maps in $\mathsf{Coalg}_l(I,TF)$ are exactly coalgebra homomorphisms. For the first direction of the main theorem, that is, that coalgebra homomorphisms are open, we need two extra axioms:

(Ax4) – describing that the order on $\mathbb{C}(X,TY)$ is point-wise. This holds for the powerset because every set is the union of its singleton subsets.

(Ax5) – describing that $\mathbb{C}(X,TY)$ admits a choice-principle. This holds for the powerset because whenever $y\in h[x]$ for a map $h\colon X\longrightarrow Y$ and $x\subseteq X$, then there is some $\{x'\}\subseteq x$ with $h(x')=y$.

Theorem 3.20. *Under the assumptions of Table 2, a coalgebra homomorphism in* $\mathsf{Coalg}_l(I,TF)$ *is* $\mathsf{Path}(I,F+1)$*-open.*

Table 2. Main assumptions on $F,T\colon \mathbb{C}\longrightarrow\mathbb{C}$, $\sqsubseteq^T$, $\mathcal{S}\subseteq \mathbf{obj}\,\mathbb{C}$

F	(Ax1)	$F+1$ admits precise factorizations, w.r.t. $\mathcal{S}$ and $I\in\mathcal{S}$
T	(Ax2)	If $(e_i\colon X_i\longrightarrow Y)_{i\in I}$ jointly epic, then $f\cdot e_i\sqsubseteq g\cdot e_i$ for all $i\in I\Rightarrow f\sqsubseteq g$.
	(Ax3)	$[\eta,\bot]\colon \mathrm{Id}+1\longrightarrow T$, with $\bot_Y\cdot !_X\sqsubseteq f$ for all $f\colon X\longrightarrow TY$
	(Ax4)	For every $f\colon X\longrightarrow TY$, $X\in\mathcal{S}$, $f=\bigsqcup\{[\eta,\bot]_Y\cdot f'\sqsubseteq f\mid f'\colon X\longrightarrow Y+1\}$
	(Ax5)	$\forall A\in\mathcal{S}$ $\quad\begin{array}{ccc} A & \xrightarrow{x} & TX \\ {\scriptstyle y}\downarrow & \sqsupseteq & \downarrow{\scriptstyle Th} \\ Y+1 & \xrightarrow{[\eta,\bot]_Y} & TY \end{array} \overset{\exists x'}{\Longrightarrow} \begin{array}{ccc} A & \overset{x}{\longrightarrow} & TX \\ {\scriptstyle x'}\searrow & \sqcup\!\lvert & \nearrow{\scriptstyle [\eta,\bot]_X} \\ {\scriptstyle y}\searrow & X+1 & \downarrow{\scriptstyle Th} \\ & \downarrow{\scriptstyle h+1} & \\ & Y+1 \xrightarrow{[\eta,\bot]_Y} & TY \end{array}$

The converse is not true in general, because intuitively, open maps reflect runs, and thus only reflect edges of reachable states, as we have seen in Sect. 2.1. The notion of a state being reached by a path is the following:

Definition 3.21. *A system* (X, ξ, x_0) *is* path-reachable *if the family of runs* $[x_k]_{k\le n}\colon J(\boldsymbol{P}_{n+1}, \boldsymbol{p}_n) \longrightarrow (X, \xi, x_0)$ *(of paths from* $\mathsf{Path}(I, F+1)$*) is jointly epic.*

Example 3.22. For LTSs, this means that every state in X is reached by a run, that is, there is a path from the initial state to every state of X.

Remark 3.23. In Definition 3.21, it is crucial that we consider $\mathsf{Path}(I, F+1)$ and not $\mathsf{Path}(I, F)$ for functors incorporating 'arities ≥ 2'. This does not affect the example of LTSs, but for $I = 1$, $FX = X \times X$ and $T = \mathcal{P}$ in Set, the coalgebra (X, ξ, x_0) on $X = \{x_0, y_1, y_2, z_1, z_2\}$ given by $\xi(x_0) = \{(y_1, y_2)\}$, $\xi(y_1) = \{(z_1, z_2)\}$, $\xi(y_2) = \xi(z_1) = \xi(z_2) = \emptyset$ is path-reachable for $\mathsf{Path}(I, F+1)$. There is no run of a length 2 path from $\mathsf{Path}(I, F)$, because y_2 has no successors, and so there is no path to z_1 or to z_2.

Theorem 3.24. *Under the assumptions of Table 2, if* (X, ξ, x_0) *is path-reachable, then an open morphism* $h\colon (X, \xi, x_0) \longrightarrow (Y, \zeta, y_0)$ *is a coalgebra homomorphism.*

3.5 Connection to Other Notions of Reachability

There is another concise notion for reachability in the coalgebraic literature [2].

Definition 3.25. *A* subcoalgebra *of* (X, ξ, x_0) *is a coalgebra homomorphism* $h\colon (Y, \zeta, y_0) \longrightarrow (X, \xi, x_0)$ *that is carried by a monomorphism* $h\colon X \rightarrowtail Y$. *Furthermore* (X, ξ, x_0) *is called* reachable *if it has no proper subcoalgebra, i.e. if any subcoalgebra* h *is an isomorphism.*

Under the following assumptions, this notion coincides with the path-based definition of reachability (Definition 3.21).

Assumption 3.26. For the present Subsect. 3.5, let $\mathbb{C}$ be cocomplete, have (epi,mono)-factorizations and wide pullbacks of monomorphisms.

The first direction follows directly from Theorem 3.20:

Proposition 3.27. *Every path-reachable* (X, ξ, x_0) *has no proper subcoalgebra.*

For the other direction it is needed that TF preserves arbitrary intersections, that is, wide pullbacks of monomorphisms. In Set, this means that for a family $(X_i \subseteq Y)_{i\in I}$ of subsets we have $\bigcap_{i\in I} TFX_i = TF \bigcap_{i\in I} X_i$ as subsets of TFY.

Proposition 3.28. *If, furthermore, for every monomorphism* $m\colon Y \longrightarrow Z$*, the function* $\mathbb{C}(-, Tm)\colon\ \mathbb{C}(X, TY) \longrightarrow \mathbb{C}(X, TZ)$ *reflects joins and if* TF *preserves arbitrary intersections, then a reachable coalgebra* (X, ξ, x_0) *is also path-reachable.*

All those technical assumptions are satisfied in the case of LTSs, and will also be satisfied in all our instances in Sect. 4.

3.6 Trace Semantics for Pointed Coalgebras

The characterization from Theorems 3.20 and 3.24 points out a natural way of defining a trace semantics for pointed coalgebras. Indeed, the paths category $\mathsf{Path}(I, F+1)$ provides a natural way of defining the runs of a system. A possible way to go from runs to trace semantics is to describe accepting runs as the subcategory $J'\colon \mathsf{Path}(I, F) \hookrightarrow \mathsf{Path}(I, F+1)$. We can define the *trace semantics* of a system (X, ξ, x_o) as the set:

$$\mathsf{tr}(X, \xi, x_0) = \{\mathsf{Comp}(\boldsymbol{P}_{n+1}, \boldsymbol{p}_n) \mid \exists \text{ run } [x_k]_{k\leq n}\colon JJ'(\boldsymbol{P}_{n+1}, \boldsymbol{p}_n) \longrightarrow (X, \xi, x_0) \\ \text{with } (\boldsymbol{P}_{n+1}, \boldsymbol{p}_n) \in \mathsf{Path}(I, F)\}$$

Since $\mathsf{Path}(I, F)$-open maps preserve and reflect runs, we have the following:

Corollary 3.29. $\mathsf{tr}\colon \mathsf{Coalg}_l(I, TF) \longrightarrow (\mathcal{P}(\mathsf{PathOrd}(I, F)), \subseteq)$ *is a functor and if* $f\colon (X, \xi, x_0) \longrightarrow (Y, \zeta, y_0)$ *is* $\mathsf{Path}(I, F+1)$*-open, then* $\mathsf{tr}(X, \xi, x_0) = \mathsf{tr}(Y, \zeta, y_0)$.

Let us look at two LTS-related examples (we will describe some others in the next section). First, for $FX = A \times X$. The usual trace semantics is given by all the words in $A^\star$ that are labelled of a run of a system. This trace semantics is obtained because $\mathsf{PathOrd}(I, F) = \coprod_{n\geq 0} A^n$ and because Comp maps every path to its underlying word. Another example is given for $FX = A \times X + \{\checkmark\}$, where $\checkmark$ marks final states. In this case, a path in $\mathsf{Path}(I, F)$ of length n is either a path that can still be extended or encodes less than n steps to an accepting state $\checkmark$. This obtains the trace semantics containing the set of accepted words, as in automata theory, plus the set of possibly infinite runs.

4 Instances

4.1 Analytic Functors and Tree Automata

In Example 3.7, we have seen that every polynomial Set-functors, in particular the functor $X \mapsto A \times X$, has precise factorizations with respect to all sets. This allowed us to see LTSs, modelled as $\{*\}$-pointed $\mathcal{P}(A \times \_)$-coalgebra, as an instance of our theory. This allowed us in particular to describe their trace semantics using our path category in Sect. 3.6. This can be extended to tree automata as follows. Assume given a signature Σ, that is, a collection $(\Sigma_n)_{n\in\mathbb{N}}$ of disjoint sets. When σ belongs to Σ_n, we say that n is the *arity of* σ or that σ is a *symbol of arity* n. A top-down non-deterministic tree automata as defined in [6] is then the same as a $\{*\}$-pointed $\mathcal{P}F$-coalgebra where F is the polynomial functor $X \mapsto \coprod_{\sigma\in\Sigma_n} X^n$. For this functor, $F^n(1)$ is the set of trees over $\Sigma \sqcup \{*(0)\}$ of depth at most $n+1$ such that a leaf is labelled by $*$ if and only if it is at depth $n+1$. Intuitively, elements of $F^n(1)$ are partial runs of length n that can possibly be extended. Then, the trace semantics of a tree automata, seen as a pointed coalgebra, is given by the set of partial runs of the automata. In particular, this contains the set of accepted finite trees as those partial runs without any $*$, and the set of accepted infinite trees, encoded as the sequence of their truncations of depth n, for every n.

In the following, we would like to extend this to other kinds of tree automata by allowing some symmetries. For example, in a tree, we may not care about the order of the children. This boils down to quotient the set X^n of n-tuples, by some permutations of the indices. This can be done generally given a subgroup G of the permutation group $\mathfrak{S}_n$ on n elements by defining X^n/G as the quotient of X^n under the equivalence relation: $(x_1, \ldots, x_n) \equiv_G (y_1, \ldots, y_n)$ iff there is $\pi \in G$ such that for all i, $x_i = y_{\pi(i)}$. Concretely, this means that we replace the polynomial functor F by a so-called *analytic functor*:

Definition 4.1 [14,15]. *An analytic* Set*-functor is a functor of the form* $FX = \coprod_{\sigma\in\Sigma_n} X^n/G_\sigma$ *where for every* $\sigma \in \Sigma_n$, *we have a subgroup* G_σ *of the permutation group* $\mathfrak{S}_n$ *on* n *elements.*

Example 4.2. Every polynomial functor is analytic. The bag-functor is analytic, with $\Sigma = (\{*\})_{n\in\mathbb{N}}$ has one operation symbol per arity and $G_\sigma = \mathfrak{S}_{\mathsf{ar}(\sigma)}$ is the full permutation group on $\mathsf{ar}(\sigma)$ elements. It is the archetype of an analytic functor, in the sense that for every analytic functor $F\colon \mathsf{Set} \longrightarrow \mathsf{Set}$, there is a natural transformation into the bag functor $\alpha\colon F \longrightarrow \mathcal{B}$. If F is given by Σ and G_σ as above, then α_X is given by

$$FX = \coprod_{\sigma\in\Sigma_n} X^n/G_\sigma \twoheadrightarrow \coprod_{\sigma\in\Sigma_n} X^n/\mathfrak{S}_n \rightarrow \coprod_{n\in\mathbb{N}} X^n/\mathfrak{S}_n = \mathcal{B}X.$$

Proposition 4.3. *For an analytic* Set*-functor* F, *the following are equivalent (1) a map* $f\colon X \longrightarrow FY$ *is* F*-precise, (2)* $\alpha_Y \cdot f$ *is* $\mathcal{B}$*-precise, (3) every element of* Y *appears precisely once in the definition of* f, *i.e. for every* $y \in Y$, *there is exactly one* x *in* X, *such that* $f(x)$ *is the equivalence class of a tuple* $(y_1, \ldots, y_n)$ *where there is an index* i, *such that* $y_i = y$; *and furthermore this index is unique. So every analytic functor has precise factorizations w.r.t.* Set.

4.2 Nominal Sets: Regular Nondeterministic Nominal Automata

We derive an open map situation from the coalgebraic situation for *regular nondeterministic nominal automata (RNNAs)* [26]. They are an extension of automata to accept *words with binders*, consisting of literals $a \in \mathbb{A}$ and binders $|_a$ for $a \in \mathbb{A}$; the latter is counted as length 1. An example of such a word of length 4 is $a|_c bc$, where the last c is bound by $|_c$. The order of binders makes difference: $|_a|_b ab \neq |_a|_b ba$. RNNAs are coalgebraically represented in the category of nominal sets [10], a formalism about atoms (e.g. variables) that sit in more complex structures (e.g. lambda terms), and gives a notion of *binding*. Because the choice principles (Ax4) and (Ax5) are not satisfied by every nominal sets, we instead use the class of *strong nominal sets* for the precise factorization (Definition 3.4).

Definition 4.4 [10,24]. *Fix a countably infinite set* $\mathbb{A}$, *called the set of* atoms. *For the group* $\mathfrak{S}_\mathsf{f}(\mathbb{A})$ *of finite permutations on the set* $\mathbb{A}$, *a* group action $(X, \cdot)$ *is a set* X *together with a group homomorphism* $\cdot\colon \mathfrak{S}_\mathsf{f}(\mathbb{A}) \longrightarrow \mathfrak{S}_\mathsf{f}(X)$, *written in* *infix notation. An element* $x \in X$ *is* supported by $S \subseteq \mathbb{A}$, *if for all* $\pi \in \mathfrak{S}_f(\mathbb{A})$ *with* $\pi(a) = a\ \forall a \in S$ *we have* $\pi \cdot x = x$. *A* nominal set *is a group action for* $\mathfrak{S}_f(\mathbb{A})$ *such that every* $x \in X$ *is finitely supported, i.e. supported by a finite* $S \subseteq \mathbb{A}$. *A map* $f\colon (X,\cdot) \longrightarrow (Y,\star)$ *is* equivariant *if for all* $x \in X$ *and* $\pi \in \mathfrak{S}_f(\mathbb{A})$ *we have* $f(\pi \cdot x) = \pi \star f(x)$. *The category of nominal sets and equivariant maps is denoted by* Nom. *A nominal set* $(X,\cdot)$ *is called* strong *if for all* $x \in X$ *and* $\pi \in \mathfrak{S}_f(\mathbb{A})$ *with* $\pi \cdot x = x$ *we have* $\pi(a) = a$ *for all* $a \in \mathsf{supp}(x)$.

Intuitively, the support of an element is the set of free literals. An equivariant map can forget some of the support of an element, but can never introduce new atoms, i.e. $\mathsf{supp}(f(x)) \subseteq \mathsf{supp}(x)$. The intuition behind strong nominal sets is that all atoms appear in a fixed order, that is, $\mathbb{A}^n$ is strong, but $\mathcal{P}_f(\mathbb{A})$ (the finite powerset) is not. We set $\mathcal{S}$ to be the class of strong nominal sets:

Example 4.5. The Nom-functor of unordered pairs admits precise factorizations w.r.t. strong nominal sets, but not w.r.t. all nominal sets.

In the application, we fix the set $I = \mathbb{A}^{\#n}$ of distinct n-tuples of atoms ($n \geq 0$) as the pointing. The hom-sets $\mathsf{Nom}(X, \mathcal{P}_{\mathsf{ufs}} Y)$ are ordered point-wise.

Proposition 4.6. *Uniformly finitely supported powerset* $\mathcal{P}_{\mathsf{ufs}}(X) = \{Y \subseteq X \mid \bigcup_{y \in Y} \mathsf{supp}(y)$ *finite*$\}$ *satisfies* (Ax2-5) *w.r.t.* $\mathcal{S}$ *the class of strong nominal sets.*[1]

As for F, we study an LTS-like functor, extended with the *binding functor* [10]:

Definition 4.7. *For a nominal set* X, *define the* α*-equivalence relation* $\sim_\alpha$ *on* $\mathbb{A} \times X$ *by:* $(a,x) \sim_\alpha (b,y) \Leftrightarrow \exists c \in \mathbb{A} \setminus \mathsf{supp}(x) \setminus \mathsf{supp}(y)$ *with* $(a\,c) \cdot x = (b\,c) \cdot y$. *Denote the quotient by* $[\mathbb{A}]X := \mathbb{A} \times X/\sim_\alpha$. *The assignment* $X \mapsto [\mathbb{A}]X$ *extends to a functor, called the* binding functor $[\mathbb{A}]\colon \mathsf{Nom} \longrightarrow \mathsf{Nom}$.

RNNA are precisely $\mathcal{P}_{\mathsf{ufs}} F$-coalgebras for $FX = \{\checkmark\} + [\mathbb{A}]X + \mathbb{A} \times X$ [26]. In this paper we additionally consider initial states for RNNAs.

Proposition 4.8. *The binding functor* $[\mathbb{A}]$ *admits precise factorizations w.r.t. strong nominal sets and so does* $FX = \{\checkmark\} + [\mathbb{A}]X + \mathbb{A} \times X$.

An element in $\mathsf{PathOrd}(\mathbb{A}^{\#n}, F)$ may be regarded as a word with binders under a context $\boldsymbol{a} \vdash w$, where $\boldsymbol{a} \in \mathbb{A}^{\#n}$, all literals in w are bound or in $\boldsymbol{a}$, and w may end with $\checkmark$. Moreover, two word-in-contexts $\boldsymbol{a} \vdash w$ and $\boldsymbol{a}' \vdash w'$ are identified if their closures are α-equivalent, that is, $|_{a_1} \cdots |_{a_n} w = |_{a_1'} \cdots |_{a_n'} w'$. The trace semantics of a RNNA T contains all the word-in-contexts corresponding to runs in T. This trace semantics distinguishes whether words are concluded by $\checkmark$.

4.3 Subsuming Arbitrary Open Morphism Situations

Lasota [19] provides a translation of a small path-category $\mathbb{P} \hookrightarrow \mathbb{M}$ into a functor $\mathbb{F}\colon \mathsf{Set}^{\mathsf{obj}\,\mathbb{P}} \longrightarrow \mathsf{Set}^{\mathsf{obj}\,\mathbb{P}}$ defined by $\mathbb{F}(X_P)_P = \left(\prod_{Q \in \mathbb{P}} (\mathcal{P}(X_Q))^{\mathbb{P}(P,Q)}\right)_{P \in \mathbb{P}}$.

[1] There are two variants of powersets discussed in [26]. The finite powerset $\mathcal{P}_f$ also fulfils the axioms. However, *finitely supported* powerset $\mathcal{P}_{\mathsf{fs}}$ does not fulfil (Ax5).

So the hom-sets $\mathsf{Set}^{\mathbf{obj}\,\mathbb{P}}(X, \mathbb{F}Y)$ have a canonical order, namely the point-wise inclusion. This admits a functor Beh from $\mathbb{M}$ to $\mathbb{F}$-coalgebras and lax coalgebra homomorphisms, and Lasota shows that $f \in \mathbb{M}(X, Y)$ is $\mathbb{P}$-open iff $\mathsf{Beh}(f)$ is a coalgebra homomorphism. In the following, we show that we can apply our framework to $\mathbb{F}$ by a suitable decomposition $\mathbb{F} = TF$ and a suitable object I for the initial state pointing. As usual in open map papers, we require that $\mathbb{P}$ and $\mathbb{M}$ have a common initial object $0_{\mathbb{P}}$. Observe that we have $\mathbb{F} = T \cdot F$ where

$$T(X_P)_{P\in\mathbb{P}} = \big(\mathcal{P}(X_P)\big)_{P\in\mathbb{P}} \quad \text{and} \quad F(X_P)_{P\in\mathbb{P}} = \big(\textstyle\coprod_{Q\in\mathbb{P}}\mathbb{P}(P, Q) \times X_Q\big)_{P\in\mathbb{P}}.$$

Lasota considers coalgebras without pointing, but one indeed has a canonical pointing as follows. For $P \in \mathbb{P}$, define the characteristic family $\chi^P \in \mathsf{Set}^{\mathbf{obj}\,\mathbb{P}}$ by $\chi^P_Q = 1$ if $P = Q$ and $\chi^P_Q = \emptyset$ if $P \neq Q$. With this, we fix the pointing $I = \chi^{0_{\mathbb{P}}}$.

Proposition 4.9. *T, F and I satisfy the axioms from Table 2, with $\mathcal{S} = \mathsf{Set}^{\mathbf{obj}\,\mathbb{P}}$.*

The path category in $\mathsf{Coalg}_l(I, TF)$ from our theory can be described as follows.

Proposition 4.10. *An object of $\mathsf{Path}(I, F)$ is a sequence of composable $\mathbb{P}$-morphisms $0_{\mathbb{P}} \xrightarrow{m_1} P_1 \xrightarrow{m_2} P_2 \cdots \xrightarrow{m_n} P_n$.*

5 Conclusions and Further Work

We proved that coalgebra homomorphisms for systems with non-deterministic branching can be seen as open maps for a canonical path-category, constructed from the computation type F. This limitation to non-deterministic systems is unsurprising: as we have proved in Sect. 4.3 on Lasota's work [19], every open map situation can been encoded as a coalgebra situation with a powerset-like functor, so with non-deterministic branching. As a future work, we would like to extend this theory of path-categories to coalgebras for further kinds of branching, especially probabilistic and weighted. This will require (1) to adapt open maps to allow those kinds of branching (2) adapt the axioms from Table 2, by replacing the "+1" part of (Ax1) to something depending on the branching type.

References

1. Adámek, J., Herrlich, H., Strecker, G.E.: Abstract and concrete categories: the joy of cats. online and enhanced edition of the book published in 1990 by John Wiley and Sons (2004). http://katmat.math.uni-bremen.de/acc/acc.pdf
2. Adámek, J., Milius, S., Moss, L.S., Sousa, L.: Well-pointed coalgebras. Logical Methods Comput. Sci. **9**(3), 1–51 (2013)
3. Awodey, S.: Category Theory, 2nd edn. Oxford University Press, Inc., New York (2010)
4. Beohar, H., Küpper, S.: On path-based coalgebras and weak notions of bisimulation. In: 7th Conference on Algebra and Coalgebra in Computer Science, CALCO 2017, Ljubljana, Slovenia, 12–16 June 2017, pp. 6:1–6:17 (2017). https://doi.org/10.4230/LIPIcs.CALCO.2017.6

5. Bonchi, F., Silva, A., Sokolova, A.: The power of convex algebras. In: Meyer, R., Nestmann, U. (eds.) 28th International Conference on Concurrency Theory (CONCUR 2017), Dagstuhl, Germany, vol. 85, pp. 23:1–23:18 (2017). https://doi.org/10.4230/LIPIcs.CONCUR.2017.23
6. Comon, H., et al.: Tree Automata Techniques and Applications (2007). http://tata.gforge.inria.fr
7. Dubut, J., Goubault, É., Goubault-Larrecq, J.: Natural homology. In: Halldórsson, M.M., Iwama, K., Kobayashi, N., Speckmann, B. (eds.) ICALP 2015, Part II. LNCS, vol. 9135, pp. 171–183. Springer, Heidelberg (2015). https://doi.org/10.1007/978-3-662-47666-6_14
8. Fahrenberg, U., Legay, A.: History-preserving bisimilarity for higher-dimensional automata via open maps. Electron. Notes Theor. Comput. Sci. **298**, 165–178 (2013)
9. Fiore, M.P., Cattani, G.L., Winskel, G.: Weak bisimulation and open maps. In: 14th Annual IEEE Symposium on Logic in Computer Science (LICS 1999), pp. 67–76 (1999)
10. Gabbay, M., Pitts, A.M.: A new approach to abstract syntax involving binders. In: Longo, G. (ed.) Proceedings of the Fourteenth Annual IEEE Symposium on Logic in Computer Science, LICS 1999, pp. 214–224. IEEE Computer Society Press (1999)
11. Hansen, H.H., Klin, B.: Pointwise extensions of GSOS-defined operations. Math. Struct. Comput. Sci. **21**(1), 321–361 (2011)
12. Hasuo, I., Jacobs, B., Sokolova, A.: Generic trace semantics via coinduction. Logical Methods Comput. Sci. **3**(4), 1–36 (2007)
13. Jacobs, B., Sokolova, A.: Traces, executions and schedulers, coalgebraically. In: Kurz, A., Lenisa, M., Tarlecki, A. (eds.) CALCO 2009. LNCS, vol. 5728, pp. 206–220. Springer, Heidelberg (2009). https://doi.org/10.1007/978-3-642-03741-2_15
14. Joyal, A.: Une théorie combinatoire des séries formelles. Adv. Math. **42**(1), 1–82 (1981)
15. Joyal, A.: Foncteurs analytiques et espèces de structures. In: Labelle, G., Leroux, P. (eds.) Combinatoire énumérative. LNM, vol. 1234, pp. 126–159. Springer, Heidelberg (1986). https://doi.org/10.1007/BFb0072514
16. Joyal, A., Nielsen, M., Winskel, G.: Bisimulation from open maps. Inf. Comput. **127**, 164–185 (1996)
17. Kozen, D., Mamouras, K., Petrişan, D., Silva, A.: Nominal Kleene coalgebra. In: Halldórsson, M.M., Iwama, K., Kobayashi, N., Speckmann, B. (eds.) ICALP 2015, Part II. LNCS, vol. 9135, pp. 286–298. Springer, Heidelberg (2015). https://doi.org/10.1007/978-3-662-47666-6_23
18. Kurz, A., Petrisan, D., Severi, P., de Vries, F.: Nominal coalgebraic data types with applications to lambda calculus. Logical Methods Comput. Sci. **9**(4) (2013). https://doi.org/10.2168/LMCS-9(4:20)2013
19. Lasota, S.: Coalgebra morphisms subsume open maps. Theor. Comput. Sci. **280**(1), 123–135 (2002)
20. Milius, S.: A sound and complete calculus for finite stream circuits. In: Proceedings of the 25th Annual Symposium on Logic in Computer Science (LICS 2010), pp. 449–458 (2010)
21. Milius, S., Pattinson, D., Schröder, L.: Generic trace semantics and graded monads. In: Moss, L.S., Sobocinski, P. (eds.) Proceedings of 6th Conference on Algebra and Coalgebra in Computer Science, CALCO 2015. Leibniz International Proceedings in Informatics, vol. 35, pp. 253–269 (2015). http://www8.cs.fau.de/_media/research:papers:traces-gm.pdf

22. Nielsen, M., Hune, T.: Bisimulation and open maps for timed transition systems. Fundam. Inform. **38**, 61–77 (1999)
23. Park, D.: Concurrency and automata on infinite sequences. Theor. Comput. Sci. **104**, 167–183 (1981)
24. Pitts, A.M.: Nominal Sets: Names and Symmetry in Computer Science. Cambridge Tracts in Theoretical Computer Science, vol. 57. Cambridge University Press, Cambridge (2013)
25. Rutten, J.: Universal coalgebra: a theory of systems. Theor. Comput. Sci. **249**(1), 3–80 (2000)
26. Schröder, L., Kozen, D., Milius, S., Wißmann, T.: Nominal automata with name binding. In: Esparza, J., Murawski, A.S. (eds.) FoSSaCS 2017. LNCS, vol. 10203, pp. 124–142. Springer, Heidelberg (2017). https://doi.org/10.1007/978-3-662-54458-7_8

Author Index